Organometallic Chemistry

Volume 13

A Specialist Periodical Report

Organometallic Chemistry
Volume 13

A Review of the Literature Published during 1983

Senior Reporters
E. W. Abel, *Department of Chemistry, University of Exeter*
F. G. A. Stone, *Department of Inorganic Chemistry, University of Bristol*

Reporters
D. A. Armitage, *Queen Elizabeth College, University of London*
B. J. Brisdon, *University of Bath*
D. A. Edwards, *University of Bath*
M. E. Fakley, *Imperial Chemical Industries, Billingham*
P. G. Harrison, *University of Nottingham*
J. A. S. Howell, *University of Keele*
K. J. Karel, *E.I. du Pont de Nemours, Wilmington, U.S.A.*
W. E. Lindsell, *Heriot-Watt University*
B. Ridge, *University of Exeter*
D. R. Russell, *University of Leicester*
A. K. Smith, *University of Liverpool*
T. R. Spalding, *University College, Cork*
J. L. Wardell, *University of Aberdeen*
P. L. Watson, *E.I. du Pont de Nemours, Wilmington, U.S.A.*
W. E. Watts, *The New University of Ulster*
J. W. Wilson, *The New University of Ulster*

The Royal Society of Chemistry
Burlington House, London W1V 0BN

ISBN 0-85186-611-5
ISSN 0301-0074

Printed in Great Britain at the Alden Press, Oxford, London and Northampton

Foreword

This Volume of the Specialist Periodical Reports 'Organometallic Chemistry' surveys the literature for the calendar year 1983. The organization is similar to that in previous Volumes, with the section on organometallic compounds in biological chemistry covering the two year period 1982–83.

To assist in the speed of publication of these Volumes and to reduce their costs, the Royal Society of Chemistry is now employing 'camera ready' production methods for this series. We believe that this will be advantageous to our readership.

E. W. Abel
F. G. A. Stone

Contents

Abbreviations

Ac	acetate ($MeCOO^-$)
acac	acetylacetonate
acacen	*NN'*-ethylenebis(acetylacetone iminate)
AIBN	azoisobutyronitrile
Ar	Aryl
arphos	1-(diphenylphosphino)-2-(diphenylarsino)ethane
ATP	adenosine triphosphate
Azb	azobenzene
9-BBN	9-borabicyclo[3.3.1] nonane
bipy	2,2'-bipyridyl
Bz	benzyl
Bzac	benzoylacetonate
cbd	cyclobutadiene
1,5,9-cdt	cyclododeca-1,5,9-triene
chd	cyclohexadiene
chpt	cycloheptatriene
[Co]	cobalamin
(Co)	cobaloxime [$Co(dmg)_2$ derivative]
cod	cyclo-octa-1,5-diene
cot	cyclo-octatetraene
Cp	η^5-cyclopentadienyl
CTTM	charge transfer to metal
Cy	cyclohexyl
dab	1,4-diazabutadiene
dba	dibenzylideneacetone
depe	1,2-bis(diethylphosphino)ethane
depm	1,2-bis(diethylphosphino)methane
diars	*o*-phenylenebis(dimethyl)arsine
diarsop	{[2,2-dimethyl-1,3-dioxolan-4,5-diyl)bis(methylene)] bis-(diphenylarsine)}
diop	{[2,2-dimethyl-1,3-dioxolan-4,5-diyl)bis(methylene)] bis-(diphenylphosphine)}
diphos	1,2-bis(diphenylphosphino)ethane
DME	dimethoxyethane
DMF	*NN*-dimethylformamide
dmg	dimethylglyoximate
$dmgH_2$	dimethylglyoxime
dmpe	1,2-bis(dimethylphosphino)ethane
DMSO	dimethyl sulphoxide
dpa	di(2-pyridyl)amine
dpae	1,2-bis(diphenylarsino)ethane
dpam	bis(diphenylarsino)methane
dppb	1,4-bis(diphenylphosphino)butane
dppe	1,2-bis(diphenylphosphino)ethane
dppm	bis(diphenylphosphino)methane
dppp	1,3-bis(diphenylphosphino)propane
en	ethylene-1,2-diamine
EXAFS	extended X-ray absorption fine structure
F_6acac	hexafluoroacetylacetonate
Fc	ferrocenyl

Fp	$Fe(CO)_2Cp$
GVB	generalized valence band
HDPG	diphenylglyoximato
hfa	hexafluoroacetone
hfacac	hexafluoroacetylacetonato
hfb	hexafluorobutyne
HMPA	hexamethyl phosphoric triamide
$Me_6[14]dieneN_4$	5,7,7,12,14,14-hexamethyl-1,4,8,11-tetra-azacyclotetradeca-4,11-diene
$Me_6[14]N_4$	5,5,7,12,12,14-hexamethyl-1,4,8,11-tetra-azacyclotetradecane
4,7-Me_2phen	4,7-dimethyl-1,10-phenanthroline
3,4,7,8-Me_4phen	3,4,7,8-tetramethyl-1,10-phenanthroline
nap	1-naphthyl
nbd	norbornadiene
NBS	*N*-bromosuccinimide
OEP	octaethylporphyrin
Pc	phthalocyanin
PMDT	pentamethylenediethylenetramine
pd	pentane-2,4-dionate
phen	1,10-phenanthroline
$[PPN]^+$	$[(Ph_3P)_2N]^+$
py	pyridine
pz	pyrazolyl
sal	salicylaldehyde
salen	*NN'*-bis-(salicylaldehydo)ethylenediamine
saloph	*NN*-bisalicylidene-*o*-phenylenediamine
SCF	self consistent field
TCNE	tetracyanoethylene
TCNQ	7,7,8,8-tetracyanoquinodimethane
terpy	2,2',2''-terpyridyl
TFA	trifluoroacetic acid
tfacac	trifluoroacetylacetonato
tfo	triflate, trifluoromethylsulphonate
THF	tetrahydrofuran
tht	tetrahydrothiophen
TMBD	*NNN'N'*-tetramethyl-2-butene-1,4-diamine
TMED	tetramethylethylenediamine
tol	tolyl
TPP	*meso*-tetraphenylporphyrin
triphos	1,1,1-tris(diphenylphosphinomethyl)ethane

1
Group I: The Alkali and Coinage Metals

BY J. L. WARDELL

1. Alkali-metal Compounds

1.1 Hydrocarbon Radical Anion and Dianion Alkali-Metal Compounds. The cations in c.i.p. acenaphthylene^{2-},2Li$^+$ (1, ArH^{2+},2M$^+$; M=Li) occupy non-equivalent positions and are differently solvated;[1] ion-pair equilibria, c.i.p.$\rightleftharpoons$ s.s.i.p. were studied for (1) using n.m.r. and u.v.-visible spectra. ^{13}C N.m.r. spectra of pyrene^{2-},2M$^+$ (M=Li or Na) in Et_2O indicate[2] that the highest charge density is at C-4. The n.m.r. and e.s.r. spectra of ArH^{2-},2M$^+$ (ArH=fused benzenoid compound) are dependent on M$^+$, solvent, and temperature; this was assumed to arise from equilibrium involving a singlet ground state and a thermally accessible triplet excited state.[3] Various annulene species were investigated:[4] (i) homo[8]annulene^{2-}, 2K$^+$ ($\Delta H_f \simeq 0$, prepared from cis-bicyclo[6.1.0]nona-2,4,6-triene and K, (ii) 1,6-methano[10]annulene$^{-\cdot}$,K$^+$ (ΔH_f=67.6 kJ mol^{-1} more negative than ΔH_f of Naph$^{-\cdot}$,K$^+$, (iii) [12]annulene$^{-\cdot}$,Li$^+$. Reaction of $PhCH_2CH_2Ph$ with a Cs/K/Na alloy at -75° in THF provides[5] $2PhCH_2Cs$ via ($\cdot^-PhCH_2CH_2Ph^{-\cdot}$),2Cs$^+$ (2). A driving force for the reaction is the transformation of electrons from a π^* orbital of (2) to non-bonding orbitals of $PhCH_2Cs$; $PhCH_2CH_2CH{=}CH_2$ is also cleaved to $PhCH_2Cs$ and $(CH_2{\cdots}CH{\cdots}CH_2)^-$,Cs$^+$. Ring opening of benzocyclobutene (BCD) occurs[6] on reaction with Li in THF via BCD$^{-\cdot}$,Li$^+$ (or BCD^{2-},2Li$^+$) to o-$LiC_6H_4CH_2CH_2Li$.

1.2 General Organolithium Compounds. The use of ^{13}C n.m.r. spectra of ^{13}C-enriched-R-^{6}Li species in determining structures of lithiated hydrocarbons and α-halolithiumcarbenoids has been further illustrated by Seebach et al.[7] Coupling constants (^{13}C-^{6}Li) in donor solvents (R_2O or R_3N) at -150°C were observed for all except c.i.p. compounds. Halocarbenoids exist as monomers or heteroatom bridged oligomers, while RLi (R=Bu, cyclo-C_3H_5, bicyclo-[1.1.0]butyl, $CH_2{=}CH$ and Ph) are planar bridged dimers.

1.3 Alkyl Alkali-metal Compounds. Polycyclic alkyl-lithiums (RLi, R=1-or 2-ademantyl, 1-diamantyl, 1-twistyl, 1-triptycyl, or 3-homo-adamantyl) were produced from RCl or RBr and a 2% Na/Li alloy in pentane at 35° or in Et_2O at -45°. The presence of crushed glass and vigorous stirring (to scour metal surface) was necessary to give high yields.[8] Crystal structures of samples of MeNa containing MeLi (ratio of Na:Li from 36:1 to 3:1) were reported.[9] All samples contained $(MeNa)_4$ units: the geometry (and bonding) of $(MeNa)_4$ is analogous to that of $(MeLi)_4$ [Na-Na 3.12(3) and 3.18(3) Å, Na-C(intra) 2.58(4) and 2.64(4) Å, Na-C(inter) 2.76(4) Å]. The arrangement of $(MeNa)_4$ units provide large cavities into which $(MeLi)_4$ units can be placed up to Na:Li = 3:1. The THF adducts (3 and 4) of compounds $(RMe_2Si)_3CLi$(R=Me and Ph, respectively) have different crystal structures: (3) is an ate complex $[Li(THF)_4]$-$[Li\{C(SiMe_3)_3\}_2]$ m.p. >300° (anions are linear, Li-C 2.16(1) and 2.20(1) Å; tetrahedral cation, Li-O 1.96(8) Å), while (4) is a covalently-bonded monomer, $(PhMe_2Si)_3CLi THF$: co-ordinated to Li are O, central C of $C(SiMe_2Ph)_3$ [Li-C 2.12(2) Å] and the ipso C of a phenyl group [Li-C 2.40(2) Å]. N.m.r. spectra of (3) and (4) were reported.[10]

The structure of $(cyclo-C_3H_5Li)_2(LiBr)_2(Et_2O)_4$ was obtained; Et_2O-co-ordinated-Li atoms form a distorted tetrahedron with the faces capped by Br and cyclo-C_3H_5 groups. Orientation of the cyclopropyl groups allows H..Li interaction.[11] The ^1H N.m.r. spectrum of 2,2-Me_2-1-Li-Ph-cyclopropane showed non-equivalent Me groups at low T; the Me signals coalesce at -45°C. Restricted rotation of the Ph group was also indicated.[12] The opening of the cyclo-propyl ring in lithiated 3-R-2,4-Ph_2-*endo*-tricyclo-$[3.2.1.0^{2,4}]$octane (R=Ph or CN) occurs at -75°C surprisingly by a disrotatory process, but not, however, in a concerted step.[13] 1,ω-Proton shifts in $Ph_2CH(CH_2)_nLi$ (5) were investigated; (5, *n*=2) and $Ph(CH_2)_3Li$ do not undergo 1,3-shifts; (5, *n*=3 or 4) do take part in 1,(*n*+1) shifts expecially in THF to give $Ph(CH_2)_nCHPhLi$ [relative ease (5, *n*=3) > (5, *n*=4)].[14]

1.4 Polylithiated alkanes: A study was made of the reactions of halocarbons with Li vapour; the flash vaporisation mass spectra of the products, $CH_{4-n}Li_n$ (6) were reported.[15] The compound (6, *n*=2) was also obtained[16] from $CH_2(HgI)_2$ and Li or Bu^tLi; $MeCHLi_2$ (7) was similarly prepared from $MeCH(HgCl)_2$ and Li in Et_2O. Compound (7), calculated (*ab-initio*, 3-21G basis set) to have a classical structure, decomposes[16] at r.t. to CH_2=CHLi and LiH. A refined calculation[17] [STO-3G/4-31G/5-21G; energy optimization 6-31G*/

6-31G set] on tetralithiotetrahedrane (8) suggested, in contrast to an earlier study, that (8) is not a minimum on the p.e. surface and that it would decompose to $(LiC{\equiv}CLi)_{\underline{n}}$ ($\underline{n}$=1 or 2). Another theoretical study (3-21G basis set level with full geometry optimization) pointed to the stability of CLi_5 ($D_{3\underline{h}}$) and CLi_6 ($O_{\underline{h}}$) towards dissociation. All Li atoms were bound to carbon to give hyperlithiated species, in which the extra electrons beyond the octet are involved with Li-Li bonding rather than with C-Li interactions.[18]

1.5 Functionalized Alkyl-alkali-metal Compounds. Lithiation of alkoxypolycycloalkanes is assisted both in terms of reactivity and selectivity by intramolecular co-ordination of RLi to the alkoxy-group.[19] Formation of $ROCH_2Li$ has been achieved[20] as shown in Scheme (1). This *in situ* process is a simpler variant on the

$$ROCH_2Br \xrightarrow{i} ROCH_2SnX_3 \xrightarrow{ii} ROCH_2Li$$

R=Me, Bu^t, $PhCH_2$, $MeOCH_2CH_2$ or Ph; X = halide

Reagents: i, $LiSnCl_2Br$, THF, ½h; ii, BuLi, -78°, 1h

Scheme 1

established process using $ROCH_2SnBu_3$ and BuLi as employed[21] in the synthesis of *cis*-*endo*-2,6-$(LiOCH_2)_2$-bicyclo[3.3.0]octa-3,7-diene. An alternative route to Bu^tOCH_2M (M=K) was described[22] using $Bu^tOMe/Bu^sLi/KOBu^t$. Successive treatments of chlorohydrins with (i) BuLi and (ii) $Naph^{-\cdot},Li^+$ at -78° provide[23] $RR'C(OLi)CH_2Li$ (R,R'=H, Me, Bu^i, Ph, $PhCH_2$, CH_2=CH-CH_2 etc). The compound, $MeOCH_2CH_2CHMeLi$, prepared from $(MeOCH_2CH_2CHMe)_2Hg$ and Li, is tetrameric in PhLi and crystallises as the chelated *meso*-tetramer;[24] Li atoms form a distorted tetrahedron, to each face of which is attached a $MeOCH_2CH_2CHMe$ group by 4-centre bonding [$(Li-C)_{av}$ 2.313 Å, $(Li-Li)_{av}$ 2.495 Å, $(Li-O)_{av}$ 1.923 Å]. A theoretical treatment (SCF in conjunction with a basis set of better than double zeta plus polarization quality) of FCH_2Li (9) was performed; as in a less-refined earlier study 3 minima (essentially $H_2CLi^+...F^-$ $H_2C...LiF$ and $H_2C...FLi$) were found though there were some structural differences.[25a] A computational investigation was also made of the reaction of (9) with CH_2=CH_2.[25b] A theoretical study of lithiation of HCONHMe (to $HCONHCH_2Li$) indicated that proton removal is best achieved from the conformation in which the anion is not in conjunction with the amide π-system.[26]

1.6 Aryl-lithiums. An issue of Tetrahedron Letters (1983, Vol. 39, No. 12) was devoted to heteroatom-directed metallations in heterocyclic synthesis; a similar theme was the subject of a review.[27] Metallations, using BuLi/TMED, of 1-Li-naphthalene and 9-Li-anthracene occur readily at adjacent peri positions; a MNDO calculation revealed both the stabilization of the doubly-bridged di-lithiated products and the enhanced acidities of replaced protons.[28] A complex is formed between Bu^sLi and 2,4,6-$Pr^i{}_3C_6H_2CONMe_2$ in cyclohexane, prior to α-metallation.[29] α-Lithiations to the following groups were among those reported:[30] (i) sec- and tert-amides [internal competition in p-$Et_2NCOC_6H_4$-CONHEt showed the sec-group to be the more powerful α director, while intermolecular competition between PhCONHR and $PhCONR_2$ (R= Et or Pr) indicated the tert-group to be the more effective],[30a] (ii) $CONMeCH_2CH_2NMe_2$ and $C(OLi)RNMeCH_2CH_2NMe_2$,[30b] (iii) OCH_2OMe [ratio of 2-:4-lithiation of 3-$Me_2NC_6H_4OCH_2OMe$ is 99:1 for Bu^tLi/Et_2O and 2:98 for BuLi/hexane, both at 0°C],[30c] (iv) $CH(OR)_2$, CH(OR)R' and $CH_2CH(OR)_2$,[30d] (v) $NPhSO_2Ph$[30e] and (vi) $OCONEt_2$ at -78°C (at 0°C, o-$LiC_6H_4OCONEt_2$ rearranges to o-$LiOC_6H_4CONEt_2$).[30f]

α-Metallation in substituted pyridines continues to attract attention, e.g. in (i) 3-$MeOCH_2O$-py (mainly at the 4- but also the 2-site),[30c] (ii) 3-R_2NSO_2-py (at C_4),[31a] (iii) py-$OC(CF_3)_2$ complex at C_2),[31b] (iv) 3-F-py (at C_2)[31c] and (v) 2- or 4-Bu^tCONH-py (at C_3) and 3-Bu^tCONH-py (at C_4).[31d] The complex,[32] (3-$MeOC_6H_4CH_2OH$)Cr-$(CO)_3$, is selectively lithiated at the 4-position, in contrast to lithiation of the free arene, at C_2. Lithiation of $(Pr^i{}_3SiOC_6H_4)Cr(CO)_3$ by Bu^tLi in THF at -78°C provides the m- and p-products in a 10:1 ratio; $[1,3-(Pr^i{}_3SiO)_2C_6H_4]Cr(CO)_3$ is lithiated at the 5-position.[33]

In the crystal of $(PhLi.Et_2O)_4$ there are two interlocking tetrahedron of 4Li and 4 C atoms in a distorted cubane framework; each Li is co-ordinated to an Et_2O molecule.[34] The framework in $(PhLi)_3.LiBr(Et_2O)_3$ is further distorted by the presence of the single Br; the Li diagonally opposite Br is unco-ordinated by Et_2O. The biaryl $(2\text{-Li-6-}MeC_6H_3C_6H_3)_2$, obtained[35] from the optically pure diiodo compound and BuLi in Et_2O, is stable towards racemisation at 0°C.

1.7 Benzylic and allylic Compounds. Klein[36] has reviewed directive effects in allylic and benzylic polymetallations. Metallations by BuLi/Bu^tOM (M=Na or K) were found (by n.m.r.) to generate organo-M species.[37] Metallations by BuLi/EtMe$_2$COK were illustrated, e.g. (i) of PhCHMe$_2$, which provided PhCKMeCH$_2$K (10), (10) calculated to have a doubly-bridged structure, decomposes via the radical anion [PhCMe=CH$_2$]$^{-\cdot}$,K$^+$, to [PhCMeKCH$_2$]$_2$, (ii) of PhCH$_2$CH$_2$Ph to give [PhCH=CHPh]$^{-\cdot}$,K$^+$, detected by e.s.r., probably via PhCHKCHKPh, and (iii) of cyclic alkenes.[38]

The compound, 2-(Me$_3$Si)$_2$CH-py, unlike (Me$_3$Si)$_2$CHPh, is lithiated by BuLi in Et$_2$O/hexane. The crystal structure of the thermally-stable product, [2-(Me$_3$Si)$_2$CLi-py]$_2$, revealed[39] that Li is not involved in electron deficient bonding but is bonded to C$_\alpha$ of one ligand and to N of a centrosymmetrically related ligand; there are Li...H(CH$_3$) contacts of 2.20, 2.23 and 2.31 Å [Li-C 2.213(7) Å, Li...Li 2.560(9) Å].

Ambident anionic character was demonstrated in reactions with electrophiles by 9-anthracenyl-CHY$^-$,Li$^+$ (Y=MeO, Me$_3$Si, H or Ar (reaction at C$_{10}$ and C$_{11}$)[40a] and 2-LiCH$_2$-3-Me-naphthalene (reaction at C$_1$ and C$_2$);[40b] a 1,4-Me$_3$Si migration occurs in 9,9-(Me$_3$Si)$_2$-10-Li-9,10-dihydroanthracene.[41] Electrophilic reactions[42] of 1-Li-2-R-tetrahydroisoquinoline [R=(R)-PhCHMe or (S,S)-Me$_3$SiOCPhHCH(CH$_2$OSi-Me$_3$)] can lead to asymmetric products (>90% e.e.).

1.8 Alkenyl, Alkynyl and Related Compounds. Metallation of (Z)-RCH(OMe)CH=CH(SBu^t)(11, R=H) by BuLi/Bu^{t}OK in PhH/THF at -78°C provides (Z)-RCH(OMe)CH=CK(SBu^t), whereas (11, R=H) with Bu^{s}Li in THF at -78°C gives (Z)-RCH(OMe)CK=CH(SBu^t). The (E)-isomers and (11, R=H) undergo allylic deprotonation[43] to [RC(OMe)$\overline{\overline{\ \ }}CH\overline{\overline{\ \ }}$CH(SBu^t)]$^-$ K$^+$. ^{13}C N.m.r. spectra and MNDO calculation were reported[44] on exo-,exo-[(PhCH$\overline{\overline{\ \ }}$)$_2C\overline{\overline{\ \ }}$CHR]$^{2-}$,2Li$^+$ (R=H or Ph), obtained from PhCH$_2$C(CH$_2$R)=CHPh or (PhCH$_2$)$_2$C=CH$_2$. In the MNDO geometry-optimized structure of bridged CH$_2$=CHCH$_2$Li, the allyl protons are distorted from planarity;[45] the $^1J(C^2H^2)$ value recently reported was reinterpreted as arising from a widening of the CCC angle.

The crystal structures of (PhC≡CLi)$_2$.[Me$_2$N(CH$_2$)$_3$NMe$_2$]$_2$ [a dimer with bridging PhC≡C groups; Li-C 2.132(8) and 2.164(12) Å] and of (PhC≡CLi)$_4$[Me$_2$N(CH$_2$)$_6$NMe$_2$]$_2$ {distorted C$_4$Li$_4$ units linked by the diamine to produce high polymer strands with helix-like structure [Li-C$_{av}$ 2.20(1) Å; Li-Li 2.72(2) Å]}. From ^{13}C-^{6}Li n.m.r. data, dimeric and tetrameric PhC≡CLi species were found to exist[46] in solution. A non-fluxional cubic tetrameric structure was indicated[6] for Me$_3$CC≡CLi in THF solution [J(^{13}C-^{6}Li) = 6Hz at <190 K].[47]

Crystal structures of $(Me_3Si)_3C_5H_2Li$ (L-L)(12) (L-L=TMED or pmdet) were obtained, both have η^5-co-ordinated cyclopentadienyl rings (Li-C 2.304(6) to 2.350(6) $\overset{o}{A}$ in (12, L-L=pmdet).[48] [1]H N.m.r. spectra of $Me_3MC_5H_4Li$ and $Me_2M(C_5H_4Li)$ [M=C, Si or Ge][49] as well as $(Me_3Si)_3C_5H_2Li^{48}$ were obtained.

Compounds, $RC{\equiv}CCH_2Li$, prepared from $RC{\equiv}CMe$, Bu^SLi in THF/ cyclohexane at 0°C, are in equilibrium with $RCLi=C=CH_2$. Addition of HMPT (5 eq) leads to the formation[50] of $RCH=C=CHLi$.

2.   Copper, Silver and Gold

2.1 π-Complexes.   E.s.r. spectra[51] were obtained for paramagnetic species produced from metal atoms and ArH, $H_2C=CH_2$ or $HC{\equiv}CH$ in matrices or in a rotating cryostat at 77 K.   Among the compounds reported were (i) (PhH)M (M=Cu, Ag or Au), (ii) $(HC{\equiv}CH)_nCu$ ($\underline{n}$ = 1 or 2) as well as σ-bonded $H\overset{\bullet}{C}=CHM$ (M=Au or Ag) and $H_2C-\overset{\bullet}{C}Au$ and (iii) $(H_2C=CH_2)_n$Au ($\underline{n}$=1 or 2) and (HC≡CH)Au.   Calculations (ab initio SCF) were conducted[52] on $(H_2C=CH_2)_n$Ag and $(HC{\equiv}CH)_n$Ag ($\underline{n}$=1 or 2). The [13]C n.m.r. spectrum of the 1:1 trans-cycloheptene-CuOTf complex   revealed that the pseudo-rotational process of the ligand is little affected on complexation.[53]   Crystal structures were determined[54] of (i) [dibenzo[$\underline{a}$,$\underline{e}$]cyclooctatetraene]$_4$(CuCl)$_4$ [8-membered CuCl ring:ligand π-bonded to Cu via 2 alkene bonds], (ii) [benzocyclooctatetraene]AgClO$_4$ (Ag is co-ordinated to 2 alkene bonds, C-C aryl bond of neighbouring ligand and 2 O) and (iii) $(COT)_n(AgNO_3)_m$ { $\underline{n}$:$\underline{m}$ = 1:1 (infinite chains linked by bridging NO_3), 1:2 (discrete moles) and 2:3 (3-D polymer)}.   Gold carbene complexes, $X_nAu[=C(NMe_2)Ph]$ ($\underline{n}$=1 or 3; X=Cl, Br or I) were obtained[55] from $(CO)_5M$ $[=C(NMe_2)Ph]$ (M=Cr, Mo or W).   The syntheses and spectra of $(\eta^5-C_5Me_5)CuL$ (13, L=CO, PR_3 or $Me_3SiC{\equiv}CSiMe_3$) were reported;   air-sensitive (13, L=CO) decomposes at 0°C to Cu, CO and decamethyl-1,1'-dihydrofulvalene.[56] Initial work was communicated on $(\eta^5-C_5Me_5)Cu$ as a metal ligand fragment 'isolobal' with methylene;   reactions of (13, L=THF) with $[W\{=C(OMe)C_6H_4Me-\underline{p}\}(CO)_5]$ etc. were reported.[57]

2.2 σ-Complexes.   Monomeric ylide complexes $(Ph_3P)_2C-MCl$ (14, M=Cu, Ag or Au)[58a] and $[Ph_3ECHR]Au(C_6F_5)_n$ [E=P or As; R=H, Me or Ph; X=Cl, Br or I, n=1 or 3][58b] were obtained;   discrete molecules of (14, M=Cu) exist [Cu-C 1.906(2) $\overset{o}{A}$, ∠ClCuC 178.2(1) ;   ylidic C is trigonal].[58]   Solution properties of Ag^1 complexes of β-disulphone carbanions were studied; the structure was determined of $K^+(R_2Ag)^-$-H_2O (RH=1,3-dithiane 1,1,3,3-tetroxide),[59] (∠CAgC 180°; Ag-C 2.147 Å). The structures of the following mesityl-metal (Mes-M) species also attracted attention: (i) $(MesCu)_5$:cyclic pentamer, 10-membered puckered ring, bridging aryl group[60a] Cu-C 1.96(2) to 2.06(2) Å; ∠CCuC 146(1)-158(1)°; Cu...Cu 2.437(8) to 2.469(9) Å; (ii) $[Cu_4Mes_4(tht)_2]$: 8-membered puckered ring: 3- and 2- co-ordination Cu with CuC_{av} 2.094(8) and 2.059(4) Å respectively;[60a] (iii) $[Cu(dppe)_2]^+[CuMes_2]^-$: tetrahedral cation;[60b] linear anion Cu-C 1.915(9) Å, and (iv) $(MesAg)_4$ (15): 8-membered ring: Ag-C 2.16(3) to 2.22(3) Å; Ag...Ag 2.733(3) Å; $∠CAgC_{av}$ 167.1(9); $∠AgCAg_{av}$ 77.1(7)°. Compound (15), soluble in most solvents (dimeric in PhH) is light-sensitive but is thermally stable.[60c] Among the nitro- and polyfluoro-phenyl metal derivatives studied were (i) $[Bu_4N][AgArX]$ (Ar=C_6F_5 or C_6Cl_5; X=Ar, AcO, CF_3CO_2 or 8-oxyquinolate),[61a] (ii) $[(PhCH_2)Ph_3P][AuArX]$ (obtained from Ar_2Hg) and ArAuL (Ar=o-, m-, or p-$NO_2C_6H_4$; X=Cl, Br or I; L=Ph_3P or Ph_3As)[61b] (iii) $Z[AuAr_2]$[Z=Ag(tht),$Au(SbPh_3)_4$ or $Au(pdma)_2$; Ar=$C_6F_3H_2$] (Mössbauer spectra);[61c] (iv) $[Ar_2ClAu(dppm)AuPPh_3]ClO_4$, $Ar_3Au(dppm)AuX$ and $Ar_2Au(dppm)ClO_4$ (crystal structure obtained)[61d] (R=C_6F_5, X=Cl or C_6F_5), and (iv) $(Ar_2AuX)_2$,$[Ar_2AuClpy]$, $Ar_2Auacac$, $[Ar_2AuL_2][Ar_2AuCl_2]$ (Ar=C_6F_5; X=Cl, Br, SCN, N_3 or CF_3CO_2).[61e] Linear $[PhC≡CAuX]^-$ (X=Cl, I or C≡CPh) species were produced.[62]

The complex $Me_2Au[HB(Pz)_3]$ (16) was obtained from Me_2AuNO_3 and $K[HB(Pz)_3]$; (16), a fluxional molecule at r.t. in solution (rapid equilibria between pyrazole ring environments), has in the solid state a square planar geometry involving co-ordination of only 2 of the 3Pz rings [Au-C 2.04(3) and 2.05(2); Au-N 2.12(2) and 2.13(1) Å].[63] Other organo-Au species produced were[64] (i) PhC=CPh-CPh=CPhAuLL' (C_4Ph_4AuLL'; L=Cl, L'=py, PPh_3 or Bu^tNC; L,L'=phen and $NEt_4[C_4Ph_4AuCl_2]$ and (ii) $[Me_2AuNH_2]_4$ (8-membered ring) and $[Me_2AuNMe_2]_2$ (4-membered ring with Au-C 2.054(6) and 2.058(6) Å).

N.m.r. spectroscopy was used to study MeCu(I)[65] and MeAg(I)[66] species in solution; e.g. MeCu(I) species were prepared from varying ratios of CuX and MeM (M=Li, MgX or ½Mg) in THF. In the presence of HMPT, [MeCuX]M and $(Me_3Cu)_2Mg$ can be obtained pure from MeMgCl and CuCl. However with X=Br it is only formed in equilibria with $(Me_2Cu)_2Mg$ and $(MeCuBr)_2Mg$. Homocuprates $(Me_2Cu)_2Mg$ were always obtained in mixtures with $(Me_3Cu_2)Mg$ and MeMgX. The use of higher order mixed organocuprates R_2CuXLi_2 was[67] variously made; e.g. with α,β-unsaturated esters, $ArOSO_2CF_3$

3. Bibliography

H. Schmidbaur and U. Deschler, Chem. Ber., 1983, 116, 1386. Phosphonium ylide complexes of alkali metals.

A.I. Meyer, W.F. Reiker, and L.M. Fuentes, J. Am. Chem. Soc., 1983, 105, 2082. Solvent effects on lithiation of $RCH_2NR'CH=NBu^t$.

T. Kauffmann, A. Mitschker, and A. Woltermann, Chem. Ber., 1983, 116, 992. Lithiation of 2-(2-thienyl)-py.

P. Ribereau and G. Queguiner, Tetrahedron, 1983, 39, 3593. Lithiation of 2-(2-furyl)-py.

A.J. Bridges, V. Fredij, and E.C. Turowksi, J. Chem. Soc. Chem. Comm., 1983, 1093. Reaction of RLi with $CH_2=C=C(SPh)SiMe_3$.

T.N. Mitchell and A. Amamria, J. Organomet. Chem., 1983, 252, 47. Reaction of $RCH=C(SnMe_3)_2$ with R'Li.

H.G. Richey,Jr, A.S. Heyn, and W.F. Erickson, J. Org. Chem., 1983, 48, 3821. Assisted addition of RLi to $PhCH=CHCH_2NMe_2$.

I.R. Butler, W.R. Cullen, J. Reglinski, and S.J. Retty, J. Organomet. Chem., 1983, 249, 183. Crystal structure of $[(\eta^5-C_5H_4Li)Fe(\eta^5-C_5H_3LiCHMeNMe_2)]_4-(LiOEt)_4(TMED)_2$.

R.M. Weinstein, W.-L. Wang and D. Seyferth, J. Org. Chem., 1983, 48, 3367; D. Seyferth, R.M. Weinstein, W.-L. Wang, and R.C. Hui, Tetrahedron Lett., 1983, 24, 4907; D. Seyferth, R.M. Weinstein and W.-L.Wang, J. Org. Chem., 1983, 48, 1144; N.S. Nudelman and A.A. Vitale, J. Organomet. Chem., 1983, 241, 143. Reactions of RLi and CO.

D. Lozach, G. Molle, P. Bauer, and J.E. Dubois, Tetrahedron Lett., 1983, 24, 4213. Mechanism of reaction of 3,5,7-Me_3-adamantyl-Li with $Bu^t{}_2CO$

P.K. Freeman and L.L. Hutchinson, J. Org. Chem., 1983, 48, 4705. Stereochemistry of reactions of $(p\text{-}Bu^tC_6H_4C_6H_4Bu^t\text{-}p)^{-\cdot},Li^+$ with RX.

B.J. Kokko and P. Beak, Tetrahedron Lett., 1983, 24, 561; G. Boche, M. Bernheim, and M. Niesner, Angew. Chem. Ind. Ed. Engl., 1983, 22, 53; N.S. Narasimhan and R. Ammanamanchi, Tetrahedron Lett., 1983, 24, 4733.

R.V. Ku and J. San Filippo, Jr, Organometallics, 1983, 2, 1360. Reaction of BuCu with peroxides.

G.M. Bancroft, T.C.S. Chan and R.J. Puddephatt, Inorg. Chem., 1983, 22, 2133. He(I) and He(II) p.e. spectra of Me_3AuPR_3.

References

1 B. Eliasson and U. Edlund, J. Chem. Soc., Perkin Trans. 2, 1983, 1837.
2 C. Tintel, J. Cornelisse and J. Lugtenburg, Rec. Trav. Chim. Pays-Bas, 1983, 102, 231.
3 A. Minsky, A.Y. Meyer, R. Poupko and M. Rabinovitz, J. Am. Chem. Soc., 1983, 105, 2164; A. Minsky, A.Y. Meyer and M. Rabinovitz, Angew. Chem. Int. Ed. Engl., 1983, 22, 45.
4 R. Concepcion, R.C. Reiter and G.R. Stevenson, J. Am. Chem. Soc., 1983, 105, 1778; G.R. Stevenson, R. Concepcion and R.C. Reiter, J. Org. Chem., 1983, 48, 2777; G.R. Stevenson and S.S. Zigler, J. Phys. Chem., 1983, 87, 895.
5 E. Grovenstein, Jr, A.M. Bhatti, D.E. Quest, D. Sengupta and D. van der Veer, J. Am. Chem. Soc., 1983, 105, 6290.
6 A. Maercker, W. Berkulin and P. Schiess, Angew. Chem., Int. Ed. Engl., 1983, 22, 246.
7 D. Seebach, R. Hässig and J. Gabriel, Helv. Chim. Acta, 1983, 66, 308.
8 G. Molle, P. Bauer and J.E. Dubois, J. Org. Chem., 1983, 48, 2975.
9 E. Weiss, G. Sauermann and G. Thirase, Chem. Ber., 1983, 116, 74.
10 C. Eaborn, P.B. Hitchcock, J.D. Smith and A.C. Sullivan, J. Chem. Soc., Chem. Comm., 1983, 827, 1390.
11 H. Schmidbaur, A. Schier, and U. Schubert, Chem. Ber., 1983, 116, 1938.
12 D. Hoell, C. Schnieder and K. Mullen, Angew. Chem., Int. Ed. Engl., 1983, 22, 243.
13 G. Boche and M. Marsch, Tetrahedron Lett., 1983, 24, 3225.
14 A. Maercker and M. Passlack, Chem. Ber., 1983, 116, 710; A. Maercker and R. Stötzel, J. Organomet. Chem., 1983, 254, 1.
15 F.J. Landro, J.A. Gurak, J.W. Chinn, Jr., and R.J. Lagow, J. Organomet. Chem., 1983, 249, 1.
16 A. Maercker, M. Theis, A.J. Kos, P.v.R. Schleyer, Angew. Chem., Int. Ed. Engl. 1983, 22, 733.
17 J.P. Ritchie, J. Am. Chem. Soc., 1983, 105, 2083, see also G. Rauscher, T. Clark, D. Poppinger and P.v.R. Schleyer, Angew. Chem., Int. Ed. Engl., 1978, 17, 276.
18 P.v.R. Schleyer, E.-U. Würthwein, E. Kauffmann and T. Clark, J. Am. Chem. Soc., 1983, 105, 5930.
19 G.W. Klumpp, M. Kool, A.H. Veefkind, M. Schakel and R.F. Schmitz, Rec. Trav. Chim. Pays-Bas., 1983, 102, 542.
20 E.J. Corey and T.M. Eckrich, Tetrahedron Lett., 1983, 24, 3163.
21 D.G. Farnum and T. Monego, Tetrahedron Lett., 1983, 24, 1361.
22 E.J. Corey and T.M. Eckrich, Tetrahedron Lett., 1983, 24, 3165.
23 J. Barluenga, J. Florez and M. Yus, J. Chem. Soc., Perkin Trans. 1, 1983, 3019.
24 G.W. Klumpp, P.J.A. Geurink, A.L. Spek and A.J.M. Duisenberg, J. Chem. Soc., Chem. Comm., 1983, 814.
25 (a) M.A. Vincent and H.F. Schaefer, III, J. Chem. Phys., 1982, 77, 6103; (b) J. Mareda, N.G. Rondan, K.N. Houk, T. Clark, and P.v.R. Schleyer, J. Am. Chem. Soc., 1983, 105, 6997.
26 R.D. Bach, M.L. Braden and G.J. Wolber, J. Org. Chem., 1983, 48, 1509.
27 N.S. Narasimhan and R.S. Mali, Synthesis, 1983. 957.
28 W. Neugebauer, T. Clark and P.v.R. Schleyer, Chem. Ber., 1983, 116, 3283.
29 M. Al-Aseer, P. Beak, D. Hay, D.J. Kempf, S. Mills and S.G. Smith, J. Am. Chem. Soc., 1983, 105, 2080.
30 (a) P. Beak, A. Tse, J. Hawkins, C-W. Chen and S. Mills, Tetrahedron, 1983, 39, 1983; (b) D.L. Comins and J.D. Brown, Tetrahedron Lett., 1983, 24, 5465; (c) R.C. Ronald and M.R. Winkle, Tetrahedron, 1983, 39, 2031; (d) E. Napolitano, E. Giannone, R. Fiaschi and A. Marsiw, J. Org. Chem., 1983, 48, 3653; (e) D. Hellwinkel, R. Lenz and F. Lämmerzahl, Tetrahedron, 1983, 39, 2073; (f) J.N. Reed and V. Snieckus, Tetrahedron Lett., 1983, 24, 3795; M.P. Sibi and V. Snieckus, J. Org. Chem., 1983, 48, 1935; see D. Hellwinkel, F. Lämmerzahl and G. Hofmann, Chem. Ber., 1983, 116, 3375.

31 (a) P. Breant, F. Marsais and G. Queguiner, Synthesis, 1983, 822; (b) S.L. Taylor, D.Y. Lee and J.C. Martin, J. Org. Chem., 1983, 48, 4156; (c) F. Marsais and G. Queguiner, Tetrahedron, 1983, 39, 2009; (d) J.A. Turner, J. Org. Chem., 1983, 48, 3401.
32 M. Uemura, N. Nishikawa, K. Take, M. Chnishi, K. Hirotsu, T. Higuchi, and Y. Hayashi, J. Org. Chem., 1983, 48, 2349.
33 N.F. Masters and D.A. Widdowson, J. Chem. Soc., Chem. Comm., 1983, 955.
34 H. Hope and P.P. Power, J. Am. Chem. Soc., 1983, 105, 5320.
35 T. Frejd and T. Klingstedt, J. Chem. Soc., Chem. Comm., 1983, 1021.
36 J. Klein, Tetrahedron, 1983, 39, 2733.
37 G. Boche and H. Etzordt, Tetrahedron Lett., 1983, 24, 5477.
38 D. Wilhelm, T. Clark, P.v.R. Schleyer, J. Chem. Soc., Chem. Comm., 1983, 211; D. Wilhelm, T. Clark, T. Friedl and P.v.R. Schleyer, Chem. Ber., 1983, 116, 751; D. Wilhelm, T. Clark, J.L. Corutneidge and A.G. Davies, J. Chem. Soc., Chem. Comm., 1983, 213.
39 R.I. Papasergio, C.L. Raston and A.H. White, J. Chem. Soc., Chem. Comm., 1983, 1419.
40 (a) M. Takagi, M. Nojima and S. Kusabayashi, J. Am. Chem. Soc., 1983, 105, 4676; T.A. Enger and H. Shechter, Tetrahedron Lett., 1983, 24, 4645; (b) R.H. Mitchell, T.W. Dingle and R.V. Williams, J. Org. Chem., 1983, 48, 903.
41 M. Daney, R. Lapouyade and H. Bouas-Laurent, J. Org. Chem., 1983, 48, 5055.
42 A.I. Meyers and L.M. Fuentes, J. Am. Chem. Soc., 1983, 105, 117.
43 C.B. Bi Ekogha, O. Ruel and S.A. Julia, Tetrahedron Lett., 1983, 24, 4825, 4829.
44 D. Wilhelm, T. Clark and P.v.R. Schleyer, Tetrahedron Lett., 1983, 24, 3985; D. Wilhelm, T. Clark, P.v.R. Schleyer, K. Buckl, and G. Boche, Chem. Ber., 1983, 116, 1669.
45 G. Decher and G. Boche, J. Organomet. Chem., 1983, 259, 31; see also T. Clark, C. Rohde and P.v.R. Schleyer, Organometallics, 1983, 2, 1344.
46 B. Schubert and E. Weiss, Chem. Ber., 1983, 116, 3212; Angew. Chem., Int. Ed. Engl., 1983, 22, 496; R. Hassig and D. Seebach, Helv. Chim. Acta, 1983, 66, 2269.
47 G. Fraenkel and P. Pramanik, J. Chem. Soc., Chem. Comm., 1983, 1527.
48 P. Jutzi, E. Schlüter, C. Krüger and S. Pohl, Angew. Chem., Int. Ed. Engl., 1983, 22, 994.
49 A. Kopf and N. Klouras, Chem. Scripta, 1983, 22, 64.
50 C. Huynh and G. Linstrumelle, J. Chem. Soc., Chem. Comm., 1983, 1133.
51 J.H.B. Chenier, J.A. Howard, B. Mile and R. Sutcliffe, J. Am. Chem. Soc., 1983, 105, 788; P.H. Kasai, ibid., 1983, 105, 6704; A.J. Buck, B. Mile and J.A. Howard, ibid., 1983, 105, 3381.
52 D. Cohen and H. Basch, J. Am. Chem. Soc., 1983, 105, 6980.
53 G.M. Wallraff, R.H. Boyd and J. Michl, J. Am. Chem. Soc., 1983, 105, 4550.
54 T.C.W. Mak, H.N.C. Wong, K.H. Sze and L. Book, J. Organomet. Chem., 1983, 255, 123; T.C.W. Mak, W.C. Ho and N.Z. Huang, ibid., 1983, 251, 413; W.C. Ho and T.C.W. Mak, ibid., 1983, 241, 131; idem., ibid., 1983, 243, 233; T.C.W. Mak, ibid., 1983, 246, 331.
55 E.O. Fischer, M. Bock and R. Aumann, Chem. Ber., 1983, 116, 3618.
56 D.W. Macomber and M.D. Rausch, J. Am. Chem. Soc., 1983, 105, 5325.
57 G.A. Carriedo, J.A.K. Howard and F.G.A. Stone, J. Organomet. Chem., 1983, 250, C28.
58 (a) H. Schmidbaur, C.E. Zybil, G. Müller and C. Kruger, Angew. Chem., Int. Ed. Engl., 1983, 22, 729; (b) R. Uson, A. Laguna, M. Laguna and A. Uson, Inorg. Chim. Acta, 1983, 73, 63.
59 J.R. DeMember, H.F. Evans, F.A. Wallace and P.A. Tariverdian, J. Am. Chem. Soc., 1983, 105, 5647.
60 (a) S. Gambarotta, C. Floriani, A. Chiesi-Villa and C. Guastini, J. Chem. Soc. Chem. Comm., 1983, 1156; (b) P. Leoni, M. Pasquali and C.A. Ghilardi, ibid. 1983, 240; (c) S. Gambarotta, C. Floriani, A. Chiesi-Villa and C. Guastini, ibid., 1983, 1087.

61 (a) R. Uson, A. Laguna, and J.A. Abad, J. Organomet. Chem., 1983, 246, 341; (b) J. Vicente, A. Arcas and M.T. Chicote, ibid., 1983, 252, 257; (c) K. Moss, R.V. Parish, A. Laguna, M. Laguna and R. Uson, J. Chem. Soc., Dalton Trans., 1983, 2071; (d) R. Uson, A. Laguna, M. Laguna, E. Fernandez, M.D. Villacampa, P.G. Jones and G.M. Sheldrick, ibid., 1983, 1679; (e) R. Uson, A. Laguna, M. Laguna and M. Abad, J. Organomet. Chem., 1983, 249, 437.

62 O.M. Abu-Salah and A.R. Al-Ohaly, J. Organomet. Chem., 1983, 255, C39; Inorg. Chim. Acta, 1983, 77, L159.

63 A.J. Canty, N.J. Minchin, J.M. Patrick and A.H. White, Aust. J. Chem., 1983, 36, 1107.

64 R. Uson, J. Vicente, M.T. Chicote, P.G. Jones and G.M. Sheldrick, J. Chem. Soc., Dalton Trans., 1983, 1131; H.N. Adams, U. Grässle, W. Hiller and J. Strähle, Z. Anorg. Allg. Chem., 1983, 504, 7.

65 H. Westmijze, A.V.E. George and P. Vermeer, Rec. Trav. Chim. Pays-Bas, 1983, 102, 322.

66 D.E. Bergbreiter, T.J. Lynch and S. Shimazu, Organometallics, 1983, 2, 1354.

67 B.H. Lipshutz, J.A. Kozlowski and R.S. Wilhelm, J. Org. Chem., 1983, 48, 546; B.H. Lipshutz, Tetrahedron Lett., 1983, 24, 127; J.E. McMurray and S. Mohanraj, ibid., 1983, 24, 2723; B.H. Lipshutz, D. Parker, J.A. Kozlowski and R.D. Miller, J. Org. Chem., 1983, 48, 3334.

2

Group II: The Alkaline Earths and Zinc and its Congeners

BY J. L. WARDELL

1. Beryllium

Phosphorus ylides, $Ph_3P=CHR$ (1, R=H or Me), react with $BeCl_2$ to give complexes containing trigonal Be, e.g. (1, R=H) provides[1] the chlorine bridged dimer $[Ph_3PCH_2MCl_2MCH_2PPh_2]^{2+}$,$2Cl^-$ (2, M=Be). Reaction of Ph_2Be with 4,4'-bipyridine (L-L) in the presence of K in THF leads[2] to the blue-coloured radical cation $[PhM-(L-L)-MPh]^{+\cdot}$ (3, M=Be). Theoretical considerations were given[3] to the mass spectrum of Me_2Be (MNDO) and to the singlet-triplet states of H_2CBe and HCBeH.

2. Magnesium

Pyrophoric anthracene^{2-}, Mg^{2+}.3THF (Anth^{2-}, Mg^{2+}.3THF) has been isolated[4] from reactions of $MgBr_2$ and AnthNa or of Mg, $MgBr_2$ and Anth. Equilibrium (1) lies to the left at 40°C and to the right at -40°C

$$MgBr_2 + Anth + Anth^{2-}, Mg^{2+} \rightleftharpoons 2Anth^{-\cdot}, MgBr^+ \qquad (1)$$
$$\text{yellow} \qquad\qquad \text{blue}$$

The carbon-halogen bond strength has a crucial influence on the competitive reactions (via radical and ion-radical mechanisms) which occur between co-condensed Mg atoms and excess RX (R=alkyl or Ph; X=F, Cl, Br or I).[5] A simple synthesis of Bu_2Mg was reported.[6] The preparation of organomagnesiums from 3-haloalkyl ethers, e.g. $Br(CH_2)_3OPh$, has been achieved at -78°C using reactive Mg (from $MgCl_2$,Li and naphthalene).[7] Complexes, (2, M=Mg)[1] and (3, M=Mg)[2] were obtained. Additions of RMgX to isolated multiple bonds in functionally-substituted compounds have been reviewed.[8] The following transition metal catalysed reactions were reported[9] (i) Bu^iMgBr to $Me_3SiC{\equiv}CCRHOH$ (catalyst Cp_2TiCl_2, product (Z)-$Me_3SiCMgBr{=}CHCRHOMgBr$, (ii) Et_2Mg to $RCH{=}CH_2$ (catalyst Cp_2ZrCl_2, product $RCHEtCH_2MgX$) and (iii) ArMgX to $RC{\equiv}CCH{=}CH_2$ [Ni(II) catalyst, cis-addition to triple bond]. Hydrometallation of 1-octene[10] using RMgX(X=R, H or halide), is catalysed by Cp_2TiCl_2.

An ab-initio (Hartree-Fock) calculation of magnesocene (4) indicated that the Mg bonding with the ring π-ring system mainly involves Mg 3s and 3p orbitals. The charge-separation in (4) is slightly larger than that found in ferrocene. Comparisons were made between the theoretical findings for (4) and X-ray and electron diffraction data.[11] Other calculations were made on MeMgCl (5) (at 6-31G* level), and on it's reaction with R_2CO, and on $MeMg_2X$ (6, X=F or Cl). Compounds (5) and (6, X=Cl) have comparable stability;[12] there is a strong Mg-Mg bond in (6).

Crystal structures of EtMgBr[(-)-sparteine] (7) [Mg-C 2.24(3) Å] and Bu^tMgCl [(-)-sparteine] (8) [Mg-C 2.19(2) Å] were determined.[13] In (7) and (8), the Mg atoms are tetrahedral with the bidentate sparteine ligands assuming all-chair conformation; (7), unlike (8), initiates asymmetric selective polymerisation of racemic methacrylates. Of interest, the geometric relationship of the alkyl and halide groups in (7) is reverse to that in (8). The mean C-Mg bond dissociation energy in monomeric $(Me_3CCH_2)_2Mg$ was determined to be 133.7±5.0 kJ mol^{-1}; the bond energy ($\bar{D}_B$) of the bridging Me_3CCH_2-Mg bonds in the dimer was found[14] to be 205±3 kJ mol^{-1}.

The configurational stability of a cyclopropyl-MgX containing a metallated 2-$HOCH_2$ group has been shown:[15] the Grignard reagent from a cis:trans mixture of 2-Br-3-$HOCH_2$-1,1-Me_2-cyclopropane provides, on deuterolysis, products having the same cis:trans-ratio. A review of electron transfer processes in RMgX reactions has been published.[16] The $R_2N\cdot$ radicals, triphenylimadazolyl and $Ph_2N\cdot$ (obtained from R_2NNR_2), react[17] with R'MgX to give R_2NMgX and R·'. Addition of crotyl-MgCl to RCHO in the presence of $AlCl_3$ leads, after hydrolysis, predominately (>70%) to (E)-RCH(OH)CH_2=CHMe.[18] The use of chiral solvents, e.g. EtCHMeOMe and 1-Pr^i-2-MeO-4-Me-cyclohexane, results in asymmetric induction in the reactions of RMgX with RCHO (R=Ph or Bu^t) and cinnamates.[19]

The following chiral transition metal compounds were used as catalysts in coupling reactions:-[20] (i) $[Ph_2PCMeHCMeHPPh_2]NiCl_2$ for RMgX with allyl phenyl ethers (up to 97% e.e.), (ii) $[Me_2NCHRCH_2PPh_3]NiCl_2$, >70% e.e. for reactions of ArMeCHMgCl and CH_2=CHBr, (iii) $[Ph_2PCH_2CHMe]_2NiCl_2$ up to 50% e.e. for reactions of ArX and Bu^sMgX, and (iv) [1-Me_2NCHMe-2-Ph_2P-ferrocene]$PdCl_2$ 18% e.e. for PhC≡CCHPhSiMe_3, from PhC≡CBr and Me_3SiCHPhMgBr.

3 Calcium and Barium

Co-condensation of ArX and Ca vapour onto cooled surfaces (77 K) produce solvent-free ArCaX (9, X=F, Cl or Br); reactions of (9), which ignites in air, with electrophiles were reported.[21] Dicumylbarium (10), $(PhCMe_2)_2Ba$, was obtained from $PhCMe_2OMe$ and finely divided Ba in THF; (10) exists in THF or THP solution as triple-ion species in equilibrium with free ions and ion-pairs.[22]

4 Zinc and Cadmium

Synthesis and n.m.r. spectra were reported[23] for $(\underline{o}\text{-}Me_2NC_6H_4CH_2)_2M$ (11, M=Zn or Cd) and $(\underline{o}\text{-}Me_2NCH_2C_6H_4)_2M$ (12, M=Zn or Cd). Compounds, (11) and (12), thermally stable but light- and air-sensitive, were produced from the appropriate organolithium and MCl_2; in addition (12, M=Zn) was obtained from (12, M=Hg) and Zn. Spectra (m.s., i.r. and ^{13}C n.m.r.) and the thermal decomposition were reported[24] for bis-3-nortricyclylzinc (R_2Zn), obtained from RMgCl and $ZnCl_2$ and its bipy complexes, R_2Zn.bipy(red) and $(R_2Zn)_2$.bipy (yellow).

Ultrasound was used[25a] in a one-pot synthesis of Ar_2Zn, from ArBr, Li wire and $ZnBr_2$ in dry Et_2O or THF. Perfluoroalkylzincs were similarly prepared,[25b] as well as from reactions of R_FI with Zn/Cu in sulpholane.[25c]

Alkenylzincs[26] have been obtained (i) by the catalysed <u>cis</u>-addition of R_2Zn to alkynes, equation (2) (>98% stereoselectivity and >70% regioselectivity) and (ii) from reaction of vinylic-Li, e.g. CH_2=C(OEt)Li, with $ZnCl_2$.

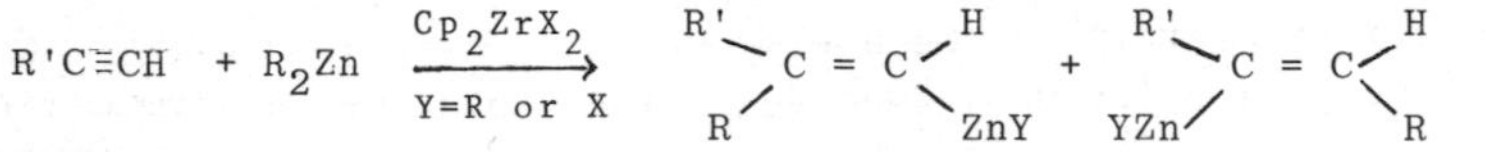

$$R'C{\equiv}CH + R_2Zn \xrightarrow[Y=R\ \text{or}\ X]{Cp_2ZrX_2} \underset{\text{major product}}{R'(R)C{=}C(H)ZnY} + R'(YZn)C{=}C(H)R \qquad (2)$$

Reactions of Et_2Zn with Bu^tN=CHCH=NBu^t and 2-(Bu^tN=CH)-pyridine have been reviewed;[28] Zn-N bonded radical cation complexes (3, M=Zn) were obtained.[2] Crystal structures were determined for the dimeric compounds (i) $(MeZnNPh_2)_2$ [bridging NPh_2 groups; $(Zn\text{-}C)_{av}$ 1.948(12) Å; $(Zn\text{-}N)_{av}$ 2.072(8) Å][29] (ii) EtZnOCMe=$CHNBu^t$-Et, obtained from Et_2Zn and t-Bu-N=CHCOMe [there is a central Zn_2O_2 ring and the ligands form N, O chelates][30] and (iii) the Reformatsky reagent, $(BrZnCH_2CO_2Bu^t.THF)_2$ (13) [tetrahedral-Zn, co-ordinated by 2 O, Br and C; bridging-$CH_2C(OBu^t)O$- groups producing a 8-membered, non-planar ring: $(Zn-C)_{av}$ 2.03 Å, $(Zn-O)_{av}$ 2.01 Å; ∠ $OZnC_{av}$ 111.0°, ∠ ZnOC 125.5°]. The dimeric structure of $BrZnCH_2CO_2R$ (R=Bu^t and Et) persists in THF, py or Me_2CO solutions.[31]

Various transition-metal catalysed coupling reactions of organozincs were studied; these included the $(Ph_3P)_4Pd$ catalysed couplings of (i) $Ph(CH_2)_2ZnCl$ to vinylic and aryl iodides, (ii) $CH_2=CH(CH_2)_2ZnCl$ to ArBr, (iii) $Me_3SiC{\equiv}C(CH_2)_2ZnCl$ to ArI ,[32a] (iv) PhZnCl to (R)-$HC{\equiv}CCPhH(OCOCF_3)$, producing PhCH=C=CHPh with anti-stereoselectivity,[32b] (v) RZnCl to $H_2C=C=C(OMe)CR^1{}_2X$ (X= OAc or O_2SMe; R=vinyl, $Me_3SiC{\equiv}C$ or Ph; R=H or Me), providing $H_2C=CRC(OMe)=CR^1{}_2$,[32c] (vi) RCH=CYZnCl to R^1X (Y=OEt, SEt or $SiMe_3$; R^1=alkenyl or aryl)[27] and (vii) RZnY (R=vinyl, aryl, alkenyl, alkynyl or $PhCH_2$) to R^1COCl.[32d]

Other catalysts used include $Ni(acac)_2$ (for reactions of ArZnX with $BrCH_2CO_2Et$[33] or α,β-enones[25a]), $Cu(acac)_2$[34] (for the reaction of Reformatsky reagents with allylic halides), and chiral [1-Me_2NCHMe-2-Ph_2P-ferrocene]$PdCl_2$ (for reactions of ArRCHZnCl with vinylic bromides (up to 86% e.e.).[35]

5 Mercury

5.1 General. The impact of solvation on the various coupling constants in the ^1H, ^{13}C and ^{19}F n.m.r. spectra of R_2Hg compounds (R=benzylic, allylic, alkenyl, alkynyl, alkyl, aryl or polyfluoroalkyl) was studied.[36]

5.2 Alkyl- and Arylmercurials. Extensive use has been made of the hydrazone route to α-acetoxyalkylmercurials,[37] Scheme 1

$$RR^1C=O \xrightarrow{i} RR^1C=NNH_2 \xrightarrow{ii} RR^1C(HgOAc)OAc$$

R,R^1=alkyl, acyl or cycloalkyl.

Scheme 1. (i) N_2H_4; (ii) HgO,$Hg(OAc)_2$,dioxane.

Compounds (11, M=Hg) and (12, M=Hg) were produced; the crystal structure determination of (12, M=Hg) revealed strong Hg-C bonds [2.10(2) Å] and weak Hg-N bonds [2.89(1) Å] with bond angles $CHgC^1$ and $NHgN^1$ 180° and CHgN 71(1)°. The bonding in (12,M=Hg) was considered to be due to either (i) d-s mixing or (ii) donation into a single acceptor orbital ($6p_x$ or $6p_y$) of electron density from two donor np orbitals.[23] The products of reaction of

4-Bu-4-Me-1-methylenecyclohexa-2,5-dienes with Hg^{II} salts are the benzylic mercurials, 3-Bu-4-$MeC_6H_3CH_2HgX$ (14) and 4-Bu-3-$MeC_6H_3CH_2HgX$ (15); the product ratio [(14):(15) = 5:1], indicates that the 1,2-shift of the Bu group occurs more readily.[38] Direct reaction of $C_6X_5CH_2Br$ with Hg leads to $(C_6X_5CH_2)_2Hg$ [16, X=Cl or Br] in Et_4NBr/DMF or to $C_6Br_5CH_2HgBr$ (17) in PhH; (16) and (17) are light-sensitive compounds having appreciable thermal stability,[39] e.g. (16, X=Cl) is sensitive to daylight and decomposes at 250-300°C, mainly to $C_6X_5CH_3$.

Isotopic chemical shifts (δ_{is}) of ^{199}Hg nuclei were measured for replacement of ^{28}Si by ^{29}Si and/or of ^{12}C by ^{13}C, directly bound to Hg in $Et_3SiHgEt$ (e.g. δ_{is} 0.07±0.02 and 0.28±0.03, respectively) and in R_2Hg [R=Me, Et, $(CF_3)_2CF$ or Et_3Si]; there is a low field displacement of $\delta\ ^{199}Hg$ on replacement with the heavier isotope,[40a] $\delta\ ^{199}Hg$ was measured for a variety of $Hg(SiRR^1R^2)_2$ species.[40b]

5.3 Oxymercuration and Related Reactions. Methoxymercuration, and subsequent demercuration, of the exocyclic double bonds of rigid 2,4-Me_2-7-methylenetetracyclo[3.3.0.0.^{2,8}0^{4,6}]oct-3-one and the *exo*- and *endo*-3-ols were studied.[41] Only for the *endo*-3-ol derivative was the MeO group stereospecifically incorporated into an *endo*-site. Amidomercuration of alkenes (and demercuration) was further reported on;[42] H_2NCO_2Et also reacts with $R^1R^2C{=}CHR^3$ and $Hg(NO_3)_2$ to give, after $NaBH_4$ treatment, $EtO_2CNHCR^1R^2CH_2R^3$. *syn*-Addition of $HgCl_2$ occurs to RCHOHC≡CCOMe (R=H or Me), equation (3); the initial products readily dehydrate to 3-furyl- or dihydrofuryl-mercurials on treatment with dilute acid or simply on attempts to recrystallise.[43] In contrast *anti*-addition

$$HOCRR^1C\ CCOR^2 + HgCl_2 \longrightarrow (Z)\text{-}HOCHR^1CCl = C(HgCl)COMe \qquad (3)$$

(18, R=H, R^1=H or Me, R^2=Me) (19)

occurs to (18, $R{=}R^1{=}R^2$=Me; $R{=}R^1$=H, R^2=OMe).[42] The RC≡CH: $Hg(OAc)_2$ mole ratio (m:n) has an influence on the products of reactions with amines R^1NH_2, in CH_2Cl_2 at r.t.; with m:n >2, $(RC{\equiv}C)_2Hg$ results while at m:n <2, $RC(CH_2HgOAc){=}NR^1$ is formed.[44] The reaction of PhC≡CH and $Hg(OAc)_2$ (5:9 mole ratio) in $CHCl_3$ at r.t. provides $PhC(OAc){=}C(HgOAc)_2$. Reactions of bicyclo[1.1.0] butane with HgX_2 (X=Cl, Br or I) at 20°C in CH_2Cl_2 produces[45] 1-X-3-XHg-cyclobutane (zero-bridge cleavage) and 1-XHg-2-XCH_2-

cyclopropane, the latter is slowly converted to $XHgCH=CH(CH_2)_2X$. The zero-bridge in 1,3-dehydroadamantane is also cleaved by $Hg(OAc)_2$ in EtOH to give, after NaCl treatment, 1-EtO-3-ClHg-adamantane; coupling constants in the n.m.r. spectrum of this conformationally rigid molecule were measured.[46]

5.4 Perfluoro-organomercurials. Mercuration[47a] of C_6F_5H occurs using HgX_4^{2-} (X=Br or I),OH^- in aq DMF at r.t. and provides $(C_6F_5)_2Hg$; $(C_6F_{13}CF=CF)_2Hg(R_2Hg)$ was prepared[47b] from RMgBr. Decarboxylation of the inner-salt of α-hydroxymercurihexafluoro-isobutyric acid $[^+HgC(CF_3)_2CO_2^-]$ in pyridine provides $[(CF_3)_2CHg]_5$.-2py.$2H_2O$ (20); (20) has a cyclic structure, in which the 5 Hg atoms in the 10-membered ring are coplanar [C-Hg 2.07(3) to 2.15(3) Å; ∠ $CHgC_{av}$ 176(1)°].[48] A force-field study of CX_3HgY (X=F or Cl, Y=Cl, Br or I) was reported.[49]

5.5 Gem Di- and Poly-mercuriated Alkanes. The following spectral studies of $CH_{4-n}(HgX)_n$ were investigated:[50a] (i) ^1H, ^{13}C and ^{199}Hg n.m.r. spectra ($1 \leqslant n \leqslant 4$; X=Cl, Br, I or CN) [increasing deshielding of all nuclei with increasing n; δ^1H and $\delta^{13}C$ but not $\delta^{199}Hg$ for the same n increase in the sequence X=I > Br > Cl > CN (i.e. with increasing *trans*-effects of X); ^1J(C-H) increases and ^2J(Hg-H) decreases with increasing n], (ii) u.v. spectra ($1 \geqslant n \geqslant 4$; X=Cl, Br, CN or SCN) [in MeOH, the bands are charge-transfer transitions from non-bonding (X-centred)HOMO's to poly-centre LUMO's between Hg atoms] and (iii) i.r. and Raman spectra (n=4; X=CN, AcO and CF_3CO_2). The crystal structure of $C(HgCl)_4$.DMSO was determined; $C\text{-}Hg_{av}$ 2.070 Å, $Hg\text{-}Cl_{av}$ 2.332 Å, ∠ $CHgCl_{av}$ 174.3°; DMSO is co-ordinated *via* O to 2 Hg atoms of one $C(HgCl)_4$ unit and to another of a neighbouring unit. There are also intermolecular Cl--Hg (strong) and S--Hg (weak) interactions.[50b] The structures of $(ClHg)_3CCHO$.DMSO and $[OHg_3CCHO]NO_3.H_2O$ [product of mercuriation of CH_3CHO (or HC≡CH) in aq $Hg(NO_3)_2$] were also determined.[51]

5.6 Alkenyl Compounds. Determinations were made of the crystal structures[52] of *trans*-$(PhCH=CH)_2Hg$ [Hg-C 2.07(4) Å; ∠ CHgC 178°] and camphene-4-mercury chloride [Hg-C 2.08 Å; ∠ CHgCl 175°]. Solvent effects[53] on the ^1H and ^{13}C n.m.r. spectra of *cis*- and *trans*-ClCH=CHHgX and isomerization of *cis*-ClCH=CHHgR were studied. σ,π-Conjugation in allyl-mercurials and related compounds produces a sharp increase in the shielding of the Hg nucleus.[54] The photo electron spectrum of CpHgCl was studied; the σ,π-interaction of the σ-MO of the C-Hg bond and the $2b_i$-π-MO results in a decrease in energy of the σ-orbital and an increase in that of the $2b_i$-π MO by 0.85eV.[55]

5.7 Other Compounds. Complexes, $[MeHg(L)]NO_3$ [e.g. L=bis(N-methylimidazol-2-yl)-methanol (tmm)] were produced. The structure of $[MeHg(tmm)]NO_3$ was determined; one N-methylimidazol-2-yl group is co-ordinated to Hg [Hg--N 2.125(7) Å; ∠ 176.1(5)°; there is in addition a weak Hg---OH interaction].[56] Studies involving Hg-S bonded compounds included; (i) determination of the solid state structures of $MeHg^+$ complexes of 1-methylimidazoline-2-thione(BH)-MeHgB [C-Hg 2.05(10) Å; S-Hg 2.338(7) Å, ∠ SHgC 176(2)°] and $[MeHg(HB)]^+NO_3^-$ [C-Hg 2.09(1) Å; S-Hg 2.382(2) Å, ∠ SHgC 176.1(4)°, Hg---O contacts][57] (ii) RHg^+ complexes of 2-thiouracil: isomeric solid state complexes with bonding to S and N; only MeHg-S bonding in solution,[58] (iii) $\delta\,^{199}Hg$ and vibrational spectra[59a] of PhHgSAr; νHgS for PhHgSPh 365(Raman) and $367cm^{-1}$(ir); (iv) $\delta\,^{19}F$ for $ArHgSC_6H_4F$-p and $ArSHgC_6H_4F$-p,[59b] (v) formation constants, β_{mlh}, for $MeHg^+$ complexes with thiols, such as $HSCH_2CH_2OH$, $HSCH_2CO_2R$, $HSCH(CO_2H)CH_2CO_2H$, L-cysteine, D,L-penicillamine, glutathione, thiocholine and 4-HS-1-Me-piperidine (equation 4)[60]; the 3 values for glutathione, $HO_2CCH(NH_2)CH_2CH_2CONHCH(CH_2SH)CONHCH_2CO_2H$,

$$mMeHg^+ + \ell RS^{n-} + hH^+ \longrightarrow [(MeHg)_m(RS)_l H_h]^{m+h-nl} \qquad (4)$$

$$\beta_{mlh} = [\{(MeHg)_m(RS)_l H_h\}^{m+h-nl}]/[MeHg^+]^m[RS^{n-}]^l[H^+]^h$$

are log β_{110} 15.99(1), log β_{111} 25.24(1) and log β_{112} 28.68(2).

5.8 Reactions. Cleavage of R_2Hg (R=Me or Ph) by F_2 to RHgF occurs in the liquid phase at a low temperature;[61a] XeF_2 also[61b] cleaves R-Hg bonds in R_2Hg (R=alkynyl, Ar or $PhCH_2$).

Reactions of R·, generated from RHgX on treatment with $NaBH_4$ or $NaBH(OMe)_3$,[37,62] (and also on photolysis)[63], especially with electron deficient alkenes, $Y^1Y^2C{=}CZ^1Z^2$ (which provide $RY^1Y^2CCZ^1Z^2H$) have attracted attention; the radicals, R·, used include (i) $\cdot CRR^1OAc$;[34] (ii) $MeCONHCHR^2CH_2\cdot$,[62a] and (iii) $ROCHR^1CHR^2\cdot$,[62b] under phase transfer conditions, equation (5).

$$R^1CH{=}CHR^2 + ROH + HgX_2 \longrightarrow ROCHR^1CHR^2CH_2CH_2Z \qquad (5)$$
$$Z{=}CN \text{ or } CO_2Me$$

Reagents; (i) NaOH, Triton X-100; (ii) $CH_2{=}CHZ, CH_2Cl_2$; (iii) $NaBH_4, OH^-$, 0°C

The β-peroxy-substituted radicals, $ROOCHR^1CR^2R^3\cdot$, generated from $ROOCHR^1CR^2R^3HgOAc$ (R^3=HO or MeCO) on reduction with $NaBH_4$ form the useful 1-R^3-epoxides.[64] Free radical chain reactions occur between RHgX and QY (Q-Y=PhS-SPh, PhSe-SePh, PhTe-TePh or $ArSO_2$-Cl) to provide[65] RY, and between $RCHOR^1CH_2HgOAc$ and $Me_2CNO_2^-, Li^+$, which gives on irradiation[66] $RCHOR^1CH_2CMe_2NO_2$. Catalysis by transition metal compounds of RHgX reactions have been well studied; the reactions include (i) carbonylation of (19), catalysed[43] by Li_2PdCl_4, (ii) coupling of ArHgX and vinylic halides, catalysed[67] by $RhCl(PPh_3)_3$, (iii) acylation of ArHgX by RCOCl, catalysed[68a] by $(Ph_3P)_2PdPhI/I^-$ or $[Rh(CO)_2Cl]_2$, (iv) demercuration[68b] of ArHgX, using $(Ph_3P)_2PdPhI/I^-$ and (v) decomposition[69] of $HOCH_2CH_2HgCl$ by $[\{Rh(C_5Me_5)\}_2(OH)_3]Cl$.

6 Bibliography

Magnesium.- W. Oppolzer, Kagaku, Zokau (Kijoto), 1983, 35 (Chem. Abstr. 1983, 99, 212552. Grignard reagents from allylic halides, without side reactions.

J.G. Duboudin, B. Jousseaume, and E. Thoumazeau, Bull. Chim. Soc. Fr., 1983, Part II, 105. Reactions of $MgCH{=}CRCH_2O$.

C. Alvarez-Ibarra, O. Arjona, R. Perez-Ossorio, A. Perez-Rubalcaba, M.L. Quiroga, and M.J. Santesmases, J. Chem. Soc., Perkin Trans. 2, 1983, 1645. Stereoselectivity in reactions of RMgX with PhCHMeCOR.

M. Tiecco, L. Testaferri, M. Imgoli, and E. Wenkert, Tetrahedron, 1983, 39, 2289. Ni^{II} catalysed coupling of RMgX to $ArSR^1$.

E. Wenkert, J.B. Fernandes, E.L. Michelotti, and C.S. Swindell, Synthesis, 1983, 701. Ni^{II} catalysed coupling of MeMgBr to ArCH=CHSMe.

M.R.H. Elmoghayer, P. Groth, and K. Undheim, Acta Chem. Scand., Part B, 1983, 37, 109; M.R.H. Elmoghayer and K. Undheim, ibid., 1983, 37, 160. Ni^{II} catalysed coupling of RMgX to halopyrimidines.

F. Babudri, S. Florio, L. Ronzini, and M. Aresta, Tetrahedron, 1983, 39, 1515. Ni^{II} catalyst coupling of RMgX to 2-X-benzothiazoles.

L. Jalander and M. Broms, Acta Chem. Scand., Part B, 1983, 37, 173. CuI catalysed coupling of RMgX to $ArCCl{=}CHCO_2Et$ (complete retention for (Z)-isomer).

M. Julia, A. Righini-Tapie, and J.-N. Vepeaux, Tetrahedron, 1983, 39, 3283. $Cu(acac)_2$ catalysed coupling of RMgX to allylic sulphones.

S. Nunomoto, Y. Kawakami, and Y. Yamashita, J. Org. Chem., 1983, 48, 1912. Li_2CuCl_4 catalysed coupling of CH_2=CH-CMgCl=CH_2 to RX.

G.A. Molander, B.J. Rahn, D.C. Saubert, and S.E. Bonds, Tetrahedron Lett., 1983, 24, 5449. Fe^{III} catalysed coupling of ArMgX to alkenyl halides.

G. Boche, M. Bernheim, and M. Niessner, Angew. Chem. Int. Ed. Engl., 1983, 22, 53. Amination of PhC≡CMgBr using Me_2NX (X=Ph_2PO or $MeSO_3$).

R.W. Hoffmann and K. Ditrich, Synthesis, 1983, 107. Formation of ArOH from ArMgBr and $(RO)_2BOOBu^t$.

F. Fujiwara, Y. Kurita, and T. Sato, Chem. Lett., 1983, 1537. RMgX → ketoximes.

Zinc.- M. Bourhis, J.J. Bosc, and R. Golse, J. Organomet. Chem., 1983, 256, 193. RZnX + $R^1OCH(CO_2Me)NEt_2$.

G. Rousseau and J. Drouin, Tetrahedron, 1983, 39, 2307. RCN+R_2C=$CHCH_2ZnBr$.

C.J. Chen and R.M. Osgood, Mater. Res. Soc. Symp. Proc., 1983, 17. Photolysis of Me_2M(M=Zn, Cd or Hg) in vapour or adsorbed phase.

A.M. Ob'edkov, B.V. Zhuk, G.A. Domrachev, V.T. Bychkov, and I.V. Lomakova, Khim. Elementoorg. Soedin, 1982, 27. Decomposition of Et_2M (M=Zn, Cd or Hg) in an electrode free high frequency discharge.

Mercury.- M.V. Garad, A. Pelter, B. Singaram and J.W. Wilson, Tetrahedron Lett., 1983, 24, 637. Preparation of $[(2,4,6\text{-}Me_3C_6H_2)_2BCH_2]_2Hg$.

J.-R. Neeser, L.D. Hall, and J.A. Balatoni, Helv. Chim. Acta, 1983, 66, 1018. Preparation of bis(5,6-didioxy-1,2-O-isopropylidene-3-O-methanesulphonyl-α-D-xylohexafuranos-6-yl)mercury.

D.N. Kravtsov, A.S. Peregudov, and V.F. Ivanov, Bull. Acad. Sci. USSR, Div. Chem. Sci., 1983, 32, 468. Equilibrium between $ArSO_2NHR$ and $ArSO_2NR^1HgPh$.

C.A. Obafemi, Phosphorus-Sulfur, 1983, 17, 91. Mass spectra of 2XHg-thiophenes.

G.K. Semin, E.V. Bryukhova, D.N. Kravtsov, and L.S. Golovchenko, Bull. Acad. Sci. USSR, Div. Chem. Sci., 1982, 31, 931. N.q.r. of $PhHgOC_6H_{5-n}X_n$.

Yu.K. Grishin, Yu.A. Ustynyuk, D.N. Kravtsov, A.S. Peregudov, and V.F. Wanov, Bull. Acad. Sci. USSR., Div. Chem. Sci., 1983, 32, 738. ^{199}Hg N.m.r. spectra of $PhHgO_2CAr$.

J.B. Lambert, E.G. Larson, and R.J. Bosch, Tetrahedron Lett., 1983, 24, 3799. Mechanism of $:CBr_2$ transfer from $PhHgCBr_3$.

References

1 Y. Yamamoto, Bull. Chem. Soc. Jpn., 1983, 56, 1772.
2 W. Kaim, J. Organomet. Chem., 1983, 241, 157.
3 J.R. Bews and C. Glidewell, Theochem., 1982, 7, 151; P.T. Luke, J.A. Pople and P.v.R. Schleyer, Chem. Phys. Lett., 1983, 97, 265.
4 P.K. Freeman and L.L. Hutchinson, J. Org. Chem., 1983, 48, 879.
5 G.B. Sergeev, V.V. Zagorsky, and F.Z. Badeev, J. Organomet. Chem., 1983, 243, 123.
6 K. Lühder, D. Nehls, and K. Madeja, J. Prakt. Chem., 1983, 325, 1027.
7 T.P. Burns and R.D. Rieke, J. Org. Chem., 1983, 48, 4141.
8 J.V.N. Varaprasad and C.N. Pillai, J. Organomet. Chem., 1983, 259, 1.
9 F. Sato and H. Katsuno, Tetrahedron Lett., 1983, 24, 1809; U.M. Dzhemilev, O.S. Vostrikova, and R.M. Sultanov, Bull. Acad. Sci. USSR, Div. Chem. Sci., 1983, 32, 193; L.M. Zubritskii, T.N. Fomina, and Kh.V. Balyan, J. Org. Chem. USSR (Engl. Transl.), 1982, 18, 1209.
10 E.C. Ashby and R.D. Ainslie, J. Organomet. Chem., 1983, 250, 1.
11 K. Faegri,Jr., J.Almlöf, and H.P. Lüthi, J. Organomet. Chem., 1983, 249, 303.
12 M. Inomoto, Setchaku, 1983, 27, 1 (Chem. Abstr., 1983, 99, 53809); P.G. Jasien and C.E. Dykstra, J. Am. Chem. Soc., 1983, 105, 2089.
13 H. Kageyama, K. Miki, N. Tanaka, N. Kasai, Y. Okamoto, and H. Yuki, Bull. Chem. Soc. Jpn., 1983, 56, 1319; H. Kageyama, K. Miki, Y. Kai, N. Kasai, Y. Okamoto, and H. Yuki, ibid., 1983, 56, 2411.
14 O.S. Akkermann, G. Schat, E.A.I.M. Evers, and F. Bickelhaupt, Recl. Trav. Chim. Pays-Bas, 1983, 102, 109.
15. H.G. Richey,Jr., and L.M. Moses, J. Org. Chem., 1983, 48, 4013.
16. T. Holm, Acta Chem. Scand., Part B, 1983, 37, 567.
17 B.S. Tanaseichuk, A.I. Belozerov, E.P. Sanaeva, and K.P. Butin, J. Org. Chem. USSR (Engl. Transl.), 1983, 19, 494.
18 Y. Yamamoto and K. Maruyama, J. Org. Chem.,1983, 48, 1564.
19 L. Jalander and R. Standberg, Acta Chem. Scand., Part B, 1983, 37, 15.
20 T. Hayashi, M. Konishi, M. Fukushima, K. Kanehira, T. Hioki, and M. Kumada, J. Org. Chem., 1983, 48, 2195; G. Consiglio, F. Morandini, and O. Piccolo, J. Chem. Soc., Chem. Commun., 1983, 112; idem., Tetrahedron, 1983, 39, 2699; T. Hayashi, Y. Okamoto, and M. Kumada, Tetrahedron Lett., 1983, 24, 807.
21 K. Mochida and H. Ogawa, J. Organomet. Chem., 1983, 243, 131.
22 L.C. Tang, C. Mathis, and B. Francois, J. Organomet. Chem., 1983, 243, 359.
23 J.L. Atwood, D.E. Berry, S.R. Stobart, and M.J. Zaworotko, Inorg. Chem., 1983, 22, 3480.
24 V. Dimitrov and K.-H. Thiele, Z. Anorg. Allg. Chem., 1982, 494, 139.
25 (a) J.L. Luche, C. Petrier, J.-P. Lansard, and A.E. Greene, J. Org. Chem., 1983, 48, 3837; (b) N. Ishikawa and T. Kitazume, Yuki Gosei Kagaku Kyokaishi, 1983, 41, 432 (Chem. Abstr., 1983, 99, 70894); (c) S. Benfice, H. Blancou, and A. Commeyras, J. Fluorine Chem., 1983, 23, 47.
26 E.-I. Negishi, D.E. van Horn, T. Yoshida, and C.L. Rand, Organometallics, 1983, 2, 563.
27 E.-I. Negishi and F.-T. Luo, J. Org. Chem., 1983, 48, 1560.
28 G. van Koten, J.T.B.H. Jastrzebski, and K. Vrieze, J. Organomet. Chem., 1983, 250, 49.
29 N.A. Bell, H.M.M. Harrison, and C.B. Spencer, Acta Crystallogr., Sect. C, 1983, 39, 1182.
30 M.R.P. van Vliet, J.T.B.H. Jastrzebski, G. van Koten, K. Vrieze, and A.L. Spek, J. Organomet. Chem., 1983, 251, C17.
31 J. Dekker, J. Boersma, and G.J.M. van der Kerk, J. Chem. Soc., Chem. Commun., 1983, 553.
32 (a) E.-I. Negishi, H. Matsushita, M. Kobayashi, and C.L. Rand, Tetrahedron Lett., 1983, 24, 3823; (b) C.J. Elsevier, P.M. Stehouwer, H. Westmijze, and P. Vermeer, J. Org. Chem., 1983, 48, 1103; (c) H. Kleijn, H. Westmijze, J. Meijer and P. Vermeer, Recl. Trav. Chim. Pays-Bas, 1983,

102, 378; (d) E.-I. Negishi, V. Bagheri, S. Chatterjee, F.-T. Luo, J.A. Miller, and A.T. Stoll, Tetrahedron Lett., 1983, 24, 5181.
33 T. Klingstedt and T. Frejd, Organometallics, 1983, 2, 598.
34 M. Gaudemar, Tetrahedron Lett., 1983, 24, 2749.
35 T. Hayashi, T. Hagihara, Y. Katsuro, and M. Kumada, Bull. Chem. Soc. Jpn., 1983, 56, 363.
36 L.A. Fedorov, J. Struct. Chem. (Engl. Transl.), 1982, 23, 673.
37 B. Giese and U. Erfort, Chem. Ber., 1983, 116, 1240.
38 V. I. Rozenberg, G.V. Gavrilova, B.I. Ginsberg, V.A. Nikanorov, and O.A. Reutov, Bull. Acad. Sci. USSR, Div. Chem. Sci., 1982, 31, 1707.
39 K.P. Butin, A.A. Ivkina, and O.A. Reutov, Bull. Acad. Sci. USSR, Div. Chem. Sci., 1983, 32, 416.
40 (a) Yu.K. Grishin and Yu.A. Ustynyuk, J. Struct. Chem., (Engl. Transl.), 1982, 23, 801; (b) M.J. Albright, T.F. Schaaf, A.K. Hovland, and J.P. Oliver, J. Organomet. Chem., 1983, 259, 37.
41 M. Nitta, A. Omata, S. Hirayama, and Y. Yajima, Bull. Chem. Soc. Jpn., 1983, 56, 514.
42 J. Barluenga, C. Jimenez, C. Najera, and M. Yus, J. Chem. Soc., Perkin Trans. 1, 1983, 591.
43 R.C. Larock and C.-L. Liu, J. Org. Chem., 1983, 48, 2151.
44 J. Barluenga, F. Aznar, R. Liz, and R. Rodes, J. Chem. Soc., Perkin Trans. 1, 1983, 1087.
45 N.M. Abramova, S.V. Zotova, and O.A. Nesmeyanova, Bull. Acad. Sci. USSR, Div. Chem. Sci., 1982, 31, 961.
46 B.E. Kogai, V.A. Sokolenko, P.V. Petrovskii, and V.I. Sokolov, Bull. Acad. Sci. USSR, Div. Chem. Sci., 1982, 31, 1464.
47 (a) G.B. Deacon and R.N.M. Smith, J. Org. Chem. USSR, (Engl. Transl.), 1982, 18, 1584; (b) N. Redwane, P. Moreau, and A. Commeyras, J. Fluorine Chem., 1982, 20, 699.
48 M.Yu. Antipin, Yu.T. Struchkov, A.Yu. Volkonskii, and E.M. Rokhlin, Bull. Acad. Sci. USSR, Div. Chem. Soc., 1983, 32, 410.
49 J. Mink and P.L. Goggin, J. Organomet. Chem., 1983, 246, 115.
50 (a) D.K. Breitinger, W. Kress, R. Sendelbeck, and K. Ishiwada, J. Organomet. Chem., 1983, 243, 245; W. Kress, D.K. Breitinger, and R. Sendelbeck, ibid., 1983, 246, 1; D.K. Breitinger and W. Kress, ibid., 1983, 256, 217; J. Mink, Z. Meic, M. Gal, and B. Korpar-Colig, ibid., 1983, 256, 203; (b) D.K. Breitinger, G. Petrikowski, G. Liehr, and R. Sendelbeck, Z. Naturforsch., Teil B, 1983, 38, 357.
51 M. Sikirica and D. Grdenic, Cryst. Struct. Commun., 1982, 11, 1571; D. Grdenic, M. Sikirica, D. Matkovic-Calogovic, and A. Nagl, J. Organomet. Chem., 1983, 253, 283.
52 B. Tecle, K.F. Siddiqui, C. Ceccarelli, and J.P. Oliver, J. Organomet. Chem. 1983, 255, 11; I.V. Shchirina-Eingorn, L.G. Kuz'mina, A.G. Makarouskaya, Yu.T. Struchkov, and I.I. Kritskaya, Bull. Acad. Sci. USSR, Div. Chem. Sci., 1983, 32, 337.
53 L.A. Fedorov, and A.K. Prokof'ev, J. Struct. Chem. (Engl. Transl.), 1982, 23, 665; N.S. Erdyneev, D.M. Mognonov, V.P. Markhaeva, and O.Y. Okhlobystin, Izv. Sib. Otd. Akad. Nauk SSSR, Ser. Khim. Nauk, 1983, 83 (Chem. Abstr., 1983, 99, 53892).
54 Yu. K. Grishin, Yu.A. Strelenko, Yu.A. Ustynyuk, A.A. Erdman, I.V. Shchirina-Eingorn, and I.I. Kritskaya, Bull. Acad. Sci. USSR, Div. Chem. Soc., 1982, 31, 921.
55 V.N. Baidin, M.M. Timoshenko, Yu.V. Chizhov, R.B. Materikova, and Yu.A. Ustynyuk, Bull. Acad. Sci. USSR, Div. Chem. Soc., 1982, 31, 427.
56 A.J. Canty, J.M. Patrick, and A.H. White, J. Chem. Soc., Dalton Trans., 1983, 1873.
57 R. Norris, S.E. Taylor, E. Buncel, F. Belanger-Gariepy, and A.L. Beauchamp, Can. J. Chem., 1983, 61, 1536.
58 G.C. Stocco, A. Tamburello, and M. Assunta Girasold, Inorg. Chim. Acta, 1983, 78, 57.

59 (a) Yu.K. Grishin, Yu.A. Strelenko, Yu.A. Ustynynuk, A.S. Peregudov, L.S. Golovchenko, E.M. Rokhlina, and D.N. Kravtsov, Bull. Acad. Sci. USSR, Div. Chem. Soc., 1982, 31, 926; L.M. Epshtein, E.S. Shubina, E.M. Rokhlina, D.N. Kravtsov, and L.A. Kazitsyna, ibid., 1982, 31, 1803; (b) S.I. Pombrik, L.S. Golovchenko, A.S. Peregudov, E.I. Fedin, and D.N. Kravtsov, ibid., 1983, 32, 732.

60 A.P. Arnold and A.J. Canty, Can. J. Chem., 1983, 61, 1428.

61 (a) D. Naumann and H. Lange, J. Fluorine Chem., 1983, 23, 37, see also M.J. Adam, J.M. Berry, L.D. Hall, B.D. Pate, and T.J. Ruth, Can. J. Chem., 1983, 61, 650; N.P. Dubin, Yu.M. Kiselev, T.V. Magdesieva, and O.A. Reutov, Bull. Acad. Sci. USSR, Div. Chem. Sci., 1982, 31, 641.

62 (a) R. Henning and H. Urbach, Tetrahedron Lett., 1983, 24, 5343; S. Danishefsky and E. Taniyama, ibid., 1983, 24, 15; S. Danishefsky, E. Taniyama, and R.R. Webb,Jr., ibid., 1983, 24, 11; A.P. Kozikowski and J. Scripko, ibid., 1983, 24, 2051; (b) J. Barluenga, J. Lopez-Prado, P.J. Campos, and G. Asenso, Tetrahedron, 1983, 39, 2863.

63 A.J. Bloodworth, A.G. Davies, R.A. Savva, and J.N. Winter, J. Organomet. Chem., 1983, 253, 1.

64 E.J. Corey, G.Schmidt, and K. Shimoji, Tetrahedron Lett., 1983, 24, 3169.

65 G.A. Russell and H. Tashtoush, J. Am. Chem. Soc., 1983, 105, 1398.

66 H. Kurosawa, H. Okada, M. Sato, and T. Hattori, J. Organomet. Chem., 1983, 250, 83.

67 R.C. Larock, K. Narayanam, and S.S. Hershberger, J. Org. Chem., 1983, 48, 4377.

68 (a) N.A. Bumagin, I.O. Kalinovskii, and I.P. Beletskaya, J. Org. Chem. USSR, Engl. Transl., 1982, 18, 1151; (b) idem., ibid., 1982, 18, 1152.

69 J. Cook and P.M. Maitlis, J. Chem. Soc.,Dalton Trans ., 1983, 1319.

3

Boron with the Exception of the Carbaboranes

BY J. W. WILSON

1 Introduction

This chapter is an attempt to give a balanced report on the significant chemistry of organoboron compounds containing at least one boron-carbon bond. As such it is not a comprehensive review of the chemistry of organic compounds of boron.

2 Books and Reviews

Two extensive treatments of organoboron chemistry have become generally available this year[1,2] in addition to the usual annual surveys (for 1981).[3] Reviews have appeared on the following topics: applications of vinylic organoboranes to pheromone syntheses,[4] new hydroborating agents,[5] chiral organoborane reagents in asymmetric syntheses,[6] the use of organoborates in organic syntheses[7] and the use of organoborane chemistry to incorporate stable and radioactive isotopes.[8]

3 Uses of Organoboranes and Organoborates in Organic Syntheses

3.1 Hydroboration and Reduction.- The mode of reaction of $(9\text{-BBN})_2$ with a range of substrates has been elucidated from mechanistic studies. With alkenes and alkynes hydroboration proceeds via a pathway which involves dissociation of the dimer followed by rapid reaction of the monomer with the unsaturated molecule. The catalytic role of Lewis bases in this reaction has also been investigated.[9] A similar mechanism operates in the reduction of aldehydes and ketones in carbon tetrachloride and it seems likely that the 9-BBN monomer is coordinated to the carbonyl oxygen during the reduction.[10] The protonolysis of representative alcohols follows a similar pathway in THF. In carbon tetrachloride the steric properties of the substrate exert an influence. With hindered alcohols the dissociation mechanism is still observed but with unhindered substrates a pathway involving direct attack of the alcohol on the dimer is also seen.[11] Complexation of $(9\text{-BBN})_2$ with a range of amines shows that both mechanisms are important, the steric requirements of the amine being the controlling factor.[12]

Hydroboration of alkenes with $Br_2BH{\cdot}SMe_2$ also follows a dissociation pathway. The reaction is catalysed by BBr_3 which facilitates the efficient hydroboration of less reactive alkenes.[13] Contrary to an earlier report 9-BBN in THF does not selectively hydroborate *cis* or *trans* isomers in a predictable way.[14] The mechanism for the thermal rearrangement of organoboranes and its relationship to that of hydroboration reactions has been probed.[15]

Hydroboration of 1-bromopropargylamines gives α-bromovinylboranes which produce aminoalkylidenecycloalkanes in high yields[16] whereas hydroboration of a range *N*-allylamines leads to the appropriate aminoalcohols.[17] The relative reactivities of alkenes and alkynes towards hydroboration by $Br_2BH{\cdot}SMe_2$ indicate that this reagent is a useful addition to the arsenal of monofunctional selective hydroborating reagents.[18] Furthermore dimesitylborane is a very efficient reagent for the rapid regioselective hydroboration of unsymmetrical alkynes. The reaction with alkenes is slow and sensitive to steric factors.[19] Haloboranes have also been used in efficient and relatively simple syntheses of alkylalkenylalkynylboranes.[20] Bromoalkylboranes are readily obtained by the catalysed redistribution of BR_3 and BBr_3.[21] The synthetic routes to alkyl substituted hydroborates have been reviewed[22] and the possible advantages of the direct method from lithium alkyls and borane adducts highlighted.[23] Lithium triethylhydroborate has been shown to be a very powerful nucleophile which facilitates the hydrogenolysis of carbon-halogen bonds in a clean and near quantitative manner.[24] Its ability to reduce epoxides together with a quantitative analytical method for its estimation have also been reported.[25] Reduction of acyclic chiral ketones proceeds *via* different stereochemical pathways when carried out by nucleophilic and electrophilic reducing agents.[26]

3.2 Chiral Syntheses.- Some α-chloro substituted (+)pinanediol boronates, made by homologation of (+)pinanediol boronic esters with (dichloromethyl) lithium, have been shown to be useful reagents for directed chiral syntheses.[27] Rearrangement of the intermediate borate complexes (1) to the possible epimeric pair of α-chloro substituted esters is catalysed by lithium and zinc chlorides[28] and this realization has led to improvements in the method to the extent that 99% chiral selectivity has been achieved as demonstrated by the synthesis of two insect pheromones.[29] In contrast (+)pinanediol dichloromethane boronate is not a useful starting material for directed chiral syntheses.[30] Pinanediol-[α-(trimethylsilyl)organyl]boronates are also of limited use in chiral syntheses due to their low diastereoselectivity. They are however readily desilylated to boronic esters by tetrabutylammonium fluoride in moist THF.[31] Chiral organoboron enolate reagents, prepared from S and R-mandelic acids, condense with achiral aldehydes to give stereospecific aldol products.[32] Two reports on reagents for asymmetric carbon-carbon bond formation have appeared. B-allyldiisopinocampheylborane condenses with aldehydes to give secondary homoallylic alcohols with enantiomeric purities in the range 83-96%[33] whereas allenylboronates of appropriate tartaric acid esters, prepared from allenylboronic acid and dialkyltartrates, provide β-acetylenic alcohols with high enantioselectivity.[34]

3.3 Boronic Acid Systems.- In addition to those mentioned above the synthetic utility of several other boronic acid derivatives has been reported. A series of pinacol α-substituted allylboronates $CH_2=CHCH(X)BO_2C_6H_{12}$ (X = Cl, Br, SEt, SBu^t, OMe) add to aldehydes to give good yields of the substituted homoallyl alcohols containing upto 97% of the Z isomer.[35] Vinyl lithium reagents have been used in syntheses of allylboronates which react stereospecifically with a range of aldehydes in reasonable yields.[36] Reaction of (dichloromethyl) lithium with a wide range of boronic esters gives the homologated α-chloro esters in good yields. These readily undergo nucleophilic displacement of chloride.[37] Methoxy(phenylthio)methyl lithium aided by mercuric (II) chloride readily homologates alkyl boronic esters giving α-methoxy derivatives which on oxidation yield the corresponding aldehydes in good yields. This method provides a novel way of introducing the aldehyde functionality into olefins in a regio and stereocontrolled manner.[38]

3.4 Boron Stabilised Carbanions.- Two routes for the synthesis of α-trimethylsilyl boronic esters have been reported. The pinacol derivatives can be deprotonated to give synthetically useful anions. These are readily alkylated and react with ketones and aldehydes to give alkenylboronates by exclusive silicon elimination. With carboxylic esters (trimethylsilyl) methyl ketones are obtained showing that with these substrates boron elimination is the major pathway. An analogous boron substituted Wittig reagent reacts with benzophenone or benzaldehyde first by boron elimination followed by phosphorus to give the corresponding allenes. Attempts to homologate alkyl boronic esters with PhSCH(X)Li (X = SPh, SMe_2^+, NMe_2^+, Cl) failed under the conditions used.[39] Carbanions stabilised by an α-dimesitylboryl group have been characterised and shown to be versatile reagents.[40] Alkylation proceeds efficiently allowing one pot one, two or three alkyl insertion reactions,[41] the anions react cleanly with ketones and aldehydes to yield alkenes[42] and the anion derived from the allyl compound is attacked by electrophiles in a regio and stereospecific manner.[43] Heteroatom substituted dimesitylboryl methanes Mes_2BCH_2X (X = $SiMe_3$, SnR_3, SPh, $PbPh_3$, $HgCH_2BMes_2$) have been synthesised, deprotonation and methylation is readily achieved.[44] A boron stabilised carbanion derived from triethylborane by electron impact reacts with carbon dioxide in the gas phase at pressures of the order 10^{-7} torr to give the diethyl borinate anion and a ketene.[45]

<u>3.5 Haloboranes</u>.- Bromodiorganoboranes afford a general and efficient method for the cleavage of acetals, ketals and methyl ethers.[46] Halodiorganoboranes and benzyldihaloboranes reversibly haloborate 1-alkynes at ambient temperatures to give predominantly <u>Z</u>-alkenes. At elevated temperatures the irreversible 1,1-organoboration is dominant.[47] Dichlorophenylborane reacts with α,β-unsaturated ortho-aryl azides under mild conditions to give the corresponding 1,2-dihydro-<u>N</u>-phenyl-1-aza-2- borabenzene derivatives in very high yields.[48]

<u>3.6 Thermal Isomerisations</u>.- Hydroboration of 2-methyl-2-butene and <u>cis</u>-3-hexene with a series of monofunctional hydroborating agents gives structurally defined organoboranes amenable to thermal isomerisation studies. Of the dialkylboryl migrating groups the bis(bicyclo[2.2.2]octyl) derivative is the most efficient giving 100% boron distribution on C(1) for the 3-hexene system and 90% equilibrium distribution on C(4) for the 2-methyl-2--butene derivative. With dihaloorganoboranes, only the diiodoboryl group shows any aptitude to migrate giving 85% distribution on C(1) with the 3-hexene derivative and 100% C(1) substitution with the 2-methyl-2-butene system.[49-51]

<u>3.7 Organoboration Reactions</u>.- Alkylboranes organoborate platinum[52] and tin[53] alkynyl compounds to give metal alkenylboranes with high stereospecificity. The tin compounds react with an addition equivalent of alkynylstannane to yield allenes. Hydroboration of alkynylstannanes with 9-BBN does not occur but the

reaction takes place via alkynyl-hydride transfer to give the alkene compound according to equation (1).

$$R'_3Sn-C\equiv CR'' + 9\text{-}BBN \rightarrow R'_3Sn(R'')C{=}C(H)BC_8H_{14} \qquad (1)$$

3.8 Reactions of Alkenyl and Alkynylboranes.- The palladium catalysed cross-coupling reaction of alkenylboranes and alkenyl-halides for the stereospecific synthesis of conjugated dienes has been used in the synthesis of bombykol[54] and the sex pheromone of the European grapevine moth.[55] A highly regioselective route to the unconjugated 3-(1-hydroxylalkyl)penta-1,4-dienes from ketones and pentadienylboranes has been reported.[56]

Alkynylboranes generated in situ react with oxiranes to produce β-hydroxy alkynes[57] and with lactams to give adducts which on reduction with lithium aluminium hydride give alkynyl--azacycloalkanes in good yields.[58] Trimethylsilyl substituted α-allenic alcohols have been synthesised from propargylorgano-boranes with high regioselectivity and in good yields.[59]

3.9 Miscellaneous.- The combination of alkylboranes with (phenylselenyl)allyl lithium produces a reagent which reacts with aldehydes to give selective syntheses of branched or linear homoallylic alcohols depending on temperature, reaction time or amount of excess alkylborane.[60] The chemistry of some tris(trimethylsilyl)methyl boranes $(Me_3Si)_3CBX_2$ (X = Me, OR, OH) has been investigated[61] as has the reaction between phenyliso-cyanate and a series of aminophenylboranes.[62] The protonolysis of triethylborane by carboxylic acids has been examined thoroughly[63] and ^{13}C labelled carboxylic acids are readily made from ^{13}C enriched carbon monoxide and functionalised organo-boranes.[64]

4 Preparations and Reactions of Organoborane Systems

4.1 Cyclic systems.- A 2:3 equilibrium mixture of 7-chloromethyl--3-allyl-3-borabicyclo[3.3.1]non-6-ene and 6-chloro-7-methylene--3-allyl-3-borabicyclo[3.3.1]nonane results from the isomerisation of the former at 120°C. Hydroboration of the mixture with diethylborane gives 1-boraadamantane and 4-chloro-1-boraadamantane.[65] The bromination of 2-isopropyl-2-boraadamantane in dichloromethane proceeds by both radical and electrophilic mechanisms to give 3α-hydroxy-7α-(2-hydroxy-2-propyl)bicyclo[3.3.1]nonane or 3α,7α-dihydroxybicyclo[3.3.1]nonane on subsequent oxidation. In

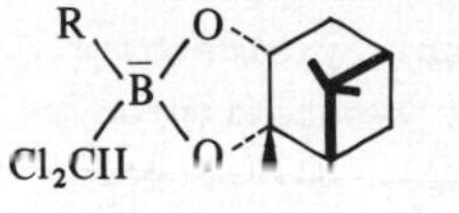

(1)

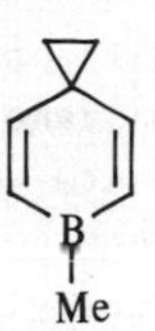

(2)

(3)

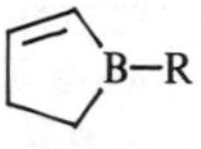

(4)

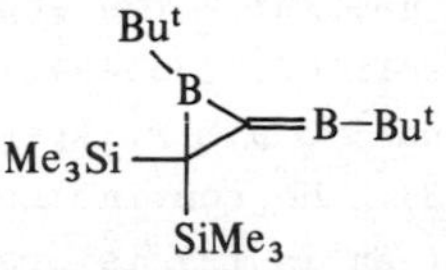

(5)

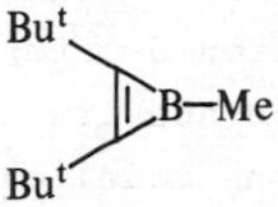

(6)

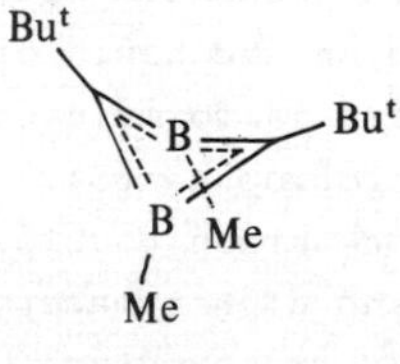

(7)

(8)

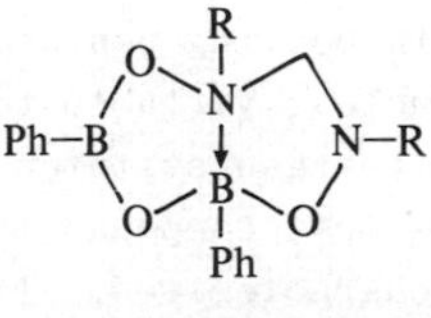

(9)

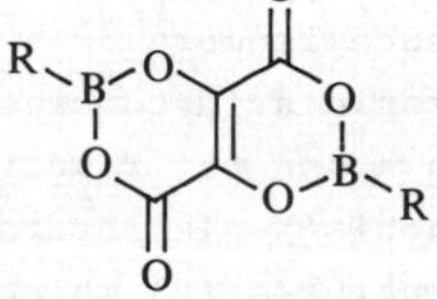

(10)

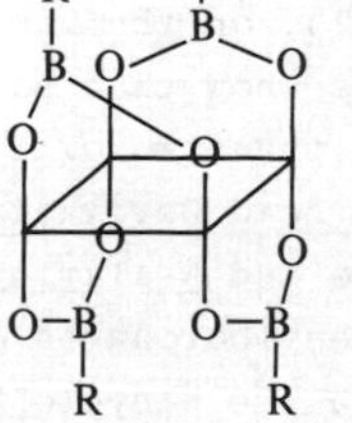

(11)

the presence of water Br^+ participates in the bromination and in thise case the oxidation product is 3-noradamantol.[66] Iodonation of 1-boraadamantane ate complexes in THF yields 3-substituted-7--iodomethyl-3-borabicyclo[3.3.1]nonanes[67] and 1-boraadamantane can be converted to 1-azaadamantane in acceptable yields by a three stage synthesis.[68] The ate complexes of 7-substituted-3--alkyl-3-borabicyclo[3.3.1]nonanes and 3-alkyl-3-borabicyclo-[3.3.1]non-6-enes react with acetyl chloride under mild conditions by an intermolecular β-hydride transfer to form 5-substituted-3--methylenecyclohex-1-ylmethyl(dialkyl)boranes.[69]

The efficient synthesis and isolation of borinane have been thoroughly investigated.[70] A neutral boron analogue of the phenonium ion 6-methyl-6-boraspiro[2,5]octa-4,7-diene (2) has been synthesised and its ^{1}H, ^{13}C and ^{11}B n.m.r. spectra compared with those of the model compound (3). No convincing evidence for extensive cyclopropyl conjugation in (2) is observed.[71] Two examples of substituted 2-borolenes (4) (R = NPr^i_2, Ph) have been isolated in good yields by the catalytic isomerisation of the corresponding 3-borolenes.[72] The reaction of 1,1-bis(tert-butyl-chloroboryl)-2,2-bis(trimethylsilyl)ethene with potassium-sodium alloy in boiling pentane gives a product which on the basis of multinuclear variable temperature n.m.r. spectra is formulated as the tetrasubstituted 2,4-diboramethylenecyclopropane (5). This is the first compound containing a formal boron-carbon double bond to be characterised. It reacts with acetone and diphenylacetylene at low temperatures to give respectively, the corresponding 1,2-oxaboretane and 2-boretene.[73] Bulky substituents are also likely to be responsible for the synthesis of the substituted borirene (6) instead of its dimer, the corresponding 1,4-dibora--2,5-cyclohexadiene. A second product is formulated as the corresponding 1,3-diboretene, with the probable puckered structure (7), on the basis of extensive calculations.[74] *Ab initio* calculations show that borepin (8) should be planar and appreciably more stable than two of its bicyclic valence isomers.[75]

4.2 Boronic Acid Derivatives.- Phenylboronic acid combines with formaldehyde and *N*-alkylhydroxylamines in a one pot reaction to give bis-phenylboronates of *N*,*N*'-methylenebis(*N*-alkylhydroxyl-amines) with the bicyclo[3.3.0]octane structures (9) which undergo topomerisation in solution.[76] A wide range of 2,6-diorgano--1,3,5,7-tetraoxa-2,6-dibora-4,8-octalindiones (10) have been prepared by a variety of routes.[77] Except for R = Bu^t or aryl these compounds show a novel and reversible thermochromic effect which is assumed to arise from inter-molecular interaction between the boron atoms of one molecule and the carbonyl oxygens of neighbouring molecules.[78] Octahydroxycyclobutane reacts with various monoorganoboranes to give tetrakis[organoboranediylbis-(oxy)]derivatives (11) in high yields. Alcoholysis of these compounds results in partial or complete deborylation depending on the substituent.[79] Ortholithiation of N,N-dimethylbenzylamines followed by reaction with trimethoxyborane and subsequent hydrolysis gives moderate yields of the corresponding boronic acids.[80] Alkyl and alkenyl boronic acids and their esters are readily obtained by the hydroboration of alkenes and alkynes with $HBBr_2 \cdot SMe_2$ followed by reaction with water or the appropriate alcohol. Efficient and almost instantaneous esterification of boronic acids with primary or secondary alcohols (the method fails with tertiary alcohols) is achieved by carrying out the reaction in pentane. Glycols give the cyclic esters in preference to the acyclic analogues. This procedure does away with the usual azeotropic distillation of a ternary mixture.[81] For derivatives that cannot be synthesised by a hydroboration route an alternative and very efficient method based on the monoalkylation of selected trialkoxyboranes with alkyl lithium reagents and reaction of the ate product with anhydrous hydrogen chloride has been developed.[82]

4.3 Heterocyclic Systems.- X-ray structure analysis of the new boron-sulphur ring system 1,4-dimethyl-1,4-dithionia-2,5-diborata-cyclohexane (12) shows that in the solid state the ring prefers the chair conformation with equitorial methyl groups.[83] Some 1-phospha and 1-arsa-4-boracyclohexa-2,5-dienes have been made and characterised by multinuclear n.m.r. spectroscopy. The chemistry of the phosphorus compound has been investigated.[84] Thermal decomposition of diarylazidoboranes Ar_2BN_3 in different inert solvents gives either the diazadiboretidines $(ArBNAr)_2$ or the borazines.[85] The iminoborane $BuB{=}NBu^t$ undergoes a 2+2 cycloaddition reaction with the iminophosphanes $R_2NP{=}NR'$ ($R = SiMe_3, Pr^i$; $R' = SiMe_3, Bu^t$) to give the corresponding 1,3,2,4-diazaphospha-boretidines (13).[86] The same iminoborane dimerises in the presence of $M(CO)_5THF$ (M = Cr, W) to give the diazadiboretidine (14) which then acts as a four electron donor to give the complexes $(BuBNBu^t)_2 \cdot MCO_4$. These can also be made directly from (14) and $M(CO)_5THF$. X-ray structure analysis of the chromium compound shows the ring to be non-planar.[87]

4.4 Transition Metal Complexes.- Triple-decker complexes of manganese and iron with a 2-ethyl-1-phenylborole (15) bridging ligand are obtained by the thermal isomerisation of 1-phenyl-4,5-dihydroborepin (16) in refluxing mesitylene in the presence of a suitable metal substrate.[88] A more general route to complexes of this ligand is available by the transition metal complex (M = Mn, Ru, Rh) catalysed isomerisation of 1-phenyl-2,5-dihydroborole (1-phenyl-3-borolene) to the 2-borolene derivative followed by dehydrogenation to give the borole complexes.[89] Tricarbonyl(1-methylborinato)chromium anions feature in a range of compounds, some of which have been studied by cyclic voltammetry.[90] Substituted cyclobutadienyl cobalt complexes containing the 1-methyl or 1-phenylborabenzene ligands have been made and the electrophilic substitutions reactions of $Co(C_5H_5BCH_3)(C_4(CH_3)_4)$ studied.[91] In addition a theoretical study of the bonding, structure and substitution reactions of bis(borabenzene)iron is reported.[92]

Reaction of $(C_5H_5)Co(C_2H_4)_2$ with alkyl derivatives of 1,3-diborolene gives $(\eta^5\text{-}C_5H_5)(\eta^5$-1,3-diborolenyl) cobalt complexes (17) which contain a five coordinated carbon atom. These compounds are readily deprotonated and react with various metal substrates to give tetra-decker complexes.[93] The tin compound is the first tetra-decker sandwich containing a main group metal and has a bent structure with the Co-Sn vectors forming a 130° angle.[94] The same ligand features as the bridging group in a series of triple-decker sandwiches whose electrochemistry and X-ray structures have been studied.[95] The preparation of carbonyl(1,4-diferrocenyl-1,4-diboracyclohexa-2,5-dienyl)transition metal complexes (M = Cr, Mo, W, Fe, Ru, Os) is reported. In all cases the 1,4-diboracyclohexadienyl moiety acts as an η^6 back-bonding ligand.[96] Several bis(1,2,dialkyl η^5-1,2-azaborolenyl)cobalt complexes (18) have been made in which the two rings are staggered.[97] The compounds are readily oxidised to the cobalticinium analogue by iodine or the ferricinium cation and pure salts can be isolated.[98] A nickel(II) complex with two dimethylmethylene phosphoranyl dimethylborato(1-)ligands has been reported and shows similar properties to the Mn(II) and Co(II) complexes with the dihydroborato(1-)congener.[99]

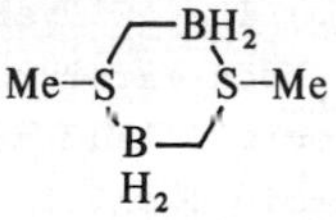

(12)

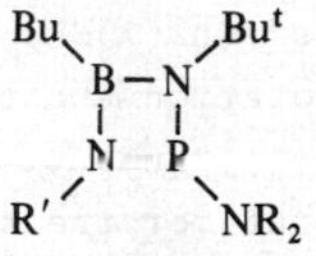

(13)

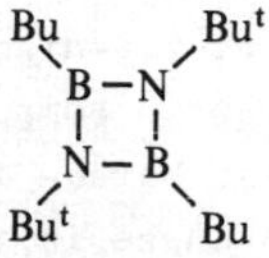

(14)

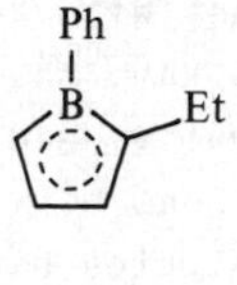

(15)

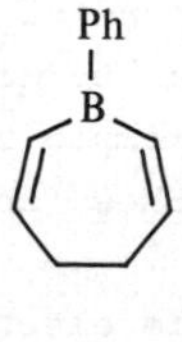

(16)

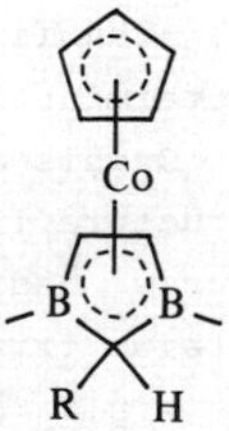

(17)

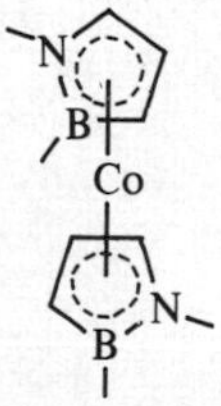

(18)

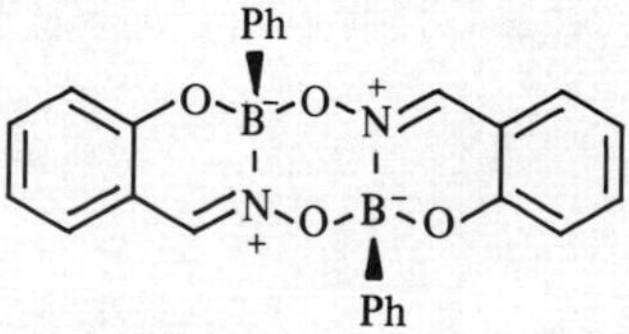

(19)

5 Physical Data

5.1 N.M.R. Spectroscopic Studies.- The synthesis and ^{13}C n.m.r. data for.a series of compounds $PhB(SEt)NR_2$ have been reported[100] and variable temperature measurements recorded for some amino-phenylboranes ($PhB(X)NR_2$; X = hal, OMe, SEt). The $\Delta G^{\ddagger}$ values for restricted rotation about the B-N bonds have been interpreted in terms of steric, inductive and mesomeric effects[101] and attempts made to account for the trends in the ^{13}C chemical shifts of the ortho, meta, para and C-1 carbon atoms of the phenyl group.[102] The variable temperature ^{13}C n.m.r. spectra of two dimesityl-organoselenoboranes have also been investigated.[103] The ^{13}C spectra of the products of mono and dihydroboration of alkynes and 1-halo-1-alkynes with 9-BBN and dicyclohexylborane have been fully assigned[104] as have the ^{13}C spectra of a dibora-2,4-diaza-1,3--benzene derivative and its corresponding borazine compound.[105] Exchange reactions between triorganoboranes and borane in THF and dimethylsulphide have been monitored by changes in their ^{11}B spectra.[106]

5.2 Crystal Structures.- The structure of the dimeric aldimino-borane $(PhCH{=}NBMe_2)_2$ shows the $(BN)_2$ ring to be planar.[107] This paper also reports a tris(ketimino)borane with a novel paddle wheel structure. The first $B_2N_2O_2$ ring has been characterised by the structure determination of bis[(salicylaldoximato(2-))phenyl-boron] (19).[108] Oxybis(dimesitylborane) $Mes_2BOBMes_2$ adopts a bent twisted structure in the solid state.[109]

5.3 Miscellaneous.- Radical products from single electron transfer reactions of lithium triethylhydroborate with N-heterocycles have been detected by e.s.r. and multinuclear ENDOR spectroscopy.[110] Analyses of published data have produced single bond enthalpy terms for a range of B-X bonds including 350±10 kJ mol^{-1} for the B-C bond.[111]

References

1 Houben-Weyl Methods in Organic Chemistry, 13/3b, Organoboron Compds., part 2, 4th Ed., 1983.
2 Comprehensive Organometallic Chemistry, vols. 1 and 7, eds. G.Wilkinson, F.G.A.Stone and E.W.Abel, Pergamon, Oxford, 1982.
3 G.W.Kabalka, J. Organomet. Chem., 1983 245, 1; K.Niedenzu, ibid., 1983, 245, 29.
4 H.C.Brown, Kagaku Zokan (Kyoto), 1983, 101.
5 H.C.Brown, Curr. Trends Org. Synth., Proc. 4th Int. Conf. ed. H.Nozaki, Pergamon, 1983.
6 C. Wang, Huaxue Shiji, 1983, 5, 210.
7 A.Suzuki, Top. Curr. Chem., 1983, 112, 67.
8 G.W.Kabalka, Chem. Abstr., 1983, 98 160760z.
9 H.C.Brown and K.K.Wang, J. Amer. Chem. Soc., 1982, 104, 7148.
10 H.C.Brown, K.K.Wang and J.Chandrasekharan, J. Amer. Chem. Soc., 1983, 105, 2340.
11 H.C.Brown, J.Chandrasekharan and K.K.Wang, J. Org. Chem., 1983, 48, 2901.
12 H.C.Brown, J.Chandrasekharan and K.K.Wang, J. Org. Chem., 1983, 48, 3689.
13 H.C.Brown and J.Chandrasekharan, Organometallics, 1983, 2, 1261.
14 H.C.Brown, D.J.Nelson and C.G.Scouten, J. Org. Chem., 1983, 48, 641.
15 S.E.Wood and B.Rickborn, J. Org. Chem., 1983, 48, 555.
16 J.-L.Torregrosa, M.Baboulene, V.Speziale and A.Lattes, J. Organomet. Chem., 1983, 244, 311.
17 M.Ferles and S.Kafka, Collect. Czech. Chem. Commun., 1982, 47, 2150.
18 H.C.Brown and J.Chandrasekharan, J. Org. Chem., 1983, 48, 644.
19 A.Pelter, H.C.Brown and S.Singaram, Tetrahedron Lett., 1983, 24, 1433.
20 H.C.Brown, D.Basavaiah and N.G.Bhat, Organometallics, 1983, 2, 1468.
21 H.C.Brown, D.Basavaiah and N.G.Bhat, Organometallics, 1983, 2, 1309.
22 H.C.Brown, B.Singaram and S.Singaram, J. Organomet. Chem., 1982, 239, 43.
23 W.Biffar, H.Nöth and D.Sedlak, Organometallics, 1983, 2, 579.
24 S.Krishnamurthy and H.C.Brown, J. Org. Chem., 1983, 48, 3085.
25 H.C.Brown, S.Narasimhan and V.Somayoji, J. Org. Chem., 1983, 48, 3091.
26 M.M.Midland and Y.C.Kwan, J. Amer. Chem. Soc., 1983, 105, 3725.
27 D.S.Matteson, R.Ray, R.R.Rocks and D.J.Tsai, Organometallics, 1983, 2, 1536.
28 D.S.Matteson and E.Erdik, Organometallics, 1983, 2, 1083.
29 D.S.Matteson and K.M.Sadhu, J. Amer. Chem. Soc., 1983, 105, 2077 and 6195.
30 D.J.S.Tsai, P.K.Jesthi and D.S.Matteson, Organometallics, 1983, 2, 1543.
31 D.J.S.Tsai and D.S.Matteson, Organometallics, 1983, 2, 236.
32 S.Masamune and W.Choy, U.S. patent, U.S. 4405802, Chem. Abstr., 1984, 100, 6828f.
33 H.C.Brown and P.K.Jadhav, J. Amer. Chem. Soc., 1983, 105, 2092.
34 R.Haruta, M.Ishiguro, N.Ikeda and H.Yamamoto, J. Amer. Chem. Soc., 1982, 104, 7667.
35 R.W.Hoffmann and B.Landmann, Tetrahedron Lett., 1983, 24, 3209.
36 P.G.M.Wuts, P.A.Thompson and G.R.Callen, J. Org. Chem., 1983, 48, 5398.
37 D.S.Matteson and D.Majumdar, Organometallics, 1983, 2, 1529.
38 H.C.Brown and T.Imai, J. Amer. Chem. Soc., 1983, 105, 6285.
39 D.S.Matteson and D.Majumdar, Organometallics, 1983, 2, 230.
40 A.Pelter, L.Williams and J.W.Wilson, Tetrahedron Lett., 1983, 24, 623.
41 A.Pelter, L.Williams and J.W.Wilson, Tetrahedron Lett., 1983, 24, 627.
42 A.Pelter, B.Singaram and J.W.Wilson, Tetrahedron Lett., 1983, 24, 635.
43 A.Pelter, B.Singaram and J.W.Wilson, Tetrahedron Lett., 1983, 24, 631.
44 M.V.Garad, A.Pelter, B.Singaram and J.W.Wilson, Tetrahedron Lett., 1983, 24, 637.
45 C.L.Johlman, C.F.Ijames, C.L.Wilkins and T.H.Morton, J. Org. Chem., 1983, 48, 2628.
46 Y.Giundon, C.Yoakim and H.E.Morton, Tetrahedron Lett., 1983, 24, 2969 and 3969.

47 R-J.Binnewirtz, H.Klingenberger, R.Welte and P.Paetzold, Chem. Ber., 1983, 116, 1271.
48 R. Leardini and P.Zanirato, J. Chem. Soc. Chem. Commun., 1983, 396.
49 H.C.Brown and U.S.Racherla, J. Organomet. Chem., 1983, 241, C37.
50 H.C.Brown and U.S.Racherla, J. Org. Chem., 1983, 48, 1389.
51 H.C.Brown and U.S.Racherla, J. Amer. Chem. Soc., 1983, 105, 6506.
52 A.Sebald and B.Wrackmeyer, J. Chem. Soc. Chem. Commun., 1983, 309.
53 B.Wrackmeyer, C.Bihlmayer and M.Schilling, Chem. Ber., 1983, 116, 3182.
54 N.Miyaura, H.Suginome and A.Suzuki, Tetrahedron Lett., 1983, 24, 1527.
55 G.Cassani, P.Massardo and P.Piccardi, Tetrahedron Lett., 1983, 24, 2513.
56 M.G.Hutchings, W.E.Paget and K.Smith, J. Chem. Res. (S), 1983, 31.
57 M.Yamaguchi and I.Hirao, Tetrahedron Lett., 1983, 24, 391.
58 M.Yamaguchi and I.Hirao, Tetrahedron Lett., 1983, 24, 1719.
59 K.K.Wang, S.S.Nickam and C.D.Ho, J. Org. Chem., 1983, 48, 5376.
60 Y.Yamamoto, Y.Saito and K.Maruyama, J. Org. Chem., 1983, 48, 5408.
61 C.Eaborn, M.N.El-Kheli, N.Retta and J.D.Smith, J. Organomet. Chem., 1983, 249, 23.
62 R.H.Cragg and T.J.Miller, J. Organomet. Chem., 1983, 255, 143.
63 H.C.Brown and N.C.Hébert, J. Organomet. Chem., 1983, 255, 135.
64 G.W.Kabalka, M.C.Delgado, U.Sastry and K.A.R.Sastry, J. Chem. Soc. Chem. Commun., 1982, 1273.
65 B.M.Mikhailov and K.L.Cherkasova, J. Organomet. Chem., 1983, 246, 9.
66 B.M.Mikhailov, T.A.Shshegoleva, E.M.Shashkova and V.G.Kiselev, J. Organomet. Chem., 1983, 250, 23.
67 B.M.Mikhailov, M.E.Gurskii and D.G.Pershin, J. Organomet. Chem., 1983, 246, 19.
68 B.M.Mikhailov and E.A.Shagova, J. Organomet. Chem., 1983, 258, 131.
69 M.E.Gurskii, S.V.Baranin, A.S.Shashkov, A.I.Lutsenko and B.M.Mikhailov, J. Organomet. Chem., 1983, 246, 129.
70 H.C.Brown and G.G.Pai, J. Organomet. Chem., 1983, 250, 13.
71 A.J.Ashe III, T.Abu-Orabi, O.Eisenstein and H.T.Sandford, J. Org. Chem., 1983, 48, 901.
72 G.E.Herberich, B.Hessner and S.Söhnen, J. Organomet. Chem., 1983, 256, C23.
73 H.Klusik and A.Berndt, Angew. Chem. Int. Ed. Engl., 1983, 22, 877.
74 S.M.van der Kerk, P.H.M.Budzelaar, A.van der Kerk-van Hoof, G.J.M. van der Ker and P.v.R.Schleyer, Angew. Chem. Int. Ed. Engl., 1983, 22, 48.
75 R.L.Disch, M.L.Sabio and J.M.Schulman, Tetrahedron Lett., 1983, 24, 1863.
76 W.Kliegel, J. Organometal. Chem., 1983, 253, 9.
77 M.Yalpani and R.Köster, Chem. Ber., 1983, 116, 3332.
78 M.Yalpani, R.Boese and D.Bläser, Chem. Ber., 1983, 116, 3338.
79 M.Yalpani, R.Köster and G.Wilke, Chem. Ber., 1983, 116, 1336.
80 M.Lauer and G.Wulff, J. Organomet. Chem., 1983, 256, 1.
81 H.C.Brown, N.G.Bhat and V.Somayaji, Organometallics, 1983, 2, 1311.
82 H.C.Brown and T.E.Cole, Organometallics, 1983, 2, 1316.
83 H.Nöth and D. Sedlak, Chem. Ber., 1983, 116, 1479.
84 H-O.Berger and H.Nöth, J. Organomet. Chem., 1983, 250, 33.
85 P. Paetzold and R.Truppat, Chem. Ber., 1983, 116, 1531.
86 P.Paetzold, C.v.Plotho, E.Niecke and R.Rüger, Chem. Ber., 1983, 116, 1678.
87 K.Delpy, D.Schmitz and P.Paetzold, Chem. Ber., 1983, 116, 2994.
88 G.E.Herberich, J.Hengesbach, G.Huttner, A.Frank and U.Schubert, J. Organomet. Chem., 1983, 246, 141.
89 G.E.Herberich, B.Hessner, W.Boveleth, H.Lüthe, R.Saive and L.Zelenka, Angew. Chem. Int. Ed. Engl., 1983, 22, 996.
90 G.E.Herberich and D.Söhnen, J. Organomet. Chem., 1983, 254, 143.
91 G.E.Herberich and A.K.Naithani, J. Organomet. Chem., 1983, 241, 1.
92 N.M.Kostíc and R.F.Fenske, Organometallics, 1983, 2, 1319.
93 J.Edwin, M.C.Böhn, N.Chester, D.M.Hoffman, R.Hoffman, H.Pritzkow, W.Siebert, K.Stumpf and H.Wadepohl, Organometallics, 1983, 2, 1666.
94 H.Wadepohl, H.Pritzkow and W.Siebert, Organometallics, 1983, 2, 1899.

95 J.Edwin, M.Bochmann, M.C.Böhm, D.E.Brennan, W.E.Geiger, C.Krüger, J.Pebler, H.Pritzkow, W.Siebert, W.Swiridorf, H.Wadepohl, J.Weiss and U.Zenneck, J. Amer. Chem. Soc., 1983, 105, 2582.
96 G.E.Herberich and M.M.Kucharska-Jansen, J. Organomet. Chem., 1983, 243, 45.
97 G.Schmid and R.Boese, Z. Naturforsch., Teil B, 1983, 38, 485.
98 G.Schmid, V.Höhner, D.Kampmann, D.Zaika and R.Boese, J. Organomet. Chem., 1983, 256, 225.
99 G.Müller, D.Neugebauer, W.Geike, F.H.Köhler, J.Pebler and H.Schmidbauer, Organometallics, 1983, 2, 257.
100 R.H.Cragg and T.J.Miller, J. Organomet. Chem., 1983, [illegible], [illegible].
101 C.Brown, R.H.Cragg, T.J.Miller and D.O'N.Smith, J. Organomet. Chem., 1983, 244, 209.
102 R.H.Cragg and T.J.Miller, J. Organomet. Chem., 1983, 241, 289.
103 R.I.Baxter, R.J.M.Sands and J.W.Wilson, J. Chem. Res. (S), 1983, 94.
104 C.D.Blue and D.J.Nelson, J. Org. Chem., 1983, 48, 4538.
105 S.Allaoud, H.Bitar, M.El Mouhtadi and B.Frange, J. Organomet. Chem., 1983, 248, 123.
106 R.Contreras and B.Wrackmeyer, Spectrochim. Acta. Part A, 1982, 38A, 941.
107 J.R.Jennings, R.Snaith, M.A.Mahmoud, S.C.Wallwork, S.J.Bryan, J.Halfpenny, E.A.Petch and K.Wade, J. Organomet. Chem., 1983, 249, C1.
108 S.J.Rettig and J.Trotter, Can. J. Chem., 1983, 61, 206.
109 C.J.Cardin, H.E.Parge and J.W.Wilson, J. Chem. Res. (S), 1983, 93.
110 W.Kaim and W.Lubitz, Angew. Chem. Suppl. Int. Ed. Engl., 1983, 1209.
111 J.B.Holbrook, B.C.Smith, C.E.Housecraft and K.Wade, Polyhedron, 1982, 1, 701.

4

Carbaboranes, including their Metal Complexes

BY T. R. SPALDING

1 Introduction, Reviews, and theoretical aspects

The layout of this Chapter is similar to previous years.[1] Section one refers to reviews, articles of general interest, and papers dealing with theoretical aspects. Section two deals with carbaborane syntheses including organic derivitatives i.e. mainly σ-bonded C,N,P,S, and halogen compounds, their characterisation and reactions. Carbaboranes with other *exo*-bonded groups are discussed in Section three. Section four concerns carbaboranes with heteroatoms incorporated into the cage. Results from physico-chemical studies including *X*-ray crystallographic and spectroscopic studies are discussed in each relevant Section.

Methods for systematically describing closed, non-closed, and capped triangulated polyhedral molecules have been discussed.[2] A new five part descriptor system including the point group symmetry and number of triangulated faces of the corresponding closed poly-borane was developed to provide unambiguous structurally definitive names. For the purposes of this current report however the common descriptors *closo-*, *nido-* will continue to be used as they were used in the original literature.

Grimes has published several reviews concerning carbaborane chemistry[3] including one specifically on carbon-rich carbaboranes and their metal derivitatives.[3a] He has also edited a book entitled "Metal Interactions with Boron Clusters" which has chapters on structural and bonding features of metallacarbaboranes, *closo*-carbaborane-metal complexes containing metal-carbon and metal -boron σ bonds, and sections on the electrochemistry of metalla-carbaboranes, and carbaboranes containing main group and transition metal hydride units.[4]

The preparation of metallacarbaborane catalysts has been discussed and described in theses.[5a-e] Another thesis concerned cobaltacarbaboranes.[5f]

A review on heteroarylcarbaboranes has appeared.[6]

Other reviews of general interest have dealt with borane synthesis,[7a] metallaboranes,[7b] and main group III general chemistry.[7c]

1.1. Theoretical Aspects.- The theory of cluster bonding based on tensor surface harmonics has been evaluated for *closo* compounds using MNDO calculations.[8] It was shown that, as with the borane anions $[B_nH_n]^{2-}$, the cluster bonding m.o.s. can be divided into a low lying, very stable, S^σ and P^σ set, and a high lying $D^\parallel$ and $F^\parallel$ set. The variation of the energy of these m.o.s. with cluster size was investigated. MNDO calculated protonated carbaborane structures and proton affinities were reported to predict B-B edge protonation for 1,6-$C_2B_4H_6$ but B-B-B face protonation for 2,4-$C_2B_5H_7$ and 1,12-$C_2B_{10}H_{12}$.[9] Protonation at carbon occurred for 2-CB_5H_9 and 1,5-$C_2B_3H_5$. Considerable care was needed in interpreting the MNDO results since the version used tended to underestimate the importance of multicentre bonding.

Stabilities of *closo*-carbaborane isomers have been correlated with the orbital overlap preferences of BH and CH units.[10] Whereas the CH group prefers bonding to three or four membered rings, the BH group prefers five membered rings. Thus 2,4-$C_2B_5H_7$ is more stable than 1,2-$C_2B_5H_7$ which is more stable than 1,7-$C_2B_5H_7$. The relationships between bond lengths in nine atom clusters including $C_2B_7H_9$, B_9Cl_9, and $[M_9]^{2-}$ (M = main group IV atom) and the form and occupancy of the frontier orbitals have been investigated.[11]

An analysis of experimental evidence from transition metal complexes containing $[C_5H_5]^-$, $[C_5Me_5]^-$, and $[C_2B_9H_{11}]^{2-}$ ligands suggests that the steric and electronic properties of the carbaborane ligand is more like $[C_5Me_5]^-$ than $[C_5H_5]^-$.[12]

2 Carbaborane Synthesis, Characterisation, and Reactions

Gas phase pyrolysis of 1,5-$C_2B_3H_5$ produced a new C_4-carbaborane, $C_4B_7H_{11}$ as well as dimeric 1:2'-(1,5-$C_2B_3H_4$)$_2$ and 2:2',1':2"- and 2:2',3':1"- isomers of trimeric (1,5-$C_2B_3H_4$)(1',5'-$C_2B_3H_3$)(1",5"-$C_2B_3H_4$).[13] From spectroscopic data and comparison with other 11-atom *nido* systems, it was suggested that the C_4-compound was 1,2,8,10-$C_4B_7H_{11}$. Other products from the pyrolysis included the isomeric 2:2'-dimer and the 2:2',3':2"-trimer. Copyrolysis of 1,5-$C_2B_3H_5$ and 1,6-$C_2B_4H_6$ gave

2:2'-(1',5'-$C_2B_3H_4$)(1,6-$C_2B_4H_5$). A new Pd-catalysed route to alkenylpentaboranes from olefins and B_5H_9 has been described.[14] On pyrolysis, these compounds give 2-CB_5H_9 derivatives. Heating 7,8-$C_2B_9H_{13}$ in benzene or toluene produced 3:8'(5',6'-$C_2B_8H_{11}$)(1,2-$C_2B_{10}H_{11}$).[15] Oxidation of C-substituted derivatives of $C_2B_{10}H_{12}$ by Tl(III) in the presence of Pd(II) gave B-B' bonded dicarbaboranes.[16]

A previously reported preparation of 1-allyl-1,2-$C_2B_{10}H_{11}$ has been questioned and a new route to this and the 1-propenyl compound was described.[17] Gas phase pyrolysis of 3-vinyl or 3-ethynyl-1,2-$C_2B_{10}H_{11}$ produced mixtures of the corresponding 2- and 4-substituted 1,7-$C_2B_{10}H_{12}$ compounds.[18] Alkyl substitution at B followed by isomerisation and some carbaborane coupling occurred when 1,2- and 1,7-$C_2B_{10}H_{12}$ were reacted with Pr^nX over $AlCl_3$.[19] Compounds based on 1-cyclopentadiene-1,2-$C_2B_{10}H_{11}$ have been prepared.[20] Chromium tricarbonyl complexes of phenyl derivatives of 1,2-$C_2B_{10}H_{12}$ were synthesised.[21]

Reaction of the carbethoxycarbene electrophile with 1,2-$C_2B_{10}H_{12}$ gave mixtures of all four possible B-CH_2CO_2Et substituted isomers.[22] The proportions corresponded approximately with the ground state charges on BH units, i.e. B(9,12) most abundant then B(8,10), B(4,5,7,11), and B(3,6). C-Substituted esters of 1,2- and 1,7-$C_2B_{10}H_{12}$ were prepared using a phase transfer catalyst.[23] Allyl and propynyl esters of 1,7- and 1,12-$C_2B_{10}H_{12}$ dicarboxylic acids have been synthesised.[24]

Some acylation and alkylation reactions of the amine group in 9-NH_2-1,7-$C_2B_{10}H_{11}$, and oxidation of NH_2 in 3-NH_2-1,2-$C_2B_{10}H_{11}$ have been studied.[25] Syntheses of isoxazole and isoxazoline derivatives of 1,2- and 1,7-$C_2B_{10}H_{12}$,[26] and a N-diethylphosphoryl methyl carbamate derivative of 1,2-$C_2B_{10}H_{12}$ have been reported.[27] Reactions of [9-phenyliodo-1,7-$C_2B_{10}H_{11}$][BF_4] with various species including pseudohalides, amido-, and benzoyl-, groups gave 9-substituted products.[28a] Similar reactions of phenyliodo - and phenylbromo-derivatives with Ph_3P gave 9-halocarbaboranes.[28b]

An unusual exchange of Cl for Br occurs when the adduct Me_3N.5-Br-2,4-$C_2B_5H_6$ is dissolved in CH_2Cl_2.[29] The products are 5-Cl-2,4-$C_2B_5H_6$ and [Me_3NCH_2Cl]Br. From control experiments and a consideration of thermodynamic factors it was suggested that [Me_3N.2,4-$C_2B_5H_6$]$^+$ was the intermediate species involved.

The C-bonded phosphazene derivative (1) polymerises on heating .[30] Both monomer (1) and polymer react with piperidine to replace all Cl atoms by $C_5H_{10}N$-groups and degrade the carbaborane unit to the corresponding *nido* $[C_2B_9H_{12}]^-$ species. Reactions of these anionic phosphazene compounds with $[RhCl(Ph_3P)_3]$ produced catalytically active rhodacarbaborane complexes. Further reaction of the anionic compounds with Na[H] gave the corresponding dianions which reacted with $[M(CO)_6]$ (M = Mo or W) to give dianionic species suggested to have one $M(CO)_3$ unit bonded to the carbaborane residue and one $M(CO)_4$ unit bonded to a phosphazene N atom. The thermal stability of carbaborane containing phosphazenes has been studied.[31]

The synthesis of carbaborane containing phosphines and arsines of type (2) and the *X*-ray crystallographic study of (2, MR = PPh) have been reported.[32] Reactions of (2) with $[M(CO)_6]$ compounds gave $[M(2)(CO)_5]$ derivatives.

The structure of $CH_2(-9-S-1,2-C_2B_{10}H_{11})_2$ has been obtained using *X*-ray crystallography.[33] Several cyclic and polymeric derivatives of $1,2-(SH)_2-1,2-C_2B_{10}H_{12}$ have been synthesised including (3) and the corresponding bis-$(7,8-C_2B_9H_{10})$ - containing dianion.[34] Treatment of 9-SCN derivatives of 1,2- or $1,7-C_2B_8H_{12}$ with $K_2[C_8H_8]$ gave 9-SH-,and 9,9'-S_2- coupled products.[35]

A number of spectroscopic studies have been reported. Analysis of $^1J(^{11}B-^1H)$ values in eleven *closo*-carbaboranes enabled an empirical equation to be developed which correlates ^1J values with the number of C-atoms bonded to the B-atom concerned and the average cage "umbrella" angle that the BH unit makes with the cage atoms to which it is attached.[36] Vibrational spectroscopy has revealed unusual splitting of the C-H vibration in the C-bonded $CH_2(1,2-C_2B_{10}H_{11})_2$.[37] I.r. studies of the proton donor ability of C-H bonds in C- and B-ethynyl derivatives of 1,2- and $1,7-C_2B_{10}H_{12}$, and in decachlorocarbaboranes interacting with O- and N- bases have been reported.[38] Raman and ^{11}B n.m.r. spectroscopy were used to study a low temperature phase transition in $1,7-C_2B_{10}H_{12}$.[39]

Phase equilibria in carbaborane systems were investigated calorimetrically.[40]

Electron capture mass spectrometry was used to identify radical anions from $C_2B_{10}H_{12}$ molecules.[41]

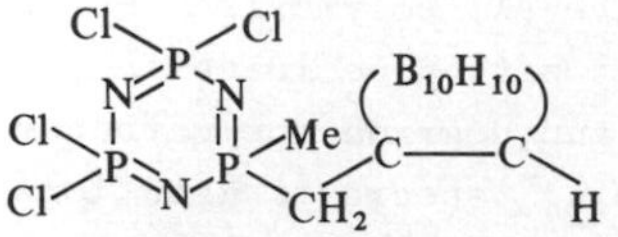

(1)

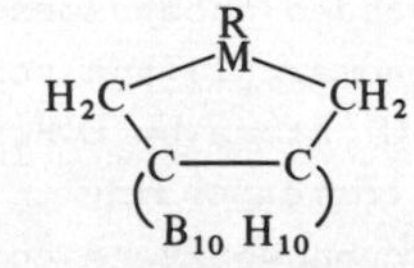

(2) M = P, As; R = Ph, Me

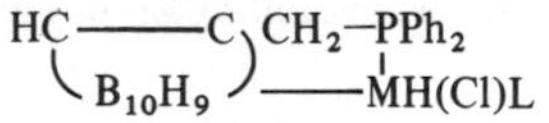

(3) M L = Rh(4-Mepy), Ir(cod)

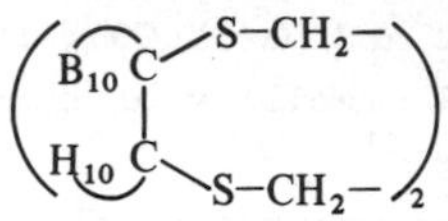

(4)

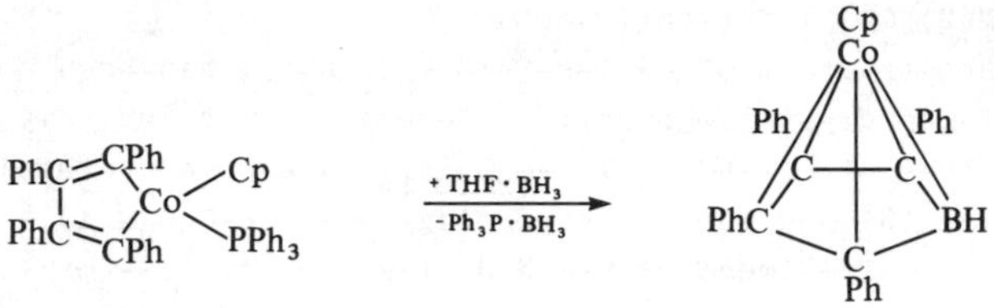

Scheme 1

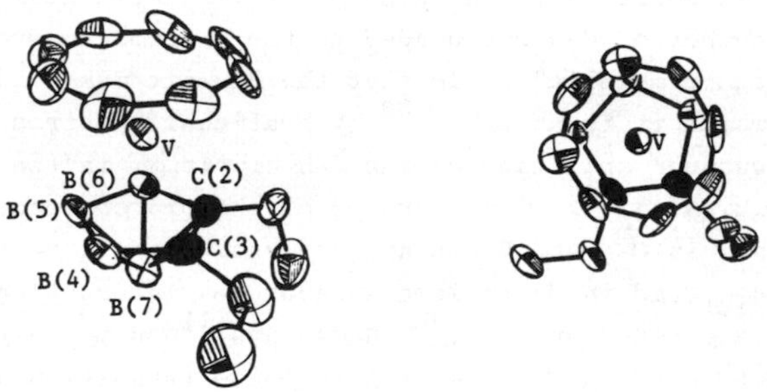

Figure 1 [$V(Et_2C_2B_4H_4)$ (η-C_8H_8)] viewed obliquely (left) and along V-B(7) axis (right)
(Reproduced with permission from *J. Am. Chem. Soc.*, 1983, *105*, 2079)

An *X*-ray crystallographic study of the 1:1 adduct of Me_2SO and B-decachlorocarbaborane was reported.[42]

Carbaborane containing organosilicon monomers and polymers have been synthesised.[43] Methods for the analysis of carbaborane-silicone polymers have been reported.[44] The use of a carbaborane-silicone polymer as the stationary phase for gas chromatography was investigated.[45]

Several reports and patents concerning carbaborane containing polymers based on polyphenylene,[46] polyester,[47] polyamide and related polymers have been published.[48] Applications of such compounds have included their addition to rubber vulcanizates and their use in the preparation of composite ceramics.[49]

Extraction of uranium from spent nuclear fuel is increased when 1,7-$C_2B_{10}H_{12}$ is added to the tributyl phosphate containing phase.[50]

Carbaborane related aspects of the boron neutron capture therapy of cancers have been discussed.[51] These include the synthesis of carbaborane derivatives of estradiol, testosterone, and other steroids,[52a,b] and the analysis of such compounds.[52c] Carbaborane containing polyphosphonates have been prepared for the treatment of calcific tumors,[53] and the use of B-decachloro-1,2-$H_2C_2B_{10}Cl_{10}$ derivatives has also been studied.[54]

3 σ-Bonded Metallacarbaboranes.

Mercurated or thalliated 9-B-derivatives of 1,2- and 1,7-$C_2B_{10}H_{12}$ continue to be used as synthetic reagents. Reactions of $Hg(C_2B_{10}H_{11})_2$ with selenium or tellurium in the presence of a solvent gave $(C_2B_9H_{11})HgM(C_2B_9H_{11})$, but heating in the solid state produced $Se_2(C_2B_9H_{11})_2$ and $Te(C_2B_9H_{11})_2$.[55a] The last compound was also obtained from the reaction of Te with $Tl(C_2B_9H_{11})(CO_2CF_3)_2$ in the presence of solvent. Reaction of $SnCl_2$ with 9-Hg derivatives of 1,2- or 1,7- $C_2B_{10}H_{10}R_2$ led to 9-$(SnCl_3)$-substituted products.[55b] Thermolysis of 9-Hg compounds gave mono- and oligo-carbaboranes as major products.[55c] Small amounts of B-B' coupled dicarbaboranes were also formed. Reaction of $Tl(CO_2CF_3)_3$ with 2-$\{Hg(CO_2CF_3)\}$-1,2-$C_2B_{10}H_{11}$ gave 2-$Tl(CO_2CF_3)_2$-1,2-$C_2B_{10}H_{11}$.[55d] Reduction of 9-TlX_2-substituted 1,2- and 1,7-$C_2B_{10}H_{12}$ by $K_2[C_8H_8]$ yielded the carbaboranes.[35]

E.s.r. was used to characterise the free radical product from the reaction of 9-{$SnCl_3$}-1,7-$C_2B_{10}H_{11}$ and 3,6-di*tert*butyl-σ-benzoquinone.[56] ^{199}Hg n.m.r. spectra of 9-HgR-derivatives (R = Me or 1,7-$C_2B_{10}H_9R_2$') of 1,7-$C_2B_{10}H_9R_2$' (R' = H,Me, or $SiMe_3$) have been recorded.[57] The structure of bis(1,2-dimethyl-1,2-dicarboran-9-yl) dimethyltin has been determined *X*-ray crystallographically.[58]

Carbon- and B-bonded carbaborane derivatives of some lanthanides have been synthesised by reaction of the elements (La, Tm,Yb) with mercurated carbaboranes, and by reaction of lanthanum trichlorides with C-lithiated carbaboranes.[59] Reactions in THF often gave solvated products, e.g. the B-bonded [$Yb(C_2B_{10}H_{11})_2$.THF] and C-bonded [$Tm(C_2B_{10}H_{10}Me)_3$.3THF].[59a] Reactions with lithiated carbaboranes gave mono-, di-, or tri-substituted products, some of which were solvated, depending on the reagents and the reaction conditions.[59b]

Carbon-bonded 1,2-$C_2B_{10}H_{12}$ derivatives of Co, Ni, and Pd have been synthesised.[60] They include complexes of the type [$M(C_2B_{10}H_{11})_2$bipy] and compounds based on 1-carboxylic acid derivatives of 1,2-$C_2B_{10}H_{12}$.

The use of 1-[Ir(CO)(PhCN)(Ph_3P)]-7-Ph-1,2-$C_2B_{10}H_{10}$ as an hydrogenation catalyst for alkenes and alkynes has been investigated.[61] The reactivity was due to the ability of Ir(I) to oxidatively add H_2 and the ease of displacement of the PhCN ligand. Synthesis of B-bonded Rh(III) and Ir(III) complexes (4) from [{MCl(L)$_2$}$_2$] and 1-Ph_2PCH_2-1,2-$C_2B_{10}H_{11}$ {in the presence of 4-methylpyridine for Rh(I)} was reported.[62] Reaction without 4-Mepy gave the phosphine complex [Rh(Cl)(CO)(1-Ph_2PCH_2-1,2-$C_2B_{10}H_{11}$)$_2$].

4 Cage Metallacarbaboranes

This area continues to produce compounds which are structurally innovative. Consequently *X*-ray crystallography plays a particularly important role in the elucidation of the structures of the more unusual compounds. Although a few established synthetic routes has been developed, the mechanism of formation of most compounds is usually not well understood. A new route, the reaction of a borane fragment with a metallacycle, has produced a carbon-rich metallacarbaborane, Scheme 1.[63] The removal of the

Ph_3P ligand by complexing to BH_3 may be a significant part of the mechanism.

The first example of a planar η^8-C_8H_8 unit bonded to V has been observed in the paramagnetic [V(Et$_2$C$_2$B$_4$H$_4$)(η^8-C$_8$H$_8$)] (Figure 1).[64] The vanadium atom was rather further (1.830 Å) from the carbaborane C_2B_3 face than expected by comparison with related complexes. The air stable compound was formed by the reaction of [VCl$_3$], K$_2$[C$_8$H$_8$], and [2,3-Et$_2$C$_2$B$_4$H$_5$]$^-$. Replacement of [VCl$_3$] by [FeCl$_2$] in the above reaction yielded a 1,3,5-cyclooctatriene complex [Fe(Et$_2$C$_2$B$_4$H$_4$)(η^6-C$_8$H$_{10}$)] as the major product and [Fe(Et$_2$C$_2$B$_4$H$_4$)(η^6-C$_6$H$_6$)] and [Fe(Et$_2$C$_2$B$_4$H$_4$)(C$_{16}$H$_{18}$)] as minor products.[65] The cyclooctatriene ligand could be replaced by η^6-arene ligands on reaction with arenes (C_6H_6, MeC_6H_5, 1,3,5-Me$_3$C$_6$H$_3$, and C_6Me_6) over $AlCl_3$.[66] An alternative route to the η^6-toluene derivative involved reaction of Fe atoms, toluene, and 2,3-Et$_2$C$_2$B$_4$H$_6$.[67] *X*-ray diffraction studies were reported for [Fe(2,3-Me$_2$C$_2$B$_4$H$_4$)(η^6-C$_8$H$_{10}$)],[65] and all the [Fe(2,3-Et$_2$C$_2$B$_4$H$_4$)(η^6-arene)] compounds mentioned.[66,67] In all cases the FeC_2B_4 systems adopted similar *closo* structures which were analogous to previously reported structures of MC_2B_4 systems.

The direct insertion of low-valent ligated metal species into carbaborane cages continues to be a fruitful approach to metallacarbaboranes. By using this approach CpFe-, (Et$_3$P)$_2$Co-, and (Et$_3$P)$_2$ Pt-units have been introduced into the dihydrido-carbaferraborane [Fe(H)$_2$(2,3-Me$_2$C$_2$B$_4$H$_4$)$_2$], Scheme 2,[68] to give "BH-wedged" complexes. An *X*-ray crystallographic study of the Fe-Co compound confirmed that the structure was basically like an earlier CpCo-derivative. However, the reduction in the number of electrons formally supplied by the (Et$_3$P)$_2$Co -unit leads to significantly longer interactions in the FeCo-containing cage of this molecule and a slightly shorter Co-B(wedge) interaction.

The eight vertex [1,2-Me$_2$-3-(η^5-C$_5$H$_5$)-3,1,2-CoC$_2$B$_5$H$_5$] system has been studied by *X*-ray diffraction methods.[69] The dodecahedral structure was analogous to related compounds. The structural characterisation of 2:3'-[1,8-(η^5-C$_5$H$_5$)$_2$-1,8,5,6-Co$_2$C$_2$B$_5$H$_6$][2',4'-C$_2$B$_5$H$_6$] has been reported.[70] The cobaltacarbaborane cage is based on a tricapped trigonal prism and is bonded by a B-B single bond to the B(3) equatorial position on the pentagonal biprismatic $C_2B_5H_6$-species.

Reaction of $Li[CB_8H_{13}]$ and $[IrCl(Ph_3P)_3]$ afforded several products including $[1-(Ph_3P)-2-H-2,2-(Ph_3P)_2-2,10-IrCB_8H_8]$ which was characterised *X*-ray crystallographically.[71] The structure was basically a distorted bicapped square antiprism. However the cage geometry was unexpected in the light of previous work on metal derivatives of the isoelectronic species $[B_9H_{14}]^-$ and $1,3-C_2B_7H_{13}$. A novel cage closing mechanism was proposed for the iridium compound.

The synthesis of $[6-(\eta^5-C_5H_5)-6,1-NiCB_8H_9]$ from $Na[CB_8H_{13}]$ and nickelocene was reported.[72] This compound rearranged under mild conditions to the 10-CpNi-isomer. Reaction of CB_8H_{14} with CpH and $[Co(H_2O)_6]Cl_2$ in alcoholic KOH gave the $[2-(\eta^5-C_5H_5)-2,1-CoCB_8H_9]^-$ anion.

Thia- and selenacarbaborane derivatives *nido*-$B_9H_9XCNH_2R$ (R = Bu^t or *cyclo*-hexyl) were prepared by reacting alkyl isocyanides with $B_9H_{11}X$.[73] From spectroscopic evidence it was suggested that X and C were adjacent on a B_3XC open face with NH_2R attached to C.

Eleven vertex *nido*-compounds $[9-(ML_n)-7,8,C_2B_8H_{11}]$ (ML_n = CpNi-,[74] or $(Et_3P)_2(H)Pt-$ [75]) have been synthesised by the direct insertion route. Reaction of the nickel compound with $[AuMe(Ph_3P)]$ gave the $\mu_{10,11}$ - $Au(Ph_3P)$-bridged product, Scheme 3. The structures of these compounds were established by *X*-ray crystallography. On thermolysis the platinum compound eliminated H_2 to produce $[9-H-9,10-(Et_3P)_2-7,8,9-C_2PtB_8H_9]$ which however did not adopt a completely closed polyhedral structure, Figure 2. Also the Pt-H unit was unexpected since the $(R_3P)_2Pt$-group has been shown to be retained in the $[8,8-(Me_3P)_2-7,8,10-CPtCB_8H_{10}]$ analogue.

Whereas reaction of $[3,3-(Ph_3P)_2-3-H-3,1,2-RhC_2B_9H_{11}]$ with B_2H_6 or BF_3 gave $[Rh(Ph_3P)C_2B_9H_{11}]_2$, with BBr_3 it afforded an ionic complex $[HPPh_3][3-(Ph_3P)-3,3-Br_2-3,1,2-RhC_2B_9H_{11}]$ (Figure 3).[76] An *X*-ray crystallographic study confirmed the icosahedral based structure of the anion and indicated strong ion pairing (Br...H distances 2.70 and 3.06 Å). This feature probably also persisted in benzene solution. Both the iodo- and chloro-analogues of the anion were isolated as $[R_4N]^+$ salts. The synthesis of anionic complexes containing a formally Rh(I) vertex has been described.[77] The 3,1,2-,2,1,7- and 2,1,12-isomers of $[\{Rh(Ph_3P)_2\} C_2B_9H_{11}]^-$ were obtained as $[R_4N]^+$ or K[18-crown-6]$^+$

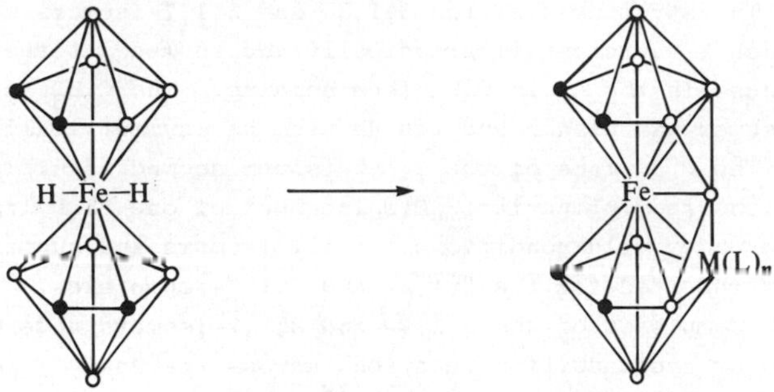

Reagents: i, [Co(Et$_3$P)$_4$] ; ii, [Pt$_2$ (μCOD) (Et$_3$P)$_4$] ; iii, [Fe(COD) (η-C$_5$H$_5$)]

Scheme 2

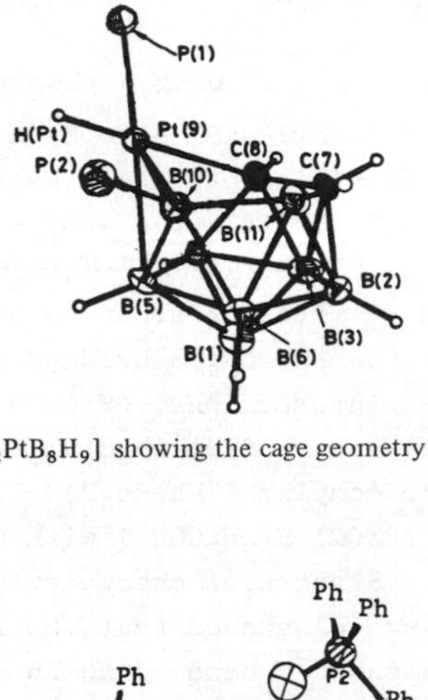

Figure 2 [9-H-9,10-(Et$_3$P)$_2$-7,8,9-C$_2$PtB$_8$H$_9$] showing the cage geometry

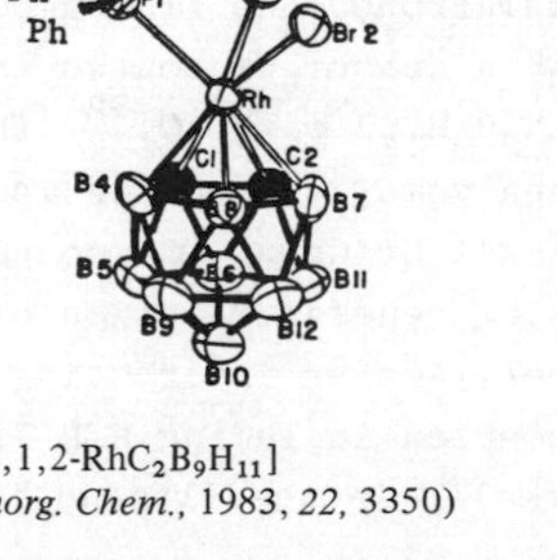

Figure 3 [HPPh$_3$] [3-(Ph$_3$P)-3,3-Br$_2$-3,1,2-RhC$_2$B$_9$H$_{11}$]
(Reproduced with permission from *Inorg. Chem.*, 1983, *22,* 3350)

salts. The structures of the 3,1,2- and 2,1,7-isomers were determined *X*-ray crystallographically and showed interesting differences in the Rh to C_2B_3 face bonding. The C_2B_3 face in the 3,1,2-isomer was planar and the Rh atom was symmetrically bonded to it. The C_2B_3 face of the 2,1,7-isomer showed significant distortions from planarity. Displacement of one Ph_3P ligand by CO occurred under mild conditions for all isomers and substitution by C_2H_4 was reported for the 3,1,2- and 2,1,7- complexes. The carbonyl complexes of the 3,1,2- and 2,1,7-isomers underwent 1,3-dipolar cycloaddition reactions across the Rh-C(O) bond when reacted with arylnitrile *N*-oxides.[78] An *X*-ray diffraction study carried out on the cycloaddition product from the 2,1,7-isomer showed the basic icosahedral structure to be retained. Reaction of the carbonyl complex of the 3,1,2-isomer with CH_2Cl_2 produced the chloro-complex [3-(Ph_3P)-3,3-Cl_2-3,1,2-$RhC_2B_9H_{11}$] mentioned previously.[76]

The syntheses of [2-$(L)_n$-7-Me-2,7,1-$MPCB_9H_{10}$] {$M(L)_n$ = $Rh(Ph_3P)_2$ and Ni(Br)Ph_3P} have been reported.[79]

Several dimeric metallacarbaboranes have been prepared. Reaction of [Rh(Cl)$(Ph_3P)_3$] with *nido* - $B_{10}H_{12}CNH_3$ in the presence of $Bu_4{}^nNOH$ gave [$Bu_4{}^nN$] [2,2-$(Ph_3P)_2$-2-H-1-(NH_2)-2,1-$RhCB_{10}H_{10}$].[80] Heating this compound in methanol produced an anionic bis-rhodocarbaborane with the two cages bridged by two NH_2-units, and with Rh-H-Rh and Rh-Rh interactions, (Figure 4). Both the rhodium hydride (1.92, 1.90 Å) and rhodium-rhodium (2.998 Å) interactions were rather long. The complex [{Rh(Et_3P)} $C_2B_9H_{10}$]$_2$ was obtained from the reaction of [Rh(Cl)Et_3P(COD)] with Cs_2[7,7'-(7',8'-$C_2B_9H_{11}$)(7,8-$C_2B_9H_{11}$)].[81] Structural characterisation by *X*-ray crystallography, (Figure 5), showed that the rhodacarbaborane cages were held together by the C-C bond, a Rh-Rh bond (2.725 Å), and two Rh-H-B interactions (Rh-H 1.92, 1.85 Å).

The dimeric nickelcarbaborane [{3-(μ-CO)-8-Ph_3P-3,1,2-$NiC_2B_9H_{11}$}$_2$] was prepared by a number of routes including reaction of [3,3-$(Ph_3P)_2$-3,1,2-$NiC_2B_9H_{11}$] with CO.[82] The structure, as determined by *X*-ray analysis, showed the icosahedra joined by a Ni-Ni bond and two nickel bridging CO-groups. The precusor complex had been prepared by the general reaction of [$Ni(Cl)_2(R_3P)_2$] with [7,8-$C_2B_9H_{11}$]$^{2-}$. Several other interesting reactions were reported for the mono-nickel complex including R_3P ligand replacement by CO and pyridine, and thermally induced ligand interchange to give

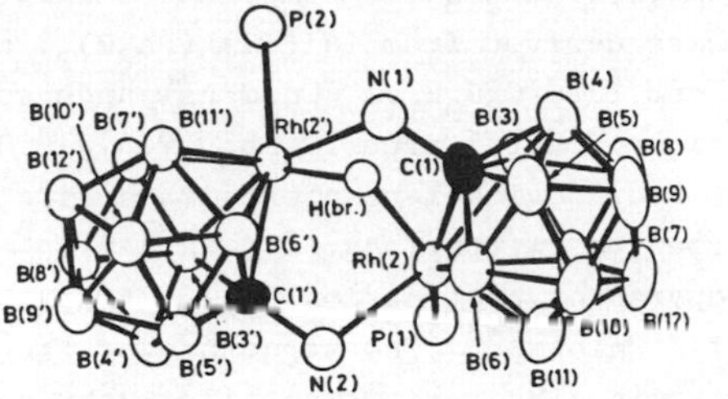

Figure 4 Cage structure of dirhodium complex anion $[\{(Ph_3P)RhCB_{10}H_{10}NH_2\}_2H]^-$

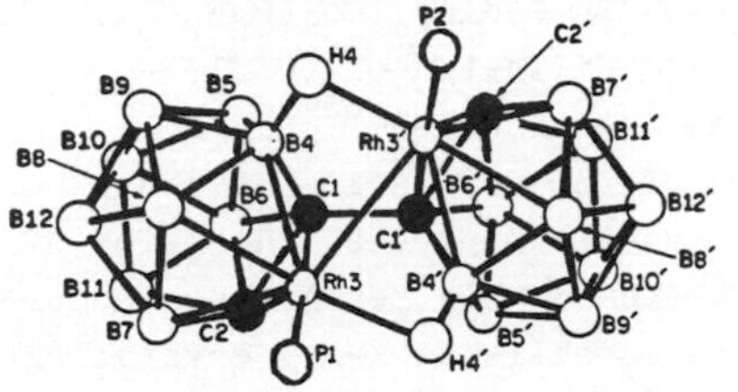

Figure 5 Cage structure of dirhodium complex $[(Et_3P)RhC_2B_9H_{10}]$
(Reproduced with permission from *Angew. Chem., Int. Ed. Engl.*, 1983, *22*, 722)

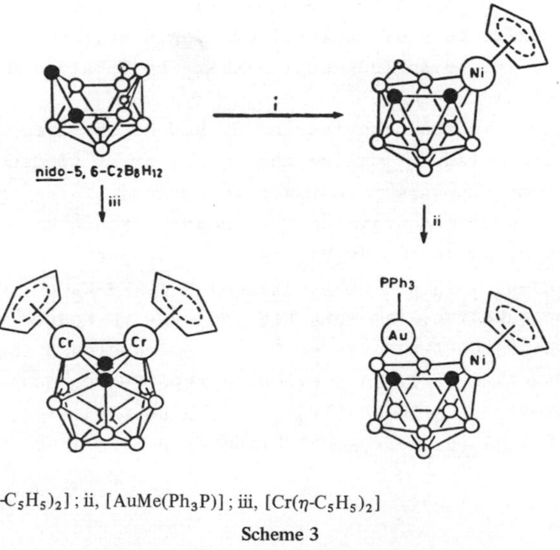

Reagents: i, $[Ni(\eta\text{-}C_5H_5)_2]$; ii, $[AuMe(Ph_3P)]$; iii, $[Cr(\eta\text{-}C_5H_5)_2]$

Scheme 3

$[3,8-(R_3P)_2-3-H-3,1,2-NiC_2B_9H_{11}]$ complexes. Surprisingly, the isomeric complexes derived from $[Ni(Cl)_2(Ph_3P)_2]$ and $[7,9-C_2B_9H_{11}]^{2-}$ or $[2,9-C_2B_9H_{11}]^{2-}$ did not undergo the ligand migration reactions. Metal free $10-Ph_3P-7,8-C_2B_9H_{11}$ was obtained on treating the hydridonickel-complex with acetic acid.

Stable carbenium zwitterions $[1-(CRR')-3-(\eta^5-C_5H_5)-3,1,2-FeC_2B_9H_{10}]$ were prepared by protonation of 1-CH(OR)R' carbinols and $1-C(R)=CH_2$ olefin anions.[83] The structure of the (R = H, R' = Ph) compound as determined by *X*-ray crystallography showed a strong interaction between the carbocationic centre and the iron atom ($Fe-C^+=2.23$ Å). Reactions between carbocationic groups and various nucleophiles were studied. The diamagnetic properties of Co polynuclear complexes including $[Co_4(C_2B_8H_{10})_3(C_2B_9H_{11})_2]^{4-}$ have been reported.[84] The diffusion of $[Co(1,2-C_2B_9H_{11})_2]^-$ through lipid membranes has been studied.[85]

Incorporation of the CpCr-unit into carbaborane cages has produced several unusual complexes. Reaction of $5,6-C_2B_8H_{12}$ with $[Cr(\eta^5-C_5H_5)_2]$ gave the double metal inserted product, Scheme 3, whose structure was established *X*-ray crystallographically.[74] Reacting $[Et_4C_4B_8H_8]^{2-}$, $Na[C_5H_5]$ and $CrCl_2$ produced two isomers each of $[Cr(\eta^5-C_5H_5)Et_4C_4B_7H_7]$ and $[Cr(\eta^5-C_5H_5)Et_4C_4B_8H_8]$.[86] The structures of the more stable isomers were determined with *X*-ray diffraction methods and the cage geometries are shown in (5) and (6). The insufficiency of skeletal bonding electrons in these structures has led to the suggestion that the Cr atoms may adopt 15-electron configurations.

Differences in the behaviour of Co and Ni sub-group $(R_3P)_2M$-units have been noted.[87] Unlike the Ni,Pd, or Pt species, the $(Et_3P)Co$-fragment appears more reactive towards *closo*-carbaboranes and can give products containing Co-H bonds. Reaction of $1-Me-1,2-C_2B_{10}H_{11}$ with $[Co(Et_3P)_4]$ gave paramagnetic $[1-Me-4-(Et_3P)-\mu_{4,7\text{ or }6}-\{Co(Et_3P)_2-\mu(H)_2\}-1,2,4-C_2CoB_{10}H_{10}]$. The distorted docosahedron framework had the cage Co bonded to a C_2B_4 face and a $(Et_3P)_2Co$-fragment *exo* to the polyhedron bridging the B(7)-Co interaction, (7). A previously reported compound $[4-(Et_3P)-1,7-Me_2-\mu_{4,8}-\{Co(H)(Et_3P)_2-\mu-(H)-\mu(Et_2P)\}-1,4,7-CCoCB_5H_4]$ had a $(Et_3P)_2(H)Co$-fragment bridging a B-Co link, (8).

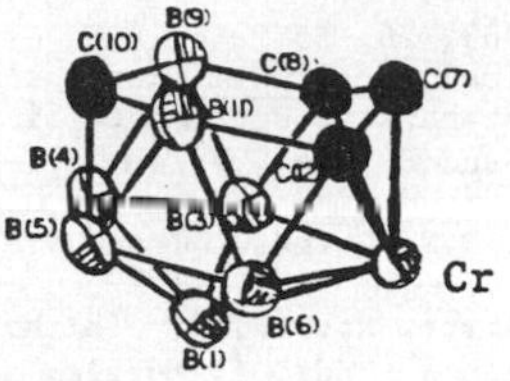
C(10)
B(9)
C(8)
C(7)
B(11)
B(4)
B(5)
Cr
B(6)
B(1)

(5)

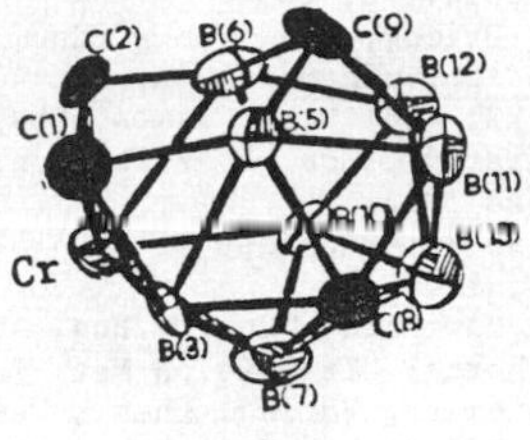
C(2)
B(6)
C(9)
B(12)
C(1)
B(5)
B(11)
Cr
B(3)
C(8)
B(7)

(6)

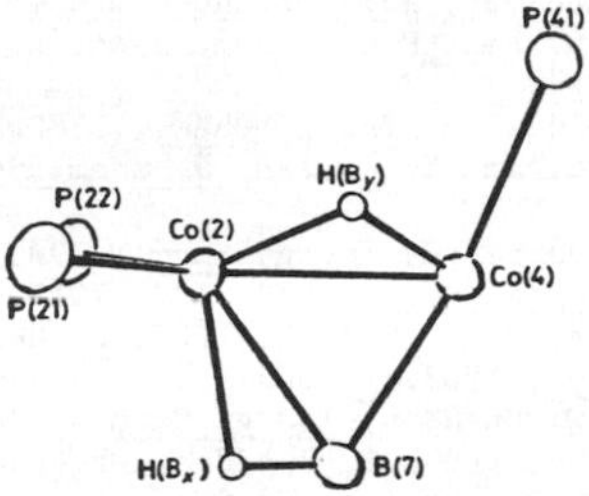
P(41)
H(B$_y$)
P(22)
Co(2)
Co(4)
P(21)
H(B$_x$)
B(7)

(7)

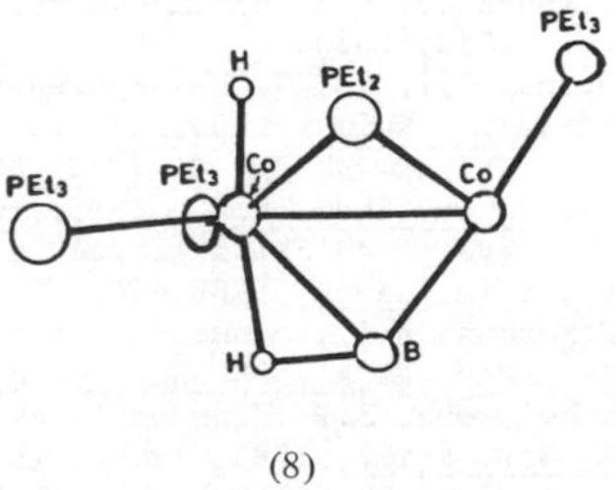
PEt$_3$
H
PEt$_2$
PEt$_3$
PEt$_3$
Co
Co
H
B

(8)

References

1 J.H. Morris, Chap. 4 in "Organometallic Chemistry", Eds. E. Abel and F.G.A. Stone, S.P.R., The Chemical Society, London, 1983.

2(a) J.B. Casey, W.J. Evans, and W.H. Powell, Inorg. Chem., 1983, 22, 2228; (b) ibid., p. 2236.

3(a) R.N. Grimes, Adv. Inorg. Chem. Radiochem., 1983, 26, 55; (b) R.N. Grimes, Pure Appl. Chem., 1982, 54, 43; (c) R.N. Grimes, Acc. Chem. Res., 1983, 16, 22; (d) R.N. Grimes, Chap. 5.5 in "Comprehensive Organometallic Chemistry", Eds. G. Wilkinson, F.G.A. Stone, and E. Abel, Pergamon, Oxford 1982.

4 "Metal Interactions with Boron Clusters", Ed. R.N. Grimes, Plenum Press, N.Y., 1982.

5(a) M.F. Hawthorne, Energy Res. Abstr.,1983, 8, Abstr. No. 28792; (b) M.F. Hawthorne, "Transition Met. Chem., Proc. Workshop", Eds. A. Mueller and E. Diemann, Verlag. Chem., Weinheim, Ger., 1981; (c) M.F. Hawthorne and M.S. Delaney, U.S.A. Pat., US 4363747A, 14th Dec. 1982.; (d) R.E. King, Diss. Abstr. Int. B., 1982, 43, 1094; (e) J. T. Petty, Diss. Abstr. Int. B, 1983, 43, 3235; (f) L. Borodinsky, Diss. Abstr., Int. B., 1983, 43, 2544.

6 O.V. Drygina and A.D. Garnovskii, Khim. Geterotsikl. Soedin., 1983, 5, 579.

7(a) S.G. Shore, Gov. Rep. Announce Index (U.S.), 1983, 83, 253; (b) C.E. Housecroft and T.P. Fehlner, Adv. Organomet. Chem., 1982, 21,57; (c) A.J. Welch, Chap. 3 in Annu. Rep. Prog. Chem. Sect. A, Inorg. Chem., 1982, 79A, 19.

8(a) A.J. Stone and M.J. Alderton, Inorg. Chem., 1982, 21, 2297; (b) P. Brint, J.P. Cronin, E. Seward, and T. Whelan, J. Chem. Soc.,Dalton Trans., 1983, 975.

9 R.L. DeKock and C.P. Jasperse, Inorg. Chem., 1983, 22, 3843.

10 E.D. Jemmis, J. Am. Chem. Soc., 1982, 104, 7017.

11 M.E. O'Neill and K. Wade, Polyhedron, 1983, 2, 963.

12 T.P. Hanusa, Polyhedron, 1982, 1, 663.

13 R.J. Astheimer and L.G. Sneddon, Inorg. Chem., 1983, 22, 1928.

14 T. Davan, E.W. Corcoran, and L.G. Sneddon, Organometallics, 1983, 2, 1693.

15 Z. Janousek, J. Plesek, B. Stibr, and S. Hermanek, Collect. Czech. Chem. Commun., 1983, 48, 228.

16 A. Ya. Usyatinskii, A.D. Ryabov, T.M. Shcherbina, V.I. Bregadze, and N.N. Godovikov, Izv. Akad. Nauk SSSR, Ser. Khim., 1983, 1637.

17 J. Plesek, B. Stibr, E. Drdakova, Z. Plzak, and S. Hermanek, Chem. Ind. (London), 1982, 778.

18 V.N. Kalinin, N.I. Kobel'kova, and L.I. Zakharkin, Izv. Akad. Nauk SSSR, Ser. Khim., 1982, 1661.

19 L.I. Zakharkin, A.I. Kovredov, V.A. Ol'shevskaya, and S.V. Vitt, Izv. Akad. Nauk SSSR, Ser. Khim., 1983, 1680.

20 O.V. Drygina, A.D. Garnovskii, Yu. V. Kolodyazhnyi, M.G. Gruntfest, and A.V. Kazantsev, Zh. Obshch. Khim., 1983, 53, 1066.

21 L.I. Zakharkin and G.G. Zhigareva, Zh. Obshch. Khim. 1983, 53, 953.

22 G. Zheng and M. Jones, J. Am. Chem. Soc., 1983, 105, 6488.

23 L.A. Churkina, T.D. Zvereva, I.A. Shingel, and Yu. A. Ol'dekop, Vestsi Akad. Navuk BSSR, Ser. Khim. Navuk, 1983, 56.

24 V.V. Korshak, N.I. Bekasova, A.I. Solomatina, T.M. Frunze, A.A. Sakharova, and O.A. Mel'nik, Isv. Akad. Nauk SSSR, Ser. Khim., 1982, 1904.

25(a) V.N. Kalinin, N.I. Kobel'kova, E.V. Krasnokutskaya, and L.I. Zakharkin, Izv. Akad. Nauk SSSR, Ser. Khim., 1983, 1200; (b) L.I. Zakharkin, G.G. Zhigareva, and E.I. Litonina, Zh. Obshch. Khim., 1982, 52, 2367.

26 V.N. Kalinin, A.V. Astakhin, A.V. Kazantsev, and L.I. Zakharkin, Zh. Obshch. Khim., 1982, 52, 1932.

27 V.A. Kolesova, N.A. Tagirova, Yu. A. Strepikheev, V.N. Kalinin, L.I. Virin, and Yu. I. Popova, Zh. Obshch. Khim., 1983, 53, 239.

28(a) V.V. Grushin, T.P. Tolstaya, and A.N. Vanchikov, Dokl. Akad. Nauk SSSR, 1982, 264, 868; (b) V.V. Grushin, T.P. Tolstaya, I.N. Lisichkina, Yu. K.

28(b) Grishin, T.M. Shcherbina, V. Ts. Kampel. V.I. Bregadze, and N.N. Godovikov, Izv. Akad. Nauk SSSR., Ser. Khim., 1983, 472.
29 K. Fuller and T. Onak, J. Organometal. Chem., 1983, 249. C6.
30 H.R. Allcock, A.G. Scopelianos, R.R. Whittle, and N.M. Tollefson, J. Am. Chem. Soc., 1983, 105, 1316.
31 L.L. Fewell, R.J. Basi, and J.A. Parker, J. Appl. Polym. Sci., 1983, 28, 2659; L.L. Fewell, R.J. Basi, and J.A. Parker, Sci. Tech. Aerosp. Rep., 1983, 21 (11), Abstr. No. N83-21034.
32 L.I. Zakharkin, M.G. Meiramov, V.A. Antonovich, A.V. Kazantsev, A.I. Yanovskii, and Yu. T. Struchkov, Zh. Obshch. Khim., 1983, 53, 90.
33 C. Novak, V. Subrtova, A. Linek, and J. Hasek, Acta Crystallogr., 1983, 39C,1393.
34 F. Teixidor and R.W. Rudolph, J. Organomet. Chem., 1983, 241, 301.
35 A. Ya. Usyatinskii, Z.V. Todres, T.M. Shcherbina, V.I. Bregadze, and N.N. Godovikov, Izv. Akad. Nauk SSSR., Ser. Khim., 1983, 1640.
36 W. Jarvis, Z.J. Abdou, and J. Onak, Polyhedron, 1983, 2, 1067.
37 L.A. Leites, L.E. Vinogradova, A.I. Kovredov, Zh. S. Shaugumbekova, and L.I. Zakharkin, Izv. Akad. Nauk SSSR, Ser. Khim., 1982, 2170.
38(a) L.E. Vinogradova, L.A. Leites, A.I. Kovredov, V.A. Ol'shevskaya, and L.I. Zakharkin, Izv. Akad. Nauk SSSR., Ser. Khim., 1982, 1663; (b) L.A. Leites, L.E. Vinogradova, J. Belloc, and A. Novak, J. Mol. Struct., 1983, 100, 379.
39 S.S. Bukalov, L.A. Leites, A.L. Blumenfeld, and E.I. Fedin, J. Raman Spectrosc., 1983, 14, 210.
40(a) S.F. Chistov, E.S. Sobolev, I.I. Skorokhodov, and E.I. Larikov, Zh. Obshch. Khim., 1983, 53, 733; (b) E.S. Sobolev, S.F. Chistov, E.I. Larikov, and I.I. Skorokhodov, *ibid*, p. 1847.
41 V.A. Mazunov, Yu. S. Nekrasov, V.I. Khvostenko, and V.I. Stanko, Izv. Akad. Nauk SSSR., Ser. Khim., 1983, 223.
42 A.I. Yanovskii, Yu. T. Struchkov, L.E. Vinogradova, and L.A. Leites, Izv. Akad. Nauk SSSR., Ser. Khim., 1982, 2257.
43(a) A.I. Yanovskii, I.L. Dubschak, V.E. Shklover, Yu. T. Struchkov, V.N. Kalinin, B.A. Ismailov, V.D. Myakushev, and L.I. Zakharkin, Zh. Strukt. Khim., 1982, 23, 88; (b) B.A. Ismailov, V.N. Kalinin, A.A. Zhdanov, and L.I. Zakharkin, Vysokomol. Soedin., Ser. A, 1983, 25, 1253.
44(a) G.D. Riska, S.A. Estes, J.O. Beyer,and P.C. Uden, Spectrochim. Acta, 1983, 38B, 407; (b) L.G. Sarto, Diss. Abstr. Int. B, 1983, 43, 2541.
45 M.M. Shchedrina, M.A. Dubrova, K.K. Prilepskaya, G.D. Kayukova, and T.A. Rudol'fi, Vysokoeffekt. Gaz. Khromatografiya, M., 1982, 124.
46(a) I.A. Khotina, A.I. Kalachev, M.M. Teplyakov, P.M. Valetskii, V.V. Korshak, S.V. Vinogradova, V.N. Kalinin, L.I. Zakharkin, V.I. Stanko, and A.I. Klimova, Otkrytiya, Isobret., Prom. Obraztsy, Tovarnye Znaki, 1982, (30), 296; (b) V.V. Korshak, M.M. Teplyakov, I.A. Khotina, A.I. Kovalev, and Kalinin, *ibid*, 1982, (38), 296; (c) M.M. Teplyakov, I.A. Khotina, A.I. Kalachev, P.M. Valetskii, S.V. Vinogradova, and V.V. Korshak, *ibid*, 1982, (26), 315, (d) V.S. Voishchev, P.M. Valetskii, V.I. Sidorenko, O.V. Voishcheva, S.V. Vinogradova, and V.V. Korshak, Vysokomol. Soedin., Ser. A, 1982, 24, 2211.
47 V.V. Korshak, N.I. Bekasova, A.I. Solomatina, Zh. P. Vagina, N.V. Klimentova, A.P. Supnin, Vysokomol. Soedin., Ser. A, 1983, 25, 989.
48(a) V.V. Korshak, N.I. Bekasova, and L.G. Komarova, Vysokomol. Soedin., Ser. A, 1982, 24, 2424; (b) S.A. Pavlova, P.N. Gribkova, T.N. Balykova, T.V. Polina, L.G. Kamarova, N.I. Bekasova, and V.V. Korshak, *ibid*, Ser. A, 1983, 25, 1270; (c) V.V. Korshak, L.G. Kormarova, P.V. Petrovskii, and N.I. Bekasova, *ibid*, Ser. B, 1983, 25, 30; (d) H. Raubach, A. Schumann, H. Oehlert, V.V. Korshak, E.S. Krongauz, and N.I. Bekasova, German (East) Pat. DD1550752, 12 May 1982.
49(a) G.G. Frenkel, Yu. K. Kirilenko, A.M. Shchetinin, V.N. Chichkova, P.M. Valetskii, A.I. Kalachev, and G.S. Bondarevskii, Otkrytiya, Izobret., Prom.

49(a) Obraztsy, Tovarnye Znaki, 1982, 34, 132; (b) V.V. Korshak, T.M. Frunze, A.A. Sakharova, O.A. Mel'nik, N. I. Bekasova, and A.I. Solomatina, *ibid*, 1982, 24, 125; (c) R. Sh. Frenkel, *ibid*, 1983, 25, 89; (d) B.E. Walker, R.W. Rice, P.F. Becher, B.A. Bender, and W.S. Coblenz, Am. Ceram. Soc. Bull., 1983, 62, 916.
50 M. Kyrs, J. Plesek, J. Rais, and E. Makrlik, Czech Pat. 211942B, 1 Nov. 1982.
51(a) "Synthesis and Applications of Isotopically Labelled Compounds", Eds. W.P. Duncan and A.B. Susan, Elsevier Co., Amsterdam, 1983; (b) E.V. Shumilina, Deposited Doc., 1981, VINITI, 5314.
52(a) F. Sweet and B.R. Samant in ref. 51(a) p. 175; (b) S.B. Kahl in ref. 51(a) p. 181; (c) I.S. Krull, S.W. Jordan, S. Kahl, and S.B. Smith, J. Chromatogr. Sci., 1982, 20, 489.
53 J.J. Benedict, Eur. Pat. EP68584 A1, 5 Jan. 1983.
54(a) D. Gabel and R. Walczyna, Z. Naurforsch, 1982, 37C, 1038; (b) D. Gabel, German (West) Pat., DE 3142442 A1, 5 May 1983.
55(a) V.I. Bregadze, V. Ts. Kampel, A. Ya. Usyatinskii, O.B. Ponomareva, and N.N. Godovikov, J. Organomet. Chem., 1982, 233, C33; (b) V.I. Bregadze, T.K. Dzhashiashvili, D.N. Sadzhaya, M.V. Petriashvili, O.B. Ponomareva, T.M. Shcherbina, V. Ts. Kampel, L.B. Kukushkina, V. Ya. Rochev, and N.N. Godovikov, Izv. Akad. Nauk SSSR., Ser. Khim., 1983, 907; (c) L.I. Zakharkin and I.V. Pisareva, Izv. Akad. Nauk SSSR., Ser. Khim., 1983, 1158; (d) V.I. Bregadze, A. Ya. Ustyatinskii, and N.N. Godovikov, Izv. Akad. Nauk SSSR., Ser. Khim., 1983, 1405.
56 Z.K. Kasymbekova, A.I. Prokof'ev, N.N. Bubnov, S.P. Solodovnikov, V.I. Bregadze, V. Ts. Kampel, M.V. Petriashvili, N.N. Godovikov, and M.I. Kabachnik, Izv. Akad. Nauk SSSR., Ser. Khim., 1983, 316.
57 Yu. K. Grishin, V.A. Roznyatovskii, Yu. A. Ustynyuk, V. Ts. Kampel,and V.I. Bregadze, Vestn. Mosk. Univ., Ser. 2: Khim., 1982, 23, 488.
58 A.I. Yanovskii, Yu. T. Struchkov, V.I. Bregadze, V. Ts. Kampel, M.V. Petriashvili, and N.N. Godovikov, Izv. Akad. Nauk SSSR., Ser. Khim., 1983, 1523.
59(a) G.Z. Suleimanov, V.I. Bregadze, N.A. Koval'chuk, and I.P. Beletskaya, J. Organomet. Chem., 1982, 235, C17; (b) V.I. Bregadze, N.A. Koval'chuk, N. N. Godovikov, G.Z. Suleimanov, and I.P. Beletskaya, J. Organomet. Chem., 1983, 241, C13.
60(a) Yu. A. Ol'dekop, N.A. Maier, A.A. Erdman, and V.P. Prokopovich, Zh. Obshch. Khim., 1982, 52, 2256; (b) *idem*, Vestsi Akad.Navuk BSSR., Ser. Khim. Navuk, 1983, 114; (c) *ibid*, 1982, 72.
61 F. Morandini, B. Longato, and S. Bresadola, J. Organomet. Chem., 1982, 239, 377.
62 V.N. Kalinin, A.V. Usator, and L.I. Zakharkin, Zh. Obshch. Khim., 1983, 53, 945.
63 D.B. Palladino and T.P. Fehlner, Organometallics, 1983, 2, 1692.
64 R.G. Swisher, E. Sinn, G.A. Brewer, and R.N. Grimes, J. Am. Chem. Soc., 1983, 105, 2079.
65 R.B. Maynard, R.G. Swisher, and R.N. Grimes, Organometallics, 1983, 2, 500.
66 R.G. Swisher, E. Sinn, and R.N. Grimes, Organometallics, 1983, 2, 506.
67 R.P. Micciche and L.G. Sneddon, Organometallics, 1983, 2, 674.
68 G.K. Barker, M.P. Garcia, M. Green, F.G.A. Stone, and A.J. Welch, J. Chem. Soc., Dalton Trans., 1982, 1679.
69 G.J. Zimmerman and L.G. Sneddon, Acta Crystallogr., 1983, 39C, 856.
70 R.P. Micciche, J.S. Plotkin, and L.G. Sneddon, Inorg. Chem., 1983, 22, 1765.
71 N.W. Alcock, J.G. Taylor, and M.G. Wallbridge, J. Chem. Soc., Chem. Commun., 1983, 1168.
72 B. Stibr, Z. Janonsek, K. Base, J. Dolansky, S. Hermanek, K.A. Solnstev, L.A. Butman, I.I. Kuznetsov, and N.T. Kuznetsov, Polyhedron, 1982, 1, 833.
73 A. Arafat, G.D. Friesen, and L.J. Todd, Inorg. Chem., 1983, 22, 3721.

74 G.K. Barker, N.R. Godfrey, M. Green, H.E. Parge, F.G.A. Stone, and A.J. Welch, J. Chem. Soc., Chem. Commun., 1983, 277.

75(a) G.K. Barker, M. Green, F.G.A. Stone, W.C. Wolsey, and A.J. Welch, J. Chem. Soc., Dalton Trans., 1983, 2063; (b) ref. 8 in (a).

76 L. Zheng, R.T. Baker, C.B. Knobler, J.A. Walker, and M.F. Hawthorne, Inorg. Chem., 1983, 22, 3350.

77. J.A. Walker, C.B. Knobler, and M.F. Hawthorne, J. Am. Chem. Soc., 1983, 105, 3368.

78 J.A. Walker, C.B. Knobler, and M.F. Hawthorne, J. Am. Chem. Soc., 1983, 105, 3370.

79 L.I. Zakharkin and G.G. Zhigareva, Zh. Obshch. Khim., 1982, 52, 2802.

80 J.A. Walker, C.A. O'Con, L. Zheng, C.B. Knobler, and M.F. Hawthorne, J. Chem. Soc., Chem. Commun., 1983, 803.

81 P.E. Behnken, C.B. Knobler, and M.F. Hawthorne, Angew. Chem., Int. Ed. Engl., 1983, 22, 722.

82 R.E.King, S.B. Miller, C.B. Knobler, and M.F. Hawthorne, Inorg. Chem., 1983, 22, 3548.

83 L.I. Zakharkin, V.V. Kobak, A.I. Yanovsky, and Yu. T. Struchkov, Tetrahedron, 1982, 38, 3515.

84 V.V. Volkov, V.N. Ikorskii, and S. Ya. Dvurechenskaya, Izv. Akad. Nauk SSSR., Ser. Khim., 1983, 252.

85 G. Amblard, B. Issaurat, B. D'Epenoux, and C. Gavach, J. Electroanal. Chem. Interfacial Electrochem., 1983, 144, 373.

86 R.B. Maynard, Z-T. Wang, E. Sinn, and R.N. Grimes, Inorg. Chem., 1983, 22, 873.

87 G.K. Barker, M.P. Garcia, M. Green, F.G.A. Stone, and A.J. Welch, J. Chem. Soc., Chem. Commun., 1983, 137.

5

Group III: Aluminium, Gallium, Indium, and Thallium

BY P. G. HARRISON

1 Synthesis, Structure, and Bonding.

Methane reacts with aluminium atoms, but not with a wide range of other transition and main group metal atoms, to afford $HAlCH_3$ as the primary initial species:

$$:Al\cdot \ + \ CH_4 \ \rightarrow \ H\overset{\cdot}{-}\overset{Al}{-}\overset{\cdot}{-}CH_3 \ \rightarrow \ H-Al-CH_3$$

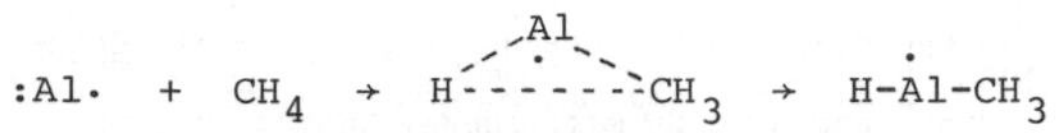

The unusual reactivity of aluminium is ascribed to its ^2P radical character, coupled with comparatively high Al-H and Al-C bond strengths (the reaction is predicted to be favourable by 34 kcal. mol^{-1}).[1] The stereochemistry, the regiochemistry and the influence of structure on reactivity has been investigated for the hydroalumination of a series of cyclic alkenes by diisobutyl-aluminium hydrides. Reactivity was observed to decrease along the series acenaphthylene>1,1-dimethyl-3-trimethyl-silylindene>1-methyl-acenaphthylene>1,1-dimethylindene>1-phenylacenaphthylene>>1,1-dimethyl-3-phenylindene. Experiments carried out in donor solvents demonstrated that cis-hydroalumination is the kinetically controlled mode of addition. Reactivity for 3-substituted-1,1-dimethylindenes reflected polar effects, whereas that for 1-substituted acenaphthylenes responded to steric effects. The regiochemistry of hydroalumination was generally determined by steric factors, except for the case of the silylindene. Isomerisation of the adducts also can occur, either by aluminium-carbon bond inversion or by reversible aluminium-hydrogen bond elimination. The former process takes place readily in the absence of Lewis bases, but the latter requires elevated temperatures and protracted times. The stable cis-hydro-alumination adducts undergo protodealumination with retention of configuration at the Al-C bond, although acenaphthenyl systems afforded significant amounts of 1,3-dihydroacenaphthylenes.[2] Triethylaluminium and methyl iodide undergo nearly quantitative alkyl group exchange on heating at 92° for 18 hours.[3]

Compounds of aluminium and gallium which contain the very bulky tris(trimethylsilyl)methyl, $(Me_3Si)_3C$ (Tsi), group have been synthesised by metathesis. $TsiAlCl_2$ is obtained in low yield from the reaction between TsiLi and $AlCl_3$ in ether/THF, whilst the compound TsiGa(OH)Me could be isolated from a reaction mixture of $GaCl_3$ and TsiLi from which residual MeLi had not been removed.[4] The silylmethylaluminium compound, $(Me_3SiCH_2)_2AlPPh_2$, which can function as an amphoteric ligand, has been synthesised from the reactions of either $(Me_3SiCH_2)_3Al$ or $(Me_3SiCH_2)_2AlH$ with PPh_2H, or of $(Me_3SiCH_2)_2AlBr$ with $KPPh_2$. In benzene solution, the phosphide exists as a monomer-dimer equilibrium, and, like the less sterically-hindered phosphides, $(Me_2AlPPh_2)_2$ and $(Et_2AlPPh_2)_2$, cleaves THF and reduce the triple bond of acetonitrile. No isolable adducts of the phosphides could be obtained from either THF or diethylether.[5] The reaction of $(Me_3SiCH_2)_2AlPPh_2$ with $Cr(CO)_5NMe_3$ in benzene leads to the formation of the complex $Cr(CO)_5[PPh_2Al(CH_2SiMe_2)_2.NMe_3]$ (<u>I</u>). In this complex, the geometry about both phosphorus and aluminium is rather distorted tetrahedral, and, although the Al-N bond distance is normal, the Al-P and Cr-P distances are unusually long. Total cleavage occurs when I is treated with anhydrous HBr and leads to the formation of $Cr(CO)_5PPh_2H$, $Br_3Al.NMe_3$, and Me_4Si, most probably via the initial scission of the long Al-P bond. No complexes with a similar Cr-P-Al sequence of atoms could be isolated from the reactions of $Cr(CO)_5L$ (L = CO, MeCN, THP) with the phosphides, R_2AlPPh_2 (R = Me, Et); rather attack on the labile ligand on chromium appeared to occur.[6] Reactions of the complexes $Cr(CO)_5PPh_2K.2$(dioxane) with the aluminium bromides AlR_2Br (R = Br, Me, Et, CH_2SiMe_3) in THF give high yields of compounds with the empirical formula $Cr(CO)_5[PPh_2(CH_2)_4OAlR_2]$, arising from cleavage of THF. No THF-Al adducts are observed. Crystals of $Cr(CO)_5[PPh_2(CH_2)_4OAl(CH_2SiMe_3)_2]$ are dimeric, the molecules lying on an inversion centre (<u>II</u>). Each octahedral $(OC)_5CrPPh_2$ fragment is linked by an $-(CH_2)_4O-$ unit to the two $Al(CH_2SiMe_3)_2$ fragments. The central $[Al_2O_2]$ ring is strictly planar, with the angle at aluminium only 79.7(5)°. Attempted small molecule elimination reactions between $Cr(CO)_5PPh_2H$ and Me_3Al or $AlMe_2H$ do not lead to compounds with Cr-P-Al linkages.[7] The similar amphoteric ligands, $R_2AlN(^tBu)PPh_2$ (R = Me, Et), react with $HMn(CO)_5$ to give the complexes <u>III</u>, resulting from a net migration

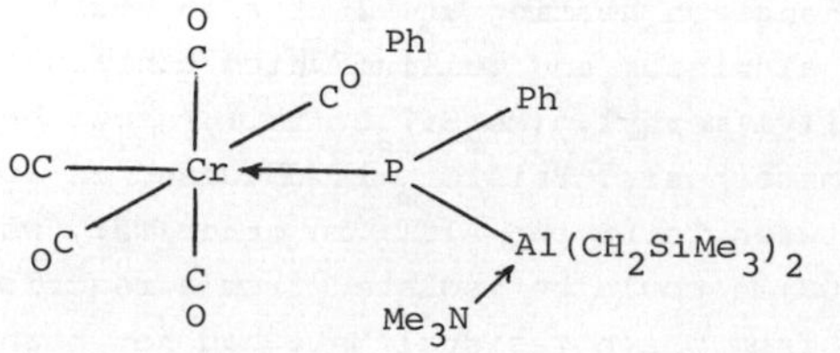
O
C
Ph
CO
Ph
OC
Cr
P
OC
Al(CH2SiMe3)2
C
O
Me3N

(I)

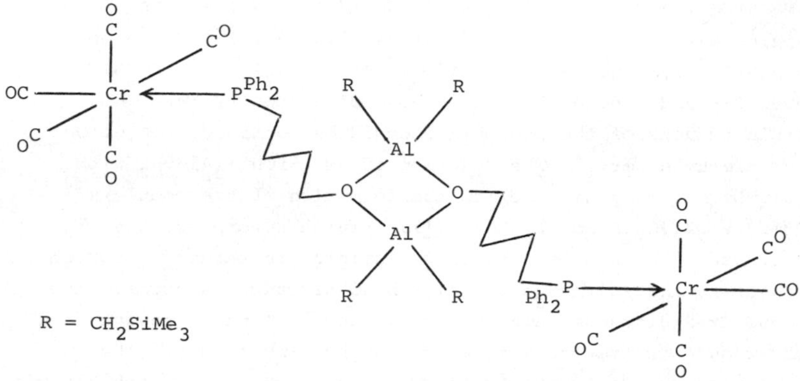
O
C
CO
OC
Cr
PPh2
OC
C
O
R
R
Al
O
O
Al
R
R
O
C
CO
Ph2
P
Cr
CO
OC
C
O
R = CH2SiMe3

(II)

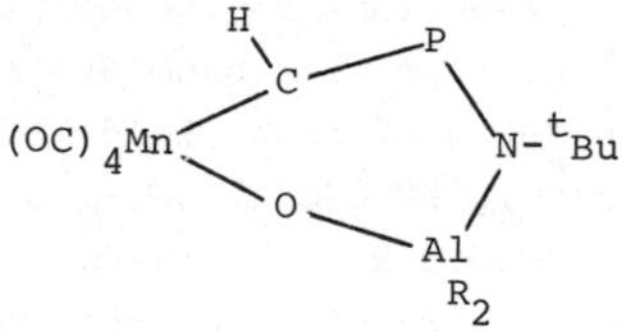
H
C
P
(OC)4Mn
N-tBu
O
Al
R2

(III)

of hydrogen from Mn to CO. However, NMR studies indicate that a direct migration does not occur, but rather proton transfer from Mn to P is the initial process. An analogue to the proposed transfer intermediate, the complex $[(C_5H_5)Mo(CO)_3][AlMe_2N(^tBu)PPh_2H]$, has been isolated from the reaction of $Ph_2PN(^tBu)AlMe_2$ with $(C_5H_5)Mo(CO)_3H$. The complexes III react with a second equivalent of amphoteric ligand. In the case of R = Me, the product was characterised by X-ray crystallography as the polycyclic heterocycle, IV, in which a C-H bond from an Al-Me group functions as one of the ligands to manganese.[8] $Ph_2PN(^tBu)AlEt_2$ induces rapid methyl group migration when treated with $(C_5H_5)(CO)_2FeMe$. The initial product of the reaction (V) has been shown to be an η^2(CO) complex of an acylphosphonium ligand, which in solution gradually rearranges to a mixture of the chelated acyl complex VI and the ethenolate complex VII, formed via a 1,2-hydrogen migration. The reaction of $(C_5H_5)Mo(CO)_3Me$ with $Ph_2PN(^iPr)AlEt_2$ affords predominantly VIII.[9]

Reaction of $(Ph_2P)_2NH$ with trimethyl- or triethylalane or -gallane yields the dimeric products, $[R_2MN(PPh_2)_2]_2$ (M = Al, Ga; R = Me, Et), with evolution of the corresponding alkane. Spectroscopic data for the products indicate that in the aluminium derivatives the ligand is unsymmetrically P,N-coordinated resulting in six-membered [AlNPAlNP] rings (IX), whilst with the gallium compounds a second symmetrical P,P-coordinate isomer with an eight-membered ring (X) appears to be part of an equilibrium in solution. The proposed structure of the dimethylaluminium compound has been confirmed by X-ray crystallography, and the six-membered ring is present in a twist-conformation. The nitrogen atoms adopt a trigonal planar configuration.[10] The unusual cobalt-dinitrogen-aluminium complex, $[(Me_3P)_3CoN_2AlMe_2]_2$, comprises a central four-membered $[Al_2N_2]$ ring (XI).[11] A similar ring system has been proposed to be present in the dimethylaluminium γ-picoline dimer (XII), formed by the reaction of the complex, $Me_2ClAl.NC_5H_4Me$ with lithium metal, on the basis of spectroscopic data.[12]

Four-membered $[Al_2O_2]$ and six-membered $[Al_3O_3]$ rings have been proposed or characterised in aluminoxane systems. Structures (XIII) and (XIV) have been proposed for tetraisobutylaluminoxane and triisobutylhydroaluminoxane, respectively, on the basis of spectroscopic data.[13] The structures of two novel aluminoxane anions have been determined by X-ray crystallography. Trimethyl-

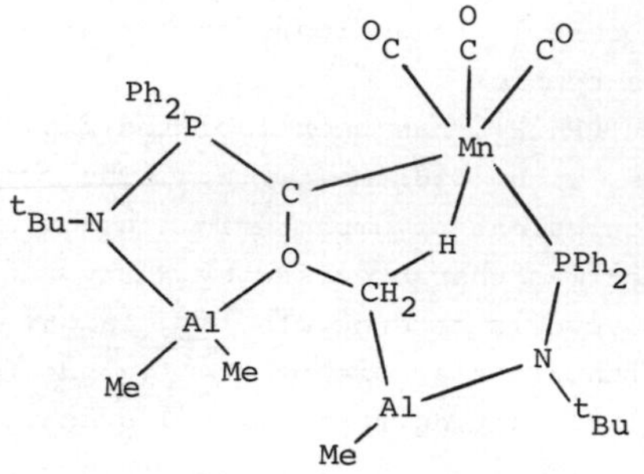
O
OC
C
CO
Ph2P
Mn
tBu-N
C
H
O
PPh2
CH2
Al
N
Me
Me
Al
tBu
Me

(IV)

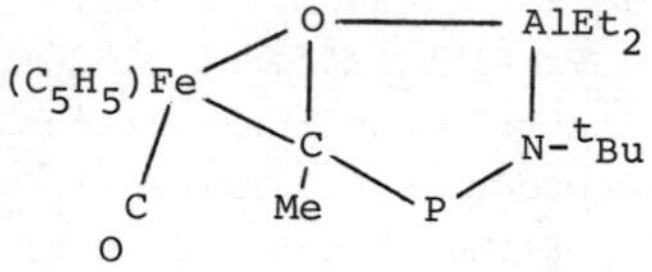
O
AlEt2
(C5H5)Fe
N-tBu
C
C
Me
P
O

(V)

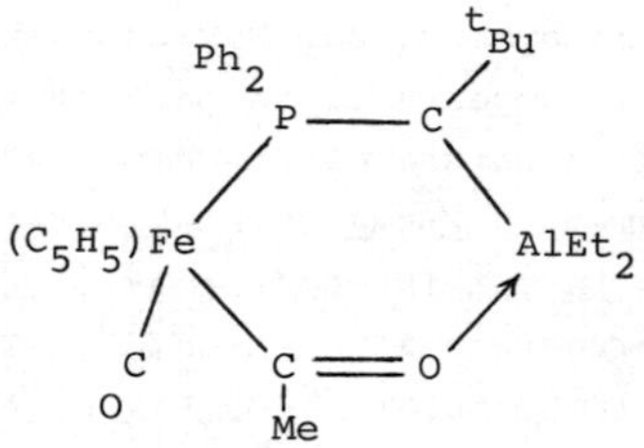
tBu
Ph2
P
C
(C5H5)Fe
AlEt2
C
C
O
O
Me

(VI)

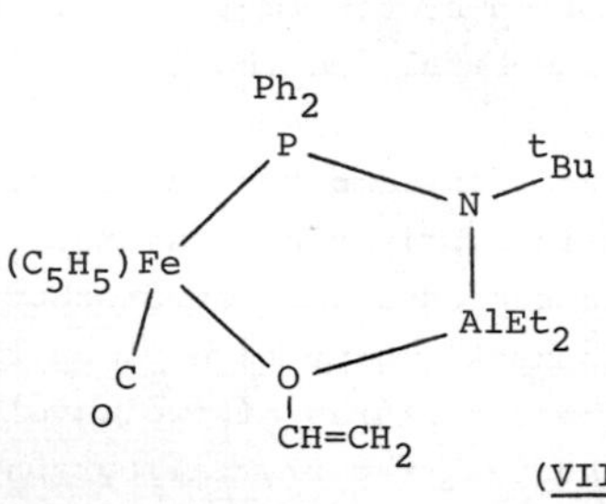
Ph2
P
tBu
N
(C5H5)Fe
AlEt2
O
C
O
CH=CH2

(VII)

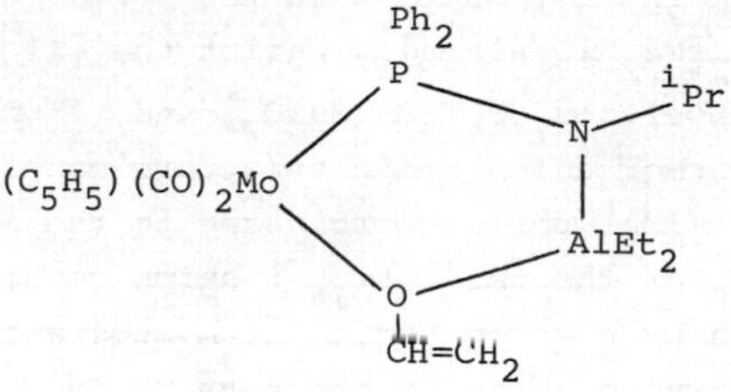

(VIII)

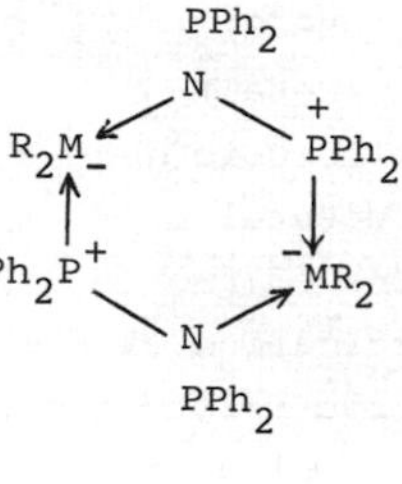

(IX)

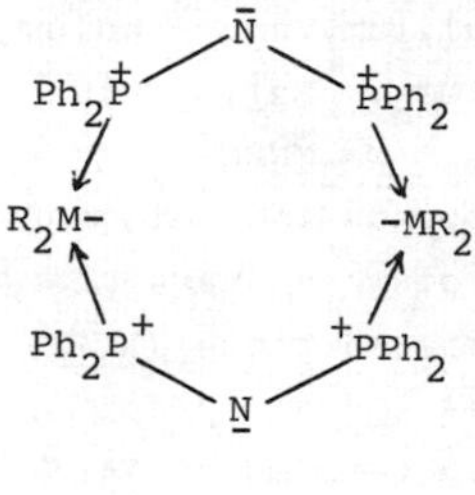

(X)

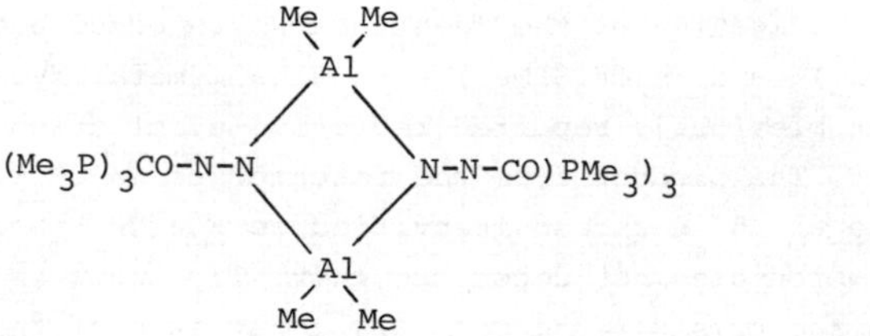

(XI)

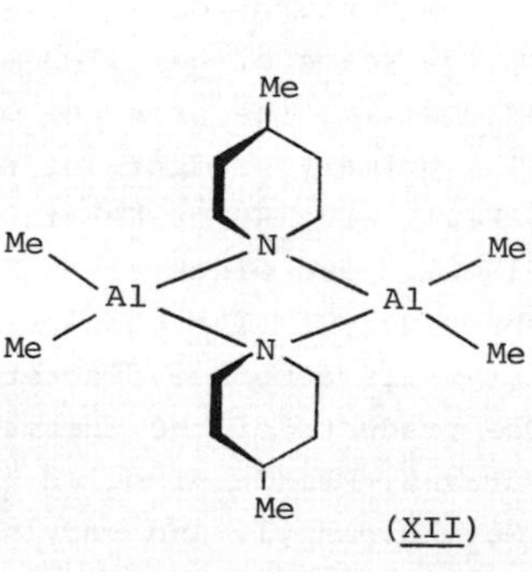

(XII)

aluminium methylates the $[Me_2AsO_2^-]$ anion with the formation of the anion compound $[Me_4As]_2[Me_2AlO.AlMe_3]_2$, whilst the $[Al_7O_6Me_{16}^-]$ ion is formed in the decomposition of $K[Al_2Me_6O_2]$ and $Cs_2[Al_2Me_6O]$.[14,15] The former anion has a structure consisting of a planar $[Al_2O_2]$ ring with $AlMe_3$ units coordinated to the oxygen atoms (XV). The structure of the $[Al_7O_6Me_{16}^-]$ anion consists of an open $[Al_6O_6]$ cage capped by a seventh aluminium atom which is bonded to three alternate oxygen atoms in the cage. Each oxygen atom is three-coordinated; the ones on the exterior (not bonded to the unique aluminium atom) bridge two aluminium atoms and are also bonded to one methyl group each. The cleavage of trimethyl-aluminium, -gallium, and -indium by dithiooxamides, $(HNRCS)_2$ (R = H, Me, $SiMe_3$, ^tBu), yields monomeric bis(dimethylmetal)-dithiooxamides, $(Me_2MNRCS)_2$ (M = Al, Ga, In). Spectroscopic data for these compounds indicate the presence of two structural isomers, both of which have planar molecular skeletons of two fused five-membered rings (XVII), but different coordination of the two metal atoms (*E*- and *Z*-forms). The centrosymmetric *E*-form has also been characterised by X-ray crystallography.[16] Molecular structures and energies for Me_3Al, the molecular cation and for all fragment ions and neutral fragments in the mass spectrum, have been calculated by MNDO with UHF wavefunctions.[17]

A reinvestigation of the reactions which occur between $Ga(CH_2SiMe_3)_3$ and $In(CH_2SiMe_3)$ with alkali metal hydrides has shown that previously reported reductive-elimination reactions do not occur. The product from the reaction of $Ga(CH_2SiMe_3)_3$ and KH in benzene at 45° has been identified as $KGa(CH_2SiMe_3)_3H$, a compound which does not decompose thermally until in excess of 200°. In the crystal, the geometry about each of the two gallium atoms is distorted tetrahedral, with Ga-H distances of 1.49 and 1.36Å. In the infrared from deuteration studies, the ν(Ga-H) is expected at the unusually low value of *ca*. 1500cm^{-1}. The reaction chemistry of the compound confirms the presence of a typically nucleophilic hydride. The primary products of reactions of the indium analogue, $In(CH_2SiMe_3)_3$ with KH or NaH in pentane at room temperature are $KIn(CH_2SiMe_3)_3H$ and mixtures of $NaIn(CH_2SiMe_3)_3H$ and $NaIn(CH_2SiMe_3)_4$, respectively. The complex $KIn(CH_2SiMe_3)_4$ has also been isolated, and crystals comprise discrete K^+ and $In(CH_2SiMe_3)_4^-$ ions. The products of the thermal decomposition of the indium hydride derivatives, $MIn(CH_2SiMe_3)_3H$ (M = K,Na), include indium metal, $SiMe_4$, biphenyl, and varying amounts of H_2

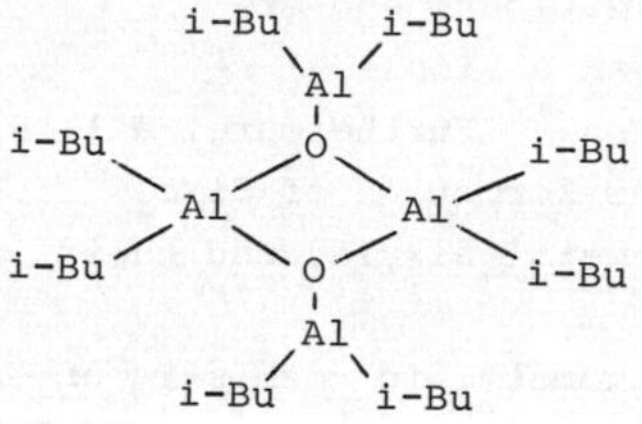

(XIII)

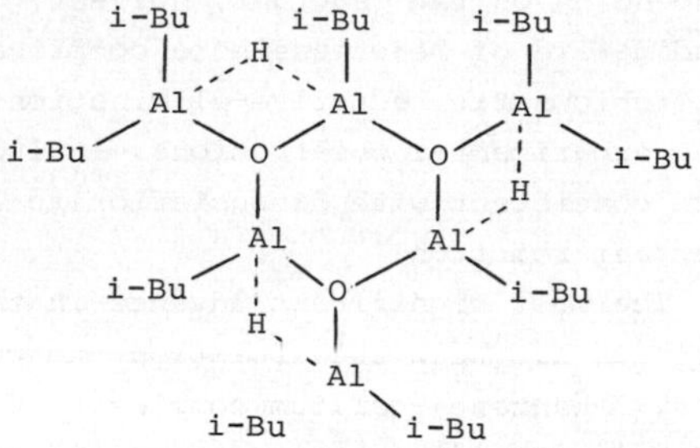

(XIV)

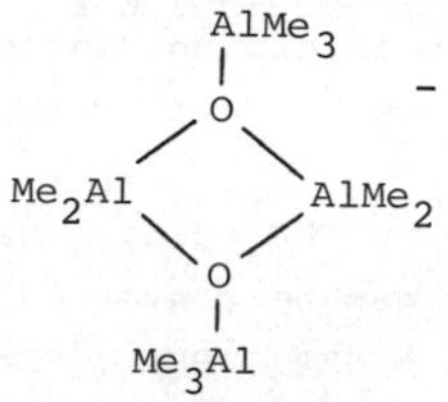

(XV)

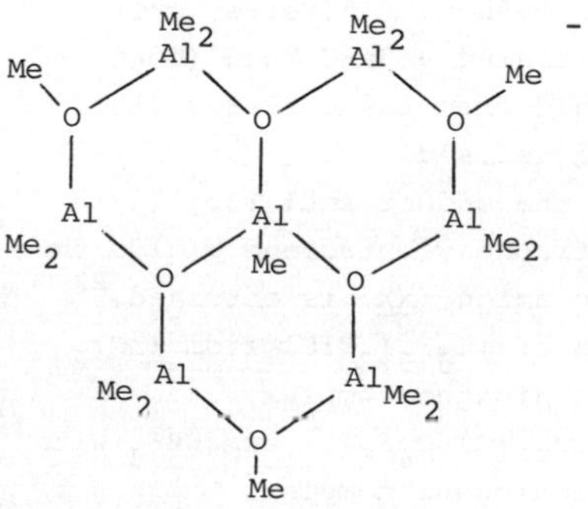

(XVI)

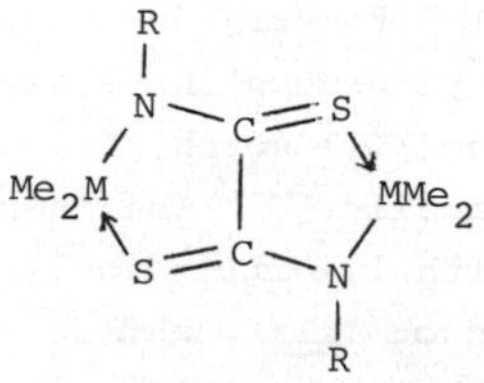

(XVII)

depending on the reactant, solvent, and conditions, and are indicative of reactions more complicated than a simple stoichiometric reductive-elimination reaction. Furthermore, all the experimental observations, including the formation of $SiMe_4$, are consistent with dissociation to alkali metal alkyls, and their further reaction.[18]

The most significant advance in the organometallic chemistry of gallium has been the isolation and characterisation of a bis(η^6-benzene)-gallium complex. Crystals of composition $(C_6H_6)_{3.5}Ga_2Cl_4$, obtained from solutions of Ga_2Cl_4 in anhydrous benzene, consist of cyclic centrosymmetrical $[(C_6H_6)_2Ga^I.Ga^{III}Cl_4]_2$ units and isolated benzene molecules. In the $[(C_6H_6)_2Ga^I]$ moieties the two η^6-benzene rings form an interplane angle of 124.4°, and pairs of gallium(I) centres are bridged by the $[GaCl_4]$ units (XVIII).[19]

The electrolysis of dimethylmagnesium in THF using a gallium anode produces $GaMe_3$.THF directly, which can be converted into the amine or phosphine complexes, $GaMe_3$.L (L = NEt_3, PMe_3, PEt_3), by the addition of excess of the ligand L. Uncomplexed $GaMe_3$ can be obtained by the reaction of a Mg-Ga alloy with methyl iodide in an ether solvent.[20] Several gallium, indium, and thallium compounds have been prepared by the organolithium method. Several novel gallium(III) porphyrin derivatives containing a Ga-C bond (both alkyl and aryl groups) have been obtained from the corresponding chlorogallium(III) porphyrin, and characterised spectroscopically.[21] The reaction of the adduct $InCl_3$.3py (py = pyridine) with 1,4-dilithio-1,2,3,4-tetraphenylbutadiene yields the pyridine adduct (XIX), whilst the spiro anion (XX) is obtained.[22] Details of the high yields of synthesis of $(C_6F_5)_2TlCl$ from $TlCl_3$ and C_6F_5Li have been described.[23] The dioxane complex, $Tl(C_6F_5)_3$.dioxane, is obtained from $[NBu_4][Tl(C_6F_5)_4]$ and HBF_4 with subsequent addition of dioxane. The analogous complex, $Tl(C_6F_3H_2$-2,4,6$)_3$.dioxane, is prepared by metathesis from $TlCl_3$ and the organolithium reagent in the presence of dioxane. The donor ligand can be replaced by other neutral donors [giving other complexes such as TlR_3.L (L = $OPPh_3$, pyridine, PPh_3 or $Ph_2PCH_2PPh_2$)], anionic ligands (to give the anions $[TlR_3X^-]$ (X = Cl^-, CN^-, NO_3^- or $CF_3CO_2^-$), or transition metal carbonyl anions (resulting in the formation of the anions $[TlR_3M^*]^-$ [M^* = $Mo(cp)(CO)_3^-$, $W(cp)(CO)_3^-$, $Co(CO)_4^-$ or $Mn(CO)_5^-$]).[24] Perfluoroarenes such as m-$H_2C_6H_4$, m-$O_2NC_6F_4H$, m- and p-BrC_6F_4H, (p-$HC_6F_4)_2$,

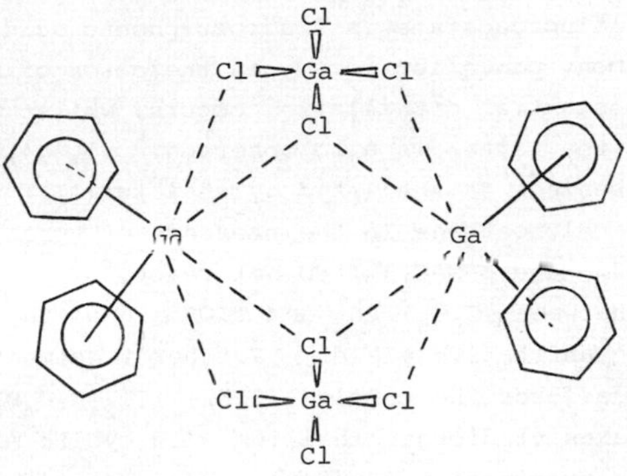

(XVIII)

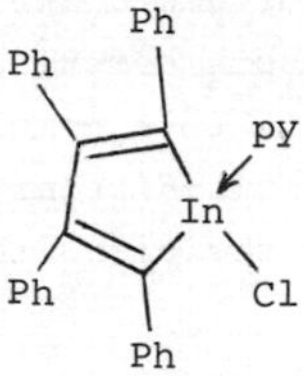

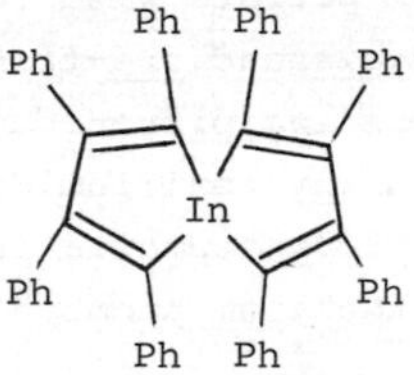

(XX)

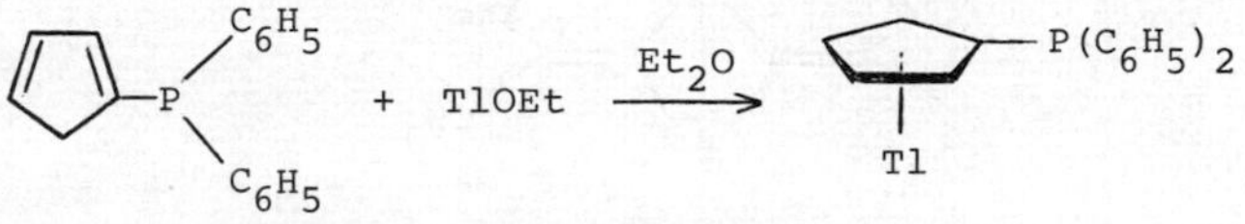

(XXI)

C_6F_5H, 1,3,5-$F_3C_6H_3$ and m-$FSO_2C_6H_4H$ undergo thallation by thallium(III) trifluoroacetate in fluorosulphonic acid in the presence of antimony pentafluoride. In the cases of m-$H_2C_6F_4$ and 1,3,5-$F_2C_6H_3$, substantial dithallation occurs, whilst some slight trithallation of the latter is also observed. 1,2,4,5-Tetra-fluoro-3-methoxybenzene is thallated by thallium(III) trifluoro-acetate in 1,2-dichloroethane in the presence of boron trifluoride etherate, but C_6F_5H and p-MeC_6F_4H do not react.[25]

The reaction between $(C_5H_5)PPh_2$ and TlOEt in diethyl ether produces XXI in quantitative yield. Further treatment of (XXI) with $(C_5H_5)TiCl_3$ affords the complex $(C_5H_5)TiCl_2[C_5H_4PPh_2]$.[26]

Several complexes of diorganothallium with cyclic multidentate ligands have been examined. Complexes of an N_6-macrocyclic ligand (XXII) have been synthesised by a Schiff base condensation of 2,6-diacetylpyridine with ethylenediamine using the thallium as a template.[27] The structures of two complexes with 18-crown-6 cyclic polyethers have been determined. Crystals of the complexes formed by dithallium picrate with two isomers of dicyclohexane-18-crown-6 (cis-cisoid-cis and cis-transoid-cis) both comprise $Me_2Tl(crown)^+$ cations and picrate anions.[28] In this complex, and in the analogous dimethylthallium(dibenzo-18-crown-6)trinitro-phenolate, the $[Me_2Tl]$ group axis is normal to the plane through the six oxygen and thallium atoms.[29]

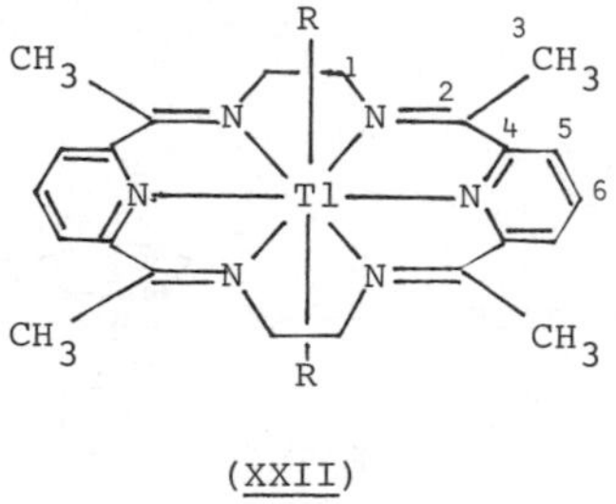

(XXII)

The organothallium dithiohydrazide derivatives, $RTlBr(S_2CNHNHPh)$; $R_2Tl(S_2CNHNHPh)$ and $RTl(S_2CNHNHPh_2)_2$, and dithiocarbamates $R_2Tl(S_2CNR'R'')$, where R = C_5H_5, MeC_5H_4, indenyl or fluorenyl, are monomeric non-electrolytes, in which the dithio group functions as a bidentate ligand.[30,31] The major factor influencing the coupling constants J(Tl-C) and J(Tl-H) in organo-thallium compounds of the types R_2TlX and $RTlX_2$ (R_2 = acyclic or alicyclic alkyl, alkenyl; X = anionic species) is the number of R groups attached to thallium, and the ratios of the analogous couplings in R_2TlX and $RTlX_2$ compounds are generally close to the value of 1:2.2 predicted on the assumption that the Fermi contact contribution dominates. The effects of solvent and anion X changes are generally minor.[32]

2 Reactions and Applications in Organic Synthesis

Interest in electron transfer reactions resulting in the formation of paramagnetic species appears to have increased. Triethylaluminium initially forms 1:1 complexes with the diimines ^tBu-N=CH-CH=N;^tBu (^{t}BuDAB) and ^tBu-N=CH-2-C_5H_4N (^{t}BuPya) (XXIII) and (XXIV), respectively, which subsequently react further to a mixture of N- and C-alkylated products (XXV) and (XXVI), respectively), (XXVII), formed by a hydride shift, and small amounts (<2%) of the stable radical (XXVIII). With ^{t}BuPya, the C-alkylated product (XXIX) is formed in nearly quantitative yield.[33] With tetracyanoethylene at -78° in toluene or polar solvents triethylaluminium undergoes 1,2-addition to the C=C double bond:

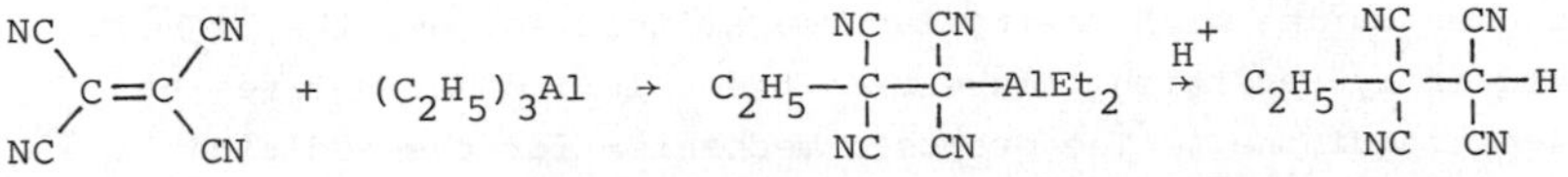

More than one Al-C bond can take part in the addition if an excess of TCNE is employed. Reaction with 7,7,8,8-tetracyanoquinodimethane results in 1,6-addition:

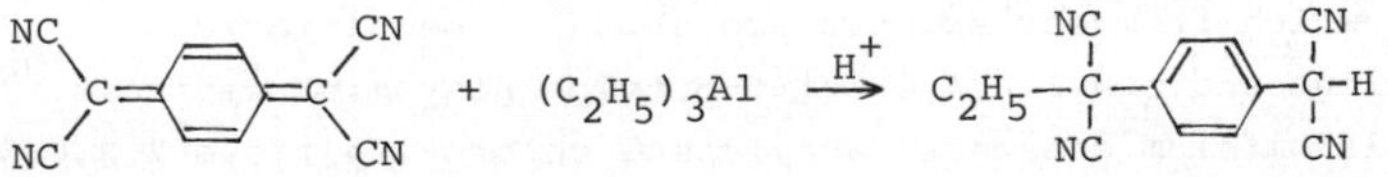

In toluene, ethylaluminium dichloride undergoes the 1,2- and 1,6-addition only at higher temperatures (ca. 0°) and in much lower yields than Et_3Al, although complexation by a polar solvent or a quaternary ammonium salt causes a marked increase in the yield.[34] Stable free-radicals have been observed in the aluminium trialkyl/-phenanthroline or 2,2'-bipyridyl, and the trimethylgallium, -indium, or -thallium/4,4'-bipyridyl systems.[35,36]

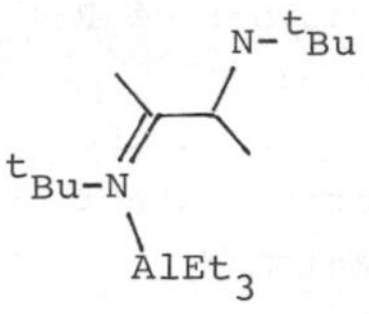

XXIII

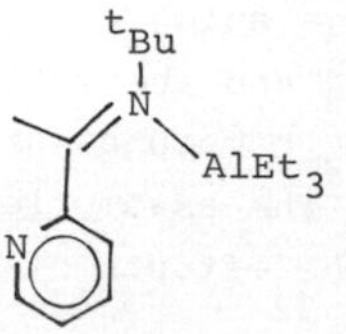

XXIV

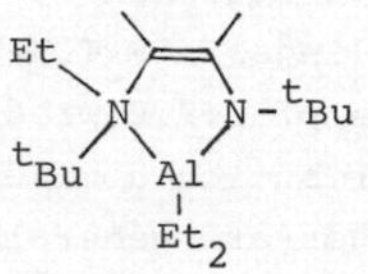

XXV

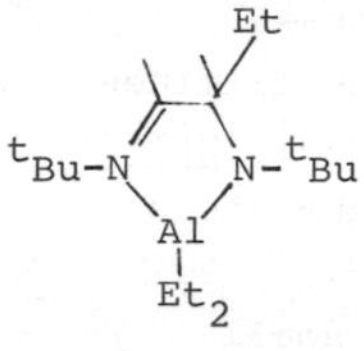

XXVI

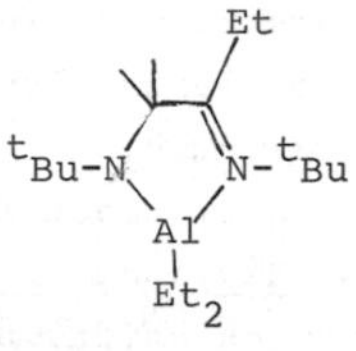

XXVII

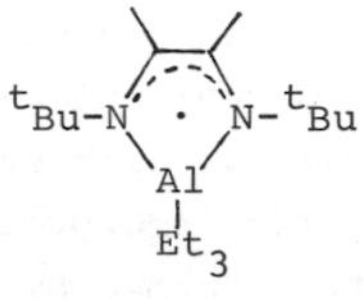

XXVIII

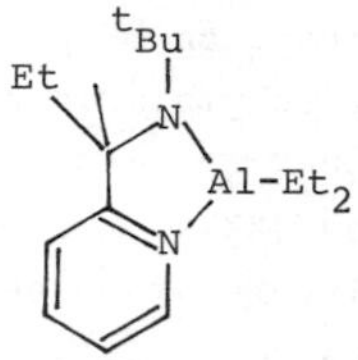

XXIX

Alkylaluminium dichlorides (except $MeAlCl_2$) undergo 1,6-addition to the conjugate O=C-C=C-C=O bond system of para-quinones. Triethylaluminium is much less reactive. The reactivities of the quinones varies with their electron affinities, and the highest yield of 1,6-addition occurs with the chlorine derivatives of 1,4-benzoquinone. The proposed mechanism for the addition involves homolytic cleavage of the Al-C bond in the donor-acceptor complex formed between the reactants, followed by a combination of alkyl radicals and aluminium derivatives of semiquinone within a solvent cage.[37] Long-lived paramagnetic products are formed when 2,4,6-tri-t-butylnitrosobenzene reacts with Me_3Al, Et_3Al, $EtAlCl_2$ and $^tBu_3Al$, which from esr spectra are shown to be nitroxide, N-alkoxyaniline and ethyl 2,4,6-tri-t-butylphenoxyamino radicals.[38]

Trimethylaluminium does not methylate 1-chloro-1-nitroso-2,2,6,6-tetramethylcyclohexane, but rather functions as an electron donor and initiator of radical reactions which lead to the formation of the oxime derivatives (XXX)-(XXXII), and the imine (XXXIII).[39]

The photolysis or thermolysis of the 18-crown-6-stabilised complex of phenyl(iodo)thallium picrate affords good yields of iodobenzene, at least partly via a phenyl radical intermediate. The corresponding chloride and bromide complexes are much more reluctant to undergo reaction, and only very small amounts of

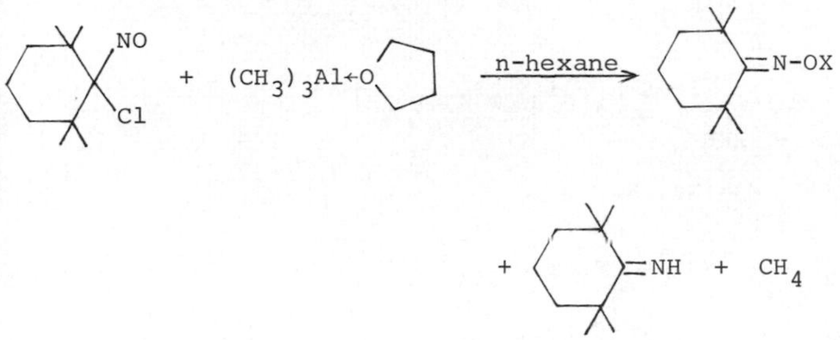

(XXXIII)

(XXX) X = 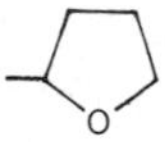

(XXXI) X = C_6H_{13}

(XXXII) X = $C(CH_3)_2(CH_2)_3C(CH_3)_2CN$

benzene and no halobenzene could be detected. Reduction of the analogous *p*-tolyl(hydroxo)thallium-crown ether complex with $P(OMe)_3$, ascorbic acid or *N*-benzyl-1,4-dihydronicotinamide gave toluene, but not, however, the di-*p*-tolylthallium species, which is formed in considerable amounts in the reduction of the crown-free *p*-tolylthallium cation.[40] The reaction of the organothallium diacetates, $RTl(OAc)_2$ (R = $PhCH(OMe)CH_2$, *p*-MeC_6H_4, *E*- and *Z*-PhCH=CH), with the anion of 2-nitropropane affords moderate-to-good yields of the coupled products, RMe_2CNO_2, probably *via* a radical non-chain substitution mechanism.[41] Decomposition of methylthallium diacetate in methanol in the presence of 2-, 3-, and 4-picoline, $PhCH_2NMe_2$, pyridine or picolinic acid gives mostly *N*-methylion. *O*-Methylation of acetate is only of minor importance. The data suggest that $MeTlOAc^+$ is the active electrophile.[42]

Mixed alkali metal alkylhydroaluminates, $MAlH_4R_{4-n}$ (M = Li, Na; R = Me, Et, ^{i}Bu; n = 1-3), can be prepared via redistribution:

$$nMAlH_4 + (4-n)MAlR_4 \rightarrow 4MAlH_4R_{4-n}$$

Al^{27} nmr studies show that the monoalkyl species $MAlH_3R$ are unstable and disproportionate in solution. The di- and trialkyl species are more stable, and the latter readily associate.[43] Other alkali metal aluminate salts havebeen obtained according to the equations:[44,45]

$$2\,{}^{i}Bu_2AlH + 2M \rightarrow M_2[H^{i}Bu_2AlAl^{i}Bu_2H]$$

$${}^{i}Bu_2AlAl^{i}Bu_2 + 2KH \rightarrow M_2[H^{i}Bu_2AlAl^{i}Bu_2H]$$

M = Na, K.

$$Me(CH_2)_4C(AlEt_2)_3 + MX \rightarrow M[Me(CH_2)_4C(AlEt_2)_3X]$$

M = Li, Na; X = H
M = K; X = F.

A Japanese patent has reported the quantitative conversion of triethylaluminium into $Et_2AlOAlEt_2$ by hydrolysis in toluene as solvent.[46] However, a more detailed study of the hydrolysis of trialkylaluminium compounds in diethyl ether has appeared. The hydrolysis takes place in several steps. The first is the formation of a complex between the trialkylaluminium, R_3Al, and water, and does not depend on the $Al:H_2O$ molar ratio. At low temperatures (<ca. -45°) the complex is stable, probably due to hydrogen-bond formation with the ether solvent as in (XXXIV). The second step, which occurs at ca. -45° involves loss of alkane and formation of dialkylaluminium hydroxide, which most probably exists in solution as a dimer (XXXV). Finally, reaction of the dialkylaluminium hydroxide with more trialkylaluminium affords the tetraalkyldialuminium oxane:

$$(R_2AlOH...OEt_2)_n + nR_3Al.OEt_2 \rightarrow nR_2AlOAlR_2.OEt_2 + nRH$$

High selectivity for aluminoxane formation can be achieved when the alkyl group is larger than methyl.[47] The hydrolysis of diisobutylalane results in the formation of the trialkylhydrido-aluminoxane, ${}^{i}Bu_2AlOAl^{i}BuH$, via the intermediate (XXXVI).[13] The mechanism of hydrolysis of the complex $Bu_3Al.MgBu_2$ proceeds via the formation of $Bu_2AlOMgBu$ and ultimately yields (XXXVII).[48] The species $[(AlO)(MeO)Bu]_{40}$, BuMgOAlBuOMgBu, $Bu_2AlOAlBuIMgBu$, $[(AlO)(MgO)Et]_{35}$ and $Et_2AlOMgEt$ have also been reported from the hydrolysis of trialkylaluminium/dialkylmagnesium phases in solvating or non-solvating solvents.[49] The partial hydrolysis of triisobutylaluminium by "cellosolves" affords (XXXVIII).[50] Several adducts of the type $Ph_2GaX.L$ and anions $Ph_2GaX_2^-$ (X = Cl, Br; L =

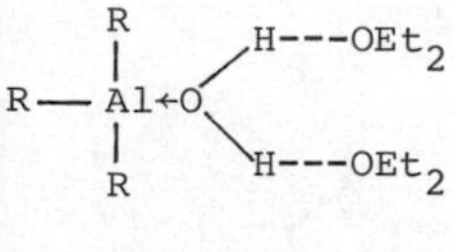

(XXXIV)

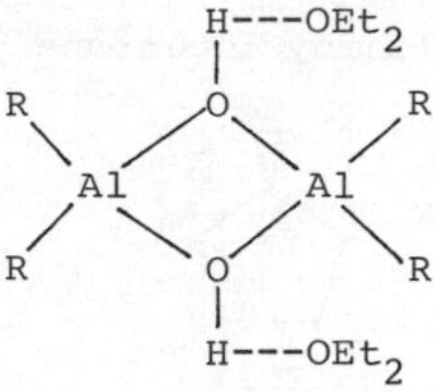

(XXXV)

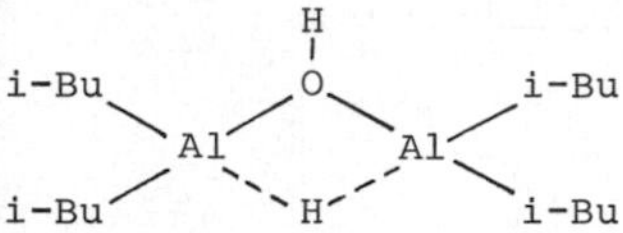

(XXXVI)

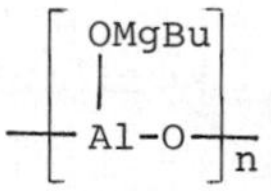

(XXXVII)

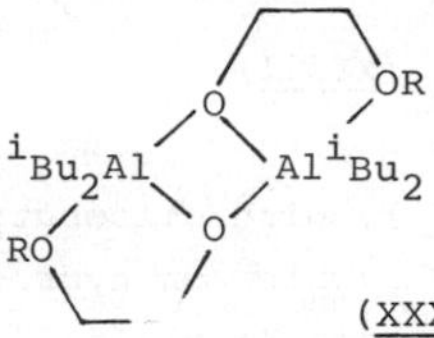

(XXXVIII)

pyridine, PPh_3) have been characterised.[51] Thallium derivatives of p-carborane have been synthesised by trans-metallation, eg:[52]

$$\underline{p}\text{-}HCB_{10}H_9(HgO_2CCF_3\text{-}2)CH + Tl(O_2CCF_3)_3$$

$$\rightarrow \underline{p}\text{-}HCB_{10}H_9[Tl(O_2CCF_3)_2\text{-}2]CH_2$$

As usual, organoaluminium compounds have featured strongly as reagents in organic synthesis. Reaction of alkynyl acetates, $RR^1C(OAc)C{\equiv}CR^2$, with trialkylalanes or lithium tetraalkylaluminates in the presence of transition metal compounds eg. $FeCl_3$ or $PdCl_2$ afford a method for the preparation of substituted allenes, $RR^1C{:}C{:}CR^2R^3$.[53] 3-Substituted-hex-1-en-5-ynes (XXXIX) undergo cyclization when treated with diethylaluminium hydride, although the nature of the product depends upon the conditions. In a hydrocarbon/ether solvent, trans-2-substituted-methyl-cyclopentane (XL) is produced, whereas in the absence of ether the major product is the 2-substituted methylenecyclopentane (XLI).[54]

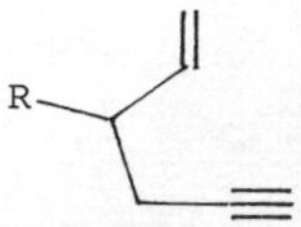

(XXXIX)

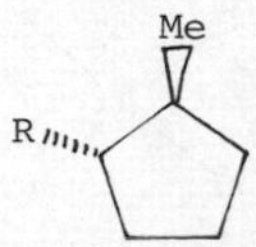

(XL)

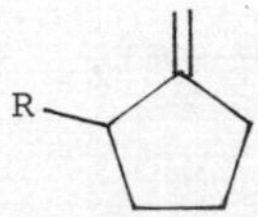

(XLI)

The reaction 1-(trimethylsilyl)-1,3-diynes with Li[AlHI^{i}Bu)$_2$^{n}Bu] proceeds regioselectively to give the enynylalanates (XLII), which is easily converted to the enynylsilanes, (XLIII):

R-C≡C-C≡C-SiMe$_3$ + Li[AlH(^{i}Bu)$_2$^{n}Bu]

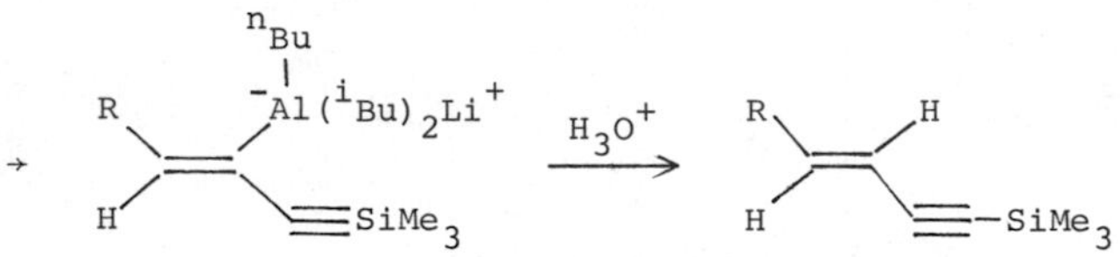

(XLII) (XLIII)

Further treatment of the silanes, (XLIII), with diisobutylaluminium hydride produces the dienylalanes (XLIV), which on hydrolysis afford the 1-(trimethylsilyl)-(1Z,3E)-dienes (XLV):[55]

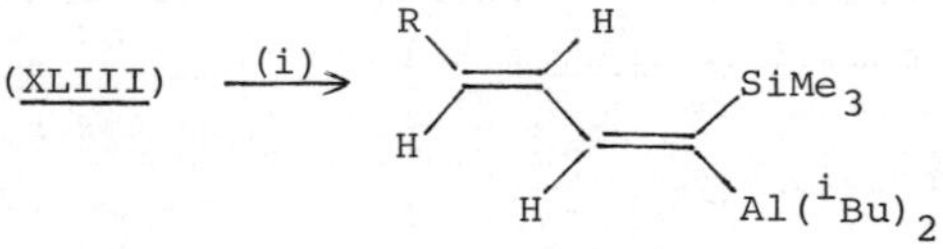

(XLIV)

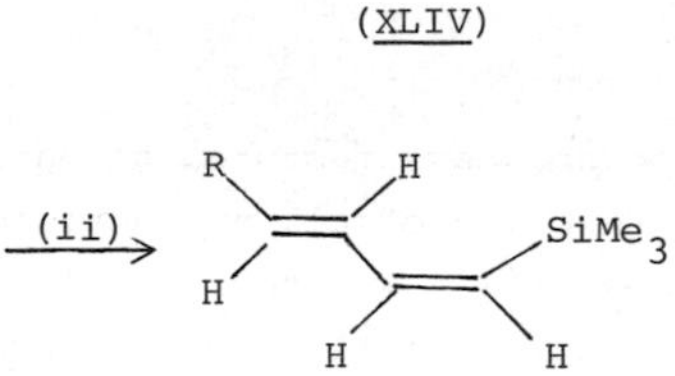

(XLV)

(i) ^{i}Bu$_2$AlH, (ii) 3N NaOH.

Allylic silanes are readily prepared in very good yield by the reaction of allylic phosphates with the silyl-aluminium reagent,

$PhMe_2Si\text{-}AlEt_2$, <u>eg</u>:

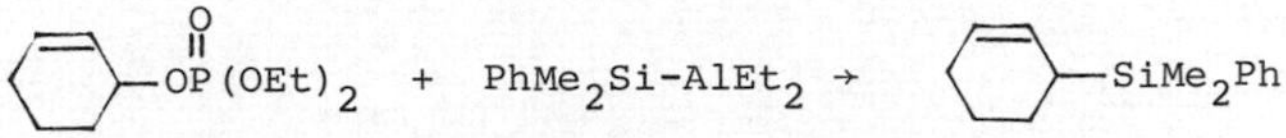

I

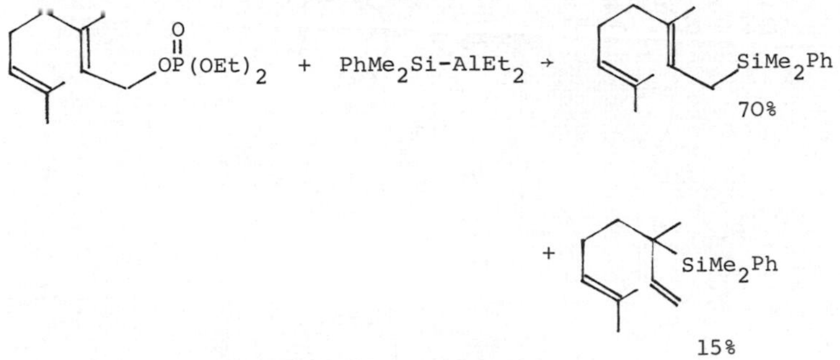

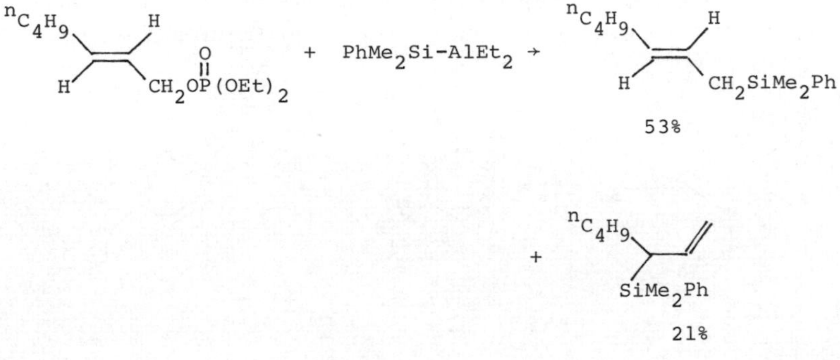

Cross-coupling with enol phosphates with the same reagent also occurs but only in the presence of palladium catalysts such as $Pd(PPh_3)_4$ and $PdCl_2(PPh_3)_2$:[56]

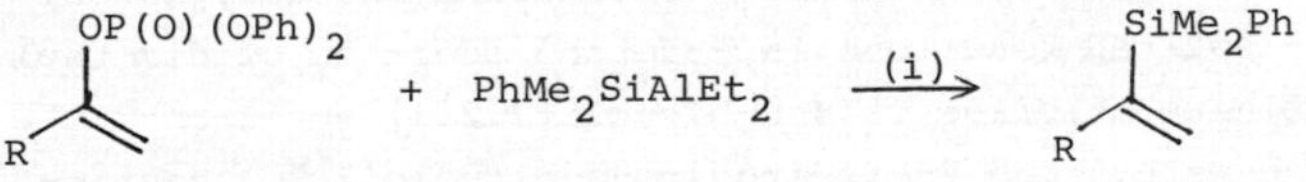

(i) Pd catalyst.

The selective coupling of enol silyl ethers with oxime sulphonates to give enaminones is promoted by $EtAlCl_2$ and Et_2AlCl,[57] eg:

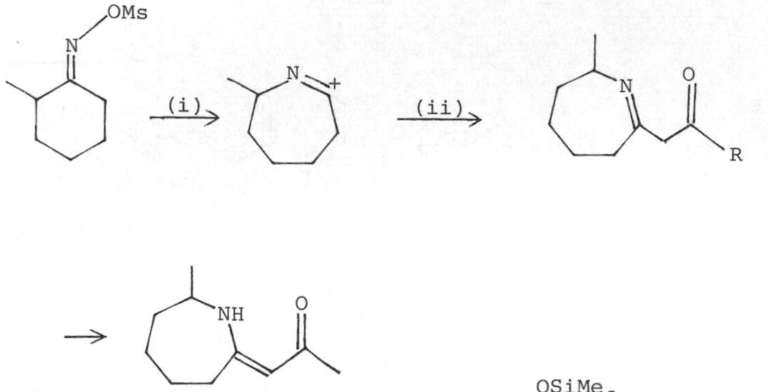

Ms = mesylate (i) Et_2AlCl (ii) 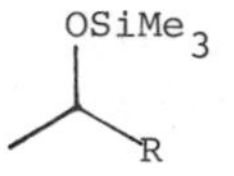

Alkyl or alkynyl groups may be introduced on the aromatic ring of arylhydroxylamine derivatives using triorganoaluminium reagents, eg:

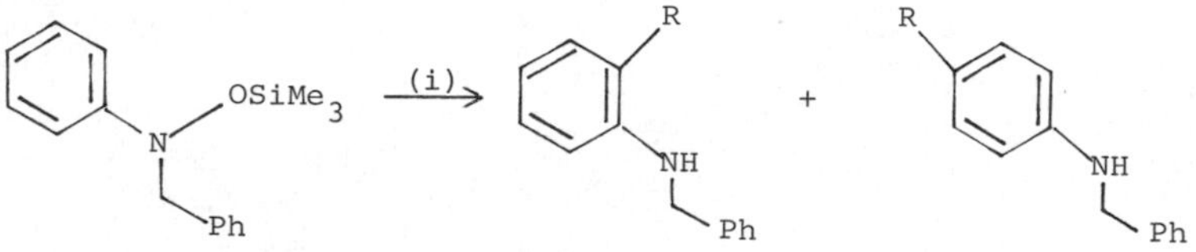

(i) R_3Al.

Overall yields of product vary from 39% to 83%. The resulting aromatic amines bearing an alkynyl in the ortho position are useful precursors for the synthesis of indoles.[58] The catalyst system $TiCl_4$-Et_2AlCl induces $[_{\pi}6s + _{\pi}2s]$ cycloadditions of cycloheptatriene to buta-1,3-diene, norbornadiene and acetylenes.[59] Other titanium-assisted reactions include the addition of dimethyl- or diethyl-aluminium chloride to syn-bicyclo[2.2.1]hept-2-en-7-ol to produce 2-exo-methyl and 2-exo-ethyl-syn-bicyclo[2.2.1]heptan-7-ol:

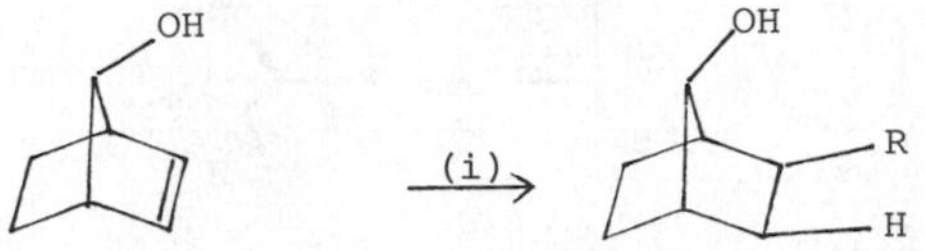

(i) R_2AlCl (R = Me, Et).

and to endo-bicyclo[2.2.1]hept-5-en-2-ol to give 5-endo-methyl- and 5-endo-ethyl-endo-bicyclo[2.2.1]heptan-2-ol:

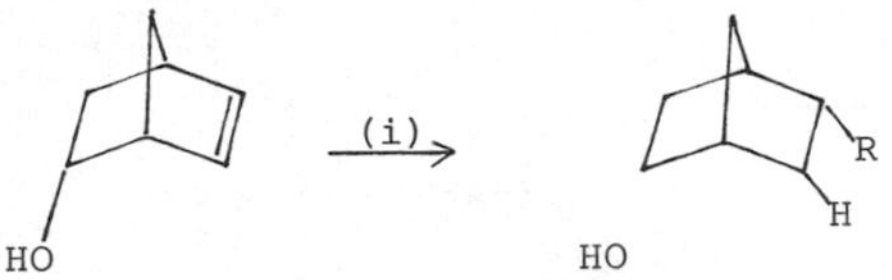

(i) R_2AlCl (R = Me, Et).

The results indicate a preference in such titanium-assisted additions to alkenols for attachment of the alkyl group to the side of the double bond near the hydroxyl group.[60] The methylation of the yne group in 5-en-3-yn-1-ols using $TiCl_4$-$AlMe_3$ proceeds smoothly to afford (3*E*)-4-methylalka-3,5-dien-1-ols <u>via</u> the scheme:[61]

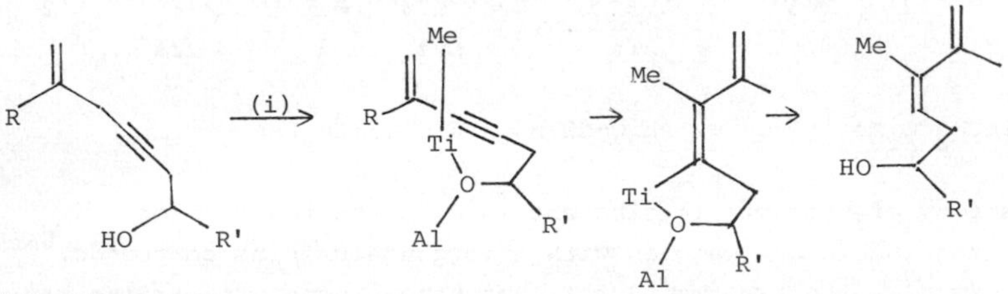

(i) $TiCl_4$-$AlMe_3$

The reaction of γ-alkoxyallylaluminium compounds, generated <u>in situ</u> with aldehydes and ketones at -78° leads to the highly diastereo-selective formation of mono-protected vicinal diols:[62]

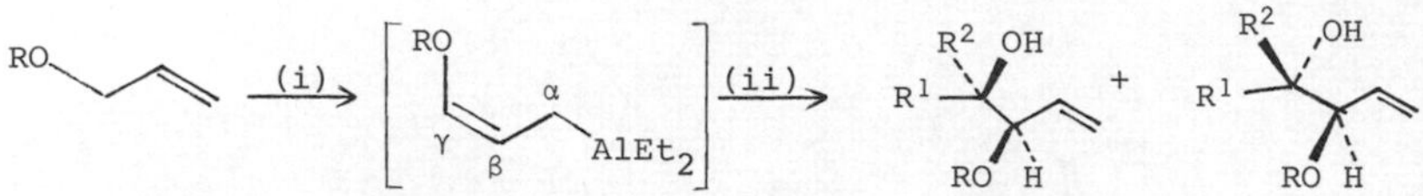

(i) Et_2AlCl, (ii) $R^1R^2C=O$

Treatment of optically active 2,3-epoxyalcohols with trialkyl-aluminium reagents followed by periodate cleavage is a convenient synthesis of α-chiral aldehyde, especially when the branching alkyl group is methyl, eg:[63]

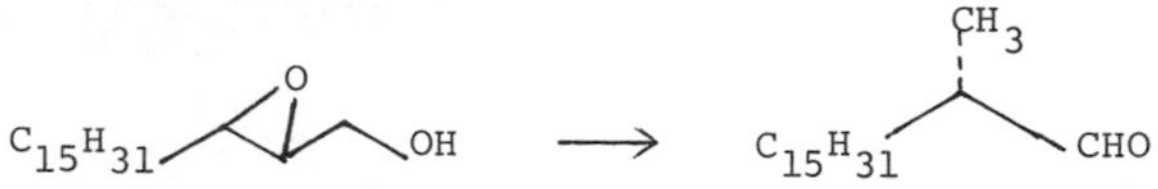

(92% ee) (94-96%; 92% ee)

(i) Me_3Al, 0°; $NaIO_4$.

The asymmetric ortho-specific hydroxylation of phenols with CCl_3CHO in the presence of chiral alkoxyaluminium chlorides gives rise to optically-active 2-(2,2,2-trichloro-1-hydroxyethyl)phenols, with good optical yields.[64] Aromatic acid anhydrides react with triphenylaluminium to yield polyarylmethanes, whereas aliphatic analogues give alkenes, ketones and tertiary alcohols, eg:[65]

$$(PhCO)_2O + Ph_3Al \rightarrow \underset{31\%}{Ph_3CH} + \underset{39\%}{Ph_2CHC_6H_4Ph\text{-}\underline{p}} + \underset{22\%}{PhCH(C_6H_4Ph\text{-}\underline{p})_2}$$

$$PhCH_2CO_2Me + Ph_3 \rightarrow Ph_2C{=}CHPh + Ph_2C(OH)CH_2Ph$$

A variety of products is also obtained in the reaction of p-benzoquinones and ketones with triorganoaluminium compounds.[66-68] For example, the treatment of 1,4-naphthoquinone with triisobutyl-aluminium affords 2,5-dimethylhexane, 1-naphtholes, 1,4-naphthalene-diols, naphthalene, and 1- and 2-alkylnaphthalenes, whilst acetophenone and triphenylaluminium give a complex mixture comprising PhCMe:CHCOPh, 5-methyl-1,3,5-triphenyl-1,3-cyclohexa-diene, $PhCO_2H$, $Ph_2C=CH_2$, Ph_2CMeOH, 1,3,5-$C_6H_3Ph_3$, Ph_2CMeCH_2COPh, and 2-benzoyl-5-methyl-1,3,5-triphenyl-1,3-cyclohexadiene. Irradiation of benzylamines complexed with trialkylaluminium compounds in benzene affords alkylbenzenes and amines:

$$PhCH_2NR'R''.AlR_3 \xrightarrow{h\nu} PhCH_3 + PhCH_2R + HNR'R''$$

1,2-Diphenylethane is also obtained as a side-product.[69]
Irradiation dramatically accelerates substitution reactions of
(tetraphenylporphinato)ethylaluminium with hindered phenols such
as 2,6-di-t-butyl-4-methylphenol and 2,6-di-t-butyl-4-methoxyphenol
to form the corresponding (porphinato)phenoxyaluminium derivative.
It would appear that upon irradiation and excitation of the porphyrin moiety, the electron density of the ethyl group bound to the
metal is increased, resulting in an enhanced nucleophilicity of the
metal-carbon bond.[70]

Reaction of triethylaluminium with $[(C_5Me_5M)_2Cl_4]$ (M = Rh or Ir)
gives the complexes $[C_5H_5M(C_2H_4)_2]$ together with some
$[C_5Me_5MCl(\eta^3-CH_2CHCHMe)]$, via the scheme:

$$[(C_5Me_5M)_2Cl_4] + 2Al_2Et_6 \rightarrow 2[C_5Me_5MEt_2.Al_2Cl_2Et_4]$$

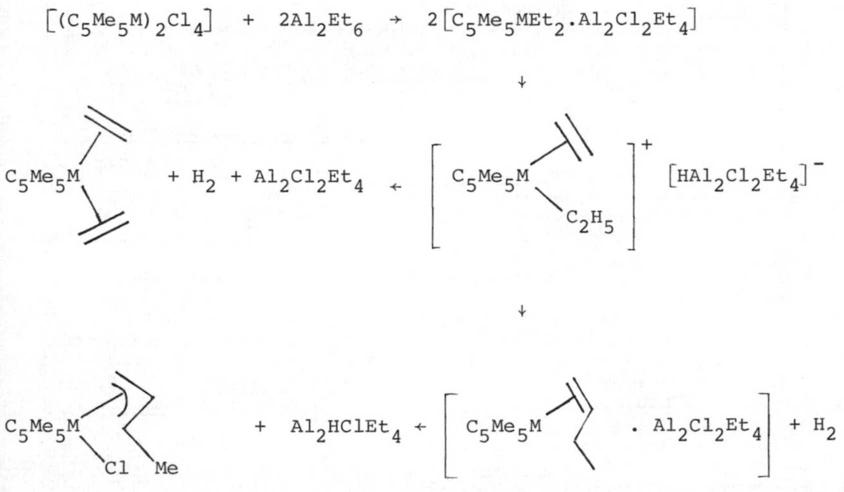

In contrast, the Rh-Et complex, $[C_5Me_5RhEt_2(PMe_3)]$ is obtained on
reaction with $[C_5Me_5RhCl_2(PMe_3)]$.[71] Because of the interest in
gallium arsenide as a semiconductor, methods of preparation of this
material by chemical vapour pyrolytic deposition have been studied.
Polycrystalline and single crystal layers have been obtained by
pyrolysis of the compound $ClMe_2Ga.AsEt_3$,[72] whilst the decomposition
of arsine in arsine-trimethylgallium mixtures has been shown to be
strongly affected by the presence of trimethylgallium.[73] Trans-

metallation occurs between arylthallium(III) trifluoroacetate and
dibroane to afford organoborane intermediates which, on hydrolysis
or oxidation give arylboronic acids or phenols, respectively:[74]

$$ArTl(OCOCF_3)_2 + B_2H_6 \rightarrow [Ar-B\langle] \begin{array}{c} \nearrow H_2O_2 \rightarrow ArOH \\ \searrow H_2O \rightarrow ArB(OH)_2 \end{array}$$

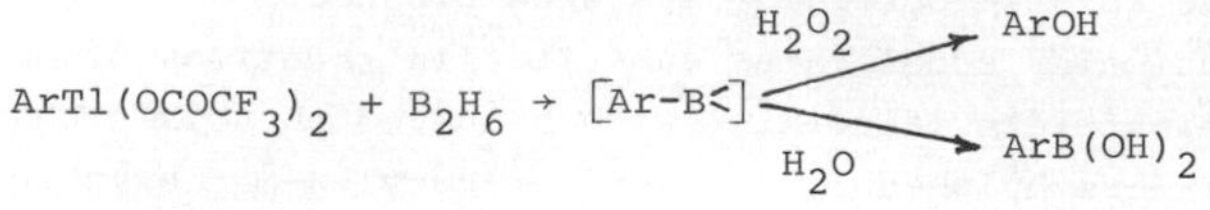

References

1   K.J.Klabunde and Y.Tanaka, J. Am. Chem. Soc., 1983, 105, 3544.
2   J.J.Eisch and K.C.Fichter, J. Organometal. Chem., 1983, 250, 63.
3   U.S. Patent 4,364,874; Chem. Abstr., 1983, 98, 107536e.
4   C.Eaborn, M.N.El-Kheli, N.Retta and J.D.Smith, J. Organometal. Chem.,
    1983, 249, 23.
5   O.T.Beachley and C.Tessler-Youngs, Organometallics, 1983, 2, 796.
6   C.Tessler-Youngs, C.Bueno, O.T.Beachley and M.R.Churchill, Inorg. Chem.,
    1983, 22, 1054.
7   C.Tessler-Youngs, W.J.Youngs, O.T.Beachley and M.R.Churchill,
    Organometallics, 1983, 2, 1128.
8   D.L.Grimmett, J.A.Labinger, J.N.Bonfiglio, S.T.Masuo, E.Shearin and J.S.
    Miller, Organometallics, 1983, 2, 1325.
9   J.A.Labinger and J.S.Miller, J. Am. Chem. Soc., 1982, 104, 6856.
10 H.Schmidbauer, S.Lauteschlager and B.Milewski-Mahria, Chem. Ber., 1983, 116,
    1403.
11 H.F.Klein, K.Ellrich and K.Ackermann, J. Chem. Soc., Chem. Commun.,
    1983, 888.
12 W.E.Dorogy and E.P.Schram, Inorg. Chim. Acta, 1983, 72, 187.
13 M.Boleslawski and J.Serwatowski, J. Organometal. Chem., 1983, 254, 159.
14 J.L.Atwood and M.J.Zaworotko, J. Chem. Soc., Chem. Commun., 1983, 302.
15 J.L.Atwood, D.C.Hrncir, R.D.Priester and R.D.Rodgers, Organometallics, 1983,
    2, 985.
16 T.Halder, W.Schwarz, J.Weidlein and P.Fischer, J. Organometal. Chem.,
    1983, 246, 29.
17 J.R.Bews and C.Glidewell, Theochem., 1982, 7, 151.
18 R.B.Hallock, O.T.Beachley, Y-J.Li, W.M.Sanders, M.R.Churchill, W.E.Hunter
    and J.L.Atwood, Inorg. Chem., 1983, 22, 3683.
19 H.Schmidbaur, U.Thewalt and T.Zafiropoulos, Organometallics, 1983, 2, 1550.
20 A.C.Jones, D.J.Cole-Hamilton, A.K.Holliday and M.M.Ahmad, J. Chem. Soc.,
    Dalton Trans., 1983, 1047.
21 A.Coutsolelos and R.Guilard, J. Organometal. Chem., 1983, 253, 273.
22 C.Peppe and D.G.Tuck, Polyhedron, 1982, 1, 549.
23 R.Uson and A.Laguna, Inorg. Synth., 1982, 21, 71.
24 R.Uson, A.Laguna, J.A.Abad and E. de Jesus, J. Chem. Soc., Dalton Trans.,
    1983, 1127.
25 G.B.Deacon and R.N.M.Smith, Austr. J. Chem., 1982, 35, 1587.
26 M.D.Rausch, B.H.Edwards, R.D.Rogers and J.L.Atwood, J. Am. Chem. Soc.,
    1983, 105, 3882.
27 Y.Kawasaki and N.Okuda, Chem. Letters, 1982, 1161.
28 D.L.Hughes and M.R.Truter, Acta Crystallogr., 1983, B39, 329.
29 J.Crowder, K.Hendrick, R.W.Matthews and B.L.Podgima, J. Chem. Research (S),
    1983, 82.
30 A.K.Sharma, B.Bhushan and N.K.Kaushik, Acta Chim. Hung., 1983, 113, 29.
31 B.Khera, A.K.Sharma and N.K.Kaushik, Synth. React. Inorg. Met.-Org. Chem.,
    1982, 12, 583.

32 F.Brady, R.W.Matthews, M.M.Thakur and D.G.Gillies, J. Organometal. Chem., 1983, 252, 1.
33 G. van Koten, J.T.B.H.Jastrzebski and K.Vrieze, J. Organometal. Chem., 1983, 49.
34 Z.Florjanczyk and U.Iwaniak, J. Organometal. Chem., 1983, 252, 275.
35 A.Yu.Fisenko, V.A.Grindin, B.A.Ershov, A.I.Kol'tsov, A.G.Boldyrev, E.B.Milovskaya and E.P.Skvovtsevich, Zh. Obshch. Khim., 1983, 53, 483.
36 W.Kaim, J. Organometal. Chem., 1983, 241, 157.
37 Z.Florjanczyk and E.Szymanska-Zachara, J. Organometal. Chem., 1983, 259, 127.
38 Z.Florjanczyk and J.Sitkowska, J. Organometal. Chem., 1983, 251, 1
39 J.Lub, M.L.Beekes and T.J. de Boer, J. Chem. Soc., Perkin Trans. II, 1983, 721.
40 H.Kurosawa, N.Okuda and Y.Kawasaki, J. Organometal. Chem., 1983, 255, 153.
41 H.Kurosawa, H.Okuda, M.Sato and T.Hattori, J. Organometal. Chem., 1983, 250, 83.
42 U.Knips and F.Huber, Z. Naturforsch., Teil B, 1983, 38, 434.
43 V.V.Gavrilenko, M.I.Vinnikov, V.A.Antonovich and L.I.Zakharkin, Izv. Akad. Nauk SSSR, Ser. Khim., 1982, 2367.
44 V.V.Gavrilenko, L.A.Chekulaeva and L.I.Zakharkin, Zh. Obshch. Khim., 1983, 53, 481.
45 V.V.Gavrilenko, L.A.Chekulaeva, V.A.Antonovich and L.I.Zakharkin, Izv. Akad. Nauk SSSR, Ser. Khim., 1983, 636.
46 Jpn Kokai Tokkyo Koho JP 57,158 792; Chem. Abstr., 1983, 98, 107534c.
47 M.Boleslawski and J.Serwatowski, J. Organometal. Chem., 1983, 255, 269.
48 S.L.Gershkokhen, N.N.Korneev and G.I.Shchetbakova, Zh. Obshch. Khim., 1982, 52, 2543.
49 N.N.Korneev, S.L.Gershkokhen and G.I.Shcherbakova, Zh. Obshch. Khim., 1982, 52, 2666.
50 A.I.Belokon and V.N.Bochkarev, Zh. Obshch. Khim., 1983, 53, 101.
51 B.N.Helliwell and M.J.Taylor, Aust. J. Chem., 1983, 36, 385.
52 V.I.Bregadze, A.Ya. Ustyatinskii and N.N.Godovikov, Izv. Akad. Nauk SSSR, Ser. Khim., 1983, 1405.
53 G.A.Tolstikov, T.Yu.Romanova and A.V.Kuchin, Izv. Akad. Nauk SSSR, Ser. Khim., 1983, 629.
54 M.J.Smith and S.E.Wilson, Tet. Letters, 1982, 23, 5013.
55 J.A.Miller and G.Zweifel, J. Am. Chem. Soc., 1983, 105, 1383.
56 Y.Okuda, M.Sato, K.Oshima and H.Nozaki, Tet. Letters, 1983, 24, 2015.
57 Y.Matsumura, J.Fujiwara, K.Maruoka and H.Yamamoto, J. Am. Chem. Soc., 1983, 105, 6312.
58 J.Fujiwara, Y.Fukutani, H.Sano, K.Maruoka and H.Yamamoto, J. Am. Chem. Soc., 1983, 105, 7177.
59 K.Mach, H.Antropiusova, P.Sedmera, V.Hanus and F.Turecek, J. Chem. Soc., Chem. Commun., 1983, 805.
60 H.G.Richey, L.M.Moses and J.J.Hangelend, Organometallics, 1983, 2, 1545.
61 T.J.Zitzelberger, M.D.Schiavelli and D.W.Thompson, J. Org. Chem., 1983, 48, 4781.
62 M.Koreeda and Y.Tanaka, J. Chem. Soc., Chem. Commun., 1982, 845.
63 W.R.Roush, M.A.Adam and S.M.Pesckes, Tet. Letters, 1983, 24, 1377.
64 F.Bigi, G.Casiraghi, G.Casnati, G.Sartori and L.Zetta, J. Chem. Soc., Chem. Commun., 1983, 1210.
65 A.Alberola, R.Pedrosa, J.L.Perez Bragado and J.F.Rodriguez Amo, An. Quim., Ser. C, 1982, 78, 159.
66 A.Alberola, A.M.Gonzalez Nogal, J.A.Martinez de Ilarduya and F.J.Putido, A. Quim., Ser. C, 1982, 78, 166.
67 A.Alberola, A.M.Gonzalez Nogal, J.A.Martinez de Ilarduya and F.J.Pulido, An. Quim., Ser. C, 1982, 78, 175.
68 A.Alberola, A.G.Ortega, R.Pedrosa, J.L.P.Bragado and J.F.R.Arno, An. Quim., Ser. C, 1982, 78, 360.
69 K.Omura and J.Furukawa, Chem. Letters, 1982, 1633.
70 H.Murayama, S.Inone and Y.Ohkatsu, Chem. Letters, 1983, 381.
71 A.V.de Miguel and P.M.Maitlis, J. Organometal. Chem., 1983, 244, C35.

72 A.Zaouk and G.Constant, J. Phys., Colloq., 1982, 421.
73 J.Nishizawa and T.Kurabayashi, J. Electrochem. Soc., 1983, 130, 413.
74 G.M.Pickles, T.Spencer, F.G.Thorpe, A.D.Ayala and J.C.Podesta, J. Chem. Soc., Perkin Trans. I, 1982, 2949.

Bibliography

Additional references, including review articles, not included in the text.

'Aluminium - Annual Survey covering the year 1981' D.J.Smith, J. Organometal. Chem., 1983, 257, 17.

'Thallium - Annual Survey covering the year 1981' H.Kurosawa, J. Organometal. Chem., 1983, 245, 1.

'Aluminium-27 NMR Spectroscopy: a probe for three-, four-, five- and six-fold coordinated aluminium atoms in organoaluminium compounds'. R.Benn,A.Rufinska, H.Lehmkuhl, E.Jannsen and C.Knieger, Angew. Chem., 1983, 95, 808.

'Terminal vs bridge bonding of methylene to metal systems: Al_2CH_2 as a model system'. D.J.Fox and H.F.Schaeffer, J. Chem. Phys., 1983, 78, 328.

'Preparation and chemical reactivity of aluminium halide sigma-bonded complexes of cyclobutadienes' H.Hogeveen , Kagaku, Zokan (Kyoto), 1983, 157.

'Reactivity of organoaluminium compounds towards Lewis bases III, Diarylaluminium amides and phosphinamides' D.Giurgui, I.Popescu and V.Hamciuc, Rev. Chim. (Bucharest), 1983, 34, 232.

'Synthesis of higher tri-n-alkyls of aluminium.' L.V.Gaponik and V.P.Mardykin, Izv. Vyssh. Unchebn. Zaved., Khim. Khim. Tekhnol., 1983, 26, 30.

'Aluminium trihydrocarbyls' U.S. Patent 4,349,483; Chem. Abstr., 1983, 98, 54194h.

'Direct synthesis of alkylaluminiums using aluminium-silicon alloys.' L.I.Zakharkin, V.V.Gavrilenko, E.J.Yacubovskii and V.P.Kiselev, Sovrem. Dostizh. v. Pr-ve i Obrab. Alyuminiya i ego Splavov L., 1981, 117; Chem. Abstr., 1983, 98, 89433n.

'Aluminium alkyls' U.S. Patent 4,364,872; Chem. Abstr., 1983, 98, 849647k.

'Aluminium alkyls' U.S. Patent 4,364,873; Chem. Abstr., 1983, 98, 849648m.

'Cyclohexylaluminium chlorides' Ger. (East) Patent 156,534; Chem. Abstr., 1983, 98, 126374S.

'Separation of aromatic hydrocarbons using $M[Al_2Et_6X]$ complexes' J.J.Harrish and J.C.Montagna, Sep. Sci. Techol., 1982, 17, 1151.

'Calorimetric study of complexes of triethylaluminium with oxygen-containing donors' V.P.Mardykin, A.V.Pavlovich, P.N.Gaponik and O.M.Baranov, Zh. Obshch. Khim., 1982, 52, 2620.

"The effect of treatment with triethylaluminium on the hydrogenation and hydrodesulphurisation activity of molydbenum, cobalt and nickel sulphide catalysts' M.Bhaduri and P.C.H.Mitchell, J. Catalysis, 1982, 77, 132.

'Regioselectivity of the syn addition of tris(trimethylsilyl)aluminium to unsymmetrical alkynes' L.Roesch, G.Altnau and G.Jas, Chem. Ztg., 1982, 106, 441.

'X-ray crystallographic study of trimethylsilyaluminium compounds' L.Roesch, G.Altnau, C.Krüger and Y.H.Tsay, Z. Naturforsch, Teil B, 1983, 38, 34.

'Reaction of tris(trimethylsilyl)aluminium with α,β-unsaturated carbonyl compounds' G.Altnau, L.Roesch and G.Jas, Tet. Letters, 1983, 24, 45.

'Tris(trimethylsilyl)aluminium and transition-metal catalysts. Silylation of allyl acetates.' B.M.Trost, J.Yoshida and M.Lautens, J. Am. Chem. Soc., 1983, 105, 4494.

'Silicon-metal bonded compounds, 17. The synthesis and crystal structure of bis[tris(trimethylsilyl)aluminium]-tetramethylethylenediamine and of bis(trimethylsilyl)magnesium-tetramethylethylenediamine' D.W.Goebel, J.L. Hencher and J.P.Oliver, Organometallics, 1983, 2, 746.

'Crystal and molecular structure of an yttrium dicyclopentadienyl chloride complex with triethylaminalane $[\eta-(C_5H_5)YCl.AlH_3.NEt_3]_2$' E.B.Labkovsky, G.L.Soloveychik, B.M.Bulychev, A.B.Erofeev, E.I.Gusev and N.I.Kirillova, J. Organometal. Chem., 1983, 254, 167.

'Synthesis and hydrolysis of a fluorosilyl-t-butylaminodichloroalane. Crystal structures of the cyclic compounds $[(Me_3SiNMe)_2SiNCMe_3]_2$ and $(Me_3CNHAlCl_2)_2$' W.Clegg, M.Haase, U.Klingebiel, J.Neemann and G.M.Sheldrick, J. Organometal. Chem., 1983, 251, 281.

'Novel inorganic ring systems. XXXIX A dimeric cycloalumdisilatriazane' U.Wannagat, T.Blumenthal, D.J.Brauer and H.Bürger, J. Organometal. Chem., 1983, 249, 33.

'Stabilisation of silicenium ylids by adduct formation with aluminium trihalides: the crystal structure of $[(Me_3C)_2SiNCMe_3]_2AlClF_2$' W.Clegg, U.Klingebiel, J.Neemann and G.M.Sheldrick, J. Organometal. Chem., 1983, 249, 47.

'Synthesis and molecular structure of tris(dimethyldithiocarbamato)aluminium, a compound with hexacoordinated aluminium' H.Nöth and P.Konrad, Chem. Ber., 1983, 116, 3552.

'The catalytic system $[(C_5H_5)_2TiCl]_2/LiAlH_4$ in aromatic solvents. I. Formation of titanocene hydride complexes; an esr study' K.Mach and H.Antropiusova, J. Organometal. Chem., 1983, 248, 287.

'Vapour-liquid equilibrium of organoaluminium compounds. I. Vapour pressure of methylaluminium chlorides' W.Ulfnalski and A.Sporzynski, J. Organometal. Chem., 1983, 244, 1.

'Multinuclear NMR studies on Al(III) complexes of ATP and related compounds' S.J.Karlik, G.A.Elgavish and G.L.Eichham, J. Am. Chem. Soc., 1983, 105, 602.

'Aluminium-27 nuclear magnetic resonance study of aluminium(III) interactions with carboxylate ligands' S.J.Karlik, E.Tarien, G.A.Elgavish and G.L.Eichham, Inorg. Chem., 1983, 32, 525.

'Heterogeneous metathesis of unsaturated esters using a rhenium-based catalyst' R.H.A.Bosma, G.C.N. van den Aardweg and J.C.Mol, J. Organometal. Chem., 1983, 255, 159.

'Molecular force fields and bonding in methyl-gallate and methylindate ions: $(R_{4-n}M^{III}Cl_n)^-$; R = CH_3 or CD_3; M^{III} = Ga or In; n = 0-4.' A.Haaland and J.Weidlein, Acta Chem. Scand., Ser. A, 1982, 36, 805.

'Kinetics of the thermal decomposition of trimethylgallium' L.S.Ivarov, N.P.Fatyushina, I.I.Nalivaiko, V.M.Timashkov and S.A.Sadof'eva, Nauch. Tr. N. -i. i Proekt. In-t Redkomet. Prom-sti Giredmet., 1982, 10; Chem. Abstr., 1983, 98, 72193m.

'Use of gas chromatography to study the thermal decomposition of organogallium peroxide compounds' I.G.Stepanova, I.V.Chikinova, G.I.Makin and Yu.A. Aleksandrov, Fizi-Khim, Metody Analiza, M, 1982, 49; Chem. Abstr., 1983, 99, 195051s.

'Matrix isolation-mass spectroscopy study of the reaction of boron trifluoride with trimethylamine-alane and trimethylamine-gallane' A.E.Shirk and J.S.Shirk, Inorg. Chem., 1983, 22, 72.

'Crystal structure of indium(III) dithizonate' J.McB.Harrowfield, C.Pakawatchai and A.H.White, J. Chem. Soc., Dalton Trans., 1983, 1109.

'Organic Metals. Introduction of indium and thallium tetrachloride anions into polyacetylene by anodic oxidation' M.Zagóvska, A.Proń, J.Przyluski, B.Krishe and G.Ahlgren, J. Chem. Soc., Chem. Commun., 1983, 1125.

'Oxidation of carboranes by thallium(III) in the presence of palladium(II) - a new route to biscarboranes' A.Ya.Usyatinskii, A.D.Ryabov, T.M.Shcherbina, V.I.Bregadze and N.N.Godovikov, Izv. Akad. Nauk, 1983, 1637.

'Some molecular adducts of phenyl-thallium bis(trifluoroacetate)' T.N.Srivastava and M.N.Pande, Indian J. Chem., 1983, 22A, 344.

'Complexes of hexamethylphosphoramide with thallium trihalides and monophenylthallium dihalides' S.Blanco, J.S.Casa, A.Sanchez, J.Sordo, J.M.F.Solis and M.Gayoso, J. Chem. Research (S), 1982, 328.

'Reactivity of thallium tris[tris(pentafluorophenyl)germyl]mercurate' G.A.Razuvaev, J. Organometal. Chem., 1983, 250, 135.

6
Group IV: The Silicon Group

BY D. A. ARMITAGE

1 Introduction

The use of organosilicon compounds in organic synthesis is recognised with the publication separately of two books by E.W.Colvin and by W.P.Weber,[1] while specific aspects reviewed include silyl enol ethers, siloxycyclopropanes, α,β-epoxysilanes, carbon-carbon bond formation, and silylation-desilylation.[2] These and other aspects are also included in 'Symposia-in-Print Number 7', with current trends in organosilicon chemistry briefly reviewed.[3] Specifically mentioned are the synthesis of photochromes and the preparation of ligands for supported catalysts, while silyl azides and phosphanes are extensively covered,[4] as are the reactions of organosilicon radicals, their cations, and the electrochemistry of silanes.[5] Proposals for catalysts for enantioselective hydrosilylation,[6] silaethenes and disilenes, multiple M-C bonded compounds,[7] and the formation of optically active silyl- and germyl-transition metal compounds are separately surveyed.[8]

Heterocyclic organogermanium species are reviewed,[9] triorganotin--sulphur compounds surveyed for Gmelin, and organotin compounds in general for Topics in Current Chemistry.[10] Reviews for tin cover alkynyl compounds, Schiff base complexes, and reactions of the silicon subgroup organoelement peroxides with electron donors.[11] The role of ion pairs in electrophilic substitution, and solvent effects on tinalkyl exchange are also covered.[12]

Annual surveys cover Si, Sn, and Pb for 1980 and Ge for 1979.[13] This article includes 800 references.

2 The Carbon-Metalloid Bond

The silaethene $Me_2Si{=}C(SiMe_3)(SiMeBu^t{}_2)$ results in THF from lithium and $Me_2FSiC(SiMe_3)(SiMeBu^t{}_2)$ by LiF elimination using Me_3SiCl. It can be recrystallised from Et_2O at -78^oC, decomposes within a few days at room temperature and adds Br_2, H_2O, BF_3, Me_2CO, and buta--1,3-diene.[14] Bu^tLi and MePhvinylSiCl generate E- and Z-$PhMeSi=CHCH_2Bu^t$ consistently in the ratio 70:30, showing stereochemical induction and, if prepared by pyrolysing the dibenzo-2--silabicyclo[2.2.2]octane, is configurationally stable to 300^oC.[15] Calculations indicate that HCl addition to silaethene proceeds through a 2- rather than a 4-centre transition state, and that silaethyne is trans-bent and more stable than isomeric silylidene. Barriers to rotation in 1- and 2-silaallenes are calculated to be 35 and 20 kcal mol^{-1} (allene 53 kcal mol^{-1}).[16]

Vacuum pyrolysis of cis-$Me_5Si_2CH=CHCMe=CH_2$ at 635^oC gives isomeric silyl hydrides through 1,5-Me_3Si migration from Si to C, to produce an intermediate 1-sila-1,3-butadiene which underwent 1,5-H migration back to doubly bonded Si. A similar mechanism is suggested for the formation of a 2-sila-2,3-dihydrofuran in good yield from the cis-enone $Me_5Si_2CH=CHCOPr^n$.[17] Model systems indicate that the pyrolysis of [3.3]silaspirocycloheptane to silacyclopentenes results through rearrangement of $H_2C=\overline{Si(CH_2)_3}$ to $:\overline{Si(CH_2)_4}$. $Me_3SiSivinyl$, generated thermally from $(Me_3Si)_2Si(OMe)vinyl$, gives 3,3-dimethyl-3,5-disilacyclopentene through silirene and silacyclopropanylidene intermediates, while alkenylsilylenes cyclise by β-CH insertion to 4-silacyclopentenes.[18] The thermally generated silylene $Me_3SiSiCH(SiMe_3)_2$ rearranges to the 1,3-disilcyclobutane $Me_3Si\overline{SiHCH(SiMe_3)SiMe_2CH_2}$ by CH insertion or to $(Me_3Si)_2Si=CHSiMe_3$ by silyl migration. Polysilylated diazomethanes generated the carbene on photolysis or pyrolysis, which rearranges via silyl migration to the silaethene, notably $(Me_3Si)_3SiCSiMe_3$ to $(Me_3Si)_2Si=C(SiMe_3)_2$.[19]

Heats of formation calculated for isomeric 2-silapropene and Me_2Si: indicate the former to be the more stable by 28 kcal mol^{-1} and, with supporting data, contradicts evidence for the equilibrium favoring the silylene. However, both are stable at 35K and both give $Me_2Si=O$ in the presence of N_2O, suggesting matrix catalysed hydrogen migration. Photolysing $Me_2\overline{Si(CH_2)_2}$ gives C_2H_4 and $Me_2Si=CH_2$ if in singlet benzene, but c-C_3H_6 and Me_2Si: in the presence of triplet benzene.[20]

Me_2Si: and Ph_2Si: add stereospecifically to cis- and trans-2-butene to give the silirane, to 1,3-dienes to give silacyclopentenes, and will deoxygenate sulphoxides.[21] The exclusive extrusion of silylenes and germylenes results on photolysing dibenzo-1,1,2,2-tetramethyl-1,2-disila- (or digerma-) cyclohexa-3,5-diene, and not the disilene (or digermene). Addition of Me_2Ge:, generated thermally, to diallenes supports a singlet state, while the first stable germirenes have been prepared by addition of R_2Ge: (R=Me,Et) to a thiacycloheptyne.[22]

The trisilylmethylsilanols $(RMe_2Si)_n(Ph_2MeSi)_{3-n}CSiMe_2OH$ (n=3, R=Me,Ph; n=2, R=Me) rearrange in MeONa/MeOH to give the bis silylmethyldisiloxanes. While methanolysis of $(Me_3Si)_3CSiMe_2Cl$ ($TsiSiMe_2Cl$) is very slow, that of $(Me_3Si)_2C(SiMe_2OMe)(SiMe_2Cl)$ is very fast, and anchimeric assistance of X for leaving Y is found in $(Me_3Si)_2C(SiMe_2X)(SiMe_2Y)$ to be in the order $OMe > OSO_2CF_3, OClO_3 > F > Cl, I > Me$.[23] The ease of nucleophilic displacement of iodide in $TsiSiMe_2I$ decreases in the order $N_3 > F > CN > NCS > NCO$ in MeOH, which reverses in MeCN. The cyanate isomerises to the isocyanate by second order kinetics, and its facile hydrolysis provides for a useful route to $(HOMe_2Si)_4C$ [from $(IMe_2Si)_4C$, AgOCN and moist ether].[24]

The rapid conversion of TsiSiPh(OH)I into TsiSiPh(OH)OMe by NaOMe is thought to occur through an Si=O intermediate, while the iodides $TsiSiR_2I$ give alkoxides in MeOH through an intermediate silicocation. Radicals formed from TsiSiHXY (X=Y=H,Me,Ph) do not rearrange and have the expected couplings constants.[25]

Adduct TsiLi.THF is an 'ate' complex $[Li(THF)_4]\ Li[C(SiMe_3)_3]_2$ with a linear anion, while $(PhMe_2Si)_3CLi(THF)$ is covalent with the third interaction of lithium to the ipso carbon atom of one of the phenyl rings. The Li and Cu^I derivatives of 2-bis(trimethylsilyl)-methylpyridine are dimeric with strong metal-metal interactions.[26]

An extensive series of Tsi derivatives of boron, aluminium, and gallium have been made while, despite steric crowding, $TsiPPh_2$ can be readily methylated, the phosphonium salt decomposing to the ylide $MePh_2P{=}C(SiMe_3)_2$.[27] The antimony alkyls $R_{\underline{n}}SbCl_{3-\underline{n}}$ [$R=(Me_3Si)_2CH$; $\underline{n}$=1-3) result from RMgCl and $SbCl_3$, R_3Sb from R_2SbCl and RLi, and $TsiSbCl_2$ from TsiLi and $SbCl_3$. R_3Bi results directly from $BiCl_3$ and RLi and, despite severe distortion in its trigonal bipyramidal structure, is thermally stable to 148°C.[28]

Reducing $(Me_3Si)_2C{=}C(BClBu^t)_2$ with Na-K gives a boracyclopropane with exocyclic boron-carbon double bond.[29] Polysilylated methyl derivatives of phosphorus,[30] arsenic and antimony can be reduced to the double bonded analogues of diazenes, stabilised through steric hindrance[31] and as transition metal complexes. $[(Me_3Si)_2CH]_2PNa$ and $(Me_3P)_2NiCl_2$ give $(Me_3P)_2Ni[(Me_3Si)_2C{=}PCH(SiMe_3)_2]$, with a NiCP triangle.[32]

A series of $(XMe_2Si)_3CH$ derivatives have been made and for X=OH, thermal dehydration gives $HC(SiMe_2OSiMe_2)_3CH$, a hexasilabicycloundecane with a manxane type structure. The hydrolysis of $TsiSnMeCl_2$ in the presence of MeOK/MeOH leads to silyl cleavage from one of the Tsi groups of the expected cyclotristannoxane $(TsiMeSnO)_3$, while $TsiGePh_2Cl$ undergoes C-Si cleavage by CsF in diglyme, giving the digermacyclobutane $(Ph_2GeCHSiMe_3)_2$. The structure of $(Ph_3SnCH_2)_3CH$ shows the phenyl groups to enclose and shield the 3^y hydrogen, accounting for its low reactivity in redox reactions.[33]

Excess TMS monomethylates $HN(SO_2Cl)_2$ to give N-chlorosulphonyl methanesulphonamide, while cleavage of the Ph-M bond (M=Si-Pb; Sn best) by F_2 is used to ^{18}F label aryl compounds. Sulphinyl chlorides RSOCl result from R_4Sn and $SOCl_2$ with ease of cleavage $Bu^n > Pr^n > Et > Me > CH_2Ph \gg p\text{-}MeC_6H_4$,[34] while Ph_4Sn gives $PhAsCl_2$ and Ph_2AsCl in good yield under appropriate conditions with $AsCl_3$. $MeSbX_2$ and Me_2SbX result similarly from Me_4M (M=Sn,Pb). The importance of work terms in the free energy relationship for charge transfer in R_4M (M=Sn,Pb) with TCNE, Br_2 and $HgCl_2$ is evaluated, and bond dissociation energies determined for Me_4Ge and Et_4Pb by very low pressure pyrolysis (83 and 54 kcal mol^{-1} respectively).[35]

Electron diffraction shows the steric interaction in $(Me_3Si)_2CH_2$ to be accommodated by a wide angle ($Si\widehat{C}Si$ 123.2°) and C-Si bonds slightly longer than normal. $(Ph_2SiCH_2)_3$ possesses a flattened twisted boat structure and $(\underline{o}\text{-tolyl})_4Sn$ has Sn-C bonds a little longer than those of Ph_4Sn.[36] The axial conformer of cyclobutylsilane has a rotation barrier potential of about 2 kcal mol^{-1}, while fundamental vibrations of XCH_2SiMe_2H (X=Me,Cl,Br,I) are assigned.[37]

A synthetic sequence stereospecific at silicon has been utilised to make $napPhMeSiCH_2CH_2X$ (X=Cl,Br) which recrystallised from hexane as pure diastereomers, the structures of which were determined. 2-Trimethylsilylethyl-4-nitrophenyl carbonate is used to protect α,ω-aminoacids, and $Me_3Si(CH_2)_{\underline{n}}OCOCH_2COMe$ ($\underline{n}$=1,2) are reduced microbiologically to the alcohol. Cyclic $\underline{E}$-β-trimethylsilyl ketoximine acetates cleaved catalytically by $Me_3SiOSO_2CF_3$ to the unsaturated nitrile with regio- and stereospecificity of the double bond, while converting β-oxidoalkylsilanes to the olefin (Peterson olefination) proceeds by an antipathway.[38]

An extensive series of silylalkyl phosphines have been prepared and the ylide $Ph_3\overset{+}{P}$-$\overset{-}{C}HOCH_2CH_2SiMe_3$ used in homologation of aldehydes and ketones. The complex $(\mu$-$H)_2Ru_3(CO)_9[\mu_3$-$PCH_2CH_2Si(OEt)_3]$, supported on oxides, is used as a catalyst, while the mass spectra of β-silylethyl derivatives yield the non-classical ethylene trimethylsilanium ion.[39]

The aldol reaction of α-stannyl ketones with aromatic and aliphatic aldehydes is highly erythro selective, but the MCl_4 (M=Sn,Ti) mediated aldol reaction of enol silyl ethers and PhCHO is distinctly different. 3^y and benzyl alcohols with a γ-R_3Sn group form cyclopropanes stereospecifically with inversion at both carbon atoms. $Ph_3Sn(CH_2)_{\underline{n}}SC_6H_4Me$-$\underline{p}$ ($\underline{n}$=3,4) can be oxidised stepwise to the sulphoxide and sulphone with MCPBA.[40] Halo(alkyl)stannylpropyl compounds undergo intramolecular coordination if terminally substituted by oxygen groups, as do $Cl_3Sn(CH_2)_3CO_2Et$ and $(ROCOCH_2CH_2)_2SnCl_2$.[41]

Silylmethyl Grignard reagents function as hydroxymethylating agents if the Si-C bond is cleaved oxidatively, and Me_3SiCCl_2 gives 1,1-dichloroalkenes with aldehydes.[42] $Me_3SiCHClLi$ induces homologation of boronates, and $(Me_3SiCH_2)_2AlPPh_2$ readily substitutes $Me_3NCr(CO)_5$, giving $(Me_3SiCH_2)_2Al(NMe_3)PPhCr(CO)_5$. KH complexes with $(Me_3SiCH_2)_3M$ (M=Ga, In) without reductive elimination , but for M=In, $KIn(CH_2SiMe_3)_4$ is formed in addition.[43]

$Me_{4-\underline{x}}M(CH_2PMe_2)_{\underline{x}}$ (M=Si,Sn) result from $LiCH_2PMe_2$ and for M=Si, $\underline{x}$=2, readily chelates to Ni. $\underline{\text{trans}}$-η^5-$cp(CO)_2(Me_3P)M{=}C(OMe)R^+$, a carbene complex, is rapidly converted to alkenyl complexes by $Me_3P{=}CHSiMe_3$, which is reduced by $(\eta^5$-$cp)_2MoH_2$ to $(Me_3SiCH_2)_2$ and $(\eta^5$-$cp)_2MoPMe_3$, and displaces THF from various organolutetium complexes.[44] Base induces C-$SiMe_3$ insertions into Fe-CO in η^5-$cpFe(CO)_2CH(SiMe_3)PMe_3$, giving the ketenyl complex, while CO inserts the W-C bond of $WCl(CH_2SiMe_{\underline{3}})_3PMe_3$.[45] Arsorane $Br(Me_3SiCH_2)_2As{=}CH_2$ and η^5-$cpM(CO)_3^-$ (M=Mo,W) give η^5-$cp(CO)_3MCH_2As(CH_2SiMe_3)_2$ which cyclise with CO loss to the first η^2-arsinomethyl complexes.[46] $(\eta^5$-$Me_5C_5)_2Th(CH_2SiMe_3)_2$ is assymmetric in the $Me_3SiCH_2ThCH_2SiMe_3$ unit, close intramolecular contact between CH_3 and CH_2 (2.3Å) providing for Me_4Si elimination and metallocycle formation. $Ir(CH_2SiMe_3)(CO)[P(OMe)_3]_3$ is fluxional with Me_3SiCH_2 axial in the solid.[47]

Disulphene $(H_2CSO_2)_2$ can be readily monosilylated at each carbon atom, Cl_2 fully chlorinating 2 of the 3 methyl groups of each Me_3Si

substituent. α-Silyl carbanions alkylidenate SO_2 to sulphines $R^1R^2C{=}S{=}O$, while sulphones result from R^1R^2CO and $Me_3SiCHSO_2Ph$. Phenyl(thiomethyl)silanes give aldehydes on oxidative hydrolysis.[48]

Formyltributyltin is the unstable intermediate in the hydrolysis of $Bu^n{}_3SnCH(Cl)OEt$, which gives $ArCH_2OEt$ and CO with ArCHO. The rearrangement of α-phenyl-1,4-dilithio-N-phenyl amide to the β-phenyl isomer proceeds through a cyclopropane intermediate, which has been established by lithiation of $PhNHCOCD_2CH_2SnBu^n{}_3$, followed by silylation.[49] $ROCH_2Cl$ readily methylates $SnX_2.LiX$, giving $ROCH_2SnX_3$, which with Bu^nLi gives $ROCH_2Li$ and $Bu^n{}_4Sn$. Me_3SnCH_2CN results from Me_3SnI and $ClMgCH_2CN$, while Me_3SnH and $H_2C{=}CHCN$ give $Me_3SnCH_2CH_2CN$. 2-Cyano-1,3-butadienes can be conveniently prepared from diethyl 2-cyano-2-trimethylsilylethanephosphonate with ketones.[50]

Me_3SiCN has been used to synthesise ArCOCN from ArCOCl, cyanhydrin silyl ethers from glyoxylic esters or α-substituted ketones using KCN/18-crown-6, β-cyanoethoxysilanes from epoxides,[51] α-cyanopyridines from pyridine-N-oxides, 2-alkoxyalkane nitriles from acetals, cyanosugars, esters of α-fluoro alcohols and the natural hydroxyamides Tembamide and Aegeline.[52] It also inserts nickel alkyne complexes as the isocyanide, to give enones on hydrolysis. Trisubstitution of $MeSi(CN)_3$, prepared in situ, by R^1R^2CO provides for β-aminoethyl alcohols on reduction.[53] The shift in the C-H stretching frequency of $CHCl_3$ for amines and nitriles show an order of base strength with $Me_3Ge(CH_2)_3NH_2 > Me_3SnCH_2CH(CH_2NH_2)(CH_2)_3NH_2 > Me_3Sn(CH_2)_3NH_2, Me_3(CH_2)_2CN$ (Sn>Ge>C) and Me_3MCN (Ge≃C>Sn).[54]

Cycloadducts of Me_3SiCNO with ethene or alkynes rearrange readily and hydrolyse to β-hydroxynitriles or β-oxonitriles, pyrrolidines are formed from olefins using $Me_3SiCH2N(CH_2CN)CH_2Ph$ and silylimmonium salts as azomethine ylide equivalent.[55] Heterocycles doubly bonded at nitrogen α-substituted by Me_3M (M=Si,Sn) are used to form acyl derivatives with RCOCl. $Me_3SnC(CN)_3.H_2O$ possesses a structure with Sn 5-coordinate and the planar $C(CN)_3$ unit bonding to Sn, and the hydrogen atoms of separate water molecules.[56]

Me_3SiCHN_2 results in good yield from $Me_3SiOSO_2CF_3$ and CH_2N_2, or from $(PhO)_2P(O)N_3$ and Me_3CH_2MgCl, while silyldiazo ketones are formed from $Me_5Si_2C(Li)N_2$ and RCOCl.[57] Polysilylated diazomethanes photolyse and pyrolyse through silyl migration to a carbene centre, the silene so formed being trapped by Bu^tOH or Me_2CO. $Me_3SiCH_2N_3$ results from Me_3SiCH_2Cl and NaN_3 in dry DMF, gives $ArNH_2$ with ArMgBr, and adds to acetylenic dipolarophiles as MeN_3 equivalent, since it can be readily desilylated.[58]

$Bu^t{}_3SiCHX_2$ (X=Cl,Br) and RLi undergo Li/X or Li/H exchange to give the carbene , which, for X=Cl, dimerises to the <u>trans</u>-disilyl alkene. This is not twisted about the double bond, but exhibits a wide angle at carbon (137.7^o) with long Si-C(sp^2) bonds (195 pm). Attempts to methylate $Bu^t{}_3SiCBrMe_2$ gave $Bu^t{}_3SiC(Me)CH_2$ through HBr elimination, and not $Bu^t{}_4Si$.[59]

A series of β-silyl-substituted divinyl ketones have been prepared from enols or ketones, and readily cyclise to cyclopentanones, trichothecane resulting similarly, while 3-silylfurans are formed from $Me_3SiC{=}CCH(R)OH$ and $Bu^iMgBr-Cp_2TiCl_2$ then R^1CN-H^+.[60] The synthesis of an estrapentaene provides the first evidence of hindered rotation in a vinyltrimethylsilane. These are also intermediate in the preparation of azacyclics, isoxazoles and pyrazolines, in intramolecular acylations to α-alkylidenecycloalkanones, and of α-chloroketones by photooxidation with $O_2/FeCl_3$.[61] $ArPd^+$ substitutes β-silylstyrene by <u>syn</u>-addition followed by <u>anti</u>-elimination, while Ni and Pd catalysts effect chemoselective and stereocontrolled silylation of vinyl and aryl halides, the latter derivatives providing a regioselective approach to cyclophanes.[62]

2-Nitrovinylsilanes give nitromethylalkylsilanes with organometallic compounds, while at -50°C, MCPBA cleaves the Si-C bond of alkenyltrifluorosilanes to give the ketone. Excess MCPBA gives aldehyde through C=C cleavage.[63] Several new ligands have been prepared from $Cl_2SiMeCH{=}CH_2$ on amination at Si and phosphine addition to the vinyl group. Lewis acids catalyse the condensation of styrylsilanes with EtO(PhS)CHMe or ethoxydithiolanes, giving allyl sulphides or styryl dithiolanes. Carbanion adducts result on cleaving 1-silyl-1-phenylthio-1-alkenes with $Bu^n{}_4NF$ in the presence of aldehydes, while $Br(Me_3Si)C{=}CHR$ behaves as a single synthon for both carbonyl cation and anion.[64]

The trends in energies of various α- and β-silylated unsaturated hydrocarbons is thought to arise through hyperconjugative interactions. Acid catalysed desilylation of $Me_2CHCMe{=}CHCH(SiMe_3)SiMe_2F$ to give the vinyl fluorosilane proceeds with 40-50% stereoselectivity, and adducts of $Me_3SiCH_2CH(Li)SOPh$ with aldehydes, ketones or epoxides eliminate PhSOH on neutralisation to give the 3-silyl allyl alcohol or 3-alken-1-ol.[65]

In the presence of $Fe_3(CO)_{12}$, $Me_3SiCH_2{=}CH_2$ and S_8 form a series of polythiacycloalkanes and the silylethylenedithiolato complex

$Me_3Si\overline{CHCH_2S}SFe_2(CO)_6$. The mass spectra of the cycloalkanes indicate migration of S to Si, and the loss of unsaturated fragments. Me_2SiS_2 and Me_2SiS are the final products of rearrangement.[66]

Vinylsilanes result stereoselectively from vinylsulphones through $Bu^n{}_3SnLi$ addition, a method used to prepare γ-lactones, while the 1,3-silyl shift from C to O in β-hydroxyvinylsilanes provides the first demonstration of migration from an sp^2 carbon atom.[67] E-3-Trimethylsilyl-2-buten-1-ol results in good yield from $MeC\equiv CCO_2Et$ and $(Me_3Si)_2CuLi$, followed by reduction. A series of β-silylated α,β-enones have been prepared. Trifluorolysis of cis-4--alkylcyclohex-2-enyl derivatives of Si, Ge, and Sn is stereospecifically γ-anti, whereas the trans-isomers are less specific.[68]

Palladium catalyses the coupling of allyl halides with vinyl and aryl derivatives of tin, and of α-oxygenated vinyltin compounds with acid chlorides (giving α-oxygenated enones and subsequently α-diketones, dienes or methyl vinyl ketones).[69] $RCH{=}C(SnMe_3)_2$ is iododestannylated by I_2 with greater specificity at lower temperatures, while addition of the stannyl cuprate $R_3Sn(L)CuLi$ adds reversibly to the alkyne $MeC\equiv CCO_2Et$, the adduct, of low pK_a ($\sim$5), being acidified to the substituted vinyl stannane.[70]

The red solution formed from $Ph_2C{=}CPhLi$ and $SnCl_2$ is thought to contain the dialkenyltin species. It undergoes oxidative addition with Bu^nBr, while with $Ph_2C{=}CPhLi$ and Bu^nBr, $(Ph_2C{=}CPh)_3SnBu^n$ results, the structure resembling that of $(Me_2C{=}CPh)_4Sn$, with considerable variation in the angles at tin.[71]

The tribenzylidenemethane dianion can be readily obtained from $(PhCH_2)_2C{=}CHPh$ with excess Bu^nLi, and gives the bis-silyl derivative with Me_3SiCl, $[Me_3Si(Ph)CH]_2C{=}CHPh$.[72] The stereochemistry of electrophilic attack of allylsilyl groups substituted on the cyclohexane moiety is anti in the allylsilyl portion but this can be offset by axial or equatorial preference in the ring. Protonation of allylsilanes containing a basic site occur with intramolecular proton transfer and concomitant remote asymmetric induction , while the stereochemical aspects of trifluoroacetolysis of allylic bis(trimethylsilyl)cyclohexenes has been studied.[73] E-Crotyl- and E-cinnamyltrimethylsilanes give erythro homoallyl alcohols with 93% selectivity using $RCHO/TiCl_4$, but lower selectivity was observed for Z-allylsilanes. Desilylation of optically active allylsilanes with Li_2PdX_4 proceeds stereospecifically to give active π-allylpalladium complexes, and allylsilane complexes of Pt hydrolyse to the allyl-Pt complex. An examination of (-)-α-napPhMeallylSi as a template for effecting chirality transfer from Si to C shows this system to be too congested and active.[74]

Allylsilanes reductively couple to pyridine-N-oxides and can be oxidised by PhI=O to conjugated enones, which add the anion of $Me_3Si(RS)CHCH=CH_2$ with γ-regioselectivity.[75] 3^y allylic and propargylic acetates react with silylcuprates cleanly anti , to give stereodefined allyl- and allenylsilanes. Activated α-chlorosulphides couple with allylsilanes, giving allylsulphides, and $allylSiMe_3/F^-$ induce conjugate addition of $allyl^-$ to Michael acceptors chemoselectively, and a terminal allylsilane function provides for the 1 step formation of a steroid nucleus from an acyclic polyene chain.[76]

The regioselective protonation and alkylation of hydroxyallylsilanes with OH appropriately sited provide a facile preparation of 3-hydroxy-1,5-hexadienes. 1,5-Dienes with the allylsilane fragment cyclise to methylenecyclohexanes (eg the terpene ±-trixagol).[77] Hydrosilation of cyclic dienes gives optically active cyclic allylsilanes which couple with ethylene oxide through Si-C insertion. Chiral α- and β-alkoxy aldehydes couple allylsilanes with diastereofacial preference, while amide-acetals and $MeCH=CHCHOHCH_2SiMe_3$ couple with little preference.[78] Both (3-pyrrolidinoallyl)- and (1-pyrrolidinoallyl)trimethylsilanes can be prepared from allylpyrrolidine. The former behaves as both allylsilane and enamine, and reacting with electrophiles to which the 1-pyrrolidinoallyl isomer is inert, though fluoride activates reaction with carbonyl compounds to give amino alcohols and furans.[79]

3-Alkenylsilanes and stannanes result from regioselective sulphenic acid or amide elimination from γ-substituted sulphoxides or imines. Allyl cations generated from the silane, germane or stannane substitute aromatic nucleophiles and nitriles. But-2-enylsilanes or stannanes, and pyruvates give the threo-isomer, and has been used to prepare cis-crobarbatic acid.[80]

The allylation of aldehydes and ketones to give homoallyl alcohols can be successfully carried out using allyl bromide and Sn with H_2O, while alkoxymethyl substituted allylstannanes allylate PhCHO to give cis-4-alkoxymethylbut-3-en-1-ol regardless of regio- or stereochemistry of the allylstannane.[81] Allylchlorostannanes couple with acyl chlorides, add to ketones and give homoallylic alcohols and substituted tetrahydropyrans with aldehydes. Radiofrequency probing (RFP) and chemically induced nuclear polarisation (CIDNP) indicated the mechanism of the reaction of allyl$SnMe_3$ with Cl_3CBr to involve radical steps. Thermal fluorine atoms and $allyl_4Ge$ give $H_2C{=}CHF$.[82]

2-Bromomethyl-3-(trimethylsilylmethyl)buta-1,3-diene behaves as a 2,2'-biallyl synthon, and results from the bis(silylmethyl)diene, itself undergoing tandem [6.5] annulations through addition to both sides of the diene. 1,4-Bis(trimethylsilyl)buta-1,3-diene adds to 2,3-dimethoxy-1,4-benzoquinone to give the naphthoquinone, and mono-silylated conjugated dienes undergo an intramolecular Diels-Alder addition to provide for model studies on nagilactone.[83] Conjugated dienes with Me_3M (M=Si,Ge,Sn) in an allylic site undergo trifluoro-acetolysis at the ϵ-position but SO_2 reacts regiospecifically at the γ-position of the stannanes. E-6,6,6-Triphenyl-6-stannahexa-1,3-diene is non-fluxional but the Z-isomer rearranges through a supra-facial [1,5]-shift.[84] $Me_3Si(PhS)C{=}C{=}CMe_2$ is cleaved at C-Si by MeLi but at C-S by Bu^tLi, the latter adding ketones to give substituted silylacetylenes, and Et_3SiCl to give the 1,1-bis-silylallene. The allenyl ether $H_2C{=}C{=}C(OMe)SiMe_3$ gives silylmethyl acetylenes on lithiation, followed by hydrolysis or ketone addition.[85]

$Me_3SiC{\equiv}CMe$ is polymerised with transition metal catalysts virtually quantitatively and shows extremely high gas permeability, while $Me_3SiC{\equiv}CCH(Me)OH$ can be <u>cis</u>-reduced by $LiAlH_4$ to <u>Z</u>-4-(trimethylsilyl)-3-buten-2-ol.[86] α,β-Unsaturated cyclopentenones have been prepared from alkynes (<u>eg</u> $Me3SiC{\equiv}CC_6H_{13}$-n) by allylzincation, and 3-substituted furans by tandem and retro Diels-Alder reactions of 4-phenyloxazole with acetylenes.[87] $Me_3C{\equiv}CH$ couples <u>N</u>-heteroaryl halides, and indoles result from <u>N</u>-benzyl[2-(trimethylsilyl)ethynyl]aniline.[88] Cobalt catalysed cocyclisation of isocyanatoalkynes with silylacetylenes gives 5-indolizinones (applied to synthesis of camptothecin), and bis(carbyne) clusters undergo unprecedented isomerisations through cage rearrangements, while the cycloaddition of N-methyl-2-pyridones with $Me_3SiC{\equiv}CPh$ gives silylated biphenyls.[89]

$Me_3SiC{\equiv}CSiMe_3$ adds cleanly to cycloheptatriene to give a bicyclic derivative rather than cyclotrimerise like other alkynes. Dienes and enynes result stereodefined on hydroalumination of $RC{\equiv}C{-}C{\equiv}CSiMe_3$, which can be conveniently synthesised by condensing $RC{\equiv}CH$ with $Me_3SiC{\equiv}CBr$, while bis-silylation of the diGrignard reagent of α,ω-diynes occurs in high yield.[90] $Me_3SiC{\equiv}C(CH_2)_2X$ (X=halogen or OTs) cyclises to the silylcyclobutene in the presence of Me_3Al/cp_2ZrCl_2, while $Me_3SiC{\equiv}CR$ undergoes [2+2] cycloaddition to dichloroketene, giving cyclobutenones. Cationic alkoxycarbene complexes of Pt^{II} are formed from $(R_3P)_2PtX(ROH)$ by $Me_3SiC{\equiv}CR$ substitution, followed by Si-C cleavage.[91]

The ketones $RC{\equiv}CCOMMe_3$ (M=Si,Ge,Sn) can be prepared from the 1,3-dioxan-2-yl acyl anion equivalent, and the anion of $Ph_2CHC{=}CPh$ gives a mixture of substituted alkynes and allenes with Me_3MCl.[92] Silyl and stannylethynylamines react with carbodiimides to give quinolines and 3-aminopropiolamidines, while 2-[$cpFe(CO)_2CO$]thiophene adds $Ph_3GeC{\equiv}CPh$ to give a cyclopenta[b]thiophen-4-one.[93] $FeCl_3$ oxidatively couples $Me_3MC{\equiv}CMe$ giving $MeC{\equiv}C{-}C{\equiv}CMe$ if M=Ge,Sn, and $Me_3SnC{\equiv}CSiMe_3$ giving $Me_3SiC{\equiv}C{-}C{\equiv}CSiMe_3$. Tin substituted alkenes and allenes are formed from $R_3SnC{\equiv}CR'$ and Et_3B, and the vibrational spectra of $Me_3SnC{\equiv}CH$ analysed.[94]

Silyl substituted cyclopropylcarbinyl trifluoroacetates are much more reactive towards ionisation than the unsubstituted system, and show inverse stereochemical effects in the mass spectra through Si---O stabilisation.[95] Tetracyclopropylsilane, which is prepared from cyclopropyl-lithium, can be monochlorinated by $GaCl_3$, while heating $Me_3SiC{=}C(SiMe_3)CHCOCHN_2$ with CuBr gives 1,5-bis(trimethylsilyl) tricyclo[$2.1.0.0^{2,5}$]pentan-3-one, which can be desilylated, the residual oil photolysing to cyclobutadiene and cyclopentadienone and not tetrahedrane.[96]

The structure of $(Me_3Si)_3C_5H_2Li(Me_2NCH_2CH_2NMeCH_2CH_2NMe_2)$ shows the ring to be η^5-coordinated. Silylation of $C_5H_5^-$ is stereoselective, and of the anion of isodicyclopentadiene occurs with predominant below plane electrophile capture.[97] Lanthanide metals form anions $[(Me_3Si)_2C_5H_3]_2LnCl_2^-$ from $LnCl_3$ or from $\{[(Me_3Si)_2C_5H_3]_2LnCl\}_2$, and for Ln=Nd, the structure is a bent sandwich. Reductive silylation of guaiazulene gives the 2,3,5,6-tetrasilyl derivative along with a little of the hexasilylated 2,5,6-dimer, coupled through the 4 positions.[98] Direct polysilylation of indoles occur without ring opening, but benzothiophen gives a hexasilyl substituted cyclohexene (with desulphuration) and benzofuran a polysilylated aryloxysilane.[99] Derivatives of C-silyl quinolizine are formed by cycloaddition of $MeO_2CC{\equiv}CCO_2Me$ to C-silylated pyridines, 2-trimethylsilylthiazole inserts aldehydes at the C-Si bond, and silylfurans photorearrange to the 4-silyl-2,3-butadienal.[100]

The adducts of 9-trimethylsilylanthracene with $MeO_2CC{\equiv}CCO_2Me$ and p-benzoquinone show temperature dependent ^1H nmr spectra, indicating slow rotation about the Si-C single bond. The 1,4-dimethoxy-9-(trimethylsilyl)triptycene has a frozen conformation. Palladium mediated ethynylation of iodoarenes coupled with cobalt catalysed alkyne cotrimerisations allows access to substituted benzo[3,4]cyclobuta[1,2-b]biphenylenes, the tetrasilyl derivative showing more electron localisation in outer unreactive silylated rings than in the inner ring.[101]

The effect of d-orbitals is more important in $PhSiF_3$ than in $PhSiH_3$ or $PhSiMe_3$, where hyperconjugative effects dominate the behavior, as they do in anionic states of p-disubstituted benzene derivatives p-$(Me_3M)_2C_6H_4$.[102] Benzyne is intermediate in the fluoride induced 1,2-elimination of o-trimethylsilylphenyl triflate, while the silylphenyl cation is generated from its o-bromo derivative at 70eV electron impact ionisation.[103] Both bromo and iodo substituted phenols can be made from the acetates of (trimethylsilyl)phenols, while 3-methoxy-6-substituted benzamides result from the 2-trimethylsilyl derivative which can be lithiated at position 6, substituted, and then desilylated by fluoride.[104] Electrophilic assistance to cleavage of substituted aryltin derivatives falls with proton availability in the Me_2SO/aq.KOH medium. The dearylation of $ArSnBu^n_3$ by lead tetraacetate is Hg^{II} catalysed, high yielding, and has been used to prepare the anti--inflammatory drug, naproxen. The aryl-lead triacetate so formed readily arylates α-hydroxymethylene ketones.[105] The mechanism of electrophilic attack by $[Fe(CO)_3(1,5\text{-}\eta\text{-dienyl})]^+$ with $XC_6H_4MMe_3$ (M=Si,Sn) is examined and the reaction used to prepare aryl derivatives of cyclohexadiene. 7-$R_3MC_7H_7Fe(CO)_3$ shows fluxional behaviour.[106]

Lithiated 2-(trimethylsilylmethyl)pyridine couples with PhCN to give E- and Z-1-phenyl-2-(2-pyridyl)-1-(trimethylsilylamino)ethene, which acidify to 2-phenacylpyridine. Benzylsilanes add ketones to give the appropriate 3^y alcohol on fluoride catalysed hydrolysis, but, rather surprisingly, photolysing o-(trimethylsilylmethyl)acetophenone gives no evidence of silyl migration to oxygen.[107]

The stannylation of lithiated bicyclo[3.2.1]octa-2,6-diene gives a mixture of two isomers, while both thiophene and furan give the 2,5-bis-stannyl derivative. (3-Thienyl)tin halides are not associated.[108] Benzthiazol-2-yl- and benzoxazol-2-yl-trimethylstannane reacts with $MeGeCl_3$ giving mono-, di-, and trisubstitution.

The mono- and disubstituted derivatives are reduced to the germane by Me_3SnH, which then decomposed by α-elimination. Lithiated benzoxazole exists as two isomers in mobile equilibrium, the one C-lithiated, the other a ring opened O-lithiated isonitrile. Silylation stabilises the latter, stannylation the former. An extensive series of lithiated heterocycles can be stannylated at the 2-position. [8-(Dimethylamino)naphthyl]trimethyltin is demethylated by Me_3SnX (X=Cl,Br), giving the 5-coordinated derivative with rigid, flat NCCCSn chelate ring. The trimethyltin derivative also methylates $CODPtCl_2$ at room temperature.[109]

An extensive series of pentamethylcyclopentadienyl derivatives of silicon and germanium have been prepared from Me_5C_5Li and $Me_{\underline{n}}MCl_{4-\underline{n}}$ ($\underline{n}$=0-3). $Me_5C_5GeMe_2Cl$ give $Me_5C_5GeMe_2X$ (X=F,I,OMe,SMe) with the appropriate nucleophile. Me_5C_5Li gives $(Me_5C_5)_2Ge$, Me_5C_5GeCl or $Me_5C_5Ge_2Cl_3$ with dioxan.$GeCl_2$, depending on the molar ratios. Decamethylstannocene $(Me_5C_5)_2Sn$ exchanges one or both rings with $C_5(CO_2Me)_5H$, while $Me_5C_5Sn^+CF_3SO_3^-$ is substituted at Sn by BI using BI_3, giving the <u>nido</u>-carbaborane cation $Me_5C_5BI^+$.[110] The structures of $(C_5H_4Me)_2M$ (M=Ge,Sn) show both to be bent sandwich complexes, as is $(Me_3SiC_5H_4)_2Sn$, prepared from dilithiated stannocene. $(C_5H_4Me)_2Sn$ functions as a dehydrating agent for the condensation of cyclic acid anhydrides with amines, to give <u>N</u>-alkyl imides. The plumbacenes $[C_5(Me_3Si)_{\underline{n}}H_{5-\underline{n}}]_2Pb$ ($\underline{n}$=1-3) result from $PbCl_2$ and increase in stability to oxidation with silylation.[111]

Heating 1-silacyclopropenes, substituted at the double bond by Me or Ph and $SiMe_2R$ (R=Me,Et,Bu^t,Ph) give products dependent upon these substituents. With Ph and R=Me or Et, an isomeric mixture of 1,4-disilacyclohexadienes result, with R=Bu^t, the disilacyclobutene results, while for R=Ph, this and two isomeric silacyclopentadienes form. The 2,5-disilyl-1-silacyclopentadiene and $PhC{\equiv}CSiMe_3$ dominate the products of pyrolysis of 1,1,2-triphenyl-3-(trimethylsilyl)-1-silacyclopropene, while the 1,2-diphenyl-1-methyl derivative gives silacyclopentadiene and 1,4-disilacyclohexadiene. Silylene insertions and coupling explain the products.[112] Silacyclopropenes with Si or H substitution at Si thermally rearrange to vinylsilylenes while the silacyclobutene results from the pyrolysis of silacyclobutane with acetylenes. The main product is the silylallene, but the 3-silylprop-1-yne also forms. FVP of methoxydisilanes gives propenylsilylenes which afford silacyclobutene derivatives and the first methylenesilacyclobutane.[113]

The thermal decomposition of silacyclobutane leads not only to silene, but to methylsilylene and to silylene, established through addition of all to buta-1,3-diene. The proportion of 1-silacyclopent-3-ene to its Me-homologue (MeHSi: reaction) plus silacyclohexene ($H_2Si=CH_2$) is about 1:5, while the proportion of this last product diminishes relative to 1-methyl-1-silacyclopent-3-ene as the pyrolysis temperature rises. Mono- and 1,3-disilacyclobutanes are readily opened by Lewis acids of B, Al, Sn, P, As, Sb, and S.[114]

Permethyl-1,2-disilyl-3,4-disilacyclobut-1-ene inserts carbene into the Si-Si bond (the first example) to give the cyclopent-4-ene, the radical anion of which shows a temperature dependent esr signal due to ring flipping. The acid promoted cleavage of both exo- and endo-2,2-dimethyl-3-neopentyl-2-silanorborn-5-ene gives clean, almost quantitative ring opening to 3-(3-cyclopenten-1-yl)-2,5,5-trimethyl-2-silahex-2-yl products.[115]

The Et_2NLi induced rearrangement of 6-oxa-3-metallabicyclo[3.1.0]hexanes provides a convenient way of synthesising cyclic allylic alcohols of Si or Ge, thereby providing the precursors to siloles and germoles. 1,1,3,4-Tetramethylsilole results from PhNCO addition to the appropriate alcohol, followed by thermolysis in CCl_4, shows no tendency to dimerise, and forms a diene complex with $Fe_2(CO)_9$. 1,1-Dimethylsilole and germole can be prepared similarly and both complex with the $Fe(CO)_3$ residue as dienes. The Diels-Alder dimer of 1,1-dimethylsilole exhibits large deshielding at the bridge Si (lower field than disilene or silene).[116]

Photo-oxygenation of 1,1-dimethyl-2,3,4,5-tetraphenylsilole gives PhCO(Ph)C=C(Ph)COPh and a novel bicyclic product with the 2,4-dioxa-3-silabicyclo[3.2.0]hept-6-ene skeleton which ring expands on further heating to the tetraphenyl dioxasilacycloheptadiene. α,ω-Bis-(1-methyl-2,5-diphenyl-1-silacyclopentadienyl)alkanes undergo reversible intramolecular [2+2] cycloaddition with higher quantum yields than detached dienes, and via a singlet excited state.[117] $Ph_2MeSiLi$ desilylates 1-methyl-1-(trimethylsilyl)dibenzosilole, giving $Ph_2MeSiSiMe_3$ and the silole anion in high yield. With 1-methyl-3,4-diphenyl-1,2,5-tris(trimethylsilyl)silole, however, $Ph_2MeSiLi$ reduces the diene to isomeric silacyclopent-3-enes. RLi reagents behave similarly but will not generate the dibenzosilole anion. Calculations suggest that the silole anion is only 25% as aromatic as $C_5H_5^-$, while silabenzene is more than 80% as aromatic as C_6H_6.[118]

Silacyclopentadienylmethylene can be generated from $(PhC)_4SiMeCHN_2$ and $(PhC)_4SiMeCHN{=}N$, and rearranges by ring expansion to silabenzene, and by a 1,2-methyl migration to 5-silafulvene, intermediates being successfully trapped by alcohols, diene or ketones. Hexamethyl-1,4-disilacyclohexa-2,5-diene and $Bu^t{}_2O_2$ react photolytically to add Bu^tO. to C=C at $-105^{\circ}C$ but to attack Si-H at $-30^{\circ}C$. The silyl radical so formed rearranges to the silacycloprop-2-yl radical. Arsabenzenes result from stannacyclohexadienes and $AsBr_3$.[119] Metallocycloalkanones are formed directly from the hydroboration of dialkenyl derivatives of Si, Ge, and Sn. The structures of a series of anthracene-like heterocyclic compounds have been used to support the presence of conjugative effects within the hetero (N,O,S), group IV substituted (Si,Ge,Sn) central ring. 10,10-Dimethylphenothiastannin has a "butterfly" structure.[120]

Attempts to synthesise stannacyclobutanes normally result in the isolation of their cyclic oligomers. With $Me_2C(CH_2MgBr)_2$, however, the cyclobutane has been isolated and characterised. It is highly volatile and, like strained stannacyclopentanes, readily reacts with oxygen. $Ph_3Sn(CH_2)_4SnPh_3$ is intermediate in the preparation of the cyclic dimer and tetramer of $Ph_2\overline{Sn(CH_2)_4}$. HBr monobrominates it to give the diphenylbromo derivative, which can be coupled with $BrMg(CH_2)_4MgBr$ to both macrocycles, while coupling first with $Cl(CH_2)_4MgBr$, then $LiPh_2Sn(CH_2)_4SnPh_2Li$ gives mainly tetramer.[121]

A comparison of the reactivity of 1-chloro-1-sila-adamantanes with 1-silabicyclo[2.2.1]heptane and [2.2.2]octane shows reactivity to increase with closeness of the geometry at silicon to trigonal bipyramidal, with the 1-sila-adamantyl system more reactive than the tetrasila-adamantyl one. 1-Methyl-1-sila-adamantane results in 67% yield from <u>cis</u>-1,3,5-tris(trimethylsilylmethyl)cyclohexane by acid catalysed redistribution.[122] $AlBr_3$ catalyses Me_4Si elimination at $20^{\circ}C$ from carbosilanes containing a chair conformation with a 3^y or 4^y carbon atom, giving condensed 1,3,5-tricyclohexanes. These condense further with CH_4 elimination, and the mechanism followed by deuteration. The structure of the dispirosilane with 4-membered 1,3-disilacyclobutane spiro with 1,3,5-trisilacyclohexane at the carbon atoms and fully methylated at Si, shows flattened cyclohexane rings, long Si-C bonds in the 4-membered ring, and short Si-C bonds at Si opposite to the 4-membered ring.[123]

The alkaloid gephyrotoxin has been prepared by using <u>o</u>-quinone methide N-alkenylimine as intermediate. These are made from o-$(Me_3SiNR)C_6H_4CH_2\overset{+}{N}Me_3$ using fluoride. The sila substituted derivatives of trihexyphenidyl, cycrimin, and procyclidin, PhR'Si(OH)-$(CH_2)_2NR_2$ (R',R_2 cyclic), 30 sila analogues of nifedipine, silylated dopamines 4-(β-aminoethyl)resorcinol and 4,4-diphenylpiperidines, and antiParkinson agents have been prepared and tested. The first report of a microbiological transformation of a silicon compound involves the reduction of trimethylsilylalkyl acetoacetates to give (+)-3(S)hydroxybutanoates with 80% enantiomeric purity.[124]

3 Catenation

Aromatic heterocyclic N-oxides are reduced by Me_6Si_2 at 20°C in the presence of F^- and γ,γ-difluoroallylsilanes, which afford gem-difluoroallyl adducts of aldehydes and ketones, result from $CF_3C(R)=CH_2$ and $Me_6Si_2/Bu^n{}_4NF$. This provides a novel route to metal free silyl anions, and promotes RCHO insertion into disilanes and trisilanes.[125] CF_2 inserts $(FMe_2Si)_2$, giving $(FMe_2Si)_2CF_2$ which can be substituted at C using Me_3SiPMe_2, the phosphine isomerising with fluorine migration to the ylide. Benzylidene di- and trichlorides can be reductively coupled to give stilbenes using Me_6Si_2 or $(ClMe_2Si)_2$ with $Pd(PPh_3)_4$. Benzylic chlorides yield $ArCH_2SiMe_nCl_{3-n}$ similarly.[126] The crystal structure of $CpFe(CO)_2SiMe_2Ph_3$ shows the Si-Si bond of 2.374Å to be 0.018Å longer then that of $Me_3SiSiPh_3$. The mass spectra of $(Me_3Si)_3SiCOR$ indicate fragmentation to $(Me_3Si)_3SiCO^+$. NaOMe catalyses the coupling of styrenes with dimethylsilylene using α,ω-dimethoxypolysilanes. Thus $MeO(SiMe_2)_2OMe$ and $PhRC=CH_2$ (R=H,Me) give 1-sila and 1,2,3-trisilacyclopentanes as isomeric mixtures, together with a 1,2,3,4-tetrasilacyclohexane.[127]

The first cyclic ethynylene polysilane results from $(Me_2SiC{\equiv}CMgBr)_2$ and $Cl(Me_2Si)_3Cl$. It loses Me_2Si: on heating to give the tetrasila and trisiladiynes, the latter, with a 7-membered ring, being the smallest cyclic diyne prepared to date. Ring strain induces a bathochromic shift in the UV spectra of these rings. The first stable C-unsubstituted 1,2-disilacyclohexa-3,5-diene results from the oxidation of the cyclohexen-4-ol and dehydration with Al_2O_3. 1,1,2,2-Tetrafluoro-1,2-disilacyclobutene reacts with conjugated aldehydes or sterically hindered ketones to give ring opening at the Si-Si bond, followed by closure or H abstraction.

Dimesitylsilylene adds to $R(Me_3Si)C=C=SiMes_2$ ($R=Me_3Si$, Ph) to give the first stable disilacyclopropanes $Mes_2SiMes_2SiC=C(SiMe_3)R$.[128]

Tetramesityldisilene adopts a trans-bent geometry with an Si-Si bond of 216pm, some 20pm shorter than the single bond length. The significant π-component suggested is supported by ^{29}Si nmr spectral evidence. The disilene thermolyses to the benzosilacyclobutene, adds HCl and Cl_2, water and alcohols, and gives cyclic adducts with ketones and ethynes.[129]

Tetra-t-butyldisilene results from the reduction of $Bu^t{}_2SiXSiXBu^t{}_2$ (X=Cl,Br,I) using alkali metals. It has been trapped as PhCOCOPh and 1,3-diene adducts. It also results from the low temperature photolysis of 2,3-benzo-7,7,8,8-tetra-t-butyl-7,8-disilabicyclo[2.2.2]octa-2,5-diene which also isomerises. Trimethylsilyltrimethyldisilene can be generated similarly, and rearranges to the disilylene. The reverse rearrangement, with silylene generated from $(Me_5Si_2)_2SiMe(OMe)$ or the appropriate sila bridged benzonorbornadiene, is equally facile.[130]

Hexa-neopentylcyclotrisilane is formed from $(Bu^tCH_2)_2SiCl_2$ and Li in THF. Lithium also reduces $(Et_2CH)_2SiXCl$ (X=H,Cl) to the cyclotrisilane while $(Pr^i{}_2Si)_3$ results from the 1,3-dichlorotrisilane. On irradiation, these cyclotrisilanes give disilene and silylene which can be trapped. 1,2-Dichlorodisilanes give cyclotetra- and pentasilanes, while methylchloropolysilanes are formed from the redistribution of methylchlorodisilanes using $Bu^n{}_4PCl$, and give fused cyclopolysilanes with seven 5- or 6-membered rings.[131]

Perethylcyclopolysilanes $(Et_2Si)_n$ (n=4-8) result from Et_2SiCl_2 and an alkali metal, which determines n. The cyclotetrasilane is strained and readily inserts alkynes, and like $(Me_2Si)_6$ is oxidised by MCPBA through consecutive oxygen atom insertion.[132] $PhMeSiCl_2$/Li/THF with $Ph_3SiSiMe_3$ gives 62% hexamer (5 isomers) and 25% pentamer. 1,1-Dichloro-1-silacyclohexane is reduced to the cyclopenta- and cyclohexasilanes with potassium, but with lithium 1,1-dichloro-1-silacyclopentane gives $[(CH_2)_4Si]_n$ n=5-12, together with a septamer with an exocyclic Si-Si bond. $(Ph_2Si)_5O$ possesses a boat structure with the angle at oxygen 146°.[133] A series of α,ω-dihalopolysilanes have been prepared by cleavege of the cyclopolysilane with X_2, and the calculated σ-orbital energies of linear polysilanes correlate closely with their photoelectron spectra.[134]

The structure of $MePh(X)SiGePh_3$ (X=H,F,Cl,OR) has been resolved and the stereochemistry of nucleophilic substitution at Si shown not to change with Ph_3Ge as substituent. Cleavage of Si-Ge by organometallic nucleophiles (X=H,OMe) gives Ph_3Ge^-, radical anions or catalysed decomposition, depending on the nature of the nucleophile. A series of digermasilanes and polygermanes have been prepared by a Wurtz type reaction. λ_{max} for UV absorption of the Ge-Ge chromophore is enhanced by phenyl or halogen substitution for ethyl, and when the chain is lengthened.[135] $(Ph_2Ge)_5$ crystallises in five phases from solution, three with solvent, and $(Ph_2Ge)_6$ is plastic at 346^oC. Calculations indicate digermene to be antipyramidal and not planar.[136] The crystal structures of $(Ph_2Sn)_6$.2toluene, $(Ph_2SnPh_2SnCH_2)_2$ and $[(Me_3Sn)_3Sn]_2(Me_2N)_4Mo_2$ all have Sn-Sn bonds of 2.78Å. Me_6Sn_2 adds to 1-alkynes catalytically to give Z-1,2-bis-(trimethylstannyl)-1-alkenes which photochemically isomerises to the E-distannylalkene.[137]

The vibrational spectra of $MeSiH_2SiH_2Me$ support rotational isomerism about the Si-Si bond, the dipole of $[(MeO)_2MeSi]_2$ is determined and the spectra of O, N, and S substituted silanes and disilanes compared.[138] The ^{29}Si nmr chemical shifts of substituted disilanes are interpreted, and the intensity of such signals found to depend on long range Si-H coupling.[139]

4 Hydrides

The reaction of gaseous B_2H_6 or B_2D_6 with alkyl and arylalkylsilanes show H-D exchange, and cleavage reactions to involve electrophilic borane insertion into Si-H, Si-C or Si-Si bonds. The kinetics of bromination of organosilicon hydrides involves a 1:1 molecular transition state with partial positive charge build-up on silicon, while acid catalysed hydrolysis of Si-H shows branching α to the reaction centre to have a greater effect than β-substituents.[140] Carbosiloxytetrahydrothiophenes result from Et_3SiH reduction of a thiophene ketal. Carbon centred radicals form during the photolysis of $Bu^t{}_2O_2$ and alkenyldimethylsilanes by H migration to give allyl radicals, or intermolecular double bond addition except with pent--4-enylsilyl radicals, where intramolecular cyclisation to the endo- rather than the exo-isomer occurs. $SnCl_4$ readily monochlorinates organosilanes without Si-C or Si-Si cleavage.[141]

Palladium catalyses the dehydrogenation of Me_3SiH to the silaethene, of silacyclobutane to the diene (only mass spectrometric evidence) and silacyclohexane to silabenzene. The reaction of MeNCO with Et_3SiH and $[HRu_3(CO)_{10}(SiEt_3)_2]^-$ (5:1:0.001) gives a [4,5]-spirocyclic derivative involving 5 isocyanate units with one of them deoxygenated by Et_3SiH and thereby providing the spiro site. $(Me_5C_5Ir)_2Cl_4$ is both reduced and silylated by Et_3SiH.[142]

Catalysed hydrosilation includes addition to carbonyl compounds to give prochiral ketones, the ring opening of furans, and the substitution of alcohols.[143] $Co_2(CO)_8$ catalyses the hydrosilation of the C=C bond of methyl acrylate in high yield, and the carbonylation-hydrosilation of alkyl acetates to (siloxymethylene)-alkanes.[144] Platinum catalyses addition to allyloxyethylboranes, and, with sonic waves, the high yield addition of X_3SiH (X=Cl,Et, EtO) to 1-hexene. Ruthenium-phosphine catalysts have been found to be more efficient catalysts for hydrosilation of 1-alkenes and vinylsilanes than other alkenes. α-Nap(o-Me_2NCH_2)$C_6H_4SiH_2$ is the first 5-coordinate non-halogenated silane and possesses very reactive Si-H bonds.[145]

Halomethyldimethylsilanes, halodimethylsilanes, and p-$(Me_3Si)_2C_6H_4$ have been vibrationally analysed, and substituent and inductive effects determined for the stretching frequency of isolated Ge-H groups.[146] A series of cycloalkylgermanes have been prepared from the trichloride using $LiAlH_4$, $(C_6H_5)_3GeH$ reacts with Et_3Tl to give Ge-Tl derivatives, while $(CF_3)_3GeH$ photolytically cleaves Mn-Mn and Co-Co bonds of carbonyls, to give the appropriate germyl-transition metal derivative. $Me_nH_{3-n}SnMn(CO)_5$ result from $Me_nH_{3-n}SnCl$ and $NaMn(CO)_5$, and show increasing thermal stability with methylation.[147]

Hydrostannylation of bicyclo[3.1.0]hex-2-ene gives an isomeric mixture of 4-methyl and 5-methyl-3-trimethylstannylcyclopentenes, while 1-stannyl-1-alkynes give 1,1- and 1,2-distannyl-1-alkenes.[148] Though a low basicity hydride, $Bu^n{}_3SnH$ induces β-hydride elimination in allyl acetates (Pd^o catalysed) to give the diene. Polymethyl-hydrosiloxane, however, gives chemoselective reduction with no β-hydride elimination. Palladium also catalyses the carbonylation of organic halides to aldehydes using $Bu^n{}_3SnH$, as does the diastereoselective synthesis of C-glycopyranosides.[149] The reduction of acyl halides by $Bu^n{}_3SnH$ is now thought not to involve a radical mechanism, though the hydrostannylation of both conjugated enynes and steroid ketones is catalysed by AIBN, supporting a radical route. The reduction of halosilanes by $Bu^n{}_3SnH$ is free radical and faster for α-halosilanes than comparable haloalkanes.[150]

5 Metal Derivatives and Radicals

Me_3SiLi in the presence of Cu^I reacts with allyl chlorides to give allylsilanes, and provides an improved preparation of E-3-trimethylsilyl-2-buten-1-ol.[151] Silylcarbene complexes result from the reaction of Ph_3SiLi with $Re_2(CO)_{10}$, and the ^{199}Hg FT NMR spectra of silyl-mercury compounds show a correlation with orbital electronegativites on silicon and the lowest energy UV absorption maxima.[152] Transition metals catalyse the silylmetallation of acetylenes, while $(Me_3Si)_2Mg.DME$ results in improved yield (80%) from $(Me_3Si)_2Hg$. It reacts with TMEDA to displace the ether, as does $[(Me_3Si)_3Al].THF$ to give $(Me_3Si)_2Mg.TMEDA$ and $[(Me_3Si)_3Al]_2.TMEDA$, with Si-Mg and Si-Al bonds of 2.63Å and 2.47Å.[153]

$(Me_3Si)_3Al$ has been used for the transition metal catalysed silylation of allyl acetates, undergoes room temperature 1,2- and low temperature 1,4-addition to α,β-unsaturated carbonyl compounds, and gives $Me_3SiCH{=}CHCH(OH)R$ with β-hydroxyalkynes. Allylic phosphates give allylsilanes with $PhMe_2SiAlEt_2$, but vinylsilanes with enol phosphates-Pd^{II}.[154] The molecular structure of $(Me_3Si)_4AlNa$ shows a chain arrangement with Na bridged tetrahedral anions, while $(Me_3Si)_4Pb$, the first organosilicon-lead compound, results from $(Me_3Si)_2Mg/PbCl_2$ as a pale yellow, light sensitive, solid decomposing above 80^oC.[155]

Acylgermanes result from amides and Et_3GeLi, while the ^{13}C nmr spectra of sodium derivatives of phenylgermanes show less anion delocalisation in the rings than Ph_2P^- or Ph_2As^- do. Thallium displacement in $[(C_6F_5)_3Ge]_3HgTl.1.5DME$ by Li or $(C_6H_6)_2Cr$ gives $\{[(C_6F_5)_3Ge]_4Hg\}[Cr(C_6H_6)_2]_2$ with the latter, and many metal halides form metal-mercury derivatives.[156]

$Bu^n{}_3SnLi$ adds to formaldehyde giving $Bu^n{}_3SnCH_2O^-$, a useful precursor for halomethyltin derivatives, while α,β-unsaturated sulphones give the Michael type adducts, β-stannyl sulphones then desuphonylated olefins.[157] Alkylation of Me_3SnNa and Ph_3MLi (M=Si,Ge) by 6-bromo-1-heptene gives (2-methylcyclopentyl)methyl derivatives, but not via 1-methyl-5-hexenyl radicals. Excellent yields of 2-(trimethylstannyl)-1-alkenes result from ω-substituted 1-alkynes with 2 equivalents of $Me_3SnCu.SMe_2$ in THF, while the electrochemical reduction of R_3SnCl involves $(R_3Sn)_2Hg$ as intermediates which disproportionate to R_6Sn_2 and Hg.[158]

Hydrogen atoms abstract H from silanes to give silyl radicals, and several, prepared from $Bu^tO.$, have been spin-trapped using $Bu^tNO..$[159] The ratio of 2- and 3-trichlorosilylpentane formed from $Cl_3Si.$ and <u>cis</u>- and <u>trans</u>-2-pentene can be interpreted by the decomposition of the adduct radical $.CHEtCH(SiCl_3)Me$, while $Cl_3Si.$ addition to ethene has been reinvestigated. Decomposition of 3,6-bis(trimethylsilyl)cyclohexadienyl radical, generated from the 1,4-cyclohexadiene, gives the disilylbenzene at 0^oC but at 130^oC, $Me_3Si.$ is eliminated.[160]

The increase in hyperfine splitting constants with size of substituents in silylalkenes can be attributed to twisting about the C-C bond in the radical cation, which then produces enhanced σ-π conjugation with the silyl substituent. The advantages of ENDOR and TRIPLE resonance techniques over ESR measurements are demonstrated with the spectra of radical cations of bicyclic silacycloalkenes fused at the double bond.[161] Silylated cyclopropanes react with $Bu^tO.$ with deprotonation of CH_2 activated by cyclopropyl or Me_3Si groups. The cyclopropylmethyl radicals so formed ring-open to give homoallylic radicals. The spectra of $RN(O.)CH_2X$ (R and/or $X=Me_3SiCH_2$ and/or $SiMe_3$) show the β-silyl group exerts an influence favoring a conformation in which Me_3Si lies in the same plane as the N-O bond.[162]

The products of the reaction of Et_3GeM (M=Na,K) with $PhCH_2Cl$ result through singlet radical intermediates, while at 80^oC in C_6H_6 allyldibutylgermyl radicals give $Bu^n{}_2Ge(allyl)_2$ and $Bu^n{}_2GeH_2$. The rate of H abstraction from $Bu^n{}_3GeH$ by the 5-hexenyl radical indicates reactivity 5% that of $Bu^n{}_3SnH$. The $Me_3Sn(CH_2)_2CHMe$ radical randomises excess internal vibrational energy, contrary to previous reports suggesting heavy metal blocking.[163]

The esr spectra of $Cl_2RSnOC_6H_2Bu^t{}_2O.$ show 5-coordinate tin with inequivalent Cl atoms at low temperature, and an extensive range of spin adducts of $R_3M.$ (M=Si,Ge,Sn) with $PhCOMPh_3$, of $Bu^tN{=}CHCH{=}NBu^t$ with $Ph_3M.$, and <u>N</u>-silylated 4,4'-bipyridine investigated.[164] $CF_3.$ exchanges with Me_4M (M=Sn,Pb) to give Me_3MCF_3 and $Me_2Pb(CF_3)_2$, the spectral shifts of $R_3M.$ (M=Si,Ge,Sn) reported, Me_4M^+ (M=Sn,Pb) shown to have $\underline{C}_{3\underline{v}}$ symmetry, and $Me_6M_2{}^+$ $\underline{D}_{3\underline{d}}$ symmetry (M=Si,Sn).[165]

6 Nitrogen Derivatives

Enthalpies of reaction of silylamines with phenol, iodine, chloroform and Me_3Al indicate thet they may not be weaker bases than the corresponding aliphatic amines. Similar work coupled with dipole moments indicates donor ability to increase with Si-N dipole, except for triaminosilanes.[166] Though Me_3SiCN exists primarily as this isomer, it coordinates to FeI_2 as the isocyanide, giving cis-$FeI_2(CNSiMe_3)_4$. Photoelectron spectral and theoretical studies of trisilyl and related amines indicate structures determined by dipole-dipole repulsions rather than by d-orbital participation, though this is important in ionisation potential determinations.[167]

The radical anion obtained by reducing the glyoxal diimines $Bu^tN{=}CHCH{=}NBu^t$ reacts with chlorosilanes to give Z-monosilylated and E-disilylated enediamines, and diazasilacyclopentenes. Silylketenimines react with aldehydes to provide a stereocontrolled synthesis of 2-methyl-2-alkene nitriles, while the aldol condensation of enamines, which result from the silylation of α-aminonitriles, or the $Fe(CO)_5$ photocatalysed isomerisation of straight chain olefinic amines, is erythro selective.[168]

Silylated carbohydroxamic acids have been prepared from acetohydroxamic acid and $(Me_3Si)_2NH$ or $(Me_3Si)_2NOSiMe_3$ and RCOCl.[169] $Ru_3(CO)_{12}$ catalyses the carboxylation of aminosilanes and disilazanes, and silylated dithiooxamides demethylate Me_3M (M=Al,Ga,In) at the amide protons. N-Halogenodisilazanes induce free radical chain halogenation of hydrocarbons.[170]

Primary amines can be monomethylated on silylation, lithiation and methylation, while a structure determination of $(Me_3Si)_2Li.OEt_2$ indicates a lithium-bridged dimer. Lithium and sodium derivatives of $(Me_3SiNH)_2SiMe_2$ are dimeric in benzene solution, but trimeric in the solid state.[171] The first cycloaluminodisiladiazane and triazane result from $AlCl_3$ and $(LiMeNSiMe_2)_2$ or $(LiMeNSiMe_2)_2NMe$ and both are dimeric as is $[(Me_3SiNMe)_2SiFBu^tAlCl_2]_2$. Reacting $RR'SiFNBu^tLi$ with $AlCl_3$ in petrol eliminates LiF to give the silicenium ylids $(RR'SiNBu^t)AlCl_3$, while the bisylid $(Bu^t{}_2SiNBu^t)_2AlClF_2$ forms in THF, and possesses bent fluoride bridges with angles of 93°.[172]

$[(Me_3Si)_2N]_2M$ (M=Sn,Pb) are V-shaped monomers in solid and vapour, and give $(PhS)_2M$ with PhSH.[173] $\overline{Ti(NBu^tSiMe_2N}Bu^t)Cl_2$ possesses a planar four-membered ring, and heating $R_2M[N(SiMe_3)_2]_2$ (R=Zr,Hf) to 60^oC under vacuum pyrolysis produces the heterocycle $[(Me_3Si)_2N\overline{MCHSiMe_2N}SiMe_3]_2$ on RH elimination, while the titanium amide is stable to 190^oC.[174] $Me_2PCH_2CH_2PMe_2$ and $R_2M[N(SiMe_3)_2]$ give 6-coordinate bis(metallocycles) which insert CO under pressure. The amides $[(R_2PCH_2SiMe_2)_2N]_2MCl_2$ (M=Zr,Hf) have trans-chloride and cis-phosphine ligands, with one phosphine uncoordinated. They are chiral in the solid and solution by virtue of the "gear" effect of the disilylamide ligands. $(Ph_2PCH_2SiMe_2)_2N^-$ derivatives of Rh and Ir have also been made, and Cr amides electrolytically reduced, while a range of $Bu^t(Me_3Si)N^-$ complexes of d^o V, Cr, Mo, and Re have been prepared.[175]

Silyl azides are intermediates in the synthesis of aroyl azides (ZnI_2 catalyst), of α alkoxy azides, $Ph_2C(N_3)_2$ and 1,5-diphenyltetrazole ($SnCl_4$ catalyst), and of B-N and S-N rings.[176] The $(Me_3Si)N$. radical, generated by pyrolysing silylated hydrazines, hydroxylamines, triazenes and tetrazenes, is more reactive than dialkylaminyl radicals, but similar to Bu^tO..[177]

A series of spirocyclic metal silylamides have been prepared from $(Me_3SiNM)_2SiMe_2$ (M=Li,Na) and $M'Cl_4$ (M'=Si,Ge,Sn,Ti,Zr,Hf), and all are thermally stable and very volatile. Bis(fluorosilyl)amines condense with dilithiated hydrazines to give four and five-membered silylhydrazine rings, the former isomerising with BuLi if the exocyclic nitrogen atom is protonated. The highly volatile $(FMe_2Si)_3N$ results as an unexpected byproduct of the condensation of $(Me_3SiNLi)_2$ with Me_2SiF_2.[178] The molecular structures of tetramethyl-N,N'-bisarylcyclodisilazanes supports those predicted by ^{13}C nmr spectra, and with the o-chlorophenyl compound, the Si-N bond distance (1.756Å) is the longest reported for such compounds, and its coplanarity is stabilised by a Cl---Si interaction. The mass spectra of $(PhNSiRR')_2$ indicate molecular ions as base peaks, supporting high stability of the cyclodisilazane ring.[179]

An extensive series of 5 and 6-membered heterocycles have been prepared, notably $R_2\overline{MNHCH_2CHRS}$ and cysteamine derivatives, cyclic silaamides,[180] diazasilacyclopentenes, which can be readily oxidised and substituted,[181] BF bridged cyclotrisilazanes, which possess a ring conformation between boat and twist, while $(Bu^tSiFNH)_3$ results from Bu^tSiF_3 and $LiNH_2$, along with condensed linear oligomers. The crystal structure of 1-trichlorosilyl-1,2,3,4-tetrahydro-1,10-phenanthroline shows Si 5-coordinate with the coordinated Si-N bond the longer by 14%.[182] Reductive silylation of aromatic N-heterocycles give compounds with unusual electronic structures stabilised by the silyl substituents, notably the 8π-electron conjugation of 1,4-disilyl-1,4-dihydropyrazines. Tri-isopropylsilylpyrrole is substituted electrophilically at position 3 due to hindrance, while hindered silylaminoboron fluorides eliminate Me_3SiF to give -B≡N-derivatives.[183]

Chelating ligands $(2\text{-}Me_2E'C_6H_4)Si(EMe_2)Me_2$ (E,E'=N,P,As) result from $PhNMe_2$ or o-ClC_6H_4Br, while $(4\text{-}XC_6H_4)SiMe(EMe_2)(CH_2CH_2E'Me_2)$ are formed from $MevinylSiCl_2$, and complex with $M(CO)_4$ residues (M=Cr,Mo,W). $Me_3SiNHPSCl_2$ results from $O(PSCl_2)_2$ and $(Me_3Si)_2NH$, and a series of disilazane derivatives of phosphorus, both mono- and disubstituted, prepared, with $(Me_3Si)_2NPRCl$ both substituted and reduced by Pr^iMgCl.[184] The diphosphene $(Me_3Si)_2NP{=}PN(SiMe_3)_2$ is formed from $(Me_3Si)_2NLi$ (with PCl_3, $LiAlH_4$ and Et_3N), $(Me_3Si)_2NPH\text{-}PClN(SiMe_3)_2$ (with $LiNHBu^tR$), or $(Me_3Si)_2NPCl_2$ (with $Cr(CO)_5^{2-}$ and $Fe(CO)_4^{2-}$), and readily dimerises to the cyclotetraphosphine. $(Me_3Si)_2NP(Cl)Bu^t$ eliminates Me_3SiCl to give $Me_3SiN{=}PBu^t$, which dimerises through phosphene addition across the P=N double bond, then gives cyclotriphosphine and phosphadiazene $Bu^tP(=NSiMe_3)_2$. The silylphosphine $Me_3SiP[N(SiMe_3)_2]_2$ results from $Me_3SiN{=}PN(SiMe_3)_2$ and $(Me_3Si)_2Hg$.[185]

A series of $Re(CO)_3$ complexes involving silylated P-N ligands result from $[Re(CO)_3(THF)Br]_2$ and $(Me_3Si)_2NP(=NSiMe_3)_2$ or $Bu^t(Me_3Si)NP{=}NBu^t$, while $Bu^t(Me_3Si)NP(S){=}NBu^t$ isomerises to the diazaphosphasiletidinesulphide $Me(S)PNBu^tSiMe_2NBu^t$ through Me migration, then complexes to $Re(CO)_3Br$ residues. $(Me_3Si)_2NNa$ converts allyl$M(CO)_4$ (M=Mn,Re) to the cyanide which with Me_3SnCl gives the neutral isonitrile complex allyl$M(CO)_3CNSnMe_3$.[186] Cyclic diars-V-azanes $[(CF_3)_2AsClN(SiMe_3)_2]_2$ result from $(CF_3)_2AsN(SiMe_3)_2$ and Cl_2 in CH_2Cl_2, and eliminate Me_3SiCl on heating to give $[(CF_3)_2AsN]_n$ (n=3,4).[187]

Cyclogermazanes, prepared from R_2GeCl_2 and a primary amine, react with bifunctional protic reagents to give germa heterocycles, insert heterocumulenes, and the 1,3-addition to nitrones provides evidence for $R_2Ge{=}NR'$ formation. A new antiplastic methyl germyl porphyrin has been shown to be highly lipophilic.[188]

Allyltin compounds add to R_fCN to give tin substituted ketimines which rearrange to the fluoroalkyl enamine. Me_3SnNR_2 ring cleaves lactones to give ω-stannoxyamides, while aromatic α-ketols and $Bu^n{}_3SnNEt_2$ at 60°C affords 1,2-bis(stannoxy)-1,2-diarylethenes which give vinylene imino- and thiocarbonates with RNCS and CS_2.[189] Imides can be readily stannylated with $(R_3Sn)_2O$ while N-trimethylstannyl succinimide is a helical polymer associated through CO---Sn interactions. ^{119}Sn nmr spectra of stannyl hydrazines show coupling constants dependent on conformation, which changes on warming.[190]

The cyclic diazastannylene $Bu^tNSiMe_2NBu^tSn$: gives a Bu^tNH_2 adduct stabilised through N---Sn coordination and hydrogen bonding. A fluxional 1:1 adduct forms with C_5H_6, but two moles of C_5H_6 gives $(C_5H_5)_2Sn$ and $Me_2Si(NHBu^t)_2$, while $PhPCl_2$ is reduced to $(PhP)_5$ and 2,4,6-tri-t-butylphenyldichlorophosphine to the diphosphene.[191]

7 Phosphorus, Arsenic, Antimony, and Bismuth Derivatives

1-Trimethylsilyl-1-phosphacyclohexane gives a spirophosphonium salt with $Cl(CH_2)_4Cl$, while $Br(CH_2)_4Br$ and 1-trimethylsilylmethyl-1-phosphacyclohexane give the phosphonium salt which forms a spiro ylid with $NaNH_2$.[192] $[(Me_2N)_2CCl]^+Cl^-$ is coupled by $(Me_3Si)_3P$ to give tetrakis(dimethylamino)phosphaallyl chloride $[(Me_2N)_2CPC(NMe_2)_2]^+Cl^-$, and $Ph(Me_3Si)C{=}PCl$ condenses with methoxymethylenecycloheptatriene to provide the first synthesis of a 2-phosphaazulene.[193] The first organodithioxophosphorane results from 2,4,6-$Bu^tC_6H_2P(SiMe_3)_2$ and S_2Cl_2, while $Bu^tP(SiMe_3)_2$ and Te give $(Me_3Si)_2Te$ and $(Bu^tP)_3Te$.[194] $Me_3Si(Bu^t)P(PBu^t)_2P(Bu^t)SiMe_3$, prepared from $K(PBu^t)_4K$ is remarkably stable to disproportionation and exists in solution below -30^oC as the erythro/d,l/erythro diastereomer of the 6 possible. It hydrolyses to a 3 diastereomer mixture.[195]

(Iminomethylidene)phosphines RP=C=NPh result from $R(Me_3Si)PC(=X)YSiMe_3$ (X=NPh,S; Y=O,NPh) through loss of $(Me_3Si)_2Z$ (Z=O,S), while $(Me_2N)_2C{=}PSiMe_3$ adds $Me_3SiCHCO$ to give phosphadienes.[196] P-Cloro-alkylidenephosphoranes and (α-chloroalkyl)phosphanes are interchangeable isomers by 1,2(C-P) chlorine shift, C-silylated compounds being determined by the P-substituent. A variety of aroyl and long chain alcoyl diphenylphosphines result from RCOCl and Me_3SiPPh_2.[197]

2,4,6-$Bu^t{}_3C_6H_2P(SiMe_3)_2$ and $COCl_2$ give the first phosphaketene stable at room temperature - Bu^tPCO, prepared from $Bu^tP(SiMe_3)_2$ and $COCl_2$, is only stable to -60^oC.[198] With excess of the silylphosphine, the cyclotetraphosphetanone $OC(PBu^t)_3PBu^t$ results, while trans-1,2-$(ClCO)_2C_6H_{10}$ and $PhP(SiMe_3)_2$ give the substituted 1,2-diphospha-3,9-cyclodecadiene through ring opening.[199] The 2,3-diphosphabuta-1,3-diene $Bu^t(Me_3SiO)C{=}P{-}P{=}C(OSiMe_3)Bu^t$ is formed from $(Me_3Si)_2PP(SiMe_3)_2$ and Bu^tCOCl, and $[Bu^t(Me_3Si)P]_2CO$ with $RPCl_2$ gives the cyclotriphosphetanone, the first cyclic phosphaurea with a 4-membered ring, $OCPBu^tPRPBu^t$, which loses CO on photolysis to give the cyclotriphosphine ($R=Bu^t$).[200]

$[(Me_3Si)_2P]_2PLi$ and Ph_2PCl give $(Me_3Si)_2PPPh_2$ and Ph_4P_2 as the main products, together with $(Me_3Si)_3P$, Me_3SiPPh_2 and a little $[(Me_3Si)_2P]_2PPPh_2$, but with Bu^tPCl_2 or PCl_3, $[(Me_3Si)_2P]_2PLi$ gives cyclotriphosphines. Bu^tLi opens P_4, silylation then giving $P_4(SiMe_3)Bu^t{}_3$, which, like $P_4(SiMe_3)_2Bu^t{}_2$, can be desilylated by Bu^nLi, the derivatives decomposing to $P_{\underline{n}}$ residues ($\underline{n}$=4 or less). With $P_4(SiMe_3)_4$, however, P_7 residues quickly form, and with MeLi in DME, P_4 gives $Li_{\underline{n}}Me_{3-\underline{n}}P_7$ ($\underline{n}$=1-3), which can be silylated with Me_3SiCl ($\underline{n}$=1-3) and Ph_3SiCl ($\underline{n}$=3), or stannylated by Me_3SnCl ($\underline{n}$=3). The Me_3Si derivatives can be substituted stepwise by Me_3SnX (X=Cl,Br). The electronic structures of the nortricyclane skeletons of $(Me_3Si)_3P_7$ and $P_4(SiMe_2)_3$ are compared and used to explain differences in reactivity with metal carbonyls.[201]

The first directed synthesis of a tricyclophosphine skeleton involves bridging the dilithio salt of 1,2,5,6-tetraphosphabicyclo-[3.3.0]octane with $RPCl_2$, while Me_2SiCl_2 gives the 9-silatricyclo-[3.3.1.0^{2,6}]nonane. $ClMe_2SiSiMe_2Cl$ and Na/K phosphide give the barralene $P(SiMe_2SiMe_2)_3P$, while $Bu^t{}_2SiFPHBu^t$ gives $(Bu^t{}_2SiPBu^t)_2$ with Bu^tLi, and $Bu^t{}_2Si(PHBu^t)_2$, the cyclosilatriphosphane $Bu^t{}_2SiPButPPhPBu^t$ with $Bu^nLi/PhPCl_2$. The tridentate phosphinosilane $MeSi(PR_2)_3$ readily caps faces of $Ru_3(CO)_{12}$, $M_4(CO)_{12}$, and $Rh_6(CO)_{16}$.[202]

Raman evidence supports an infinite linear chain of Sb atoms in $(Me_3Si)_2SbSb(SiMe_3)_2$, and such a structure is observed for the dibismuthine with shortened intermolecular contacts. With $(Me_3Si)_2BiLi(DME)$, a screw chain of alternating Bi and Li atoms provides the polymer skeleton.[203]

Reacting $(KBu^tP)_2$ with R_2GeCl_2 (R=Et,Ph) in pentane at -40°C gives the first P_2Ge rings, together with P_2Ge_2, P_3Ge, P_4Ge, and P_4Ge_2 rings as byproducts. The four membered rings, again as first examples, are better prepared using Et_2O at room temperature, $(Bu^tP)_2(GePh_2)_2$ possessing a non-planar ring. The first P_2Sn ring can be made similarly, using $Bu^t{}_2SnCl_2$, and is stable at room temperature. The P_2Sn_2 and P_4Sn_2 rings also form, while Et_2SnCl_2 gives $(P_2Sn)_2$, P_3Sn, and P_3Sn_2 rings.[204]

8 Oxygen Derivatives

Calculations indicate that silanone $H_2Si=O$ is more reactive than

$H_2Si=CH_2$ or $H_2Si=SiH_2$, but that appropriate substituents could stabilise it. 2-Silapyrans result from $Me_5Si_2CH=CHCH=CHOMe$ on pyrolysis, and extrude $Me_2Si=O$ under Diels-Alder conditions, $CF_3C\equiv CCF_3$ giving o-bis(trifluoromethyl)benzene.[205] The silenes $(Me_3SiO)RC=Si(SiMe_3)_2$ ($R=Ph, Bu^t$) add to cyclopentadienones to give 1,2-silaoxetanes which readily rearrange, while that reported to be derived from 7-norbornone has been shown to be a 6-membered cyclic ketene acetal.[206] Dehydrogenation of $RR'SiHCH_2OH$ gives the 6-membered $\overline{SiCOSiCO}$ ring, while base cyclises $Me_3SiO(CH_2)_nSiMe_n(OEt)_{3-n}$ to give 7- and 8-membered oxasilacycloalkanes.[207]

The crown ethers containing n oxygen atoms (n=3-6) have been prepared with Me_2Si for C_2H_4 from $RR'Si(OEt)_2$ and $HO(CH_2CH_2O)_{n-1}H$, and new crown 9,10-anthraconaphanes, prepared from 9,10-bis(trimethylsiloxy)anthracene and $ICH_2(CH_2OCH_2)_nCH_2I$, combine cation complexing with photochemical properties. Alcoholysis equilibria of $Me_2Si(OR)_2$ are catalysed by I_2 and IBr, the latter readily promoting 3^y alcohol exchange.[208]

Carboxylic acids can be readily esterified via an intermediate silyl ester, the range of acids RCO_2H encompassing R=hydroxyalkyl and alkyl, as well many amino acids. Secondary trimethylsilyl ethers are oxidised to ketones with CrO_3, and epoxides reduced to alcohols with Zn/HCl.[209]

(Acyloxymethyl)diorganylsilanes R_2SiHCH_2OCOR rearrange thermally through 1,2-H/acyloxy exchange intramolecularly to give the acyloxy(methyl)silane, the acyloxy migration also being radical initiated. Siloxybenzamides undergo anion induced O---C silyl rearrangement to salicylamides silylated ortho to the amide group.[210] O-Trimethylsilylated cyanohydrins readily add RMgX to give acyloins on hydrolysis, but while enaminones unsubstituted at nitrogen are N-silylated, those N-monosubstituted give O-silylated derivatives. 1-Triethylsilylallyliminium salts undergo desilylation on irradiation, to give C-vinylazomethine ylids.[211] Silylated carbamates readily silylate alcohols, phenols and carboxylic acids, while silylated oxycarbamic acid readily cracks thermally to give the siloxyisocyanate $Me_3SiONCO$ and disiloxane.[212] Silyl nitronates have been used in the synthesis of prostaglandin intermediates, 3(2H)furanones, functionalised butadienes, and 2,5-substituted furans.[213]

A series of alkoxy and alkoxysiloxysilanols have been prepared, together with the Si analogue of α-terpineol, dimethyl(4-methyl-3-cyclohexenyl)silanol, which smells of lily of the valley.[214] The silanol endo-3-methyl-exo-3-hydroxy-3-silabicyclo[3.2.1]octane possesses the silacyclohexane ring in the chair form substantially flattened at the Si end. Reacting $Ru_3(CO)_{12}$ with $Ph_2PCH_2SiMe_2H$ affords the dimer $Ru_2(\mu\text{-}Ph_2PCH_2SiMe_2)_2(CO)_6$ which with CF_3CO_2H gives a mononuclear silanol complex in which the phosphinomethylsilanol ligand chelates to Ru in an apparently unique manner. The length of the Si-O bond is 1.692Å and the O---O contact distance with solvating diethyl ether of 2.64Å is typical of a short hydrogen bond.[215]

Disiloxanes $(HRMeSi)_2O$ react with low valent Pt and Ir complexes to form cyclometallodisiloxanes as isomeric mixtures. Both $(C_5H_5SiMe_2)_2O$ and $(C_5H_5SiMeO)_n$ give titanocene complexes if reduced and then reacted with $Ti(py)_2Cl_4$.[216] Electron diffraction shows $(MeH_2Si)_2O$ present as a mixture of at least 2 conformers in the gas phase, one dominating, but $(Me_2HSi)_2O$ is present only as one conformer (Si-O ~1.64Å). $(HOPr^i{}_2Si)_2O$ shows the siloxane bond to be similar, the silanol one to be 1.619Å, and SiOSi to be 164°.[217] Sugars protected by the 1,1,3,3-tetraisopropyldisiloxane 1,3-diyl group can be selectively allylated, while $(ClMe_2Si)_2O$ with acetamide and benzamides give disiladioxazines, that of acetamide giving a tautomeric equilibrium involving N-acetylcyclosiloxane with a 4-membered ring. At -78°C, alkyl-lithium reagents substitute only one chlorine atom of $(ClMe_2Si)_2O$, to provide a route for the selective preparation of 1-alkyl-3-chlorodisiloxanes and thence siloxysilanols by hydrolysis. The second chlorine atom can be alkylated at 0°C, and in $(ClMe_2SiO)_2SiMe2$ both react competitively at -78°C.[218]

Allylsilsesquioxanes have been prepared and tested as new radiation sensitive substances, and the structure of $(Me_2SiO)_8$ has been shown to closely resemble that of $(Me_2PN)_8$. The synthesis, thermal decomposition and rearrangement of cross-linked and branched methylcyclosiloxanes shows the larger rings to be the more stable.[219] Organosiloxane-silicate and silica supported poly-γ-cyano (or amino)propylsiloxane provide support for Rh and Pd catalysts. The mass spectra of $(MeRSiO)_n$ show silanone loss and ring contraction (R=H, n=4-6), C_6H_6 loss (R=Ph, n=3,4), Br loss but no fragmentation (R=Br, n=3,4), and CO loss ($R=Co(CO)_4$, n=4,5).[220]

Siloxycyclopropanes are oxidatively coupled by Ag or Cu tetrafluoroborate to give the 1,6-diketones, and 1,2-bis(trimethylsiloxy)-cyclobutenes oxidatively ring contract to the siloxycyclopropyl carboxylate as major product.[221] A series of 1,3-dioxa-6-aza-2-silacyclooctanes have been prepared from diethanolamine, and shown to be present in solution in conformational equilibrium with the boat-boat conformer possessing an Si---N interaction much weaker than that of silatranes.[222] Atranes of the type $Me_2SiCH_2MMe_2OCH_2CH_2NMeCH_2CH_2O$ (M=Si,Sn) result similarly, while silatranyl derivatives of urea and thiourea form from $X{=}C[NH(CH_2)_3Si(OEt)_3]_2$ and triethanolamine. Further references are listed as abstracts.[223]

Diastereoisomeric amides made from $MeCH{=}CH(OH)CHMeCH_2OSiBu^tMe_2$ have been used to establish the stereostructure of the molecular fossil 13,16-dimethyloctacosane-1,28-diol, and an improved synthesis found for 4-t-butyldimethylsiloxy-2(2'-acetoxyethylcyclopentene-1--one.[224] Tubercidin is trisilylated by excess Bu^tMe_2SiCl at 2'-,3'- and 5'-positions, while $\underline{N}^2$-benzoyl and $\underline{N}^2$-isobutyrylguanosin gives mono-, di, and trisilyl derivatives. The synthesis of anthacycline antitumour compounds and tetracycline antibiotics utilises protective silylation and the advantages of $Bu^t{}_2Si$ as a protecting group for 1,2- and 1,3-diols are assessed. Base induces migration of Bu^tMe_2Si between trans-diaxial OH groups, and it assists in the monitoring of hydroxy fatty acids.[225] 1-Aminoalkane acids result from silyl phosphites and RN=CHPh, while $(EtO)_2POSiMe_3$ and epoxides give alkanephosphonates.[226]

The bridgehead sulphenic acid derived from triptycene gives a Me_3Si ester, and $Me_3SiOSO_2CF_3$, prepared most conveniently from CF_3SO_3H and 3-(trimethylsilyl)-2-oxazolidinone, is intermediate in the preparation of ketene acetals.[227] It also silylates metal complexes of the chelating O(2),C(4)-trimethylsiloxybutenone ligand, and $Me_5C_5(CO)_2FeC(O)CH{=}PMe_3$ at the ylidic carbon atom to give iron acyl phosphonium salts, and catalyses the reaction of $Me_3SiO_2CHRCO_2SiMe_3$ with aldehydes and ketones, giving 1,2,4-trioxan-5-ones, the crucial structural element of Qinghaosu, which possesses potent antimalarial activity.[228]

Silyl peroxides insert SO_3, the esters deflagrating at -10^oC to give aldehydes, and ketones are oxidised to esters by $Me_3SiOOSiMe_3$.[229] Terminal dicarboxylic acids result from the metal ion promoted ring opening of $\underline{O}$-silylated cyclic enones by H_2O_2, while $PhCOO_2SiMe_3$ [from $(PhCO_2)_2$ and $(Me_3Si)_2NH$] epoxidise cholestene derivatives[230] and $Me_3SiOOMMe_3$ (M=C,Si), allyloxy and homoallyloxysilanes or 1^y allylic alcohols using $VO(acac)_2$ as catalyst.[231]

Pyridinium dichromate/$Me_3SiOOSiMe_3$ is used to oxidise alcohols to aldehydes and ketones, while with $RuCl_2(PPh_3)_2$, primary alcohols are selectively oxidised in the presence of secondary alcohols. $(Ph_3SiO)_2CrO_2$ oxidises alkyl and alkenyl ferrocenes to ketone derivatives, while $Me_3SiOCrO_2Cl$ converts both alcohols and oxines to carbonyl compounds, and thiols to disulphides. $(Me_3Si)_2O$ silylates SeO_2F_2 giving first the fluoroselenate then $(Me_3SiO)_2SeO_2$.[232]

Silylenol ethers react with H_2O_2 to give α-siloxy hydroperoxides, which also result from $^1O_2$ and siloxybuta-1,3-dienes. $PhSCH_2Cl$ thiomethylates silyl dienol ethers both α and γ, and Me_3SiCl silylates AcCH=CHOMe, giving the silyl dienol ether.[233] A series of synthetic routes to silyl enol ethers are devised[234] with 1,2-bis(trimethylsiloxy)alkenes prepared from α-diketones, and β-diketones with R_2SiCl_2 used to prepare cyclic dienolates.[235] Silyl enol ethers,[236] variously substituted silyl acetals,[237] and silylated enones[238] are widely used in organic synthesis, notably to prepare α-stannyl and α-plumbyl ketones.[239]

$Bu^t{}_2GeCl_2$ hydrolyses to the diol which is associated through hydrogen bonding to give continuous double chains. It dehydrates to the planar $(Bu^t{}_2GeO)_3$. Hydrolysing Bu^tGeCl_3 gives $(Bu^tGe)_6O_9$ as the smallest example of a group IV sesquichalcogenide, containing 2 bridged 6 membered rings. $Ph_3GeOCOCF_3$ shows weak intramolecular 5 coordination.[240] Germatranes and their derivatives have been further studied.[241]

1,3,2-Dioxastannolans $Bu^n{}_2Sn(OCR_2)_2$ result from the diol and di-n-butyltin oxide, insert $Bu^n{}_2SnO$ units and dimerise. They give ethylene carbonates with $COCl_2$ and oxalyl chloride (on decarbonylation), and complex with pyridine and oxygen ligands.[242] There is temperature dependent diastereoselectivity in the aldol reaction of stannyl enolates with PhCHO, and regioselective mono-O-alkylation of disaccharide glycosides through $Bu^n{}_2Sn$ derivatives.[243] 1,3,2-Dioxastannolens are formed from acyloins or their enediol carbonates, associate in concentrated solution or as solids, and are thought to eliminate the stannylene on heating.[244]

ArMgBr (Ar=2,6-diethylphenyl) and $(acac)_2SnCl_2$ gives a rare example of a stannoxane unassociated through 5 coordination. $(Ar_2SnO)_3$ has a planar ring and with conc. HCl, gives Ar_2SnCl_2 which can be reduced to the first cyclotristannane, possessing an isosceles triangle of tin atoms (*cf* the Si_3 analogue).[245] $(MeOCO)_5C_5H$ and $(Bu^n{}_3Sn)_2O$ gives the complex $[Bu^n{}_3Sn(OH_2)_2]^+[C_5(CO_2Me)_5]^-$ with the cation apically hydrated. The siloxy bridged derivative of Sn^{II} $[(Me_3SiO)_2Sn]_2$ results from $(C_5H_5)_2Sn$, and coordinates to $Ni(CO)_4$. The unusual $(Ph_3Sn)_2(CO)_3WSnPh_2ORSnPh_2$, with a 4 membered ring, results from $W(CO)_3$(PMTA) and Ph_3SnCl through Ph-Sn cleavage and alkoxide addition.[246]

The "ladder-type" structure is exemplified by $(R_2ClSnOSnR_2OH)_2$ ($R=Pr^i$ or Me_3SiCH_2) and $(ClMe_2SnOSnMe_2Cl)_2$, while methylstannatrane is a hydrated linear trimer with O-bridges.[247]

9 Sulphur, Selenium, and Tellurium Derivatives

The first thiasilacyclopropanes result from the reaction of dimesitylsilylene and 1,1,3,3-tetramethyl-2-indanethione or adamantanethione, and both are stable to oxygen, water and heat.[248] Silylsulphenyl halides add sulphur to C=C and phosphites, and thiosilanes are cleaved by gallium halides.[249] The crystal structure of $(Me_2SiS)_2$ shows a planar ring with an Si-S bond of 2.152A, and angles at sulphur (82.5^{o}) less than those at silicon. This is the only Si-S product resulting from the low pressure pyrolysis of $(Me_2SiS)_3$ or $Me_2SiSCH_2CH_2$ though $Me_2Si=CH_2$ and $H_2C=S$ could both be isolated in matrix.[250] The mass spectra of $Me_{4-n}Si(SMe)_n$ and $Me_2SiS(CH_2)_2S$ are reported, $ClCH_2SiMe_2(CH_2)_nCH=CH_2$ ($\underline{n}$=0,1) readily cyclise with KSH, to give the 1,3-thiacyclopentane and hexane. Warming $(Et_2Si)_4$ with sulphur at 50°C inserts one sulphur atom, but with more sulphur at 190°C, the rings Si_3S_2 (1,3-) and Si_4S_2 (1,4- and 1,3-) result.2[51]

3-Germathietanes result from bis(iodomethyl)germanes and H_2S, their UV spectra supporting a transannular S---Ge interaction. Electron impact mass spectrometry generates germaethene and germathione ($Me_2Ge=S$) ions, the latter, along with $Me_2Si=S$ having been detected spectroscopically for the first time by pyrolysing the trimer. Diallylgermylenes insert oxiranes and thiirane to give unstable germacyclobutanes, decomposing to $R_2Ge=X$ (X=O,S) which insert the oxirane or thiirane to give the 5 membered diox(or dithi)olane.[252] The structures of $(c\text{-}C_6H_{11})_3GeSH$ and $[(c\text{-}C_6H_{11})_3Ge]_2S_3$ are determined, while vibrational spectroscopy shows the π-acidity of Me_3MSPh (M=Si,Ge,Sn) to the $W(CO)_5$ residue to decrease with increasing at.wt. of M.[253]

Thiosulphinates RSS(O)R' result from R'S(O)Cl and $Bu^n{}_3SnSR$, while PhCOSX (X=Br,I) can be made from $Ph_2Sn(SCOPh)_2$ and NBS or I_2.[254] The rings $Bu^t{}_8Sn_4X$ (X=S,Se,Te) result from terminal diiodide and

H_2X, and are almost planar, with long Sn-Sn bonds. The non-planarity of Sn_3Y_3 in the spiro derivatives $R_8Sn_5Y_6$ (Y=S,Se) gives conformational isomerism dependent on R (Bu^t,D_2; Pr^i,S_4).[255] $Me_2\overline{SnS(CH_2)_2NMe(CH_2)_2S}$ and $(CO)_5Cr\overline{SnS(CH_2)_2NBu^t(CH_2)_2S}$ both exhibit transannular N---Sn interactions, but no association is noted for $(Ph_3Sn)_2Te$, unlike $Me_2\overline{SnS(CH_2)_2S}$.[256] $Ph_3SnCS_2^-$, prepared from Ph_3SnLi and CS_2, complexes with transition metals as the anion or thioester.[257] $Ph_3PbSeCN$ associates through N-bridging, unlike the isothiocyanate.[258]

10 Halogen Derivatives

The conductivity of Me_3SiCl with pyridine derivatives (D) in $PhNO_2$ supports the formation of $(Me_3SiD)^+(Me_3SiCl_2)^-$ and Me_3SiClD. The pyridine complexes of Me_3SiX (X=Br,I) show Me_3Sipy^+ tetrahedral, Si-N 1.86A and no Si---X interaction.[259] Dioxolanes result from ketones and diols in the presence of Me_3SiCl, and by the electrolysis of ketones with $(Me_3SiOCH_2)_2$.[260] $(Me_3Si)_2O$ and Me_2SiCl_2 exchange catalytically, and ^{29}Si and ^{13}C nmr spectroscopy used to study silica surfaces silylated with chlorosilanes.[261]

The strength of the Si-F bond is used to generate enolates from silyl enol ethers using fluoride generated from the equilibrium of $(Et_2N)_3S^+Me_3SiF_2^-$ with $(Et_2N)_3S^+F^-$ and Me_3SiF. While aroyloxymethyl trifluorosilanes are 5 coordinated through carbonyl donation to silicon, the di- and monofluorosilanes exhibit an equilibrium involving 4- and 5-coordinate silicon.[262]

R_3SiI (R=Me,Et) result in good yield from the chloride and lithium iodide, while trimethylsilyl polyphosphate with sodium iodide readily converts alcohols to iodides. IBr and ICl catalyse the cleavage of C-O bonds normally unreactive to Me_3SiX (X=Br,Cl).[263] Treating sulphoxides with Me_3SiI in the presence of a sterically hindered 3^y amine provides a high yield dehydration route to give vinyl sulphides. Aldoximes can be dehydrated similarly, and ketoximes rearrange.[264] With 2-alkenoic acids, 1,4-addition of Me_3SiI occurs smoothly, but with aldehydes, 1,2-addition results.[265] The conjugate addition of furans to enones is mediated by Me_3SiI, which cleaves 2,5-dimethoxydihydrofurans to give γ-ketoesters.[266] 2,5-Dimethoxytetrahydrofurans and 2,6-dimethoxytetrahydropyrans give α,ω-dihalodimethoxyalkanes quantitatively with 2 moles of Me_3SiX (X=Br,I) and in very good yield for X=Cl. Quassinoid synthesis utilises Me_3SiI to promote an intramolecular Reformatsky-type reaction, and it catalyses erythro selective condensation of silyl enol ethers with acetals, while Me_3SiI-$Me_3SiOCOCF_3$ activate selectively chlorine in α-chloroethers, thereby giving homoallyl ethers with allylsilanes.[267]

Vibrational spectra of $MeMCl_3$ (M=Si,Ge), and ^{19}F nmr spectra of trifluoromethyl germanium halides and mixed polyhalides, are assigned and rationalised in terms of substituent effects.[268] $ArMCl_3$ (M=Ge,Sn) results in good yield from $ArMeSiCl_2$ and MCl_4. A number of halides Me_2RSnX and MeR_2SnX (R=Bu^tCH_2 or 1-methylcyclohexyl) result from RMgCl and Me_3SnCl or Me_2SnCl_2. $HgCl_2$ cleaves the Me-Sn bond.[269]

Organotin^{II} halides are reported as intermediates in the direct synthesis of methyltin chlorides. The ideal conditions for reacting excess BuNa with $SnCl_4$ use hydrocarbon solvents and slow addition at 0^oC.[270] UV light degrades methyltin chlorides in CCl_4 to give $SnCl_4$ via $Me_{\underline{n}}SnCl_{4-\underline{n}}$ ($\underline{n}$=2,1), and in aqueous solution to give hydrated and hydroxylated tin^{IV} cations, while 400 MHz nmr spectroscopy establishes the presence of $MeSn(OH)Cl_2.2H_2O$, $MeSn(OH)_2Cl.\underline{n}H_2O$ and $[MeSn(OH)(H_2O)_4]^{2+}$ on hydrolysis of $MeSnCl_3$.[271] The Co^{III}--Co^{II} reduction potential of model vitamin B_{12} complexes with $MeSnCl_3$ and Me_2SnCl_2 show Co-C cleavage to be influenced mainly by the electron donor ability of the equatorial ligand and the nature of the tin centre. Crown ethers (L) bridge tin centres in the complexes $(Ph_3SnCl)_2.L.2H_2O$ and $(R_2SnCl_2)_2.L.2H_2O$, while $(Ph_3PMe)^+(Ph_2SnCl_3)^-$ results from Ph_2SnCl_2, $Ph_3MeP^+I^-$ and HCl/EtOH.[272] $Bu^n{}_2SnCl_2$ incorporated into a PVC matrix is thought to break down from a 6-coordinate polymer unit to a 5-coordinate dimer. R_3SnI forms a charge transfer complex with I_2. A series of mixed haloanions $(Ph_2MXYZ)^-$, $(Ph_2MXYZ_2)^{2-}$, and $(Ph_2MX_3Y)^{2-}$ (M=Sn,Pb; X,Y,Z=Cl,Br,I,N_3,NCS) result from Ph_2MXY and Z^-.[273]

References

1 E.W.Colvin, 'Silicon in Organic Synthesis", Butterworths, London 1981; W.P.Weber, 'Silicon Reagents for Organic Synthesis', Springer-Verlag, Berlin, 1983.

2 P.Brownbridge, Synthesis, 1983, 1 and 85; S.Murai, I.Ryu, and N.Sonoda, J. Organomet. Chem., 1983, **250**, 121; Y.Ohshiro and T.Hirao, Yukagaku, 1983, **32**, 355 (Chem. Abs., 1983, **99**, 212543); P.Magnus in 'Reviews of Silicon, Germanium, Tin, and Lead Compounds', ed. M.Gielen, Freund, Tel Aviv, 1982, Vol.6, p.37; J.Dunogues, Ann. Chim. (Paris), 1983, **8**, 135.

3 Tetrahedron, 1983, **39**, No.6; M.Kumada, Yuki Gosei Kagaku Kyokai Sh, 1982, **40**, 462 (Chem. Abs., 1983, **98**, 34608).

4 C.Biran, Ann. Chim. (Paris), 1983, 8, 151; R.M.Pike and N.Sobinski, J. Organomet. Chem., 1983, **253**, 183; G.Fritz, Comments on Inorg. Chem., 1982, **1**, 329.
5 J.W.Wilt in 'React. Intermed.', ed. M.Jones,jr. and R.A.Moss, Plenum, 1983, **3**, 113; H.Bock and W.Kaim, Acc. Chem. Res., 1982, **15**, 9; E.M.Genies and F.El Omar, Electrochim Acta, 1983, **28**, 541.
6 H.Brunner, Angew. Chem. Int. Ed. Engl., 1983, **22**, 897.
7 H.Sakurai, Kagaku (Kyoto), 1982, **37**, 925 (Chem. Abs., 1983, **98**, 107332); R.Aumann, Nachr. Chem. Tech. Lab., 1982, **30**, 771 Chem. Abs., (1983, **98**, 34609).
8 R.J.P.Corriu and E.Colomer, Ann. Chim. (Paris), 1983, **8**, 121.
9 J.Satge, Bull. Soc. Chim. Belg., 1982, **91**, 1019.
10 H.Schumann and I.Schumann, Gmelin Handbook of Inorganic Chemistry, 8th Edition, Tin-Organotin Compounds, Part 9, Springer-Verlag, Berlin, 1982; Topics in Current Chemistry, ed. F.L.Boschke, Springer-Verlag, Berlin, 1982, Vol.104.
11 B.Wrackmeyer in 'Reviews of Silicon, Germanium, Tin, and Lead Compounds', 1982, Vol.6, p.75; R.C.Mehrotra, G.Srivastava, and B.S.Saraswat, ibid, p.171; Yu.A.Aleksandrov and N.V.Yablokova, ibid, p.1.
12 O.A.Reutov, J. Organomet. Chem., 1983, **250**, 145; V.S.Petrosyan, ibid, p.157.
13 J. Organomet. Chem. Library, ed. D.Seyferth and R.B.King, Elsevier, Amsterdam, 1982, Vol. 13.
14 N.Wiberg and G.Wagner, Angew. Chem. Int. Ed. Engl., 1983, **22**, 1005.
15 P.R.Jones, M.E.Lee, and L.T.Lin, Organometallics, 1983, **2**, 1039; P.R.Jones and M.E.Lee, J. Am. Chem. Soc., 1983, **105**, 6725.
16 S.Nagase and T.Kudo, J. Chem. Soc., Chem. Commun., 1983, 363; M.R.Hoffmann, Y.Yoshioka, and H.F.Schaeffer III, J. Am. Chem. Soc., 1983, **105**, 1084; K.Krogh-Jespersen, J. Comput. Chem., 1982, **3**, 571.
17 T.J.Barton, W.D.Wulff, and S.A.Burns, Organometallics, 1983, **2**, 4; T.J.Barton, G.T.Burns, and D.Gschneidner, ibid, p.8.
18 T.J.Barton and G.T.Burns, Organometallics, 1983, **2**, 1 and Tetrahedron Lett., 1983, **24**, 159.
19 A.Sekiguchi and W.Ando, Tetrahedron Lett., 1983, **24**, 2791 and Chem. Lett., 1983, 871.
20 C.F.Pau, W.J.Pietro, and W.J.Hehre, J. Am. Chem. Soc., 1983, **105**, 16; C.A.Arrington, R.West, and J.Michl, ibid, p.6176; C.George and R.D.Koob, Organometallics, 1983, **2**, 39.
21 V.J.Tortorelli, M.Jones,jr., S.Wu, and Z.Li, Organometallics, 1983, **2**, 759; T.J.Barton, S.A.Burns, P.P.Gaspar, and Y.-S.Chen, Synth. React. Inorg. Metal-Org. Chem., 1983, **13**, 881; H.Sakurai, Y.Kobayashi, R.Sato, and Y.Nakadaira, Chem. Lett., 1983, 1197; I.S.Alnaimi and W.P.Weber, J. Organomet. Chem., 1983, **241**, 171.
22 M.Kira, K.Sakamoto, and H.Sakurai, J. Am. Chem. Soc., 1983, **105**, 7469; M.Schriewer and W.P.Neumann, ibid, p.897; A.Krebs and J.Berndt, Tetrahedron Lett., 1983, **23**, 4083.
23 R.Damrauer, C.Eaborn, D.A.R.Happer, and A.I.Mansour, J. Chem. Soc., Chem. Commun., 1983, 348; C.Eaborn and D.E.Reed, ibid, p.495.
24 S.A.I.Al-Shali and C.Eaborn, J. Organomet. Chem., 1983, **246**, C34; C.Eaborn, Y.Y.El-Kaddar, and P.D.Lickiss, J. Chem. Soc., Chem. Commun., 1983, 1450.
25 Z.H.Aiube, J.Chojnowski, C.Eaborn, and W.A.Stanczyk, J. Chem. Soc., Chem. Commun., 1983, 493; C.Eaborn and A.I.Mansour, J. Organomet. Chem., 1983, **254**, 273; A.I.Al-Wassil, C.Eaborn, A.Hudson, and R.A,Jackson, ibid, 1983, **258**, 271.
26 C.Eaborn, P.B.Hitchcock, J.D.Smith, and A.C.Sullivan, J. Chem.

Soc., Chem. Commun., 1983, 827 and 1390; R.I.Papasergio, C.L.Raston, and A.H.White, ibid, p.1419.
27 C.Eaborn, M.N.El-Kheli, N.Retta, and J.D.Smith, J. Organomet. Chem., 1983, **249**, 23; C.Eaborn, N.Retta, and J.D.Smith, J. Chem. Soc., Dalton Trans., 1983, 905.
28 H.J.Breunig, W.Kanig, and A.Soltani-Neshan, Polyhedron, 1983, **2**, 291; B.Murray, J.Hvoslef, H.Hope, and P.P.Power, Inorg. Chem., 1983, **22**, 3421.
29 H.Klusik and A.Berndt, Angew. Chem. Int. Ed. Engl., 1983, **22**, 877.
30 A.H.Cowley, J.E.Kilduff, M.Pakulski, and C.A.Stewart, J. Am. Chem. Soc., 1983, **105**, 1655; A.H.Cowley, J.E.Kilduff, N.C.Norman, M.Pakulski, J.L.Atwood, and W.E.Hunter, ibid, p.4845; J.Jaud, C.Couret, and J.Escudie, J. Organomet. Chem., 1983, **249**, C25; H.Schmidt, C.Wirkner, and K.Issleib, Z. Chem., 1983, **23**, 67; A.H.Cowley, J.E.Kilduff, S.K.Mehrotra, N.C.Norman, and M.Pakulski, J. Chem. Soc., Chem. Commun., 1983, 528.
31 A.H.Cowley, J.G.Lasch, N.C.Norman, M.Pakulski, and B.R.Whittlesey, J. Chem. Soc., Chem. Commun., 1983, 881; A.H.Cowley, J.G.Lasch, N.C.Norman, and M.Palulski, J. Am. Chem. Soc., 1983, **105**, 5506; C.Couret, J.Escudie, Y.Madaule, H.Ranaivonjatovo, and J.-G.Wolf, Tetrahedron Lett., 1983, **24**, 2769; J.Escudie, C.Couret, H.Ranaivonjatovo, and J.G.Wolf, ibid, p.3625.
32 K.M.Flynn, M.M.Olmstead, and P.P.Power, J. Am. Chem. Soc., 1983, **105**, 2085; A.H.Cowley, R.A.Jones, C.A.Stewart, A.L.Stuart, J.L.Atwood, W.E.Hunter, and H.-M.Zhang, ibid, p.3737.
33 C.Eaborn, P.B.Hitchcock, and P.D.Lickiss, J. Organomet. Chem., 1983, **252**, 281; V.K.Belsky, N.N.Zemlyansky, I.V.Borisova, N.D.Kolosova, and I.P.Beletskaya, ibid, 1983, **254**, 189 and Izv.Akad. Nauk SSSR, Ser. Khim., 1983, 959 (Chem. Abs., 1983, **99**, 38572); A.L.Beauchamp, S.Latour, M.J.Olivier, and J.D.Wuest, J.Organomet. Chem., 1983, **254**, 283.
34 A.Blaschette and G.Seurig, Z. Naturforsch., Teil B, 1983, **38**, 793; M.J.Adam, J.M.Berry, L.D.Hall, B.D.Pate, and T.J.Ruth, Can. J. Chem., 1983, **61**, 658; S.P.Narula, R.K.Sharma, S.Lata, and R.Walia, Indian J. Chem., Sect.A, 1983, **22**, 246 (Chem. Abs., 1983, **99**, 70882).
35 L.Silaghi-Dumitrescu and I.Haiduc, Synth. React. Inorg. Metal-Org. Chem., 1983, **13**, 475; M.Wieber, D.Wirth, and I.Fetzer, Z. Anorg. Allg. Chem., 1983, **505**, 134; S.Fukuzumi and J.K.Kochi, Bull. Chem. Soc. Jpn., 1983, **56**, 969; G.P.Smith and R.Patrick, Int. J. Chem. Kinet., 1983, **15**, 167 (Chem. Abs., 1983, **98**, 126281).
36 T.Fjeldberg, R.Seip, M.F.Lappert, and A.J.Thorne, J. Mol. Struct., 1983, **99**, 295; K.Peters and H.-G.von Schnering, Z. Anorg. Allg. Chem., 1983, **502**, 55; V.K.Belsky, A.A.Simonenko, V.O.Reikhsfeld, and I.E.Saratov, J. Organomet. Chem., 1983, **244**, 125.
37 A.Wurstner-Rueck and H.D.Rudolph, J. Mol. Struct., 1983, **97**, 327; K.Ohno, K.Suehiro, and H.Murata, ibid, 1983, **98**, 251.
38 G.L.Larson, S.Sandoval, F.Cartledge, and F.R.Fronezek, Organometallics, 1983, **2**, 810; A.Rosowsky and J.E.Wright, J. Org. Chem., 1983, **48**, 1539, R.Tacke, H.Linoh, B.Stumpf, W.R.Abraham, K.Kieslich, and L.Ernst, Z. Naturforsch., Teil B, 1983, **38**, 616; H.Nishiyama, K.Sakuta, N.Osaka, and K.Itoh, Tetrahedron Lett., 1983, **24**, 4021; K.Yamamoto and Y.Tomo, ibid, p.1997.
39 R.D.Holmes-Smith, R.D.Osei, and S.R.Stobart, J. Chem. Soc., Perkin Trans. 1, 1983, 861; K.Schonauer and E.Zbiral,

Tetrahedron Lett.,1983, **24**, 573; S.L.Cook and J.Evans, J. Chem. Soc., Chem. Commun., 1983, 713; B.Ciommer and H.Schwarz, J. Organomet. Chem., 1983, **244**, 319.

40 E.Nakamura and I.Kuwajima, Tetrahedron Lett., 1983, **24**, 3347; I.Fleming and C.J.Urch, ibid, p.4591; J.L.Wardell and J.McM.Wigzell, J. Organomet. Chem., 1983, **244**, 225.

41 H.G.Kuivila, T.J.Karol, and K.Swani, Organometallics, 1983, 2, 909; R.A.Howie, E.S.Paterson, J.L.Wardell, and J.W.Burley, J. Organomet. Chem., 1983, **259**, 71; H.-S.Cheng, T.-L.Hwang, and C.-S.Liu, ibid, 1983, **254**, 43.

42 K.Tameo, N.Ishida, and M.Kumada, J. Org. Chem., 1983, **48**, 2120; A.Hosomi, M.Inaba, and H.Sakurai, Tetrahedron Lett., 1983, **24**, 4727.

43 D.S.Matteson and D.Majumdar, Organometallics, 1983, 2, 230; O.T.Beachley,jr. and C.Tessier-Youngs, ibid, p.796; C.Tessier-Youngs, C.Bueno, O.T.Beachley,jr., and M.R.Churchill, Inorg. Chem., 1983, **22**, 1054; R.B.Hallock, O.T.Beachley,jr., Y.-J.Li, W.M.Sanders, M.R.Churchill, W.E.Hunter, and J.L.Atwood, ibid, p.3683.

44 H.H.Karsch and A.Appelt, Z. Naturforsch., Teil B, 1983, **38**, 1399; G.Grotsch and W.Malisch, J. Organomet. Chem., 1983, **258**, 297; K.Fiederling, I.Grob, and W.Malisch, ibid, 1983, **255**, 299; H.Schumann, F.W.Reier, and M.Dettlaff, ibid, p.305.

45 S.Voran and W.Malisch, Angew. Chem. Int. Ed. Engl., 1983, **22**, 151; E.Carmona, J.M.Marin, M.L.Poveda, L.Sanchez, R.D.Rogers, and J.L.Atwood, J. Chem. Soc., Dalton Trans., 1983, 1003.

46 A.Meyer, A.Hartl, and W.Malisch, Chem. Ber., 1983, **116**, 348.

47 J.W.Bruno, T.J.Marks, and L.R.Morss, J. Am. Chem. Soc., 1983, **105**, 6824; J.W.Bruno and T.J.Marks, J. Organomet. Chem., 1983, **250**, 237; L.Dahlenburg and F.Mirzaei, ibid, 1983, **251**, 123.

48 U.Rheude and W.Sundermeyer, Chem. Ber., 1983, **116**, 1285; P.A.T.W.Porskamp, M.van der Leij, B.H.M.Lammerink, and B.Zwanenburg, Recl. Trav. Chim. Pays-Bas, 1983, **102**, 400; S.V.Ley and N.S.Simpkins, J. Chem. Soc., Chem. Commun., 1983, 1281; D.J.Ager, J. Chem. Soc., Perkin Trans. 1, 1983, 1131.

49 J.-P.Quintard, B.Elissondo, and D.Mouko-Mpegna, J. Organomet. Chem., 1983, **251**, 175; R.Goswami and D.E.Corcoran, J. Am. Chem. Soc., 1983, **105**, 7182.

50 E.J.Corey and T.M.Eckrich, Tetrahedron Lett., 1983, **24**, 3163; L.D'Ornelas and J.Cadenas, Acta Cient. Venez., 1982, **33**, 312 (Chem. Abs., 1983, **99**, 38560); M.Nakano and Y.Okamoto, Synthesis, 1983, 917.

51 G.A.Olah, M.Arvanaghi, and G.K.S.Prakesh, Synth. Commun., 1983, 636; T.Mukaiyama, T.Oriyama, and M.Murakami, Chem. Lett., 1983, 985; W.J.Greenlee and D.G.Hangauer, Tetrahedron Lett., 1983, **24**, 4559; G.O.Spessard, A.R.Ritter, D.M.Johnson, and A.M.Montgomery, ibid, p.655.

52 H.Vorbruggen and K.Krolikiewicz, Synthesis, 1983, 316; S.Kirchmeyer, A.Merteus, M.Arvanaghi, and G.A.Olah, ibid, p.498; F.G.de las Heras, A.S.Felix, and D.Fernandez-Resa, Tetrahedron, 1983, **39**, 1617; W.A.Vinson, K.S.Prickett, B.Spahic, and P.R.Ortiz de Montellano, J. Org. Chem., 1983, **48**, 4661; R.Somanathan, H.R.Aguilar, G.R.Ventura, and K.M.Smith, Synth. Commun., 1983, **13**, 273.

53 J.J.Eisch, A.A.Aradi, and K.I.Han, Tetrahedron Lett., 1983, **24**, 2073; F.Duboudin, P.Cazeau, O.Babot, and F.Moulines, ibid, p.4335.

54 L.D'Ornelas and J.Cadenas, Acta Cient. Venez., 1982, **33**, 317 (Chem. Abs., 1983, **99**, 38561).

55 F.De Sarlo, A.Brandi, A.Goti, A.Guarna, and P.Rovero, Heterocycles, 1983, **20**, 511; A.Padwa and Y.-Y.Chen, Tetrahedron

Lett., 1983, **24**, 3447; A.Padwa, G.Haffmanns, and M.Tomas, ibid, p.4303.
56 P.Jutzi and U.Gilge, J. Heterocycl. Chem., 1983, 1011; D.Britton and Y.M.Chow, Acta Crystallogr., 1983, **C39**, 1539.
57 M.Martin, Synth. Commun., 1983, **13**, 809; S.Mori, I.Sakai, T.Aoyama, and T.Shioiri, Chem. Pharm. Bull., 1982, **30**, 3380 (Chem. Abs., 1983, **98**, 34639); A.Sekiguchi, T.Sato, and W.Ando, Chem. Lett., 1983, 1083.
58 A.Sekiguchi and W.Ando, Chem. Lett., 1983, 871; K.Nishiyama and N.Tanaka, J. Chem. Soc., Chem. Commun., 1983, 1322; O.Tsuge S.Kanemasa, and K.Matsuda, Chem. Lett., 1983, 1131.
59 M.Weidenbruch, H.Flott, B.Ralle, K.Peters, and H.G.von Schnering, Z. Naturforsch., Teil B, 1983, 38, 1062.
60 T.K.Jones and S.E.Denmark, Helv. Chim. Acta, 1983, **66**, 2377 and 2397; E.Nakamura, K.Fukuzaki, and I.Kuwajima, J. Chem. Soc., Chem. Commun., 1983, 499; F.Sato and H.Katsuno, Tetrahedron Lett., 1983, **24**, 1809.
61 J.-C.Clinet, E.Dunach, and K.P.C.Vollhardt, J. Am. Chem. Soc., 1983, **105**, 6710; L.E.Overman, T.C.Malone, and G.P.Meier, ibid, p.6993; A.Padwa and J.G.Macdonald, J. Org. Chem., 1983, **48**, 3189; K.Mikami, N.Kishi, and T.Nakai, Tetrahedron Lett., 1983, **24**, 795; T.Sato, T.Tonegawa, K.Naoi, and E.Murayama, Bull. Chem. Soc. Jpn., 1983, **56**, 285.
62 K.Kikukawa, K.Ikenaga, F.Wada, and T.Matsuda, Chem. Lett., 1983, 1337; B.M.Trost and J.-i.Yoshida, Tetrahedron Lett., 1983, **24**, 4895.
63 T.Hayama, S.Tomoda, Y.Takeuchi, and Y.Nomura, Tetrahedron Lett., 1983, **24**, 2795; K.Tameo, M.Akita, and M.Kumada, J. Organomet. Chem.,1983, **254**, 13.
64 P.Aslanidis and J.Grobe, Z. Naturforsch., Teil B, 1983, **38**, 289; T.Hirao, S.Kohno, Y.Ohshiro, and T.Agawa, Bull. Chem. Soc. Jpn., 1983, **56**, 1569; H.Oda, M.Sato, Y.Morizawa, K.Oshima, and H.Nozaki, Tetrahedron Lett., 1983, **24**, 2877; R.B.Miller, M.I.Al-Hassan, and G.McGarvey. Synth. Commun., 1983, **13**, 969.
65 J.C.Giordan, J. Am. Chem. Soc., 1983, **105**, 6544; H.Wetter and P.Scherer, Helv. Chim. Acta, 1983, **66**, 118; C.-N.Hsiao and H.Scherer, Tetrahedron Lett., 1983, **24**, 2371.
66 E.A.Chernyshev, O.V.Kuz'min, A.V.Lebedev, A.I.Guzev, N.I.Kirillova, N.S.Nametkin, V.D.Tyurin, A.M.Krapivin, and N.A.Kubasova, J. Organomet. Chem.,1983, **252**, 133; E.A.Chernyshev, O.V.Kuz'min, A.V.Lebedev, A.I.Gusev, M.G.Los', N.V.Alekseev, N.S.Nametkin, V.D.Tyurin, A.M.Krapivin, N.A.Kubasova, and V.G.Zaikin, ibid, p.143; A.I.Mikaya, E.A.Trusova, V.G.Zaikin, V.D.Tyurin, O.V.Kuz'min, and A.V.Lebedev, ibid, 1983, **256**, 97.
67 M.Ochiai, T.Ukita, and E.Fujita, Chem. Lett., 1983, 1457 and Tetrahedron Lett., 1983, **24**, 4025; F.Sato, Y.Tanaka, and M.Sato, J. Chem. Soc., Chem. Commun., 1983, 165; S.R.Wilson and G.M.Georgiadis, J. Org. Chem., 1983, **48**, 4143.
68 J.E.Audia and J.A.Marshall, Synth. Commun., 1983, **13**, 531; J.Otera, T.Mandai, M.Shiba, T.Saito, K.Shimohata, K.Takemori, and Y.Kawasaki, Organometallics, 1983, **2**, 332; D.Young, W.Kitching, and G.Wickham, Tetrahedron Lett., 1983, **24**, 5789.
69 F.K.Sheffy and J.K.Stille, J. Am. Chem. Soc., 1983, **105**, 7173; J.A.Soderquist and W.W.-H.Leong, Tetrahedron Lett., 1983, **24**, 2361.
70 T.N.Mitchell and A.Amamria, J. Organomet. Chem.,1983, **256**, 37; S.D.Cox and F.Wudl, Organometallics, 1983, **2**, 184.
71 C.J.Cardin, D.J.Cardin, R.J.Norton, H.E.Parge, and K.W.Muir, J. Chem. Soc., Dalton Trans., 1983, 665; C.J.Cardin, D.J.Cardin, J.M.Kelly, R.J.Norton, A.Ray, B.J.Hathaway, and T.J.King, ibid,

p.671.
72 D.Wilhelm, T.Clark, P.von Rague Schleyer, K.Buckl, and G.Boche, Chem. Ber., 1983, **116**, 1669.
73 I.Fleming and N.K.Terrett, Tetrahedron Lett., 1983, **24**, 4153; S.R.Wilson and M.F.Price, ibid, p.569; G.Wickham and W.Kitching, Organometallics, 1983, **2**, 541.
74 T.Hayashi, K.Kabeta, I.Hamachi, and M.Kumada, Tetrahedron Lett., 1983, **24**, 2865; T.Hayashi, M.Konishi, and M.Kumada, J. Org. Chem., 1983, **48**, 281 and J. Chem. Soc., Chem. Commun., 1983, 736; J.W.Fitch, M.Brown, N.H.Hall, P.P.Mebe, K.A.Owens, and M.R.Roesch, J. Organomet. Chem.,1983, **244**, 201; S.J.Hathaway and L.A.Paquette, J. Org. Chem., 1983, **48**, 3351.
75 H.Vorbruggen and K.Krolikiewicz, Tetrahedron Lett., 1983, **24**, 889; M.Ochiai and E.Fujita, ibid, p.777; K.S.Kyler, M.A.Netzel, S.Arseniyadisand, and D.S.Watt, J. Org. Chem., 1983, **48**, 383.
76 I.Fleming and N.K.Terrett, Tetrahedron Lett., 1983, **24**, 4151; M.Wadd, T.Shigehasi, and K.-Y.Akiba, ibid, p.1711; G.Majetich, A.M.Casares, D.Chapman, and M.Behnke, ibid, p.1909; G.Majetich, R.Desmond, and A.M.Casares, ibid, p.1913; W.S.Johnson, Y.-Q.Chen, and M.S.Kellogg, J. Am. Chem. Soc., 1983, **105**, 6653.
77 H.Urabe and I.Kuwajima, Tetrahedron Lett., 1983, **24**, 4241; R.J.Armstrong and L.Weiler, Can. J. Chem., 1983, **61**, 2530.
78 T.Hayashi, K.Kabeta, T.Yamamoto, K.Tamao, and M.Kumada, Tetrahedron Lett., 1983, **24**, 5661; S.-i.Kiyooka and C.H.Heathcock, ibid, p.4765; E.H.Smith and N.D.Tyrell, J. Chem. Soc., Chem. Commun., 1983, 285.
79 R.J.P.Corriu, V.Huynh, and J.J.E.Moreau, J. Organomet. Chem., 1983, **259**, 283.
80 A.Hosomi, M.Inaba, and H.Sakurai, Chem. Lett., 1983, 1763; M.Ochiai, E.Fujita, M.Arimoto, and H.Yamaguchi, Chem. Pharm. Bull., 1982, **30**, 3994 and 1983, **31**, 86; M.Ochiai, S.Tada, M.Arimoto, and E.Fujita, ibid, 1982, **30**, 2836; Y.Yamamoto, T.Komatsu, and K.Maruyama, J. Chem. Soc., Chem. Commun., 1983, 191.
81 J.Nokami, J.Otera, T.Sudo, and R.Okawara, Organometallics, 1983, **2**, 191; Y.Naruta and K.Maruyama, J. Chem. Soc., Chem. Commun., 1983, 1264.
82 A.Gambaro, V.Peruzzo, and D.Marton, J. Organomet. Chem.,1983, **258**, 291; A.Gambaro, A.Boaretto, D.Marton, and G.Tagliavini, ibid, 1983, **254**, 293; T.V.Leshina, R.Z.Sagdeev, N.E.Polyakov, M.B.Taraban, V.I.Valyaev, V.I.Rakhlin, R.G.Mirskov, S.Kh.Khangazheev, and M.G.Voronkov, ibid, 1983, **259**, 295; P.J.Rogers, J.I.Selco, and F.S.Rowland, Chem. Phys. Lett., 1983, **97**, 313.
83 B.M.Trost and R.Remuson, Tetrahedron Lett., 1983, **24**, 1129; B.M.Trost and M.Shimizu, J. Am. Chem. Soc., 1983, **105**, 6757; L.Birkofer and V.Foremny, Z. Chem., 1983, **23**, 250; S.D.Burke, S.M.S.Strickland, and T.H.Powner, J. Org. Chem., 1983, **48**, 454.
84 M.Jones and W.Kitching, J. Organomet. Chem.,1983, **247**, C5; M.J.Hails, B.E.Mann, and C.M.Spencer, J. Chem. Soc., Dalton Trans., 1983, 729.
85 A.J.Bridges, V.Fedij, and E.C.Turowski, J. Chem. Soc., Chem. Commun., 1983, 1093; I.Kuwajima, S.Sugahara, and J.Enda, Tetrahedron Lett., 1983, **24**, 1061.
86 T.Masuda, E.Isobe, and T.Higashimura, J. Am. Chem. Soc., 1983, **105**, 7473; M.L.Mancini and J.F.Honek, Tetrahedron Lett., 1983, **24**, 4295.
87 E.-i.Negishi and J.A.Miller, J. Am. Chem. Soc., 1983, **105**, 6761; D.Liotta, M.Saindane, and W.Ott, Tetrahedron Lett., 1983, **24**, 2473.
88 T.Sakamoto, M.Shiraiwa, Y.Kondo, and H.Yamanaka, Synthesis,

1983, 312; J.Fujiwara, Y.Fukutani, H.Sano, K.Maruoka, and H.Yamamoto, J. Am. Chem. Soc., 1983, **105**, 7177.

89 R.A.Earl and K.P.C.Vollhardt, J. Am. Chem. Soc., 1983, **105**, 6991; N.T.Allison, J.R.Fritsch, K.P.C.Vollhardt, and E.C.Walborsky, ibid, p.1384; L.Birkofer and B.Wahle, Chem. Ber., 1983, **116**, 3309.

90 K.Mach, H.Antropiusova, P.Sedmera, V.Hanus, and F.Turecek, J. Chem. Soc., Chem. Commun., 1983, 805; J.A.Miller and G.Zweifel, J. Am. Chem. Soc., 1983, **105**, 1383 and Synthesis, 1983, 128; P.C.Anderset and A.S.Dreiding, Synth. Commun., 1983, **13**, 881.

91 E.-i.Negishi, L.D.Boardman, J.M.Tour, H.Sawada, and C.L.Rand, J. Am. Chem. Soc., 1983, **105**, 6344; R.L.Danheiser and H.Sard, Tetrahedron Lett., 1983, **24**, 23; H.C.Clark, V.K.Jain, and G.S.Rao, J. Organomet. Chem.,1983, **259**, 275.

92 K.J.H.Kruithof, R.F.Schmitz, and G.W.Klumpp, J. Chem. Soc., Chem. Commun., 1983, 239; P.I.Dem'yanov, I.B.Fedot'eva, E.V.Babaev, V.S.Petrosyan, and O.A.Reutov, Dokl. Akad. Nauk SSSR, 1983, **268**, 1403 (Chem. Abs., 1983, **99**, 195033).

93 G.Himbert and W.Schwickerath, Liebigs Ann. Chem., 1983, 1185; L.V.Goncharenko, N.E.Kolobova, L.G.Kuz'mina and Yu.T.Struchtov, Izv. Akad. Nauk SSSR, Ser. Khim., 1983, 1162 (Chem. Abs., 1983, **99**, 105387).

94 J.Braun and B.K.Trung, Bull. Soc. Chim. Fr., 1983, 16; B.Wrackmeyer, C.Bihlmayer, and M.Schilling, Chem. Ber., 1983, **116**, 3182; A.V.Belyakov, E.T.Bogoradovskii, V.S.Zavgorodnii, G.M.Apal'kova, V.S.Nikitin, and L.S.Khaikin, J. Mol. Struct., 1983, **98**, 27.

95 G.DeLucca and L.A.Paquette, Tetrahedron Lett., 1983, **24**, 4931; K.Vekey, J.Tamas, G.Czira, and I.E.Dolgy, Org. Mass Spectrom., 1982, **17**, 620 (Chem. Abs., 1983, **99**, 5674).

96 H.Schmidbaur and A.Schier, Synthesis, 1983, 372; G.Maier, M.Hoppe, and H.P.Reisenauer, Angew. Chem. Int. Ed. Engl., 1983, **22**, 990.

97 P.Jutzi, E.Schluter, C.Kruger, and S.Pohl, Angew. Chem. Int. Ed. Engl., 1983, **22**, 994; H.Yasuda, T.Nishi, K.Lee, and A.Nakamura, Organometallics, 1983, **2**, 21; L.A.Paquette, P.Charumilind, and J.C.Gallucci, J. Am. Chem. Soc., 1983, **105**, 7364.

98 M.F.Lappert, A.Singh, J.L.Atwood, W.E.Hunter, and H.-M.Zhang, J. Chem. Soc., Chem. Commun., 1983, 69; G.Felix, J.Dunogues, M.Petraud, and B.Barbe, J. Organomet. Chem., 1983, **258**, C49.

99 C.Biran, B.Efendene, and J.Dunogues, J. Organomet. Chem., 1983, **253**, C13.

100 L.Birkofer and B.Wahle, Chem. Ber., 1983, **116**, 2564; A.Medici, G.Fantin, M.Fogagnolo, and A.Dondoni, Tetrahedron Lett., 1983, **24**, 2901; T.J.Barton and G.P.Hussmann, J. Am. Chem. Soc., 1983, **105**, 6316.

101 N.Nakamura, M.Kohno, and M.Oki, Chem. Lett., 1983, 1809; B.C.Berris, G.H.Hovakeemian, and K.P.C.Vollhardt, J. Chem. Soc., Chem. Commun., 1983, 502; G.H.Hovakeemian and K.P.C.Vollhardt, Angew. Chem. Int. Ed. Engl., 1983, **22**, 994.

102 T.Veszpremi and J.Nagy, J. Organomet. Chem., 1983, **255**, 41; J.C.Giordan and J.H.Moore, J. Am. Chem. Soc., 1983, **105**, 6541.

103 Y.Himeshima, T.Sonoda, and H.Kobayashi, Chem. Lett., 1983, 1211; G.Depke, M.Hanack, W.Hummer, and H.Schwarz, Angew. Chem. Int. Ed. Engl., 1983, **22**, 786.

104 D.S.Wilbur, W.E.Stone, and K.W.Anderson, J. Org. Chem., 1983, **48**, 1542; R.J.Mills and V.Snieckus, ibid, p.1565.

105 P.Dembech, G.Seconi, and C.Eaborn, J. Chem. Soc., Perkin Trans. 2, 1983, 301; R.P.Kozyrod and J.T.Pinhey, Tetrahedron Lett., 1983, **24**, 1301; J.T.Pinhey and B.A.Rowe, Aust. J. Chem., 1983, **36**, 789.

106 G.R.John, L.A.P.Kane-Maguire, T.I.Odiaka, and C.Eaborn, J. Chem. Soc., Dalton Trans., 1983, 1721; L.K.K.Li Shing Man, J.G.A.Reuvers, J.Takats, and G.Deganello, Organometallics, 1983, **2** , 28.
107 T.Konakahara and K.Sato, Bull. Chem. Soc. Jpn., 1983, **56**, 1241; B.Bennetau and J.Dunogues, Tetrahedron Lett., 1983, **24**, 4217; W.R.Bergmark, M.Meador, J.Isaacs, and M.Thiim, Tetrahedron, 1983, **39**, 1109.
108 F.H.Kohler and N.Hertkorn, Z. Naturforsch., Teil B, 1983, **38**, 407; D.E.Seitz, S.-H.Lee, R.N.Hanson, and J.C.Bottaro, Synth. Commun., 1983, **13**, 121; D.W.Allen, D.J.Derbyshire, J.S.Brooks, and P.J.Smith, J. Organomet. Chem., 1983, **251**, 45.
109 P.Jutzi and U.Gilge, J. Organomet. Chem., 1983, **244**, 355 and 1983, **246**, 159 and 163; J.T.B.H.Jastrzebski, C.T.Knaap, and G.van Koten, ibid, 1983, **255**, 287.
110 P.Jutzi, H.Saleske, D.Buhl, and H.Grohe, J. Organomet. Chem., 1983, **252**, 29; F.X.Kohl and P.Jutzi, ibid, 1983, **243**, 31; F.X.Kohl, E.Schluter, and P.Jutzi, ibid, p.C37; Z.X.Kohl and P.Jutzi, Angew. Chem. Int. Ed. Engl., 1983, **22**, 56.
111 J.Almlof, L.Fernholt, K.Faegri, jr., A.Haaland, B.E.R.Schilling, R.Seip, and K.Taugboel, Acta Chem. Scand., Ser.A, 1983, **37**, 131; A.H.Cowley, J.G.Lasch, N.C.Norman, C.A.Stewart, and T.C.Wright, Organometallics, 1983, **2**, 1691; T.Mukaiyama, J.Ichikawa, M.Toba, and M.Asami, Chem. Lett., 1983, 879; P.Jutzi and E.Schluter, J. Organomet. Chem., 1983, **253**, 313.
112 M.Ishikawa, H.Sugisawa, M.Kumada, H.Kawakami, and M.Yamada, Organometallics, 1983, **2**, 974.
113 T.J.Barton, S.A.Burns, and G.T.Burns, Organometallics, 1983, **2**, 199; R.T.Conlin, Y.-W.Kwak, and H.B.Huffaker, ibid, p.343; G.T.Burns and T.J.Barton, J. Am. Chem. Soc., 1983, **105**, 2006.
114 R.T.Conlin and R.S.Gill, J. Am. Chem. Soc., 1983, **105**, 618; N.Auner and J.Grobe, Z. Anorg. Allg. Chem., 1983, **500**, 132.
115 H.Sakurai, Y.Nakadaira, and H.Tobita, Chem. Lett., 1983, 207; P.R.Jones, R.A.Pierce, and A.H.B.Cheng, Organometallics, 1983, **2**, 12.
116 G.Manuel, G.Bertrand, and F.E.Auba, Organometallics, 1983, **2**, 391; A.Laporterie, H.Iloughmane, and J.Dubac, Tetrahedron Lett., 1983, **24**, 3521 and J. Organomet. Chem., 1983, **244**, C12; H.Sakurai, Y.Nakadaira, T.Koyama, and H.Sakaba, Chem. Lett., 1983, 213.
117 Y.Nakadaira, T.Nomura, S.Kanouchi, R.Sato, C.Kabuto, and H.Sakurai, Chem. Lett., 1983, 209; H.Sakurai, A.Nakamura, and Y.Nakadaira, Organometallics, 1983, **2**, 1814.
118 M.Ishikawa, T.Tabohashi, H.Ohashi, M.Kumada, and J.Iyoda, Organometallics, 1983, **2**, 351; M.Ishikawa, T.Tabohashi, H.Sugisawa, K.Nishimura, and M.Kumada, J. Organomet. Chem., 1983, **250**, 109; M.S.Gordon, P.Boudjouk, and F.Anwari, J. Am. Chem. Soc., 1983, **105**, 4972.
119 W.Ando, H.Tanikawa, and A.Sekiguchi, Tetrahedron Lett., 1983, **24**, 4245; J.D.Rich and R.West, J. Am. Chem. Soc., 1983, **105**, 5211; A.J.Ashe III and S.T.Abu-Orabi, J. Org. Chem., 1983, **48**, 767.
120 J.A.Soderquist and A.Hassner, J. Org. Chem., 1983, **48**, 1801; V.K.Belsky, I.E.Saratov, V.O.Reikhsfeld, and A.A.Simonenko, J. Organomet. Chem., 1983, **258**, 283; W.T.Pennington, A.W.Cordes, J.C.Graham, and Y.W.Jung, Acta Crystallogr., 1983, **C39**, 712.
121 J.W.F.L.Seetz, G.Schat, O.S.Akkerman, and F.Bickelhaupt, J. Am. Chem. Soc., 1983, **105**, 3336; N.Newcomb, Y.Azuma, and A.R.Courtney, Organometallics, 1983, **2**, 175.
122 P.Boudjouk, C.A.Kapfer, and R.F.Cunico, Organometallics, 1983, **2**, 336.

123 G.Fritz, J.Neutzner, and H.Volk, Z. Anorg. Allg. Chem., 1983, **497**, 21; G.Fritz, H.Volk, K.Peters, E.-M.Peters, and H.G.von Schnering, ibid, p.119; G.Fritz and G.Brauch, ibid, p.134; K.Peters, E.-M.Peters, and H.G.von Schnering, ibid, 1983, **502**, 61.

124 Y.Ito, E.Nakajo, Y.Morizawa, K.Oshima, and H.Nozaki, Tetrahedron Lett., 1983, **24**, 2881; R.Tacke, M.Strecker, G.Lambrecht, U.Moser, and E.Mutschler, Liebigs Ann. Chem., 1983, 922; R.Tacke, A.Bentlage, R.Towart, and E.Moeller, Eur. J. Med. Chem.-Chim. Ther., 1983, **18**, 155 (Chem. Abs., 1983, **99**, 38524); M.Gerlach, P.Jutzi, J. P.Stasch, and H.Przuntek, Z. Naturforsch., Teil B, 1983, **38**, 237; R.Krebs, R.Bartsch, and K.Ruhlmann, Pharmazie, 1982, **37**, 483 (Chem. Abs., 1983, **98**, 72203); R.Tacke, H.Linoh, B.Stumpf, W.R.Abraham, K.Kieslich, and L.Ernst, Z. Naturforsch., Teil B, 1983, **38**, 616.

125 H.Vorbruggen and K.Krolikiewicz, Tetrahedron Lett., 1983, **24**, 5337; T.Hiyama, M.Obayashi, and M.Sawahata, ibid, p.4113; T.Hiyama, M.Obayashi, J.Mori, and H.Nozaki, J. Org. Chem., 1983, **48**, 912; T.Hiyama and M.Obayashi, Tetrahedron Lett., 1983, **24**, 4109.

126 G.Fritz and H.Bauer, Angew. Chem. Int. Ed. Engl., 1983, **22**, 730; H.Matsumoto, T.Arai, M.Takahashi, T.Ashizawa, T.Nakano, and Y.Nagai, Bull. Chem. Soc. Jpn., 1983, **56**, 3009; H.Matsumoto, M.Kasahara, I.Matsubara, M.Takahashi, T.Arai, M.Hasegawa, T.Nakano, and Y.Nagai, J. Organomet. Chem., 1983, **250**, 99.

127 L.Parkanyi, K.H.Pannell, and C.Hernandez, J. Organomet. Chem., 1983, **252**, 127; A.G.Brook, A.G.Harrison, and R.K.M.R.Kallury, Org. Mass Spectrom., 1982, **17**, 360 (Chem. Abs., 1983, **98**, 107365); H.Watanabe, J.Inose, T.Muraoka, M.Saito, and Y.Nagai, J. Organomet. Chem., 1983, **244**, 329.

128 H.Sakurai, Y.Nakadaira, A.Hosomi, Y.Eriyama, and C.Kabuto, J. Am. Chem. Soc., 1983, **105**, 3359; A.Laporterie, M.Joanny, H.Iloughmane, and J.Dubac, Nouv. J. Chim., 1983, 7, 225; Y.-B.Lu and C.-S.Liu, J. Organomet. Chem., 1983, **243**, 393; M.Ishikawa, H.Sugisawa, M.Kumada, T.Higuchi, K.Matsui, K.Hirotsu, and J.Iyoda, Organometallics, 1983, **2**, 174.

129 M.J.Fink, M.J.Michalczyk, K.J.Haller, R.West, and J.Michl, J. Chem. Soc., Chem. Commun., 1983, 1010; K.W.Zilm, D.M.Grant, J.Michl, M.J.Fink, and R.West, Organometallics, 1983, **2**, 193; M.J.Fink, D.J.De Young, R.West, and J.Michl, J. Am. Chem. Soc., 1983, **105**, 1070.

130 M.Weidenbruch, A.Schafer, and K.L.Thom, Z. Naturforsch., Teil B, 1983, **38**, 1695; S.Masamune, S.Murakami, and H.Tobita, Organometallics, 1983, **2**, 1464; H.Sakurai, Y.Nakadaira, and H.Sakaba, ibid, p.1484.

131 H.Watanabe, T.Okawa, M.Kato, and Y.Nagai, J. Chem. Soc., Chem. Commun., 1983, 781; S.Masamune, H.Tobita, and S.Murakami, J. Am. Chem. Soc., 1983, **105**, 6524; H.Watanabe, J.Inose, K.Fukushima, Y.Konga, and Y.Nagai, Chem. Lett., 1983, 1711; R.H.Baney, J.H.Gaul,jr., and T.K.Hilty, Organometallics, 1983, **2**, 859.

132 C.W.Carlson and R.West, Organometallics, 1983, **2**, 1792 and 1801; I.S.Alnaimi and W.P.Weber, ibid, p.903.

133 S.-M.Chen, L.D.David, K.J.Haller, C.L.Wadsworth, and R.West, Organometallics, 1983, **2**, 409; C.W.Carlson, X.-H.Zhang, and R.West, ibid, p.453; L.Parkanyi, E.Hengge, and H.Stuger, J. Organomet. Chem., 1983, **251**, 167.

134 H.G.Schuster and E.Hengge, Monatsh. Chem., 1983, **114**, 1305; A.Herman, B.Dreczewski. and W.Wojnowski, J. Organomet. Chem., 1983, **251**, 7.

135 R.J.P.Corriu, S.Ould-Kada, and G.Lanneau, J. Organomet. Chem., 1983, **248**, 23 and 39; A.Castel, P.Riviere, B.Saint-Roch,

J.Satge, and J.P.Malrieu, ibid, 1983, **247**, 149.
136 L.Ross and M.Drager, Z. Naturforsch., Teil B, 1983, **38**, 665; S.Nagase and T.Kudo, J. Mol. Struct., 1983, **103**, 35.
137 M.Drager, B.Mathiasch, L.Ross, and M.Ross, Z. Anorg. Allg. Chem., 1983, **506**, 99; J.Meunier-Piret, M.van Meerssche, M.Gielen, and K.Jurkschat, J. Organomet. Chem., 1983, **252**, 289; M.J.Chetcuti, M.H.Chisholm, H.T.Chiu, and J.C.Huffman, J. Am. Chem. Soc., 1983, **105**, 1060; T.N.Mitchell, A.Amamria, H.Killing, and D.Rutschow, J. Organomet. Chem., 1983, **241**, C45.
138 K.Ohno, M.Hayashi, and H.Murata, Spectrochim. Acta, Part A, 1983, **39**, 373; J.Nagy, G.Zsombok, E.Hengge, and W.Veigl, J. Mol. Struct., 1982, **84**, 97; R.Steudel, E.Popowski, and H.Kelling, Z. Anorg. Allg. Chem., 1983, **506**, 195.
139 H.Sollradl and E.Hengge, J. Organomet. Chem., 1983, **243**, 257; R.Radeglia and G.Engelhardt, ibid, 1983, **254**, Cl; R.Radeglia and A.Porzel, Z. Chem., 1983, **23**, 253.
140 G.A.Olah, L.D.Feld, K.Lammertsma, D.Paquin, and K.Summerman, Nouv. J. Chim., 1983, **7**, 279; N.M.K.El-Durini and R.A.Jackson, J. Chem. Soc., Perkin Trans. 2, 1983, 1275; F.K.Cartledge, Organometallics, 1983, **2**, 425.
141 G.Roy, Synth. Commun., 1983, **13**, 459; C.Chatgilialoglu, H.Woynar, K.U.Ingold, and A.G.Davies, J. Chem. Soc., Perkin Trans. 2, 1983, 555; N.S.Hosmane, S.Cradock, and E.A.V.Ebsworth, Inorg. Chim. Acta, 1983, **72**, 181.
142 T.M.Gentle and E.L.Muetterties, J. Am. Chem. Soc., 1983, **105**, 304; G.Suss-Fink, G.Hermann, and U.Thewalt, Angew. Chem. Int. Ed. Engl., 1983, **22**, 880; M.-J.Fernandez and P.M.Maitlis, Organometallics, 1983, **2**, 165.
143 R.L.Yates, J. Catal., 1982, **78**, 111 (Chem. Abs., 1983, **98**, 72220); D.Wang and T.H.Chan, Tetrahedron Lett., 1983, **24**, 1573; H.Brunner, G.Riepl, and H.Weitzer, Angew. Chem. Int. Ed. Engl., 1983, **22**, 331; T.Murai, Y.Hatayama, S.Murai, and N.Sonoda, Organometallics, 1983, 2, 1883; U.Ohmichen and H.Singer, J. Organomet. Chem., 1983, **243**, 199.
144 K.Takeshita, Y.Seki, K.Kawamoto, S.Murai, and N.Sonoda, J. Chem. Soc., Chem. Commun., 1983, 1193; N.Chatani, S.Murai, and N.Sonoda, J. Am. Chem. Soc., 1983, **105**, 1370.
145 M.G.Veliev, M.M.Gusimov, A.M.Garamanov, and I.A.Khudoyarov, Synthesis, 1983, 205; B.-H.Han and P.Boudjouk, Organometallics, 1983, **2**, 769; L.-Z.Wang and Y.-Y.Jiang, J. Organomet. Chem., 1983, **251**, 39; B.Marciniec and J.Gulinski, ibid, 1983, **253**, 349; B.J.Helmer, R.West, R.J.P.Corriu, M.Poirier, G.Royo, and A.de Saxce, ibid, 1983, **251**, 295.
146 K.Ohno, K.Suehiro, and H.Murata, J. Mol. Struct., 1983, **98**, 251; A.J.F.Clark, J.E.Drake, R.T.Hemmings, and Q.Shen, Spectrochim. Acta, Part A, 1983, **39**, 127; G.Sbrana, N.Neto, M.Muniz-Miranda, and M.Nocentini, ibid, p.295; D.C.McKean, I.Torto, M.W.MacKenzie, and A.R.Morrison, ibid, p.387; D.C.McKean, I.Torto, and M.W.Mackenzie, ibid, p.399.
147 M.Dakkouri and H.Kehrer, Chem. Ber., 1983, **116**, 2041; M.N.Bochkarev, T.A.Basalgina, G.S.Kalinina, and G.A.Razuvaev, J. Organomet. Chem., 1983, **243**, 405; D.J.Brauer and R.Eujen, Organometallics, 1983, **2**, 263; S.P.Foster and K.M.Mackay, J. Organomet. Chem., 1983, **247**, 21.
148 G.Dumartin, J.-P.Quintard, and M.Pereyre, J. Organomet. Chem., 1983, **252**, 37; T.N.Mitchell and A.Amamria, ibid, p.47.
149 E.Keinan and N.Greenspoon, J. Org. Chem., 1983, **48**, 3544; V.P.Baillargeon and J.K.Stille, J. Am. Chem. Soc., 1983, **105**, 7175; B.Giese and J.Dupius, Angew. Chem. Int. Ed. Engl., 1983, **22**, 622.
150 J.Lusztyk, E.Lusztyk, B.Maillard, L.Lunazzi, and K.U.Ingold, J.

Am. Chem. Soc., 1983, **105**, 4475; J.C.Cochran, A.J.Leusink, and J.G.Noltes, Organometallics, 1983, **2**, 1099; H.Laurent, P.Esperling, and G.Baude, Liebigs Ann. Chem., 1983, 1996; J.W.Wilt, F.G.Belmonte, and P.A.Zieske. J. Am. Chem. Soc., 1983, **105**, 5665.

151 J.G.Smith, N.R.Quinn, and M.Viswanathan, Synth. Commun., 1983, **13**, 1 and 773; B.Laycock, W.Kitching, and G.Wickham, Tetrahedron Lett., 1983, **24**, 5785; J.E.Audia and J.A.Marshall, Synth. Commun., 1983, **13**, 531.

152 E.O.Fischer, P.Rustemeyer, O.Orama, D.Neugebauer, and U.Schubert, J. Organomet. Chem., 1983, **247**, 7; M.J.Albright, T.F.Schaaf, A.K.Hovland, and J.P.Oliver, ibid, 1983, **259**, 37.

153 H.Hayami, M.Sato, S.Kanemoto, Y.Morizawa, K.Oshima, and H.Nozaki, J. Am. Chem. Soc., 1983, **105**, 4491; L.Rosch and U.Starke, Z. Naturforsch., Teil B, 1983, **38**, 1292; D.W.Goebel,jr.,J.L.Hencher, and J.P.Oliver, Organometallics, 1983, **2**, 746.

154 B.M.Trost, J.-i.Yoshida, and M.Lautens, J. Am. Chem. Soc., 1983, **105**, 4494; G.Altnau and L.Rosch, Tetrahedron Lett., 1983, **24**, 45; L.Rosch, G.Altnau, and G.Jas, Chem.-Ztg., 1983, **107**, 128; Y.Okuda, M.Sato, K.Oshima, and H.Nozaki, Tetrahedron Lett., 1983, **24**, 2015.

155 L.Rosch, G.Altnau, C.Kruger, and Y.-H.Tsay, Z. Naturforsch., Teil B, 1983, **38**, 34; L.Rosch and U.Starke, Angew. Chem. Int. Ed. Engl., 1983, **22**, 557.

156 D.A.Bravo-Zhivotovskii, S.D.Pigarev, I.D.Kalikhman, O.A.Vyazankina, and N.S.Vyazankin, J. Organomet. Chem., 1983, **248**, 51; R.J.Batchelor and T.Birchall, J. Am. Chem. Soc., 1983, **105**, 3848; G.A.Razuvaev, M.N.Bochkarev, and L.V.Pankrato, J. Organomet. Chem., 1983, **250**, 135.

157 D.E.Seitz, J.J.Carroll, C.P.Cartaya M, S.H.Lee, and A.Zapata, Synth. Commun., 1983, **13**, 129; M.Ochiai, T.Ukita, and E.Fujita, J. Chem. Soc., Chem. Commun., 1983, 619.

158 K.W.Lee and J.San Filippo jr., Organometallics, 1983, **2**, 906; C.Feasson and M.Devaud, Bull. Soc. Chim. Fr., 1983, 40; E.Piers and J.M.Chong, J. Chem. Soc., Chem. Commun., 1983, 934.

159 K.Woersdorfer, B.Reimann, and P.Potzinger, Z. Naturforsch., Teil B, 1983, **38**, 896; J.Planinic, Radiochem. Radioanal. Lett., 1983, **56**, 205 (Chem. Abs., 1983, **99**, 88260); H.Chandra, I.M.T.Davidson, and M.C.R.Symons, J. Chem. Soc., Faraday Trans. 1, 1983, **79**, 2705.

160 T.Dohmaru and Y.Nagata, Bull. Chem. Soc. Jpn., 1983, **56**, 1847 and 2387; H.Sakurai, M.Kira, and H.Sugiyama, Chem. Lett., 1983, 599 and J. Am. Chem. Soc., 1983, **105**, 6436.

161 M.Kira, H.Nakazawa, and H.Sakurai, J. Am. Chem. Soc., 1983, **105**, 6983; H.Bock, B.Hierholzer, H.Kurreck, and W.Lubitz, Angew. Chem. Int. Ed. Engl., 1983, **22**, 787.

162 A.G.Davies, J.A.-A.Hawari, M.Grignon-Dubois, and M.Pereyre, J. Organomet. Chem., 1983, **255**, 29; M.Kira, H.Osawa, and H.Sakurai, ibid, 1983, **259**, 51.

163 M.B.Taraban, T.V.Leshina, K.M.Salikhov, R.Z.Sagdeev, Yu.N.Molin, O.I.Margorskaya, and N.S.Vyazankin, J. Organomet. Chem., 1983, **256**, 31; K.Mochida and I.Miyagawa, Bull. Chem. Soc. Jpn., 1983, **56**, 1875; J.Lusztyk, B.Maillard, D.A.Lindsay, and K.U.Ingold, J. Am. Chem. Soc., 1983, **105**, 3578; S.P.Wrigley and B.S.Rabinovitch, Chem. Phys. Lett., 1983, **98**, 386.

164 A.G.Davies and J.A.-A.Hawari, J. Organomet. Chem., 1983, **251**, 53; A.Alberti, G.Seconi, G.F.Pedulli, and A.Degl'Innocenti, ibid, 1983, **253**, 291; A.Alberti and A.Hudson, ibid, 1983, **241**, 313; W.Kaim, ibid, p.157.

165 M.A.Guerra, R.L.Armstrong, W.I.Bailey,jr., and R.J.Lagow, J.

Organomet. Chem., 1983, **254**, 53; C.Chatgilialoglu, K.U.Ingold, J.Lusztyk, A.S.Nazran, and J.C.Scaiano, Organometallics, 1983, **2**, 1332; B.W.Walther, F.Williams, W.Lall, and J.K.Kochi, ibid, p.688; C.Glidewell, J. Chem. Res. (S), 1983, 22.
166 K.J.Fisher and C.E.Ezeani, Polyhedron, 1983, **2**, 393; Yu.V.Kolodyazhnyi, M.G.Gruntfest, N.I.Sizova, A.P.Sadimenko, L.I.Ol'khovskaya, and N.V.Komarov, Zh. Obshch. Khim., 1983, **53**, 678 (Chem. Abs., 1983, **99**, 70805).
167 R.A.Jones and M.H.Seeberger, J. Chem. Soc., Dalton Trans., 1983, 181; P.Livant, M.L.McKee, and S.D.Worley, Inorg. Chem., 1983, **22**, 895.
168 H.T.Dieck, B.Bruder, and K.-D.Franz, Chem. Ber., 1983, **116**, 136; H.Okada, I.Matsuda, and Y.Izumi, Chem. Lett., 1983, 97; W.Ando and H.Tsumaki, ibid, p.1409; H.Ahlbrecht and E.O.Duber, Synthesis, 1983, 56; R.J.P.Corriu, V.Huynh, J.J.E.Moreau, and M.Pataud-Sat, J. Organomet. Chem., 1983, **255**, 359.
169 W.Heuchel, M.Boldhaus, and C.Bliefert, Chem.-Ztg., 1983, **107**, 69; W.Ando and H.Tsumaki, Synth. Commun., 1983, **13**, 1053.
170 M.T.Zoeckler and R.M.Laine, J. Org. Chem., 1983, **48**, 2539; J.R.Bowser, P.J.Williams, and K.Kurz, ibid, p.4111; T.Halder, W.Schwarz, J.Weidlein, and P.Fischer, J. Organomet. Chem., 1983, **246**, 29; M.D.Cook, B.P.Roberts, and K.Singh, J. Chem. Soc., Perkin Trans. 2, 1983, 635.
171 M.J.Calverley, Synth. Commun., 1983, **13**, 601; M.F.Lappert, M.J.Slade, and A.Singh, J. Am. Chem. Soc., 1983, **105**, 302; L.M.Engelhardt, A.S.May, C.L.Raston, and A.H.White, J. Chem. Soc., Dalton Trans., 1983, 1671; D.J.Brauer, H.Burger, W.Geschwandtner, G.R.Liewald, and C.Kruger, J. Organomet. Chem., 1983, **248**, 1.
172 T.Blumenthal, U.Wannagat, D.J.Brauer, and H.Burger, Monatsh. Chem., 1983, **114**, 1271 and J. Organomet. Chem., 1983, **249**, 33; W.Clegg, M.Haase, U.Klingebiel, J.Neemann, and G.M.Sheldrick, ibid, 1983, **251**, 281; W.Clegg, U.Klingebiel, J.Neemann, and G.M.Sheldrick, ibid, 1983, **249**, 47.
173 T.Fjeldberg, H.Hope, M.F.Lappert, P.P.Power, and A.J.Thorne, J. Chem. Soc., Chem. Commun., 1983, 639; P.B.Hitchcock, M.F.Lappert, B.J.Samways, and E.L.Weinberg, ibid, p.1492.
174 R.A.Jones, M.H.Seeberger, J.L.Atwood, and W.E.Hunter, J. Organomet. Chem., 1983, **247**, 1; R.P.Planalp, R.A.Andersen, and A.Zalkin, Organometallics, 1983, **2**, 16 and 1675.
175 M.D.Fryzuk, H.D.Williams, and S.J.Rettig, Inorg. Chem., 1983, **22**, 863; M.D.Fryzuk and P.A.MacNeil, Organometallics, 1983, 2, 355; D.C.Bradley and M.Ahmed, Polyhedron, 1983, **2**, 87; W.A.Nugent, Inorg. Chem., 1983, **22**, 965.
176 G.K.S.Prakash, P.S.Iyer, M.Arvanaghi. and G.A.Olah, J. Org. Chem., 1983, **48**, 3358; S.Kirchmeyer, A.Mertens, and G.A.Olah, Synthesis, 1983, 500; P.Paetzold and R.Truppat, Chem. Ber., 1983, **116**, 1531; D.Bielefeldt and A.Haas, ibid, p.1257.
177 J.C.Brand, B.P.Roberts, and J.N.Winter, J. Chem. Soc., Perkin Trans. 2, 1983, 261.
178 H.Burger, W.Geschwandtner, and G.R.Liewald, J. Organomet. Chem., 1983, **259**, 145; W.Clegg, M.Haase, H.Hluchy, M.Klingebiel, and G.M.Sheldrick, Chem. Ber., 1983, **116**, 290.
179 A.Szollosy, L.Parkanyi, L.Bihatsi, and P.Hencsei, J. Organomet. Chem., 1983, **251**, 159; L.Parkanyi, A.Szollosy, L.Bihatsi, P.Hencsei, and P.Nagy, ibid, 1983, **256**, 235; T.Muller and L.Bihatsi, Acta Chim. Hung., 1983, **112**, 99 (Chem. Abs., 1983, **99**, 70826).
180 J.Satge, A.Cazes, M.Bouchaut, M.Fatome, H.Sentenae-Roumanou, and C.Lion, Eur. J. Med. Chem.-Chim. Ther., 1982, **17**, 433 (Chem. Abs., 1983, **98**, 54061); W.Maringgele, Z. Naturforsch., Teil B,

1983, **38**, 71.

181 O.Graalmann, M.Hesse, U.Klingebiel, W.Clegg, M.Haase, and G.M.Sheldrick, Angew. Chem. Int. Ed. Engl., 1983, **22**, 621; M.Hesse, U.Klingebiel, J.Heinze, and J.Mortensen, Z. Naturforsch., Teil B, 1983, **38**, 953; M.Hesse and U.Klingebiel, Z. Anorg. Allg. Chem., 1983, **501**, 57; W.Clegg, O.Graalmann, M.Haase, W.Klingebiel, G.M.Sheldrick, P.Werner, G.Henkel, and B.Krebs, Chem. Ber., 1983, **116**, 282.

182 W.Clegg, Acta Crystallogr., 1983, **C39**, 387; U.Klingebiel and N.Vater, Chem. Ber., 1983, **116**, 3277; G.Klebe, J.W.Bats, and K.Hensen, Z. Naturforsch., Teil B, 1983, **38**, 825.

183 W.Kaim, J. Am. Chem. Soc., 1983, **105**, 707; J.M.Muchowski and D.R.Solas, Tetrahedron Lett., 1983, **24**, 3455; H.Noth and S.Weber, Z. Naturforsch., Teil B, 1983, **38**, 1460.

184 P.Aslanidis and J.Grobe, Z. Naturforsch., Teil B, 1983, **38**, 280, 289 and J. Organomet. Chem., 1983, **249**, 103; K.Dostal, J.Vesela, M.Meisel, and C.Donath, Z. Chem., 1983, **23**, 185; B.-L.Li, J.S.Engenito,jr., R.H.Neilson, and P.Wisian-Neilson, Inorg. Chem., 1983, **22**, 575; H.R.O'Neal and R.H.Neilson, ibid, p.814.

185 E.Niecke, R.Ruger, M.Lysek, S.Pohl, and W.Schoeller, Angew. Chem. Int. Ed. Engl., 1983, **22**, 486; E.Niecke and R.Ruger, ibid, p.155; K.M.Flynn, B.D.Murray, M.M.Olmstead, and P.P.Power, J. Am. Chem. Soc., 1983, **105**, 7460; E.Niecke, R.Ruger, B.Krebs, and M.Dartmann, Angew. Chem. Int. Ed. Engl., 1983, **22**, 552; V.D.Romanenko, V.F.Shul'gin, V.V.Scopenko, and L.N.Markovski, J. Chem. Soc., Chem. Commun., 1983, 808.

186 O.J.Scherer, J.Kerth, and M.L.Ziegler, Angew. Chem. Int. Ed. Engl., 1983, **22**, 503; O.J.Scherer, J.Kerth, R.Anselmann, and W.S.Sheldrick, ibid, p.988; O.J.Scherer and J.Kerth, J. Organomet. Chem., 1983, **243**, C33; M.Moll, H.Behrens, H.-J.Seibold, and P.Merbach, ibid, 1983, **248**, 329.

187 H.W.Roesky, R.Bohra, and W.S.Sheldrick, J. Fluorine Chem., 1983, **22**, 199; R.Bohra, H.W.Roesky, J.Lucas, M.Noltemeyer, and G.M.Sheldrick, J. Chem. Soc., Dalton Trans., 1983, 1011.

188 G.Lacrampe, H.Lavayssiere, M.Riviere-Baudet, and J.Satge, Recl. Trav. Chim. Pays-Bas, 1983, **102**, 21; M.Riviere-Baudet, A.Castel, G.Lacrampe, and J.Satge, ibid, p.65; T.K.Miyamoto, N.Sugita, Y.Matsumoto, Y.Sasaki, and M.Kouno, Chem. Lett., 1983, 1695.

189 W.A.Nugent and B.E.Smart, J. Organomet. Chem., 1983, **256**, C9; A.Ricci, M.N.Romanelli, M.Taddai, G.Seconi, and A.Shanzer, Synthesis, 1983, 319; S.Sakai, M.Murata, N.Wada, and T.Fujinami, Bull. Chem. Soc. Jpn., 1983, **56**, 1873.

190 G.A.Razuvaev, V.I.Shcherbakov, and N.E.Stolyarova, Synth. React. Inorg. Metal-Org. Chem., 1983, **13**, 59; F.E.Hahn, T.S.Dory, C.L.Barnes, M.B.Hossain, D.van der Helm, and J.J.Zuckerman, Organometallics, 1983, **2**, 969; T.Gasparis-Ebeling, H.Noth, and B.Wrackmeyer, J. Chem. Soc., Dalton Trans., 1983, 97.

191 M.Veith, G.Schlemmer, and M.L.Sommer, Z. Anorg. Allg. Chem., 1983, **497**, 157; M Veith and F.Tollner, J. Organomet. Chem., 1983, **246**, 219; M.Veith, V.Huch, J.-P.Majoral, G.Bertrand, and G.Manuel, Tetrahedron Lett., 1983, **24**, 4219.

192 H.Schmidbaur and A.Mortl, Z. Chem., 1983, **23**, 249.

193 A.Schmidpeter, S.Lochschmidt, and A.Wallhalm, Angew. Chem. Int. Ed. Engl., 1983, **22**, 545; G.Markl, E.Seidl, and I.Trotsch, ibid, p.879.

194 R.Appel, F.Knoch, and H.Kunze, Angew. Chem. Int. Ed. Engl., 1983, **22**, 1004; W.W.du Mont, T.Severengiz, and B.Meyer, ibid, p.983.

195 M.Baudler, G.Reuschenbach, and J.Hahn, Chem. Ber., 1983, **116**, 847; M.Baudler, G.Reuschenbach, J.Hellmann, and J.Hahn, Z. Anorg. Allg. Chem., 1983, **499**, 89.

196 C.Wentrup, H.Briehl, G.Becker, G.Uhl, H.-J.Wessely, A.Maquestiau, and R.Flammang, J. Am. Chem. Soc., 1983, **105**, 7194; L.N.Markovskii, V.D.Romanenko, and T.V.Pidvarko, Zh. Obshch. Khim., 1983, **53**, 1672 (Chem. Abs., 1983, **99**, 195100).
197 R.Appel, M.Huppertz, and A.Westerhaus, Chem. Ber., 1983, **116**, 114; E.Lindner and D.Hubner, ibid, p.2574.
198 R.Appel and W.Paulen, Angew. Chem. Int. Ed. Engl., 1983, **22**, 785 and Tetrahedron Lett., 1983, **24**, 2639.
199 R.Appel and W.Paulen, Chem. Ber., 1983, **116**, 109; R.Appel, J.Hunerbein, and F.Knoch, Angew. Chem. Int. Ed. Engl., 1983, **22**, 61.
200 R.Appel, V.Barth, and F.Knoth, Chem. Ber., 1983, **116**, 938; R. Appel and W.Paulen, ibid, p.2371.
201 G.Fritz and J.Harer, Z. Anorg. Allg. Chem., 1983, **500**, 14; G.Fritz, J.Harer, and K.Stoll, ibid, 1983, **504**, 47; G.Fritz and J.Harer, ibid, p.23; G.Fritz, K.D.Hoppe, W.Honle, D.Weber, C.Mujica, V.Manriquez, and H.G.von Schnering, J. Organomet. Chem., 1983, **249**, 63; R.Gleiter, M.C.Bohm, M.Eckert-Maksic, W.Schaefer, M.Baudler, Y.Aktalay, G.Fritz, R.Hoppe, and D.Klaus, Chem. Ber., 1983, **116**, 2972.
202 M.Baudler and S.Esat, Chem. Ber., 1983, **116**, 2711; K.Hassler, J. Organomet. Chem., 1983, **246**, C31; W.Clegg, M.Haase, U.Klingebiel, and G.M.Sheldrick, Chem. Ber., 1983, **116**, 146; D.F.Foster, B.S.Nicholls, and A.K.Smith, J. Organomet. Chem., 1983, **244**, 159.
203 H.Burger, R.Eujen, G.Becker, O.Mundt, M.Westerhausen, and C.Witthauer, J. Mol. Struct., 1983, **98**, 265; O.Mundt, G.Becker, M.Rossler, and C.Witthauer, Z. Anorg. Allg. Chem., 1983, **506**, 42.
204 M.Baudler and H.Suchomel, Z. Anorg. Allg. Chem., 1983, 503, 7 and 1983, **506**, 22; K.-F.Tebbe and R.Frohlich, ibid, p.27; M.Baudler and H.Suchomel, ibid, 1983, **505**, 39.
205 T.Kudo and S.Nagase, J. Organomet. Chem., 1983, **253**, C23; G.Hussmann, W.D.Wulff, and T.J.Barton, J. Am. Chem. Soc., 1983, **105**, 1263.
206 G.Markl and M.Horn, Tetrahedron Lett., 1983, **24**, 1477; T.J.Barton and G.P.Hussmann, Organometallics, 1983, **2**, 692.
207 R.Tacke, H.Lange, A.Bentlage, W.S.Sheldrick, and L.Ernst, Z. Naturforsch., Teil B, 1983, **38**, 190; V.Chvalovsky and W.S.El-Hamouly, Tetrahedron, 1983, **39**, 1195.
208 B.Arkles, K.King, R.Anderson, and W.Peterson, Organometallics, 1983, **2**, 454; A.Castellan, M.Daney, J.-P.Desvergne, M.-H.Riffaud, and H.Bouas-Laurent, Tetrahedron Lett., 1983, **24**, 5215; K.Ito and T.Ibaraki, Bull. Chem. Soc. Jpn., 1983, **56**, 295.
209 M.A.Brook and T.H.Chan, Synthesis, 1983, 201; R.Baker, V.B.Rao, P.D.Ravenscroft, and C.J.Swain, ibid, p.572; Y.D.Vandar, P.S.Arya, and C.T.Rao, Synth. Commun., 1983, **13**, 869.
210 R.Tacke and H.Lange, Chem. Ber., 1983, **116**, 3685; J.W.Wilt and S.M.Keller, J. Am. Chem. Soc., 1983, **105**, 1395; R.J.Billedeau, M.P.Sibi, and V.Snieckus, Tetrahedron Lett., 1983, **24**, 4515.
211 L.R.Krepski, S.M.Heilmann, and J.K.Rasmussen, Tetrahedron Lett., 1983, **24**, 4075; T.Proll and W.Walter, Chem. Ber., 1983, **116**, 1564; S.-f.Chen, J.W.Ullrich, and P.S.Mariano, J. Am. Chem. Soc., 1983, **116**, 6160.
212 D.Knausz, A.Meszticzky, L.Szakacs, B.Csakvari, and K.Ujszaszy, J. Organomet. Chem., 1983, **256**, 11; V.D.Sheludyakov, A.B.Dmitrieva, A.D.Kirilin, and E.A.Chernyshev, Zh. Obshch. Khim., 1983, **53**, 706 (Chem. Abs., 1983, **99**, 70806).
213 N.B.Das and K.B.G.Torssell, Tetrahedron, 1983, **39**, 2227 and 2247; S.H.Andersen, K.K.Sharma, and K.B.G.Torssell, ibid, p.2241.

214 H.-J.Holdt, G.Schott, E.Popowski, and H.Kelling, Z. Chem., 1983, **23**, 252; D.Wrobel and U.Wannagat, Liebigs Ann. Chem., 1983, 211.

215 M.-u.Haque, W.Horne, S.E.Cremer, and C.S.Blankenship, J. Chem. Soc., Perkin Trans. 2, 1983, 395; M.J.Auburn, R.D.Holmes-Smith, S.R.Stobart, M.J.Zaworotko, T.S.Cameron, and A.Kumari, J. Chem. Soc., Chem. Commun., 1983, 1523.

216 L.G.Bell, W.A.Gustavson, S.Thanedar, and M.D.Curtis, Organometallics, 1983, **2**, 740; M.D.Curtis, J.J.D'Errics, D.N.Duffy, P.S.Epstein, and L.G.Bell, ibid, p.1806.

217 D.W.H.Rankin and H.E.Robertson, J. Chem. Soc., Dalton Trans., 1983, 265; W.Clegg, Acta Crystallogr., 1983, **C39**, 901.

218 J.J.Oltvoort, M.Kloostermans, and J.H.van Boom, Recl. Trav. Chim. Pays-Bas, 1983, **102**, 501; B.Dejak and Z.Lasocki, J. Organomet. Chem., 1983, **246**, 151; S.A.Kazoura and W.P.Weber, ibid, 1983, **243**, 149.

219 T.N.Martynova, V.P.Korchkov, and P.P.Semyannikov, J. Organomet. Chem., 1983, **258**, 277; N.L.Paddock, S.J.Rettig, and J.Trotter, Can. J. Chem., 1983, **61**, 541; N.N.Makarova, M.Blazso, and E.Jakab, Polyhedron, 1983, **2**, 257 and 455.

220 B.Marciniec and W.Urbaniak, J. Mol. Catal., 1983, **18**, 49; Y.-Z.Zhou and Y.-Y.Jiang, J. Organomet. Chem., 1983, **251**, 31; E.Pelletier and J.F.Harrod, Can. J. Chem., 1983, **61**, 762.

221 I.Ryu, M.Ando, A.Ogawa, S.Murai, and N.Sonada, J. Am. Chem. Soc., 1983, **105**, 7192; H.-G.Heine, W.Hartmann, H.-J.Knops, and U.Priesnitz, Tetrahedron Lett., 1983, **24**, 2635.

222 E.Kupce, E.Liepins, and E.Lukevics, J. Organomet. Chem., 1983, **248**, 131; P.Hegyes, S.Foldeak, P.Hencsei, G.Zsombok, and J.Nagy, ibid, 1983, **251**, 289; E.Liepins, I.Birgele, G.Zelcans, I.Urtane, and E.Lukevics, Zh. Obshch. Khim., 1983, **53**, 1076 (Chem. Abs., 1983, **99**, 105328).

223 J.Grobe and N.Voulgarakis, Z. Naturforsch., Teil B, 1983, **38**, 269; M.G.Voronkov, A.E.Pestunovich, E.I.Kositsyna, B.Z.Sterenberg, T.A.Pusechkina, and N.N.Vlasova, Z. Chem., 1983, **23**, 248; Chem. Abs., 1983, **98**, 89440, 126195, 160800, 179459, 215667, and 215699; Chem. Abs., 1983, **99**, 105325, 105333, 105335, 175934, and 212600.

224 C.H.Heathcock and B.L.Finkelstein, J. Chem. Soc., Chem. Commun., 1983, 919; G.F.Cooper, N.L.McClure, A.R.van Horn, and D.Wren, Synth. Commun., 1983, **13**, 225.

225 F.Seela, E.Hissmann, and J.Ott, Liebigs Ann. Chem., 1983, 1169; D.Flockerzi, W.Schlosser, and W.Pfleiderer, Helv. Chim. Acta, 1983, **66**, 2069; B.M.Trost, C.G.Caldwell, E.Murayama, and D.Heissler, J. Org. Chem., 1983, **48**, 3252; C.A.A.van Boeckel, S.F.van Aelst, and T.Beetz, Recl. Trav. Chim. Pays-Bas, 1983, **102**, 415; P.M.Woollard, Biomed. Mass Spectrom., 1983, **10**, 143 (Chem. Abs., 1983, **99**, 122534).

226 K.Issleib, A.Balszeiweit, H.-J.Richter, and W.Tonk, Z. Chem., 1983, **23**, 434; T.Azuhata and Y.Okamoto, Synthesis, 1983, 916.

227 N.Nakamura, J. Am. Chem. Soc., 1983, **105**, 7172; M.Ballester and A.L.Palomo, Synthesis, 1983, 571; H.Emde and G.Simchen, Liebigs Ann. Chem., 1983, 816.

228 G.Grotsch, W.Malisch, and H.Blau, J. Organomet. Chem., 1983, **252**, C19; A.Stansunik and W.Malisch, ibid, 1983, **247**, C47; C.W.Jefford, J.-G.Rossier, and G.D.Richardson, J. Chem. Soc., Chem. Commun., 1983, 1064.

229 A.Blaschette and H.Safari, Phosphorus Sulphur, 1983, **17**, 57; S.Matsubara, K.Takai, and H.Nozaki, Bull. Chem. Soc. Jpn., 1983, **56**, 2029.

230 I.Saito, R.Nagata, K.Yuba, and T.Matsuura, Tetrahedron Lett., 1983, **24**, 4439; R.N.Baruah, R.P.Sharma, and J.N.Baruah, Chem. Ind. (London), 1983, 825.

231 T.Hiyama and M.Obayashi, Tetrahedron Lett., 1983, **24**, 395; S.Matsubara, K.Takai, and H.Nozaki, ibid, p.3741.
232 S.Kanemoto, K.Oshima, S.Matsubara, K.Takai, and H.Nozaki, Tetrahedron Lett., 1983, **24**, 2185; J.Holecek, K.Handlir, and M.Nadvornik, J. Prakt. Chem., 1983, **325**, 341; J.M.Aizpurna and C.Palomo, Tetrahedron Lett., 1983, **24**, 4367; K.Dostal and J.Sikola, Z. Chem., 1983, 23, 185.
233 I.Saito, R.Nagata, K.Yuba, and T.Matsuura, Tetrahedron Lett., 1983, **24**, 1737; E.L.Clennan and R.P.L'Esperance, ibid, p.4291; I.Fleming and J.Iqbal, ibid, p.2913; S.Danishefsky, T.Kitahara, and P.F.Schuda, Org. Synth., 1983, **61**, 147.
234 I.Kuwajima, E.Nakamura, and K.Hashimoto, Org. Synth., 1983, **61**, 122; M.E.Krafft and R.A.Holton, Tetrahedron Lett., 1983, **24**, 1345; I.Matsuda, S.Sato, and Y.Izumi, ibid, p.2787 and 3855; J.Hooz and J.Oudenes, ibid, p.5695.
235 J.K.Rasmussen, L.R.Krepski, S.M.Heilmann, H.K.Smith II, and M.L.Tumey, Synthesis, 1983, 457; J.A.Cella and T.D.Mitchell, J. Organomet. Chem., 1983, **244**, C5.
236 E.Piers, C.K.Lau, and I.Nagakura, Can. J. Chem., 1983, **61**, 288; R.N.Renard, D.Berube, and C.J.Stephens, ibid, p.1379; M.Wada, M.Kakatani, and K.Akiba, Chem. Lett., 1983, 39; T.Hosokawa, S.Inui, and S.-i.Murahashi, ibid, p.1081; Y.Matsumura, J.Fujiwara, K.Maruoka, and H.Yamamoto, J. Am. Chem. Soc., 1983, **105**, 6312; Y.Yamamoto, K.Murayama, and K.Matsumoto, ibid, p.6963; E.Nakamura, Y.Horiguchi, J.-i.Shimada, and I.Kuwajima, J. Chem. Soc., Chem. Commun., 1983, 796; G.M.Rubottom and H.D.Juve,jr., J. Org. Chem., 1983, **48**, 422; E.Nakamura, M.Shimizu, I.Kuwajima, J.Sakata, K.Yokoyama, and R.Noyori, ibid, p.932; T.V.Lee and J.O.Okonkwo, Tetrahedron Lett., 1983, **24**, 323; I.Fleming and J.Iqbal, ibid, p.327; D.J.Ager, ibid, p.419; M.A.Ibragimov and W.A.Smit, ibid, p.961; A.Koskinen and M.Lounasmaa, ibid, p.1951; M.T.Reetz and M.Sauerwald, ibid, p.2837; K.-y.Akiba, Y.Nishihara, and M.Wada, ibid, p.5269; E.Akguen and U.Pindur, Chem.-Ztg., 1983, **107**, 236 and 237.
237 T.Oida, S.Tanimoto, H.Ikehira, and M.Okano, Bull. Chem. Soc. Jpn., 1983, **56**, 645; K.Yamamoto and Y.Tomo, Chem. Lett., 1983, 531; O.W.Webster, W.R.Hertler, D.Y.Sogah, W.B.Farnham, and T.V.RajanBabu, J. Am. Chem. Soc., 1983, **105**, 5706; P.J.Cowan and M.W.Rathke, Synth. Commun., 1983, 13, 183; G.Wenke, E.N.Jacobsen, G.E.Totten, A.C.Karydas, and Y.E.Rhodes, ibid, p.449; K.Ikeda, K.Achiwa, and M.Sekiya, Tetrahedron Lett., 1983, **24**, 913; Y.Kita, H.Yasuda, Y.Sugiyama, F.Fukata, J.Haruta, and Y.Tamura, ibid, p.1273; S.K.Patel and I.Paterson, ibid, p.1315; R.A.Bunce, M.F.Schlecht, W.G.Dauben, and C.H.Heathcock, ibid, p.4943.
238 P.Brownbridge, T.H.Chan, M.A.Brook, and G.J.Kang, Can. J. Chem., 1983, **61**, 688; P.R.Hamann and P.L.Fuchs, J. Org. Chem., 1983, **48**, 914; K.Hirai and I.Ojima, Tetrahedron Lett., 1983, **24**, 785.
239 E.Nakamura and I.Kuwajima, Chem. Lett., 1983, 59; H.Urabe, Y.Takano, and I.Kuwajima, J. Am. Chem. Soc., 1983, **105**, 5703; H.Urabe and I.Kuwajima, Tetrahedron Lett., 1983, **24**, 5001; B.H.Craig, J.T.Pinhey, and S.Sternhell, Aust. J. Chem., 1982, **35**, 2237.
240 H.Puff, S.Franken, W.Schuh, and W.Schwab, J. Organomet. Chem., 1983, **254**, 33 and 1983, **244**, C41; H.Puff, S.Franken, and W.Schuh, ibid, 1983, **256**, 23; C.Glidewell and D.C.Liles, ibid, 1983, **243**, 291.
241 E.Kupce, E.Liepins, A.Lapsina, G.Zelchan, and E.Lukevics, J. Organomet. Chem., 1983, **251**, 15; Chem. Abs., 1983, **98**, 54057, 89527, and 198371; Chem. Abs., 1983, **99**, 122564, 175933, and 195118.

242 C.Auchisi, A.Maccioni, A.M.Maccioni, and G.Podda, Gazz. Chim. Ital., 1983, **113**, 73; A.G.Davies, J.A.-A.Hawari, and P.Hua-De, J. Organomet. Chem., 1983, **251**, 203 and **256**, 251; A.G.Davies and A.J.Price, ibid, 1983, **258**, 7.
243 K.Kobayashi, M.Kawanisi, T.Hitomi, and S.Kozima, Chem. Lett., 1983, 851; J.Alais, A.Maranduba, and A.Veyrieres, Tetrahedron Lett., 1983, **24**, 2383.
244 A.G.Davies and J.A.-A.Hawari, J. Chem. Soc., Perkin Trans. 1, 1983, 875.
245 S.Masamune, L.R.Sita, and D.J.Williams, J. Am. Chem. Soc., 1983, **105**, 630.
246 A.G.Davies, J.P.Goddard. M.B.Hursthouse, and N.P.C.Walker, J. Chem. Soc., Chem. Commun., 1983, 597; W.W.du Mont and M.Grenz, Z. Naturforsch., Teil B, 1983, **38**, 113 and J. Organomet. Chem., 1983, **241**, C5; G.L.Rochfort and J.E.Ellis, ibid, 1983, **250**, 265 and 277.
247 H.Puff, I.Bung, E.Friedrichs, and A.Jansen, J. Organomet. Chem., 1983, **254**, 23; R.Graziani, U.Casellato, and G.Plazzegna, Acta Crystallogr., 1983, **C39**, 1188; R.G.Swisher, R.O.Day, and R.R.Holmes, Inorg. Chem., 1983, **22**, 3692.
248 W.Ando, Y.Hamada, and A.Sekiguchi, Tetrahedron Lett., 1983, **24**, 4033.
249 J.Kowalski, J.Chojnowski, and J.Michalski, J. Organomet. Chem., 1983, **258**, 1; G.G.Hoffmann, Chem. Ber., 1983, **116**, 3858.
250 W.E.Shklover, Yu.T.Struchkov, L.E.Gusel'nikov, W.W.Volkova, and W.G.Avakyan, Z. Anorg. Allg. Chem., 1983, **501**, 153; L.E.Gusel'nikov, V.V.Volkova, V.G.Avakyan, N.S.Nametkin, M.G.Voronkov, S.V.Kirpichenko, and E.N.Suslova, J. Organomet. Chem., 1983, **254**, 173.
251 L.A.Naessens, E.G.Claeys, S.Hoste, and G.P.van der Kelen, Bull. Soc. Chim. Belg., 1983, **92**, 753; M.G.Voronkov, S.V.Kirpichenko, E.N.Suslova, V.V.Keiko, and A.I.Albanov, J. Organomet. Chem., 1983, **243**, 271; C.W.Carlson and R.West, Organometallics, 1983, **2**, 1798.
252 J.Barrau, G.Rima, and J.Satge, J. Organomet. Chem., 1983, **252**, C73; C.Guimon, G.Pfister-Guillouzo, H.Lavayssiere, G.Dousse, J.Barrau, and J.Satge, ibid, 1983, **249**, C17; J.Barrau, M.Bouchaut, H.Lavayssiere, G.Dousse, and J.Satge, ibid, 1983, **243**, 281; J.Barrau, G.Rima, H.Lavayssiere, G.Dousse, and J.Satge, ibid, 1983, **246**, 227.
253 F.Brisse, F.Belanger-Gariepy, B.Zacharie, Y.Gareau, and K.Steliou, Nouv. J. Chim., 1983, **7**, 391; F,Brisse, M.Vanier, M.J.Olivier, Y.Gareau, and K.Steliou, Organometallics, 1983, 2, 878; C.R.Lucas, Can. J. Chem., 1983, **61**, 1096.
254 D.N.Harpp, T.Aida, and T.H.Chan, Tetrahedron Lett., 1983, **24**, 5173; S.Kato, K.Itoh, K.Miyagawa, and M.Ishida, Synthesis, 1983, 814.
255 H.Puff, A.Bongartz, W.Schuh, and R.Zimmer, J. Organomet. Chem., 1983, **248**, 61; H.Puff, E.Friedrichs, R.Hundt, and R.Zimmer, ibid, 1983, **259**, 79.
256 M.Drager, J. Organomet. Chem., 1983, **251**, 209; A.Tzschach, K.Jurkschat, M.Scheer, J.Meunier-Piret, and M.van Meerssche, ibid, 1983, **259**, 165; F.W.B.Einstein, C.H.W.Jones, T.Jones, and R.D.Sharma, Can. J. Chem., 1983, **61**, 2611; A.S.Secco and J.Trotter, Acta Crystallogr., 1983, **C39**, 451.
257 T.Hattich and U.Kunze, Z. Naturforsch., Teil B, 1983, **38**, 655 and Chem. Ber., 1983, **116**, 3071; U.Kunze and P.-R.Bolz, Z. Anorg. Allg. Chem., 1983, **498**, 41; B.Mathiasch and U.Kunze, Inorg. Chim. Acta, 1983, **75**, 209; S.W.Carr, R.Colton, D.Dakternieks, B.F.Hoskins, and R.J.Steen, Inorg. Chem., 1983, **22**, 3700.

258 R.Wojtowski, I.Wharf, and M.Onyszchuk, Can. J. Chem., 1983, **61**, 743; M.Onyszchuk and I.Wharf, J. Organomet. Chem., 1983, **249**, C9.
259 K.Lal, Monatsh. Chem., 1983, **114**, 33; K.Hensen, T.Zengerly, P.Pickel, and G.Klebe, Angew. Chem. Int. Ed. Engl., 1983, **22**, 725.
260 T.H.Chan, M.A.Brook, and T.Chaly, Synthesis, 1983, 203; S.Torii and T.Inokuchi, Chem. Lett., 1983, 1349.
261 L.Engelbrecht and G.Sonnek, Plaste Kautsch., 1983, **30**, 362 (Chem. Abs., 1983, **99**, 158522); D.W.Sindorf and G.E.Maciel, J. Am. Chem. Soc., 1983, **105**, 3767.
262 R.Noyori, I.Nishida, and J.Sakata, J. Am. Chem. Soc., 1983, **105**, 1598; A.I.Albanov, L.I.Gubanova, M.F.Larin, V.A.Pestunovich, and M.G.Voronkov, J. Organomet. Chem., 1983, **244**, 5; Yu.L.Frolov, M.G.Voronkov, G.A.Gavrilova, N.N.Chipanina, L.I.Gubanova, and V.M.D'Yakov, ibid, p.107.
263 M.Lissel and K.Drechsler, Synthesis, 1983, 459; T.Imamoto, T.Matsumoto, and M.Yokoyama, ibid, p.460; E.C.Friedrich and G.DeLucea, J. Org. Chem., 1983, **48**, 1678.
264 R.D.Miller and D.R.McKean, Tetrahedron Lett., 1983, **24**, 2619; M.E.Jung and Z.Long-Mei, ibid, p.4533.
265 T.Azuhata and Y.Okamoto, Synthesis, 1983, 461; M.E.Jung and P.K.Lewis, Synth. Commun., 1983, **13**, 213; M.G.Voronkov, V.G.Komarov, and E.I.Dubinskaya, Izv. Akad. Nauk SSSR, Ser. Khim., 1982, 2182 (Chem. Abs., 1983, **98**, 72211).
266 G.A.Kraus and P.Gottschalk, Tetrahedron Lett., 1983, **24**, 2727; B.L.Feringa and W.Dannenberg, Synth. Commun., 1983, **13**, 509.
267 T.H.Chan and S.D.Lee, Tetrahedron Lett., 1983, **24**, 1225; M.Voyle, N.K.Dunlap, D.S.Watt, and O.P.Anderson, J. Org. Chem., 1983, **48**, 3242; H.Sakurai, K.Sasaki, and A.Hosomi, Bull. Chem. Soc. Jpn., 1983, **56**, 3195; H.Sakurai, Y.Sakata, and A.Hosomi, Chem. Lett., 1983, 409.
268 R.Eujen and R.Mellies, J. Fluorine Chem., 1983, **22**, 263; M.S.Soliman, M.A.Khattab, and A.G.El-Kourashy, Spectrochim. Acta, Part A, 1983, **39**, 621.
269 G.V.Motsarev, E.A.Chernyshev, V.R.Rosenberg, V.T.Inshakova, and V.I.Zetkin, Zh. Obshch. Khim., 1983, **53**, 1111 (Chem. Abs., 1983, **99**, 195119); A.F.El-Farargy and W.P.Neumann, J. Organomet. Chem., 1983, **258**, 15.
270 V.D.Pomeshchikov and V.V.Pozdeev, Zh. Obshch. Khim., 1982, **52**, 2583 (Chem. Abs., 1983, **98**, 72307); A.von Rumohr, W.Sundermeyer, and W.Towae, Z. Anorg. Allg. Chem., 1983, **499**, 75; D.W.Owen and R.C.Poller, J. Organomet. Chem., 1983, **255**, 173.
271 S.J.Blunden, J. Organomet. Chem., 1983, **248**, 149; S.J.Blunden, P.J.Smith, and D.G.Gillies, Inorg. Chim. Acta, 1983, **60**, 105.
272 M.H.Darbieu and G.Cros, J. Organomet. Chem., 1983, **252**, 327; P.J.Smith and B.N.Patel, ibid, 1983, **243**, C73; M.A.Wassef and S.Hessin, Commun. Fac. Sci. Univ. Ankara, Ser.B, 1981, **27**, 141 (Chem. Abs., 1983, **98**, 54056).
273 J.S.Brooks, R.W.Clarkson, and D.W.Allen, J. Organomet. Chem., 1983, **243**, 411; S.Hoste, G.G.Herman, F.F.Roelandt, W.Lippens, L.Verdonck, and G.P.van der Kelen, Spectrochim. Acta, Part A, 1983, **39**, 959; S.K.Misra, K.Singhal, F.S.Siddiqui, A.Ranjan, and P.Raj, Indian J. Phys. Nat. Sci., 1983, **3**, 16 (Chem. Abs., 1983, **99**, 195125).

7

Group V: Arsenic, Antimony, and Bismuth

BY J. L. WARDELL

1. Tervalent Compounds

1.1 Metal-Metal Bonded Species.- Compounds containing Group V element-element double bonds have been prepared[1,2] [equations (1) and (2)]. Bulky organic groups appear essential for the stability

$$2,4,6\text{-}Bu^t{}_3C_6H_2MH_2 + (Me_3Si)_2CHM^1Cl_2 \xrightarrow[DBU]{THF} 2,4,6\text{-}Bu^t{}_3C_6H_2M{=}M^1CH(SiMe_3)_2 \qquad (1)$$

$(M{=}As,\ M^1{=}As$ or P; $M{=}P,\ M^1{=}As$ or $Sb)$

$$(Me_3Si)_3CPCl_2 + (Me_3Si)_3CAsCl_2 \xrightarrow{Bu^tLi} (Me_3Si_3)_3CM{=}M^1C(SiMe_3)_3 \qquad (2)$$

$(2,\ M,M^1{=}P,\ As)$

of $RM{=}M^1R$ compounds. Characterization of (1) and (2) was generally achieved spectroscopically (e.g. using m.s., n.m.r., u.v.) while in addition crystal structures were determined for (1, M=P and As; M^1=As).[1] Both compounds have trans-planar arrangements about the central $CM{=}M^1C$ fragments; some details are for (1, $M{=}M^1$=As) As=As 2.224 Å, ∠ CAsAs 99.9(3)° and 93.6(3)° and for (1, M=P, M^1=As) As=P 2.124(2) Å, ∠CAsP 101.2(2)° and ∠ AsPC 96.4(2)°. The Sb analogue is the least stable in solution. Sulphur adds across the double bond of (2). The arsenic-arsenic double bond length in (1; $M{=}M^1$=As) can be compared with the single bond lengths in cyclo-$(AsPh)_6$ (3) (av. 2.459 Å)[3] and in non-planar PhC=CPhAsPhAsPh (4) (2.471 Å)[4]; (4) was prepared from (3) and PhC≡CPh at 160°C. The crystal structure of 2,2',5,5'-tetramethylbiarsolyl (5) was also reported;[5] (5) was prepared from 2,5-Me_2-1-Ph-arsole on successive treatments with Li and I_2. Tetra(1-alkenyl)diantimony compounds, R_4Sb_2, were obtained in two-step syntheses from R_3Sb, e.g. see Scheme (1). As with many R_2MMR_2, (6) exhibits a thermochronic effect - it is a yellow liquid cooling to a violet solid. Ready cleavage of the Sb-Sb bond in R_4Sb_2 occurs on reaction with electrophiles, e.g. Me_4Sb_2 is cleaved by

$$(CH_2{=}CH)_3Sb \xrightarrow{i} (CH_2{=}CH)_2SbNa \xrightarrow{ii} (CH_2{=}CH)_2SbSb(CH{=}CH_2)_2$$

Reagents: i, Na,liq.NH_3; $ClCH_2CH_2Cl$

Scheme 1

Br_2, I_2, MeI and p-$MeC_6H_4TeTeC_6H_4Me$-p, the latter reagent providing the first reported Sb-Te bonded compound, $Me_2SbTeC_6H_4Me$-p.[7] The new cyclic compound $(Bu^tSb)_5$ was characterized[8] as a product of the reduction of Bu^tSbCl_2 by Mg in THF. Compounds, R_4Bi_2, [R=Me (7), Et, Pr^i, CH_2=CMe (8), Me_2C=CH, Ph etc.] were obtained analogously to R_4Sb_2 (Scheme 2). All are red liquids but (7), (8) and 1,1'-bibismolane are blue in the solid state.[9] The methylene protons in Et_4Bi_2 and the Me groups in $Pr^i{}_4Bi_2$ are diastereotopic.[9] A staggered trans-conformation was found[10] for Ph_4Bi_2 in the solid state with Bi-Bi 2.990(2) Å, ∠CBiBi 90.9(5)° and 91.6(5)°, ∠CBiB 98.3(8)°; Ph_4Bi_2, (an orange solid at RT, yellow at 77K), is stable to 100° when pure.[10] Generally, thermolysis of air-sensitive R_4Bi_2 provides R_3Bi and Bi; cleavages of R_4Bi_2 occurred on reaction with I_2, HCl, $PhCH_2Br$, BuLi and $BrCCl_3$.[9,10] The first compound containing a P_2Sb ring (Bu^tP-$SbBu^t$-PBu^t) was obtained as a product of reaction of KBu^tPPBu^tK with Bu^tSbCl_2 in pentane at -78°C; other products had P_2Sb_2 and P_3Sb cycles.[11]

1.2 Other Compounds:- Series of 4-YCH_2 arsabenzene (9; R=CO_2Et, CN, CH_2Br, CH_2CO_2Et, CH_2CN and CHMeBr) and 2-Ph-arsabenzenes have been obtained respectively by the reaction of $AsBr_3$ with 4-YCH_2-1,1-Bu_2-stannacyclohexa-2,5-dienes and by acid treatment of 1-Ph-4-HO-arsacyclohexa-2,5-dienes.[12] Chemical transformations of (9) were possible, e.g. the side chain in (9, Y=CH_2Br) was converted to $CH{=}CH_2$, CH_2CH_2CN or $CH_2CH_2NH_2$ and (9, Y=$CHNH_2CO_2$Et) was synthesised from (9, Y=CH_2CO_2Et).

Base-catalysed addition of $PhAsH_2$ to appropriate 1,3-diynes provides 2,5-R'_2-1-R-arsoles (10, R=Ph).[13] Radical anions of (10, R=Ph), produced on treatment with 1 equiv. of Li, K, or RLi (R=Ph or Bu^t), react with R^2X to give (10, R=R^2). Dianions were also produced. Arsoles (10) are oxidised by $PhICl_2$ or Bu^tOCl. The As(V) oxidation products are thermally labile, e.g. 1,1-Cl_2-2,5-$R^1{}_2$-1-R-λ^5-arsoles (11), from reaction with PhICl provides (10, R=Cl). Compound (10, R=Cl, R^1=Ph) reacts with 2,5-Ph_2-arsolyllithium to give 2,2^1,5,5^1-tetraphenyldiarsolyl.[14]

1-H-1,3-Benzazarsole (12) is a planar compound with C-As

1.822(9) and 1.920(7)°. In contrast to arsabenzenes, derivatives of (12) are alkylated at As by RLi as well as undergoing lithiation at N or C-2, without addition to the As-C π bond.[14] Cyclo-addition of CH_2=CMeCMe=CH_2 or o-tetrachlorobenzoquinone occurs across the As=C bond in 1,3-benzo arsoles.[15]

Among the triorganometallic species produced were (i) (6-Me-quinolin 8 yl)$_n$ AsPh_{3-n} (n=1 or 2),[16a] (ii) o-$Me_2(Me_2M)SiC_6H_4AsMe_2$ (M^1=P or As) and p-$XC_6H_4SiMe(MMe_2)CH_2CH_2M^1Me_2$ (X=F or Cl; M=N, P or As; M^1=As),[16b] (iii) (-)[2-((S)-$Me_2NCHMe)C_6H_4AsPh_2$] and (o-Y-R-$C_6H_3)_nMPh_{3-n}$ (Y=MeO, EtO, etc. or NMe_2; R=H or 3,5 or 6-Y; M=As, Sb or Bi; n=1, 2 or 3),[16c] (iv) $Me_2AsCH_2CH_2OR$ (e.g. R=H, COR^1, C(O)NHPh and $MeAs(CH_2CH_2OH)(CH_2CH_2X)$ (e.g. X=OH, CH_2OH, CO_2R, CN) as well as $MeAsHCH_2CH_2CO_2H$ and As(V) analogues,[16d] (v) $N(CH_2CH_2AsPh_2)_3$ (13),[16e] and (vi) ArAr'Ar"Bi,[16f] e.g. (o-MeC_6H_4)(o-$Pr^iC_6H_4$)(p-ClC_6H_4)Bi (diastereotopic effects observed in nmr).

Compound (13) reacts with HI to give $[HN(CH_2CH_2AsI_2)_3]I$ from which the cryptands (or spherands) $[N(CH_2CH_2)_3]_8(As_4X_4)_6$ (14, X=O or S) are obtained on treatment with H_2O or H_2S; a significant feature of structure of (14, X=O) is the (As_4O_4) rings. Cleavage of $(Ph_2Sb)_2CH_2$ by PhLi provides the useful intermediate, Ph_2SbCH_2Li, from which were obtained $Ph_2SbCH_2CRR'OH$ (from RR'CO), $Ph_2SbCH_2MPh_2$ (M=P or As) and Ph_2SbCH_2R (via Ph_2SbCH_2Cu).[17]

The crystal structure of Na[18-cyclopenta-2,4-dienyl-4,8,12-cyclopenta-2,4-diene-1,1,2-triyl-3a,8a-epistibino-tricyclopenta [1,4,7]-tristiboninide].3THF (prepared from $SbCl_3$ and CpNa in THF) consists of $3(\mu_3-C_5H_3)$ units with 4 Sb atoms built-up to a tetra-stibaadmantane framework.[18] Additional ligands are a $\sigma-C_5H_5$ group and a $(\mu_2-C_5H_3)^-$ unit. The THF co-ordinated Na^+ resides above the anionic ring.[18] Structures were also determined[19] for (i) Me_3As (As-C 1.968(3) Å, ∠CAsC 96.1(5)°) and Me_3Sb(Sb-C 2.163(3) Å, ∠CSbC 94.1(5)° (both by electron diffraction), (ii) Ph_3As (As-C 1.957(8) Å, ∠CAsC 100.1(4)°, (iii) thermally-stable $[(Me_3Si)_2CH]_3Bi$ (Bi-C_{av} 2.328(13) Å, ∠CBiC_{av} 102.9(5)°, and (iv) (p-$MeC_6H_4)_3Bi$ (Bi-C_{av} 2.25 Å, ∠CBiC_{av} 94.7°.

Important features of the crystal structure of $(Me_5C_5)_2M^+$,BF_4^- (15, M=As) are non-planar rings (angle 36.5°) bound in a di- and tri-fashion to As; the Sb analogue was also obtained. Both compounds are fluxional molecules in solution.[20] The exocyclic Cl in 10-chlorophenothiantimonin is in an equatorial position in contrast to its axial siting in the As analogue.[21] The compounds,

$R_2AsSP(S)Ph_2$ (R=Me or Ph) are formed from Ph_2PCl and $R_2As(S)SNa$ as a result of S migration from As to P.[22] Xanthates, $RSb(S_2COR')_2$ and R_2BiS_2COR' have been obtained by metathetic reactions from the halides and by CS_2 insertion into a M-OR' bond; $MeSb(S_2COEt)_2$ has a pentagonal bipyramidal structure (inter- and intra-molecular bridging S_2COEt units) with the Me group and the lone pair in axial sites.[23]

2. <u>Quinquevalent Compounds</u>

The syntheses were reported of R_4Sb-oxinates (16) (hexa-co-ordinate) and anthranilates (penta-co-ordinate), (R=Me or Ph), from reaction of R_5Sb with 8-HO-quinoline or o-$HO_2CC_6H_4NH_2$. Mössbauer spectra were reported for these compounds, for $Me_4Sbacac$ (hexa-co-ordinate) and for ionic $Me_4SbH(OCOR)_2$.[24] The crystal structures of monomeric (16, R=Me)[24] and hexa-co-cordinate Ph_3BiCl (2-Me-8-quinolinolate)[25] (17) were obtained; for (16, R=Me) $(Sb\text{-}O)_{equat}$ 2.187(8) Å, $(Sb\text{-}N)_{equat}$ 2.463(9) Å, and for (17) $(Bi\text{-}C)_{av}$ 2.19(2) Å $(Bi\text{-}N)_{equat}$ 2.71(2) Å. The arsonyl radical $[Ph_3AsBr]\cdot$ was detected on X-irradiation of $[Ph_3AsMe]^+Br^{-}\cdot$; e.s.r. analysis suggested a structure between trigonal bipyramid and C_{3v} symmetry.[26] Compounds, $(cyclo\text{-}C_6H_{11})_3Sb(OH)Y$ (Y-Cl, Br, AcO or NO_3) (ν_{OH} ca. 3630 cm^{-1}, ν_{SbO} 565-513 cm^{-1}) have been obtained from $[(cyclo\text{-}C_6H_{11})_3SbY]_2O$ (18) on hydrolysis or by reaction with AgY in aqueous CH_3COCH_3. Dehydration back to (18) occurs readily for Y=Cl or Br but compounds having $Y=NO_3$ or AcO are stable.[27] Thermolysis of $(Me_3SiCH_2)_3AsBr_2$ at 170° at 10^{-4}Torr provides $(Me_3SiCH_2)_2AsCH_2Br$ *via* $(Me_3SiCH_2)_2As(Br)=CH_2$.[28] Structures of $(Me_2Cl_2Sb)_2CH_2.CH_2Cl_2$[29a] (axial Cl) $Ph_3SbCl_2.SbCl_3$ (19)[29b] and $[Ph_3SbCl]^+[SbCl_6]^-$ (20)[29b] have been determined. The trigonal geometry of Ph_3SbCl_2 is only slightly perturbed on complexation in (19). The cation geometry in (20) is intermediate between trigonal bipyramidal and tetrahedral; there is some anion-cation interaction.

The synthesis of interesting heterocycles $[(CF_3)_2AsClN(SiMe_3)]_2$ and $[(CF_3)_2AsN]_n$ (21, n=3 or 4) have been reported (Scheme 2).[30]

$$(CF_3)_2AsN(SiMe_3)_2 \xrightarrow{i} (CF_3)_2\overline{As(Cl)\text{-}NSiMe_3\text{-}As(CF_3)_2Cl\text{-}N}SiMe_3 \xrightarrow{ii} (21)$$

Reagents: i, Cl_2; ii, hexane, Δ

Scheme 2

There is only a slight alteration in the As-N bond lengths in (19, n=4) 1.716(7) and 1.732(9) Å.

3. Bibliography

F.B. Whitfield, D.J. Freeman, and K.J. Shaw, Chem. Ind. (London), 1983, 786. Me_3As as off-flavour component in prawns.

R.H. Fish, R.S. Tannous, W. Walker, C.S. Weiss, and F.E. Brinckman, J. Chem. Soc., Chem. Commun., 1983, 490. Organoarsenicals in oil shales.

J.S. Edmonds and K.A. Francesconi, J. Chem. Soc. Perkin Trans. 1, 1983, 2375. Organoarsenic containing ribofuranosides, isolated from brown kelp.

N.L.M. Dereu and R.A. Zingaro, Bull. Soc. Chim. Belg., 1982, 91, 685. Preparation of 5'-dioxy-2'-Me_2AsCH_2-thymidine.

H.J. Breunig, W. Kanig, and A. Soltani-Neshan, Polyhedron, 1983, 2, 291. Synthesis of $[(Me_3Si)_2CH]_nSbCl_{3-n}$ (n=1, 2 or 3) and $(Me_3Si)_3CSbCl_2$.

G.M. Bodner, C. Gagnon, and D.N. Whittern, J. Organomet. Chem., 1983, 243, 305. ^{13}C Nmr of R_3M (M=As, Sb or Bi) and Ph_nAsR_{3-n} (R=Cl, Me, Et or Bu).

R. Gleiter, W.D. Goodmann, W. Schäfer, J. Grobe, and J. Apel. Chem. Ber., 1983, 116, 3745. P.e.s. of $Me_nM(CF_3)_{3-n}$ (M=As or Sb).

I.A. Litvinov, Yu.T. Struchkov, B.A. Arbuzov, E.N. Dianova, and E.Ya. Zabotina, Dokl. Akad. Nauk. SSSR, 1983, 268, 885. Synthesis and crystal structure of 2,5-Ph_2-4-Ph_2CH-1,2,3-diazaarsole.

M. Nunn, D.B. Sowerby, and D.M. Wesolek, J. Organomet. Chem., 1983, 251, C45; M. Wieber, D. Wirth and I. Fetzer, Z. Anorg. Allg. Chem., 1983, 505, 134. Mild syntheses of R_2SbX and $RSbX_2$ (R=Ph or Me; X=Cl or Br).

L. Silaghi-Dumitrescu and I. Haiduc, Synth. React. Inorg. Metal-Org. Chem., 1983, 13, 475. Synthesis of Ph_nAsCl_{3-n} from $AsCl_3$ and Ph_4Sn.

M. Wieber and H.G. Rüdling, Z. Anorg. Allg. Chem., 1983, 505, 147. Synthesis of Me_2BiSR (R=nitrogen heteroaryl).

G. Alonzo, Inorg. Chem. Acta, 1983, 73, 141. Synthesis and spectra of $PhM(SC_6H_4NH_2$-o$)_2$ (M=Sb or Bi)

M. Wieber, D. Wirth and K. Hess, Z. Anorg. Allg. Chem., 1983, 505, 138. $MeM(OCOR)_2$ (M=Sb or Bi).

D.G. Allen and S.B. Wild, Organometallics, 1982, 2, 394. Benzylidene transfer from chiral arsonium ylides.

Z. Xia and Z. Zhang, Huaxue Xuebao, 1983, 41, 577; Chem. Abstr., 1983, 99, 113979. Crystal structure of $Ph_3\overset{+}{A}s$-$\bar{C}(COPh)COCF_2CF_3$

G.S. Harris, D. Lloyd, W.A. MacDonald, and I. Gosney, Tetrahedron, 1983, 39, 297. Triarylarsonium cyclopentadienylides.

J.B. Ousset, C. Mioskowski, and G. Solladie, Tetrahedron Lett., 1983, 24, 4419. Uses of R^1R^2C=$CH\bar{C}H$=$\overset{+}{A}sPh_3$.

References

1 A.H. Cowley, J.G. Lasch, N.C. Norman, and M. Pakulski, J. Am. Chem. Soc., 1983, 105, 5506; A.H. Cowley, J.G. Lasch, N.C. Norman, M. Pakulski, and B.R. Whittlesey, J. Chem. Soc., Chem. Commun., 1983, 881.
2 J. Escudie, C. Couret, H. Ranaivonjatovo and J. G. Wolf, Tetrahedron Lett., 1983, 24, 3625; C. Couret, J. Escudie, Y. Madule, H. Ranaivonjatovo, and J.G. Wolf, ibid., 1983, 24, 2769.
3 A.L. Rheingold and P.J. Sullivan, Organometallics, 1983, 2, 327.
4 G. Sennyey, F. Mathey, J. Fischer, and A. Mitschler, Organometallics, 1983, 2, 298.
5 A.J. Ashe,III, W.M. Butler, and T.R. Diephouse, Organometallics, 1983, 2, 1005.
6 A.J. Ashe,III, E.G. Ludwig,Jr, and H. Pommerening, Organometallics, 1983, 2, 1573.
7 H.J. Breunig and H. Jawad, J. Organomet. Chem., 1983, 243, 417; W.W.du Mont, T. Severengiz, and H.J. Breunig, Z. Naturforsch., Teil B, 1983, 38, 1306.
8 H.J. Breunig and H. Kischkel, Z. Anorg. Allg. Chem., 1983, 502, 175.
9 A.J. Ashe,III, E.G. Ludwig,Jr, and J. Oleksyszyn, Organometallics, 1983, 2, 1859; H.J. Breunig and D. Müller, Z. Naturforsch., Teil B, 1983, 38, 125; J. Organomet. Chem., 1983, 253, C21.
10 F. Calderazzo, A. Morvillo, G. Pelizzi, and R. Poli, J. Chem. Soc., Chem. Commun., 1983, 507.
11 M. Baudler and S. Klautke, Z. Naturforsch., Teil B, 1983, 38, 121.
12 A.J. Ashe,III and S.T. Abu-Orabi, J. Org. Chem., 1983, 48, 767; G. Märkl, A. Bergbauer, and J.B. Rampal, Tetrahedron Lett., 1983, 24, 4079.
13 G. Märkl, H. Hauptmann, and A. Merz, J. Organomet. Chem., 1983, 249, 335.
14 G. Märkl and H. Hauptmann, J. Organomet. Chem., 1983, 248, 269.
15 R. Richter, J. Sieler, A. Richter, J. Heinicke, A. Tzschach, and O. Lindquist, Z. Anorg. Allg. Chem., 1983, 501, 146; J. Heinicke, A. Petrasch, and A. Tzschach, J. Organomet. Chem., 1983, 258, 257; J. Heinicke and A. Tzschach, Tetrahedron Lett., 1983, 24, 5481.
16 (a) G.L. Roberts, B.W. Shelton, A.H. White, and S.B. Wild, Aust. J. Chem., 1982, 35, 2193; (b) P. Aslanidis and J. Grobe, Z. Naturforsch., Teil B, 1983, 38, 280, 289; (c) L. Horner and G. Simons, Phosphorus-Sulfur, 1983, 14, 253; (d) P.B. Chi and F. Kober, Z. Anorg. Allg. Chem., 1983, 501, 89; (e) J. Ellermann, A. Veit, E. Lindner, and S. Hoehne, J. Organomet. Chem., 1983, 252, 153; (f) P. Bras, A. van der Gen, and J. Wolters, J. Organomet. Chem., 1983, 256, C1.
17 T. Kauffmann, R. Joussen, N. Klas, and A. Vahrenhorst, Chem. Ber., 1983, 116, 473.
18 O. Mundt and G. Becker, Z. Anorg. Allg. Chem., 1983, 496, 58.
19 R. Blom, A. Haaland, and R. Seip, Acta Chem. Scand., Ser A., 1983, 37, 595; A.N. Sobolev, V.K. Bel'sky, N.Yu. Chernikova, and F. Yu. Akhmadulina, J. Organomet. Chem., 1983, 244, 129; B. Murray, J. Hvoslef, H. Hope, and P.P. Power, Inorg. Chem., 1983, 22, 3421; A.N. Sobolev, V.K. Bel'sky, and I.P. Romm, Koord. Khim., 1983, 9, 262 (Chem. Abstr., 1983, 98, 179526).
20 P. Jutzi, T. Wippermann, C. Krüger, and H.J. Kraus, Angew. Chem., Int. Ed. Engl., 1983, 22, 250.
21 W.T. Pennington, A.W. Cordes, J.C. Graham, and Y.W. Jung, Acta Crystallogr., Sect. C, 1983, 39, 709, 1010.
22 L. Silaghi-Dumitrescu and I. Haiduc, J. Organomet. Chem., 1983, 252, 295.
23 M. Wieber, D. Wirth and Ch. Burschka, Z. Anorg. Allg. Chem., 1983, 505, 141; R.K. Gupta, V.K. Jain, A.K. Rai, R.C. Mehrotra, Indian J. Chem., Sect. A, 1983, 22, 708; M. Wieber and H.G. Rüdling, Z. Anorg. Allg. Chem., 1983, 505, 150.
24 H. Schmidbaur, B. Milewski-Mahrla and F.E. Wagner, Z. Naturforsch., Teil B, 1983, 38, 1477.

25 G. Faraglia, R. Graziani, L. Volponi, and U. Casellato, J. Organomet. Chem., 1983, 253, 317.
26 M. Geoffroy and A. Llinares, Helv. Chim. Acta, 1983, 66, 76.
27 Y. Kawasaki, Y. Yamamoto, and M. Wada, Bull. Chem. Soc. Jpn., 1983, 56, 145.
28 A. Meyer, A. Hartl, and W. Malisch, Chem. Ber., 1983, 116, 348.
29 (a) W. Kolondra, W. Schwarz, and J. Weidlein, Z. Anorg. Allg. Chem., 1983, 501, 137; (b) M. Hall and D.B. Sowerby, J. Chem. Soc. Dalton Trans., 1983, 1095.
30 H.W. Roesky, R. Bohra, and W.S. Sheldrick, J. Fluorine Chem., 1983, 22, 199; R. Bohra, H.W. Roesky, J. Lucas, M. Noltemeyer, and G.M. Sheldrick, J. Chem. Soc., Dalton Trans., 1983, 1011.

8
Metal Carbonyls

BY B. J. BRISDON

1 Introduction

The format of this report follows that used last year, but space limitations have restricted the number of references which can be included. Consequently, in a year which has seen a further increase in the total number of relevant publications on metal carbonyl chemistry, some significant contributions have had to be omitted entirely.

Several interesting reviews concerning aspects of metal carbonyl and nitrosyl chemistry have appeared in 1983. These include articles on the synthetic and catalytic applications of metal carbonyls,[1] activation and reduction of CO,[2] hydrido complexes of transition metals,[3] ligand substitution reactions,[4] kinetic behaviour of M-M bonded carbonyls,[5] mechanisms for reactions of electrophiles with polynuclear metal carbonyls,[6] and aspects of transition metal nitrosyl and thionitrosyl chemistry.[7]

Additional relevant articles are to be found in Volumes 249 and 250 of the *Journal of Organometallic Chemistry* which feature invited contributions to mark Professor H.J. Emeléus' 80th birthday, and the publication of the 250th Volume of this Journal respectively. The former issue contains a most authorative account of the deca-, hexa- and penta-nuclear carbido cluster carbonyls of Ru and Os,[8] which together with reviews of the chemistry of carbidocarbonyl clusters[9] and basic metal cluster reactions,[10] provides a very comprehensive coverage of this area of metal carbonyl chemistry.

Studies in the primary journals which in the opinion of this author are of particular significance include an interesting series of papers on highly reduced organometallics of the early transition metals. Alkali metal reductions of $[M(CO)_6]^-$ (M = Nb or Ta) yield $[M(CO)_5]^{3-}$ anions,[11] and by carrying out similar reductions on di- and tri-substituted Group VI carbonyls containing non-π-acceptor ligands, the substituent ligands are preferentially labilised, so producing binary anionic carbonylates. In this way routes to $[M(CO)_4]^{4-}$ and to formal derivatives of $[M(CO)_3]^{6-}$ (M = Cr, Mo or W) have been developed.[12] By using K(*sec*-Bu_3BH) as reductant, the hydride anions $[M_2(CO)_8H_2]^{2-}$ (M = Cr, Mo or W) and $[M(CO)_3H]_4{}^{4-}$ (M = Mo or W) have been prepared.[13] Reduction of $[Mn(CO)_4NO]^-$ again results in preferential loss of the weaker π-acceptor ligand, and the formation of $[Mn(CO)_3NO]^{2-}$.[14]

The first CO adduct of a first-row transition metal in the +3 oxidation state has been reported.[15] The complex [Mn(salen)CO]ClO_4 was isolated in almost quantitative yield from the aerial oxidation of a basic, alcoholic solution of [Mn(salen)$(H_2O)_2$] containing perchlorate ions. A dimeric CO bridged structure was proposed for the cation.

Conversion of CO to a product containing two or more carbon atoms, either by selective reduction of CO, or by carbonylation of a carbido or related complex, continues to attract much interest. An electrophilic methylidene ligand has been shown to pick up exogeneous CO under extremely mild conditions to give a stable (η^2-C,C)ketene complex. This ligand is readily transformed into a carbomethoxymethyl group, representing a novel synthesis of a C_2 alkyl ligand from CO.[16] [$Os_3(CO)_{12}${η^2(C,C)μ-CHO}] was formed on carbonylation of [$Os_3(CO)_{11}(\mu$-$CH_2)$],[17] but the same starting material loses CO on heating to yield [$Os_3(\mu$-H$)_2$-$(CO)_9(\mu_3$-CCO)]. This product contains a linear CCO group perpendicular to the M_3 plane, a feature not found in the isoelectronic [$Fe_3(CO)_9(\mu_3$-CCO)]$^{2-}$ ion.[18] The first example of a borylidyne containing carbonyl cluster, formed by insertion of boron into a M-CO bond of the unsaturated cluster [$Os_3(\mu$-H$)_2(CO)_{10}$], has been reported.[19] The structure of [$Os_3(\mu$-H$)_3(CO)_9(\mu_3$-BCO)] is analogous to that of [$Os_3(\mu$-H$)_3(CO)_9(\mu_3$-CCO)]$^+$. Another osmium cluster complex, [$Os_3Pt(\mu$-H$)_2$-$(CO)_{10}P(C_6H_{11})_3$], provides the first well authenticated example of reversible reactivity towards both H_2 and CO. The parent *closo*-Os_3Pt 58-electron cluster reacts with H_2 to give a *closo*-Os_3Pt tetrahydrido-60-electron species, whereas a 60-electron "butterfly" complex is formed on reaction with CO.[20]

The ^{17}O spin-lattice relaxation times of several metallocarbonyls have been measured and related to variations in the correlation times for molecular reorientation and to changes in the ^{17}O electric quadrupole coupling constants (QCC). *Intra*-molecular variations in $\underline{T}_1$ values can be ascribed mainly to variations in QCC values which show stereochemical dependence. Consequently such measurements may be of diagnostic value in future structural studies on carbonyls.[21] In another development, non-bonded Os···Os distances have been identified from the EXAFS of Os L(111) edge spectra of a series of polynuclear osmium carbonyl cluster compounds. These results demonstrate that for highly backscattering metals at least, EXAFS can provide strong evidence about the skeletal geometry of a high nuclearity cluster.[22]

Professor R. Hoffmann's Nobel lecture published in 1982 in *Angewandte Chemie* focussed attention on the increasing impact of theoretical studies on organotransition metal chemistry. Two such studies are highlighted here from the 1983 literature. The synthetic strategy required for coupling CO ligands has been defined from theoretical models,[23] and a general procedure for evaluating the number of cluster valence molecular orbitals in molecules with condensed polyhedral geometries has been derived from MO calculations, and shown to be widely applicable.[24]

2 Theoretical and Mechanistic Studies

Theoretical calculations, often coupled with PES measurements, have been carried out on many metal carbonyls and their derivatives. The role of bridging carbonyls in metal cluster compounds has been analysed using a fragment MO approach, and structural differences between light- and heavy-atom carbonyl clusters have been shown to be at least in part electronic in origin.[25] In a complementary paper, the electronic structures of $Rh_4(CO)_{12}$ and $Ir_4(CO)_{12}$ have been studied by the EHMO method,[26] and MO calculations on the binuclear complexes $[Pd_2Cl_4(CO)_2]^{n-}$ ($\underline{n}$ = 0 or 2) have been used to explain the observed change from Cl bridges ($\underline{n}$ = 0) to CO bridges ($\underline{n}$ = 2).[27]

In calculations carried out on mononuclear carbonyl derivatives, the importance of including correlation effects in a quantitative description of the electronic structure of such compounds has been demonstrated,[28] and reasons for the colour of solid $V(CO)_6$ sought.[29] The Mössbauer parameters of iron carbonyls have been correlated with the type of iron site in the complex,[30] and PE spectra, together with MO calculations, have been reported for sulphur-containing derivatives of Fe, Co, Ru and Os carbonyls, and for hydrogen- and halogen-bridged triosmium carbonyl clusters.[31]

A detailed analysis of the kinetics of electrocatalytic ligand substitution reactions in metal carbonyls has been carried out,[32] and the mechanism of the Lewis base induced disproportionation of $V(CO)_6$ shown to involve an isocarbonyl bridged intermediate.[33] The photofragmentation dynamics of gaseous $Cr(CO)_6$ have been examined,[34] and the multiphoton dissociation and ionisation of metal carbonyl complexes used to produce highly unsaturated metal-containing molecular ions.[35] Several papers devoted to the mechanism of $M_2(CO)_{10}$ (M = Mn or Re) substitution reactions under varying conditions have appeared, and the relative importance of M-M bond scission and CO dissociation processes evaluated.[36]

3 Chemistry of Metal Carbonyls

3.1 Hydride Attack on Co-ordinated CO.- Aspects of metal carbonyl chemistry which are even distantly related to the conversion of CO and H_2 to organic products in the presence of homogeneous catalysts continue to attract much attention, and several further examples of mononuclear complexes containing CH_xO_y ligands prepared by hydride attack on co-ordinated CO have been reported. Thus $NaBH_4$ reduction of $[Fe(\eta-C_5Me_5)(CO)_3]^+$ gives high yields of $[Fe(\eta-C_5Me_5)-(CO)_2R]$ (R = H, CH_2OH or CH_3) depending upon reaction conditions,[37] and in related studies, the hydride reduction of both $[Fecp(diphos)(CO)]^+$ and $[Fe(\eta-C_5Me_5)(CO)_2(PR_3)]^+$ have been investigated.[38] For the former complex, hydride addition occurs regioselectively at CO at -78° C, while at elevated temperatures *exo* hydride attack on the cp ligand competes with attack at CO. Reduction of $[Os(CO)_2(diphos)_2]^{2+}$ by $KBH(OPr)_3$ yields *trans*-$[Os(CO)(CHO)(diphos)_2]^+$ which has been characterised by *X*-ray analysis.[39] In an interesting approach which avoids the use of powerful, non-regenerable hydride donors such as borohydrides, the metal formyl complex [Recp(CO)(CHO)(NO)] was generated from $[Recp(CO)_2NO]^+$ by hydride transfer from the $[Ru(CO)_4H]^-$ ion, which is plausibly regenerated from H_2 and CO in a catalytic system.[40] The bimetallic formyl complex anions $[M_2(CO)_9$-(CHO)] (M = Mn or Re) have been isolated from the reaction of $LiBHEt_3$ with $M_2(CO)_{10}$,[41] and the Re complex has been the subject of a mechanistic investigation which showed that the Re-CHO to Re-H conversion involves a free radical chain mechanism.[42] The generality of this process for other metal-formyl complexes remains to be determined.

To date, definite evidence for an *intra*-molecular migration step generating a C-H bond from ligated CO and H ligands has been obtained in very few instances. The reaction of Ph_2PN-*t*-$BuAlR_2$ with $Mn(CO)_5H$ does yield a product resulting from net migration of H from Mn to CO, but nmr studies reveal that proton transfer occurs indirectly *via* the P atom in this case.[43] In some other complexes, oxidatively induced migration of H from metal to CO does appear to parallel the well-known Lewis acid induced alkyl migration reactions of $Mn(CO)_5Me$, although other possible mechanisms have not been definitely excluded as yet.[44]

3.2 Mononuclear Carbonyl Derivatives.- Further details of the preparations of $[M(CO)_6]^-$ (M = Nb or Ta) by reductive carbonylation of M_2Cl_{10} at atmospheric pressure and ambient temperature have been published, and the structures of the PPN salts of both hexacarbonylate anions were shown to be isostructural with the corresponding vanadium derivative.[45] Various low temperature species including $[Cr(CO)_{6-n}L_n]$ (L = Xe, n = 1; L = N_2, n = 1-5) and $[Ni(CO)_3(N_2)]$ have been identified in doped liquid noble gas solutions,[46] and the effects of high intensity ultrasound on substitution reactions of metal carbonyls observed.[47] The rates of sonochemical ligand substitution were shown to be first order in carbonyl concentration and independent of ligand concentration, as expected for a dissociative mechanism in which co-ordinatively unsaturated species are produced by the cavitation process.

Nucleophilic nitrosylations of a wide range of metal carbonyls using $PPN(NO_2)$ in dipolar aprotic solvents gave high yields of nitrosyl products with few of the problems found with other nitrosylating reagents.[48] Further papers describing the chemical or structural features of $PPN[Fe(CO)_3NO]$, $[V(CO)_5NO]$ and $[Re(CO)_5$-$(NS)](AsF_6)_2$ have appeared.[49]

3.3 Binuclear Carbonyl Derivatives.- The structure of $Co_2(CO)_8$ has been redetermined at low temperatures, and significant differences noted between axial and equatorial Co-C bond lengths which were not apparent at room temperature.

Further evidence supporting strong through ligand metal-metal interactions rather than a direct Co-Co link is presented.[50] In the complex $[(OC)_5Os\text{-}Os(CO)_3(GaCl_3)\text{-}Cl]$, the $Os(CO)_5$ moiety acts as a donor ligand to the second osmium centre *via* an unsupported donor-acceptor Os-Os bond. The Os-Os distances [range 2.916(2) - 2.931(1) Å in the three crystallographically independent molecules] are somewhat longer than usual unbridged Os-Os bond lengths.[51] A quantitative study of the equilibrium between $Rh_2(CO)_8$ and $Rh_4(CO)_{12}$ under CO pressure has been carried out and thermodynamic parameters reported.[52] Further quantitative data, this time on enthalpies of formation, have been determined for $Re_2(CO)_{10}$ and its derivatives.[53]

3.4 Polynuclear Carbonyl Derivatives.- In a systematic study of the thermal activity of rhodium carbonyl clusters, the ease of formation of complexes of high nuclearity was, in general, highly counterion dependent.[54] A redox reaction between a rhodium carbonyl anion and a substituted ammonium cation is implicated in the thermal growth process. Controlled oxidation of $[Rh_7(CO)_{16}]^{3-}$ yields the new $[Rh_{11}(CO)_{23}]^{3-}$ ion, which has a novel metal skeleton (of ideal D_{3h} symmetry) derived from a face-to-face condensation of three octahedral sub-units.[55] The largest mixed-metal cluster anions as yet known also involve Rh.[56] Controlled pyrolysis of $[PtRh_5(CO)_{15}]^-$ yields $[Pt_2Rh_{11}(CO)_{24}]^{3-}$ and $[PtRh_{12}(CO)_{24}]^{4-}$ containing a hexagonal close-packed arrangement of metal atoms similar to that found in the isoelectronic series $[Rh_{13}(CO)_{24}H_{5-\underline{n}}]^{\underline{n}-}$ ($\underline{n}$ = 2, 3 or 4). Other mixed metal clusters, including those containing main group metals, have been prepared from cobalt and iron carbonyls.[57]

The synthesis of the highly charged ruthenium carbonyl cluster anions $[Ru_4(CO)_{11}]^{6-}$, $[Ru_6(CO)_{17}]^{4-}$ and $[Ru_6(CO)_{16}]^{6-}$ has been achieved by reduction of $Ru_3(CO)_{12}$ using alkali-metal benzophenone in THF,[58] and some sixty known and new derivatives of $Ru_3(CO)_{12}$ and $[Ru_4(CO)_{12}H_4]$ prepared by sodium diphenylketyl initiated substitution reactions.[59]

4 Cluster Carbonyls containing C, N, O or S

The chemistry of metal carbonyl clusters containing exposed, non-metallic elements such as C, N or O, or derivatives thereof, continues to attract considerable interest in the belief that such species provide the most reliable models for certain reactions on metal surfaces. In a continuation of previous chemical studies on the $[Rh_6C(CO)_{15}]^{2-}$ anion, reaction with propan-2-ol at 70° C in the presence of H_2SO_4 yields the new anion $[Rh_{12}C_2(CO)_{24}]^{2-}$ which contains isolated C atoms in prismatic cavities in the metal framework.[60] Reaction of the same starting material with $AgBF_4$ in acetone, gives a series of oligomers containing Rh_6 trigonal prismatic units bridged by Ag atoms, including the $[Ag\{Rh_6(CO)_{15}C\}_2]^{3-}$ which has been characterised by *X*-ray crystallography.[61]

$[Ru_5C(CO)_{15}]$ has been synthesised by the carbonylation of $[Ru_6C(CO)_{17}]$, and contains an exposed carbido atom lying 0.11 Å below the basal plane of the square pyramidal Ru_5 unit. Reaction with MeCN or with halide ions gives 1:1 adducts containing bridged butterfly arrangements of metal atoms,[62] and carbonylation of $[M_5C(CO)_{15}]$ (where M = Ru or Os) yields $[M_5C(CO)_{16}]$.[63] A far larger undeca-osmium cluster species $[Os_{11}C(CO)_{27}]$ was isolated as a low yield product of the vacuum pyrolysis of $Os_3(CO)_{12}$.[64] The novel metal core geometry is not readily explained by the Wade skeletal electron counting method, but can be rationalised using the Mingos approach.[24]

The synthesis of NO containing clusters is currently receiving much attention, partly because of the greater reactivity of such derivatives compared with many saturated carbonyl clusters. Thus the reaction of $[Ru_6C(CO)_{17}]$ with $PPN(NO_2)$ in THF leads to the formation of the first hexaruthenium nitrosyl cluster species $PPN[Ru_6C(CO)_{15}NO]$ and $[Ru_6C(CO)_{14}(NO)_2]$. The latter contains terminal NO ligands attached to opposite vertices of a carbido-centred octahedral Ru_6 core.[65]

Despite being one of the most useful nitrosylating reagents, the reactivity of NO^+ with metal carbonyls is frequently difficult to control, and the products difficult to predict. This is well demonstrated in a paper summarising the reactions of mono- and poly-nuclear carbonyl anions with NO^+, which gave $[CoRu_3N(CO)_{12}]$ and $[FeCo_2(NH)(CO)_9]$ as just two of the products.[66] Other nitrido cluster carbonyls were isolated from the reaction of $[M_4(CO)_{12}H_3]^-$ (M = Ru or Os) with NO^+, and the structures of $[Ru_4(\mu\text{-}H)_3(CO)_{11}(\mu_4\text{-}N)]$ and $[Os_4(CO)_{12}(\mu_4\text{-}N)]^-$ were established by *X*-ray analysis.[67] The interaction of $Ru_3(CO)_{12}$ with $PPN(N_3)$ gave the isocyanto carbonyl clusters $[Ru_3(NCO)(CO)_{11}]^-$ and $[Ru_3(NCO)(CO)_{10}]^-$, which slowly form $[Ru_4(NCO)(CO)_{13}]^-$ at room temperature and $[Ru_6N(CO)_{16}]^-$ in refluxing THF.[68]

Although the first low-valent metal cluster containing an exposed oxygen atom was characterised more than a decade ago, few other examples have been reported in the interim period. In 1983 two additional such complexes have been synthesised and structurally characterised. $[Fe_3(CO)_9(\mu_3\text{-}O)]^{2-}$ is formed in almost quantitative yield from the reaction of $[Fe_3(CO)_{11}]^{2-}$ with O_2,[69] and a new "raft" complex, $[Os_6(\mu_3\text{-}CO)(CO)_{18}(\mu_3\text{-}O)]$ results from the reaction of $Os_6(CO)_{20}$ with O_2.[70] Several new Os and Ru cluster carbonyls containing 3, 4, 6 and 7 metal atoms and μ_3- or μ_4-S ligands have been reported,[71] and some interesting observations made on the key role played by the sulphur atoms in the initial linking of the low nuclearity clusters, and in subsequent structural re-arrangements of the products. A μ_4-S atom is also crucial in the formation of the first structurally characterised chromium carbonyl cluster complex.[72] De-protonation of $[Cr_2(CO)_{10}SH]^-$ results in enhanced nucleophilicity for the product $[Cr_2(CO)_{10}S]^{2-}$, which gives $[(\mu_2\text{-}CO)_3(CO)_9Cr_3(\mu_4\text{-}S)Cr(CO)_5]^{2-}$ on treatment with

$Cr(CO)_5THF$. The anion contains a Cr_3 triangle [Cr-Cr, 2.850(9) Å] capped by an $SCr(CO)_5$ group.[72]

5 Metal Carbonyl Hydrides

Matrix isolation studies have provided further evidence that hydrogen bears a negative charge on some metal carbonyl hydrides,[73] in spite of the acidity of the molecules in polar solvents.[74]   The catalytic effect of base on the formation of $Co(CO)_4H$ from $Co_2(CO)_8$ has been examined,[75] and the pronounced increase in yield of $Co(CO)_4H$ noted when reductive carbonylation of basic cobalt carbonate is carried out in the presence of butyric or acetic acid.[76]

Several reports on polynuclear rhenium carbonyl hydrides have appeared.[77] The unsaturated cluster anion $[Re_3(\mu-H)_4(CO)_{10}]^-$ reacts with I_2 in donor solvents such as ethanol to form $[Re_3(\mu-H)_3(\mu-I)(CO)_{10}]^-$ only, but in CH_2Cl_2 stepwise degradation occurs leading to $[Re_2(\mu-I)_3(CO)_6]^-$ and $[Re_2(\mu-I)_2(CO)_8]$ via $[Re_3(\mu-H)_2(\mu-I)_2(CO)_{10}]^-$ and $[Re_2(\mu-H)(\mu-I)_2(CO)_6]^-$, both of which have been characterised by X-ray analysis.   Thermal decomposition of $[Re(CO)_4H_2]^{2-}$ salts gives moderate yields of $[Re_6(\mu_3-H)_2C(CO)_{18}]^{2-}$, which appears to be the first octahedral carbonyl cluster of rhenium.[78]   Structures of several new Ru and Os hydrido carbonyl clusters have been reported, including the first non-carbido decaosmium cluster $[Os_{10}H_4(CO)_{24}]^{2-}$ which can be prepared in low yields by the thermal decomposition of several Os_3 cluster carbonyls.[79]

6 Metal Carbonyl Halides

Several new and convenient routes to carbonyl halides and their anions have been reported.   Reaction of SO_2Cl_2 with $M_2(CO)_{10}$ (M = Mn or Re) in inert solvents gives $M(CO)_5Cl$ in very high yields.[80] $Re(CO)_5Cl$ is also formed on high pressure carbonylation of $[Re_2Cl_8]^{2-}$ in MeCN, and the X-ray structure of the product has been reported.[81]   An improved synthesis of the $[Pt_2(CO)_2X_4]^{2-}$ anions (X = Cl, Br or I) has been found and some detailed spectroscopic studies carried out,[82] and the quantitative conversion of cis-$[Os(CO)_4Cl_2]$ to $[Os(CO)_3Cl_2]_2$ observed, following the preparation of the former from $OsCl_3$.[83]   The mixed metal cluster compounds $[MOs_3(\mu-H)_2(\mu-Cl)(CO)_2]$ (M = Ir or Rh) were formed on reaction of $[MCl(CO)_2(NH_2C_6H_4Me-4)]$ with $[Os_3(\mu-H)_2(CO)_{10}]$, and the structure of the Ir complex determined.[84]   Reaction of the monoanion $[Os_8H(CO)_{22}]^-$ with I_2 to afford $[Os_8H(CO)_{22}I]$, results in an unprecedented change in overall cluster geometry which cannot be explained by current skeletal electron counting procedures.[85]   The lack of an interstitial atom to stabilise the metal core geometry may account for the great flexibility observed in these Os_8 systems compared to carbide containing Os_{10} systems.

References

1 E.L.Muetterties and M.J.Krause, Angew. Chem. Int. Ed. Engl., 1983, 22, 135; R.D. Adams, Acc. Chem. Res., 1983, 16, 67; M.Franck-Neumann, Pure Appl. Chem., 1983, 55, 1715.
2 D.F.Schriver, Chem. Brit., 1983, 19, 482; L.C.Costa, Catal. Rev., 1983, 25, 325.
3 D.S.Moore and S.D.Robinson, Chem. Soc. Rev., 1983, 12, 415.
4 J.D.Atwood, M.J.Wovkulich and D.C.Sonnenberger, Acc. Chem. Res., 1983 16, 350; J.A.S.Howell and P.M.Burkinshaw, Chem. Rev., 1983, 83, 557.
5 A.Pöe, Chem. Brit., 1983, 19, 997.
6 A.R.Manning, Coord. Chem. Rev., 1983, 51, 41.
7 K.K.Pandey, Coord. Chem. Rev., 1983, 51, 69; H.W.Roesky and K.K.Pandey, Adv. Inorg. Chem. Radiochem., 1983, 26, 337.
8 B.F.G.Johnson, J.Lewis, W.J.H.Nelson, J.N.Nicholls and M.D.Vargas, J. Organomet. Chem., 1983, 249, 255.
9 J.S.Bradley, Adv. Organomet. Chem., 1983, 22, 1.
10 H.Vahrenkamp, Adv. Organomet. Chem., 1983, 22, 169.
11 G.F.P.Warnock, J.Sprague, L.Kristi and J.E.Ellis, J. Am. Chem. Soc., 1983, 105, 672.
12 J.T.Lin, G.P.Hagen and J.E.Ellis, J. Am. Chem. Soc., 1983, 105, 2296; idem, Organometallics, 1983, 2, 1145; G.L.Rochfort and J.E.Ellis, J. Organomet. Chem., 1983, 250, 265 and 277.
13 J.T.Lin and J.E.Ellis, J. Am. Chem. Soc., 1983, 105, 6252.
14 Y-S.Chen and J.E.Ellis, J. Am. Chem. Soc., 1983, 105, 1689.
15 F.M.Ashmawy, C.A.McAuliffe, K.L.Mintin, P.V.Parish and J.Tames, J. Chem. Soc., Chem. Commun., 1983, 436.
16 T.W.Bodnar and A.R.Cutler, J. Am. Chem. Soc., 1983, 105, 5926.
17 E.D.Morrison, G.R.Stemmetz, G.L.Geoffroy, W.C.Fultz and A.L.Rheingold, J. Am. Chem. Soc., 1983, 105, 4104; J.R.Shapley, D.S.Strickland, G.M.St.George, M.R.Churchill and C.Bueno, Organometallics, 1983, 2, 185.
18 J.W.Kolis, E.M.Holt and D.F.Shriver, J. Am. Chem. Soc., 1983, 105, 7307.
19 S.G.Shore, D-Y.Jan, L-Y.Hsu and W-L.Hsu, J. Am. Chem. Soc., 1983, 105, 5923.
20 L.J.Farrugia, M.Green, D.R.Hankey, G.A.Orpen and F.G.A.Stone, J. Chem. Soc., Chem. Commun., 1983, 310.
21 S.Aime, R.Gobetto, D.Osella, L.Milone, G.E.Hawkes and E.W.Randall, J. Chem. Soc., Chem. Commun., 1983, 794.
22 S.L.Cook, J.Evans, G.N.Greaves, B.F.G.Johnson, J.Lewis, P.R.Raithby, P.B.Wells and P.Worthington, J. Chem. Soc., Chem. Commun., 1983, 777.
23 R.Hoffman, C.N.Wilker and S.J.Lippard, J. Amer. Chem. Soc., 1983, 105, 146.
24 D.M.P.Mingos, J. Chem. Soc., Chem. Commun., 1983, 706; D.M.P.Mingos and D.G.Evans, J. Organomet. Chem., 1983, 251, C13.
25 D.G.Evans, J. Chem. Soc., Chem. Commun., 1983, 675.
26 H.Miessner, Z. Anorg. Allg. Chem., 1983, 505, 187.
27 N.M.Koskić and R.F.Fenske, Inorg. Chem., 1983, 22, 666.
28 D.Moncrieff, P.C.Ford, I.H.Hillier and V.R.Saunders, J. Chem. Soc., Chem. Commun., 1983, 1108.
29 G.F.Holland, M.C.Manning, D.E.Ellis and W.C.Trogler, J. Am. Chem. Soc., 1983, 105, 2308.
30 R.P.Brint, M.P.Collins, T.R.Spalding and F.A.Deeney, J. Organomet. Chem., 1983, 258, C57; R.P.Brint, K.O'Cuill, T.R.Spalding and F.A.Deeney, ibid., 247, 61.

31 P.T.Chesky and M.B.Hall, Inorg. Chem., 1983, 22, 2102, 2998 and 3327; G.Granozzi, R.Benoni, E.Tondello, M.Casarin, S.Aime and D.Osella, ibid., 3899.
32 J.W.Hershberger, C.Amatore and J.K.Kochi, J. Organomet. Chem., 1983, 250, 345.
33 T.C.Richmond, Q-Z.Shi, W.C.Trogler and F.Bazolo, J. Chem. Soc., Chem. Commun., 1983, 650.
34 T.R.Fletcher and R.N.Rosenfeld, J. Am. Chem. Soc., 1983, 105, 6358.
35 D.G.Leopold and V.Vaida, J. Am. Chem. Soc., 1983, 105, 6809.
36 A.M.Stolzenberg and E.L.Muetterties, J. Am. Chem. Soc., 1983, 105, 822; N.J.Coville, A.M.Stolzenberg and E.L.Muetterties, ibid., 2499; A.F.Hepp and M.S.Wrighton, ibid., 5934; H.Yesaka, T.Kobayashi, K.Yasufuku and S.Nagakura, ibid., 6249; C.Busetto, A.M.Mattucci, E.M.Cernia and R.Bertani, J. Organomet. Chem., 1983, 246, 183; S.W.Lee, L.F.Wang and C.P.Cheng, ibid., 248, 189.
37 C.Lapinte and D.Astruc, J. Chem. Soc., Chem. Commun., 1983, 430.
38 S.G.Davies, J.Hibberd and S.J.Simpson, J. Organomet. Chem., 1983, 246, C16; S.G.Davies and S.J.Simpson, ibid., 254, C29.
39 G.Smith, D.J.Cole-Hamilton, M.Thornton-Pett and M.B.Hursthouse, Polyhedron, 1983, 2, 1241.
40 D.B.Dombek and A.M.Harrison, J. Am. Chem. Soc., 1983, 105, 2485.
41 W.Tam, M.Marsi and J.A.Gladysz, Inorg. Chem., 1983, 22, 1413.
42 B.A.Narayanan, C.Amatore, C.P.Casey and J.K.Kochi, J. Am. Chem. Soc., 1983, 105, 6351.
43 D.L.Grimmett, J.A.Labinger, J.N.Bonfiglio, S.T.Masuo, E.Shearin and J.S.Miller, Organometallics, 1983, 2, 1325.
44 A.Cameron, V.H.Smith and M.Baird, Organometallics, 1983, 2, 465.
45 F.Calderazzo, U.Englert, G.Pampaloni, G.Pelizzi and R.Zamboni, Inorg. Chem., 1983, 22, 1865.
46 M.B.Simpson, M.Poliakoff, J.J.Turner, W.B.Maier II and J.G.McLaughlin, J. Chem. Soc., Chem. Commun., 1983, 1355; J.J.Turner, M.B.Simpson, M.Poliakoff, W.B.Maier II and M.A. Graham, Inorg. Chem., 1983, 22, 911; J.J.Turner, M.B.Simpson, M.Poliakoff and W.B.Maier II, J. Am. Chem. Soc., 1983, 105, 3898.
47 K.S.Suslick, J.W.Goodale, P.F.Schubert and H.H.Wang, J. Am. Chem. Soc., 1983, 105, 5781.
48 R.E.Stevens and W.L.Gladfelter, Inorg. Chem., 1983, 22, 2034.
49 K.H.Pannell, Y-S.Chen, K.Belknap, C.C.Wu, I.Bernal, M.W.Creswick and H.N.Huang, Inorg. Chem., 1983, 22, 418; K.L.Kjare and J.E.Ellis, J. Am. Chem. Soc., 1983, 105, 2303; R.Mews and C.Liu, Angew. Chem., Int. Ed. Engl., 1983, 22, 162.
50 P.C.Leung and P.Coppens, Acta Crystallogr., 1983, B39, 535.
51 F.W.B.Einstein, R.K.Pomeroy, P.Rushman and A.C.Willis, J. Chem. Soc., Chem. Commun., 1983, 854.
52 F.Oldani and G.Bor, J. Organomet. Chem., 1983, 246, 309.
53 G.Al-Takhin, J.A.Connor and H.A.Skinner, J. Organomet. Chem., 1983, 259, 313.
54 J.L.Vidal and R.C.Schoening, J. Organomet. Chem., 1983, 241, 395.
55 A.Fumagalli, S.Martinengo, G.Ciani and A.Sironi, J. Chem. Soc., Chem. Commun., 1983, 453.
56 A.Fumagalli, S.Martinengo and G.Ciani, J. Chem. Soc., Chem. Commun., 1983, 1381.
57 B.A.Sosinsky, R.G.Shong, B.J.Fitzgerald, N.Norem and C.O'Rourke, Inorg. Chem., 1983, 22, 3124; R.B.Petersen, J.M.Rogosta, G.E.Whitwell and J.M.Burlitch, ibid., 3407; P.Braunstein, J.Rose. Y.Dusausoy and J-P.Mangeot, J. Organomet. Chem., 1983, 256, 125.
58 A.A.Bhattacharyya, C.C.Nagel and S.G.Shore, Organometallics, 1983, 2, 1187.

59 M.I.Bruce, J.G.Matisons and B.K.Nicholson, J. Organomet. Chem., 1983, 247, 321.

60 V.G.Albano, D.Braga, P.Chini, D.Strumolo and S.Martinengo, J. Chem. Soc., Dalton Trans., 1983, 249; D.Strumolo, C.Seregni, S.Seregni, V.G.Albano and D.Braga, J. Organomet. Chem., 1983, 252, C93.

61 B.T.Heaton, L.Strona, S.Martinengo, D.Strumolo, V.G.Albano and D.Braga, J. Chem. Soc., Dalton Trans., 1983, 2175.

62 B.F.G.Johnson, J.Lewis, J.N.Nicholls, J.Puga, P.R.Raithby, M.J.Rosales, M.McPartlin and W.Clegg, J. Chem. Soc., Dalton Trans., 1983, 277.

63 B.F.G.Johnson, J.Lewis, W.J.H.Nelson, J.N.Nicholls, J.Puga, P.R.Raithby, M.J.Rosales, H.Schröder and M.D.Vargas, J. Chem. Soc., Dalton Trans., 1983, 2447.

64 D.Braga, K.Henrick, B.F.G.Johnson, J.Lewis, M.McPartlin, W.J.H.Nelson, A.Sironi and M.D.Vargas, J. Chem. Soc., Chem. Commun., 1983, 1131.

65 B.F.G.Johnson, J.Lewis, W.J.H.Nelson, J.Puga and P.R.Raithby, J. Organomet. Chem., 1983, 243, C13; B.F.G.Johnson, J.Lewis, W.J.H.Nelson, J.Puga, M.McPartlin and A.Sironi, ibid., 253, C5.

66 D.E.Fjare, D.G.Keyes and W.L.Gladfelter, J. Organomet. Chem., 1983, 250, 383.

67 M.A.Collins, B.F.G.Johnson, J.Lewis, J.H.Mace, J.Morris, M.McPartlin, W.J.H.Nelson, J.Puga and P.R.Raithby, J. Chem. Soc., Chem. Commun., 1983, 689.

68 D.E.Fjare, J.A.Jensen and W.L.Gladfelter, Inorg. Chem., 1983, 22, 1774.

69 A.Ceriotti, L.Resconi, F.Demartin, G.Longoni, M.Manassero and M.Sansoni, J. Organomet. Chem., 1983, 249, C35.

70 R.J.Gondsmit, B.F.G.Johnson, J.Lewis, P.R.Raithby and K.H.Whitmire, J. Chem. Soc., Chem. Commun., 1983, 246.

71 R.D.Adams, I.T.Horvarth and L-W.Yang, J. Am. Chem. Soc., 1983, 105, 1533; idem, Organometallics, 1983, 2, 1257; R.D.Adams, D.Männig and B.E.Segmüller, Organometallics, 1983, 2, 149; R.D.Adams, I.T.Horvath, P.Mathur, B.E.Segmüller and L-W.Yang, ibid., 1078; R.D.Adams and L-W.Yang, J. Am. Chem. Soc., 1983, 105, 235.

72 M.Hoefler, K-F.Tebbe, H.Veit and N.E.Weiler, J. Am. Chem. Soc., 1983, 105, 6338.

73 R.L.Sweany and J.W.Owens, J. Organomet. Chem., 1983, 255, 327.

74 H.W.Walker, R.G.Pearson and P.C.Ford, J. Amer. Chem. Soc., 1983, 105, 1179.

75 A.Sisak, F.Ungvary and L.Marko, Organometallics, 1983, 2, 1244.

76 N.S.Imyanitov and B.E.Kuvaev, Zh. Obsch. Khim., 1983, 53, 1686.

77 G.Ciani, G.D'Alfonso, P.Romiti, A.Sironi and M.Freni, Inorg. Chem., 1983, 22, 3115; idem, J. Organomet. Chem., 1983, 254, C37.

78 G.Ciani, G.D'Alfonso, P.Romiti, A.Sironi and M.Freni, J. Organomet. Chem., 1983, 244, C27.

79 D.Braga, J.Lewis, B.F.G.Johnson, M.McPartlin, W.J.H.Nelson and M.D.Vargas, J. Chem. Soc., Chem. Commun., 1983, 241; J.A.Jensen, D.E.Fjare and W.L.Gladfelter, Inorg. Chem., 1983, 22, 1250.

80 A.R.Manning, G.McNally, R.Davis and C.C.Rowland, J. Organomet. Chem., 1983, 259, C15.

81 F.A.Cotton, L.M.Daniels and C.D.Schmulbach, Inorg. Chim. Acta, 1983, 75, 163; F.A.Cotton and L.M.Daniels, Acta Crystallogr., 1983, C39, 1495.

82 N.M.Boag, P.L.Goggin, R.J. Goodfellow and I.R.Herbert, J. Chem. Soc., Dalton Trans., 1983, 1101.

83 R.Psaro and C.Dossi, Inorg. Chim. Acta, 1983, 77, L255.

84 L.J.Farrugia, G.A.Orpen and F.G.A.Stone, Polyhedron, 1983, 1, 171.

85 B.F.G.Johnson, J.Lewis, W.J.H.Nelson, M.Vargas, D.Braga and M.McPartlin, J. Organomet. Chem., 1983, 249, C21.

9

Organometallic Compounds containing Metal–Metal Bonds

BY W. E. LINDSELL

1 Introduction

The continued expansion in the literature of metal-metal bonded organometallic compounds has necessitated some selection of the material included in this chapter. A short bibliography of some additional papers is appended.

1.1 Reviews.- General accounts, primarily of binuclear compounds, include discussions of multiple metal-metal bonds[1] and of metal-metal and metal-carbon bonds in the chemistry of molybdenum and tungsten alkoxides.[2] Volume 22 of Advances in Organometallic Chemistry includes reviews by Vahrenkamp on basic metal cluster reactions[3a] and by Bradley on carbidocarbonyl clusters[3b] as well as accounts of vinylidene and propadienylidene (allenylidene) metal complexes[3c] and of isocyanide complexes[3d] which encompass bi- and poly-nuclear examples. An outline of rational approaches to cluster chemistry has appeared[5] and more specialised reviews cover the topics: (a) alkyne substituted homo- and hetero-metallic clusters of Fe, Co and Ni triads[5]; (b) carbido clusters of Ru and Os[6]; (c) chemistry of triosmium carbonyl clusters.[7] Metal-metal bonded compounds are also discussed in reviews of bridged hydrocarbyl or hydrocarbon binuclear transition metal complexes[8], of methylene bridged organometallics[9] and of complexes of dppm.[10] Surveys have appeared of heteronuclear metal-metal bonded derivatives reported in 1980-81.[11]

A brief account of the kinetic behaviour of metal-metal bonded carbonyls[12] has been published and, also, a mechanistic review of reactions of electrophiles with polynuclear carbonyl complexes.[13]

1.2 Theoretical Studies.- The nature of metal-metal bonds in bridged bi- and poly-nuclear systems has continued to be a subject of study. Dimers doubly bridged by π-acceptor ligands may be classified into three geometrically distinct groups and it is concluded that the major factor determining the separation of metals is bonding through the bridging unit, with direct metal-metal interaction being negligible or anti-bonding.[14] Fenske-Hall type calculations on $[Pd_2Cl_4(\mu\text{-}CO)_2]^{2-}$ also indicate no genuine Pd-Pd bond.[15] An electronic, rather than steric, rationalisation for the occurrence of μ_2- and μ_3-CO groups in clusters invokes the use of 2 or 3 σ-donor orbitals by CO in its interaction with metals.[16]

Calculations on triply bonded d^3 species $[M_2Me_6]$ give M≡M bond energies 440 or 276 $kJmol^{-1}$ for M = Mo or Cr, respectively.[17] In $[Fe_2(CO)_6(\mu\text{-}S_2)]$, new calculations support only the presence of a single "bent" Fe-Fe bond.[18]

Further theoretical and U.V. photoelectron (P.E.) spectroscopic studies of clusters have been reported. Investigations of clusters $[Co_{3-\underline{n}}Fe_{\underline{n}}(CO)_9(\mu\text{-}H)_{\underline{n}-1}(\mu_3\text{-}S)]$ ($\underline{n}$ = 0-3), $[M_3(CO)_9(\mu\text{-}H)_2(\mu_3\text{-}S)]$ (M = Ru, Os) and $[Os_3(CO)_{10}(\mu\text{-}X)(\mu\text{-}Y)]$ (Y = H, Y = Cl, Br, I, X = Y = Cl, Br I) conclude that (μ-H) groups substantially disrupt direct M-M bonds and that, in the $Os(\mu\text{-}Y)_2Os$ systems, the direct Os-Os interaction is weakly antibonding;[19] P.E. spectra also indicate localisation of valence electrons on ionisation of $Co_{3-n}Fe_n$ clusters. Related studies have appeared on alkyne complexes $[Ni_2Cp_2(\mu\text{-}C_2R_2)]$,[20a] $[Co_2(CO)_6(\mu\text{-}C_2R_2)]$,[21] $[Fe_3(CO)_9(\mu_3\text{-}C_2Et_2)]$,[20b] $[Ru_4(CO)_{12}(\mu_4\text{-}C_2R_2)]$,[20c] $[Co_4(CO)_{10}(\mu_4\text{-}C_2R_2)]$,[21] on $[Co_3(CO)_9(\mu_3\text{-}CR)]$[21] and on alkynyl complexes $[M_3(CO)_9(\mu\text{-}H)(\mu_3\text{-}C_2R)]$ (M = Ru, Os).[22] The bonding of the methyne fragment in "butterfly" cluster $[Fe_4(CO)_{14}(\mu\text{-}H)(\mu\text{-}\eta^2\text{-}CH)]$ has also been theoretically examined.[23]

The polyhedral sketalal electron pair theory has been extended to provide rules which predict geometries of lower symmetry clusters incorporating non-conical $M(CO)_4$ fragments.[24] Also, this theory has been generalised to apply to condensed clusters joined by vertex, edge or triangular face,[25a] including condensed clusters of platinum.[25b] A simple model involving a conductive surface, topologically equivalent to a superaromatic polyhedral system, can also extend the electron counting scheme to some condensed clusters.[26] The two isomers of $[Fe_2Mo_2(CO)_8Cp_2(\mu_3\text{-}S)_2]$ can be rationalised by the electron counting procedure as two Mo_2FeS tetrahedra sharing common Mo-Mo edges.[27] Close packed, high nuclearity clusters have been shown to obey an extension of the Hume-Rothery rule for metallic lattices.[28]

1.3 Physical Studies.- Characterisation of high nuclearity osmium clusters by EXAFS using Os L (111) edge X-ray absorptions requires the determination of bonded and nonbonded Os-Os distances;[29] this technique has also been applied to surface supported Os_3 clusters.[30] Analyses of vibrational spectra of $[M_4(CO)_{12}(\mu\text{-}H)_4]$ (M = Ru, Os),[31] $[Os_3(CO)_{10}(\mu\text{-}H)(\mu\text{-}CHCH_2)]$,[32] $[Os_3(CO)_9(\mu\text{-}H)_2(\mu_3\text{-}CCH_2)]$[32] and $[Os_3(CO)_9(\mu\text{-}CO)(\mu_3\text{-}C_2H_2)]$[33] have been reported. X-Ray and neutron diffraction studies on $[Ru_4(CO)_8\{P(OMe)_3\}_4(\mu\text{-}H)_4]$ characterise the vibrational behavior of H atoms as being consistent with a symmetric 3 centre 2 electron Ru-H-Ru interaction.[34] ^{2}H.N.m.r. of $[Co_3Fe(CO)_{12}(\mu_3\text{-}^2H)]$ has demonstrated the use of this technique to study hydrides of metals possessing quadrupolar nuclei,[35] and direct ^{103}Rh and $^{13}C\{^{103}Rh\}$ n.m.r. has been employed in studying Rh_4 clusters.[36] Iron containing carbonyl clusters give ^{57}Fe Mössbauer spectra for which quadrupole splittings correlate with type of $Fe(CO)_{\underline{n}}$ unit present.[37]

1.4 Surface bound species.- Surface bound clusters of catalytic importance have been studied by IR and UV spectroscopy, by EXAFS in some cases, and by chemical techniques, including: $[Fe_3(CO)_{10}H(OAl\lessdot)]$ on γ-Al_2O_3;[38] $[Ru_3(CO)_{10}H(OSi\lesseqgtr)]$ on SiO_2;[39] $[Fe_4Cp_4(\mu_3\text{-}CO)_4]$, $[Ni_3Cp_3(\mu_3\text{-}CO)_2]$ or $[Ni_2Cp_2(\mu\text{-}CO)_2]$ bound through carbonyl O atom to Al_2O_3;[40] $Al^+[RuOs_3(CO)_{12}H_3]^-$ formed from $[RuOs_3(CO)_{13}(\mu\text{-}H)_2]$ and Al_2O_3.[41] Specific tethering to oxide surfaces via phosphine ligand {L = $PPh_2(CH_2)_2Si(OEt)_3$} is reported for $[Ru_4(CO)_8H_4L_4]$ on SiO_2,[42] $[Ru_3(CO)_{11}L]$, $[Ru_4(CO)_{11}H_4L]$, $[Ru_5(CO)_{14}(C)L]$ or $[Ru_6(CO)_{16}(C)L]$ on various oxides[43] and $[M_4(CO)_8\{(PPh_2)_3CH\}L]$ (M = Rh, Ir) on SiO_2.[44] Acid-base reactions between hydridic clusters and aminated silica have afforded ionic materials such as $\{SIL(CH_2)_3NH_3{}^+[FeCo_3(CO)_{12}]^-\}$.[45] $[Pt_3(PPh_3)_4(\mu\text{-}CO)_3]$ transforms into $[Pt_5(CO)(PPh_3)_4(\mu\text{-}CO)_5]$ upon impregnation into inorganic oxide supports.[46]

Other papers on surface bond clusters may be found in the proceedings of the 4th international symposium on the relationship between homogeneous and heterogeneous catalysis (Asilomar, California, October 1983).[47]

2 Compounds Containing Homonuclear Bonds between Transition Metals

2.1 Vanadium, Niobium, and Tantalum.- Reactions of $[V_2(CO)_5Cp_2]$ with alkynes include the coupling of electron poor alkynes to form mononuclear η-cyclobutadiene complexes.[48] Complex $[V_2Cp_2(\mu\text{-}\eta^5,\eta^5\text{-cot})]$ possesses a short V-V bond {2.439(1)Å} and is reduced to an anion-radical characterised by e.s.r. spectroscopy.[49] The fluxional, benzene complex $[V_2Cp_2(\mu\text{-}H)_2(\mu\text{-}\eta^4,\eta^4\text{-}C_6H_6)]$ {V-V = 2.425(1)Å} is formed from $K[VCp_2]$ and 1,3-chd.[50] X-Ray analysis of $[V_2(\eta\text{-}C_5H_4\text{-}Pr^i)_2(\mu\text{-}S)_2(\mu\text{-}S_2)]$ indicates the presence of an $(\mu\text{-}\eta^1,\eta^1\text{-}S_2)$ group whereas the hexafluorobut-2-yne adduct $[V_2Cp_2\{\mu\text{-}S_2C_2(CF_3)_2\}(\mu\text{-}S_2)]$ has an $(\mu\text{-}\eta^2,\eta^2\text{-}S_2)$ bridge.[51]

Fulvalene niobium dimers $[Nb_2Cp_2(\mu\text{-}C_{10}H_8)(\mu\text{-}X)]$ (X = NR, Br, SCN) have been prepared.[52]

Reactions of $[Ta_2Cl_4Cp'_2(\mu\text{-}H)_2]$ ($Cp' = C_5Me_4Et$) with MeCN forms $[Ta_2Cl_3Cp'_2(\mu\text{-}H)(\mu\text{-}NCHMe)(\mu\text{-}Cl)]$, characterised by X-ray diffraction;[53] details of the chemistry of $[Ta_2Cl_4Cp'_2(\mu\text{-}H)(\mu\text{-}CHO)]$, formed by reaction of the former complex with CO, are also presented.[54]

2.2 Chromium, Molybdenum, and Tungsten.- The cyclooctatetraene species $[Cr_2Cp_2(\mu\text{-}\eta^5,\eta^5\text{-cot})]$, an analogue of the vanadium compound above, has a formal Cr=Cr bond {2.390(2)Å}.[55] Activation of the CH bond of the metallacyclononatetraene in $[Mo_2Cp_2(\mu\text{-}C_8Me_8)]$ via 2e oxidation with CF_3SO_3H or $FeCp_2{}^+$ gives (1), whereas oxidation with $SbCl_5$ produces (2).[56]

Matrix isolation studies (12-77K) present evidence for primary CO dissociation on photolysis of dimers $[M_2(CO)_6Cp_2]$ (M = Mo, W) and IR spectra suggest an intermediate $[M_2(CO)_4Cp_2(\mu\text{-}\eta^1,\eta^2\text{-}CO)]$.[57] Photolysis of $[WH(CO)_3Cp]$ produces dimeric $[W_2(CO)_4Cp_2(\mu\text{-}H)_2]$, and a related molybdenum species may be formed from $[Mo_2(CO)_4(\eta\text{-}C_5Me_5)_2]$ and dihydrogen.[58] The stable paramagnetic monomer $[Mo(CO)_3$-(HBpz)] dimerises thermally to $[Mo_2(CO)_4(HBpz)_2]$ (Mo≡Mo = 2.507(1)Å)[59] in which the formal triple bond is longer than in $[Mo_2(CO)_4Cp'_2]$ (Cp′ = Cp or C_5Me_5[60]).

The μ-allylidene ligand in $[Mo_2(CO)_4Cp_2(\mu\text{-}CHCHCMe_2)]$ (from $[Mo_2(CO)_4Cp_2]$ and 3,3-dimethylcyclopropene) isomerises on heating to an alkyne or isopropene group in $[Mo_2(CO)_4Cp_2(\mu\text{-}HC_2Pr^i)]$ or $[Mo_2(CO)_2(\mu\text{-}\eta^4\text{-}CH_2CHCMeCH_2)]$, respectively, and also forms trinuclear complex (3).[61] Reaction of the μ-allylidene complex with alkynes has also been studied: HC_2Bu^t affords an unusually bound μ-cyclohexadienone ligand.[62] The structure of cation $[Mo_2(CO)_4(OH_2)Cp_2(\mu\text{-}CHCHPh)]^+$, obtained by protonation of $[Mo_2(CO)_4Cp_2(\mu\text{-}HC_2Ph)]$, is reported.[63]

Many reactions of compounds $[M_2(CO)_4Cp'_2]$ (M = Mo, W; Cp′ = Cp, C_5Me_5) have been described. 9-Diazafluorene reacts to form alkylidene bridged species $[Mo_2(CO)_4Cp_2(\mu\text{-}\overline{CC_6H_4C_6H_4})]$ {Mo–Mo = 2.798(1)Å},[64] a dimetallacyclopropene. Reactions with diazoalkanes and organic azides are reported by two groups: 2-diazopropane affords $[Mo_2(CO)_4(\eta\text{-}C_5R_5)_2(\mu\text{-}N_2CMe_2)]$ (R = H) which exists in two interconverting isomeric forms;[65] the analogous derivative (R = Me) thermally isomerises to isocyanate $[M_2(NCO)(CO)_3(\eta\text{-}C_5Me_5)_2(\mu\text{-}NCMe_2)]$ (M = Mo, Mo=Mo = 2.745(2)Å)[66a] and the tungsten isocyanato-complex (M = W) is formed directly from Me_2CN_2 at < 0°C.[66b] Aryl azides form a labile 1:1 adduct[65] with $[M_2(CO)_4Cp_2]$ but further reaction produces nitrene compounds $[M_2(CO)_2(NR)Cp_2(\mu\text{-}CONRN_2)]$, characterised by X-ray diffraction (M = Mo) when Cp′ = Cp, R = 4-$Bu^tC_6H_4$[65] or Cp′ = C_5Me_5, R = Et.[67] One product from reaction between MeNCS and $[Mo_2(CO)_4$-$(\eta\text{-}C_5Me_5)_2]$ is the fluxional, bridged isocyanide $[Mo_2(CO)_4(\eta\text{-}C_5Me_5)_2(\mu\text{-}\eta^2\text{-}CNMe)]$ {Mo–Mo = 3.240(3)Å}[68a] and a similar derivative accompanies asymmetrically bridged product, $[Mo_2(CO)_2Cp_2(\mu\text{-}RNCNR)]$, from reactions with carbodiimides.[68b] Other products from $[Mo_2(CO)_4Cp_2]$, characterised by X-ray diffraction, include $[Mo_2(CO)_4Cp_2(\mu\text{-}PCBu^t)]$,[69] $[Mo_2(CO)_4Cp_2\{\mu\text{-}SCO(CH_2)_3\}]$[70] and $[Mo_2(CO)_4Cp_2(\mu\text{-}SPC_6H_4$-OMe-4)][71] formed from Bu^tCP, thiobutyrolactone and $[4\text{-}MeOC_6H_4P(S)S]_2$ (Lawesson's reagent), respectively.

Electrophilic addition of halogens to $[Mo_2(CO)_4Cp_2]$ gives products $[Mo_2(CO)_4$-$Cp_2(\mu\text{-}X)(\mu\text{-}Y)]$ (e.g. X = Y = I) with no Mo–Mo bond whereas hydrogen halides give adducts with shorter Mo–Mo distances {e.g. X = H, Y = I: 3.310(2)Å}.[72] Complete decarbonylation of $[Mo_2(CO)_4Cp_2]$ occurs on reaction with aromatic nitro or nitroso compounds to form $[Mo_2O_2Cp_2(\mu\text{-}NAr)(\mu\text{-}O)]$ {e.g. R = p-tol; Mo–Mo = 2.6497(5)Å}.[73] The W≡C triple bond of $[W(CR)(CO)_2Cp]$ reacts with $[M_2(CO)_nCp_2]$ (n = 4 or 6; M = Mo, W) to form homo- or hetero-nuclear clusters $[M_2W(CO)_6Cp_3$-$(\mu_3\text{-}CR)]$ (R = p-tol).[74]

Reduction of species $[M(CO)_3(Me_5dien)]$ (M = Mo, W) yields coordinatively unsaturated anions $[M_4H_4(CO)_{12}]^{4-}$ for which tetrahedral geometry has been established (M = Mo).[75]

Alcoholysis of organometallic amides $[Mo_2R_2(NMe_2)_4]$ gives dialkyl alkoxides when R = Me, e.g. $[Mo_2Me_2(OBu^t)_4py_2]$ {Mo≡Mo = 2.256(1)Å}, but when R = Et or Pr^n cooperative β-H atom elimination occurs across the bimetallic system to yield $[Mo_2R(OR')_5]$.[76] Thermolysis of $[Mo_2(CH_2SiMe_3)_4(OAr)_2]$ (Ar = 2,6-$Me_2C_6H_3$) in the presence of pyridine causes elimination of $SiMe_4$, via α-abstraction, and second α-H addition to the Mo-Mo bond to produce $[Mo_2(CH_2SiMe_3)_2(OAr)_2py_2(\mu\text{-}H)\text{-}(\mu\text{-}CSiMe_3)]$ {Mo=Mo: 2.380(2)Å}.[79]

Novel products are formed on carbonylation of $[W_2(OPr^i)_6py_2]$; a structurally characterised species (4) comprises two binuclear units bonded by O atoms of two $(\mu_2\text{-}CO)$ groups.[78] Several products arise from reactions of $[W_2(OR)_6]$ with alkynes $C_2R'_2$ depending on the nature of R and R': Ph_2C_2 with $[W_2(OBu^t)_6]$ at 70° yields dimers $[W_2(OBu^t)_4(\mu\text{-}C_2Ph_2)_2]$ and $[W_2(OBu^t)_4(\mu\text{-}CPh)_2]$;[79] Et_2C_2 at 75-80° gives $[\{W_3(OBu^t)_5(\mu\text{-}CEt)(\mu\text{-}O)\}_2(\mu\text{-}O)_2]$, shown by X-ray to have structure (5).[80] Alkynes R'_2C_2 (R' = H, Me) react with $[W_2(OR)_6]$ in the presence of pyridine to form $[W_2(OR)_6py_2(\mu\text{-}C_2R'_2)]$, $[W_2(OR)_6(\eta\text{-}C_2R'_2)(\mu\text{-}C_4R'_4)]$ or $[W(CR')(OR)_3]$ depending on R,R' and reaction conditions.[81]

2.3 Manganese and Rhenium.- There is good evidence for homolytic fission of the Mn-Mn bonds of $[Mn_2(CO)_8L_2]$ (L = PCy_3, PPh_3) in thermally activated reactions.[82] However, further evidence using isotopic labelling establishes that no significant M-M bond cleavage occurs in thermal substitution reactions of $[M_2(CO)_{10}]$ (M = Mn, Re) at temperatures up to 150°C;[83] the Mn-Mn bond energy must be greater than earlier estimates of 92 $kJmol^{-1}$.[83] In microcalorimetric measurements on reactions of $[Re_2(CO)_{10}]$ a value of 187 $kJmol^{-1}$ was taken for D(Re-Re).[84] Metal-metal bond cleavage does occur on photolysis of $[M_2(CO)_{10}]$ (M = Mn, Re) in solution.[83,85]

Substitution or alkyne insertion reactions of $[Mn_2(CO)_8(\mu\text{-}H)(\mu\text{-}PPh_2)]$ yield products $[Mn_2(CO)_{8-n}L_n(\mu\text{-}H)(\mu\text{-}PPh_2)]$ (n = 1,2) or $[Mn_2(CO)_7(\mu\text{-}CRCHR)(\mu\text{-}PPh_2)]$, characterised by X-ray diffraction for L = $CNBu^t$ or R = H, respectively.[86] Reaction of $[Re_2(CO)_8(\mu\text{-}H)(\mu\text{-}CHCHEt)]$ with 3,3-dimethylcyclopropene affords the μ-carbene complex $[Re_2(CO)_8(\mu\text{-}\eta^1,\eta^3\text{-}CHCHCMe_2)]$ which undergoes Re-Re bond cleavage on addition of CO.[87] The structures and substitution reactions of complexes $[Re_2(CO)_7L(\mu\text{-}H)(\mu\text{-}NC_5H_4)]$ (L = CO, Me_3NO; NC_5H_4 = α-pyridyl) are described.[88]

Diazomethane adds to a carbonyl ligand in $[Mn_2(CO)_4(\mu\text{-}CO)(\mu\text{-}dppm)_2]$ to form $[Mn_2(CO)_4(\mu\text{-}COCH_2N_2)(\mu\text{-}dppm)_2]$ and this reaction is reversed on photolysis or thermolysis.[89] A formally unsaturated dimer $[Mn_2(CO)_4(\mu\text{-}H)_2(\mu\text{-}dppm)_2]$ undergoes insertion reactions with MeCHO and MeNC;[90] a dimer with a formal Mn≡Mn bond $[Mn_2(\eta\text{-}C_5Me_5)_2(\mu\text{-}CO)_3]$ is also described.[91]

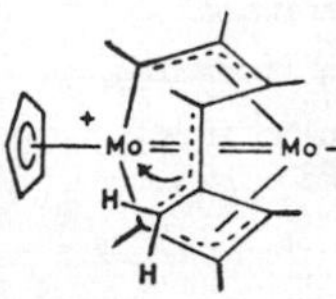

Mo
Mo
H
H

(1)

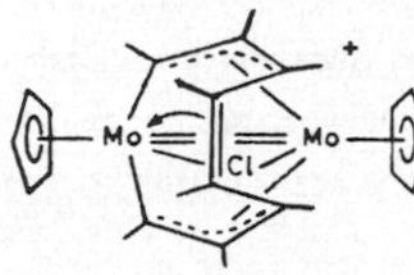

Mo
Mo
Cl

(2)

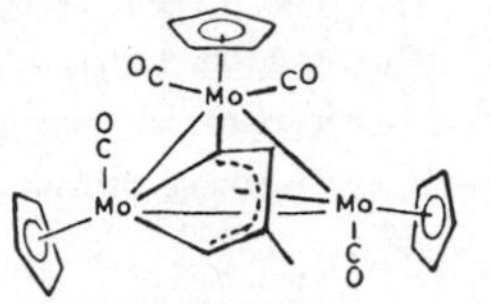

OC
Mo
CO
O
C
Mo
Mo
C
O

(3)

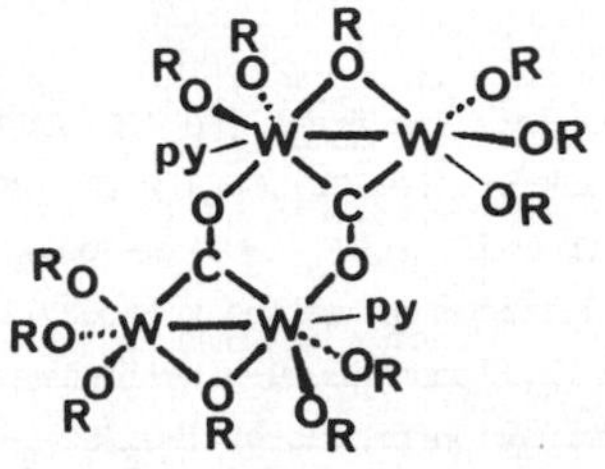

R
O
R
O
OR
RO
py
W
W
OR
O
C
OR
RO
C
O
RO
W
W
py
OR
R
O
O
R
OR

(4)

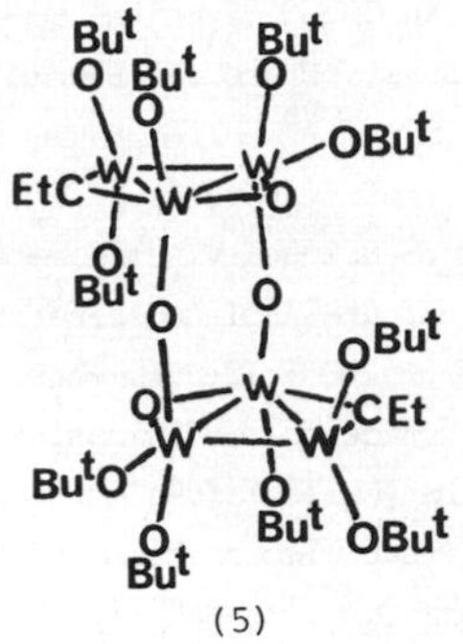

Bu^t
O
Bu^t
O
Bu^t
O
OBu^t
EtC
W
W
W
O
O
Bu^t
O
O
OBu^t
W
CEt
W
W
Bu^tO
O
Bu^t
OBu^t
O
Bu^t

(5)

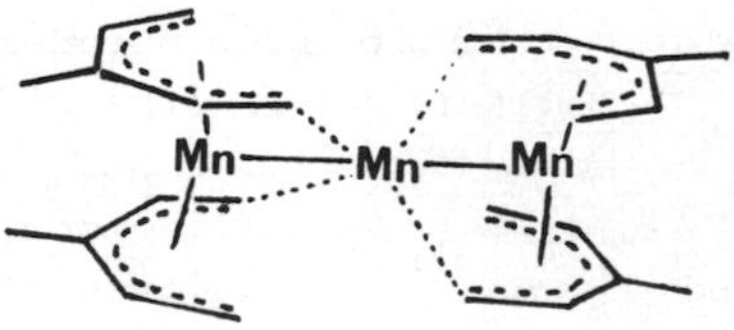

Mn
Mn
Mn

(6)

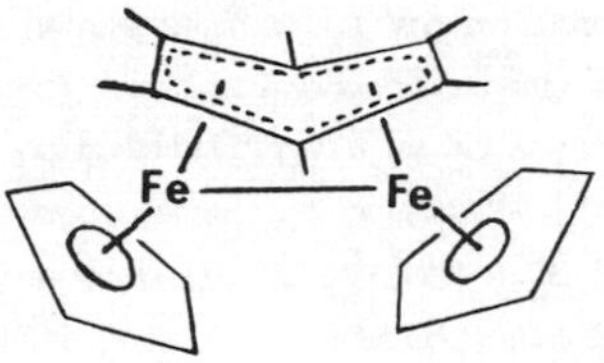

Fe
Fe

(7)

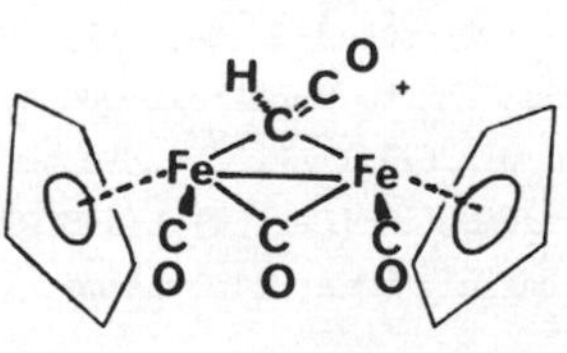

H
C
O
C
+
Fe
Fe
C
O
C
O
C
O

(8)

In $[Mn_3(\eta-C_5H_6Me-3)_4]$ (6) the central Mn atom interacts with a terminal C atom of each ligand and, although Mn atoms may exhibit weak covalent coupling, strong electrostatic forces between them must exist;[92] only weak Mn-Mn interactions are present in the alkyl derivatives $[Mn_3(C_6H_2Me_3-2,4,6)_6]$[93a] and $[Mn_2R_4(PMe_3)_2]$.[93b]

The stepwise degradation of $[Re_3(CO)_{10}(\mu-H)_4]^-$ with I_2 has been monitored, with intermediate anions $[Re_3(CO)_{10}(\mu-H)_2(\mu-I)_2]^-$ and $[Re_2(CO)_6(\mu-H)(\mu-I)_2]^-$ being structurally characterised;[94] oxidation of the same tetrahydride anion with CF_3SO_3H in CH_3CN gives the reactive diaxial-trans-$[Re_3(CO)_{10}(NCMe)_2(\mu-H)_3]$.[95]

The first octahedral rhenium cluster $[Re_6(CO)_{18}(\mu_3-H)_2(\mu_6-C)]^{2-}$ has been characterised.[96]

2.4 Iron.- Bridged vinylidene di-iron complexes, cis and trans $[Fe_2(CO)_2Cp(\mu-CO)-(\mu-CCH_2)]$, are isolated after nucleophilic attack of $[Fe_2(CO)_4Cp_2]$ with LiMe whereas $[Fe_2(CO)_3(\mu-H)(\mu-Ph_2PCHPPh_2)]$ results from $[Fe_2(CO)_6(\mu-CO)(\mu-dppm)]$ under similar reaction conditions.[97] Nucleophilic additions of amines to acetylide ligand in $[Fe_2(CO)_6(\mu-\eta^2-C_2R)(\mu-PPh_2)]$ give μ-alkylidene and μ-vinylidene products[98a] but also combined carbonylation and amination products, $[Fe_2(CO)_6-\{\mu-OCCHCPh(NR_2)\}(\mu-PPh_2)]$,[98b] with ruthenium analogues characterised by X-ray diffraction. Hydride addition to $[Fe_2(CO)_6(\mu-PPh_2)_2]$ forms $[Fe_2(CO)_5(PPh_2H)-(\mu-CO)(\mu-PPh_2)]^-$ for which further reactions are reported and the product of methylation, $[Fe_2(CO)_5(PPh_2Me)(\mu-COMe)(\mu-PPh_2)]$, is structurally characterised.[99] Deprotonation of $[Fe_2(CO)_6(\mu-PBu^tH)_2]$ followed by reaction with 1,2-dibromoethane affords the diphosphene bridged $[Fe_2(CO)_6(\mu-\eta^2,\eta^2-P_2Bu^t_2)]$ {Fe-Fe = 2.740(1)Å}.[100]

Addition of an alkynyllithium to $[Fe_2(CO)_6(\mu-S_2)]$ initially forms $[Fe_2(CO)_6-(\mu-SC_2R)(\mu-SLi)]$ which intramolecularly adds S^- to the α- or β-C atom of coordinated SC_2R to give bridging organic disulphide ligands;[101] deprotonation of related disulphur-bridges, e.g. in $[Fe_2(CO)_6(\mu-L)]$ (L = SCH_2S, SCHSMe), gives anions which react with organic halides to form unexpected products, including $[Fe_2(CO)_6(\mu-L')]$ (L′ = $MeCOCH_2CHS_2$ and $\eta^2-SCSMe_2$) characterised by X-ray diffraction.[102] The thioketene compound $[Fe_2(CO)_6\{\mu-\eta^2-SC(Me_4Cy)\}]$ (Me_4Cy = 2,2,6,6-tetramethylcyclohexyl) reversibly adds PPh_3 to form half-opened, unbridged $[Fe_2(CO)_6(PPh_3)\{\eta^2-SC(Me_4Cy)\}]$ (Fe-Fe = 2.675(1)Å).[103] Electron transfer catalysis promotes regioselective substitution in $[Fe_2(CO)_5L(\mu-SCOR)(\mu-SMe)]$ (L = CO, R = adamantyl) by $P(OMe)_3$ to produce mainly equatorial isomer (L = $P(OMe)_3$, at C bonded Fe) with Fe-Fe = 2.623(2)Å.[104]

Spectroscopic evidence is presented for formation of $[Fe_2Cp_2(\mu-CO)_3]$ on photolysis of $[Fe(CO)_2Cp_2(\mu-CO)_2]$ in matrices,[105] whereas in photosubstitution of the latter dimer with $P(OPr^i)_3$ the intermediate $[Fe_2(CO)_3\{P(OPr^i)_3\}Cp_2(\mu-CO)]$, without Fe-Fe bond, is proposed.[106] In CH_3CN, oxidative cleavage of $[Fe_2(CO)_2-Cp_2(\mu-CO)_2]$ occurs with strong acids, HX, to give $[Fe(CO)_2(NCMe)Cp]X$ and/or

$[FeX(CO)_2Cp]$;[107a] in CH_2Cl_2, clean protonation by $HBF_4{\cdot}OMe_2$ yields $[Fe_2(CO)_4HCp_2]^+$ although cleavage products are observed from $[Cr_2(NO)_4Cp_2]$ and $[Mn_2(CO)_2(NO)_2Cp_2]$.[107b]

The extremely unstable anion $[Fe_2(CO)_{\underline{n}}(\mu\text{-}C_7H_7)]^-$ ($\underline{n}$ = 6) loses CO to form a fluxional, structurally characterised product ($\underline{n}$ = 5) with μ-η^3,η^4-cycloheptatrienyl ligand.[108] The crystal structure of fluxional, bridged hexamethylbenzene complex $[Fe_2Cp_2(\mu\text{-}C_6Me_6)]$ (7) is reported.[109]

Protonation of bridging ligands in $[Fe_2(CO)Cp_2(\mu\text{-}CO)(\mu\text{-}COCRCR')]$ (and of Ru analogues) yields μ-vinyl cations $[M_2(CO)_2Cp_2(\mu\text{-}CO)(\mu\text{-}CRCHR')]^+$ which are converted into μ-alkylidene species $[M_2(CO)_2Cp_2(\mu\text{-}CO)(\mu\text{-}CRCH_2R')]$ on reaction with $NaBH_4$; cis and trans isomers of both μ-vinyl and μ-alkylidene products are observed, and X-ray determined structures reported for cis species (M = Fe, R = R' = H).[110] μ-Alkenyl (vinyl) cations are also formed on thermal rearrangement of μ-alkylidyne cations $[Fe_2(CO)_2Cp_2(\mu\text{-}CO)(\mu\text{-}CR)]^+$ (e.g. R = Bu^n, but not Me); also, the novel μ-methylidyne cation (R = H) inserts ethylene to produce homologous μ-alkylidyne (R = Et) or adds CO to form the structurally characterised acylium derivative $[Fe_2(CO)_2Cp_2(\mu\text{-}CO)(\mu\text{-}CHCO)]^+$, (8).[111] Reaction of $[U(CHPMe_2Ph)Cp_3]$ with $[Fe_2(CO)_2Cp_2(\mu\text{-}CO)_2]$ causes coupling of bridging and terminal CO groups to form an μ-η^1,η^3-allyl ligand in $[UCp_2\{Fe_2(CO)Cp_2(\mu\text{-}CO)(\mu\text{-}COCOCHPMe_2Ph)\}]$.[112]

Properties and chemistry of the paramagnetic anion $[Fe_2Cp_2(\mu\text{-}NO)_2]^-$ have been investigated.[113] The unexpected $[Fe_2Cp_2(\mu\text{-}S_2C_{10}Cl_4S_2)(\mu\text{-}SC_{10}Cl_4S_2)]$, with a single Fe-Fe bond {2.649(4)Å}, results from desulphurisation of 2,3,6,7-tetrachloro-1,4,5,8-tetrathionaphthalene, in contrast to tetrathiolene products for M = Co and Ni, $[M_2Cp_2(\mu\text{-}S_2C_{10}Cl_4S_2)]$.[114]

The isocyanide dimers $[Fe_2(CNR)_2Cp_2(\mu\text{-}CNR)_2]$ form a mixture of isomers, non interconverting at R.T.[115] The isocyanide substituted cluster $[Fe_3(CO)_9(CNBu^t)(\mu\text{-}CO)_2]$ thermally converts into $[Fe_3(CO)_9(\mu_3\text{-}\eta^2\text{-}CNBu^t)]$, both clusters being characterised by X-ray diffraction.[116] Nitrile substituents in $[Fe_3(CO)_7L(NCR)(\mu_3\text{-}PPh)_2]$ (L = CO, NCR; R = Me, Et) are replaced by phosphorus and arsenic ligands, as in related $[Co_4(CO)_9(NCR)(\mu_4\text{-}PPh)_2]$.[117]

The structurally characterised ketenylidene anion, $[Fe_3(CO)_9(\mu_3\text{-}CCO)]^{2-}$, on protonation or methylation undergoes CO migration to form $[Fe_3(CO)_9(\mu\text{-}CO)(\mu_3\text{-}CR)]^-$ (R = H or Me).[118] Ethylene inserts into $[Fe_3(CO)_{10}(\mu\text{-}H)(\mu\text{-}CO)]^-$ to produce $[Fe_3(CO)_9(\mu_3\text{-}\eta^2\text{-}OCEt)]^-$ and the reactivity of this μ_3-propionyl anion has been studied; CO and PPh_3 cause complete cluster rupture.[119] Electrophilic attack (H^+ or Me^+) on μ_3-acetyl cluster $[Fe_3(CO)_9(\mu_3\text{-}\eta^2\text{-}OCMe)]^-$ yields a number of products, crystal structures being reported for $[Fe_3(CO)_9(\mu\text{-}H)(\mu_3\text{-}OCMe)]$, $[Fe_3(CO)_9(\mu_3\text{-}CMe)(\mu_3\text{-}OMe)]$ and $[Fe_3(CO)_9(\mu_3\text{-}CMe)(\mu_3\text{-}COMe)]$.[120] Reduction of $[Fe_3(CO)_9(\mu_3\text{-}CMe)(\mu_3\text{-}COEt)]$ produces $[Fe_3(CO)_{10}(\mu_3\text{-}CMe)]^-$ which, on reaction with methanol, is transformed into $[Fe_3(CO)_9(\mu_3\text{-}CCMe)]$ by coupling of ethylidyne with carbide produced by C-O bond cleavage.[121] The μ_2-carbyne cluster $[Fe_3(CO)_{10}(\mu\text{-}H)(\mu\text{-}COMe)]$ (also Ru and Os analogues) is reversibly hydrogenated to form $[Fe_3(CO)_9(\mu\text{-}H)_3(\mu_3\text{-}COMe)]$.[122] Complex $[Fe_3(CO)_6Cp(\mu\text{-}CO)_2(\mu_3\text{-}CMe)]$. (characterised by X-ray diffraction) is formed from $[Fe_2(CO)_2Cp_2(\mu\text{-}CO)(\mu\text{-}CCH_2)]$ and $[Fe_2(CO)_9]$; its further reaction with cyclopentadiene affords $[Fe_3Cp_3(\mu\text{-}CO)_3\text{-}(\mu_3\text{-}CMe)]$.[123]

Oxidation of $[Fe_3(CO)_9(\mu\text{-}CO)_2]^{2-}$ with O_2 quantitatively gives $[Fe_3(CO)_9\text{-}(\mu_3\text{-}O)]^{2-}$, a model for O atoms absorbed at an iron surface;[124] alkylation forms μ_3-alkoxy clusters. The μ_3-sulphur ligand clusters, $[Fe_3(CO)_9(\mu_3\text{-}ER)\{\mu_3\text{-}SM\text{-}(CO)_5\}]$ (E = P, As; M = Cr, W)[125] and $[Fe_3(CO)_9(\mu\text{-}X)(\mu_3\text{-}SR)]$ (X = PR'_2, AsR'_2, SbR'_2, SR', SeR')[126] possess an opened triangle of iron atoms with one non-bonded Fe-Fe separation (ca. 3.5Å in three examples studied by X-ray diffraction). A linear trimer $[Fe_3(CO)_6(\mu_3\text{-}SC_6H_4NN)_2]$ has been structurally characterised.[127]

The boron analogue of a protonated carbido cluster $[Fe_4(CO)_{12}(\mu\text{-}H)(\mu_4\text{-}BH_2)]$ (9)[128] has been characterised by X-ray diffraction, as has unsaturated cluster $[Fe_4(CO)_{10}(\mu\text{-}CO)\{\mu_4\text{-}P(tol\text{-}p)\}_2]$ which, with CO, is reversibly converted into $[Fe_4(CO)_{12}\{\mu_4\text{-}P(tol\text{-}p)\}_2]$.[129]

2.5 Ruthenium and Osmium.- The μ-fulvalene complex $[Ru_2(CO)_4(\mu\text{-}C_{10}H_8)]$ {Ru-Ru = 2.821(1)Å} undergoes a thermally reversible photoisomerisation to $[Ru_2(CO)_4\text{-}(\mu\text{-}\eta^1,\eta^5\text{-}C_5H_4)_2]$ involving both C-C and Ru-Ru bond fission.[130]

Alkyne oligomerisation occurs on μ-alkylidene Ru_2 complexes and steric control of the stereochemistry is observed; X-ray determined structures of products $[Ru_2(CO)Cp_2(\mu\text{-}CO)(\mu\text{-}C_4H_4CMe_2)]$ and $[Ru_2Cp_2(\mu\text{-}CO)\{C_4(CO_2Me)_4CH_2\}]$ are reported.[131] Alkylidyne-alkyne coupling occurs on a Ru_3 cluster to form C_3 ligands in $[Ru_3(CO)_9(\mu\text{-}H)(\mu\text{-}\eta^3\text{-}CRCR'CR')]$ (R = Me, Ph, OMe).[132] Allenes displace alkyne from $[Ru_2(CO)Cp_2(\mu\text{-}CO)(\mu\text{-}COC_2Ph_2)]$ forming allyl complexes which, after protonation and reaction with $NaBH_4$, give μ-alkylidene and μ-alkenyl products, e.g. $[Ru_2(CO)_2Cp_2(\mu\text{-}CO)(\mu\text{-}CMe_2)]$ and $[Ru_2(CO)_2Cp_2(\mu\text{-}CO)(\mu\text{-}CHMeCH)]$, respectively, for both of which isomers exist.[133] Molecular structures and chemistry of μ-vinylidene and μ-ethylidyne complexes cis-$[Ru_2(CO)_2Cp_2(\mu\text{-}CO)(\mu\text{-}CCH_2)]$ and cis-$[Ru_2\text{-}(CO)_2Cp_2(\mu\text{-}CO)(\mu\text{-}CMe)]^+$ are reported in detail.[134] Photolysis with dihydrogen converts $[Ru_2(CO)_2Cp_2(\mu\text{-}CO)(\mu\text{-}CHR)]$ into $[Ru_3(CO)_3Cp_3(\mu\text{-}H)_3]$ for which the structure and some reactions are reported.[135]

Application of radical-anion initiated substitution using $Na[OCPh_2]$ has afforded more than 60 derivatives of $[Ru_3(CO)_{12}]$ and $[Ru_4(CO)_{12}(\mu\text{-}H)_4]$:[136] products characterised by X-ray diffraction and obtained by this technique, include chiral cluster $[Ru_4(CO)_9(PMe_2Ph)\{P(tol\text{-}p)_3\}\{P(OCH_2)_3CEt\}(\mu\text{-}H)_4]$,[137] equatorially substituted $[Ru_3(CO)_{11}(PCy_3)]$, $[Ru_3(CO)_{10}\{P(OMe)_3\}_2]$ and $[Ru_3(CO)_9\text{-}(PMe_3)_3]$,[138] and axially substituted $[Ru_3(CO)_{12-n}(CNBu^t)_n]$ (n = 1, 2).[139] Details for synthesis of anions $[Ru_3(CO)_{11}]^{2-}$, $[Ru_4(CO)_{12}]^{4-}$, $[Ru_4(CO)_{13}]^{2-}$ and

$[Ru_6(CO)_{18}]^{2-}$ by stoichiometric reactions of $Ph_2CO^{\bullet -}$ with $[Ru_3(CO)_{12}]$ have appeared[140a] and further reductions to highly charged anions $[Ru_4(CO)_{11}]^{6-}$, $[Ru_6(CO)_{17}]^{4-}$ and $[Ru_6(CO)_{16}]^{6-}$ are also reported.[140b] Reversible substitutive oligomerisation of $[Ru_3(CO)_9(\mu\text{-}H)_2(\mu_3\text{-}S)]$ yields cyclic trimer $[Ru_3(CO)_8(\mu\text{-}H)_2\text{-}(\mu_4\text{-}S)]_3$.[141]

Salts of bis(triphenylphosphine)iminium cations have found a number of applications in Ru and Os cluster chemistry. Anions $[Ru_3(CO)_7(\mu\text{-}CO)_3(\mu\text{-}O_2CR)]^-$ {R = H(X-ray), Me} are synthesised from [PPN]$[O_2CR]$.[142] The salt [PPN]$[NO_2]$ is a nucleophilic nitrosylating agent forming anions $[M_3(CO)_{10}(NO)]^-$ (M = Ru, Os)[143] and $[Ru_6C(CO)_{15}(NO)]^-$[143,144] from parent carbonyls: the structure of the protonated product of the latter anion, $[Ru_6(CO)_{13}(NO)(\mu\text{-}H)(\mu\text{-}CO)_2(\mu_6\text{-}C)]$ is described,[144] as is that of $[Ru_6(CO)_{10}(NO)_2(\mu\text{-}CO)_4(\mu_6\text{-}C)]$ formed by further reaction with NO^+.[145] The azide PPN$[N_3]$ nucleophilically attacks a CO group in $[Ru_3(CO)_{12}]$ to form isocyanate;[146] at R.T., Ru_3 intermediates are observed but the product is $[Ru_4(NCO)(CO)_{13}]^-$ which may be converted into Ru_4 and Ru_3 derivatives including $[Ru_4(CO)_{12}(\mu\text{-}H)_3(\mu\text{-}NCO)]$[146] and $[Ru_4(CO)_{12}(\mu\text{-}H)(\mu\text{-}CO)]^-$,[147] both studied by X-ray diffraction; in refluxing THF, nitride $[Ru_6(CO)_{16}(\mu_6\text{-}N)]^-$ is produced which on carbonylation is converted into square pyramidal $[Ru_5(CO)_{13}\text{-}(\mu\text{-}CO)(\mu_5\text{-}N)]^-$.[148]

Structures and chemistry of η^2-alkene ($CF_3CH{=}CHCF_3$) and bromine containing clusters $[Os_3Br(CO)_{10}(\eta\text{-}alkene)]^-$, $[Os_3(CO)_9(\eta\text{-}alkene)(\mu\text{-}Br)]^-$ and $[Os_3(CO)_9\text{-}(\eta\text{-}alkene)(\mu\text{-}H)(\mu\text{-}Br)]$ are reported.[149] Reaction of ethylene and CO with $[Ru_3(CO)_{10}(\mu\text{-}H)(\mu\text{-}X)]$ affords $[Ru_3(CO)_{10}(\mu\text{-}OCEt)(\mu\text{-}X)]$ (X = Cl, Br, I) and further reaction of a related product (X = H) causes fragmentation to $[Ru_2(CO)_6\text{-}(\mu\text{-}OCEt)_2]$ containing a head-to-head arrangement of μ-propionyl groups.[150] Investigations of Fischer-type carbenes of $[Ru_3(CO)_{10}(\mu\text{-}H)(\mu\text{-}OCMe)]$ lead to products $[Ru_3(CO)_9(\mu\text{-}H)_2\{\mu_3\text{-}\eta^2\text{-}C(OEt)CH\}]$ and $[Ru_3(CO)_9(\mu\text{-}H)_2\{\mu_3\text{-}\eta^2\text{-}C(OEt)N(Me)\text{-}CH\}]$.[151] In addition to deprotonation, the reaction of base with $[Ru_3(CO)_{10}\text{-}(\mu\text{-}H)(\mu\text{-}OCNMe_2)]$ yields the unusual 92e hexanuclear cluster $[Ru_6H(CO)_{14}(\mu\text{-}CO)_4\text{-}(\mu\text{-}OCNMe_2)_2]^-$ with open "raft" structure.[152] The μ-η^2,P-2-styrylphosphine cluster $[Ru_3(CO)_{10}(\mu\text{-}sp)]$ undergoes thermolysis to form several products depending on the temperature; X-ray diffraction indicated the structures of resultant alkyne complexes $[Ru_3(CO)_8(\mu\text{-}H)_2(\mu_3\text{-}HC_2C_6H_4PPh_2)]$,[153] $[Ru_4(CO)_{11}(\mu_4\text{-}HC_2C_6H_4\text{-}PPh_2)]$[154] and alkylidene complex, $[Ru_2(CO)_6(\mu\text{-}MeCC_6H_4PPh_2)]$.[155]

The nitrile bond of PhCN is activated in the formation of $[Ru_3(CO)_{10}(\mu\text{-}H)\text{-}(\mu\text{-}NHCH_2Ph)]$ from $[Ru_3(CO)_{12}]$ and acetic acid.[156] Oxidative addition of PhC≡CCl to $[Ru_3(CO)_{12}]$ produces $[Ru_3(CO)_9(\mu\text{-}Cl)(\mu_3\text{-}\eta^2\text{-}C_2Ph)]$ in which a Ru-Ru bond has broken and a Ru_3Cl "butterfly" structure is formed.[157] Photolytic activation of the P-H bond in $[Ru_3(CO)_9(PHPh_2)_3]$ affords $[Ru_3(CO)_7(\mu\text{-}H)(\mu\text{-}PPh_2)_3]$ among other products;[158] thermolysis of 46e cluster $[Ru_3(CO)_9(\mu\text{-}H)(\mu\text{-}PPh_2)]$ causes elimination of benzene and one product (15%) is $[Ru_5(CO)_{16}(\mu\text{-}PPh_2)(\mu_5\text{-}P)]$ with partially encapsulated P atom.[159] Square pyramidal cluster $[Ru_5(CO)_{\underline{n}}(\mu\text{-}PPh_2)(\mu_4\text{-}\eta^2\text{-}C_2Ph)]$ ($\underline{n}$ = 13) undergoes facile reversible cleavage of a basal Ru-Ru bond on addition of CO to form species ($\underline{n}$ = 14).[160] Pyrolysis of $[Ru_5(CO)_{14}(CNBu^t)(\mu_5\text{-}CNBu^t)]$ produces octahedral $[Ru_6(CO)_{15}(CNBu^t)(\mu\text{-}CO)(\mu_6\text{-}C)]$ with carbide atom originating from coordinated isocyanide.[161]

X-Ray diffraction shows $[Os_3(CO)_{10}(\mu\text{-}H)(\mu\text{-}CH)]$ to possess a semi-triply bridging methylidyne ligand and this undergoes facile reactions with nucleophiles;[162] in refluxing hexane the cluster rearranges to $[Os_3(CO)_9(\mu\text{-}H)_2\text{-}(\mu_3\text{-}\eta^1\text{-}CCO)]$ by coupling of methylidyne carbon with CO.[163] A structurally related boron compound $[Os_3(CO)_9(\mu\text{-}H)_3(\mu_3\text{-}\eta^1\text{-}BCO)]$ is also reported.[164] The μ-methylene cluster $[Os_3(CO)_{10}(\mu\text{-}CO)(\mu\text{-}CH_2)]$ undergoes reversible C-C coupling on carbonylation to form reactive ketene complex $[Os_3(CO)_{12}(\mu\text{-}CH_2CO)]$ with an opened Os-Os bond.[165] Structurally characterised benzyne cluster $[Os_3(CO)_9(\mu\text{-}H)_2\text{-}(\mu_3\text{-}C_6H_4)]$ is obtained in good yield from $[Os_3(CO)_{10}(NCMe)_2]$ and refluxing benzene[166a] (structures of other benzyne clusters of Ru_3 and Os_3 have also been reported[166b]). Oxidative addition of Ph_3SiH to $[Os_3(CO)_{10}(\mu\text{-}H)_2]$ yields $[Os_3\text{-}(CO)_9(SiPh_3)(\mu\text{-}H)_3]$ containing a formally unsaturated $Os(\mu\text{-}H)_2Os$ unit {Os-Os = 2.7079(4) Å}.[167]

Amidines form derivatives $[Os_3(CO)_{\underline{n}}(\mu\text{-}H)(NRCR'NH)]$ ($\underline{n}$ = 9, 10) and $\underline{X}$-ray diffraction when $\underline{n}$ = 9, R = R' = Ph establishes the presence of 5e (μ_3-NPhCPhNH) ligand.[168] Cyclic thioamides form deprotonated thiolato bridges in clusters $[Os_3(CO)_{10}(\mu\text{-}H)(\mu\text{-}L)]$ ($\underline{X}$-ray structure for L = $\mu\text{-}SCNCH_2CH_2S$) which lose CO to give $[Os_3(CO)_9(\mu\text{-}H)(\mu_3\text{-}L)]$.[169] Competition between N-H and C-H activation occurs on thermolysis of $[Os_3(CO)_{11}\{NHC(CH_2)_4CH_2\}]$: C-H cleavage gives $[Os_3(CO)_{10}\text{-}(\mu\text{-}H)_2\{\mu_3\text{-}NHCC(CH_2)_3CH_2\}]$ (11%) and N-H cleavage $[Os_3(CO)_{10}(\mu\text{-}H)\{\mu_3\text{-}NC(CH_2)_4CH_2\}]$ (34%).[170] Reaction of $S(NPBu_2)_2$ with $[Os_3(CO)_{11}(NCMe)]$ or $[Os_3(CO)_{10}(\mu\text{-}H)_2]$ yields $[Os_3(CO)_{11}PBu_2(NH_2)]$ or $[Os_3(CO)_9(\mu\text{-}H)(\mu_3\text{-}SNHPBu_2)]$, respectively, after N=S bond fission.[171] Addition of isocyanide to acetylide ligand ($\mu_3\text{-}\eta^2\text{-}C_2Ph$) affords structurally characterised $[Ru_3(CO)_9(\mu\text{-}H)\{\mu_3\text{-}C(CNBu^t)CPh\}]$ or, after animation with $BuNH_2$, $[Os_3(CO)_9(\mu\text{-}PPh_2)(\mu_3\text{-}C\{C(NHBu^n)(NHBu^t)\}CPh)]$.[172]

Several products of reaction of $[NO]^+X^-$ (X = BF_4 or PF_6) with $[M_4(CO)_{12}\text{-}(\mu\text{-}H)_3]^-$ (M = Ru, Os) are reported:[173] in acetonitrile the reactive "butterfly" cluster $[Os_4(CO)_{12}(NCMe)_2(\mu\text{-}H)_3]^+$ is formed;[173a] with traces of water (X = PF_6) oxygen bridged clusters $[Os_4(CO)_{12}(\mu\text{-}H)_3(\mu\text{-}Y)]$ {Y = OH, $OPO(OH)_2$} are isolated;[173b] from CH_2Cl_2 solution, the nitrido clusters $[Ru_4(CO)_{11}(\mu\text{-}H)_3(\mu_4\text{-}N)]$ or (after deprotonation) $[Os_4(CO)_{12}(\mu_4\text{-}N)]^-$ are obtained.[173c]

Chemistry of square pyramidal clusters $[M_5(CO)_{15}(\mu_5\text{-}C)]$ (M = Ru, Os) has been studied.[174] $[Ru_5(CO)_{15}(\mu_5\text{-}C)]$, obtained by carbonylation of $[Ru_6(CO)_{16}(\mu\text{-}CO)\text{-}(\mu_6\text{-}C)]$, forms adducts with MeCN or halide ions, is substituted by phosphine ligands,[174a] and with RSH (R = H, Me, Et) or H_2Se forms opened square pyramidal species $[Ru_5(CO)_{14}(\mu\text{-}H)(\mu\text{-}XR)(\mu_5\text{-}C)]$ (X = S, Se)—reactions of the species with

X = S, R = Et are described.[174b] Carbonylation of $[M_5(CO)_n(\mu_5\text{-}C)]$ (M = Ru, Os; $\underline{n}$ = 15) under pressure affords adduct ($\underline{n}$ = 16) with a "bridged butterfly" skeleton ($M_5 = Os_5$) and reductions give $[M_5(CO)_{14}(\mu_5\text{-}C)]^{2-}$, square pyramidal with one μ_2-CO ligand bridging basal Os atoms (M = Os).[174c] Opened clusters $[Os_5(CO)_nL\text{-}$ $(\mu\text{-}H)_2]$ {$\underline{n}$ = 15; L = PPh_3, $PEt_3P(OMe)_3$} are thermally decarbonylated to trigonal bipyramidal derivatives ($\underline{n}$ = 14) and this skeletal geometry is retained on substitution by more phosphine.[175] The complex $[Os_5(CO)_{17}(\mu_3\text{-}C_2H_2)]$ has a "bow-tie" configuration (**10**).[176] The "raft" complex $[Os_6(CO)_{18}(\mu_3\text{-}CO)(\mu_3\text{-}O)]$ (**11**) formed by oxygenation of $[Os_6(CO)_{20}]$ possesses a puckered plane of fused Os_3 triangles capped centrally on each side by O and CO ligands.[177]

A range of osmium-sulphur clusters has been reported by Adams and coworkers.[178-180] Preparations included thermolyses or photolysis of Os_3 species,[178b,e] additions of H_2S[178c] or CO[178d] to sulphur containing clusters and condensations of lower clusters with $[Os_3(CO)_{10}(NCMe)_2]$;[178f-h] compounds include $[Os_3(CO)_8L(\mu_3\text{-}S)_2]$ (L = CO, PMe_2Ph),[178a] $[Os_4(CO)_n(\mu_3\text{-}S)_m]$ ($\underline{n}$ = 12, 13; $\underline{m}$ = 1, 2),[178b-d] $[Os_4(CO)_{12}(\mu\text{-}H)_2(\mu_3\text{-}S)_2]$,[178c] $[Os_5(CO)_{15}(\mu_4\text{-}S)]$,[178b,c] $[Os_5(CO)_{14}\text{-}$ $(\mu\text{-}H)_2(\mu_3\text{-}S)_2]$,[178c] $[Os_6(CO)_{17}(\mu_4\text{-}S)_2]$,[178f] $[Os_6(CO)_{16}(\mu_3\text{-}S)(\mu_4\text{-}S)]$,[178f] $[Os_6\text{-}$ $(CO)_n(\mu\text{-}H)_2(\mu_3\text{-}S)(\mu_4\text{-}S)]$ ($\underline{n}$ = 18, 17),[178e] and $[Os_7(CO)_n(\mu_4\text{-}S)_m]$ ($\underline{n}$ = 19, $\underline{m}$ = 1; $\underline{n}$ = 20, $\underline{m}$ = 2).[178g,h] Isothiocyanates ArNCS with $[Os_3(CO)_{10}(\mu\text{-}H)_2]$ form Os_3 clusters with thioformamido ligands and also cluster $[Os_3(CO)_9(\mu\text{-}H)(\mu\text{-}HCNAr)\text{-}$ $(\mu_3\text{-}S)]$, product of C-S bond fission;[179] thermolyses of the latter cluster produces Os_6 species with $(\mu_4\text{-}S)$ and, in some cases, $(\mu_3\text{-}S)$ ligands.[180] Structures of most new products have been established by X-ray diffraction.

Reaction of iodine with $[HOs_8(CO)_{22}]^-$ causes a major change in geometry to form $[Os_8(CO)_{22}(\mu\text{-}I)(\mu_x\text{-}H)]$ ($\underline{x}$ = 3?) with I bridging a "butterfly" Os_4 unit.[181] The first Os_9 species $[Os_9(CO)_{21}(\mu_4\text{-}CHCRCH)]^-$ containing allylic unit (R = Me, Et) and Os_9 core (**12**),[182] the first non-carbido Os_{10} cluster $[Os_{10}(CO)_{24}\text{-}$ $(\mu_x\text{-}H)_4]^{2-}$ with tetra-capped octahedral framework[183] and the first Os_{11} cluster, $[Os_{11}(CO)_{27}(\mu_6\text{-}C)]^{2-}$[184] are reported.

2.6 Cobalt.
Electron transfer catalysed substitutions of $[Co_2(CO)_5L\{\mu\text{-}C_2(CF_3)_2\}]$ (L = CO) quantitatively form species where L = Lewis base; when L = MeCN the ligand is equatorial whereas phosphorus ligands occupy an axial site.[185] Solution dynamics of fluxional complexes $[Co_2(CO)_4(\mu\text{-}RC_2R')(\mu\text{-}dppm)]$ are reported.[186] In new derivatives $[Co_2(CO)_4(\mu\text{-}L)(\mu\text{-}CHR)(\mu\text{-}dppm)]$ (L = CO or CH_2; R = H or CO_2Et) dynamic n.m.r. studies indicate a rapid bridge/terminal transformation of coordinated CH_2.[187] The synthesis and reactivity of dimers $[Co_2(CO)_2Cp'_2(\mu\text{-}CCR')]$ are described by two group and crystal structures are reported for R = R' = H, Cp' = C_5H_4Me[188a] or C_5Me_5.[188b] The radical ions $[Co_2Cp'_2(\mu\text{-}CO)_2]^-$ (Cp' = Cp or C_5H_4Me) react with organic halides and the dimetallacyclohexene derivative $[Co_2\text{-}$ $(\eta\text{-}C_5H_4Me)_2(\mu\text{-}CO)\{\mu\text{-}(CH_2)_2C_6H_4\text{-}2\}]$ is characterised; the retro-Diels Alder

elimination of o-xylylene from this dimetallacycle has been studied kinetically.[189] The bridged, syn-butadiene complex $[Co_2Cp_2(\mu\text{-}CO)(\mu\text{-}\eta^2,\eta^2\text{-}C_4H_6)]$ is structurally characterised[190a] and hexatriene forms isomeric complexes $[Co_2Cp_2(\mu\text{-}\eta^3,\eta^2\text{-}CH(CH)_3CHMe)]$ interconvertible by thermal and photochemical vinyl-H shifts.[190b]

The syntheses, electrochemistry and reactivity of nitrosyl species $[Co_2Cp'_2(\mu\text{-}NO)(\mu\text{-}L)]^z$ (Cp′ = Cp or C_5Me_5; L = NO, CO) are reported.[191] Complexes $[Co_2(CO)_2L_2(\mu\text{-}PR_2)_2]$ (L = CO, PMe_3; R = Bu^t [192] and L = PEt_2Ph, R = Ph[193]), characterised by X-ray diffraction, have Co=Co double bonds and planar $Co_2(\mu\text{-}P)_2$ systems; their transformation to Co_3 species is described.[193]

Coordinated ethylene of $[Co(\mu\text{-}C_5Me_5)(\mu\text{-}C_2H_4)_2]$ is converted on thermolysis into ethylidyne in $[Co_3(\mu\text{-}C_5Me_5)_3(\mu_3\text{-}CMe)_2]$[194] and alkylidyne-alkylidyne coupling occurs to form dithio- or diseleno-lene complexes on reaction of $[Co_3Cp'_3(\mu_3\text{-}CR)_2]$ with sulphur or selenium, respectively.[195] $[Co_3Cp_3(\mu_3\text{-}CNCy)(\mu_3\text{-}S)]$ contains a triply bridging isocyanide.[196] Also, reactions of $[Co_3Cp_3(\mu_3\text{-}CS)(\mu_3\text{-}S)]$ are described[197] and previously reported $[Co_3(CO)_9(CS)]_2$ has been identified by X-ray analysis as $[\{Co_3(CO)_9(\mu_3\text{-}C)\}_2SCO]$.[198]

The paramagnetic 46e cluster $[Co_3(CO)_6(PPh_3)_3(\mu_x\text{-}H)]$ (x = 3?) is characterised.[199] Chiral clusters $[Co_3(CO)_7(\mu\text{-}L\text{-}X)(\mu_3\text{-}S)]$ are formed from thioamides with $[Co_2(CO)_8]$, and the structure for L-X = {1,3-NHC(Me)S} is described.[200]

E.s.r. and/or electrochemical studies are reported on clusters $[Co_3Cp_3(\mu\text{-}CPh)_2]$,[201] $[(R)FcCCo_3(CO)_{9-n}(PR_3)_n]$[202] and $[Co_4(CO)_{12-2n}(dppm)_n]$.[203] The tetranuclear $[Co_4(CO)_8(\mu\text{-}PC_4H_2Me_2)_2]$, formed from 3,3′,4,4′-tetramethyl-1,1′-biphospholyl, has a bent Co_4 chain structure.[204] $[Co_6(CO)_{14}(\mu\text{-}CO)_2(\mu_6\text{-}P)]$ contains an open array of Co atoms with a semi-interstitial P atom[205] and $[Co_8(CO)_{16}(\mu_4\text{-}AsPh)_2(\mu_4\text{-}As)(\mu_6\text{-}As)]_2$ comprises two linked systems of four distorted $AsCo_3$ and As_2Co_2 tetrahedra.[206]

2.7 Rhodium and Iridium.

- Diphosphine catalysts $[Rh_2(CO)_2(P\text{-}P)_2(\mu\text{-}CO)_2]$, including the crystal structure for P-P = dppp, have been described.[207] In asymmetric $[Rh_2Cp_2(PPh_3)(\mu\text{-}CO)_2]$ there is a polar Rh-Rh bond {2.673(1)Å}.[208] Additions to the formal double bond of $[Rh_2(\mu\text{-}C_5Me_5)_2(\mu\text{-}CO)_2]$ {Rh-Rh = 2.564(1)Å[209]} have been reported: the ketocarbene derived from 2-diazodimedone forms $[Rh_2(\eta\text{-}C_5Me_5)_2(\mu\text{-}CO)(\mu\text{-}COCCOCH_2CMe_2CH_2CO)]$ by coupling with a CO ligand;[210] alkynes, C_2R_2, also couple with CO to form a metallacyclic bridge in $[Rh_2(\mu\text{-}C_5Me_5)_2(\mu\text{-}CO)(\mu\text{-}CR{=}CR\text{-}CO)]$ but when R = H this interconverts with a μ-ketenylmethylene system (μ-CH-CH=C=O);[211] clean additions of S_8, Se, Te, SO_2 and ArCl also occur.[212] $[Rh_2(CO)_2Cp_2(\mu\text{-}CO)]$ forms an adduct with $HgCl_2$, is cleaved by TCNE and by I_2 and is converted into a new modification of $[Rh_3Cp_3(\mu\text{-}CO)_3]$ by $PdCl_2$ or $PtCl_2$.[213] A dinuclear $\{Rh_2(\mu\text{-}CO)_3\}^{2+}$ unit (Rh-Rh = 2.58Å) may be complexed within a ditopic macrocyclic ligand $\{(NHCH_2CH_2)_3OCH_2CH_2\}_2$.[214]

X-Ray diffraction has established the structure of cis-$[Rh_2Me_2(\eta-C_5Me_5)_2(\mu-CH_2)_2]$ and this may be converted into the trans-form by action of Lewis acids.[215] Reaction of acids in the presence of donors (L = MeCN, py, etc.) with trans-$[Rh_2Me_2(\eta-C_5Me)_2(\mu-CH_2)_2]$ gives cis plus trans cations $[Rh_2LMe(\eta-C_5Me_5)_2(\mu-CH_2)_2]^+$ which equilibrate at higher temperatures by intramolecular Me migration between Rh atoms.[216] The crystal structure of $[Rh_2(\eta-C_5Me_5)_2(\mu-CH_2)_2(\mu-O_2COH)]^+$ shows Rh-Rh = 2.591(1)Å; the bicarbonato-bridge is broken by acid to liberate CO_2 and form an aquo complex.[217]

Phosphido-bridged dimers of Rh(I) can have planar/planar, planar/tetrahedral or tetrahedral/tetrahedral configurations at the two metal centres: complex $[Rh_2(CO)_4(\mu-PBu^t_2)_2]$ crystallises in each of the two former geometries, the first possessing no Rh-Rh bond but the second, Rh-Rh = 2.7609(9)Å, is assigned a single bond;[218] both isomers interconvert in solution. ^{31}P N.m.r. studies on related planar/tetrahedral complexes $[Rh_2(cod)(PR'_3)_n(\mu-PPh_2)_2]$ (n = 1, 2) are also reported.[219] Tetrahedral/tetrahedral complex $[Rh_2(CO)_2(PMe_3)_2(\mu-PBu^t_2)_2]$ has Rh-Rh = 2.550(1)Å, indicative of a double bond.[220] The complex $[Rh_2(\mu-C_5Me_5)_2(\mu-PMe_2)_2]$ reacts with O_2 and its congeners either by insertion into one or two Rh-P bonds or by addition to the Rh-Rh bond.[221]

Two centre oxidative addition of $CF_3C{\equiv}CCF_3$ to $[Ir_2(cod)_2(\mu-pz)(\mu-fpz)]$ {fpz = 3,5-$(CF_3)_2$pz} causes bridge elimination via H transfer to form unsymmetrical $[Ir_2(cod)(1\text{-}3,5,6\text{-}\eta-C_8H_{11})\{\mu-C_2(CF_3)_2\}(\mu-pz)]$.[222]

The 50e triangular compound $[Rh_3(CO)_3(\mu-CO)(\mu-Cl)_2(\mu-PPh_2)_3]$ has Rh-Rh distances consistent with single bonds; the Cl atoms may be removed as radicals with NEt_2H in presence of CO to form $[Rh_3(CO)_5(\mu-PPh_2)_3]$ or as Cl^- using $AgSbF_6$.[223] The crystal structure of $[Ir_3Ph(CO)_6(\mu-dppm)(\mu_3-PPh)]$ is reported.[224] Tri- and tetra-nuclear complexes $[Rh_3(CO)_2(\mu-CO)(\mu-I)_2\{\mu_3-PPh(CH_2PPh_2)_2\}_2]$ and $[Rh_4Cl_2(CO)_2(\mu-CO)(\mu-Cl)_2\{\mu_3-py(PPh_2)_2\}_2]$ both possess metal chains with only one short Rh-Rh separation indicative of a bond.[225] In $[Rh_4(cod)_4(\mu_4-PPh)_2]$ a square planar array of Rh_4 atoms is capped on each side by μ_4-PPh groups.[226]

N.m.r. studies at low temperatures in CH_2Cl_2 establish an equilibrium between two forms of $[Ir_4(CO)_8(PEt_3)(\mu-CO)_3]$ and that the minor isomer with PEt_3 in the plane of three (μ-CO) groups is kinetically more important in the CO scrambling process.[227] Structures of $[Ir_6(CO)_7\{P(OMe)_3\}_5(\mu-CO)(\mu_3-CO)_3]$,[228] $[Ir_6(COEt)(CO)_{11}(\mu-CO)_4]^-$, with an η^1-acyl ligand,[229] and of $[Rh_6(CO)_6(\mu-CO)_9(\mu_6-N)]$, with all edges of a Rh_6 trigonal prism bridged by CO,[230] have been reported.

The new cluster $[Rh_{11}(CO)_{11}(\mu-CO)_{12}]^{3-}$ comprises three octahedral subunits connected by face-to-face condensation.[231] $[Rh_{12}(CO)_{16}(\mu-CO)_8(\mu_8-C)_2]^{2-}$ has a Rh_{12} polyhedron with square-rhomb-square sequence and C atoms encapsulated in two isolated prismatic cavities;[232] reduction forms $[Rh_{12}(CO)_{23}C]^{n-}$ (n = 3, 4) and the paramagnetic species (n = 3), $[Rh_{12}(CO)_{13}(\mu-CO)_{10}(\mu_8-C)_2]^{3-}$, has the same framework as the parent dianion.[233] The effect of counter ions in the growth and decomposition of Rh cluster anions has been discussed;[234] higher nuclearity species are more readily formed with NR_4^+ than with Cs^+. V.T. N.m.r. studies of CO and framework fluxional processes of some Rh_9 - Rh_{12} clusters are reported.[235]

<u>2.8 Nickel</u>.- The anion $[Ni_2(\eta-C_2H_4)_4(\mu-H)]^-$ has Ni-Ni bond {2.596(1)Å}[236] and in $[Ni(CO)_2L(\mu-PBu^t_2)_2]$ Ni-Ni separations are 2.414(2)Å (L = CO) or 2.446(2)Å (L = PMe_3).[237] Reaction of $[Ni(\eta\text{-allyl})(SH)]_2$ with $LiBu^n$ gives sulphur cluster $[Ni_3(\eta-C_3H_5)_3(\mu_3-S)_2]^-$; a palladium analogue and also polynuclear derivatives $[M_2(\eta-C_3H_4Me)_2S]_x$ (M = Ni, Pd, Pt) are also reported.[238] Anions $[Ni_{12}(CO)_{21}H_{4-n}]^{n-}$ (n = 2, 3) contain interstitial H atoms and are related by reversible protonation; species (n = 4) is only formed under severe conditions; all anions are rapidly degraded by CO and possess limited thermal stability.[239]

<u>2.9 Palladium and Platinum</u>.- Paramagnetic dinuclear ions, $[Pd_2(\eta-C_5Ph_5)_2(\mu-\eta^2-C_2Ph_2)]^z$ are formed electrochemically (z = -1, +1, +2) or chemically (z = +1) from the neutral dimer; the C_5Ph_5 ligands appear to stabilise the system to oxidation.[240] In complex $[Pt_2Br_2(PPh_3)_3(\mu-CO)]$ the CO forms an unsymmetrical bridge, probably causing partial valence disproportionation.[241] Fragmentation of CS_2 in $[Pt(\eta^2-CS_2)(dppe)]$ by Pt(O) gives Pt-Pt bonded dimer $[Pt_2(CS)(PPh_3)(dppe)(\mu-S)]$.[242] The crystal structure of $[Pt_2(dppe)_2(\mu-H)(\mu-CO)]^+$ shows Pt-Pt = 2.716(1)Å.[243] Reactions using $[MgCp_2]$ have produced $[Pt_2Cp_2(\mu-C_{10}H_{10})]$ and $[Pt_2(CO)_2Cp_2]$ in good yield; the latter with alkynes C_2R_2 (R = Bu^t or Ph) gives $[Pt_2Cp_2(\mu-\eta^2-C_2Bu^t_2)]$ or $[Pt_2Cp_2(\mu\text{-CPh=CPh-CO})]$ {Pt-Pt = 2.590(1)Å}, respectively.[244] The formation of donor-acceptor Pt-Pt bonds in reactions of Pt(II) A-frame complexes has been proposed[245] and Pt-Pt bonds are present in several derivatives of type $[Pt_2X(Y)(\mu\text{-dppm})_2]^{z+}$.[246]

The triangular cation $[Pd_3(\mu\text{-dppm})_3(\mu_3-CO)]^{2+}$ is formed in high yield from $[Pd(OAc)_2]$.[247] $[Pt_7(CNR)_6(\mu-CNR)_5(\mu_3-CNR)]$ (R = 2,6-$Me_2C_6H_3$) has Pt_7 skeleton (**13**) and a 4e μ_3-CNR ligand.[248] Syntheses of clusters $[Pd_{10}(CO)_{18-n}(PR_3)_n]$ (n = 4, 6) may be achieved under mild conditions,[249a] and the structure for (n = 4, R = Bu^n), i.e. $[Pd_{10}(CO)_2(PBu^n_3)_4(\mu-CO)_8(\mu_3-CO)_4]$, with tetracapped octahedral Pd_{10} skeleton is discussed in terms of ligand close packing.[249b]

<u>2.10 Copper and Gold</u>.- Cyclic pentamer $[Cu_5(mes)_5]$ and its ring contracted adduct $[Cu_4(mes)_4(SC_4H_8)_2]$ (mes = mesityl; SC_4H_8 = tetrahydrothiophene) show relatively long Cu-Cu separations with little interaction between metals[250] but $[Au_5(mes)_5]$, with 5-pointed star Au_5C_5 arrangement, has Au-Au distances (2.692-2.710Å) supporting direct Au-Au bonding.[251] Cation $[Au_9(CNPr^i)_2(PPh_3)_6]^{3+}$ has one central Au atom with terminal ligands bonded to eight peripheral Au atoms.[252]

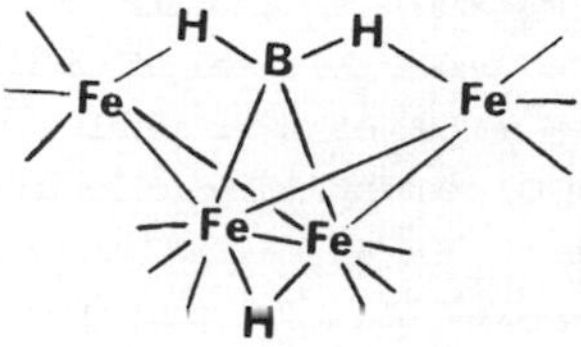

(9)

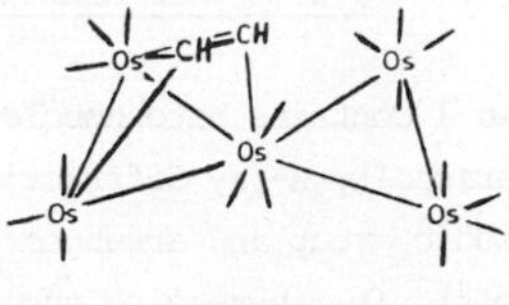

(10)

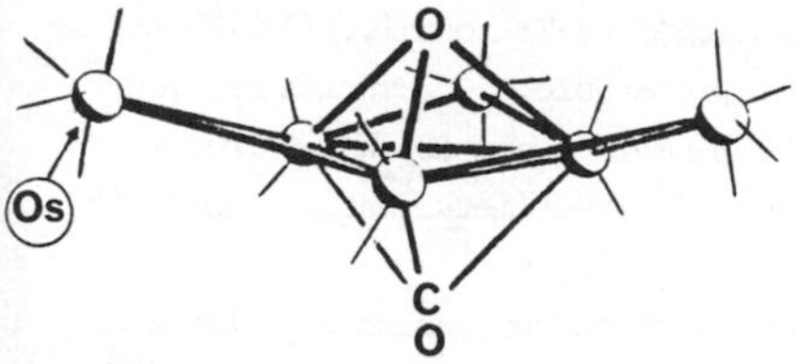

(11)

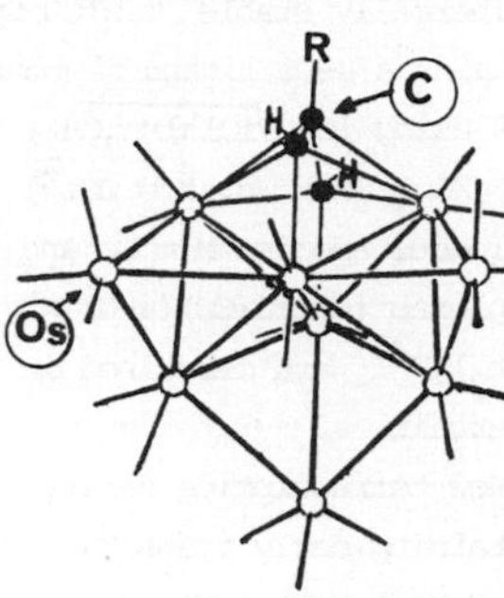

(12)

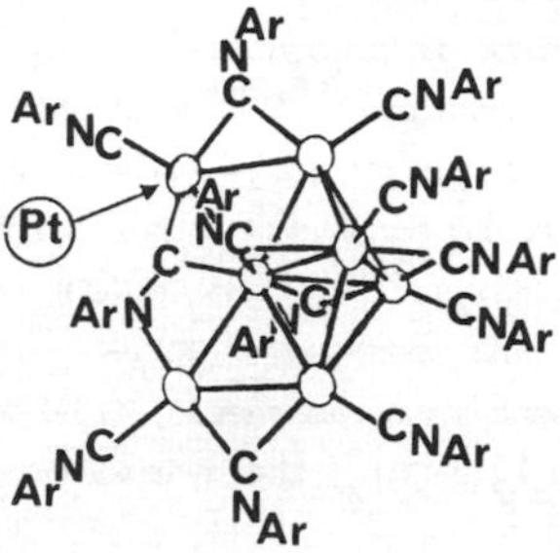

(13)

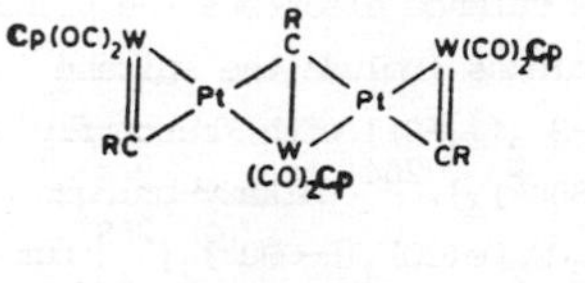

(14)

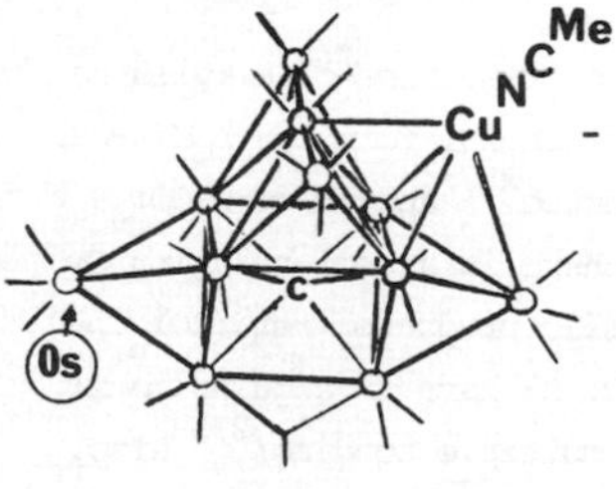

(15)

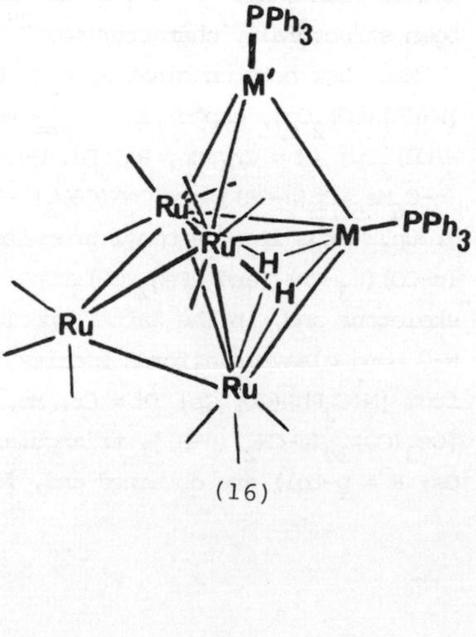

(16)

3 Compounds Containing Heteronuclear Bonds between Transition Metals

Table 1 contains heteronuclear metal-metal bonded compounds structurally characterised by X-ray diffraction; compounds are listed under the metal of earliest periodic group and arranged in (a) increasing group number of other metal atom(s), (b) increasing nuclearity, (c) decreasing CO content. The following discussion broadly follows the order of this Table.

Thermally stable [TiW(CO)Cp_3(μ-CO)(μ-CR)] (R = p-tol) has a ZrW analogue and contains a μ-CO ligand η^2-bonded to Ti;[254] methylene transfer to W≡C of this species using [Cp_2TiClAl$Me_2$$CH_2$] gives [TiW(CO)Cp(μ-CO)(μ-CR=CH_2)].[253] In [MZr-$(CO)_2$XCp_3] (M = Ru, X = OBu^t) there is an unbridged Ru-Zr bond {2.910(1)Å} and analogous species M = Fe and X = Me are also preparable.[255] Hydride bridged complexes [Cp_2Ta(CO)(μ-H)(ML_n)] (M = Cr, Mo, W, Mn)[335] and [Cp_2W$(\mu$-H$)_2$PtPh-$(PEt_3)]^+$[287] are described in which the extent of direct metal-metal bonding is uncertain.

New tetrasulphide derivatives with cubane or distorted cubane structures and containing early transition metals include [$(\eta$-$C_5H_4Me)_2V_2Fe_2(NO)_2(\mu_3$-$S)_4$],[256] [$(\eta$-$C_5Me_5)_2M_2Fe_2(NO)_2(\mu_3$-$S)_4$] (M = Cr, Mo),[336] [$Cp_3Cr_3Fe(O_2CBu^t)(\mu_3$-$S)_4$][259] and [$Cp_2Mo_2Ni_2(CO)_2(\mu_3$-$S)_4$];[267] a general synthetic method is presented.[336] In [$(\eta$-$C_5H_4Me)_2Mo_2Fe_2(CO)_6(\mu_3$-$S)_4$] the metal atoms are coplanar[266,267] with only two bonding Mo-Fe distances, whereas in [$Cp_2Mo_2Fe_2(CO)_6(\mu$-$CO)_2(\mu_3$-$S)_2$] with two less valence electrons the planar isomer has four Mo-Fe bonds.[265] Other sulphur complexes include the spirane [$Cp_4Cr_4Co(\mu$-$SBu^t)_2(\mu_3$-$S)_4$],[260] [$Cp_2Mo_2Co_2(CO)_4$-$(\mu_3$-$S)_2(\mu_4$-$S)$] with "butterfly" metal framework[267] and open [$MoFe_2(CO)_8$-$(\mu$-$SBu^t)_4$].[264] Alkoxy-bridged Cr_2 nuclei are present in paramagnetic, triangular [$Cp_2Cr_2Fe(CO)_4(\mu$-$OBu^t)_2$][258] but in [$FeRu_2(CO)_8(PPh_3)_2(\mu$-$OH)_2$] the hydroxy-bridged Ru_2 atoms are non-bonded.[296]

[$MoW(CO)_4Cp_2$] is produced by disproportionation of the homonuclear analogues and is fluxional.[337] The binuclear tungsten alkylidyne [(PhC)$(CO)_4$WCo$(CO)_4$] has been structurally characterised.[282]

Many new heteronuclear species have been formed by reactions of alkylidyne [W(CR)$(CO)_2$Cp], especially R = p-tol: compounds of general formula [L_nM(μ-CR)-W$(CO)_2$Cp] (M = Cr, Mn, Re, Fe, Co, Rh, Ir) are reported[257] and in [CrW$(CO)_4$-$(\eta$-$C_6Me_6)$Cp(μ-CR)] or [CoW$(CO)_3(\eta$-$C_5Me_5)$Cp(μ-CR)] there is a semi-bridging CO group. From iron carbonyl precursors, the trimetallic products [$Fe_2W(CO)_8$Cp-(μ-CO)$(\mu_3$-CR)] and [$FeW_2(CO)_5LCp_2(\mu_3$-$C_2R_2)$] (L = CO, O) have triangular metal skeletons and, in the latter species, possible W-W multiple bonding;[275] also, W≡C bond cleavage affords species [$Fe_2W(CO)_8Cp(\mu_3$-$C_2R)$],[277] analogous to products from [M(C_2Ph)$(CO)_3$Cp] (M = Cr, Mo, W) and [$Fe_2(CO)_9$].[276] Using [$Ru_3(CO)_{12}$] or [$Os_3(CO)_{10}(\mu$-$CH_2)(\mu$-H)], triangular derivatives [$MW_2(CO)_7Cp_2(\mu_3$-$C_2R_2)$] (M = Ru, Os; R = p-tol) are obtained and, for M = Os, exist as two isomers, fluxional in

Table 1. X-Ray determined structures of heteronuclear metal-metal bonded compounds

Titanium

$[\mathrm{TiW(CO)Cp_3(\mu{-}CO)\{\mu{-}C(tol{-}\underline{p})CH_2\}}]$ [253]
$[\mathrm{TiW(CO)Cp_3(\mu{-}CO)\{\mu{-}C(tol{-}\underline{p})\}}]$ [254]

Zirconium

$[\mathrm{ZrRu(CO)_2(OBu^t)Cp_3}]$ [255]

Vanadium

$[\mathrm{V_2Fe_2(NO)_2(\eta{-}C_5H_4Me)_2(\mu_3{-}S)_4}]$ [256]

Chromium

$[\mathrm{CrW(CO)_4(\eta{-}C_6Me_6)Cp\{\mu{-}C(tol{-}\underline{p})\}}]$ [257]
$[\mathrm{Cr_2Fe(CO)_4Cp_2(\mu{-}OBu^t)_2}]$ [258]
$[\mathrm{Cr_3Fe(O_2CBu^t)Cp_3(\mu_3{-}S)_4}]$ [259]
$[\mathrm{Cr_4CoCp_4(\mu{-}SBu^t)_2(\mu_3{-}S)_4}]$ [260]
$[\mathrm{CrRh(CO)_2(\eta{-}C_5Me_5)(\eta{-}C_6H_6)(\mu{-}CO)_2}]$ [261]
$[\mathrm{M_2Pd_2(PEt_3)_2Cp_2(\mu{-}CO)_4(\mu_3{-}CO)_2}]$
(M = Cr, Mo, W) [262]

Molybdenum

$[\mathrm{MoWFeCo(CO)_7Cp_2(\mu{-}AsMe_2)(\mu_3{-}S)}]$ [263]
$[\mathrm{MoFe_2(CO)_8(\mu{-}SBu^t)_4}]$ [264]
$[\mathrm{Mo_2Fe_2(CO)_6(\eta{-}C_5H_4Me)_2(\mu{-}CO)_2(\mu_3{-}S)_2}]$ [265]
$[\mathrm{Mo_2Fe_2(CO)_6(\eta{-}C_5H_4Me)_2(\mu_3{-}S)_4}]$ [266,267]

Molybdenum (cont.)

$[\mathrm{MoFeCo(CO)_8Cp(\mu_3{-}PMe)}]$ [268]
$[\mathrm{MoFeCo_2(CO)_8Cp(\mu{-}AsMe_2)(\mu_3{-}S)}]$ [263]
$[\mathrm{MoRuCo_2(CO)_8Cp(\mu{-}AsMe_2)(\mu_3{-}S)}]$ [263]
$[\mathrm{MoCo_2(CO)_6Cp(\mu{-}CMeNCy)(\mu_3{-}S)}]$ [269]
$[\mathrm{MoCo_2(CO)_6Cp(\mu{-}AsMe_2)(\mu_3{-}PBu^t)}]$ [268]
$[\mathrm{Mo_2Co_2(CO)_4Cp_2(\mu_3{-}S)_2(\mu_4{-}S)}]$ [267]
$[\mathrm{MoRhMe(CO)_3I(\eta{-}C_5Me_5)(\mu{-}PMe_2)_2}]$ [270]
$[\mathrm{MoRh_2(CO)_5(\eta{-}C_5Me_5)_2(\mu{-}CO)_2}]$ [271]
$[\mathrm{Mo_2Ni_2(CO)_2Cp_2(\mu_3{-}S)_4}]$ [267]
$[\mathrm{MoPt(CO)_2Cl_2(\mu{-}CO)(\mu{-}2{-}PPh_2py)_2}]$ [272]
$[\mathrm{Mo_2Pt_2(CO)(dppe)Cp_2(\mu{-}CO)_5}]$ [273]
$[\mathrm{MoCu(CO)_2(4{-}Mepy)Cp(\mu{-}OAc)_2}]$ [274]

Tungsten

$[\mathrm{WFe_2(CO)_8Cp(\mu{-}CO)\{\mu_3{-}C(tol{-}\underline{p})\}}]$ [275]
$[\mathrm{WFe_2(CO)_8Cp(\mu_3{-}C_2R)}]$
(R = Ph, [276] $\underline{p}$-tol, [277])
$[\mathrm{W_2Fe(CO)_5LCp_2\{\mu_3{-}C_2(tol{-}\underline{p})_2\}}]$
(L = CO, O) [275]
$[\mathrm{W_2Os(CO)_7Cp_2\{\mu_3{-}C_2(tol{-}\underline{p})_2\}}]$ [278]
$[\mathrm{WOs_3(CO)_{12}(PMe_2Ph)_n(\mu_3{-}S)_2}]$
(n = 1,2) [279]
$[\mathrm{WOs_3(CO)_{11}Cp\{\mu_3{-}C(tol{-}\underline{p})\}}]$ [278]

Tungsten (cont.)

$[\mathrm{WOs_3(CO)_{11}Cp\{\mu_3{-}OCCH_2(tol{-}\underline{p})\}}]$ [280]
$[\mathrm{WOs_3(CO)_{10}Cp(\mu{-}H)\{\mu_3{-}C_2(tol{-}\underline{p})_2\}}]$ [281]
$[\mathrm{WOs_3H(CO)_9Cp\{\mu_3{-}C(tol{-}\underline{p})\}_2}]$ [281]
$[\mathrm{WCo(CO)_8(CPh)}]$ [282]
$[\mathrm{WCo(CO)_3Cp(\eta{-}C_5Me_5)\{\mu{-}C(tol{-}\underline{p})\}}]$ [257]
$[\mathrm{WIrH(CO)_5(PPh_3)(\mu{-}PPh_2)_2}]$ [283]
$[\mathrm{WPt(CO)_5(dppm)(\mu{-}CCH_2)}]$ [284]
$[\mathrm{WPt(CO)_5\{\mu{-}C(OMe)(tol{-}\underline{p})\}(\mu{-}dppm)}]$ [285]
$[\mathrm{WPt(CO)_2(PMe_3)_2Cp\{\mu{-}C(tol{-}\underline{p})CH_2\}}]$ [253]
$[\mathrm{WPt(\eta{-}C_2H_4)(PEt_3)_2Cp(\mu{-}CO)_2}]^+$ [286]
$[\mathrm{WPtPh(PEt_3)Cp_2(\mu{-}H)_2}]^+$ [287]
$[\mathrm{W_2Pt_3(CO)_4(cod)_2Cp_2\{\mu_3{-}C(tol{-}\underline{p})\}_2}]$ [288]
$[\mathrm{W_3Pt_2(CO)_6Cp_3\{\mu{-}C(tol{-}\underline{p})\}_2{-}}$
$\mathrm{\{\mu_3{-}C(tol{-}\underline{p})\}}]$ [288]
$[\mathrm{WPtCu(CO)_2(PMe_3)_2Cp(\eta{-}C_5Me_5){-}}$
$\mathrm{\{\mu_3{-}C(tol{-}\underline{p})\}}]$ [289]
$[\mathrm{WCu(CO)_3(PPh_3)_2Cp}]$ [290]
$[\mathrm{W_2Au(CO)_4Cp_2\{\mu{-}C(tol{-}\underline{p})\}_2}]^+$ [288]

Manganese

$[\mathrm{MnFeCo(CO)_9(\mu{-}CO)(\mu_3{-}PBu^t)}]$ [291]
$[\mathrm{MnPd(CO)_3Br(\mu{-}dppm)_2}]$ [292]
$[\mathrm{Mn_2Pd_2(CO)_7(\mu{-}CO)(\mu{-}dppm)_2}]$ [293]

Table 1. (continued)

Iron	**Ruthenium**	**Osmium** (cont.)
$[FeRu(CO)_6(\mu-C_{16}H_{22})]$ [294]	$[RuCo_2(CO)_9(\mu_3-C_2Ph_2)]$ [310]	$[Os_3Ni(CO)_9Cp(\mu-H)(\mu_4-CCHBu^t)]$ [323]
$[FeRu(CO)_2Cp_2(\mu-CO)\{\mu-P(NPF_2)_2N\}]$ [295]	$[RuCo_3(CO)_9(\mu-CO)_3]^-$ [311]	$[Os_3Pt(CO)_{11}(PCy_3)(\mu-H)_2]$ [324]
$[FeRu_2(CO)_8(PPh_3)_2(\mu-OH)_2]$ [296]	$[RuCo_3(CO)_8(\mu-CO)_2(\mu_4-C_2Ph_2)]$ [310]	$[Os_3Pt(CO)_{10}(PCy_3)(\mu-H)_4]$ [324]
$[FeCo(CO)_3Cp\{\mu-CPhC(COOMe)C(COOMe)CPh\}]$ [297]	$[Ru_2Co_2(CO)_9(\mu-H)(\mu-CO)_3(\mu_3-H)]$ [312]	$[Os_{10}Cu(CO)_{24}(NCMe)(\mu_6-C)]^-$ [325]
$[FeCo_2(CO)_9(\mu_3-NH)]$ [298]	$[Ru_2Co_2(CO)_9(\mu-CO)_2(\mu_4-C_2Ph_2)]$ [312]	$[Os_{11}Cu(CO)_{26}(NCMe)(\mu-CO)(\mu_6-C)]^-$ [184]
$[FeCo_2(CO)_3Cp_2(\mu_3-CPh)_2]$ [299]	$[Ru_2Co_2Au_2(CO)_{10}(PPh_3)_2(\mu-CO)_2]$ [308]	$[Os_3Au(CO)_8(PPh_3)_2(\mu-2-NHpy)]$ [326]
$[Fe_2Co(CO)_9(\mu-PHPh)(\mu_3-PPh)]$ [129]	$[Ru_3Ni(CO)_9Cp(\mu-H)(\mu_4-CCHPr^i)]$ [313]	$[Os_3Au_2(CO)_{10}(PEt_3)_2]$ [327]
$[Fe_2Co(CO)_6Cp(\mu-CO)(\mu_3-PBu^t)]$ [291]	$[Ru_4MM'(CO)_{12}(PPh_3)_2(\mu_3-H)_2]$	$[Os_5Au_2(CO)_{14}(PPh_3)_2(\mu_5-C)]$ [174c]
$[FeCo_3(CO)_7(\mu-CO)_2(\mu-CPhCHPh)-$	(M = M′ = Cu,Ag: M = Cu, M′ = Ag) [314]	$[Os_8Au_2(CO)_{22}(PPh_3)_2]$ [328]
$(\mu_4-C_2Ph_2)]$ [300]	$[Ru_3Au(CO)_{10}(PPh_3)(\mu-COMe)]$ [315]	$[Os_{10}Au(CO)_{24}(PPh_3)(\mu_6-C)]^-$ [325]
$[FeCo_3(CO)_7Cp(\mu-CO)_2(\mu_4-CCH_2)]$ [123]	$[Ru_3Au(CO)_9(PPh_3)(\mu-H)_2(\mu_3-COMe)]$ [315]	**Cobalt**
$[Fe_2Co_2(CO)_{10}(\mu-CO)(\mu_4-PPh)_2]$ [129]	$[Ru_3Au_2(CO)_9(PPh_3)_2(\mu-H)(\mu_3-COMe)]$ [316]	$[CoRh(\eta-C_5Me_5)_2(\mu-CO)_2]$ [209]
$[FeCoNi(CO)_6Cp(\mu_3-PBu^t)]$ [301]	$[Ru_3Au_2(CO)_8(PPh_3)_3(\mu_3-S)]$ [316]	$[Co_2Pd_2(CO)_6(\mu-dppm)_2(\mu_3-CO)]$ [273]
$[Fe_2CoNi(CO)_8Cp(\mu_4-PMe)_2]$ [301]	$[Ru_3Au_3(CO)_9(PPh_3)_3(\mu_3-COMe)]$ [315]	$[Co_2Au_6(CO)_8(PPh_3)_4]$ [329]
$[Fe_2Rh_2(CO)_6(\mu-CO)_2(\mu-PPh_2)_4]$ [302]	$[Ru_4Au_3(CO)_{12}(PPh_3)_3(\mu-H)]$ [317]	**Rhodium**
$[Fe_5Rh(CO)_{12}(\mu-CO)_4(\mu_6-C)]^-$ [303]	$[Ru_5AuCl(CO)_{15}(PPh_3)(\mu_5-C)]$ [318]	$[Rh_2Ir_2(CO)_4(\eta-C_5Me_5)_2(\mu-CO)(\mu_3-CO)_2]$ [321]
$[FeIr(CO)_5(PPh_3)_2(\mu-PPh_2)]$ [304]	$[Ru_5Au(CO)_{14}(PPh_3)(\mu-Br)(\mu_5-C)]$ [318]	$[RhPdCl_3(CO)(\mu-2-PPh_2py)_2]$ [330a]
$[FePd(PPh_3)(\mu-\eta-C_5H_4S)_2]$ [305]	$[Ru_5Au(CO)_{13}(PPh_3)_2(\mu-I)(\mu_5-C)]$ [174b]	$[Rh_{11}Pt_2(CO)_{12}(\mu-CO)_{12}]^{3-}$ [331]
$[Fe_2Ag(CO)_6(\mu-CHCPhNHMe)(\mu-PPh_2)]^+$ [306]	$[Ru_6Au(CO)_{13}(NO)(PPh_3)(\mu-CO)_2(\mu_6-C)]$ [145]	$[Rh_{12}Pt(CO)_{12}(\mu-CO)_{12}]^{4-}$ [331]
$[Fe_3Au(CO)_9(PPh_3)(\mu_3-HCNBu^t)]$ [307]	**Osmium**	$[Rh_2Cu(\eta-C_5Me_5)_3(\mu-CO)_2]$ [289]
$[Fe_3Au_2(CO)_9(PPh_3)_2(\mu_3-S)]$ [308]	$[Os_3Co(CO)_9Cp(\mu-H)_3]$ [319]	$[Rh_2Ag(CO)_2(PPh_3)_2Cp_2]^+$ [332]
$[Fe_2Au_4(CO)_8(\mu-dppm)_2]$ [309]	$[Os_3Rh(CO)_9(\eta-C_6H_5Me)(\mu-H)_3]$ [320]	$[Rh_{12}Ag(CO)_{12}(\mu-CO)_{18}(\mu_6-C)_2]^{3-}$ [333]
$[\{FeAu_2(CO)_4\}_2(\mu-dppe)_2]$ [309]	$[Os_3Ir(CO)_{12}(\mu-H)_2(\mu-Cl)]$ [321]	**Iridium**
	$[Os_3Ni(CO)_9Cp(\mu-H)_3]$ [319,322a]	$[IrCuCl(C_2Ph)(CO)(\mu-dppm)_2]$ [334]

solution.[278] Tetrahedral $[Os_3W(CO)_{11}Cp(\mu_3\text{-}CR)]$[278] and the triangulated, rhomboidal acyl cluster $[Os_3W(CO)_{11}Cp(\mu_3\text{-}COCH_2R)]$[280] are also produced using $[W(CR)(CO)_2Cp]$ (R = p-tol). Stepwise syntheses of W-Pt or W-Au bonds using $[W(CR)(CO)_2Cp]$ yield compounds with alternating metal chains and products involving up to 5 metals are characterised by X-ray diffraction, e.g. $[Pt_2W_3(CO)_6Cp_3(\mu\text{-}CR)_2(\mu_3\text{-}CR)]$, (**14**).[288] The combination of $[Mn(CO)_2LCp]$ {L = C=CH(CO_2Me) or η-CH≡C(CO_2Me)} with $[W(CO)_5(THF)]$ affords $[MnW(CO)_6Cp\{\mu\text{-}CCH(CO_2Me)\}]$.[338]

The μ-ethylidyne group in $[PtW(CO)_2(PEt_3)_2Cp(\mu\text{-}CMe)]$ can be transformed by proton transfer reactions into a η-ethylene, μ-vinyl or μ-ethylidene ligand.[286]

$[(\eta\text{-}C_5Me_5)MeRh(\mu\text{-}PMe_2)_2Mo(CO)_3I]$ is the oxidative addition product of MeI across a polar Rh-Mo bond.[270] Nucleophilic alkylation of CO ligands at W in $[(OC)_4W(\mu\text{-}PPh_2)_2IrH(CO)(PPh_3)]$ leads to stable acyl-hydride and carbene-hydride complexes.[283] Derivatives $[(R_3P)(OC)_3Fe(\mu\text{-}PPh_2)M(CO)_n(PR)_3]$ (M = Rh, Ir; n = 1, 2) are described; the weak donor-acceptor Fe→Ir bond (M = Ir, n = 2) is cleaved by addition of CO to give n = 3.[304] Other phosphido bridged complexes include straight chain species $[Fe_2Rh_2(CO)_6(\mu\text{-}CO)_2(\mu\text{-}PPh_2)_4]$[302] and binuclear $[MnRu(CO)_6(PPh_3)_2(\mu\text{-}PPh_2)]$.[339]

Using bridging ligands ($\eta\text{-}C_5H_4PPh_2$), (2-PPh_2py) or ($\eta\text{-}C_5H_4S$), the binuclear derivatives $[MoM(CO)_3L_n(\mu\text{-}C_5H_4PPh_2)]$ (M = Rh, Ir, L = CO, PR_3, n = 2; M = Mn, Re, L = CO, n = 4),[340] $[MoPt(CO)_2Cl_2(\mu\text{-}CO)(\mu\text{-}PPh_2py)_2]$,[272] (c.f. related Rh-Pd and Rh-Pt species)[330] and $[FePd(PPh_3)(\mu\text{-}C_5H_4S)_2]$,[305] with dative Fe→Pd bond, respectively, have been characterised and studied.

Addition of $M(CO)_5$ (M = Cr, Mo, W) fragments to $[RhM'(\eta\text{-}C_5Me_5)_2(\mu\text{-}CO)_2]$ (M' = Rh, Co) affords triangular clusters $[MRhM'(CO)_5(\eta\text{-}C_5Me_5)_2(\mu\text{-}CO)_2]$ and when M = Mo, V.T. n.m.r. studies indicate alkene-like rotation of the RhM'[261,271] unit. From $[Rh(CO)_2Cp]$, binuclear species $[CrRh(CO)_2(\eta\text{-}C_5Me_5)(\eta\text{-}arene)(\mu\text{-}CO)_2]$ with formal Cr→Rh bonds are obtained.[261]

Structures of tetrametallic, planar, triangulated clusters $[M_2M'_2(PR_3)_2Cp_2(\mu\text{-}CO)_4(\mu_3\text{-}CO)_2]$ (M = Cr, Mo, W; M' = Pd; R = Et) are reported[262] and electrochemical studies on the Pd and Pt analogues (M = Cr, Mo, W; M' = Pd, Pt; R = Ph) show an irreversible 2e reduction leading to cluster rupture.[341] Other heteronuclear clusters of Pd and Pt with earlier metals include $[Mo_2Pt_2(CO)(dppe)Cp_2(\mu\text{-}CO)_5]$[273], $[MPd_2ClCp(\mu\text{-}CO)_2(\mu\text{-}dppm)_2]$ (M = Mo, W)[273], $[MM'Pt(PPr^i_3)_2(\mu\text{-}CO)_2(\mu\text{-}C_3H_4Me)(\mu_3\text{-}CO)]$ (M = Cr, Mo, W; M' = Pd, Pt)[342] and related triangular clusters of MPd_2 or MPdPt (M = Mn,[293] Fe,[343], Co[273]). Binuclear PtW[284,285] and MnPd[292] complexes, containing dppm ligands, are also described.

Copper containing dinuclear complexes with two semi-bridging carbonyls $[MCu(CO)_3(PPh_3)_2Cp]$ (M = Mo, W)[290] and $[MoCu(CO)_2(4\text{-}Mepy)Cp(\mu\text{-}OAc)_2]$[274] are characterised and two modifications of the former (M = W) exhibit different $W(CO)_2Cu$ interactions. The fragment $Cu(\eta\text{-}C_5Me_5)$, isolobal with CH_2, can be incorporated into species $[(\eta\text{-}C_5Me_5)CuPtW(CO)_2(PMe_3)_2Cp\{\mu\text{-}C(tol\text{-}p)\}]$ and also

$[CuRh_2(\eta-C_5Me_5)_3(\mu-CO)_2]$.[289]

Deliberate, stepwise syntheses of heteroclusters, including chiral examples, have been developed. Trinuclear systems with cores $\{MM'M''(\mu_3-PR)\}$ (M = Fe, M' = Co, etc., M'' = various) are formed from monomers with chlorophosphine ligands,[291] e.g. PMe(H)Cl, $PMe(NMe_2)Cl$, or with PRH_2[344] ligands and from dimers with μ-PRH bridges.[129] Addition of 4e fragments to $[Co_2Ru(CO)_{10}(\mu-CO)]$ produces $[Co_2Ru(CO)_9(\mu_3-E)]$ (E = S, Se, PR, AsR, C_2R_2) and expanded clusters (E = W(CR)(CO)Cp, $Fe(C_2R)(CO)Cp$, $Ru(CO)_4$, $Co(CO)_4^-$);[345] aggregation of the former products (E = S, PR), and related M_3E species, via initial substitution with $[M(AsR_2)(CO)_3Cp]$ (M = Cr, Mo, W) affords tetranuclear clusters, including the mixed systems $[CoMMoW(CO)_7Cp_2(\mu-AsMe_2)(\mu_3-S)]$ (M = Fe, Ru).[263] Similarly $[Fe_3(CO)_9(\mu_3-Te)_2]$, via $[Fe_2(CO)_6(\mu-Te)_2]$, is a precursor to heteroclusters, e.g. $[Fe_2Rh(CO)_n(\mu_3-Te)_2]$ (n = 6, 7).[346] Aggregation of μ_3-acetylide bridged M_3 clusters with fragment CpNi (from $[Ni_2Cp_2(\mu-CO)_2]$) gives $[M_3Ni(CO)_9Cp(\mu-H)(\mu_4-CCHR)]$ (M = Ru, Os).[313,323] Metal exchange is another synthetic route to heteroclusters: thus, $[Co_3(CO)_7LX(\mu_3-S)]$ may be converted into optically active clusters $[Co_2Mo(CO)_6LX(Cp)(\mu_3-S)]$ by reaction with $[Mo(CO)_3Cp]^-$[269] and electron transfer catalysed exchange in $[Co_3(CO)_9(\mu_3-CR)]$ produces species $[Co_2M(CO)_8Cp(\mu_3-CR)]$.[347] Metal exchange yields chiral clusters with tetrahedral [CoFeMP] cores (M = Mo, W).[268] Direct metal exchange for Ni can be achieved in $[Co_2Fe(CO)_9(\mu_3-PR)]$ with $[Ni_2Cp_2(\mu-CO)_2]$ but other products are also obtained[301] and it is noted the $[NiCp_2]$ can cause metal and/or ligand exchange with Co_3 clusters.[348]

Reversible opening of one heterometallic bond is observed on addition of many 2e ligands to $[MnFe_2(CO)_8Cp(\mu_3-PR)]$[349] or of CO to $[Os_3Pt(CO)_{10}(PCy_3)(\mu-H)_4]$.[324] Cluster $[Os_3W(CO)_{12}(PMe_2Ph)(\mu_3-S)_2]$ is partially opened on formation of adduct with PMe_2Ph.[279]

$[Co_2Ru_2(CO)_{13}]$ reacts with H_2 at a Ru atom but with C_2Ph_2 at the Co-Co bond;[312] this alkyne, C_2Ph_2, reacts with $[HFeCo_3(CO)_{12}]$ to form $[FeCo_3(CO)_7(\mu-CO)_2(\mu-CPhCHPh)(\mu_4-C_2Ph_2)]$.[300] Coordinated μ_3-alkynes undergo thermal scission into two μ_3-CR ligands on an Os_3W cluster[281] and are isomerised into μ_3-vinylidene ligand on a Co_2Ru cluster.[350] Thermolysis of $[FeCo(CO)_7(\mu-CHCH_2)]$ converts the μ-vinyl group into μ_3-ethylidyne or μ_3-vinylidene ligand in trinuclear products.[351]

Tetrahedral osmium clusters with Os_3M cores (M = Co,[319] Rh,[320] Ni[319,322]) have been structurally characterised.

The simple binuclear species $[Ph_3PAuOs(PMe_3)CNR(\eta-C_6H_6)]^+$ are reported.[352] The range of Fe, Ru and Os clusters containing $Au(PR_3)$ groups, which formally replace H ligands, has expanded (e.g. see Table 1). Most syntheses have involved reactions of cluster anions with $[AuCl(PR_3)]$ (R = Ph, etc.) (e.g. refs. 145, 307, 308, 325-328) but $[\{Au(PPh_3)\}_3O]BF_4$ has also been used[317] as have reactions of $[AuMe(PPh_3)]$ with hydrido clusters[315,316] and oxidative addition of Au-X (X = Cl,

Br) to $[Ru_5(CO)_{15}(\mu_5\text{-}C)]$.[318] The in situ synthesis of isocyanato-gold triosmium clusters from $[Os_3(CO)_{12}]$ with azide ions and $[AuCl(PR_3)]$ is also described.[353] $Au(PPh_3)$ groups normally cap 2 or, less commonly, 3[328] heterometal atoms, although at higher incorporation numbers Au atoms may become more completely involved in the metal core[308,315-317] with Au-Au interactions being present in such derivatives, e.g. $[(Ph_3P)_3Au_3Ru_3(CO)_9(\mu_3\text{-}COMe)]$[315] and $[(Ph_3P)_3Au_3Ru_4(CO)_{12}\text{-}(\mu\text{-}H)]$.[317] The fluxionality of trigonal bipyramidal $[Au_2Ru_3(CO)_8(PPh_3)_3(\mu_3\text{-}S)]$ in solution may arise through facile Au-Ru bond rupture.[316] Clusters in which Au-Au bonding is important are $[(Ph_3P)_4Au_6Co_2(CO)_8]$[329] and $[Au_4Fe_2(CO)_8\text{-}(\mu\text{-}dppm)_2]$[309]; the rhomboidal core of the latter and the separate linked Au_2Fe triangles of $[\{Au_2Fe(CO)_4\}_2(\mu\text{-}dppe)_2]$ exemplify the structural effects of the "bite" of the diphosphine ligands.[309]

Capping of a triangular face by Cu is observed in $[(MeCN)CuOs_{10}(CO)_{24}\text{-}(\mu_6\text{-}C)]^-$[325] but a fourth weaker Cu-Os interaction is present in the first structurally characterised Os_{11} cluster $[(MeCN)CuOs_{11}(CO)_{26}(\mu\text{-}CO)(\mu_6\text{-}C)]^-$ (15), with carbide in a central trigonal prismatic cavity.[184] In fluxional, bicapped tetrahedral clusters $[(Ph_3P)_2MM'Ru_4(CO)_{12}(\mu_3\text{-}H)_2]$ (16) (M,M′ = variously Cu, Ag, Au) the coinage metals occupy two distinct sites, one bonded to four and the other to three metal atoms.[314]

Unsaturated complexes $[CoRh(\eta\text{-}C_5Me_5)_2(\mu\text{-}CO)_2]$[209] and $[CpCoIr(\eta\text{-}C_5Me_5)\text{-}(\mu\text{-}CO)_2]$[354] are described; addition of low valent metal complexes to the CoRh bond of the former yields tri- and tetra-nuclear clusters.[209] Other dinuclear CoRh complexes are reported[188a,209] and binuclear complexes of Rh or Ir with Cu, Ag or Au bridged by dppm include species with formal dative metal-metal bonds.[334]

High nuclearity anions $[Pt_nRh_{13-n}(CO)_{12}(\mu\text{-}CO)_{12}]^{(5-n)-}$ (n = 1, 2) have h.c.p. metal frameworks with a central Pt atom.[331] Double and triple decker sandwich anions $[Ag_n\{Rh_6(CO)_{15}C\}_2]^{(4-n)-}$ (n = 1, 3) and $[Ag_n\{Rh_6(CO)_{15}C\}_3]^{(6-n)-}$ (n = 2, 4) and other adducts of $[Rh_6(CO)_{15}C]^{2-}$ with Cu(I), Ag(I) and Au(I) have been studied.[333]

4 Compounds Containing Bonds between Transition Metals and Elements of Groups lA and II–IVB

4.1 Group lA.– The structure of a trimeric, hydrocarbon soluble lithium carrier, $[LiCo(\mu\text{-}C_2H_4)(PMe_3)_3]_3$ is reported.[355]

4.2 Group IIB.– Binuclear $[Me_2N(CH_2)_3ZnW(CO)_3Cp]$ has bond length W-Zn = 2.685(3)Å.[356] Anions of general formula $[M'(ML_n)_3]^-$ (M′ = Zn, Cd, Hg; M = Cr, W, Fe, Co) can be prepared and a crystal structure (M′ = Zn; ML_n = $Fe(CO)_2Cp$) has been determined;[357] similar anions $[M'\{Fe(CO)_4\}_2]^{2-}$ are also described including the structural characterisation for M′ = Hg.[358] The first Zn containing hetero cluster is $[Ni_2Zn_4Cp_6]$ with a compressed octahedral skeleton.[359]

Trans-$[Mo(HgCN)(CO)_2(AsMe_2Ph)Cp]$ has a Mo-Hg bond {2.654(1)Å}.[360] In clusters $[Hg\{Ru_3(CO)_9(\mu_3\text{-}C_2Bu^t)\}_2]$ and $[Hg\{Ru_3(CO)_9(\mu_3\text{-}C_2Bu^t)\}\{Mo(CO)_3Cp\}]$ the Hg atoms connect the two metal fragments.[361] In the heterocluster $[Hg_2Pd_4Br_2(PEt_3)_4\text{-}(\mu\text{-}CO)_4]$ two triangular faces of a Pd_4 butterfly unit are capped by HgBr.[362]

4.3 Group IIIB.- Reaction of $InCl_3$ with $[Nb(BH_4)Cp'_2]$ forms $[In_2Nb_3H_3(\eta\text{-}C_5H_4Me)_6\text{-}(\mu\text{-}Cl)_2]$ containing an internal In_2Nb triangle.[363] A series of thallium anions $[TlR_3(ML_n)]^-$ (ML_n = $Mo(CO)_3Cp$, $W(CO)_3Cp$, $Mn(CO)_5$, $Co(CO)_4$) has been prepared.[364]

4.4 Group IVB.- The crystal structure of $[Fe(SiMe_2SiMe_3)(CO)_2Cp]$ has been reported.[365] The chelating ligand, $Ph_2PCH_2SiMe_2H$, forms dinuclear $[Ru_2(CO)_6\text{-}(\mu\text{-}PPh_2CH_2SiMe_2)_2]$ from $[Ru_3(CO)_{12}]$.[366] Under CO pressure, $[Os_3(CO)_{12}]$ and $SiHCl_3$ react to form $[Os_3(CO)_{12}(SiCl_3)_2]$ with a linear $SiOs_3Si$ backbone, unlike $[Os_3(CO)_9(SiCl_3)_3(\mu\text{-}H)_3]$ with triangular Os_3 core, formed in absence of CO.[367] Other interesting silicon species include $[M_2(CO)_8\{\mu\text{-}SiIM(CO)_5\}_2]$ (M = Mn, Re)[368] and Ir(V) derivatives such as $[IrH_2(SiEt_3)_2(\mu\text{-}C_5Me_5)]$.[369]

A Ge atom links four Fe atoms in $[Ge\{Fe(CO)_4\}_4]$[370] and related complex $[Ge\{Fe(CO)_4\}_2\{FeMn(CO)_6(\mu\text{-}C_5H_4Me)\}]$ has also been structurally characterised.[370a] Reactions of halogermylenes yield several transition metal products including $[FGeCo(CO)_4]$, the first germylene with a transition metal bond.[371] Compounds $[Mn(CO)_5Ge(CF_3)_3]$ and $[Co(CO)_4Ge(CF_3)_3]$ have been prepared and studied;[372] related 16e species $[Co(PMe_3)_3MPh_3]$ {M = Ge and Sn (X-ray data)} add small 2e ligands.[373] In the linear Os_2Ge heterometallic $[Os_2(GeCl_3)(CO)_8Cl]$ there is an unbridged donor-acceptor interaction between two Os atoms {Os-Os = 2.931(1)Å; Os-Ge = 2.430(3)Å}.[374]

Anions $[Nb(CO)_3(MR_3)Cp]^-$ (M = Ge, Sn, Pb) are reported[375] and crystal structures have been determined for species $[Nb(CO)(SnX_3)Cp_2]$ (X = Ph, Cl),[376] $[Cr(CO)_5(SnS(CH_2)_2NBu^t(CH_2)_2S)]$,[377] and $[W(CO)_3(SnPh_3)_2\{(SnPh_2)_2OPr^i\}]$.[378] Tin containing anions of Cr, Mo and W in highly reduced oxidation states can be prepared, i.e. $[M(CO)_4(SnPh_3)_n]^{(4-n)-}$ (n = 2, 3).[379]

Branched heterometallic chains incorporating M≡M bonds are present in complexes $[M_2\{Sn(SnMe_3)_3\}_2(NMe_2)_4]$ {M = Mo (X-ray), W}.[380] Crystal structures are reported for $[Br_2Sn\{Re(CO)_4PPh_3\}_2]$[381] and for diastereomeric $[Ru(SnCl_3)\{(R)\text{-}Ph_2P\text{-}CHMeCH_2PPh_2\}Cp]$, prepared with retention of stereochemistry by reaction of $SnCl_2$ with a Ru-Cl bond.[382]

References

1 F.A.Cotton, Chem. Soc. Rev., 1983, 12, 35.
2 M.H. Chisholm, Polyhedron, 1983, 2, 681.
3 Adv. Organomet. Chem., 1983, 22: (a) H.Vahrenkamp, p. 169; (b) J.S.Bradley, p. 1; (c) M.I.Bruce and A.G.Swincer, p. 60; (d) E.Singleton and

H.E.Oosthuizen, p. 209.
4 M.J.McGlinchey, M.Mlekuz, P.Bougeard, B.G.Sayer, A.Marinetti, J.Y.Saillard and G.Jauoen, Can. J. Chem., 1983, 61, 1319.
5 E.Sappa, A.Tiripicchio and P.Braunstein, Chem. Rev., 1983, 83, 203.
6 B.F.G.Johnson, J.Lewis, W.J.H.Nelson, J.N.Nicholls and M.D.Vargas, J. Organomet. Chem., 1983, 249, 255.
7 R.D.Adams, Acc. Chem. Res., 1983, 16, 67.
8 J.Holton, M.F.Lappert, R.Pearce and P.I.W.Yarrow, Chem. Rev., 1983, 83, 135.
9 W.A.Herrmann, J. Organomet. Chem., 1983, 250, 319.
10 R.J.Puddephatt, Chem. Soc. Rev., 1983, 12, 99.
11 M.I.Bruce, J. Organomet. Chem., 1983, 242, 147 and 257, 417.
12 A.Poë, Chem. Br., 1983, 19, 997.
13 A.R.Manning, Coord. Chem. Rev., 1983, 51, 41.
14 F.Bottomley, Inorg. Chem., 1983, 22, 2656.
15 N.M.Kostic and R.F.Fenske, Inorg. Chem., 1983, 22, 666.
16 D.G.Evans, J. Chem. Soc., Chem. Commun., 1983, 675.
17 T.Ziegler, J. Am. Chem. Soc., 1983, 105, 7543.
18 R.L.DeKock, E.J.Baerends and A.Oskam, Inorg. Chem., 1983, 22, 4158.
19 P.T.Chesky and M.B.Hall, Inorg. Chem., 1983, 22, 2102, 2998 and 3327.
20 (a) M.Casarin, D.Ajo, G.Granozzi, E.Tondello and S.Aime, J. Chem. Soc., Dalton Trans., 1983, 869; (b) G.Granozzi, E.Tondello, M.Casarin, S.Aime and D.Osella, Organometallics, 1983, 2, 430; (c) G.Granozzi, R.Bertoncello, M.Acampora, D.Ajo, D.Osella and S.Aime, J. Organomet. Chem., 1983, 244, 383.
21 R.L.DeKock, P.Deshmukh, T.K.Dutta, T.P.Fehlner, C.E.Housecroft and J.L.S. Hwang, Organometallics, 1983, 2, 1108.
22 G.Granozzi, E.Tondello, R.Bertoncello, S.Aime and D.Osella, Inorg. Chem., 1983, 22, 744.
23 C.E.Housecroft and T.P.Fehler, Organometallics, 1983, 2, 690.
24 D.G.Evans and D.M.P.Mingos, Organometallics, 1983, 2, 435.
25 (a) D.M.P.Mingos, J. Chem. Soc., Chem. Commun., 1983, 706; (b) D.M.P.Mingos and D.G.Evans, J. Organomet. Chem., 1983, 251, C13.
26 Yu.L.Slovokhotov and Yu.T.Struchkov, J. Organomet. Chem., 1983, 258, 47.
27 T.P.Fehlner, C.E.Housecroft and K.Wade, Organometallics, 1983, 2, 1426.
28 B.K.Teo, J. Chem. Soc., Chem. Commun., 1983, 1362.
29 S.L.Cook, J.Evans, G.N.Greaves, B.F.G.Johnson, J.Lewis, P.R.Raithby, P.B. Walls and P.Worthington, J. Chem. Soc., Chem. Commun., 1983, 777.
30 S.L.Cook, J.Evans and G.N.Greaves, J. Chem. Soc., Chem. Commun., 1983, 1287.
31 S.F.A.Kettle, R.Rossetti and P.Stanghellini, Inorg. Chem., 1983, 22, 661.
32 J.Evans and G.S.McNulty, J. Chem. Soc., Dalton Trans., 1983, 639.
33 C.E.Anson, B.T.Keiller, I.A.Oxton, D.B.Powell and N.Sheppard, J. Chem. Soc., Chem. Commun., 1983, 470.
34 A.G.Orpen and R.K.McMullan, J. Chem. Soc., Dalton Trans., 1983, 463.
35 G.E.Hawkes, E.W.Randall, S.Aime, R.Gobetto and D.Osella, Polyhedron, 1983, 2, 1235.
36 B.T.Heaton, L.Strona, R.D.Pergola, L.Garlaschelli, U.Sartorelli and I.H. Sadler, J. Chem. Soc., Dalton Trans., 1983, 173.
37 R.P.Brint, K.O'Cuill, T.R.Spalding and F.A.Deeney, J. Organomet. Chem., 1983, 247, 61; R.P.Brint, M.P.Collins, T.R.Spalding and F.A.Deeney, ibid., 1983, 258, C57.
38 Y.Iwasawa, M.Yamada, S.Ogasawara, Y.Sato and H.Kuroda, Chem. Lett., 1983, 621.
39 A.Theolier, A.Choplin, L.D'Ornelas, J.M.Basset, G.Zanderighi and C. Sourisseau, Polyhedron, 1983, 2, 119.
40 C.Tessier-Youngs, F.Correa, D.Pioch, R.L.Burwell and D.F.Shriver, Organometallics, 1983, 2, 898.
41 J.R.Budge, J.P.Scott and B.C.Gates, J. Chem. Soc., Chem. Commun., 1983, 342.
42 Y.Doi, K.Soga, K.Yano and Y.Ono, J. Mol. Catal., 1983, 19, 359.
43 J.Evans and B.P.Gracey, J. Chem. Soc., Chem. Commun., 1983, 247.
44 D.F.Forster, J.Harrison, B.S.Nicholls and A.K.Smith, J. Organomet. Chem., 1983, 248, C29.
45 R.Hemmerich, W.Keim and M.Röper, J. Chem. Soc., Chem. Commun., 1983, 428.

46 R.Bender and P.Braunstein, J. Chem. Soc., Chem. Commun., 1983, 334.
47 J. Mol. Catal., 1983, 21; pp 95, 189, 331 and 389.
48 L.N.Lewis and K.G.Caulton, J. Organomet. Chem., 1983, 252, 57.
49 C.Elschenbroich, J.Heck, W.Massa, E.Nun and R.Schmidt, J. Am. Chem. Soc., 1983, 105, 2905.
50 K.Jonas, V.Wiskamp, Y.H.Tsay and C.Krüger, J. Am. Chem. Soc., 1983, 105, 5480.
51 C.M.Bolinger, T.B.Rauchfuss and A.L.Rheingold, J. Am. Chem. Soc., 1983, 105, 6321.
52 D.A.Lemenovskii, K.Romanenkova and E.G.Perevalova, Koord. Khim., 1983, 9, 903; D.A.Lemenovskii, M.V.Tsikalova, S.A.Konde and E.G.Perevalova, ibid., p. 1060.
53 M.R.Churchill and H.J.Wasserman, J. Organomet. Chem., 1983, 247, 175.
54 P.A.Belmonte, F.G.N.Cloke and R.R.Schrock, J. Am. Chem. Soc., 1983, 105, 2643.
55 C.Elschenbroich, J.Heck, W.Massa and R.Schmidt, Angew. Chem., Int. Ed. Engl., 1983, 22, 330.
56 S.G.Bott, N.G.Connelly, M.Green, N.C.Norman, A.G.Orpen, J.F.Paxton and C.J. Schaverien, J. Chem. Soc., Chem. Commun., 1983, 378.
57 R.H.Hooker, K.A.Mahmoud and A.J.Rest, J. Organomet. Chem., 1983, 254, C25.
58 H.G.Alt, K.A.Mahmoud and A.J.Rest, Angew. Chem., Int. Ed. Engl., 1983, 22, 544.
59 K.B.Shiu, M.D.Curtis and J.C.Huffman, Organometallics, 1983, 2, 936.
60 J.S.Huang and L.F.Dahl, J. Organomet. Chem., 1983, 243, 57.
61 M.Green, A.G.Orpen, C.J.Schaverien and I.D.Williams, J. Chem. Soc., Chem. Commun., 1983, 583.
62 M.Green, A.G.Orpen, C.J.Schaverien and I.D.Williams, J. Chem. Soc., Chem. Commun., 1983, 181.
63 R.F.Gerlach, D.N.Duffy and M.D.Curtis, Organometallics, 1983, 2, 1172.
64 J.J.D'Errico and M.D.Curtis, J. Am. Chem. Soc., 1983, 105, 4479.
65 J.J.D'Errico, L.Messerle and M.D.Curtis, Inorg. Chem., 1983, 22, 849 and 2222.
66 (a) W.A.Herrmann, L.K.Bell, M.L.Ziegler, H.Pfisterer and C.Pahl, J. Organomet. Chem., 1983, 247, 39; (b) W.A.Herrmann and G.Ihl, ibid., 1983, 251, C1.
67 W.A.Herrmann, G.W.Kriechbaum, R.Dammel, H.Bock, M.L.Ziegler and H.Pfisterer, J. Organomet. Chem., 1983, 254, 219.
68 (a) H.Brunner, H.Buchner, J.Wachter, I.Bernal and W.H.Ries, J. Organomet. Chem., 1983, 244, 247; (b) H.Brunner, B.Hoffmann and J.Wachter, ibid., 1983, 252, C35.
69 G.Becker, W.A.Herrmann, W.Kalcher, G.W.Kriechbaum, C.Pahl, C.T.Wagner and M.L.Ziegler, Angew. Chem., Int. Ed. Engl., 1983, 22, 413.
70 H.Alper, F.W.B.Einstein, R.Nagai, J.F.Petrignani and A.C.Willis, Organometallics, 1983, 2, 1291.
71 H.Alper, F.W.B.Einstein, J.F.Petrignani and A.C.Willis, Organometallics, 1983, 2, 1422.
72 M.D.Curtis, N.A.Fotinos, K.R.Han and W.M.Butler, J. Am. Chem. Soc., 1983, 105, 2686.
73 H.Alper, J.F.Petrignani, F.W.B.Einstein and A.C.Willis, J. Am. Chem. Soc., 1983, 105, 1701; J.F.Petrignani and H.Alper, Inorg. Chim. Acta, 1983, 77, L243.
74 M.Green, S.J.Porter and F.G.A.Stone, J. Chem. Soc., Dalton Trans., 1983, 513.
75 J.T.Lin and J.E.Ellis, J. Am. Chem. Soc., 1983, 105, 6252.
76 M.H.Chisholm, J.C.Huffman and R.J.Tatz, J. Am. Chem. Soc., 1983, 105, 2075.
77 T.W.Coffindaffer, I.P.Rothwell and J.C.Huffman, J. Chem. Soc., Chem. Commun. Commun., 1983, 1249.
78 F.A.Cotton and W.Schwotzer, J. Am. Chem. Soc., 1983, 105, 4955 and 5639.
79 F.A.Cotton, W.Schwotzer and E.S.Schamshoum, Organometallics, 1983, 2, 1167.
80 F.A.Cotton, W.Schwotzer and E.S.Schamshoum, Organometallics, 1983, 2, 1340.
81 M.H.Chisholm, K.Folting, D.M.Hoffman, J.C.Huffman and J.Leonelli, J. Chem. Soc., Chem. Commun., 1983, 589.

82 A.Poë and C.Sekhar, J. Chem. Soc., Chem. Commun., 1983, 566.
83 (a) N.J.Coville, A.M.Stolzenberg and E.L.Muetterties, J. Am. Chem. Soc., 1983, 105, 2499; (b) A.Stolzenberg and E.L.Muetterties, ibid., p. 822.
84 G.Al-Takhin, J.A.Connor and H.A.Skinner, J. Organomet. Chem., 1983, 259, 313.
85 H.Yesaka, T.Kobayashi, K.Yasufuku and S.Nagakura, J. Am. Chem. Soc., 1983, 105, 6249; A.F.Hepp and M.S.Wrighton, ibid., p. 5934.
86 J.A.Iggo, M.J.Mays, P.R.Raithby and K.Hendrick, J. Chem. Soc., Dalton Trans., 1983, 205.
87 M.Green, A.G.Orpen, C.J.Schaverien and I.D.Williams, J. Chem. Soc., Chem. Commun., 1983, 1399.
88 P.O.Nubel, S.R.Wilson and T.L.Brown, Organometallics, 1983, 2, 515.
89 G.Ferguson, W.J.Laws, M.Parvez and R.J.Puddephatt, Organometallics, 1983, 2, 276.
90 H.C.Aspinall and A.J.Deeming, J. Chem. Soc., Chem. Commun., 1983, 838.
91 W.A.Herrmann, R.Serrano and J.Weichmann, J. Organomet. Chem., 1983, 246, C57.
92 M.C.Böhm, R.D.Ernst, R.Gleiter and D.R.Wilson, Inorg. Chem., 1983, 22, 3815.
93 (a) S.Gambarotta, C.Floriani, A.Chiesi-Villa and C.Guastini, J. Chem. Soc., Chem. Commun., 1983, 1128; (b) C.G.Howard, G.Wilkinson, M.Thornton-Pett and M.B.Hursthouse, J. Chem. Soc., Dalton Trans., 1983, 2025.
94 G.Ciani, G.D'Alfonso, P.Romiti, A.Sironi and M.Freni, Inorg. Chem., 1983, 22, 3115.
95 G.Ciani, A.Sironi, G.D'Alfonso, P.Romiti and M.Freni, J. Organomet. Chem., 1983, 254, C37.
96 G.Ciani, G.D'Alfonso, P.Romiti, A.Sironi and M.Freni, J. Organomet. Chem., 1983, 244, C27.
97 G.M.Dawkins, M.Green, J.C.Jeffery, C.Sambale and F.G.A.Stone, J. Chem. Soc., Dalton Trans., 1983, 499.
98 (a) G.N.Mott and A.J.Carty, Inorg. Chem., 1983, 22, 2726; (b) G.N.Mott, R.Granby, S.A.MacLaughlin, N.J.Taylor and A.J.Carty, Organometallics, 1983, 2, 189.
99 Y.F.Yu, J.Gallucci and A.Wojcicki, J. Am. Chem. Soc., 1983, 105, 4826.
100 H.Vahrenkamp and D.Wolters, Angew. Chem., Int. Ed. Engl., 1983, 22, 154.
101 D.Seyferth, G.B.Womack and L.C.Song, Organometallics, 1983, 2, 776.
102 D.Seyferth, G.B.Womack, L.C.Song, M.Cowie and B.W.Hames, Organometallics, 1983, 2, 928; D.Seyferth, G.B.Womack, M.Cowie and B.W.Hames, ibid., p. 1696.
103 H.Umland, F.Edelmann, D.Wormsbächer and U.Behrens, Angew. Chem., Int. Ed. Engl., 1983, 22, 153.
104 E.K.Lhadi, H.Patin, A.Benoit, J.Y.LeMarouille and A.Darchen, J. Organomet. Chem., 1983, 259, 321; A.Darchen, E.K.Lhadi and H.Patin, ibid., p. 189.
105 R.H.Hooker, K.A.Mahmoud and A.J.Rest, J. Chem. Soc., Chem. Commun., 1983, 1022.
106 D.R.Tyler, M.A.Schmidt and H.B.Gray, J. Am. Chem. Soc., 1983, 105, 6018.
107 (a) B.Callan and A.R.Manning, J. Organomet. Chem., 1983, 252, C81; (b) P.Legzdins, D.T.Martin, C.R.Nurse and B.Wassink, Organometallics, 1983, 2, 1238.
108 M.Moll, H.Behrens, W.Popp, G.Liehr and W.P.Fehlhammer, Z. Naturforsch., Teil B, 1983, 38, 1446.
109 K.Jonas, G.Koepe, L.Schieferstein, R.Mynott, C.Krüger and Y.H.Tsay, Angew. Chem., Int. Ed. Engl., 1983, 22, 620.
110 A.F.Dyke, S.A.R.Knox, M.J.Morris and P.J.Naish, J. Chem. Soc., Dalton Trans., 1983, 1417; A.G.Orpen, ibid., p. 1427.
111 (a) C.P.Casey, P.J.Fagan, W.H.Miles and S.R.Marder, J. Mol. Catal., 1983, 21, 173; (b) C.P.Casey, S.R.Marder and P.J.Fagan, J. Am. Chem. Soc., 1983, 105, 7197.
112 R.E.Cramer, K.T.Higa, S.L.Pruskin and J.W.Gilje, J. Am. Chem. Soc., 1983, 105, 6749.
113 M.D.Seidler and R.G.Bergman, Organometallics, 1983, 2, 1897.
114 B.K.Teo, V.Bakirtzis and P.A.Snyder-Robinson, J. Am. Chem. Soc., 1983, 105, 6330.

115 G.McNally, P.T.Murray and A.R.Manning, J. Organomet. Chem., 1983, 243, C87.
116 M.I.Bruce, T.W.Hambley and B.K.Nicholson, J. Chem. Soc., Dalton Trans., 1983, 2385.
117 J.K.Kouba, E.L.Muetterties, M.R.Thompson and V.W.Day, Organometallics, 1983, 2, 1065.
118 J.W.Kolis, E.M.Holt and D.F.Shriver, J. Am. Chem. Soc., 1983, 105, 7307.
119 M.Lourdichi, R.Pince, F.Dahan and R.Mathieu, Organometallics, 1983, 2, 1417.
120 W.K.Wong, K.W.Chiu, G.Wilkinson, A.M.R.Galas, M.Thornton-Pett and M.B. Hursthouse, J. Chem. Soc., Dalton Trans., 1983, 1557.
121 D.deMontauzon and R.Mathieu, J. Organomet. Chem., 1983, 252, C83.
122 J.B.Keister, M.W.Payne and M.J.Muscatella, Organometallics, 1983, 2, 219.
123 P.Brun, G.M.Dawkins, M.Green, R.M.Mills, J-Y.Salaün, F.G.A.Stone and P.Woodward, J. Chem. Soc., Dalton Trans., 1983, 1357.
124 A.Ceriotti, L.Resconi, F.Demartin, G.Longoni, M.Manassero and M.Sansoni, J. Organomet. Chem., 1983, 249, C35.
125 A.Winter, I.Jibril and G.Huttner, J. Organomet. Chem., 1983, 247, 259.
126 A.Winter, L.Zsolnai and G.Huttner, J. Organomet. Chem., 1983, 250, 409.
127 K.H.Pannell, A.J.Mayr and D.VanDerveer, J. Am. Chem. Soc., 1983, 105, 6186.
128 T.P.Fehlner, C.E.Housecroft, W.R.Scheidt and K.S.Wong, Organometallics, 1983, 2, 825.
129 H.Vahrenkamp, E.J.Wucherer and D.Wolters, Chem. Ber., 1983, 116, 1219.
130 K.P.C.Vollhardt and T.W.Weidman, J. Am. Chem. Soc., 1983, 105, 1676.
131 P.Q.Adams, D.L.Davies, A.F.Dyke, S.A.R.Knox, K.A.Mead and P.Woodward, J. Chem. Soc., Chem. Commun., 1983, 222.
132 L.R.Beanan, Z.A.Rahman and J.B.Keister, Organometallics, 1983, 2, 1062.
133 R.E.Colborn, A.F.Dyke, S.A.R.Knox, K.A.Mead and P.Woodward, J. Chem. Soc., Dalton Trans., 1983, 2099.
134 R.E.Colborn, D.L.Davies, A.F.Dyke, A.Endesfelder, S.A.R.Knox, A.G.Orpen and D.Plaas, J. Chem. Soc., Dalton Trans., 1983, 2661.
135 N.J.Forrow, S.A.R.Knox, M.J.Morris and A.G.Orpen, J. Chem. Soc., Chem. Commun., 1983, 234.
136 M.I.Bruce, J.G.Matisons and B.K.Nicholson, J. Organomet. Chem., 1983, 247, 321.
137 M.I.Bruce, B.K.Nicholson, J.M.Patrick and A.H.White, J. Organomet. Chem., 1983, 254, 361.
138 M.I.Bruce, J.G.Matisons, B.W.Skelton and A.H.White, J. Chem. Soc., Dalton Trans., 1983, 2375.
139 M.I.Bruce, J.G.Matisons, R.C.Wallis, J.M.Patrick, B.W.Skelton and A.H.White, J. Chem. Soc., Dalton Trans., 1983, 2365.
140 (a) A.A.Bhattacharyya, C.C.Nagel and S.G.Shore, Organometallics, 1983, 2, 1187; (b) A.A.Bhattacharyya and S.G.Shore, ibid., p. 1251.
141 R.D.Adams, D.Männig and B.E.Segmüller, Organometallics, 1983, 2, 149.
142 D.J.Darensbourg, M.Pala and J.Waller, Organometallics, 1983, 2, 1285.
143 R.E.Stevens and W.L.Gladfelter, Inorg. Chem., 1983, 22, 2034.
144 B.F.G.Johnson, J.Lewis, W.J.H.Nelson, J.Puga, M.McPartlin and A.Sironi, J. Organomet. Chem., 1983, 253, C5.
145 B.F.G.Johnson, J.Lewis, W.J.H.Nelson, J.Puga, P.R.Raithby, D.Braga, M.McPartlin and W.Clegg, J. Organomet. Chem., 1983, 243, C13.
146 D.E.Fjare, J.A.Jensen and W.L.Gladfelter, Inorg. Chem., 1983, 22, 1774.
147 J.A.Jensen, D.E.Fjare and W.L.Gladfelter, Inorg. Chem., 1983, 22, 1250.
148 M.L.Blohm, D.E.Fjare and W.L.Gladfelter, Inorg. Chem., 1983, 22, 1004.
149 Z.Dawoodi, M.J.Mays, P.R.Raithby, K.Henrick, W.Clegg and G.Weber, J. Organomet. Chem., 1983, 249, 149.
150 C.E.Kampe, N.M.Boag and H.D.Kaesz, J. Mol. Catal., 1983, 21, 297; idem., J. Am. Chem. Soc., 1983, 105, 2896.
151 C.M.Jensen and H.D.Kaesz, J. Am. Chem. Soc., 1983, 105, 6969.
152 N.M.Boag, C.B.Knobler and H.D.Kaesz, Angew. Chem., Int. Ed. Engl., 1983, 22, 249.
153 M.I.Bruce, B.K.Nicholson and M.L.Williams, J. Organomet. Chem., 1983, 243, 69.
154 M.I.Bruce, E.Horn, M.R.Snow and M.L.Williams, J. Organomet. Chem., 1983, 255, 255.

155 M.I.Bruce, M.L.Williams and B.K.Nicholson, J. Organomet. Chem., 1983, 258, 63.

156 P.M.Lausarot, G.A.Vaglio, M.Valle, A.Tiripicchio and M.T.Camellini, J. Chem. Soc., Chem. Commun., 1983, 1391.

157 S.Aime, D.Osella, A.J.Deeming, A.M.M.Lanfredi and A.Tiripicchio, J. Organomet. Chem., 1983, 244, C47.

158 R.P.Rosen, G.L.Geoffroy, C.Bueno, M.R.Churchill and R.B.Ortega, J. Organomet. Chem., 1983, 254, 89.

159 S.A.MacLaughlin, N.J.Taylor and A.J.Carty, Inorg. Chem., 1983, 22, 1409.

160 S.A.MacLaughlin, N.J.Taylor and A.J.Carty, Organometallics, 1983, 2, 1194.

161 R.D.Adams, P.Mathur and B.E.Segmüller, Organometallics, 1983, 2, 1250.

162 J.R.Shapley, M.E.Cree-Uchiyama, G.M.St.George, M.R.Churchill and C.Bueno, J. Am. Chem. Soc., 1983, 105, 140.

163 J.R.Shapley, D.S.Strickland, G.M.St.George, M.R.Churchill and C.Bueno, Organometallics, 1983, 2, 185.

164 S.G.Shore, D.Y.Jan, L.Y.Hsu and W.L.Hsu, J. Am. Chem. Soc., 1983, 105, 5923.

165 E.D.Morrison, G.R.Steinmetz, G.L.Geoffroy, W.C.Fultz and A.L.Rheingold, J. Am. Chem. Soc., 1983, 105, 4104.

166 (a) R.J.Goudsmit, B.F.G.Johnson, J.Lewis, P.R.Raithby and M.J.Rosales, J. Chem. Soc., Dalton Trans., 1983, 2257; (b) M.I.Bruce, J.M.Guss, R.Mason, B.W.Skelton and A.H.White, J. Organomet. Chem., 1983, 251, 261.

167 A.C.Willis, F.W.B.Einstein, R.M.Ramadan and P.K.Pomeroy, Organometallics, 1983, 2, 935.

168 K.Burgess, H.D.Holden, B.F.G.Johnson and J.Lewis, J. Chem. Soc., Dalton Trans., 1983, 1199.

169 A.M.Brodie, H.D.Holden, J.Lewis and M.J.Taylor, J. Organomet. Chem., 1983, 253, C1.

170 G.Süss-Fink and P.R.Raithby, Inorg. Chim. Acta, 1983, 71, 109.

171 W.Ehrenreich, M.Herberhold, G.Süss-Fink, H.P.Klein and U.Thewalt, J. Organomet. Chem., 1983, 248, 171.

172 S.A.MacLaughlin, J.P.Johnson, N.J.Taylor, A.J.Carty and E.Sappa, Organometallics, 1983, 2, 352.

173 (a) B.F.G.Johnson, J.Lewis, W.J.H.Nelson, J.Puga, P.R.Raithby and K.H. Whitmire, J. Chem. Soc., Dalton Trans., 1983, 1339; (b) B.F.G.Johnson, J.Lewis, W.J.H.Nelson, J.Puga, K.Hendrick and M.McPartlin, ibid., p. 1203; (c) M.A.Collins, B.F.G.Johnson, J.Lewis, J.M.Mace, J.Morris, M.McPartlin, W.J.H.Nelson, J.Puga, and P.R.Raithby, J. Chem. Soc., Chem. Commun., 1983, 689.

174 (a) B.F.G.Johnson, J.Lewis, J.N.Nicholls, J.Puga, P.R.Raithby, M.J.Rosales, M.McPartlin and W.Clegg, J. Chem. Soc., Dalton Trans., 1983, 277; (b) A.G. Cowie, B.F.G.Johnson, J.Lewis, J.N.Nicholls, P.R.Raithby and M.J.Rosales, ibid., p. 2311; (c) B.F.G.Johnson, J.Lewis, W.J.H.Nelson, J.N.Nicholls, J.Puga, P.R.Raithby, M.J.Rosales, M.Schröder and M.D.Vargas, ibid., p. 2447.

175 B.F.G.Johnson, J.Lewis, P.R.Raithby and M.J.Rosales, J. Organomet. Chem., 1983, 259, C9.

176 B.F.G.Johnson, J.Lewis, P.R.Raithby and M.J.Rosales, J. Chem. Soc., Dalton Trans., 1983, 2645.

177 R.J.Goudsmit, B.F.G.Johnson, J.Lewis, P.R.Raithby and K.H.Whitmire, J. Chem. Soc., Chem. Commun., 1983, 246.

178 (a) R.D.Adams, I.T.Horvath, B.E.Segmüller and L.W.Yang, Organometallics, 1983, 2, 144; (b) idem., ibid., p. 1301; (c) R.D.Adams, I.T.Horvath and L.W.Yang, ibid., p. 1257; (d) R.D.Adams and L.W.Yang, J. Am. Chem. Soc., 1983, 105, 235; (e) R.D.Adams, I.T.Horvath, P.Mathur and B.E.Segmüller, Organometallics, 1983, 2, 996; (f) R.D.Adams, I.T.Horvath and L.W.Yang, J. Am. Chem. Soc., 1983, 105, 1533; (g) R.D.Adams, D.F.Foust and P.Mathur, Organometallics, 1983, 2, 990; (h) R.D.Adams, I.T.Horvath, P.Mathur, B.E. Segmüller and L.W.Yang, ibid., p. 1078.

179 R.D.Adams, Z.Dawoodi, D.F.Foust and B.E.Segmüller, Organometallics, 1983, 2, 315.

180 (a) R.D.Adams, Z.Dawoodi, D.F.Foust and B.E.Segmüller, J. Am. Chem. Soc., 1983, 105, 831; (b) R.D.Adams and D.F.Foust, Organometallics, 1983, 2, 323;

(c) R.D.Adams, D.F.Foust and B.E.Segmüller, ibid., p. 308.
181 B.F.G.Johnson, J.Lewis, W.J.H.Nelson, M.Vargas, D.Braga and M.McPartlin, J. Organomet. Chem., 1983, 249, C21.
182 B.F.G.Johnson, J.Lewis, M.McPartlin, W.J.H.Nelson, P.R.Raithby, A.Sironi and M.D.Vargas, J. Chem. Soc., Chem. Commun., 1983, 1476.
183 D.Braga, J.Lewis, B.F.G.Johnson, M.McPartlin, W.J.H.Nelson and M.D.Vargas, J. Chem. Soc., Chem. Commun., 1983, 241.
184 D.Braga, K.Hendrick, B.F.G.Johnson, J.Lewis, M.McPartlin, W.J.H.Nelson, A.Sironi and M.D.Vargas, J. Chem. Soc., Chem. Commun., 1983, 1131.
185 M.Arewgoda, B.H.Robinson and J.Simpson, J. Am. Chem. Soc., 1983, 105, 1893.
186 B.E.Hanson and J.S.Mancini, Organometallics, 1983, 2, 126.
187 W.J.Laws and R.J.Puddephatt, J. Chem. Soc., Chem. Commun., 1983, 1020.
188 (a) K.H.Theopold and R.G.Bergman, J. Am. Chem. Soc., 1983, 105, 464; (b) W.A.Herrmann, C.Bauer, J.M.Huggins, H.Pfisterer and M.L.Ziegler, J. Organomet. Chem., 1983, 258, 81.
189 W.H.Hersh, F.J.Hollander and R.G.Bergman, J. Am. Chem. Soc., 1983, 105, 5834; W.H.Hersh and R.G.Bergman, ibid., p. 5846.
190 (a) J.A.King and K.P.C.Vollhardt, Organometallics, 1983, 2, 684; (b) idem., J. Am. Chem. Soc., 1983, 105, 4846.
191 N.G.Connelly, J.D.Payne and W.E.Geiger, J. Chem. Soc., Dalton Trans., 1983, 295.
192 R.A.Jones, A.L.Stuart, J.L.Atwood and W.E.Hunter, Organometallics, 1983, 2, 1437.
193 A.D.Harley, R.R.Whittle and G.L.Geoffroy, Organometallics, 1983, 2, 60; A.D.Harley, G.J.Guskey and G.L.Geoffroy, ibid., p. 53.
194 R.B.A.Pardy, G.W.Smith and M.E.Vickers, J. Organomet. Chem., 1983, 252, 341.
195 K.P.C.Vollhardt and E.C.Walborsky, J. Am. Chem. Soc., 1983, 105, 5507.
196 J.Fortune, A.R.Manning and F.S.Stephens, J. Chem. Soc., Chem. Commun., 1983, 1071.
197 J.Fortune and A.R.Manning, Organometallics, 1983, 2, 1719.
198 G.Gervasio, R.Rossetti, P.L.Stanghellini and G.Bor, J. Chem. Soc., Dalton Trans., 1983, 1613.
199 P.Bradamante, P.Pino, A.Stefani, G.Fachinetti and P.F.Zanazzi, J. Organomet. Chem., 1983, 251, C47.
200 A.Benoit, A.Darchen, J.Y.LeMarouille, C.Mahé and H.Patin, Organometallics, 1983, 2, 555.
201 S.Enoki, T.Kawamura and T.Yonezawa, Inorg. Chem., 1983, 22, 3821.
202 S.B.Colbran, B.H.Robinson and J.Simpson, Organometallics, 1983, 2, 943 and 952.
203 J.Rimmelin, P.Lemoine, M.Gross and D.deMontauzon, Nouveau J. Chim., 1983, 7, 453.
204 S.Holand, F.Mathey, J.Fischer and A.Mitschler, Organometallics, 1983, 2, 1234.
205 G.Ciani and A.Sironi, J. Organomet. Chem., 1983, 241, 385.
206 A.L.Rheingold and P.J.Sullivan, J. Chem. Soc., Chem. Commun., 1983, 39.
207 B.R.James, D.Mahajan, S.J.Rettig and G.M.Williams, Organometallics, 1983, 2, 1452.
208 F.Faraone, G.Bruno, S.L.Schiavo, P.Piraino and G.Bombieri, J. Chem. Soc., Dalton Trans., 1983, 1819.
209 M.Green, D.R.Hankey, J.A.K.Howard, P.Louca and F.G.A.Stone, J. Chem. Soc., Chem. Commun., 1983, 757.
210 W.A.Herrmann, C.Bauer, M.L.Ziegler and H.Pfisterer, J. Organomet. Chem., 1983, 243, C54.
211 W.A.Herrmann, C.Bauer and A.Schäfer, J. Organomet. Chem., 1983, 256, 147.
212 W.A.Herrmann, C.Bauer and J.Weichmann, J. Organomet. Chem., 1983, 243, C21.
213 F.Faraone, S.L.Schiavo, G.Bruno, P.Piraino and G.Bombieri, J. Chem. Soc., Dalton Trans., 1983, 1813.
214 J.P.Lecompte, J.M.Lehn, D.Parker, J.Guilhem and C.Pascard, J. Chem. Soc., Chem. Commun., 1983, 296.
215 K.Isobe, A.V.DeMiguel, P.M.Bailey, S.Okeya and P.M.Maitlis, J. Chem. Soc., Dalton Trans., 1983, 1441.

216 S.Okeya, B.F.Taylor and P.M.Maitlis, J. Chem. Soc., Chem. Commun., 1983, 971.
217 N.J.Meanwell, A.J.Smith, H.Adams, S.Okeya and P.M.Maitlis, Organometallics, 1983, 2, 1705.
218 R.A.Jones, T.C.Wright, J.L.Atwood and W.E.Hunter, Organometallics, 1983, 2, 470.
219 P.E.Kreter and D.W.Meek, Inorg. Chem., 1983, 22, 319.
220 R.A.Jones and T.C.Wright, Organometallics, 1983, 2, 1842.
221 B.Klingert and H.Werner, J. Organomet. Chem., 1983, 252, C47.
222 G.W.Bushnell, D.O.K.Fjeldsted, S.R.Stobart and M.J.Zaworotko, J. Chem. Soc., Chem. Commun., 1983, 580.
223 R.J.Haines, N.D.C.T.Steen and R.B.English, J. Chem. Soc., Dalton Trans., 1983, 2229.
224 M.M.Harding, B.S.Nicholls and A.K.Smith, J. Chem. Soc., Dalton Trans., 1983, 1479.
225 M.M.Olmstead, R.R.Guimerans and A.L.Balch, Inorg. Chem., 1983, 22, 2473; F.E.Wood, M.M.Olmstead and A.L.Balch, J. Am. Chem. Soc., 1983, 105, 6332.
226 E.W.Burkhardt, W.C.Mercer, G.L.Geoffroy, A.L.Rheingold and W.C.Fultz, J. Chem. Soc., Chem. Commun., 1983, 1251.
227 B.E.Mann, C.M.Spencer and A.K.Smith, J. Organomet. Chem., 1983, 244, C17.
228 F.Demartin, M.Manassero, M.Sansoni, L.Garlaschelli, M.C.Malatesta and U.Sartorelli, J. Organomet. Chem., 1983, 248, C17.
229 F.Demartin, M.Manassero, M.Sansoni, L.Garlaschelli, C.Raimondi and S.Martinengo, J. Organomet. Chem., 1983, 243, C10.
230 R.Bonfichi, G.Ciani, A.Sironi and S.Martinengo, J. Chem. Soc., Dalton Trans., 1983, 253.
231 A.Fumagalli, S.Martinengo, G.Ciani and A.Sironi, J. Chem. Soc., Chem. Commun., 1983, 453.
232 V.G.Albano, D.Braga, P.Chini, D.Strumolo and S.Martinengo, J. Chem. Soc., Dalton Trans., 1983, 249.
233 D.Strumolo, C.Seregni, S.Martinengo, V.G.Albano and D.Braga, J. Organomet. Chem., 1983, 252, C93.
234 J.L.Vidal and R.C.Schoening, J. Organomet. Chem., 1983, 241, 395.
235 B.T.Heaton, L.Strona, R.D.Pergola, J.L.Vidal and R.C.Schoening, J. Chem. Soc., Dalton Trans., 1983, 1941.
236 K.R.Pörschke, W.Kleimann, G.Wilke, K.H.Calus and C.Krüger, Angew. Chem., Int. Ed. Engl., 1983, 22, 991.
237 R.A.Jones, A.L.Stuart, J.L.Atwood and W.E.Hunter, Organometallics, 1983, 2, 874.
238 R.Benn, B.Bogdanović, P.Göttsch and M.Rubach, Z. Naturforsch., Teil B, 1983, 38, 604; B.Bogdanović, P.Göttsch and M.Rubach, ibid., p. 599.
239 A.Ceriotti, P.Chini, R.D.Pergola and G.Longoni, Inorg. Chem., 1983, 22, 1595.
240 K.Broadley, G.A.Lane, N.G.Connelly and W.E.Geiger, J. Am. Chem. Soc., 1983, 105, 2486.
241 R.J.Goodfellow, I.R.Herbert and A.G.Orpen, J. Chem. Soc., Chem. Commun., 1983, 1386.
242 W.M.Hawling, A.Walker and M.A.Woitzik, J. Chem. Soc., Chem. Commun., 1983, 11.
243 G.Minghetti, A.L.Bandini, G.Banditelli, F.Bonati, R.Szostak, C.E.Strouse, C.B.Knobler and H.D.Kaesz, Inorg. Chem., 1983, 22, 2332.
244 N.M.Boag, R.J.Goodfellow, M.Green, B.Hessner, J.A.K.Howard and F.G.A.Stone, J. Chem. Soc., Dalton Trans., 1983, 2585.
245 R.J.Puddephatt, K.A.Azam, R.H.Hill, M.P.Brown, C.D.Nelson, R.P.Moulding, K.R.Seddon and M.C.Grossel, J. Am. Chem. Soc., 1983, 105, 5642; R.H.Hill and R.J.Puddephatt, ibid., p. 5797; M.C.Grossel, R.P.Moulding and K.R.Seddon, J. Organomet. Chem., 1983, 247, C32.
246 (a) A.T.Hutton, B.Shabanzadeh and B.L.Shaw, J. Chem. Soc., Chem. Commun., 1983, 1053; (b) S.Muralidharan and J.H.Espenson, Inorg. Chem., 1983, 22, 2786; (c) K.R.Grundy and K.N.Robertson, Organometallics, 1983, 2, 1736.

247 L.Manojlović-Muir, K.W.Muir, B.R.Lloyd and R.J.Puddephatt, J. Chem. Soc., Chem. Commun., 1983, 1336.

248 Y.Yamamoto, K.Aoki and H.Yamazaki, Organometallics, 1983, 2, 1377.

249 (a) E.G.Mednikov and N.K.Eremenko, Koord. Khim., 1983, 9, 243; (b) E.G. Mednikov, N.K.Eremenko, Yu.L.Slovokhotov, Yu.T.Struchkov and S.P.Gubin, J. Organomet. Chem., 1983, 258, 247.

250 S.Gambarotta, C.Floriani, A.Chiesi-Villa and C.Guastini, J. Chem. Soc., Chem. Commun., 1983, 1156.

251 S.Gambarotta, C.Floriani, A.Chiesi-Villa and C.Guastini, J. Chem. Soc., Chem. Commun., 1983, 1304.

252 W.Bos, J.J.Bour, J.W.A.vanderVelden, J.J.Steggerda, A.L.Casalnuovo and L.H. Pignolet, J. Organomet. Chem., 1983, 253, C64.

253 R.D.Barr, M.Green, J.A.K.Howard, T.B.Marder, I.Moore and F.G.A.Stone, J. Chem. Soc., Chem. Commun., 1983, 746.

254 G.M.Dawkins, M.Green, K.A.Mead, J.Y.Salaün, F.G.A.Stone and P.Woodward, J. Chem. Soc., Dalton Trans., 1983, 527.

255 C.P.Casey, R.F.Jordan and A.L.Rheingold, J. Am. Chem. Soc., 1983, 105, 665.

256 T.B.Rauchfuss, T.D.Weatherill, S.R.Wilson and J.P.Zebrowski, J. Am. Chem. Soc., 1983, 105, 6508.

257 J.A.Abad, L.W.Bateman, J.C.Jeffery, K.A.Mead, H.Razay, F.G.A.Stone and P.Woodward, J. Chem. Soc., Dalton Trans., 1983, 2075.

258 I.L.Eremenko, A.A.Pasynskii, Yu.V.Rakitin, O.G.Ellert, V.M.Novotortsev, V.T.Kalinnikov, V.E.Shklover and Yu.T.Struchkov, J. Organomet. Chem., 1983, 256, 291.

259 I.L.Eremenko, A.A.Pasynskii, B.Orazsakhatov, O.G.Ellert, V.M.Novotortsev, V.T.Kalinnikov, M.A.Porai-Koshits, A.S.Antsyshkina, L.N.Dikareva, V.N. Ostrikova, Yu.T.Struchkov and R.G.Gerr, Inorg. Chim. Acta, 1983, 73, 225.

260 A.A.Pasynskii, I.L.Eremenko, G.Sh.Gasanov, O.G.Ellert, V.M.Novotortsev, V.T.Kalinnikov, Yu.T.Struchkov and V.E.Shklover, Bull. Acad. Sci. USSR, Div. Chem. Sci., 1983, 32, 1315.

261 R.D.Barr, M.Green, K.Marsden, F.G.A.Stone and P.Woodward, J. Chem. Soc., Dalton Trans., 1983, 507.

262 R.Bender, P.Braunstein, J.M.Jud and Y.Dusausoy, Inorg. Chem., 1983, 22, 3394.

263 F.Richter, H.Beurich, M.Müller, N.Gärtner and H.Vahrenkamp, Chem. Ber., 1983, 116, 3774.

264 S.Lu, N.Okura, T.Yoshida, S.Otsuka, K.Hirotsu and T.Higuchi, J. Am. Chem. Soc., 1983, 105, 7470.

265 P.D.Williams, M.D.Curtis, D.N.Duffy and W.M.Butler, Organometallics, 1983, 2, 165.

266 B.Cowans, J.Noordik and M.R.Dubois, Organometallics, 1983, 2, 931.

267 M.D.Curtis and P.D.Williams, Inorg. Chem., 1983, 22, 2661.

268 M.Müller and H.Vahrenkamp, Chem. Ber., 1983, 116, 2748.

269 C.Mahé, H.Patin, J.Y.LeMarouille and A.Benoit, Organometallics, 1983, 2, 1051.

270 R.G.Finke, G.Gaughan, C.Pierpont and J.H.Noordik, Organometallics, 1983, 2, 1481.

271 R.D.Barr, M.Green, J.A.K.Howard, T.B.Marder and F.G.A.Stone, J. Chem. Soc., Chem. Commun., 1983, 759.

272 J.P.Farr, M.M.Olmstead, N.M.Rutherford, F.E.Wood and A.L.Balch, Organometallics, 1983, 2, 1758.

273 P.Braunstein, J.M.Jud, Y.Dusausoy and J.Fischer, Organometallics, 1983, 2, 180.

274 H.Werner, J.Roll, K.Linse and M.L.Ziegler, Angew. Chem., Int. Ed. Engl., 1983, 22, 982.

275 L.Busetto, J.C.Jeffery, R.M.Mills, F.G.A.Stone, M.J.Went and P.Woodward, J. Chem. Soc., Dalton Trans., 1983, 101.

276 N.A.Ustynyuk, V.N.Vinogradova, V.N.Korneva, Yu.L.Slovokhotov and Yu.T. Struchkov, Koord. Khim., 1983, 9, 631.

277 M.Green, K.Marsden, I.D.Salter, F.G.A.Stone and P.Woodward, J. Chem. Soc., Chem. Commun., 1983, 446.

278 L.Busetto, M.Green, B.Hessner, J.A.K.Howard, J.C.Jeffery and F.G.A.Stone, J. Chem. Soc., Dalton Trans., 1983, 519.
279 R.D.Adams, I.T.Horvath and P.Mathur, J. Am. Chem. Soc., 1983, 105, 7202.
280 J.T.Park, J.R.Shapley, M.R.Churchill and C.Bueno, Inorg. Chem., 1983, 22, 1579.
281 J.T.Park, J.R.Shapley, M.R.Churchill and C.Bueno, J. Am. Chem. Soc., 1983, 105, 6182.
282 E.O.Fischer, P.Friedrich, T.L.Lindner, D.Neugebauer, F.R.Kreissl, W.Uedelhoven, N.Q.Dao and G.Huttner, J. Organomet. Chem., 1983, 247, 239.
283 M.J.Breen, G.L.Geoffroy, A.L.Rheingold and W.C.Fultz, J. Am. Chem. Soc., 1983, 105, 1069.
284 M.R.Awang, J.C.Jeffery and F.G.A.Stone, J. Chem. Soc., Dalton Trans., 1983, 2091.
285 K.A.Mead, I.Moore, F.G.A.Stone and P.Woodward, J. Chem. Soc., Dalton Trans., 1983, 2083.
286 M.R.Awang, J.C.Jeffery and F.G.A.Stone, J. Chem. Soc., Chem. Commun., 1983, 1426.
287 A.Albinati, R.Naegeli, A. Togni and L.M.Venanzi, Organometallics, 1983, 2, 926.
288 M.R.Awang, G.A.Carriedo, J.A.K.Howard, K.A.Mead, I.Moore, C.M.Nunn and F.G. F.G.A.Stone, J. Chem. Soc., Chem. Commun., 1983, 964.
289 G.A.Carriedo, J.A.K.Howard and F.G.A.Stone, J. Organomet. Chem., 1983, 250, C28.
290 L.Carlton, W.E.Lindsell, K.J.M.McCullough and P.N.Preston, J. Chem. Soc., Chem. Commun., 1983, 216.
291 M.Müller and H.Vahrenkamp, Chem. Ber., 1983, 116, 2322.
292 B.F.Hoskins, R.J.Steen and T.W.Turney, Inorg. Chim. Acta, 1983, 77, L69.
293 P.Braunstein, J.M.Jud and J.Fischer, J. Chem. Soc., Chem. Commun., 1983, 5.
294 I.Noda, H.Yasuda and A.Nakamura, J. Organomet. Chem., 1983, 250, 447.
295 H.R.Allcock, L.J.Wagner and M.L.Levin, J. Am. Chem. Soc., 1983, 105, 1321.
296 D.F.Jones, P.H.Dixneuf, A.Benoit and J.Y.LeMarouille, Inorg. Chem., 1983, 22, 29.
297 H.Yamazaki, K.Yasufuku and Y.Wakatsuki, Organometallics, 1983, 2, 726.
298 D.E.Fjare, D.G.Keyes and W.L.Gladfelter, J. Organomet. Chem., 1983, 250, 383.
299 B.H.Freeland, N.C.Payne, M.A.Stalteri and H.VanLeeuwen, Acta Crystallogr., Sect. C, 1983, 39, 1533.
300 S.Aime, D.Osella, L.Milone, A.M.M.Lanfredi and A.Tiripicchio, Inorg. Chim. Acta, 1983, 71, 141.
301 M.Müller and H.Vahrenkamp, Chem. Ber., 1983, 116, 2765.
302 R.J.Haines, N.D.C.T.Steen and R.B.English, J. Chem. Soc., Dalton Trans., 1983, 1607.
303 V.E.Lopatin, N.M.Mikova and S.P.Gubin, Bull. Acad. Sci. USSR, Div. Chem. Sci., 1983, 32, 1276.
304 D.A.Roberts, G.R.Steinmetz, M.J.Breen, P.M.Shulman, E.D.Morrison, M.R.Duttera, C.W.DeBrosse, R.R.Whittle and G.L.Geoffroy, Organometallics, 1983, 2, 846.
305 D.Seyferth, B.W.Hames, T.G.Rucker, M.Cowie and R.S.Dickson, Organometallics, 1983, 2, 472.
306 G.N.Mott, N.J.Taylor and A.J.Carty, Organometallics, 1983, 2, 447.
307 M.I.Bruce and B.K.Nicholson, J. Organomet. Chem., 1983, 250, 627.
308 E.Roland, K.Fischer and H.Vahrenkamp, Angew. Chem., Int. Ed. Engl., 1983, 22, 326.
309 C.E.Briant, K.P.Hall and D.M.P.Mingos, J. Chem. Soc., Chem. Commun., 1983, 843.
310 P.Braunstein, J.Rose and O.Bars, J. Organomet. Chem., 1983, 252, C101.
311 M.Hidai, M.Orisaku, M.Ue, Y.Koyasu, T.Kodama and Y.Uchida, Organometallics, 1983, 2, 292.
312 E.Roland and H.Vahrenkamp, Organometallics, 1983, 2, 183.
313 A.J.Carty, N.J.Taylor, E.Sappa and A.Tiripicchio, Inorg. Chem., 1983, 22, 1871.

314 M.J.Freeman, M.Green, A.G.Orpen, I.D.Salter and F.G.A.Stone, J. Chem. Soc., Chem. Commun., 1983, 1332.
315 L.W.Bateman, M.Green, K.A.Mead, R.M.Mills, I.D.Salter, F.G.A.Stone and P.Woodward, J. Chem. Soc., Dalton Trans., 1983, 2599.
316 L.J.Farrugia, M.J.Freeman, M.Green, A.G.Orpen, F.G.A.Stone and I.D.Salter, J. Organomet. Chem., 1983, 249, 273.
317 M.I.Bruce and B.K.Nicholson, J. Organomet. Chem., 1983, 252, 243.
318 B.F.G.Johnson, J.Lewis, J.N.Nicholls, J.Puga and K.H.Whitmire, J. Chem. Soc., Dalton Trans., 1983, 787.
319 M.R.Churchill and C.Bueno, Inorg. Chem., 1983, 22, 1510.
320 S.G.Shore, W.L.Hsu, M.R.Churchill and C.Bueno, J. Am. Chem. Soc., 1983, 105, 655; M.R.Churchill and C.Bueno, J. Organomet. Chem., 1983, 256, 357.
321 L.J.Farrugia, A.G.Orpen and F.G.A.Stone, Polyhedron, 1983, 2, 171.
322 (a) G.Lavigne, F.Papageorgiou, C.Bergounhou and J.J.Bonnet, Inorg. Chem., 1983, 22, 2485; (b) M.Castiglioni, E.Sappa, M.Valle, M.Lanfranchi and A.Tiripicchio, J. Organomet. Chem., 1983, 241, 99.
323 E.Sappa, A.Tiripicchio and M.T.Camellini, J. Organomet. Chem., 1983, 246, 287.
324 L.J.Farrugia, M.Green, D.R.Hankey, A.G.Orpen and F.G.A.Stone, J. Chem. Soc., Chem. Commun., 1983, 310.
325 B.F.G.Johnson, J.Lewis, W.J.H.Nelson, M.D.Vargas, D.Braga and M.McPartlin, J. Organomet. Chem., 1983, 246, C69.
326 K.Burgess, B.F.G.Johnson, J.Lewis and P.R.Raithby, J. Chem. Soc., Dalton Trans., 1983, 1661.
327 K.Burgess, B.F.G.Johnson, D.A.Kaner, J.Lewis, P.R.Raithby, S.N.Azman and B.Syed-Mustaffa, J. Chem. Soc., Chem. Commun., 1983, 455.
328 B.F.G.Johnson, J.Lewis, W.J.H.Nelson, P.R.Raithby and M.D.Vargas, J. Chem. Soc., Chem. Commun., 1983, 608.
329 J.W.A.vanderVelden, J.J.Bour, W.P.Bosman and J.H.Noordik, Inorg. Chem., 1983, 22, 1913.
330 (a) J.P.Farr, M.M.Olmstead and A.L.Balch, Inorg. Chem., 1983, 22, 1229; (b) J.P.Farr, M.M.Olmstead, F.E.Wood and A.L.Balch, J. Am. Chem. Soc., 1983, 105, 792.
331 A.Fumagalli, S.Martinengo and G.Ciani, J. Chem. Soc., Chem. Commun., 1983, 1381.
332 N.G.Connelly, A.R.Lucy, J.D.Payne, A.M.R.Galas and W.E.Geiger, J. Chem. Soc., Dalton Trans., 1983, 1879.
333 B.T.Heaton, L.Strona, S.Martinengo, D.Strumolo, V.G.Albano and D.Braga, J. Chem. Soc., Dalton Trans., 1983, 2175.
334 A.T.Hutton, P.G.Pringle and B.L.Shaw, Organometallics, 1983, 2, 1889.
335 J.C.Leblanc, J.F.Reynoud and C.Moise, J. Organomet. Chem., 1983, 244, C24.
336 H.Brunner, H.Kauermann and J.Wachter, Angew. Chem., Int. Ed. Engl., 1983, 22, 549.
337 M.D.Curtis, N.A.Fortinos, L.Messerle and A.P.Sattelberger, Inorg. Chem., 1983, 22, 1559.
338 N.E.Kolobova, L.L.Ivanov and O.S.Zhvanko, Bull. Acad. Sci. USSR, Div. Chem. Sci., 1983, 32, 866.
339 S.Sabo, B.Chaudret and D.Gervais, J. Organomet. Chem., 1983, 258, C19.
340 (a) C.P.Casey, R.M.Bullock and F.Nief, J. Am. Chem. Soc., 1983, 105, 7574; (b) C.P.Casey and R.M.Bullock, J. Organomet. Chem., 1983, 251, 245.
341 R.Jund, P.Lemoine, M.Gross, P.Bender and P.Braunstein, J. Chem. Soc., Chem. Commun., 1983, 86.
342 P.Thometsek and H.Werner, J. Organomet. Chem., 1983, 252, C29.
343 M.C.Grossel, R.P.Moulding and K.R.Seddon, J. Organomet. Chem., 1983, 253, C50.
344 M.Müller and H.Vahrenkamp, Chem. Ber., 1983, 116, 2311.
345 E.Roland and H.Vahrenkamp, Organometallics, 1983, 2, 1048.
346 (a) D.A.Lesch and T.B.Rauchfuss, Inorg. Chem., 1983, 22, 1854; (b) L.E. Bogan, D.A.Lesch and T.B.Rauchfuss, J. Organomet. Chem., 1983, 250, 429.
347 S.Jensen, B.H.Robinson and J.Simpson, J. Chem. Soc., Chem. Commun., 1983, 1081.

348 M.Mlekuz, B.Bougeard, M.J.McGlinchey and G.Jaouen, J. Organomet. Chem., 1983, 253, 117.
349 J.Schneider and G.Huttner, Chem. Ber., 1983, 116, 917.
350 E.Roland and H.Vahrenkamp, J. Mol. Catal., 1983, 21, 233.
351 J.Ros and R.Mathieu, Organometallics, 1983, 2, 771.
352 H.Werner and R.Weinand, Z. Naturforsch., Teil B, 1983, 38, 1518.
353 K.Burgess, B.F.G.Johnson and J.Lewis, J. Chem. Soc., Dalton Trans., 1983, 1179.
354 W.A.Herrmann, C.E.Barnes, R.Serrano and B.Koumbouris, J. Organomet. Chem., 1983, 256, C30.
355 H.F.Klein, H.Witty and U.Schubert, J. Chem. Soc., Chem. Commun., 1983, 231.
356 P.H.M.Budzelaar, H.J.Alberts-Jansen, K.Mollema, J.Boersma, G.J.M.vanderKerk, A.L.Spek and A.J.M.Duisenberg, J. Organomet. Chem., 1983, 243, 137.
357 R.B.Petersen, J.M.Ragosta, G.E.Whitwell and J.M.Burlitch, Inorg. Chem., 1983, 22, 3407.
358 B.A.Sosinsky, R.G.Shong, B.J.Fitzgerald, N.Norem and C.O'Rourke, Inorg. Chem., 1983, 22, 3124.
359 P.H.M.Budzelaar, J.Boersma and G.J.M.vanderKerk, Angew. Chem., Int. Ed. Engl., 1983, 22, 329.
360 G.Salem, B.W.Skelton, A.H.White and S.B.Wild, J. Chem. Soc., Dalton Trans., 1983, 2117.
361 S.Ermer, K.King, K.I.Hardcastle, E.Rosenberg, A.M.M.Lanfredi, A.Tiripicchio and M.T.Camellini, Inorg. Chem., 1983, 22, 1339.
362 E.G.Mednikov, V.V.Bashilov, V.I.Sokolov, Yu.L.Slovokhotov and Yu.T.Struchkov, Polyhedron, 1983, 2, 141.
363 Yu.V.Skripkin, I.L.Eremenko, A.A.Pasynskii, O.G.Volkov, M.A.Porai-Koshits, A.S.Antsyshkina, L.M.Dikareva, V.N.Ostrikova and Yu.T.Struchkov, Izv. Akad. Nauk. SSSR, Ser. Khim., 1983, 1927.
364 R.Usón, A.Laguna, J.A.Abad and E.deJesús, J. Chem. Soc., Dalton Trans., 1983, 1127.
365 L.Párkányi, K.H.Pannell and C.Hernandez, J. Organomet. Chem., 1983, 252, 127.
366 M.J.Auburn, R.D.Holmes-Smith, S.R.Stobart, M.J.Zaworotko, T.S.Cameron and A.Kumari, J. Chem. Soc., Chem. Commun., 1983, 1523.
367 A.C.Willis, G.N.VanBuuren, R.K.Pomeroy and F.W.B.Einstein, Inorg. Chem., 1983, 22, 1162.
368 B.J.Aylett and M.T.Taghipour, J. Organomet. Chem., 1983, 249, 55.
369 M.J.Fernandez and P.M.Maitlis, Organometallics, 1983, 2, 164.
370 (a) D.Melzer and E.Weiss, J. Organomet. Chem., 1983, 255, 335; (b) A.S. Batsanov, L.V.Rybin, M.I.Rybinskaya, Yu.T.Struchkov, I.M.Salimgareeva and N.G.Bogatova, ibid., 1983, 249, 319.
371 A.Castel, P.Riviere, J.Satgé, J.J.E.Moreau and R.J.P.Corriu, Organometallics, 1983, 2, 1498.
372 D.J.Brauer and R.Eujen, Organometallics, 1983, 2, 263.
373 H.F.Klein, K.Ellrich, D.Neugebauer, O.Orama and K.Krüger, Z. Naturforsch., Teil B, 1983, 38, 303.
374 F.W.B.Einstein, R.K.Pomeroy, P.Rushman and A.C.Willis, J. Chem. Soc., Chem. Commun., 1983, 854.
375 I.Pforr, F.Näumann and D.Rehder, J. Organomet. Chem., 1983, 258, 189.
376 A.A.Pasynskii, Yu.V.Skripkin, O.G.Volkov, V.T.Kalinnikov, M.A.Porai-Koshits, A.S.Antsyshkina, L.M.Dikareva and V.N.Ostrikova, Bull. Acad. Sci. USSR, Div. Chem. Sci., 1983, 32, 1093.
377 A.Tzschach, K.Jurkschat, M.Scheer, J.Meunier-Piret and M.VanMeerssche, J. Organomet. Chem., 1983, 259, 165.
378 G.L.Rochfort and J.E.Ellis, J. Organomet. Chem., 1983, 250, 277.
379 J.T.Lin, G.P.Hagen and J.E.Ellis, Organometallics, 1983, 2, 1145.
380 M.J.Chetcuti, M.H.Chisholm, H.T.Chiu and J.C.Huffman, J. Am. Chem. Soc., 1983, 105, 1060.
381 H.Preut and H.J.Haupt, Acta Crystallogr., Sect. C, 1983, 39, 981.
382 G.Consiglio, F.Morandini, G.Ciani, A.Sironi and M.Kretschmer, J. Am. Chem. Soc., 1983, 105, 1391.

Bibliography

Other published work relevant to this chapter includes:

M.Hoefler, K.F.Tebbe, H.Veit and N.E.Weiler, J. Am. Chem. Soc., 1983, 105, 6338. Synthesis and structure of $[Cr_3(CO)_9(\mu-CO)_3(\mu_4-S)\{Cr(CO)_5\}]^{2-}$.

A.A.Pasynskii, I.L.Eremenko, Yu.V.Rakitin, V.M.Novotortsev, O.G.Ellert, V.T. Kalinnikov, V.E.Shklover, Yu.T.Struchkov, S.V.Lindeman, T.Kh.Kurbanov, and G.Sh.Gasanov, J. Organomet. Chem., 1983, 248, 309. Synthesis and structures of $[Cr_4Cp'_4(\mu_3-S)_4]$ and $[Cr_4Cp'_4(\mu_3-O)(\mu_3-S)_3CuBr_2]$ from reactions of $[Cr_2Cp'_2-(\mu-S)(\mu-SBu^t)_2]$ ($Cp' = C_5H_4Me$).

A.E.Stiegman, M.Stieglitz and D.R.Tyler, J. Am. Chem. Soc., 1983, 105, 6032. Photochemical disproportionation of $[Mo_2(CO)_6Cp'_2]$.

M.J.Chetcuti, M.H.Chisholm, K.Folting, D.A.Haitko, J.C.Huffman and J.Janos, J. Am. Chem. Soc., 1983, 105, 1163. Structural investigations of complexes $[M_2R_2(NMe_2)_4]$ (M≡M) with M = Mo, (W); R = Bnz, Ar.

J.A.Bandy, C.E.Davis, J.C.Green, M.L.H.Green, K.Prout and D.P.S.Rodgers, J. Chem. Soc., Chem. Commun., 1983, 1395. Synthesis, structure and bonding of $[Mo_4Cp'_4(\mu_3-S)_4]^{z+}$ (z = 0, 1, 2).

A.Winter, O.Scheidsteger and G.Huttner, Z. Naturforsch., Teil B, 1983, 38, 1525. Crystal structures of $[W_2(CO)_8(\mu-SMe)_2]$ and $[W_3(CO)_{10}(\mu-SPh)_4]$.

J.L.LeQueré, F.Y.Pétillon, J.E.Guerchais, L.Manojlović-Muir, K.W.Muir and D.W.A. Sharp, J. Organomet. Chem., 1983, 249, 127. Structure of $[Cp(CO)_2W(\mu-I)(\mu-SMe)WI(CO)_3]$ and related species.

B.F.Hoskins and R.J.Steen, Aust. J. Chem., 1983, 36, 683. Crystal structure of $[Mn_2(CO)_8(\mu-Ph_2AsCH_2CH_2AsPh_2)]$.

W.Tam, M.Marsi and J.A.Gladysz, Inorg. Chem., 1983, 22, 1413. Properties of formyl complexes, $[MM'(CO)_9(CHO)]^-$ (M,M' = Mn, Re).

T.J.Barder, S.M.Tetrick, R.A.Walton, F.A.Cotton and G.L.Powell, J. Am. Chem. Soc., 1983, 105, 4090. Ortho-metallation at Re≡Re centre; structure of $[Re_2Cl_3\{\mu-(C_6H_4)PhPpy\}(\mu-Ph_2Ppy)]$.

N.J.Coville, M.O.Albers and E.Singleton, J. Chem. Soc., Dalton Trans., 1983, 947. $[Fe_2(CO)_4Cp_2]$ as catalyst in carbonyl substitution reactions.

R.L.De, J. Organomet. Chem., 1983, 243, 331. Crystal structures of $[Fe_2(CO)_6-(\mu-PClPh)(\mu-Cl)]$ and $[Fe_2(CO)_6(\mu-PClPh)_2]$.

B.Deppisch and H.Schäfer, Acta Crystallogr., Sect. C, 1983, 39, 975. Crystal structure of $[Fe_2(CO)_2Cp(\mu-CO)(\mu-OPOH)]$.

H. Lang, L.Zsolnai and G.Huttner, Angew. Chem., Int. Ed. Engl., 1983, 22, 976. Preparation and structures of $[\{CpMn(CO)_2P\}_2Fe_2(CO)_6]$ and $[\{Cr(CO)_5\}_2P_2Co_2-(CO)_6]$.

E.Lindner, G.A.Weiss, W.Hiller and R.Fawzi, J. Organomet. Chem., 1983, 255, 245. Synthesis and structure of $[Fe_3(CO)_9(\mu\text{-PAr})(\mu_3\text{-S})]$ (2 Fe-Fe).

S.Bhaduri, K.S.Gopalkrishnan, G.M.Sheldrick, W.Clegg and D.Stalke, J. Chem. Soc., Dalton Trans., 1983, 2339. Reactions of $PhNO_2$ with $[Ru_3(CO)_{12}]$ and structure of $[Ru_3(CO)_9(\mu_3\text{-CO})(\mu_3\text{-NPh})]$.

M.R.Churchill, L.R.Beanan, H.J.Wasserman, C.Bueno, Z.A.Rahman and J.B.Keister, Organometallics, 1983, 2, 1179. Synthesis of clusters $[Ru_3(CO)_8(chd)(\mu\text{-H})\text{-}(\mu_3\text{-CX})]$ and crystal structure for X = OMe.

M.R.Burke, J.Takats, F.W.Grevels and J.G.A.Reuvers, J. Am. Chem. Soc., 1983, 105, 4092. Structure of $[(OC)_8Os_2CH_2CHCO_2Me]$, a 1,2-diosmacyclobutane.

E.E.Sutton, M.L.Niven and J.R.Moss, Inorg. Chim. Acta, 1983, 70, 207. Structure of $[Os_2(CO)_6(\mu\text{-I})_2]$.

F.W.B.Einstein, S.Nussbaum, D.Sutton and A.C.Willis, Organometallics, 1983, 2, 1259. Structures of complexes $[Os_3(CO)_n(\mu\text{-MeO}_2CN_2CO_2Me)]$ (n = 10, 12).

E.Sappa, A.Tiripicchio and A.M.M.Lanfredi, J. Organomet. Chem., 1983, 249, 391. Synthesis and structure of $[Os_3(CO)_{10}(\mu\text{-H})(\mu\text{-}\eta^2\text{-CHCHBu}^t)]$.

M.R.Churchill and H.J.Wasserman, J. Organomet. Chem., 1983, 248, 365. Crystal structure of $[Os_3(CO)_{10}(\mu\text{-CO})(\mu\text{-CHSiMe}_3)]$.

S.Aime and A.J.Deeming, J. Chem. Soc., Dalton Trans., 1983, 1807. Intramolecular nucleophilic attack of an μ_3-alkynyl ligand of an Os_3 cluster.

M.R.Churchill and J.R.Missert, J. Organomet. Chem., 1983, 256, 349. Crystal structure of $[Os_3(CO)_{10}(\mu\text{-H})(\mu\text{-NCCHNHCMe})]$.

H.D.Holden, B.F.G.Johnson, J.Lewis, P.R.Raithby and G.Uden, Acta Crystallogr., Sect. C, 1983, 39, 1197, 1200 and 1203. Crystal structures of carbonyl Os_3 clusters with sulphur ligands.

P.Braunstein, I.Pruskil, G.Predieri and A.Tiripicchio, J. Organomet. Chem., 1983, 247, 227. Crystal structure of $[Co_2(CO)_4(dppe)(\mu\text{-CO})_2]$.

M.P.Anderson and L.H.Pignolet, Organometallics, 1983, 2, 1246. Structure of unsymmetrical complex $[Rh_2(CO)\{\mu\text{-(PPh}_2)_2CHpy\}_2]$.

J.S.Najdzionek and B.D.Santarsiero, Acta Crystallogr., Sect. C, 1983, 39, 577. Crystal structure of $[Rh_2\{\mu\text{-CNCMe}_2(CH_2)_2CMe_2NC\}_3\{\mu\text{-NCCMe}_2(CH_2)_2CMe_2\}]^+$.

J.T.Mague, Inorg. Chem., 1983, 22, 1158. Dimetallated alkene isocyanide complex, $[Rh_2(CNBu^t)_4(\mu\text{-CF}_3C_2CF_3)(\mu\text{-dppm})_2]$.

M.P.Brown, A.Yavari, L.Manojlović-Muir and K.W.Muir, J. Organomet. Chem., 1983, 256, C19. Deprotonation of bound dppm to form $[Pt_2(PPh_3)_2(\mu\text{-Ph}_2PCHPPh_2)\text{-}(\mu\text{-dppm})]^+$, a PtCPPt dimetallacycle.

N.M.Boag, P.L.Goggin, R.J.Goodfellow and I.R.Herbert, J. Chem. Soc., Dalton Trans., 1983, 1101. ^{13}C and ^{195}Pt N.m.r. studies of $[Pt_2X_4(CO)_2]^{2-}$ (X = Cl, Br, I).

N.K.Eremenko, E.G.Mednikov, S.S.Kurasov and S.P.Gubin, Bull. Acad. Sci. USSR, Div. Chem. Sci., 1983, 32, 620. Synthesis of carbonyl trialkylphosphine clusters of Pt(O).

J.L.Davidson, J. Chem. Soc., Dalton Trans., 1983, 1667. Chemistry of hetero-binuclear complexes of MoCo and WCo.

A.Ceriotti, G.Longoni, R.D.Pergola, B.T.Heaton and D.O.Smith, J. Chem. Soc., Dalton Trans., 1983, 1433. Syntheses and studies (^{13}C, ^{103}Rh and $^{13}C\{^{103}Rh\}$ n.m.r.) of anionic iron-rhodium carbonyl clusters.

M.Kretschmer, P.S.Pregosin, P.Favre and C.W.Schlaepfer, J. Organomet. Chem., 1983, 253, 17. Synthesis, Raman and n.m.r. studies of $[IrCl(SnCl_3)(HgCl)(CO)(PPh_3)_2]$.

Yu.K.Grishin, V.A.Roznyatovsky, Yu.A.Ustynyuk, S.N.Titova, G.A.Domrachev and G.A.Razuvaev, Polyhedron, 1983, 2, 895. ^{31}P, ^{195}Pt and ^{199}Hg N.m.r. studies of $[Pt(HgGePh_3)(GePh_3)(PPh_3)_2]$.

C.Lewis and M.S.Wrighton, J. Am. Chem. Soc., 1983, 105, 7768. Photochemical synthesis of $[Cp(CO)_2HWSiMe_2CH_2]$.

P.Gusbeth and H.Vahrenkamp, J. Organomet. Chem., 1983, 247, C53. Formation of $[Co_3(CO)_9(\mu_3\text{-}GeR)]$, $[Co_4(CO)_{10}(\mu\text{-}CO)(\mu_3\text{-}GeR)_2]$ and $[Co_3Mo(CO)_9Cp(\mu\text{-}CO)(\mu_3\text{-}GeR)_2]$.

M.Kretschmer, P.S.Pregosin and M.Garralda, J. Organomet. Chem., 1983, 244, 175. Synthesis and ^{119}Sn and ^{31}P n.m.r. studies of complexes $[M(SnCl_3)_2(diene)(PR_3)]^-$ (M = Rh, Ir).

S.Carr, R.Colton and D.Dakternieks, J. Organomet. Chem., 1983, 249, 327. ^{31}P, ^{119}Sn and ^{195}Pt N.m.r. studies of a series of complexes $[PtPh(PPh_3)_2(SnX_2Y)]$.

10

Substitution Reactions of Metal and Organometal Carbonyls with Group V and VI Ligands

BY D. A. EDWARDS

1 Introduction and Reviews

The scope of this Chapter is essentially unchanged from that in previous Volumes of the Series. Substitution reactions which either involve or yield mononuclear carbonyl complexes are reviewed as fully as possible, but space limitations allow only a highly selective consideration of metal-metal bonded complexes and organometal carbonyl complexes such as cyclopentadienyls. These are reviewed in depth elsewhere in this Volume.

Three excellent reviews on ligand substitution reactions of metal carbonyls have appeared. The first focusses attention on the stabilization of electron-deficient transition states formed in dissociative processes.[1] The second provides a comprehensive review of the kinetics and mechanisms of thermal substitution reactions of 18-electron complexes.[2] The final review considers mechanisms of ligand substitution reactions of octahedral metal carbonyl complexes, emphasis being placed on stereochemical features.[3] Complexes containing α-diimine[4] and $Ph_2PCH_2PPh_2$ ligands[5] have been reviewed.

2 Papers of General Interest

The effect of high intensity ultrasound on $M(CO)_6$ (M = Cr, Mo, or W), $Fe(CO)_5$, $Fe_3(CO)_{12}$, and $M'_2(CO)_{10}$ (M'_2 = Mn_2, MnRe, or Re_2) has been studied. Ultrasonic irradiation in the presence of additional ligands such as phosphines and phosphites may effect mono- or di-substitution. The rates of such sonochemical substitutions are first order in metal carbonyl, the results being consistent with a carbonyl dissociative mechanism.[6]

Further studies on electrochemically induced ligand substitution of metal carbonyls have been reported. Reactions of $[Mo(CO)_4L_2]$, $[(\eta\text{-}C_5H_4Me)Mn(CO)_2L]$, and $[CpRe(CO)_2L]$ with ligands

L' (L = py, Bu^tNC, THF, MeCN, PPh_3, DMSO, or Et_2NH; L' = phosphines, phosphites, $AsPh_3$, $SbPh_3$, py, or MeCN) take place *via* an electron-transfer chain mechanism. A 17-electron cation $[LM(CO)_{\underline{n}}]^+$ is produced at the anode and undergoes rapid exchange to produce $[L'M(CO)_{\underline{n}}]^+$. Electron transfer with $[LM(CO)_{\underline{n}}]$ follows to give the product $[L'M(CO)_{\underline{n}}]$ and regeneration of $[LM(CO)_{\underline{n}}]^+$ completes the chain propagation. Enhanced rates of substitution are also found when manganese carbonyl cations *e.g.* $[Mn(CO)_3(NCMe)_3]^+$ undergo cathodic reduction in the presence of added ligands such as PMe_2Ph or PPh_3. The products are the hydrides $[HMn(CO)_3L_2]$. The proposed mechanism involves Mn(O) radicals as reactive intermediates.[7]

Nucleophilic nitrosylations of metal carbonyls can be achieved using $[(Ph_3P)_2N][NO_2]$ in dipolar aprotic solvents. The attack of NO_2^- on a co-ordinated carbonyl ligand is thought to be the rate-determining step, an atom-transfer redox leading to the liberation of CO_2 and the co-ordination of NO. The known species $[Fe(CO)_3(NO)]^-$, $[Mn(CO)_4(NO)]$, and $[Co(CO)_3(NO)]$, the new anions $[Mn(NO)_2(CO)_2]^-$ and $[M_3(CO)_{10}(NO)]^-$ (M = Ru or Os) and the ligand substituted complexes $[Fe(NO)_2(CO)_{\underline{n}}(PPh_3)_{2-\underline{n}}]$ ($\underline{n}$ = 0 or 1), and $[Co(NO)(CO)_2(PPh_3)]$ were isolated.[8]

Thermal and photochemical activations of nitrile solutions of $[Fe_3(\mu_3\text{-}PPh)_2(CO)_9]$ and $[Co_4(\mu_4\text{-}PPh)_2(CO)_{10}]$ yield mono- and di-substituted nitrile complexes from which phosphine, phosphite and arsine complexes can be prepared by nitrile displacement.[9]

Substitution of a CO ligand in $[(\eta^5\text{-}C_5Me_5)M(CO)_2]$ (M = Co or Rh) by phosphines or phosphites, and in $[(\eta^5\text{-}C_9H_7)Rh(CO)_2]$, $[(\eta^5\text{-}C_9H_7)Mn(CO)_3]$ and $[(\eta^5\text{-}C_{13}H_9)Mn(CO)_3]$ by PPh_3, proceeds by the associative formation of η^3-allyl-ene intermediates followed by CO loss.[10]

The bulky secondary phosphine $[(Me_3Si)_2CH]_2PH$ reacts with $Fe_2(CO)_9$ and $Co_2(CO)_8$ to give $[Fe(CO)_4L]$ and $[Co_2(CO)_7L]$, respectively.[11] The chelate complexes *cis*-$\{M(CO)_2[Ph_2PN(H)PPh_2]_2\}$ (M = Cr or Mo) and $\{Co(NO)(CO)_2[Ph_2PN(H)PPh_2]\}$ and the ligand-bridged complexes $\{[Fe(NO)_2(CO)]_2Ph_2PN(H)PPh_2\}$ and $\{Fe(NO)_2[Ph_2PN(H)PPh_2]\}_2$ result from reactions of $M(CO)_6$, $Co(NO)(CO)_3$, and $Fe(NO)_2(CO)_2$ with bis(diphenylphosphino)amine.[12] Bicyclic MeNC(O)N(Me)PN(Me)C(O)N(Me)P acts as a *PP*-bridging ligand in $\{[Cr(CO)_5]_2L\}$, $[M(CO)_4L]_2$ (M = Cr or Mo), $\{[Mo(CO)_3]_2L_3\}$ and $[Rh(CO)ClL]_2$ but as a unidentate ligand in $[Mo(CO)_3L_3]$. Its *P*-monoxide, L', merely acts as a unidentate phosphine in forming $[Cr(CO)_{6-\underline{n}}L'_{\underline{n}}]$ ($\underline{n}$ = 1 or 2) and $[Fe(CO)_4L']$.[13] In $\{Ru_3(CO)_9[MeSi(PR_2)_3]\}$ (R = Et or Pr), $\{Rh_6(CO)_{13}[MeSi(PBu^n_2)_3]\}$ and $\{M_4(CO)_9[MeSi(PR_2)_3]\}$ (M = Os, R = Et, Pr, or Bu^n; M = Rh or Ir, R = Et) the phosphine ligand caps a triangular face of the metal atom framework. The related $\{M_4(CO)_9[HC(PPh_2)_3]\}$ (M = Rh or Ir) complexes react with PPh_3 or $Ph_2P(CH_2)_2Si(OEt)_3$ to give $\{M_4(CO)_8L[HC(PPh_2)_3]\}$ products. Immobilization of these complexes to give unique supported clusters can be achieved either by treating the nonacarbonyls with phosphinated silica or phosphinated polystyrene-divinylbenzene or by reacting $\{M_4(CO)_8[Ph_2P(CH_2)_2Si(OEt)_3][HC(PPh_2)_3]\}$ with silica.[14] The poly[(aryloxy)phosphazenes] $[NP(OC_6H_4PPh_2)_{\underline{x}}(OPh)_{2-\underline{x}}]_{\underline{n}}$ ($\underline{x}$ = *ca.* 0.3 or 0.6; $\underline{n}$ = *ca.* 7000) and the model trimers $N_3P_3(OPh)_5(OC_6H_4PPh_2)$ and $N_3P_3(OC_6H_4PPh_2)_6$ can act as *P*-donor ligands in reactions with $H_2Os_3(CO)_{10}$, $CpMn(CO)_3$, $\{Fe(CO)_3[PhCH{=}CHC(O)Me]\}$, and $[RhCl(CO)_2]_2$.[15]

Reactions between $[Cr(CO)_5]^{2-}$ or $[Fe(CO)_4]^{2-}$ and ECl_2R [E = P or As; R = $N(SiMe_3)_2$] are far from simple. The phosphinidene and arsinidene complexes $[(ML_{\underline{n}})_2(\mu\text{-}ER)]$, [$ML_{\underline{n}}$ = $Cr(CO)_5$, E = P or As; $ML_{\underline{n}}$ = $Fe(CO)_4$, E = As] and $[(OC)_5Cr(\mu\text{-}PR)Cr(CO)_4(\mu\text{-}PR)Cr(CO)_5]$ containing multiple MP or MAs bonds have been characterized as products, together with the diphosphene complexes $[RP{=}P(R)Cr(CO)_5]$ and $[(OC)_4Fe(R)P{=}P(R)Fe(CO)_4]$. Other diphosphene, phosphaarsene, or diarsene complexes *e.g.* $[R'P{=}P(R')Fe(CO)_4]$, $[R'P{=}P(R')Ni(CO)_3]$, $[R'P{=}E(R'')Fe(CO)_4]$, $[R''P{=}P(R'')Cr(CO)_5]$, $[(OC)_4Fe(R'')P{=}P(R'')Fe(CO)_4]$, $[R'As{=}As(R'')Cr(CO)_5]$ and $[(OC)_5Cr(Ph)P{=}P(Ph)Cr(CO)_5]$, [R' = 2,4,6-$Bu^t_3C_6H_2$, R" = $(Me_3Si)_2CH$, E = P or As] have been isolated either from reactions of metal carbonyls with diphosphenes and diarsenes or from carbonylmetallate dianions with monosubstituted phosphorus dichlorides.[16]

Reactions of metal carbonyls with $Me_2As(CH_2)_2OH$ have been investigated. Products characterized include $[W(CO)_5L]$, $[Mo(CO)_4L_2]$, $[Fe(CO)_4L]$, $[Fe(CO)_3L_2]$ and $[Ni(CO)_3L]$ in which L = unidentate *As*-bonded dimethylarsinoethanol. Subsequent reactions of some of these products with $Me_2Si(NMe_2)_2$ or alternatively reactions of metal carbonyls with $Me_2Si(OCH_2CH_2AsMe_2)_2$ lead to chelate complexes such as {$L_nM[(Me_2AsCH_2CH_2O)_2SiMe_2]$}, [$ML_{\underline{n}}$ = $Mo(CO)_4$, $Fe(CO)_3$, or $Ni(CO)_2$] or ligand-bridged {$[(OC)_5W]_2[(Me_2AsCH_2CH_2O)_2SiMe_2]$}.[17]

Trialkyl- or triaryl-tin dithiocarboxylic esters act as unidentate ligands in {$Cp(OC)_2Mn[S{=}C(SMe)SnCy_3]$}, {$Cp(OC)_2Mn[S{=}C(SR)SnAr_3]$}, {$(OC)_5W[S{=}C(SR)SnAr_3]$} and {$Re_2(CO)_9[S{=}C(SMe)SnPh_3]$}, (Ar = Ph, *o*-tol, or *p*-tol; R = Me, Et, Pr^i, or $CH_2CH{=}CH_2$). Related triorganotin dithiocarboxylate complexes may involve either unidentate anions *e.g.* {$Mn(CO)_3[P(OPh)_3]_2[SC(S)SnPh_3]$} and {$Re(CO)_4L[SC(S)SnPh_3]$} (L = CO or PPh_3) or chelating bidentate anions *e.g.* [$Mn(CO)_{4-\underline{n}}L_{\underline{n}}(S_2CSnPh_3)$] (L = phosphines, phosphites, $AsPh_3$, or $SbPh_3$; $\underline{n}$ = 1 or 2) and [$Re(CO)_3L(S_2CSnR_3)$] (R = Me, Bz, or Ph, L = CO; R = Ph, L = PPh_3).[18]

3 Groups IV and V

The *NN*-bonded [$Cp_2Ti(RN{=}NR)$] (R = Ph or *p*-tol)[19] and [$Cp_2Ti(S_3N_4)$] containing an eight-membered TiS_3N_4 ring[20] result from ligand substitution reactions of $Cp_2Ti(CO)_2$. Photolytic or thermal reactions in the presence of phosphines afford [$Cp_2Ti(CO)(PR_3)$] (R = F, Et, or Ph) and [$Cp_2Ti(PF_3)_2$].[21]

In the presence of bases *e.g.* py, THF, Et_2O, or MeCN , $V(CO)_6$ undergoes disproportionation according to a second-order rate law. The isocarbonyl-bridged intermediate [$(OC)_5VCOV(py)_4OCV(CO)_5$] has been detected by F.T. i.r. spectroscopy in a reaction leading to [$V(py)_6$][$V(CO)_6$]$_2$.[22] Reactions of Na[$V(CO)_4(dppe)$] with acidic aqueous solutions of sulphite, selenite or tellurite yield {$[V(CO)_3(dppe)]_2(\mu\text{-}E)$} (E = S, Se, or Te) complexes containing multiple V-E bonds.[23] Thermally unstable $V(CO)_5(NO)$ has been prepared from [Et_4N][$V(CO)_6$] and $NOBF_4$ in CH_2Cl_2 at -40°C. Subsequent reactions give products such as [$V(CO)_3(NO)(dppe)$], {$[V(CO)_4(NO)]_2(\mu\text{-dppe})$} , and *trans*-[$V(CO)_4(NO)L$] (L = NMe_3, PMe_3, or PPh_3).[24]

The M(-III) compounds $Na_3[M(CO)_5]$ (M = Nb or Ta) are oxidised by NH_4Cl in liquid NH_3 producing [$M(CO)_5(NH_3)$]$^-$ anions, isolated as [Ph_4As]$^+$ salts. The ammine anions react with phosphines, phosphites, and isocyanides giving [$M(CO)_5L$]$^-$ anions.[25] Photolysis of $CpNb(CO)_4$ in the presence of [$Ph_2P(CH_2)_2$]$_2$PR, L, (R = Ph or Cy) gives the chelate complexes [$CpNb(CO)_2L$] containing an unco-ordinated Ph_2P-group. These products will react with

[CpNb(CO)$_3$(THF)] yielding [CpNb(CO)$_2$(μ-L)Nb(CO)$_3$Cp] species.[26]

4 Group VI

4.1 Carbonyl Complexes of Cr^0, Mo^0, and W^0.- The nitrogen donor ligand substituted metal carbonyls [W(CO)$_5$(C$_5$H$_{10}$NH)], {[M(CO)$_5$]$_2$(μ-pyrazine)}, and [M(CO)$_4$(chelate)], (M = Cr, Mo, or W, chelate = substituted 2,2'-bipyridyls, substituted 1,10-phenanthrolines, and 2,2'-bipyrimidine) have been prepared in order to examine their photochemical and photophysical properties.[27] The complexes [W(CO)$_3$(NCR)$_3$] (R = Et or Pr^n) are claimed to be superior synthetic reagents to [W(CO)$_3$(NCMe)$_3$].[28] Many *cis*-[M(CO)$_4$(4,4'-X$_2$-2,2'-bipy)] chelate complexes (M = Cr, Mo, or W; X = H, NMe$_2$, NH$_2$, OMe, Bu^t, Me, Ph, CH=CHPh, Cl, CO$_2$H, CO$_2$Me, or NO$_2$) have been isolated and variations in their electronic, vibrational and n.m.r. spectroscopic properties correlated with the nature of X.[29] Solid [M(CO)$_4$(di-2-pyridylamine)] (M = Mo or W) chelate complexes have short intermolecular contacts between unco-ordinated NH groups and carbonyl oxygen atoms in adjoining molecules suggesting the presence of weak NH---OC hydrogen bonds.[30] Tri-2-pyridylamine, tpa, reacts with M(CO)$_6$ (M = Mo or W) in refluxing hydrocarbon solvents to give [M(CO)$_3$(tpa)] complexes which may be oxidised by halogens to [WCl$_4$(tpa)], [MCl$_3$(tpa)] and [MoBr$_3$(tpa)].[31] The compounds PhP(2-XC$_5$H$_4$N)$_2$ (X = CH$_2$, O, or NH) also act as terdentate ligands in forming {M(CO)$_3$[PhP(2-XC$_5$H$_4$N)$_2$]} complexes (M = Cr, Mo, or W).[32] The preparations and structures of [W(CO)$_4$(μ-L)]$_2$,{[W(CO)$_5$]$_2$(μ-L)} and{[Cr(CO)$_3$]$_2$(μ-L)$_3$}, (L = Me$_2\overline{\mathrm{CN{=}NC}}Me_2$) have been reported.[33] Cycloalkeno-1,2,3-selenadiazoles and - thiadiazoles react with [M(CO)$_5$(THF)] to give [M(CO)$_5$L] species (M = Cr, Mo, or W) in which the ligand L is bonded through the nitrogen in the 2-position. The complexes *cis*-[Mo(CO)$_4$L$_2$] and{[Mo(CO)$_3$]$_2$(μ-L)$_3$} (L = cyclohexeno-1,2,3-selenadiazole bridging *via* its two nitrogen atoms) were also characterized.[34]

Many studies of phosphine-substituted metal carbonyls have been reported. Photolysis of a Cr(CO)$_6$ - P(C≡CPh)$_3$ mixture in THF yields {Cr(CO)$_5$[P(C≡CPh)$_3$]}, the phosphine ligand being a good σ -donor but only a moderate π-acceptor.[35] The *P*-donor complexes [Mo(CO)$_5$(PR$_3$)] (R = *o*-, *m*-, or *p*-tol), the η^6-arene complexes [Mo(CO)$_3$(η^6-RPR$_2$)] (R = *p*-tol or mesityl) and

$\{(OC)_5Mo[P(p\text{-tol})_2(\eta^6\text{-}p\text{-tol})]Mo(CO)_3\}$ have been characterized.[36] Two types of complexes, $[M(CO)_4L]$ (M = Cr, Mo, or W) and $[M(CO)_2L]$ (M = Cr or Mo) in which L = $(o\text{-tol})_2PCH_2P(o\text{-tol})_2$ have been isolated. In the former type the bis-phosphine acts in a conventional *PP*-chelating manner whereas in the latter type co-ordination occurs *via* one phosphorus atom and an η^6-*o*-tolyl group.[37] While studying the magnetic triple resonance spectra of carbonyl - phosphine species, the new complexes $\{Cr(CO)_4[(Ph_2P)_2C{=}CH_2]\}$ and $\{M(CO)_4[(Ph_2P)_2NH]\}$ (M = Cr, Mo, or W) were prepared by direct reaction of the metal hexacarbonyls with the ligands.[38] Both $\{M(CO)_4[Ph_2P(CH_2)_2SMe]\}$ (M = Mo or W) and the cations $\{M(CO)_4[Ph_2P(CH_2)_2SMe_2]\}^+$ contain *PS*-chelating ligands. The cations are converted into *P*-unidentate bonded $\{M(CO)_4L[Ph_2P(CH_2)_2SMe_2]\}^+$ cations on reaction with additional ligands [L = CO, MeCN, or $P(OMe)_3$].[39] The diphosphine-bridged complexes $\{[M(CO)_5]_2[\mu\text{-}R(X)PP(X)R]\}$ (M = Cr or W; R = Ph or 4-$MeOC_6H_4$; X = Br or I) have been prepared from $[M(CO)_5(THF)]$ and the diphosphines. Further reactions of $\{[Cr(CO)_5]_2[\mu\text{-}Ph(Br)PP(Br)Ph]\}$ with $LiAlH_4$ or $LiBu^n$ lead to replacement of both bromine atoms by hydrogen atoms or butyl groups, respectively.[40] The unco-ordinated central phosphorus atom of the chelate complex $\{Mo(CO)_4[Ph_2PCH_2P(Ph)CH_2PPh_2]\}$ acts as a donor on reaction with $[RhCl(CO)_2]_2$. The product is *trans*-$\{RhCl(CO)[PhP(CH_2PPh_2)_2Mo(CO)_4]_2\}$.[41] Routes to $\{(OC)_5WPPh[(CH_2)_2PPh_2Mo(CO)_5][(CH_2)_2PPh_2Cr(CO)_5]\}$ and $\{(OC)_5CrPPh[(CH_2)_2PPh_2Mo(CO)_5][(CH_2)_2PPh_2W(CO)_5]\}$, the first complexes containing all three Group VI metals, have been devised.[42] The bicyclic geometry of the bis-phosphine $P(OCH_2)_3P$ allows each bridgehead phosphorus to co-ordinate to separate metal carbonyl residues. Thus reactions with $[M(CO)_4(nbd)]$ (M = Cr or W) or $[Mo(CO)_4(tmen)]$ allow syntheses of *cis*-$\{M(CO)_4[P(OCH_2)_3P]_2\}$, in which the ligands are unidentate, as well as $\{M(CO)_4[P(OCH_2)_3P]\}_4$ (M = Cr, Mo, or W) and $\{M(CO)_4[P(OCH_2)_3P]_2M'(CO)_4\}_2$ (M = Cr or W; M' = Mo) tetramers.[43] The reaction of $Mo(CO)_6$ with $(Pr^i{}_2N)_2P(O)H$ occurs with deamination and the formation of complex (1) containing a novel tetradentate phosphoxane ligand that chelates the two metals in an adamantane-like cage structure.[44] Strategies for the syntheses of the phosphine complexes *cis*-$\{M(CO)_4[Ph_2POP(O)PhPPh_2]\}$, (M = Cr, Mo, W), *cis*-$\{Mo(CO)_4[(Ph_2PO)_2PPh]M(CO)_5\}$ (M = Cr or Mo), and $\{[Mo(CO)_5]_2[(Ph_2PO)_2PPh]\}$ have been proposed and confirmed

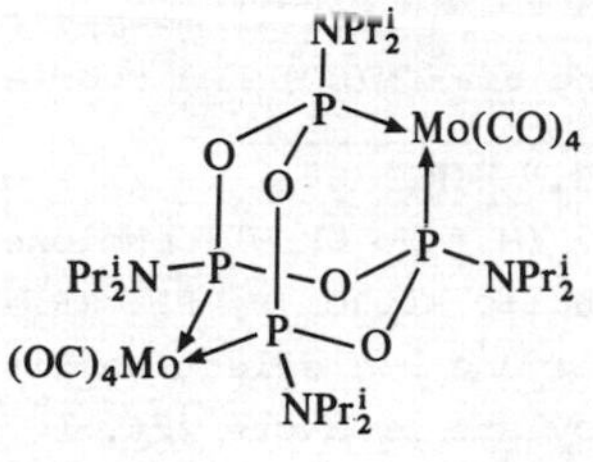

(1)

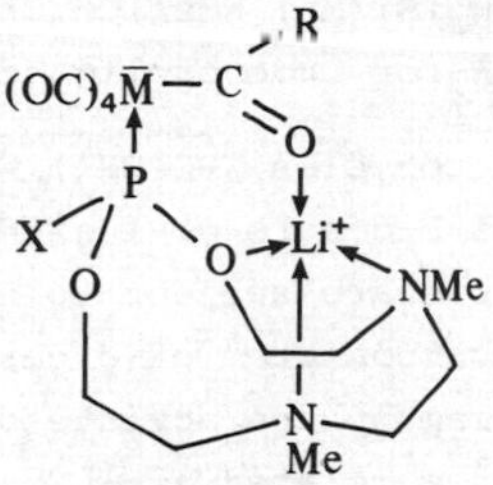

(2) X = OEt or NEt_2

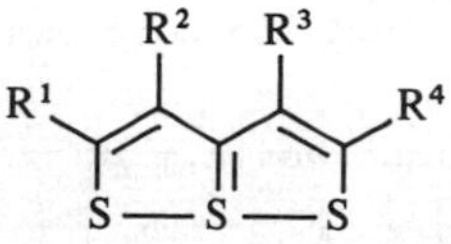

(3) $R^1 = R^4 = H, R^2, R^3 = -(CH_2)_3-$;
$R^1 = R^4 = Me, R^2 = R^3 = H$;
$R^1 = Bu^t, R^2 = R^3 = R^4 = H$

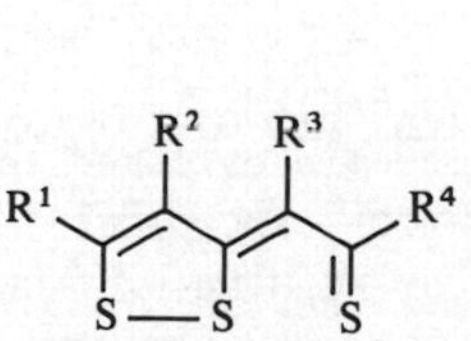

(4) R^1—R^4 as for (3)

experimentally. The products contain co-ordination-stabilized tautomers of the unknown bis(phosphoryl)phosphine $(Ph_2P{=}O)_2PPh$.[45] Reactions of $[M(CO)_4(nbd)]$ (M = Cr, Mo, or W) with $[(2\text{-}Me_2E'C_6H_4)Si(EMe_2)Me_2]$ (E = N, P, or As; E' = P or As) yield $[M(CO)_4(chelate)]$ complexes. However, $[(2\text{-}Me_2NC_6H_4)Si(PMe_2)Me_2]$ and $[(4\text{-}XC_6H_4)Si(Me)(NMe_2)CH_2CH_2E'Me_2]$ (X = F or Cl) merely act as unidentate *P*- or *As*-donor ligands giving *cis*-$[M(CO)_4L_2]$ complexes.[46]

The macrocycles $\overline{XPO(CH_2)_2N(Me)(CH_2)_2N(Me)(CH_2)_2O}$ (X = OEt or NEt_2) act as *P*-donors giving $[M(CO)_5L]$ (M = Mo or W) complexes. These products are subject to nucleophilic attack by RLi (R = Me or Ph) at the carbon atom of a carbonyl ligand *via* selective lithium cation binding by the acylate or benzoylate products (2). Similarly, the related *cis*-$[Mo(CO)_4(bidentate)]$ complexes [bidentate = $Ph_2POCH_2CH_2N(CH_2CH_2OMe)CH_2CH_2OPPh_2$, $R_2PO(CH_2)_3O(CH_2)_3$-OPR_2, $Ph_2NH(CH_2)_3NMe(CH_2)_3NHPPh_2$, or $Ph_2PNH(CH_2)_4NHPPh_2$; R = Ph or $-N(Me)CH_2CH_2(Me)N-$] possess carbonyl ligands which can be activated by reaction with lithium alkyls or aryls. Acylate or benzoylate formation and lithium ion complexation as in (2) is involved in each case.[47]

The syntheses and multinuclear n.m.r. spectroscopic features of many *cis*-$[Mo(CO)_4(PPh_2X)_2]$, (X = *e.g.* Cl, NH_2, NHMe, OMe, SEt, $ESiMe_3$, or EC_6H_4Me-*p*; E = NH or O), *cis*-$[Mo(CO)_4(PPh_2YPPh_2)]$, [Y = *e.g.* ESi(R)(R')E; E = NH or O; R, R' = Me or Ph] and $[Mo(CO)_5(PPh_2ER)]$ (E = NH or O; R = 1-4 carbon alkyls) complexes have been reported.[48] The *As*-donor ligand complexes $\{M(CO)_5[Me_{3-\underline{x}}As(NMe_2)_{\underline{x}}]\}$ (M = Cr, Mo, or W; $\underline{x}$ = 1 - 3) are prepared by reactions of $[Et_4N][M(CO)_5I]$ with the appropriate ligand in THF. Cleavage of the As-N bonds of the co-ordinated ligands by reaction with HX (X = Cl, OEt, SEt, or OH) is a route to the complexes $[M(CO)_5(Me_{3-\underline{x}}AsX_{\underline{x}})]$.[49]

I.r. and n.m.r. spectroscopic evidence indicates that the ylide is carbon-bonded in $\{W(CO)_5[PhC(O)\bar{C}H\overset{+}{Z}]\}$ (Z = SMe_2), but oxygen-bonded when Z = PPh_3 or NC_5H_4Me-*p*. Chelation *via* the C=O oxygen and the pyridyl nitrogen occurs in the 2-pyridyl analogues $\{W(CO)_4[C_5H_4NC(O)\bar{C}H\overset{+}{Z}]\}$.[50] The anion $\{CpCo[P(O)(OEt)_2]_3\}^-$ acts as a novel tripodal ligand bonding *via* the P=O oxygen atoms in forming $[M(CO)_3(tripod)]^-$ (M = Cr, Mo, or W) anions on reaction with $[M(CO)_3(NCMe)_3]$.[51] Preparations of *mer, trans*-$[M(CO)_3(PR_3)_2(SO_2)]$ (M = Mo or W; R = Ph, Cy, or Pr^i) and *cis, trans*-[Mo(CO)$_2$(PPh$_3$)$_2$(SO)$_2$L] (L = MeCN, py, or isonitriles) have been reported. Depending on the nature of L, the SO$_2$ is either η^1-*S*- or η^2-*OS*-co-ordinated. Both isomers are present in the solid state where L = CyNC or Bu^{t}NC. Syntheses of *fac*-[Mo(CO)$_3$(η^2-SO$_2$)(L$_2$)] (L$_2$ = dppe, bipy, phen, or 2 py) have also been developed and the dppe complex shown to isomerize to *mer*-[Mo(CO)$_3$(η^1-SO$_2$)(dppe)]. The dimer [Mo(CO)$_2$(py)(PPh$_3$)(μ-SO$_2$)]$_2$, in which all three atoms of the SO$_2$ ligands are co-ordinated, has also been characterized. Other electron-rich SO$_2$ complexes isolated include *fac*-[M(CO)$_3$(η^2-SO$_2$)(dmpe)], *mer*- [M(CO)$_3$(η^1-SO$_2$)(dmpe)], *mer,trans*-[M(CO)$_3$(η^1-SO$_2$)L$_2$], [M = Cr, Mo, or W; L = PPh$_2$Me, P(OMe)$_3$, or P(OPr^i)$_3$], *trans*-[Mo(CO)$_2$(dppe)(η^1-SO$_2$)L] (L = PMe$_3$, PPh$_2$Me, or PPr$^i_3$), *fac*-[Cr(CO)$_3$(bipy)(η^2-SO$_2$)] and *fac*-[Mo(CO)$_3$(NCMe)$_2$(η^2-SO$_2$)].[52] During a study of the vibrational spectra of [W(CO)$_5$L] complexes (L = main group IV organometallic sulphide), new complexes with L = PhSER$_3$ (R = Me or Ph; E = Si, Ge, or Sn) have been prepared.[53] The photochemical reactions of RSH and R$_2$S$_2$ with W(CO)$_6$ lead to dinuclear [W(CO)$_4$(μ-SR)$_2$]$_2$ and open-chain [(OC)$_4$W(μ-SR')$_2$W(CO)$_2$(μ-SR')$_3$W(CO)$_4$] metal-metal bonded complexes (R = R', Bu^t, *p*-BrC$_6$H$_4$, or *p*-MeOC$_6$H$_4$; R' = Me, Et, Pr^i, or Ph).[54] The 1,3-intramolecular shifts of the M(CO)$_5$ unit in [M(CO)$_5$L], [M = Cr or W; L = SCH$_2$SCH$_2$SCH$_2$, SCH$_2$SCH$_2$CH$_2$CH$_2$, SCH(Me)SCH$_2$CH$_2$CH$_2$, or SCH(Me)SCH(Me)SCH(Me)] have been studied using variable temperature n.m.r. spectroscopy. For [M(CO)$_5$(SCH$_2$SCH$_2$SCH$_2$SCH$_2$)] two intramolecular processes are observed, pyramidal atomic inversion about the sulphur atoms, and commutation of the M(CO)$_5$ unit between co-ordination sites on different sulphur atoms.[55]

Reactions of [Et$_4$N][W(CO)$_5$I] with the trithiapentalenes (3) in the presence of AgNO$_3$ yield W(CO)$_5$-complexes of the thioaldehyde and thioketone valence isomers of the trithiapentalenes (4), the symmetrically substituted complexes being fluxional in solution.[56]

4.2 Carbonyl Complexes of Mo^{II} and W^{II}.- Reactions of [pyH][M(CO)$_4$X$_3$] (M = Mo or W; X = Br or I) with 2,2'-bipyridyl afford [M(CO)$_3$(bipy)X$_2$] species.[57] The co-ordinatively unsaturated complexes [WBr$_2$(CO)(diolefin)L] or the co-ordinatively saturated complexes [WBr$_2$(CO)(diolefin)L$_2$] and [WBr$_2$(CO)$_2$L$_3$], [diolefin = cod, cot, or nbd; L = CNBu^t, PMe$_2$Ph or P(OMe)$_3$] result from reactions of

$[WBr_2(CO)_2(diolefin)]_2$ with ligands.[58] The sterically hindered thiols RSH (R = 2,6-$Pr^i{}_2C_6H_3$ or 2,4,6-$Pr^i{}_3C_6H_2$) react with $MBr_2(CO)_4$ (M = Mo or W) to give trigonal bipyramidal $[M(CO)_2(SR)_3]^-$ anions.[59] The seven-co-ordinate species $[Mo(CO)_3(SC_6H_4SMe)_2]$ and $\{Mo(CO)_{\underline{x}}[m\text{-}SC_6H_4(SCH_2CH_2)_{\underline{n}}SC_6H_4S\text{-}m]\}$ ($\underline{x}$ = 3, $\underline{n}$ = 1; $\underline{x}$ = 2, $\underline{n}$ = 2; $\underline{x}$ = 1, $\underline{n}$ = 3) have been prepared from either $MoCl_2(CO)_4$ or $[MoCl_2(CO)_3(PPh_3)_2]$ and the appropriate thiol.[60]

4.3 Cyclopentadienyl, Arene and Other Complexes.- The wavelength-dependent photochemical disproportionation reactions of $[(\eta\text{-}RC_5H_4)Mo(CO)_3]_2$ (R = H or Me) in the presence of added ligands lead to $[(\eta\text{-}RC_5H_4)Mo(CO)_2L_2]^+$ cations and $[(\eta\text{-}RC_5H_4)Mo(CO)_3]^-$ anions. If the added ligand is a weak donor or is sterically bulky, substitution rather than disproportionation occurs giving $[(\eta\text{-}RC_5H_4)_2Mo_2(CO)_{6-\underline{n}}L_{\underline{n}}]$ ($\underline{n}$ = 1 or 2).[61] Reaction of $[CpMCl(CO)_3]$ (M = Mo or W) with the optically active esters $(Ph_2P)_2NCH(R)CO_2Me$ (R = H, Me, Pr^i, or Bz) gives the diastereomeric chelate complexes $\{CpMCl(CO)[(Ph_2P)_2NCH(R)CO_2Me]\}$.[62] Ring-opening of *cyclo*-$(MeAs)_5$ occurs on reaction with $[CpMo(CO)_3]_2$, the product being $\{[CpMo(CO)_2]_2\mu$-*catena*-$(MeAs)_5\}$ in which the terminal arsenic atoms of the chain bridge both metal atoms.[63] Photolysis of $[CpW(CO)_3]_2$ in the presence of $Me_2NC(S)S_2C(S)NMe_2$ affords $\{CpW(CO)_3[SC(S)NMe_2]\}$ which can be transformed into the chelate $[CpW(CO)_2(S_2CNMe_2)]$.[64] The related $[(\eta\text{-}C_5Me_5)Mo(CO)_2(S_2CNHMe)]$ is one product of the reaction of $[(\eta\text{-}C_5Me_5)Mo(CO)_2]_2$ with MeNCS.[65]

Electrochemical oxidation of $[(\eta\text{-mesitylene})Cr(CO)_3]$ in the presence of $P(OEt)_3$ followed by electrochemical reduction produces a mixture of $\{(\eta\text{-mesitylene})Cr(CO)_{3-\underline{n}}[P(OEt)_3]_{\underline{n}}\}$ ($\underline{n}$ = 1 and 2).[66] The related complexes $[(\eta\text{-}2,6\text{-}Me_2C_5H_3N)Cr(CO)_{3-\underline{n}}L_{\underline{n}}]$, [L = PPh_3, $\underline{n}$ = 1; L = $P(OMe)_3$, $\underline{n}$ = 1 or 2] have been prepared by photochemical substitution.[67] A variety of η^3-allyl complexes *e.g.* $[(\eta^3\text{-}C_3H_5)Mo(CO)_2(anion)L]$ (anion = S_2COMe, S_2CNHEt, or S_2CNEt_2; L = PPh_3, bipy, or phen), $\{[(\eta^3\text{-}C_3H_5)Mo(CO)_2(anion)]_2\mu\text{-}L\}$ (L = dppe, 4,4'-bipy, or 4,4'-dithiobipyridyl), and $\{[(\eta^3\text{-}C_3H_5)Mo(anion)(dppe)]_2(\mu\text{-}CO)_2\}$ have been synthesized from $[(\eta^3\text{-}C_3H_5)Mo(CO)_2(anion)py]$.[68]

The chelate complexes (5) arise from reactions of the aryl-carbene complexes $[Cr(CO)_5C(p\text{-}C_6H_4R)OMe]$ (R = H, Me, or CF_3) with $Ph_2PC{\equiv}CPPh_2$. The intermediates $\{[Cr(CO)_4C(p\text{-}C_6H_4R)OMe]_2\mu\text{-}$

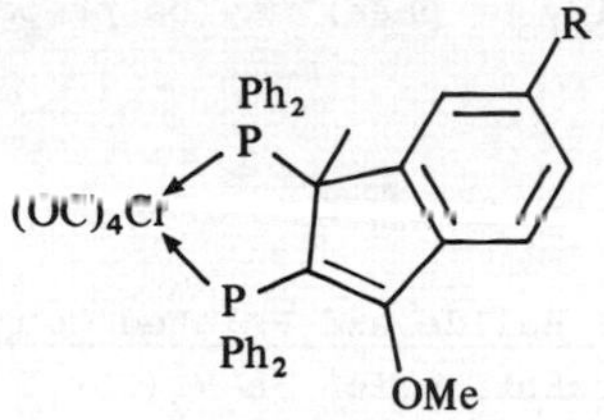

(5) R = H, Me, or CF_3

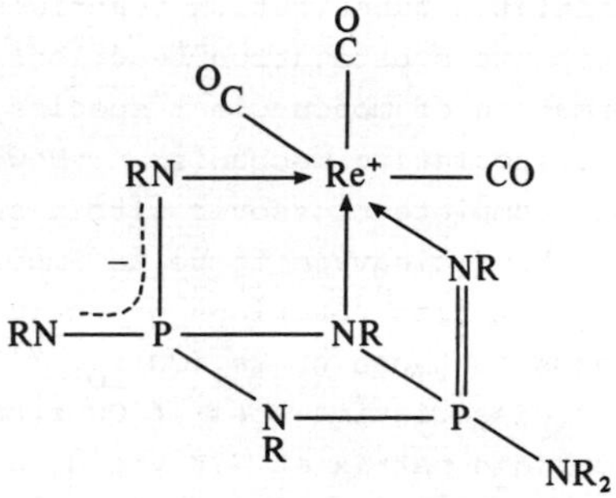

(6) R = Me_3Si

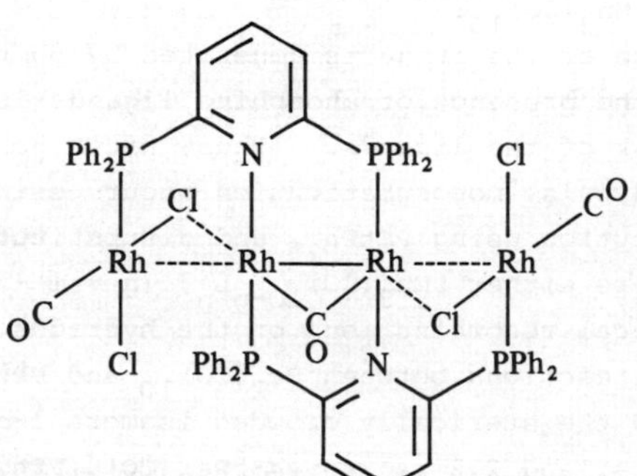

(7)

$Ph_2PC{\equiv}CPPh_2$} are involved in these reactions.[69] The thermally stable carbyne complexes [$MBr(CO)_2({\equiv}CR)L_2$] (M = Cr, Mo, or W; R = Ph or Me; L_2 = bipy or phen) may be prepared from *trans*-[$MBr(CO)_4({\equiv}CR)$].[70]

5 Group VII

5.1 Carbonyl, Carbonyl Halide and Related Complexes.- The mechanisms of ligand substitution in $M_2(CO)_{10}$ (M = Mn or Re) have been the subject of much discussion over recent years. Crossover experiments employing $^{185}Re_2(CO)_{10}$ and $^{187}Re_2(CO)_{10}$ have now been used to test for fragmentation to mononuclear species in thermal and photochemically initiated substitution reactions. It was found that thermal phosphine substitution reactions proceed without detectable formation of mononuclear species, the results supporting a carbonyl dissociative mechanism. However, in photochemical reactions complete crossover within short reaction times shows metal-metal bond cleavage to be an important mechanistic step. Substitution reactions of $Mn_2(CO)_{10}$ are mechanistically analogous to those of $Re_2(CO)_{10}$.[71] Other workers have shown that dissociative loss of CO from photo-excited $Mn_2(CO)_{10}$ in a rigid matrix at 77K yields a carbonyl-bridged $Mn_2(CO)_9$ species that adds unidentate ligands to give [$Mn_2(CO)_9L$] complexes. The rigid matrix apparently precludes cleavage of the metal-metal bond. Both metal-metal and metal-carbonyl bond scissions are primary processes in the laser flash photolysis of $Mn_2(CO)_{10}$.[72]

The nature of the products generated by photolysis of $Mn_2(CO)_{10}$ in the presence of phosphine ligands is influenced by the steric bulk of the ligands. Thus, after generation of [$Mn(CO)_5$]$^{\bullet}$ radicals, monosubstitution occurs using PBu^t_3, mono- and di-substitution using $PEtPh_2$, and disubstitution using PBu^n_3. The products are either [$Mn_2(CO)_{10-\underline{n}}L_{\underline{n}}$] ($\underline{n}$ = 1 - 3) species formed by radical recombinations or the hydrides [$HMn(CO)_3L_2$].[73] Photochemical reactions between $Re_2(CO)_{10}$ and PPh_3 carried out *in vacuo* yield the sterically crowded isomers 1-*ax*-2',4'-*dieq*-, 1-*ax*-2,2'-*dieq*-, and 2,2',4'-*trieq*-[$Re_2(CO)_7(PPh_3)_3$].[74] Reaction of $Mn_2(CO)_{10}$ with PPh_2H produces [$Mn_2(\mu\text{-}H)(\mu\text{-}PPh_2)(CO)_8$] which undergoes substitution reactions leading to the derivatives [$Mn_2(\mu\text{-}H)(\mu\text{-}PPh_2)(CO)_{8-\underline{n}}L_{\underline{n}}$] [$\underline{n}$ = 1, L = MeCN, PhCN, PPh_3, $P(OMe)_3$, or $P(OEt)_3$; $\underline{n}$ = 2, L_2 = $(EtO)_2POP(OEt)_2$; $\underline{n}$ = 4, L_4 = $2(EtO)_2POP(OEt)_2$].[75]

Paramagnetic metal carbonyl radicals have been much studied, mainly by e.s.r. spectroscopy. Examples include $[M(CO)_4(Bu^tNCHCHNBu^t)]^\bullet$ (M = Mn or Re), $[Mn(CO)_3L_2]^\bullet$ (L = phosphines or phosphites), $[Mn_4(CO)_8(asq)_4]$ (asq = 2-acetyl-1,4-benzosemiquinone), $[Re(CO)_3(PCy_3)_2]^\bullet$, $[Re(CO)_{4-\underline{n}}(dtbq)L_{\underline{n}}]^\bullet$ (dtbq = 3,5-di-*t*-butyl-1,2-benzoquinone, L = Group V donor ligand, $\underline{n}$ = 1 or 2), and various optically active $[Re(CO)_3(quinone)L]^\bullet$ radicals.[76] The nitroxyl radical 2,2,6,6-tetramethylpiperidinyl-1-oxo, tmpo, reacts with $M_2(CO)_{10}$ (M = Mn or Re) to give $[M(CO)_3(tmpo)]$ 16-electron complexes containing *ON*-bonded tmpo monoanions.[77]

Amine-*N*-oxide-assisted reactions of $Re_2(CO)_{10}$ with alkyl nitriles produce $[Re_2(CO)_{10-\underline{n}}(NCR)_{\underline{n}}]$ complexes ($\underline{n}$ = 1 or 2).[78] Cations prepared from $[Mn(CO)_5(OClO_3)]$ include $[Mn(CO)_5L]^+$, *fac*-$[Mn(CO)_3L_3]^+$, $[Mn(CO)_5(LL)]^+$, and $\{[Mn(CO)_3]_2(\mu\text{-}LL)_3\}^{2+}$ [L = RCN, R = Et, Ph, Bz or allyl; LL = $NC(CH_2)_{\underline{n}}CN$ or 1,2-$C_6H_4(CN)_2$, $\underline{n}$ = 1 - 3].[79] The bidentate ligands *o*-$NCC_6H_4O(CH_2)_{\underline{n}}OC_6H_4CN$-*o* ($\underline{n}$ = 2 - 4) react with $MnBr(CO)_5$ to give $[MnBr(CO)_3(dinitrile)]$ chelate complexes. The novel trinitrile, 1,3,5-*tris*(2-cyanophenyl)benzene, L, acts as a bidentate ligand in $[MBr(CO)_3L]$ or as a terdentate ligand in $[M(CO)_3L][PF_6]$ (M = Mn or Re).[80] Substitution of the SO_2 ligand of $[Re(CO)_5(SO_2)][AsF_6]$ by *N*-donor ligands is a route to $[Re(CO)_5L]^+$, [L = NSR, HNSO, or $NSON(CF_3)_2$; R = F, NMe_2, or $N(Me)SiMe_3$] and $[Re(CO)_5(NS)]^{2+}$ cations.[81]

Several methods for the preparation of the *NN*-chelate amidino-complexes $\{Re(CO)_4[R'NC(R)NR']\}$ and $\{Re(CO)_3L[R'NC(R)NR']\}$ (R = Me or Ph; R' = Ph or *p*-tol; L = PPh_3 or $AsPh_3$) have been reported. Amidines react directly with $ReX(CO)_5$ (X = Cl or Br) or $[ReBr(CO)_4]_2$ to produce the *o*-metallated derivatives $\{\overline{Re(CO)_3[R'NC(R)NHC}_6H_3R''\text{-}p][R'NC(R)NHR']\}$ (R = H, Me, or Ph; R' = Ph or *p*-tol; R" = H or Me).[82] The *NO*-chelate aryl-formamido-complexes $[Re(CO)_2(ArNCHO)(PPh_3)_2]$ (Ar = Ph, *p*-XC_6H_4 or *o*-tol; X = Me, Cl, MeO, or NO_2) result from reactions of $[ReCl(CO)_3(PPh_3)_2]$ with lithium arylformamidates.[83] Stepwise substitution of up to three carbonyl groups of the μ-pyridyl complex $[Re_2(CO)_8(\mu\text{-}H)(\mu\text{-}NC_5H_4)]$ by unidentate ligands *e.g.* Me_3NO, Me_3N, py, MeCN, PPh_3, or $P(OPh)_3$ can be achieved using Me_3NO-assisted, thermal, or photolytic reactions.[84] Other nitrogen-donor ligand substituted rhenium carbonyls reported include $[ReBr(CO)_3(Hgly)L]$ and $[Re(CO)_3(gly)L]$ (Hgly = glycylglycine, gly = bidentate glycylglycinate, L = dioxane),[85] $[Re(CO)_3(bipy)L]^{2+}$ (L = *N*-methyl- or *N*-phenyl-4,4'-bipyridinium ions),[86] and $[ReBr(CO)_3(bpnp)]$, [bpnp = bidentate 2,7-bis(2'-pyridyl)-1,8-naphthyridine].[87] On reaction with $ReBr(CO)_5$, the amino(imino)thiophosphorane $Bu^t(Me_3Si)NP(S){=}NBu^t$ isomerizes to the cyclic diazaphosphasiletidine sulphide $Bu^t\overline{NSi(Me)_2N(Bu^t)P}(S)Me$ which then acts as a *NN'*-bridging ligand producing $[ReBr(CO)_3\mu\text{-}Bu^t\overline{NSi(Me)_2N(Bu^t)P}(S)Me]_2$.[88] The compound $(Me_3Si)_2NP({=}NSiMe_3)_2$, which can be regarded as the first stable derivative of monomeric metaphosphoric acid, $HOPO_2$, reacts with $[ReBr(CO)_3(THF)]_2$ to afford the novel complex (6) arranged as a distorted cube with a missing corner.[89]

Treatment of the dithioacetic acid derivative $Ph_3P{=}C(CN)CS_2H$ with $MnX(CO)_5$ (X = Cl, Br, or I) gives the chelate complex $\{Mn(CO)_4[S_2CC(CN)PPh_3]\}$ whereas reaction of the lithium salt with $ReCl(CO)_5$ or $[ReCl(CO)_4]_2$ produces the dimer $\{Re_2(CO)_6[\mu\text{-}S_2CC(CN)PPh_3]_2\}$.[90]

The amino(imino)phosphine *trans*-$Me_3Si(Bu^t)NP{=}NBu^t$ isomerizes to the *cis*-form which then acts as a *PN*-bridging ligand on reaction with $[ReBr(CO)_3(THF)]_2$, the product being $\{Re_2(\mu\text{-}Br)_2(CO)_6[\mu\text{-}Me_3Si(Bu^t)NP{=}NBu^t]\}$.[91] Other products of similar structure isolated include $[Mn_2(\mu\text{-}Br)_2(CO)_6(\mu\text{-}E_2Ph_2)]$ (E = Se or Te),[92] and $[Re_2(\mu\text{-}X)_2(CO)_6(\mu\text{-}E_2R_2)]$ (X = Cl or Br; E = S or Se; R = $PhCH_2$ or Me_3SiCH_2). The synchronous pyramidal inversions at the co-ordinated S or Se atoms of the rhenium complexes have been studied using variable temperature ^1H n.m.r. spectroscopy.[93]

<u>5.2 Cyclopentadienyl Complexes.</u>- Reaction of $CpRe(CO)_3$ with PMe_3 produces *fac*-$[(\eta^1\text{-}C_5H_5)Re(CO)_3(PMe_3)_2]$. The reaction is reversible, but long reaction times convert the equilibrium mixture to $[CpRe(CO)_2(PMe_3)]$.[94] The new ligands $RC(O)PPh_2$ react with $[CpMn(CO)_2(THF)]$ to give the *P*-donor complexes $\{CpMn(CO)_2[Ph_2PC(O)R]\}$ (R = $4\text{-}ClC_6H_4$, $4\text{-}MeOC_6H_4$, or $3,5\text{-}Cl_2C_6H_3$).[95] Spectroscopic evidence points to strong acceptor properties for

S_2 and Se_2 in [CpRe(CO)$_2$(η^2-S_2)] and [(η-C_5Me_5)Mn(CO)$_2$(η^2-Se_2)]. The former complex, together with [Cp_2Re_2(CO)$_4$(μ-S)] and [Cp_2Re_2(CO)$_2$(μ-S_2)(μ-S_3)], is formed from CpRe(CO)$_3$ and S_8 or COS in ether.[96]

6 Group VIII : Iron, Ruthenium, and Osmium

6.1 Iron Carbonyl Complexes.- Evidence has been presented suggesting that the formation of [Fe_2(CO)$_7$(bipy)] and [Fe_2(CO)$_7$(phen)] from $Fe_2(CO)_9$ in THF occurs *via* [Fe(CO)$_3$(chelate)] intermediates rather than by direct substitution. These [Fe(CO)$_3$(chelate)] complexes can be isolated by displacement of diene from [Fe(CO)$_3$(PhCH=CHCH=CH_2)].[97] Several routes to the η^2-*NS*-sulphinylaniline complexes [Fe(CO)$_2$(PR_3)$_2$(ArNSO)] have been reported. The most useful is the reaction of $Fe_2(CO)_9$ with PR_3 and ArNSO in THF (R = *p*-XC_6H_4 or Bz, X = H, Cl, Me, or OMe; Ar = *p*-YC_6H_4, Y = H, F, Cl, Br, NO_2, Me, or OMe). In the absence of PR_3 the products are probably of the type [Fe(CO)$_4$(η^1-ArNSO)].[98] The dissociation of L from [Fe(CO)$_3L_2$] has been investigated by reaction with CO at elevated temperatures. The reactions obey first-order kinetics, the ease of generation of the 16-electron intermediates [Fe(CO)$_3$L] being in the order $AsPh_3 > PPh_3 > P(OPh)_3$.[99] The complexes [Fe(CO)$_4$(η^1-dppe)] and *trans*-[Fe(CO)$_3$(η^1-dppe)$_2$] have been prepared by base-promoted addition of PPh_2H to [Fe(CO)$_{5-n}$(Ph_2PCH=CH_2)$_n$] (n = 1 or 2).[100] In an extension of previous work on fluorophosphine complexes {Fe(CO)$_4$[η^1-MeN(PF_2)$_2$]}, {Fe_2(CO)$_2$(μ-CO)[μ-MeN(PF_2)$_2$]$_3$} and {(OC)$_4$Fe[P{N(Ph)PF_2}$_3$]Mo(CO)$_3$} have been isolated.[101] The fluorophosphines $PF_{3-\underline{n}}$[OC(CF_3)$_2$CN]$_{\underline{n}}$ ($\underline{n}$ = 1 - 3), containing both bulky and highly electronegative groups on phosphorus, react with $Fe_2(CO)_9$ to give axial and equatorial isomeric mixtures of [Fe(CO)$_4$L] species, the equatorial isomer being favoured as n increases.[102] The ferrocenylphosphine complex [Fe(CO)$_4$(Fc_2PCl)], prepared from $Fe_2(CO)_9$, can be converted into the phosphenium ion complex [Fe(CO)$_4$(PFc_2)][$AlCl_4$] on treatment with $AlCl_3$.[103] Reaction of $Fe_2(CO)_9$ with $PhPCl_2$ produces the compounds [Fe(CO)$_4$($PPhCl_2$)], [Fe_2(CO)$_6$(μ-Cl)(μ-PPhCl)] and [Fe_2(CO)$_6$(μ-PPhCl)$_2$].[104]

A series of trigonal bipyramidal [Fe(CO)$_2L_2$(η^1-SO_2)] complexes have been prepared either by photolysis of [Fe(CO)$_3L_2$] (L = phosphite) followed by addition of SO_2, or by displacement of CS_2 in $[Fe(CO)_2L_2(CS_2)]$ by SO_2 (L = phosphine or phosphite). Two isomers may be formed, the ligands L being arranged either axially [L = *e.g.* $P(OPh)_3$] or equatorially [L = *e.g.* $P(O\text{-tol-}m)_3$].[105] The isothiocyanate complexes $[Fe(CO)_2(PPh_3)_2(RNCS)]$ (R = Ph or Me), prepared from $[Fe(CO)_2(PPh_3)_3]$ and RNCS, decompose spontaneously even in the solid state.[106]

There have been several reports of substitution reactions of iron carbonyl nitrosyls. The bicyclic aminophosphorane $PhHP(OCH_2CH_2)_2N$ reacts with $[Fe(CO)_2(NO)_2]$ to give $[Fe(CO)_{2-n}(NO)_2L_n]$ (n = 1 or 2) complexes in which L is the *P*-bonded monocyclic open isomer $PhP(OCH_2CH_2)_2NH$.[107] The complexes $[Fe(CO)(NO)_2L]$ [L = $PBu^t_{3-n}(EMe_3)_n$; n = 0 - 3, E = Si, Ge, or Sn] have also been prepared from $[Fe(CO)_2(NO)_2]$.[108] Protonation of the anion $[Fe(CO)_3(NO)]^-$ in the presence of PPh_3 leads to $[FeH(CO)_{3-n}(NO)(PPh_3)_n]$ (n = 1 or 2) or $[Fe(CO)_2(NO)(PPh_3)_2]^+$, depending on reaction conditions.[109]

6.2 Ruthenium and Osmium Carbonyl Complexes.- Over sixty complexes of the types $[Ru_3(CO)_{12-n}L_n]$ (L = unidentate phosphines, phosphites, $AsPh_3$, or $SbPh_3$; n = 1 - 4), $[Ru_3(CO)_{12-2n}(LL)_n]$ (LL = bidentate diphosphines or diarsines; n = 1 or 2), $\{[Ru_3(CO)_{11}]_2(\mu\text{-dppm})\}$ and $[H_4Ru_4(CO)_{12-n}L_n]$ (L = unidentate phosphines or phosphites; n = 1 - 3) have been prepared by sodium diphenylketyl-initiated substitution reactions of $Ru_3(CO)_{12}$ and $H_4Ru_4(CO)_{12}$.[110] The thermal reaction of $Ru_3(CO)_{12}$ with PPh_2H produces $[Ru_3(CO)_9(PPh_2H)_3]$ which on photolysis affords the complexes $[Ru_2(CO)_6(\mu\text{-}PPh_2)_2]$, $[Ru_3(CO)_7(\mu\text{-}H)_2(\mu\text{-}PPh_2)_2L]$ (L = CO or PPh_2H), and $[Ru_3(CO)_7(\mu\text{-}H)(\mu\text{-}PPh_2)_3]$.[111] Cluster disruption occurs when $Ru_3(CO)_{12}$ reacts with the triphosphines $PhP(CH_2CH_2CH_2PR_2)_2$ (R = Ph or Cy), the products being of the type $[Ru(CO)_2(\text{triphosphine})]$.[112] Variations in the molar ratios of reactants and reaction conditions influence the nature of the products formed in the $[Ru_2(CO)_4(O_2CMe)_2]_n$ - PBu^n_3 - $MeCO_2H$ system. Complexes isolated include $[Ru(CO)_2(O_2CMe)_2(PBu^n_3)_2]$, $[Ru_2(CO)_4(O_2CMe)_2(PBu^n_3)L]$ (L = CO or PBu^n_3) and $[Ru_4(CO)_8(O_2CMe)_4(PBu^n_3)_2]$. Some dicarboxylate analogues *e.g.* $\{Ru_2(CO)_4[O_2C(CH_2)_xCO_2]\}_n$ and $\{Ru_4(CO)_8[O_2C(CH_2)_xCO_2]_2(PBu^n_3)_4\}$ (x = 0 - 4), have also been isolated.[113]

Thermal reaction between $Ru_3(CO)_{12}$ and $Ph_2PCH_2SiMe_2H$ yields the symmetrically-bridged complex $[Ru_2(CO)_6(\mu\text{-}Ph_2PCH_2SiMe_2)_2]$ which on treatment with CF_3CO_2H affords the *OP*-bonded chelate silanol complex $[Ru(CO)_2(Ph_2PCH_2SiMe_2OH)(O_2CCF_3)_2.Et_2O]$.[114] In the solid state trigonal bipyramidal $[M(CO)_4(EPh_3)]$ (M = Fe, E = P, As, or Sb; M = Ru or Os, E = P or As) have the EPh_3 ligand occupying an axial site, whereas in $[Ru(CO)_4(SbPh_3)]$ and $[Os(CO)_4(SbPh_3)]$ the $SbPh_3$ ligands are equatorial. In solution, $[M(CO)_4(EPh_3)]$ (M = Ru or Os; E = As or Sb) exist as a mixture of axial and equatorial isomers.[115]

<u>6.3 Cyclopentadienyl Complexes.</u>- Photolysis of cyclohexane solutions of $[CpFe(CO)_2]_2$ in the presence of PR_3 (R = Ph or OPr^i) proceeds *via* $[Cp(CO)_2Fe(\mu\text{-}CO)Fe(CO)(PR_3)Cp]$ intermediates yielding $[Cp_2Fe_2(CO)_3(PR_3)]$.[116] Carbonyl substitution by *P*-donors in $[CpFe(CO)_2(\eta^1\text{-}C_5H_5)]$ occurs by a radical chain mechanism in which $[CpFe(CO)_2]^{\cdot}$ is initially formed. Substitution then occurs to give $[CpFe(CO)L]^{\cdot}$ followed by C_5H_5 group transfer from $[CpFe(CO)_2(\eta^1\text{-}C_5H_5)]$ to produce $[CpFe(CO)(L)(\eta^1\text{-}C_5H_5)]$ and regenerate $[CpFe(CO)L]^{\cdot}$. The $[CpFe(CO)(L)(\eta^1\text{-}C_5H_5)]$ products are stable when L = $P(OPh)_3$ and $P(OPr^i)_3$, but for L = $P(OMe)_3$ and $P(OEt)_3$ a subsequent Arbuzov-like rearrangement leads to $\{CpFe(CO)[P(OR)_3][P(O)(OR)_2]\}$ while for phosphines a redox action with the chlorinated solvents employed leads to $[CpFe(CO)(PR_3)_2]Cl$.[117] Photolysis of $[CpFe(CO)_2I]$ in the presence of $P(OCH_2CF_3)_3$ produces $\{CpFe[P(OCH_2CF_3)_3]_2I\}$.[118] The cations $[CpFeL_{\underline{n}}(NCMe)_{3-\underline{n}}]^+$ arise from photolysis of $[CpFe(CO)_3]^+$ - ligand mixtures in MeCN ($\underline{n}$ = 1, L = CO or PPh_3; $\underline{n}$ = 2, L_2 = dppe; $\underline{n}$ = 3, L = a phosphole). Similarly, the cations $[(\eta\text{-}C_5Me_5)Fe(CO)_{3-\underline{n}}L_{\underline{n}}]^+$ ($\underline{n}$ = 1, L = PPh_3, PMe_3, or PEt_3; n = 2, L_2 = dppe or $2PMe_3$) are produced by treatment of $[(\eta\text{-}C_5Me_5)Fe(CO)_3]^+$ with phosphines.[119] Reaction of $[CpFe(CO)_2]_2$ with $(PhO)_2P(CH_2)_2P(OPh)_2$, L_2, yields $[Cp_2Fe_2(\mu\text{-}CO)_2(\mu\text{-}L_2)]$, oxidized by iodine to a mixture of diastereoisomers of $[Cp_2Fe_2(CO)_2I_2(\mu\text{-}L_2)]$. Irradiation of $Me_2Si[(\eta^5\text{-}C_5H_4)Fe(CO)_2]_2$ or $Me_2Si[(\eta^5\text{-}C_5H_4)Fe(CO)_2I]_2$ in the presence of added ligands affords the products $\{Me_2Si[(\eta^5\text{-}C_5H_4)Fe(\mu\text{-}CO)]_2(\mu\text{-}L'_2)\}$, $\{[Me_2Si(\eta^5\text{-}C_5H_4)_2Fe_2(CO)(\mu\text{-}CO)_2]_2(\mu\text{-}L'_2)\}$ (L'_2 = dppm, dppe, or

dppp) and {$Me_2Si[(\eta^5-C_5H_4Fe(CO)(L'')I]_2$}, [L" = PPh_3 or $P(OPh)_3$].[120] Photochemical and thermal reactions of [CpFe$(CO)_2$Me] with the crown ether $PhPCH_2N(CH_2CH_2O)_4CH_2CH_2$ yield *P*-bonded [CpFe(CO)Me(crown ether)] and [CpFe(CO)(COMe)(crown ether)] complexes, respectively. Both complexes bind alkali and alkaline-earth metal cations *via* co-ordinated crown ether oxygen- and nitrogen-cation interactions, the binding ability of the acetyl complex being superior because of additional acyl oxygen-cation interaction.[121]

7 Group VIII : Cobalt, Rhodium, and Iridium

7.1 Cobalt Complexes.- Photochemical reaction of $Co_2(CO)_8$ with $[Co_2(CO)_6L_2]$ (L = a phosphine) affords the three $[Co_2(CO)_{8-\underline{n}}L_{\underline{n}}]$ ($\underline{n}$ = 0 - 2) species in equilibrium. The kinetics are consistent with Co-Co bond cleavage in the dimers, followed by bimolecular recombinations of the radicals.[122]

Metal-metal double bonded $[Co(CO)_2(\mu-PBu^t{}_2)]_2$ arises from a reaction of $Co(CO)_4I$, (prepared *in situ* from $Co_2(CO)_8$ and I_2 in THF), with $Bu^t{}_2PLi$. The analogue $[Co(CO)(PEt_2Ph)(\mu-PPh_2)]_2$ can be formed from oligomeric $[Co(CO)_3(\mu-PPh_2)]_{\underline{n}}$ and PEt_2Ph.[123] The complexes $[(C_6F_5)Co(CO)_{4-\underline{n}}L_{\underline{n}}]$ ($\underline{n}$ = 1 or 2; L = phosphine or phosphite) are the products of reactions of $[(C_6F_5)Co(CO)_4]$ with *P*-donor ligands.[124]

7.2 Rhodium and Iridium Complexes.- Various nucleobases and nucleosides react with $[RhCl(CO)_2]_2$ to give $[RhCl(CO)_2L]$ species.[125] Other nucleobase complexes of the type $[Rh(CO)(PPh_3)_2L][PF_6]$ can be formed from $[Rh(CO)_3(PPh_3)_2][PF_6]$ (L = *e.g.* guanosine, inosine, purine, adenine, or adenosine).[126] Binuclear{$[RhCl(CO)_2]_2\mu$-L} (L = pyrazine or phenazine) and mononuclear $[MCl(CO)_2L]$ (M = Rh or Ir, L = phenazine; M = Rh, L = phenazine-*N*-oxide) complexes result from reactions of the ligands with $[RhCl(CO)_2]_2$ or $[IrCl(CO)_3]_{\underline{n}}$.[127] Solid $[RhCl(CO)_2(pyrazole)]$ has a columnar chain structure, the square planar units being linked by weak intermetallic bonding.[128] The 24-atom macrocycle {$[NH(CH_2)_2]_3O(CH_2)_2[NH(CH_2)_2]_3O(CH_2)_2$} reacts with $[RhCl(CO)_2]_2$ to give the cryptate-like cation $[Rh_2(\mu-CO)_3(macrocycle)]^{2+}$, isolated as its PF_6^- salt. The metal-metal bonded $Rh_2(\mu\text{-}CO)_3$ unit is held inside the ligand cavity each metal being bonded to three nitrogen atoms in addition to the bridging carbonyl groups.[129]

Several complexes containing phosphinopyridine ligands have been characterized. Addition of $[RhCl(CO)_2]_2$ to *cis*-$[PtCl_2(2\text{-}Ph_2PC_5H_4N)_2]$ initially gives the salt $[PtCl(2\text{-}Ph_2PC_5H_4N)_2]$ $[RhCl_2(CO)_2]$, which on heating rearranges to $[Cl_2(CO)Rh(\mu\text{-}2\text{-}Ph_2PC_5H_4N)_2PtCl]$. Direct reaction of $[RhCl(CO)_2]_2$ with $2,6\text{-}(Ph_2P)_2C_5H_3N$ in refluxing benzene gives complex (7), which on carbonylation followed by addition of NH_4PF_6 breaks down to binuclear $\{(MeOH)(CO)Rh[\mu\text{-}2,6\text{-}(Ph_2P)_2C_5H_3N]_2RhCl(CO)\}\{PF_6\}$.[130] The complex (8), in which the phosphorus atoms are arranged *trans* on one metal and *cis* on the other, can be prepared from $Li[RhCl_2(CO)_2]$ and $2\text{-}(Ph_2P)_2CHC_5H_4N$ in the presence of $NaBPh_4$. The related complexes $\{Rh_2(CO)_2[\mu\text{-}2\text{-}(Ph_2P)_2CHC_5H_4N]_2\}\{PF_6\}_2$ and $\{Ir_2(CO)_2(\mu\text{-}CO)[\mu\text{-}2\text{-}(Ph_2P)_2CHC_5H_4N]_2\}\{BF_4\}_2$ have also been characterized. The rhodium cation reacts with Cl^- to produce $\{RhCl(CO)[2\text{-}(Ph_2P)_2CHC_5H_4N]\}$, in which the potentially terdentate phosphinopyridine is only bidentate, and with S^{2-} to afford the 'A-frame' complex $\{Rh_2(CO)_2(\mu\text{-}S)[\mu\text{-}2\text{-}(Ph_2P)_2CHC_5H_4N]_2\}$ in which the pyridyl nitrogen atoms are not co-ordinated.[131] The complexes *trans*-$[RhCl(CO)L_2]$, [L = Ph_2PR; R = *e.g.* $(CH_2)_2CH{=}CH_2$, $p\text{-}C_6H_4(CH_2)_{\underline{n}}CH{=}CH_2$, $p\text{-}C_6H_4CH_2C(Me){=}CH_2$, or $p\text{-}C_6H_4CH{=}CHMe$; $\underline{n}$ = 0 - 3] have free olefin groups which are thereby available for formation of polymer-supported metal complex catalysts.[132] Reaction of $[RhCl(CO)_2]_2$ with $PhP(CH_2PPh_2)_2$ gives cation (9) isolated as Cl^-, PF_6^-, and BPh_4^- salts. The BPh_4^- salt reacts with NaI to produce $\{Rh_3(CO)_2(\mu\text{-}CO)(\mu\text{-}I)_2[\mu\text{-}PhP(CH_2PPh_2)_2]_2\}\{BPh_4\}$ in which the Rh_3P_6 unit distorts to accommodate bridging iodide and carbonyl ligands.[133] 2,5-Bis(diphenylphosphino)furan, dpf, acts as a *PP*-bridging ligand in $[RhCl(CO)(\mu\text{-}dpf)]_2$, $[Cl(CO)Rh(\mu\text{-}dpf)_2Rh(CO)_3]^+$ and $[Rh(CO)_2(\mu\text{-}dpf)]_2^{2+}$.[134] Depending on the molar ratio of the reactants, either $[RhCl(CO)(PPh_2H)_2]$ or $[Rh_3(\mu\text{-}Cl)_2(\mu\text{-}CO)(CO)_3(\mu\text{-}PPh_2)_3]$ can be isolated from the reaction of $[RhCl(CO)_2]_2$ with PPh_2H in benzene. Removal of chloride from the trinuclear complex in the presence of CO affords $[Rh_3(CO)_5(\mu\text{-}PPh_2)_3]$.[135] Reaction of $[RhCl(CO)_2]_2$ with $Bu^t{}_2PLi$ in THF at -78^oC produces a mixture of $[Rh(CO)_2(\mu\text{-}PBu^t{}_2)]_2$ and the new trimer $[Rh(CO)(\mu\text{-}PBu^t{}_2)]_3$ containing a co-ordinatively

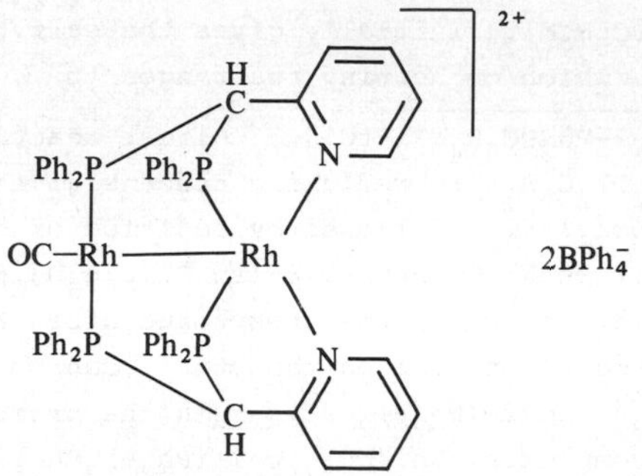

(8)

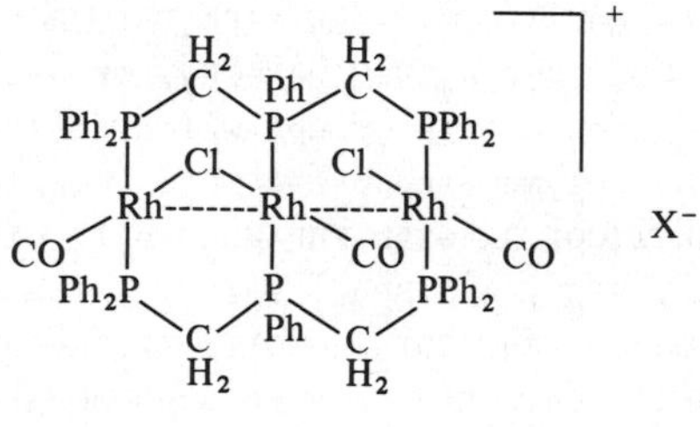

(9) X = Cl, PF_6, or BPh_4

unsaturated metal-metal bonded Rh_3P_3 core. The dimer reacts
with PMe_3 to give the metal-metal double bonded complex
$[Rh(CO)(\mu-PBu^t_2)(PMe_3)]_2$.[136] Starting from $[RhCl(CO)_2]_2$, the
PS-bonded chelate complexes $[RhCl(CO)L_2]$ and the *P*-unidentate
complexes *trans*-$[RhCl(CO)(L_2)_2]$ ($L_2 = Ph_2P(CH_2)_nSPh$, $n = 1$ or 2)
can be prepared by direct reaction with the ligands.[137]
Luminescent anionic maleonitriledithiolate complexes $[IrL_2(mnt)]$,
$[L = P(OPh)_3$ or $\frac{1}{2}$ dppe] and $[Ir(CO)(PPh_3)(mnt)]^-$, isolated as
$Bu^n_4N^+$ or $(Ph_3P)_2N^+$ salts, are formed by reaction of
$[Ir(CO)_2(mnt)]^-$ with the appropriate ligand. The $[Ir(CO)_2(mnt)]^-$
anion arises from the reaction of $[Bu^n_4N][Ir(CO)_2Br_2]$ with
$Na_2[mnt]$.[138]

7.3 Cyclopentadienyl Complexes.-

Stepwise reaction of
$[Cp_2Rh_2(CO)_2(\mu-CO)]$ with PPh_3 produces $[CpR\overline{h(\mu-CO)_2R}h(PPh_3)Cp]$
and $[CpRh(CO)(PPh_3)]$.[139] Photolysis of $[(\eta-C_5Me_5)Co(CO)]_2$ in
the presence of $R_2P(CH_2)_2PR_2$ ($R = Me$ or Ph) gives the chelate
complexes $\{(\eta-C_5Me_5)Co[R_2P(CH_2)_2PR_2]\}$.[140] The complexes
$[CpRh(CO)(\eta^1-dppm)]$, $[CpRh(dppe)]$, and $[CpRh(dppp)]$ result from
reactions of $[CpRh(CO)_2]$ with the diphosphines.[141]

8 Group VIII : Nickel, Palladium, and Platinum

A wide variety of $[Ni(CO)_3L]$ complexes ($L = P$, As, Sb, or Bi
donor ligands) have been studied using [13]C n.m.r. spectroscopy.[142]
The phosphaalkene $2,4,6-Me_3C_6H_2P=CPh_2$ reacts with $Ni(CO)_4$ to give
successively the unstable *P*-bonded complexes $[Ni(CO)_3L]$ and
$[Ni(CO)_2L_2]$.[143] Another *P*-bonded complex prepared by
substitution from $Ni(CO)_4$ is $\{Ni(CO)_3[P_4(NMe)_6S_3]\}$.[144]
Carbonylation of $[Ni(\mu-PBu^t_2)(PR_3)]_2$ ($R = Me$ or Et) produces
$[Ni_2(\mu-PBu^t_2)_2(CO)_2(PR_3)]$, one PR_3 having been replaced by two
CO ligands. Photolysis of $[Ni(\mu-PBu^t_2)(PMe_3)]_2$ in the
presence of $Mn_2(CO)_{10}$ affords $[Ni_2(\mu-PBu^t_2)_2(CO)_3]$ which then
reacts with PMe_3 to give $[Ni_2(\mu-PBu^t_2)_2(CO)_2(PMe_3)]$. These
nickel(I) complexes possess metal-metal bonds.[145]
Carbonylation of Na_2PtCl_4 in aqueous acetone gives $[Pt(CO)_2]_n$
in situ which then reacts with PEt_3 affording the clusters
$[Pt_5(CO)_6(PEt_3)_4]$, $[Pt_4(CO)_5(PEt_3)_4]$, and $[Pt_3(CO)_3(PEt_3)_4]$.[146]
The metal-metal bonded platinum(I) complex

$[Et_4N]_2[Pt_2Cl_4(\mu\text{-}Ph_2PCH_2NEt_2)_2]$, prepared from $[Et_4N]_2[Pt_2Cl_4(CO)_2]$, reacts reversibly with CO to give $[Et_4N]_2[Pt_2Cl_4(\mu\text{-}CO)(\mu\text{-}Ph_2PCH_2NEt_2)_2]$.[147] Reactions of $[Pt_2X_4(CO)_2]^{2-}$ (X = Cl or Br) with PPh_3 sequentially yield the metal-metal bonded platinum(I) species $[Pt_2X_3(CO)_2(PPh_3)]^-$, $[Pt_2X_2(CO)_2(PPh_3)_2]$ and $[Pt_2X(CO)_2(PPh_3)_3]^+$. This cation reacts with additional halide ion to give $[Pt_2X_2(\mu\text{-}CO)(PPh_3)_3]$ containing an unsymmetrical bridging carbonyl ligand.[148]

Carbonylation of palladium(II) carboxylates in non-aqueous media using various reaction conditions is a route to several clusters, for example, $[Pd(CO)(O_2CR)]_4$ (R = alkyl or Ph), $[Pd_4(CO)_5(PR'_3)_4]$, $[Pd_{10}(CO)_{12}(PR'_3)_4]$, $[Pd_{10}(CO)_{12}(PR'_3)_6]$, (R' = Et or Bu^n) and $[Pd_{10}(CO)_{14}(PBu^n_3)_4]$.[149] Reaction of $Pd(O_2CMe)_2$ with dppm under CO in aqueous acetone containing CF_3CO_2H gives the first dicationic cluster of palladium, $[Pd_3(\mu_3\text{-}CO)(\mu\text{-}dppm)_3][CF_3COO]_2$.[150]

References

1 J.D. Atwood, M.J. Wovkulich, and D.C. Sonnenberger, Acc. Chem. Res., 1983, 16, 350.

2 J.A.S. Howell and P.M. Burkinshaw, Chem. Rev., 1983, 83, 557.

3 D.J. Darensbourg, Adv. Organomet. Chem., 1982, 21, 113.

4 G. van Koten and K. Vrieze, Adv. Organomet. Chem., 1982, 21, 152.

5 R.J. Puddephatt, Chem. Soc. Rev., 1983, 12, 99.

6 K.S. Suslick, J.W. Goodale, P.F. Schubert, and H.H. Wang, J. Am. Chem. Soc., 1983, 105, 5781; K.S. Suslick and P.F. Schubert, ibid., p. 6042.

7 J.W. Hershberger, C. Amatore, and J.K. Kochi, J. Organomet. Chem., 1983, 250, 345; J.W. Hershberger, R.J. Klinger, and J.K. Kochi, J. Am. Chem. Soc., 1983, 105, 61; B.A. Narayanan, C. Amatore, and J.K. Kochi, J. Chem. Soc., Chem. Commun., 1983, 397.

8 R.E. Stevens and W.L. Gladfelter, Inorg. Chem., 1983, 22, 2034; F.J. Lalor and L.H. Brookes, J. Organomet. Chem., 1983, 251, 327.

9 J.K. Kouba, E.L. Muetterties, M.R. Thompson, and V.W. Day, Organometallics, 1983, 2, 1065.

10 M.E. Rerek, Liang-Nian Ji, and F. Basolo, J. Chem. Soc., Chem. Commun., 1983, 1208; M.E. Rerek and F. Basolo, Organometallics, 1983, 2, 372.

11 A.H. Cowley and R.A. Kemp, Inorg. Chem., 1983, 22, 547.

12 J. Ellermann and W. Wend, J. Organomet. Chem., 1983, 258, 21.

13 W.S. Sheldrick, H.W. Roesky, and D. Amirzadeh-Asl, Phosphorus Sulfur, 1983, 14, 161; H.W. Roesky, D. Amirzadeh-Asl, W. Clegg, M. Noltemeyer, and G.M. Sheldrick, J. Chem. Soc., Dalton Trans., 1983, 855; H.W. Roesky and D. Amirzadeh-Asl, Z. Naturforsch., Teil B, 1983, 38, 460.

14 D.F. Foster, B.S. Nicholls, and A.K. Smith, J. Organomet. Chem., 1983, 244, 159; D.F. Foster, J. Harrison, B.S. Nicholls, and A.K. Smith, ibid., 248, C29.

15 H.R. Allcock, K.D. Lavin, N.M. Tollefson, and T.L. Evans, Organometallics, 1983, 2, 267.

16 K.M. Flynn, B.D. Murray, M.M. Olmstead, and P.P. Power, J. Am. Chem. Soc., 1983, 105, 7460; A.H. Cowley, J.E. Kilduff, J.G. Lasch, N.C. Norman, M. Pakulski, F. Ando, and T.C. Wright, ibid, p. 7751; K.M. Flynn, H. Hope,

B.D. Murray, M.M. Olmstead, and P.P. Power, ibid., p. 7750; K.M. Flynn, M.M. Olmstead, and P.P. Power, ibid., p. 2085; A.H. Cowley, J.G. Lasch, N.C. Norman, and M. Pakulski, Angew. Chem., Int. Ed. Engl., 1983, 22, 978; J. Borm, L. Zsolnai, and G. Huttner, ibid., p. 977.

17 Pham Ba Chi and F. Kober, Z. Anorg. Allg. Chem., 1983, 498, 64.

18 U. Kunze and P.-R. Bolz, Z. Anorg. Allg. Chem., 1983, 498, 41, 50; B. Mathiasch and U. Kunze, Inorg. Chim. Acta, 1983, 75, 209; T. Hättich and U. Kunze, Z. Naturforsch., Teil B, 1983, 38, 655; U. Kunze and T.Hättich, Chem. Ber., 1983, 116, 3071.

19 G. Fochi, C. Floriani, J.C.J. Bart and G. Giunchi, J. Chem. Soc., Dalton Trans., 1983, 1515.

20 C.G. Marcellus, R.T. Oakley, A.W Cordes, and W.T. Pennington, J.Chem. Soc., Chem. Commun., 1983, 1451.

21 B H. Edwards, R.D. Rogers, D.J. Sikora, J.L. Atwood, and M.D. Rausch, J. Am. Chem. Soc., 1983, 105, 416.

22 T.G. Richmond, Qi-Zhen Shi, W.C. Trogler, and F. Basolo, J. Chem. Soc., Chem. Commun., 1983, 650.

23 J. Schiemann, P. Hübener, and E. Weiss, Angew. Chem., Int. Ed. Engl., 1983, 22, 980.

24 K.L. Fjare and J.E. Ellis, J. Am. Chem. Soc., 1983, 105, 2303.

25 G.F.P. Warnock, J. Sprague, K.L. Fjare, and J.E. Ellis, J. Am. Chem. Soc., 1983, 105, 672.

26 M. Hoch and D. Rehder, Z. Naturforsch., Teil B, 1983, 38, 446.

27 S.A. Wallin and A.F. Schreiner, Inorg. Chem., 1983, 22, 1964; S. Chun, D.C. Palmer, E.F. Mattimore, and A.J. Lees, Inorg. Chim. Acta, 1983, 77, L119; D.M. Manuta and A.J. Lees, Inorg. Chem., 1983, 22, 572, 3825; J.M. Kelly, C. Long, C.M. O'Connell, J.G. Vos, and A.H.A. Tinnemans, ibid., p. 2818; K.J. Moore and J.D. Petersen, Polyhedron, 1983, 2, 279.

28 G.J. Kubas, Inorg. Chem., 1983, 22, 692.

29 J.A. Connor and C. Overton, J. Organomet. Chem., 1983, 249, 165.

30 R.A. Howie, G. Izquierdo, and G.P. McQuillan, Inorg. Chim. Acta, 1983, 72, 165.

31 G. Cote, J. Chem. Res. (M), 1983, 1360; (S), 142.

32 E. Lindner, H. Rauleder, and W. Hiller, Z. Naturforsch., Teil B, 1983, 38, 417.

33 B. Bovio, J. Organomet. Chem., 1983, 241, 363; 244, 257; 252, 71.

34 K.H. Pannell, A.J. Mayr, R. Hoggard, J.S. McKennis, and J.C. Dawson, Chem. Ber., 1983, 116, 230.

35 A. Hengefeld, J. Kopf, and D. Rehder, Organometallics, 1983, 2, 114.

36 E.C. Alyea, G. Ferguson, and A. Somogyvari, Organometallics, 1983, 2, 668.

37 G.R. Clark and P.W. Clark, J. Organomet. Chem., 1983, 255, 205.

38 G.T. Andrews, I.J. Colquhoun, and W. McFarlane, Polyhedron, 1983, 2, 783.

39 R.D. Adams, C. Blankenship, B.E. Segmüller, and M. Shiralian, J. Am. Chem. Soc., 1983, 105, 4319.

40 A.-M. Hinke, A. Hinke, and W. Kuchen, J. Organomet. Chem., 1983, 258, 307.

41 R.R. Guimerans, M.M. Olmstead, and A.L. Balch, Inorg. Chem., 1983, 22, 2223.

42 R.L. Keiter, R.D. Borger, M.J. Madigan, S.L. Kaiser, and D.L. Rowley, Inorg. Chim. Acta, 1983, 76, L5.

43 P.M. Stricklen, E.J. Volcko, and J.G. Verkade, J. Am. Chem. Soc., 1983, 105, 2494.

44 E.H. Wong, M.M. Turnbull, E.J. Gabe, F.L. Lee, and Y. Le Page, J. Chem. Soc., Chem. Commun., 1983, 776.

45 E.H. Wong, F.C. Bradley, and E.J. Gabe, J. Organomet. Chem., 1983, 244, 235.

46 P. Aslanidis and J. Grobe, J. Organomet. Chem., 1983, 249, 103.

47 J. Powell, K.S. Ng, W.W. Ng, and S.C. Nyburg, J. Organomet. Chem., 1983, 243, C1; J. Powell, S.C. Nyburg, and S.J. Smith, Inorg. Chim. Acta, 1983, 76, L75; J. Powell, M. Gregg, A. Kuksis, and P. Meindl, J. Am. Chem. Soc., 1983, 105, 1064.

48 G.M. Gray and C.S. Kraihanzel, Inorg. Chem., 1983, 22, 2959; J. Organomet. Chem., 1983, 241, 201; G.M. Gray and R.J. Gray, Organometallics, 1983, 2, 1026.

49 F. Kober and M. Kerber, Z. Anorg. Allg. Chem., 1983, 507, 119.

50 I. Kawafune and G.-E. Matsubayashi, Inorg. Chim. Acta, 1983, 70, 1.

51 W. Kläui, A. Müller, and M. Scotti, J. Organomet. Chem., 1983, 253, 45.

52 G.J. Kubas, G.D. Jarvinen, and R.R. Ryan, J. Am. Chem. Soc., 1983, 105, 1883; W.A. Schenk and F.-E. Baumann, J. Organomet. Chem., 1983, 256, 261.

53 C.R. Lucas, Can. J. Chem., 1983, 61, 1096.

54 A. Winter, O. Scheidsteger, and G. Huttner, Z. Naturforsch., Teil B, 1983, 38, 1525.

55 E.W. Abel, G.D. King, K.G. Orrell, G.M. Pring, V. Sik, and T.S. Cameron, Polyhedron, 1983, 2, 1117; E.W. Abel, G.D. King, K.G. Orrell, and V. Sik, ibid., p. 1363.

56 P.J. Pogorzelec and D.H. Reid, J. Chem. Soc., Chem. Commun., 1983, 289; C. Glidewell, D.C. Liles, and P.J. Pogorzelec, Acta Crystallogr., Sect. C, 1983, 39, 542.

57 J.R. Moss and B.J. Smith, S. Afr. J. Chem., 1983, 36, 32.

58 J.L. Davidson and G. Vasapollo, J. Organomet. Chem., 1983, 241, C24.

59 J.R. Dilworth, J. Hutchinson, and J.A. Zubieta, J. Chem. Soc., Chem. Commun., 1983, 1034.

60 D. Sellmann and J. Schwarz, J. Organomet. Chem., 1983, 241, 343; D. Sellmann and U. Kleine-Kleffmann, ibid., 258, 315.

61 A.E. Stiegman, M. Stieglitz, and D.R. Tyler, J. Am. Chem. Soc., 1983, 105, 6032.

62 H.-G. Fick and W. Beck, J. Organomet. Chem., 1983, 252, 83.

63 A.L. Rheingold and M.R. Churchill, J. Organomet. Chem., 1983, 243, 165.

64 H.B. Abrahamson and M.L. Freeman, Organometallics, 1983, 2, 679.

65 H. Brunner, H. Buchner, J. Wachter, I. Bernal, and W.H. Ries, J. Organomet. Chem., 1983, 244, 247.

66 M.G. Peterleitner, M.V. Tolstaya, V.V. Krivykh, L.I. Denisovitch, and M.I. Rybinskaya, J. Organomet. Chem., 1983, 254, 313.

67 H.W. Choi and M.S. Sollberger, J. Organomet. Chem., 1983, 243, C39.

68 M.F. Perpinan, L. Ballester, and A. Santos, J. Organomet. Chem., 1983, 241, 215.

69 K.H. Dötz, I. Pruskil, U. Schubert, and K. Ackermann, Chem. Ber., 1983, 116, 2337.

70 A. Filippou and E.O. Fischer, Z. Naturforsch., Teil B, 1983, 38, 587.

71 A.M. Stolzenberg and E.L. Muetterties, J. Am. Chem. Soc., 1983, 105, 822; N.J. Coville, A.M. Stolzenberg, and E.L. Muetterties, ibid., p. 2499.

72 A.F. Hepp and M.S. Wrighton, J. Am. Chem. Soc., 1983, 105, 5934; H. Yesaka, T. Kobayashi, K. Yasufuku, and S. Nagakura, ibid., p. 6249.

73 C. Busetto, A.M. Mattucci, E.M. Cernia, and R. Bertani, J. Organomet. Chem., 1983, 246, 183.

74 S.W. Lee, L.F. Wang, and C.P. Cheng, J. Organomet. Chem., 1983, 248, 189.

75 J.A. Iggo, M.J. Mays, P.R. Raithby, and K. Hendrick, J. Chem. Soc., Dalton Trans., 1983, 205.

76 A. Alberti and A. Hudson, J. Organomet. Chem., 1983, 248, 199; Chiu Ping Cheng, Aina H. Cheng, and Pao Jen Shyong, Proc. Nat. Sci. Counc., Repub. China, Part B, 1982, 6, 298; P. Mathur and G.C. Dismukes, J. Am. Chem. Soc., 1983, 105, 7093; H.W. Walker, G.B. Rattinger, R.L. Belford, and T.L. Brown, Organometallics, 1983, 2, 775; K.A.M. Creber and J.K.S. Wan, Transition Met. Chem., 1983, 8,

253; Can. J. Chem., 1983, 61, 1017.

77 P. Jaitner, W. Huber, G. Huttner, and O. Scheidsteger, J. Organomet. Chem., 1983, 259, C1.

78 L.K. Peterson, R.S. Dhami, and F. Wada, Synth. React. Inorg. Met.-Org. Chem., 1983, 13, 291.

79 F.J. Garcia Alonso and V. Riera, Polyhedron, 1983, 2, 1103.

80 D.T. Plummer, G.A. Kraus, and R.J. Angelici, Inorg. Chem., 1983, 22, 3492.

81 R. Mews and C. Liu, Angew. Chem., Int. Ed. Engl., 1983, 22, 162; G. Hartmann, R. Mews, and G.M. Sheldrick, ibid., p. 723; J. Organomet. Chem., 1983, 252, 195; G. Hartmann, R. Hoppenheit, and R. Mews, Inorg. Chim. Acta, 1983, 76, L201.

82 J.A. Clark and M. Kilner, J. Chem. Soc., Dalton Trans., 1983, 2613.

83 R. Rossi, A. Duatti, L. Magon, U. Casellato, R. Graziani, and L. Toniolo, Inorg. Chim. Acta, 1983, 75, 77.

84 P.O. Nubel, S.R. Wilson,and T.L. Brown, Organometallics, 1983, 2, 515.

85 A.A. Ioganson, Yu. G. Kovalev, and E.D. Korniets, Bull. Acad. Sci. USSR, Div. Chem. Sci., 1982, 31, 1466.

86 T.D. Westmoreland, H. Le Bozec, R.W. Murray, and T.J. Meyer, J. Am. Chem. Soc., 1983, 105, 5952.

87 W. Tikkanen, W.C. Kaska, S. Moya, T. Layman, R. Kane, and C. Krüger, Inorg. Chim. Acta, 1983, 76, L29.

88 O.J. Scherer and J. Kerth, J. Organomet. Chem., 1983, 243, C33.

89 O.J. Scherer, J. Kerth, and M.L. Ziegler, Angew. Chem., Int. Ed. Engl., 1983, 22, 503.

90 U. Kunze, R. Merkel, and M. Moll, J. Organomet. Chem., 1983, 248, 205.

91 O.J. Scherer, J. Kerth, R. Anselmann, and W.S. Sheldrick, Angew. Chem., Int. Ed. Engl., 1983, 22, 984.

92 J.L. Atwood, I. Bernal, F. Calderazzo, L.G. Canada, R. Poli, R.D. Rogers, C.A. Veracini, and D. Vitali, Inorg. Chem., 1983, 22, 1797.

93 E.W. Abel, S.K. Bhargava, M.M. Bhatti, M.A. Mazid, K.G. Orrell, V. Sik, M.B. Hursthouse, and K.M.A. Malik, J. Organomet. Chem., 1983, 250, 373.

94 C.P. Casey, J.M. O'Connor, W.D. Jones and K.J. Haller, Organometallics, 1983, 2, 535.

95 E. Lindner and D. Hübner, Chem. Ber., 1983, 116, 2574.

96 M. Herberhold, D. Reiner, and U. Thewalt, Angew. Chem., Int. Ed. Engl., 1983, 22, 1000.

97 H.-W. Frühauf, J. Chem. Res.(M), 1983, 2023, 2035; (S), 216, 218.

98 H.C. Ashton and A.R. Manning, Inorg. Chem., 1983, 22, 1440.

99 S.P. Modi and J.D. Atwood, Inorg. Chem., 1983, 22, 26.

100 R.L. Keiter, A.L. Rheingold, J.J. Hamerski, and C.K. Castle, Organometallics, 1983, 2, 1635.

101 R.B. King, T.W. Lee, and J.H. Kim, Inorg. Chem., 1983, 22, 2964; R.B. King and M. Shimura, J. Organomet. Chem., 1983, 256, 71.

102 D.P. Bauer and J.K. Ruff, Inorg. Chem., 1983, 22, 1686.

103 S.G. Baxter, R.L. Collins, A.H. Cowley, and S.F. Sena, Inorg. Chem., 1983, 22, 3475.

104 R. Lal De, J. von Seyerl, and G. Huttner, J. Organomet. Chem., 1983, 243, 331.

105 P. Conway, S.M. Grant, A.R. Manning, and F.S. Stephens, Inorg. Chem., 1983, 22, 3714.

106 H.C. Ashton and A.R. Manning, Inorg. Chim. Acta, 1983, 71, 163.

107 L. Mordenti, J.-L.A. Roustan, and J.G. Riess, Organometallics, 1983, 2, 843.

108 H. Schumann, K.-H. Köhricht, and M. Meissner, Z. Naturforsch., Teil B, 1983, 38, 705.

109 J.-L.A. Roustan, A. Forgues, J.-Y. Merour, N.D. Venayak, and B.A. Morrow, Can. J. Chem., 1983, 61, 1339; M. Cygler, F.R. Ahmed, A. Forgues, and J.-L.A. Roustan, Inorg. Chem., 1983, 22, 1026.

110 M.I. Bruce, J.G. Matisons, and B.K. Nicholson, J. Organomet. Chem., 1983, 247, 321.
111 R.P. Rosen, G.L. Geoffroy, C. Bueno, M.R. Churchill, and R.B. Ortega, J. Organomet. Chem., 1983, 254, 89.
112 J.B. Letts, T.J. Mazanec, and D.W. Meek, Organometallics, 1983, 2, 695.
113 M. Bianchi, P. Frediani, U. Matteoli, G. Menchi, F. Piacenti, and G. Petrucci, J. Organomet. Chem., 1983, 259, 207.
114 M.J. Auburn, R.D. Holmes-Smith, S.R. Stobart, M.J. Zaworotko, T.S. Cameron, and A. Kumari, J. Chem. Soc., Chem. Commun., 1983, 1523.
115 L.R. Martin, F.W.B. Einstein, and R.K. Pomeroy, Inorg. Chem., 1983, 22, 1959.
116 D.R. Tyler, M.A. Schmidt, and H.B. Gray, J. Am. Chem. Soc., 1983, 105, 6018.
117 B.D. Fabian and J.A. Labinger, Organometallics, 1983, 2, 659.
118 J.M. Shreeve and S.M. Williamson, J. Organomet. Chem., 1983, 249, C13.
119 D. Catheline and D. Astruc, J. Organomet. Chem., 1983, 248, C9; A. Stasunik and W. Malisch, ibid., 247, C47; S.G. Davies, S.J. Simpson, and S.E. Thomas, ibid., 254, C29.
120 V.W. Day, M.R. Thompson, G.O. Nelson, and M.E. Wright, Organometallics, 1983, 2, 494; M.E. Wright, T.M. Mezza, G.O. Nelson, N.R. Armstrong, V.W. Day, and M.R. Thompson, ibid., p. 1711.
121 S.J. McLain, J. Am. Chem. Soc., 1983, 105, 6355.
122 R.W. Wegman and T.L. Brown, Inorg. Chem., 1983, 22, 183.
123 R.A. Jones, A.L. Stuart, J.L. Atwood, and W.E. Hunter, Organometallics, 1983, 2, 1437; A.D. Harley, G.J. Guskey, and G.L. Geoffroy, ibid., p. 53; A.D. Harley, R.R. Whittle, and G.L. Geoffroy, ibid., p. 60.
124 N. Espana, P. Gomez, P. Royo, and A.V. de Miguel, J. Organomet. Chem., 1983, 256, 141.
125 M.M. Singh, Y. Rosopulos, and W. Beck, Chem. Ber., 1983, 116, 1364.
126 D.W. Abbott and C. Woods, Inorg. Chem., 1983, 22, 597, 2918.
127 A.R. Siedle, G. Filipovich, P.E. Toren, F.J. Palensky, E. Cook, R.A. Newmark, W.L. Stebbings, K. Melancon, and H.E. Mishmash, J. Organomet. Chem., 1983, 246, 83.
128 M.J. Decker, D.O.K. Fjeldsted, S.R. Stobart, and M.J. Zaworotko, J. Chem. Soc., Chem. Commun., 1983, 1525.
129 J.-P. Lecomte, J.-M. Lehn, D. Parker, J. Guilhem, and C. Pascard, J. Chem. Soc., Chem. Commun., 1983, 296.
130 J.P. Farr, M.M. Olmstead, F.E. Wood, and A.L. Balch, J. Am. Chem. Soc., 1983, 105, 792; F.E. Wood, M.M. Olmstead, and A.L. Balch, ibid., p. 6332; F.E. Wood, J. Hvoslef, and A.L. Balch, ibid., p. 6986.
131 M.P. Anderson and L.H. Pignolet, Organometallics, 1983, 2, 1246; M.P. Anderson, C.C. Tso, B.M. Mattson, and L.H. Pignolet, Inorg. Chem., 1983, 22, 3267.
132 F.R. Hartley, S.G. Murray, and P.N. Nicholson, Inorg. Chim. Acta, 1983, 76, L51.
133 R.R. Guimerans, M.M. Olmstead, and A.L. Balch, J. Am. Chem. Soc., 1983, 105, 1677; M.M. Olmstead, R.R. Guimerans, and A.L. Balch, Inorg. Chem., 1983, 22, 2473.
134 J.M. Brown and L.R. Canning, J. Chem. Soc., Chem. Commun., 1983, 460.
135 R.J. Haines, N.D.C.T. Steen, and R.B. English, J. Chem. Soc., Dalton Trans., 1983, 2229.
136 J.L. Atwood, W.E. Hunter, R.A. Jones, and T.C. Wright, Inorg. Chem., 1983, 22, 993; R.A. Jones and T.C. Wright, Organometallics, 1983, 2, 1842.
137 A.R. Sanger, Can. J. Chem., 1983, 61, 2214.
138 C.E. Johnson, R. Eisenberg, T.R. Evans, and M.S. Burberry, J. Am. Chem. Soc., 1983, 105, 1795.

139 F. Faraone, G. Bruno, S. Lo Schiavo, P. Piraino, and G. Bombieri, J. Chem. Soc., Dalton Trans., 1983, 1819.

140 N. Dudeney, J.C. Green, P. Grebenik, and O.N. Kirchner, J. Organomet. Chem., 1983, 252, 221.

141 F. Faraone, G. Bruno, S. Lo Schiavo, G. Tresoldi, and G. Bombieri, J. Chem. Soc., Dalton Trans., 1983, 433.

142 G.M. Bodner, C. Gagnon, and D.N. Whittern, J. Organomet. Chem., 1983, 245, 305.

143 T.A. van der Knaap, L.W. Jenneskens, H.J. Meeuwissen, F. Bickelhaupt, D. Walther, E. Dinjus, E. Uhlig, and A.L. Spek, J. Organomet. Chem., 1983, 254, C33.

144 F. Casabianca and J.G. Riess, Synth. React. Inorg. Met.-Org. Chem., 1983, 13, 799.

145 R.A. Jones, A.L. Stuart, J.L. Atwood, and W.E. Hunter, Organometallics, 1983, 2, 874.

146 N.K. Eremenko, E.G. Mednikov, S.S. Kurasov, and S.P. Gubin, Bull. Acad. Sci. USSR, Div. Chem. Sci., 1983, 32, 620.

147 P. Dagnac, R. Turpin, and R. Poilblanc, J. Organomet. Chem., 1983, 253, 123.

148 R.J. Goodfellow, I.R. Herbert, and A.G. Orpen, J. Chem. Soc., Chem. Commun., 1983, 1386.

149 T.A. Stromnova, M.N. Vargaftik, and I.I. Moiseev, J. Organomet. Chem., 1983, 252, 113; E.G. Mednikov, N.K. Eremenko, Yu. L. Slovokhotov, Yu. T. Struchkov, and S.P. Gubin, ibid., 258, 247; E.G. Mednikov and N.K. Eremenko, Bull. Acad. Sci. USSR, Div. Chem. Sci., 1982, 31, 2240; E.G. Mednikov and N.K. Eremenko, Koord. Khim., 1983, 9, 243.

150 L. Manojlovic-Muir, K.W. Muir, B.R. Lloyd, and R.J. Puddephatt, J. Chem. Soc., Chem. Commun., 1983, 1336.

11

Complexes Containing Metal–Carbon σ-Bonds of the Groups Scandium to Manganese

BY K. J. KAREL AND P. L. WATSON

1. Introduction

The format of this chapter follows that of previous years. Papers are organized in Sections for each of Groups IIIA-VIIA of the Periodic Table, ordered into topics covering the synthesis and reactivity of classes of complexes having similar ligand systems, including carbene and carbyne complexes.

Several general papers have appeared which are relevant to this chapter. Aspects of the chemistry of binuclear transition metal complexes with bridging hydrocarbyl or hydrocarbon ligands[1], synthetic routes to organometallic methylene bridges[2], evidence for the role of C-H bonds as ligands in transition metal complexes[3], and the syntheses and properties of dinuclear compounds containing early and late transition metals[4] are topics that have been reviewed. Stone[5] discussed new routes to organometallic compounds relevant to catalysis and Cotton[6] outlined some organometallic chemistry of metal-metal bonded complexes with early transition metals. Carbon-13 NMR chemical shifts of carbenes, carbynes and related complexes (and their correlation with reaction chemistry) was interpreted by Fenske.[7]

In this chapter the abbreviation Cp is used for η^5-C_5H_5, Cp* for η^5-C_5Me_5, and Cp' for η^5-C_5H_4Me.

2. Group IIIA, Lanthanides and Actinides

A review on homoleptic organometallic compounds of the lanthanide metals appeared,[8] as has a general update on organo-lanthanide chemistry.[9]

Divalent organofluorine ytterbium complexes YbR_2 ($R=C_6F_5$, p-HC_6F_4) underwent R for halide exchange with a variety of Group VIII transition and main group metal complexes e.g. with $NiCl_2(PPh_3)_2$ to give trans-$Ni(C_6F_5)_2(PPh_3)_2$. CO_2 inserted into the R-Yb bonds of the YbR_2 complexes and acidification studies showed fluoride elimination from the o-position of $C_6F_5CO_2Yb$ species can be extensive.[10] Deacon also reported the X-ray crystal structure of divalent complexes $YbCp_2(\eta^2$-1,2-dimethoxyethane).[11] Both divalent aryl species PhMI (M=Yb, Eu) underwent fluorine for carbon exchange in reacting with $PhCF{=}CF_2$ to give PhCF=CFPh.[12] However reaction of ArYbI ($Ar{=}C_6H_5$, C_6F_5) with $CH_2{=}CHBr$ required the presence of $CoCl_2$ to give high yields (90%) of the coupled organic product $CH_2{=}CHAr$.[13]

$Cp^*{}_2Sm(THF)_2$ was reported to give a black Sm(III) dimer $Cp^*{}_2SmC(Ph){=}C(Ph)SmCp^*{}_2$ on treatment with PhC≡CPh. The dimer reacted with H_2 giving $[Cp^*{}_2SmH]_2$ which was characterized by X-ray crystallography.[14] Yttrium hydrides $[(C_5H_4R)_2YH(THF)]_2$ (R=H, Me) underwent insertion reactions with Me_3CNC. The dimeric products $[(C_5H_4R)_2Y(HC{=}NCMe_3)]_2$ contained bridging formimidoyl C atoms.[15] Synthesis of a Sm(III) complex with electron-deficient alkynyl bridges, (1) ($Cp'{=}C_5H_4Me$) was achieved from reaction of $Cp'_2SmCl(THF)$ with $LiC{\equiv}CCME_3$. $HPPh_2$ reacted with (1) giving monomeric Cp'_2SmPPh_2 and $HC{\equiv}CCMe_3$.[16]

The lutetium dimers $[Cp^*{}_2LuR]_2$ (R=H, Me) e.g. (2) dissociated in hydrocarbon solution to the electron-deficient monomers $Cp^*{}_2LuR$ which at 20-50°C were extremely reactive toward the C-H bonds of relatively inert hydrocarbons (equation 1).[17] Further,

$$Cp^*{}_2Lu\text{-}R + R'H \longrightarrow Cp^*{}_2Lu\text{-}R' + RH \qquad (1)$$

$$R = H, CH_3;\ R'H = py,\ SiMe_4,\ C_6H_6,\ CH_2{=}PPH_3,\ {}^{13}CH_4.$$

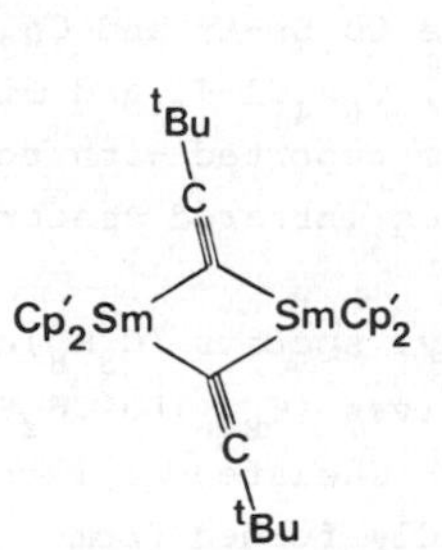

(1)

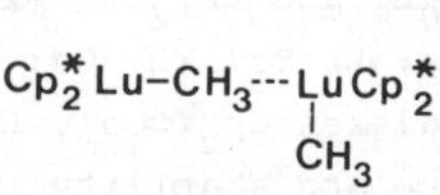

(2)

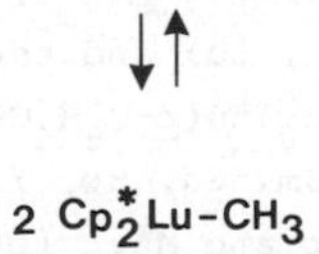

using $^{13}CH_4$ even methane was shown by NMR undergo facile C-H activation, reacting predominantly in a kinetically bimolecular process.[18] This was the first well-characterized reaction of methane with an organometallic complex.

Fluorenyl lanthanide anions (3) (Ln=La, Nd, Sm, Ho, Ln) were reportedly prepared in reasonable yields from fluorenylLi and $LnCl_3$ in THF.[19] Novel carboranyl complexes, assigned structure (4) of divalent lanthanides Eu and Yb were synthesized, though characterized only by reactions with electrophiles e.g. as shown in Scheme (1).

$$Ph-\underset{\underset{B_{10}H_{10}}{|\quad\backslash}}{C-C}-I + Ln(0) \longrightarrow Ph-\underset{\underset{B_{10}H_{10}}{|\quad\backslash}}{C-C}-Ln-I$$

(4) Ln=Eu, Yb

Cl-SiMe$_3$

$$PhC-\underset{\underset{B_{10}\;H_{10}}{|\quad\backslash}}{C}-SiMe_3$$

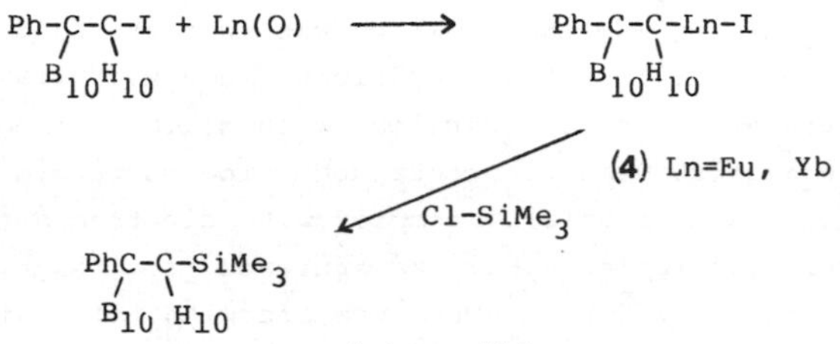

Scheme (1)

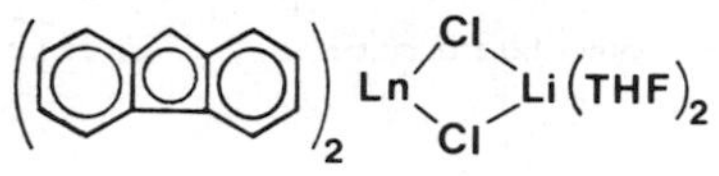

(3)

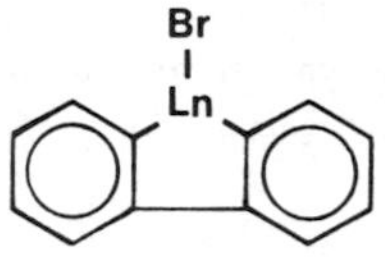

(5)

For Ln=Sm, Eu, Yb complexes (4) were independently made from $Ph-C_2B_{10}H_{10}Li$ and LnI_2.[20] Aryl complexes Cp_2Ln-Ar and $Cp_2'Ln$-Ar ($Cp'=C_5H_4Me$; Ln=Er, Yb, Gd; Ar=C_6H_4Me-4, C_6H_4Cl-4) and mixed ligand complexes Cp_2YbCp', Cp'_2ErCp were reported with comments on structure and stability of these, plus infrared spectra of a variety of Cp_2LnCl compounds.[21]

Wayda prepared new cyclo-octatetraenyl species $(C_8H_8)LnCl$-$(THF)_x$(Ln=Er, Lu) and the alkyl derivatives $(C_8H_8)Lu(CH_2SiMe_3)$-$(THF)_2$, $(C_8H_8)Lu(\underline{o}-C_6H_4CH_2NMe_2)(THF)$.[22] Chelated aryl complexes (5) (M=Pr, Sm, Gd, Ho, Yb) were reportedly formed from $\underline{o}$-LiC_6H_4Li-$\underline{o}$ and MBr_3 (ether-THF, 20°C).[23]

Stable anionic uranium(III) complexes were formed by reduction of uranium(IV) alkyls according to equation (2) with R=Me, CHMe$_2$,

CH_2CHMe_2 and R'=Me, Bu.[24] Reduction of N_2 to NH_3 (1 atm, 20°C) was observed in the presence of $[Cp_3UMe]Li$ and a strong reducing agent e.g. lithium napthalenide in THF.[25]

$$Cp_3UR + \overset{\text{excess}}{R'Li} \xrightarrow[30-50°]{THF} [Cp_3UR]Li(THF)_n + R'H \qquad (2)$$

The thorium η^2-acyl compound, $Cp^*_2Th(\eta^2\text{-}COCH_2CMe_3)Cl$, reacted with CO forming dionedienolate (6), ($R=CH_2CMe_3$) in a process kinetically first order in each reagent. Addition of PR_3 (R=Me, Ph) caused no change in kinetics, but the isolated product was then the ylide (7) rather than (6), suggesting a common ketene-like intermediate (8).[26]

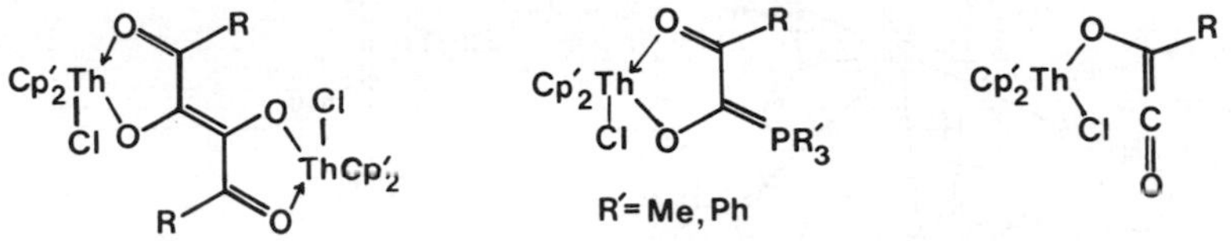

(6) (7) (8)

Analysis of $CH_{4-n}D_n$ products from reaction of alumina, or deuterated alumina, and the organo-actinides Cp_2MR_2 (M=Th, U; $R=CH_3$, CD_3) revealed several methane-evolving reaction pathways. One was proposed by Marks to result in concomitant formation of alumina-stabilized actinide alkylidene species.[27]

Anerobic alcoholysis of a related series of thorium complexes Cp_2ThR_2 (R=alkyl, aryl, metallacycle, hydride, dialkylamide) was performed calorimetrically to yield bond dissociation enthalpies for the Th-R bonds which were in the range of 60-90 kcal/mol. Effects of ancillary ligands and strain energies were addressed.[28]

Relativistic Xα-SW MO calculations were performed on Cp_2UR_2 (R=Cl, Me) and reconciled with PES studies on related Cp_2UR_2 species.[29] PES spectra of $(indenyl)_3M\text{-}R$ (M=Th,U; R=Cl, Br, OMe, Me) complexes were recorded.[30]

3. Group IVA

Reger has described[31] preparation of the 3,5-dimethylpyrazolylborate (=$HBpz_3$) complexes $Zr(HPpz_3)(OCMe_3)R_2$ (R=Cl, Me, CH_2Ph, C=CMe), in which σ-donation from the alkoxy oxygen to electron-deficient Zr can be quite important. Facile reaction of $CNCMe_3$ with the methyl complex gave the iminoacyl (9). Similarly, insertion of $CNCH_3$ into the Ti-Cl bond of $TiCl_4$ in CH_2Cl_2 gave the formimidoyl chloride (10), which adds to carbonyl groups of ketones or aldehydes forming adducts (11). Seebach utilized this methodology for a general synthesis of α-hydroxy carboxamides.[32] Preparation of alkoxy/alkyl complexes $TiR_{4-n}(OR')_n$ (R=C_6H_{11}, CH_2Ph; R'=Me, Et, C_6H_{11}, CH_2Ph) via alcoholysis of the homoleptic alkyls TiR_4 was surveyed.[33]

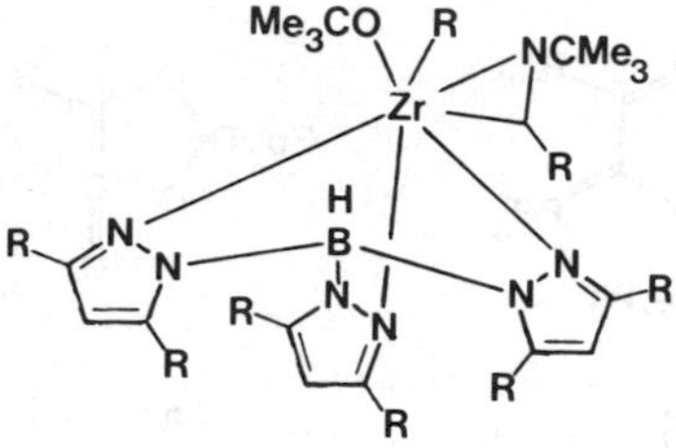

(9) R = Me

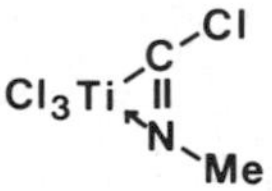

(10)

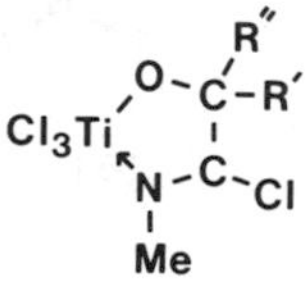

(11)

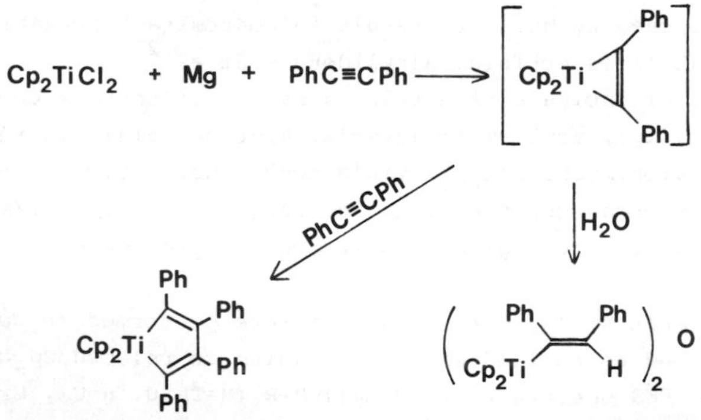

Scheme (2)

Formation of μ-oxo complexes was reported in two instances. The structure of $[Cp_2Ti(CPh{=}CHPh)]_2O$ was proved by X-ray crystallography; the complex was made according to Scheme (2).[34] Cleavage of CO rather than H_2O gave the zirconium oxocomplex (12), (shown in Scheme (3)), with concomitant formation of the zirconium enolate (13).[35]

$$[(Me_3Si)_2N]_2ZrMe_2 \xrightarrow[\text{1–2 atm, pentane}]{\text{CO}} [[(Me_3Si)_2N]_2ZrMe]_2O \;\; \textbf{(12)}$$

$$+\; [(Me_3Si)_2N]_2Zr(OCMe{=}CMe_2)Me \;\; \textbf{(13)}$$

Scheme (3)

As shown in Scheme (4) the same zirconium dialkyls $[(Me_3Si)_2N]_2ZrR_2$ (R=Me, Et, CH_2SiMe_3) decompose thermally (60°C) with alkane elimination giving bridging carbene species (14).[36] Ti and Hf complexes similar to (14) were also made. Dmpe reacts readily with (14) giving mononuclear (15).[37]

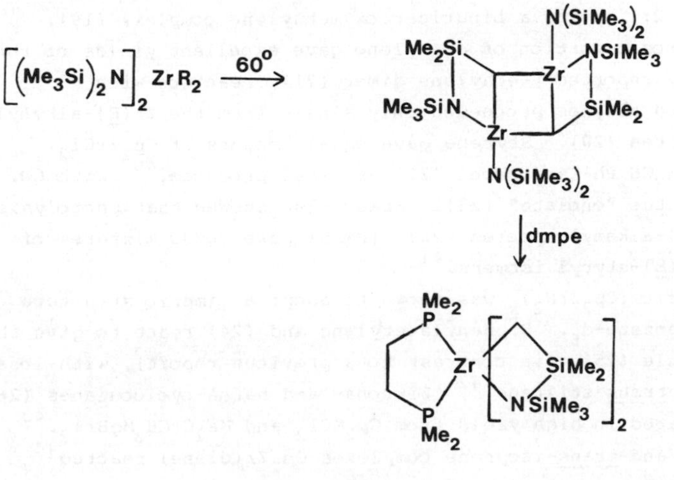

(14)

(15)

Scheme (4)

Anionic (16) and neutral (17) ketene complexes of zirconium were prepared by the general route of Scheme (5).[38] Reaction with H_2 gave enolate hydrides (e.g. (18)) with cis-enolate geometry, in a model for the reduction of CO by $Cp^*_2ZrH_2$.

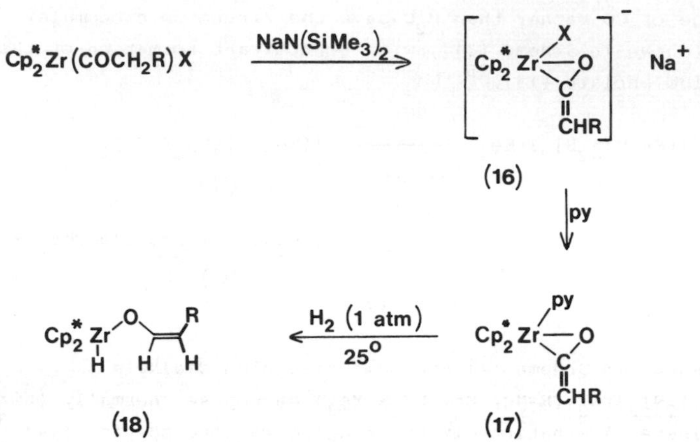

Scheme (5)

Floriani[39] showed the reaction of Cp_2ZrHCl with CO occurred in the ratio 2:1 giving a binuclear oxymethylene complex, (19). While hydrozirconation of acetylene gave excellent yields of the previously reported μ-ethylene dimer (21), reaction with substituted alkynes proceeded only slowly from the Zr(E)-alkenyl intermediates (20). Styrene gave equal amounts of Cp_2ZrCl_2, $Cp_2ZrCl(CH_2CH_2Ph)$ and novel (22) as final products.[40] With CO, (22) gave the "enolate" (23). Erker also showed that photolysis of the (E)-alkenyl species (20), (R=Ph) gave 70/30 mixtures of (Z)- and (E)-styryl isomers.[41]

Polymeric $(Cp_2ZrH_2)_x$ was shown to adopt a dimeric structure (24) in benzene-d_6. Diphenylacetylene and (24) react to give the metallocycle (25), (in contrast to a previous report), with loss of H_2 and trans-stilbene.[42] Zircona- and hafna-cyclobutanes (26) were prepared in high yield from Cp_2MCl_2 and $Me_2C(CH_2MgBr)_2$.[43] Both cis- and trans-isoprene complexes Cp_2Zr(diene) reacted[44] thermally with ketone Me_3CCOCH_3 giving the insertion product (27). Reactions of alkynes with Cp_2Zr(s-trans-η^4-butadiene) and other zirconocene complexes were also reported.[45] A structural study of titana-cyclopropane $Cp_2Ti(\eta^2-CH_2CH_2)$ was completed.[46] Oxymetallocycles (28)[47], (29)[48], (30)[49] were described.

$$(Cp_2ZrHCl)_x + \tfrac{1}{2}\,CO \longrightarrow \quad (19)$$

$$(Cp_2ZrHCl)_x \xrightarrow{HC\equiv CR} \quad (20) \xrightarrow[(Cp_2ZrHCl)_x]{H=H} Cp_2ClZr-CH_2CH_2-ZrCp_2Cl \quad (21)$$

$$(22) \xrightarrow{CO} (23)$$

$$(24) \xrightarrow{PhC\equiv CPh} (25) \qquad (26)$$

$$\text{Cp}_2\text{Zr} \quad \text{or} \quad \text{Cp}_2\text{Zr}-\text{Me} \xrightarrow{Me_3CCOMe} (27)$$

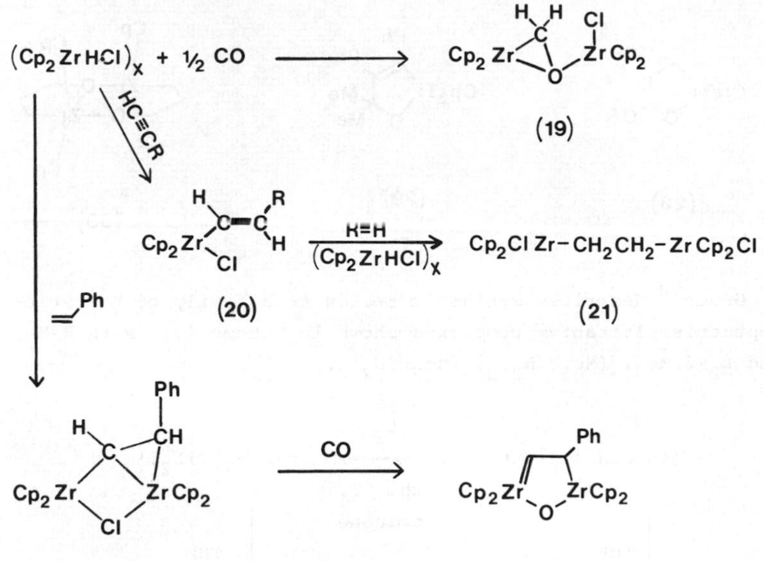

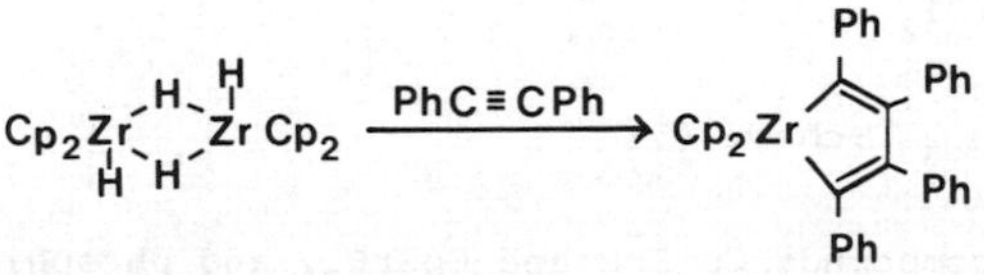

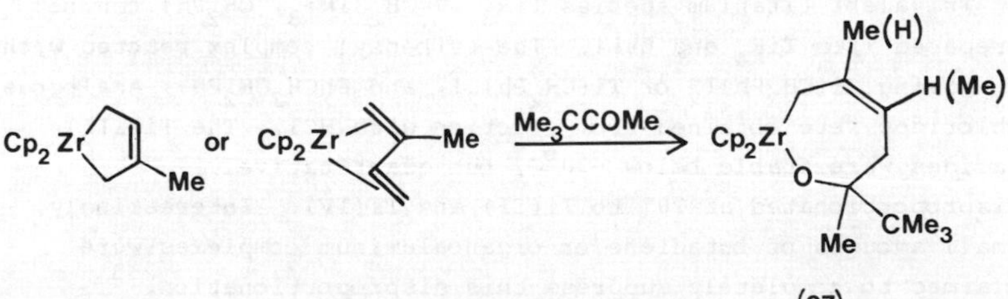

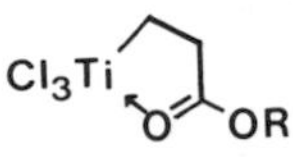

$$Cl_3Ti \underset{O}{\overset{}{\diagdown}} OR$$

$$Cp_2Ti \overset{Ph}{\underset{O}{\diagdown}} \overset{Ph}{\underset{Me}{\diagup}} Me$$

(28)   **(29)**

$$Cp_2Zr \overset{O}{\underset{O}{\diagup}} ZrCp_2$$

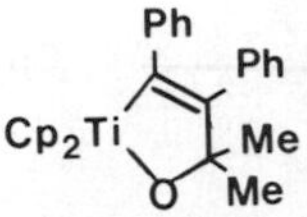

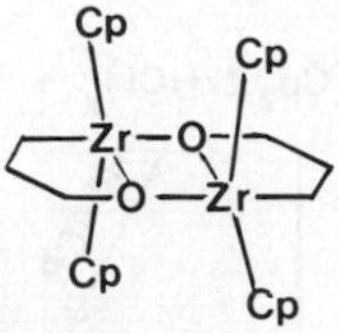

(30)

Green[50] described synthetic routes to a family of η^7-cyclo-heptatrienyltitanium complexes shown in Scheme (5), with R=Me, Et and L_2=2PMe$_3$, (Me$_2$NCH$_2$)$_2$, (Me$_2$PCH$_2$)$_2$.

$$[(\eta\text{-}C_7H_7)Ti(THF)Cl]_2 \xrightarrow[\substack{2h, RT, \\ toluene}]{L_2} (\eta\text{-}C_7H_7)TiL_2Cl$$

$\uparrow$ THF	$\downarrow$ THF RMgCl

$(\eta\text{-toluene})_2$Ti $(\eta\text{-}C_7H_7)TiL_2R$

+ C$_7$H$_8$ + EtAlCl$_2$

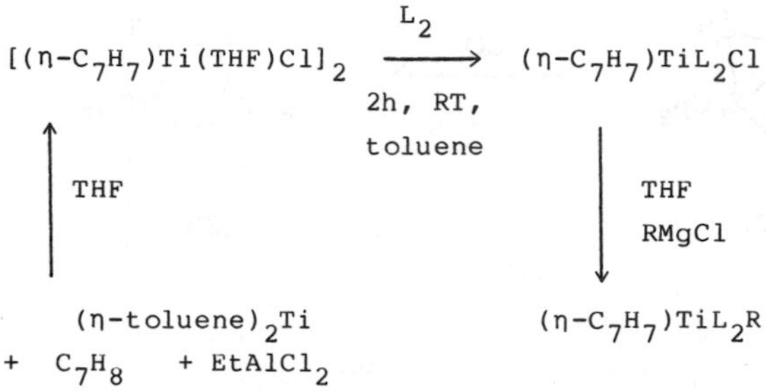

Scheme (5)

Trivalent zirconium compounds Cp$_2$ZrR and CpZrR$_2$, and phosphine adducts thereof, were generated by photolysis (at low temperatures) of Cp$_2$ZrR$_2$ (R=Ph, CH$_2$Ph, CH$_2$SiMe$_3$). ESR parameters of these unstable species were recorded.[51] A more persistent Zr(III) derivative Cp$_2$ZrCH$_2$PPh$_2$, generated by Na/Hg reduction of Cp$_2$Zr(Cl)(CH$_2$PPh$_2$) in THF, was isolated by Schore[52] as a pink powder which could be redissolved, and which was a monomer in dimethylether.

Trivalent titanium species TiR$_3$ (R=CH$_2$SiMe$_3$, CH$_2$Ph) can be prepared from TiR$_4$ and EtLi. The tribenzyl complex reacted with I$_2$ giving Ti(CH$_2$Ph)I$_2$ or Ti(CH$_2$Ph)$_2$I, and PhCH$_2$CH$_2$Ph. Analogous chlorides were obtained from reaction with HCl. The Ti(III) halides were stable below -30°C, but quantitatively disproportionated at 20° to Ti(II) and Ti(IV). Interestingly, small amounts of butadiene or organoaluminum complexes were claimed to completely suppress this disproportionation.[53]

Diazoalkanes, R_2CN_2, inserted into the Zr-R' bonds of $Cp_2ZrR'_2$ (R=Ph, CO_2Et; R'=H, Me, CH_2Ph) in a manner indiscriminate of the natures of R or R'. The X-ray structure of a typical product, $Cp_2Zr(Me)(NMe\text{-}N{=}CPh_2)$ was obtained and showed the hydrazonato ligand to be dihapto with considerable N-N double bond character.[54]

Reviews describing organometallic compounds of Ti and Zr[55] and organic synthesis using alkyl titanium reagents[56] appeared. Synthetic applications of the Tebbe reagent $Cp_2TiCH_2AlMe_2Cl$ were developed.[57] Similar chemistry allowed use of the vinylidene daughter (31) from metallacycle (32) in the synthesis of substituted allenes from (32) and carbonyl compounds.[58]

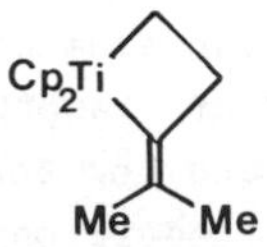

(31)

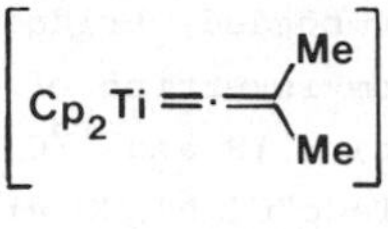

(32)

4. Group V

The related series $[Cp_2NbR_2]^{+1,0,-1}$ ($R_2=(o\text{-}CH_2C_6H_4)_2{}^{2-}$) (33), was examined by X-ray crystallography and found to vary considerably in Nb-CH_2-C and CH_2-Nb-CH_2 angles as well as torsion along the biphenyl axes.[59] ESR measurements of forty d^1 V(IV) and Nb(IV) complexes of type $[Cp_2ML_2]^{n+}$, e.g. where L_2 is a

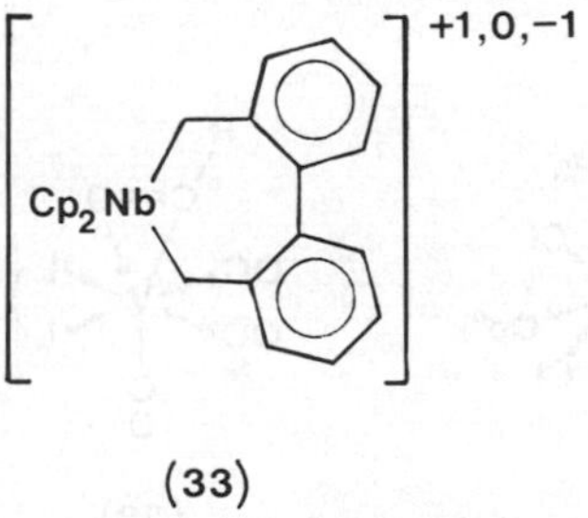

(33)

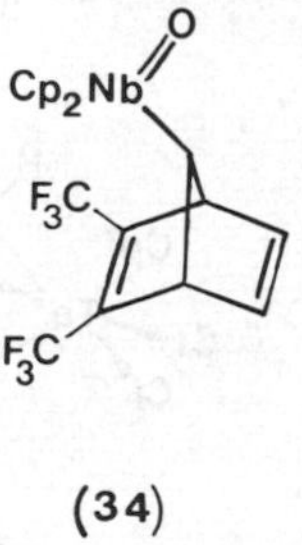

(34)

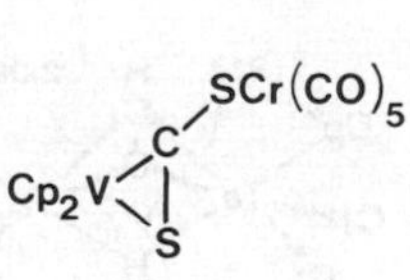

(35)

dithiolate ligand, allowed calculation of the usual ESR parameters, relation of these to the interbond L-M-L angle and assignment of the unpaired electron to an orbital of $\underline{a}_1$ symmetry.[60] Diels-Alder addition of alkyne to a η^1-C_5H_5 group occurred on mixing $CF_3C{\equiv}CCF_3$ and $Cp_2Nb(\eta^1$-$Cp)_2$. The product (34) contained an oxygen atom of unknown origin.[61] Metallotropic η^1-η^5 rearrangements and intermolecular ligand exchange in $(RC_5H_4)_3V$ (R=H, Me) were shown by ^1H and ^{13}C NMR.[62] Donor effects of alkyls in Cp_2VR (R=Me, CH_2SiMe_3, CH_2CMe_3, C_6F_5, Ph, tolyl, xylyl) were probed by PES.[63] Basicity of the exocyclic sulfur atom in $Cp_2V(\eta^2$-$CS_2)$ was demonstrated by addition of several coordinately unsaturated transition metal carbonyls, such as $Cr(CO)_5$ to give (35).[64]

A side-on bonded, bridging formyl group in tantalum dimer (36) resulted from insertion of CO into a Ta-H bond of $[TaCp''Cl_2H]_2$ ($Cp''=C_5Me_4Et$). IR and ^{13}C NMR evidence was found for formation of adduct $[TaCp''Cl_2H]_2CO$ at -78°C in solution. Addition of nucleophile PMe_3 occured at the formyl C atom of (36) causing C-O bond cleavage and formation of (37).[65]

Photochemical reaction of $[V(CO)_4L_2]^-$ anions (L_2=chelating ditertiary phosphines or arsines, e.g. dppe) with substituted benzoyl chlorides and cyclopropanyl carbonyl chloride gave neutral η^2-acyl complexes (38).[66]

Alkylidene hydride complexes such as $Me_3CCH{=}Ta(H)(dmpe)_2Cl^+$, $Me_3CCH{=}Ta(H)(PMe_3)_3Cl_2$, $Me_3CHC{=}Ta(H)Cp^*(PMe_3)Cl$, and $Me_3CCH{=}Ta(H)(PMe_3)_3X_2$ ($X=O_3SCF_3$, I) were prepared. Protons of the RCH=TaH moieties exchange readily (magnetization transfer ^1H NMR studies) at 320 K. The last (diiodo) complex polymerized ethylene although $EtTa(C_2H_4)(PMe_3)_2I_2$ did not, suggesting that a classical insertion mechanism was not operative.[67]

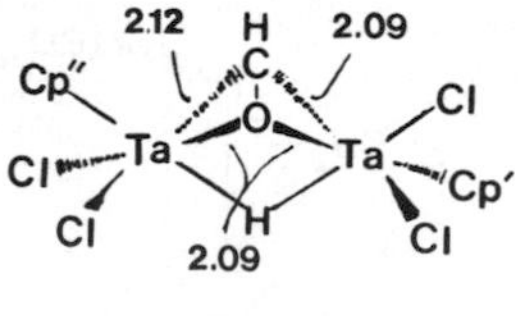

(36)

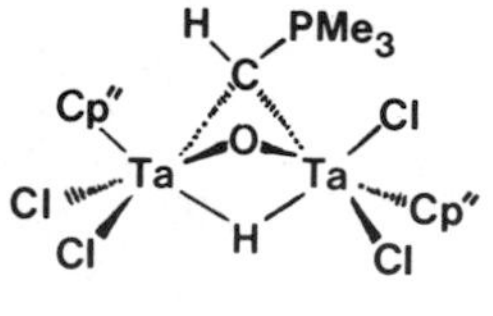

(37)

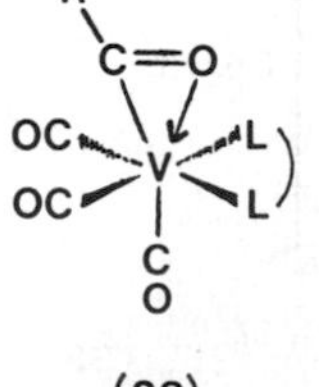

(38)

Use of PMe_3 as a solvent for reduction of $TaCl_5$ with sodium sand gave the green ligand-metallated complex $Ta(PMe_3)_3(\eta^2CH_2-PMe_2)(\eta^2CH-PMe2)$, (39).[68]

[Dihydrobis(3,5-dimethyl-1-pyrazolyl)borato]$TaClMe_3$ (40) was shown by X-ray to adopt a capped octahedral structure in the solid, with a 3-center 2-electron B-H-Ta bond.[69]

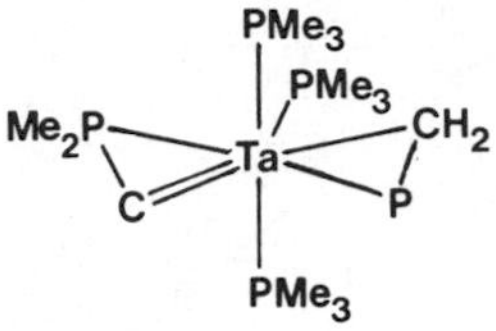

(39)

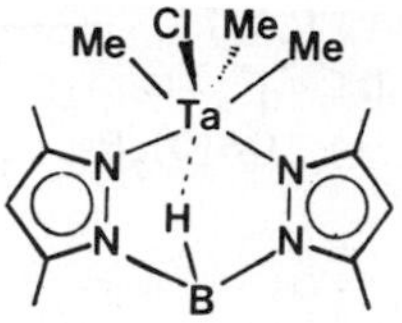

(40)

5. Group VIA

The first complexes with metal-arsenic double bonds were formed when CO was thermally eliminated (60°C, benzene) from complexes $Cp(CO)_3MAs(CMe_3)_2$ (M=Mo, W) to give $Cp(CO)_2)MAs(CMe_3)_2$[70], (41). Treatment of tungsten complex (41) with PMe_3 generated coordinately-saturated $Cp(CO)_2(Me_3P)WAs(CMe_3)_2$.

A diastereomeric ligand was used to investigate the stereochemistry of reactions at tungsten, as shown in Scheme (6). Erythro-$PhCHDCHDO_3SC_6H_4Me$-p with $Na[CpW(CO)_3]$ yielded $Cp-W(CO)_3CHDCHDPh$-threo (2) in >95% diastereomeric purity, showing the reaction to proceed with inversion of configuration at the α-carbon.[71] Insertion of SO_2 into the W-C bond of (42) proceeds with inversion of configuration, while reaction of (42) with I_2 giving threo-PhCHDCHDI and $CpW(CO)_2I_3$ proceeds with retention. The latter reaction is assumed to involve electrophilic attack of I^+ on W, followed by reductive elimination of alkyl iodide. β-Hydrogen elimination in (42) is not observed thermally but is observed upon photolysis (cyclohexane, 350nm) affording η^3-benzylic complex $CpW(CO)_2(\eta^3\text{-CHMePh})$ as the stable product. Apparently loss of CO gives $CpW(CO)_2CH_2CH_2Ph$ (43) which then β-hydrogen eliminates to $CpW(CO)_2(H)$(styrene), followed by insertion of styrene into the W-H bond to give the η^3-benzyl complex. This process is faster than recapture of CO by (43) to give (42) even under 1 atm CO.

β-Hydrogen elimination from a similar $16e^-$ intermediate is

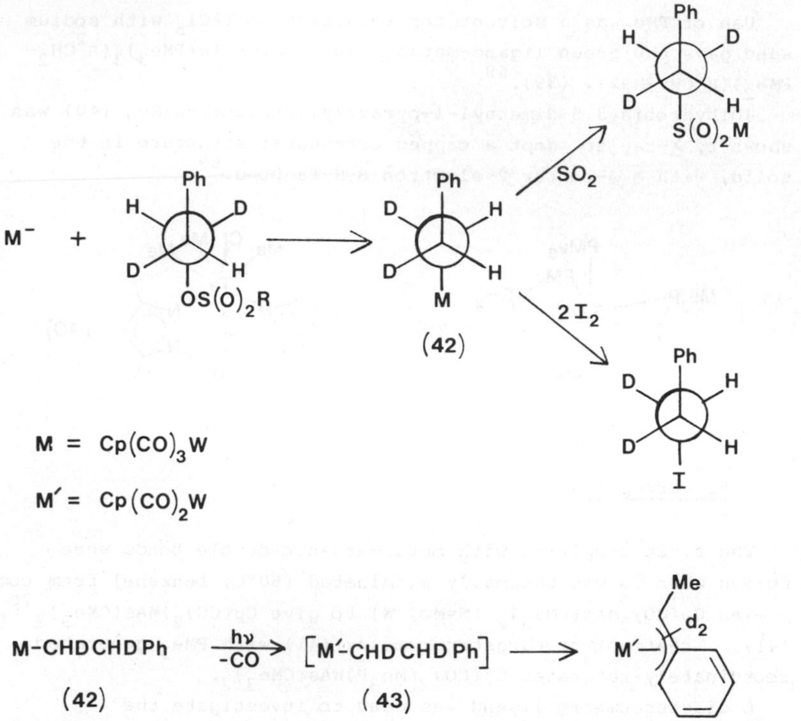

Scheme (6)

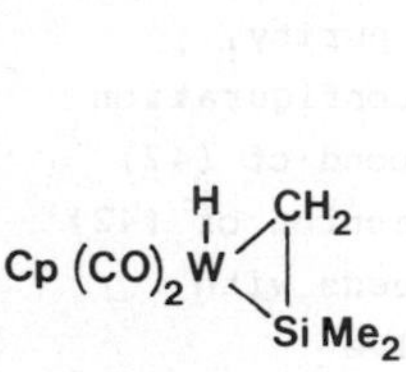

(44)

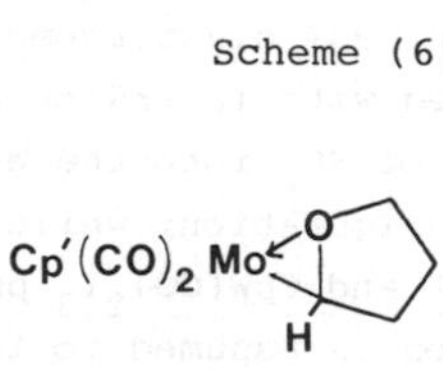

(45)

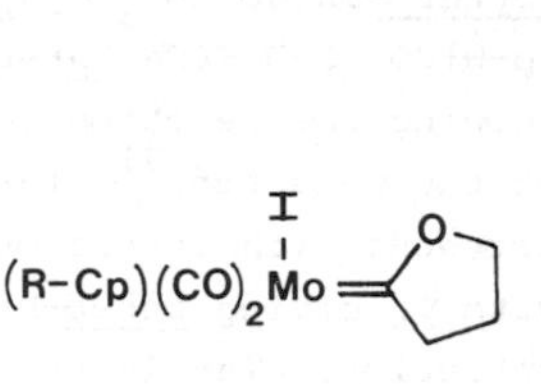

(46) (R = H, Me)

presumed in low temperature photolysis of $CpW(CO)_3CH_2Si(CH_3)_2H$ (77 or 19.6 K in hydrocarbon) which gives complex cis-(44).[72] Intermediate $CpW(CO)_2CH_2Si(CH_3)_2H$ is not observed. While fully characterized at 200 K, (4) yields significant amounts of $CpW(CO)_3SiMe_3$ upon warming with CO.

Insertion of $CF_3C{\equiv}CCF_3$ into the W-H bond of $CpW(CO)_3H$ gave $CpW(CO)_3[\eta^1\text{-}C(CF_3){=}(CF_3)H]$ which undergoes CO migration forming the η^3-acryloyl complex $CpW(CO)_2[\eta^3\text{-}COC(CF_3){=}C(CF_3)H]$, structure shown by X-ray.[73]

An unusual metallaoxapropane (45) was formed on reaction of $Cp(CO)_3Mo(CH_2)_3Br$ with $LiEt_3BH$.[74] η^2-Bonding of the α-tetrahydrofuranyl unit was proved by X-ray crystallography. Analogous to the possible intermediate, alkylidene (46)[75] did react with $LiBEt_3H$ to give (45). Tungstaphosphabicyclobutanone (47), (structure confirmed by X-ray), was produced from reaction of $ClPMe_2$ with ketene adduct $Cp(CO)(PMe_3)W(\eta^2MeC{=}C{=}O)$.[76]

The triple-bonded dimolybdenum complex $[Cp'_2Mo_2(CO)_4]$ reacted with diazofluorene to lose N_2 and form the μ-alkylidene complex (48) verified by X-ray.[77] A triple-bridged μ-alkylidyne complex $[W_3(OCMe_3)_5(\mu\text{-}O)(\mu\text{-}CEt)O]_2$ was formed from $W_2(OCMe_3)_6$ and $EtC{\equiv}CEt$ and the X-ray structure detected.[78] The bridging alkylidyne complex $Mo_2(\mu\text{-}H)(\mu\text{-}CSiMe_3)(CH_2SiMe_3)(OAr)_2py_2$ was formed by extrusion of $SiMe_4$ from $Mo_2(OAr)_2(CH_2SiMe_3)_4$ (OAr=2,6-dimethylphenoxide) in pyridine.[79]

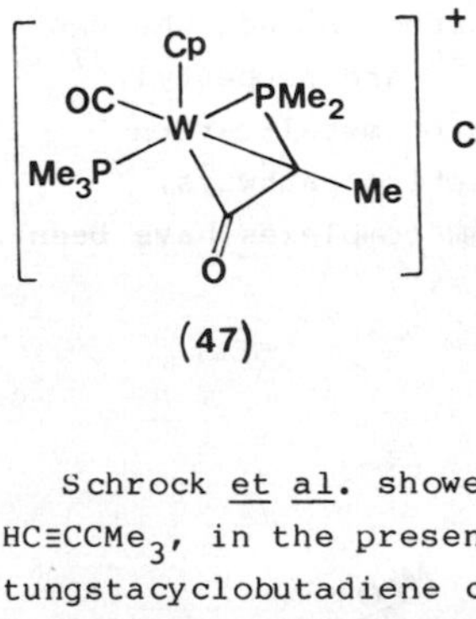

(47)

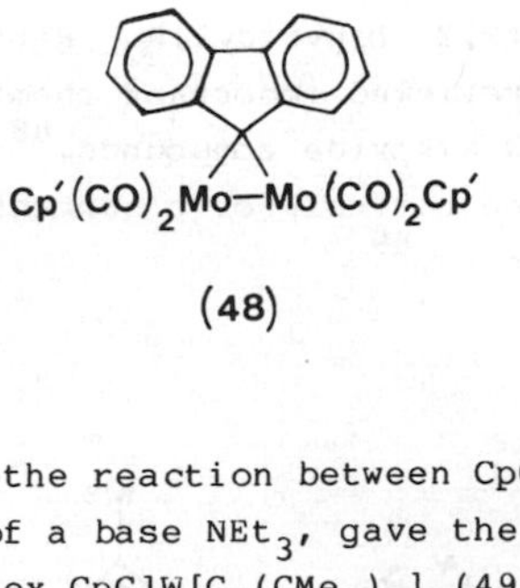

(48)

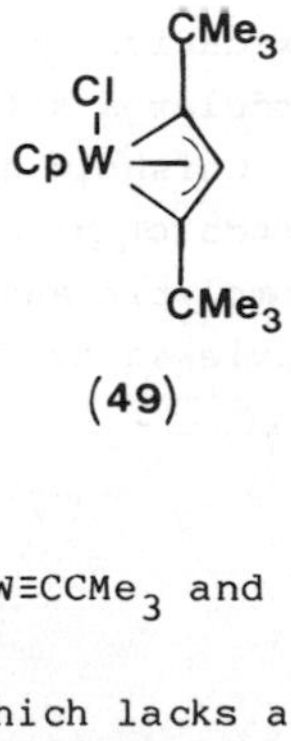

(49)

Schrock et al. showed[80] the reaction between $CpCl_3W{\equiv}CCMe_3$ and $HC{\equiv}CCMe_3$, in the presence of a base NEt_3, gave the tungstacyclobutadiene complex $CpClW[C_3(CMe_3)_2]$ (49) which lacks a β-C atom substituent. Such deprotonated but stable products were suggested to be a general reason why terminal alkynes would not metathesize. A similar metallocycle $W(OCMe_3)_2(C_3Ph_2)py_2$, resulted from reaction of $W{\equiv}CPh(OCMe_3)_3$ and $HC{\equiv}CPh$ in the presence of pyridine.

Isolation of complexes containing both coordinated alkyne and carbene ligands was achieved by Geoffroy et al.[81], one example being $W(CO)_4(Ph\text{-}C{\equiv}C\text{-}Ph)({=}C(OMe)Ph)$. Decomposition of the compounds (>25°C) gave only polyacetylenes for terminal alkyne ligands whereas internal alkynes gave products from coupling with the carbene carbon.

Matrix isolation IR studies (including ^{13}CO labelling) of the photolysis of $CpCr(CO)_3Me$ gave evidence for reversible formation of the α-CH elimination product, $CpCr(CO)_2(H)(=CH_2)$, following CO loss.[82]

Terminal alkyne ligands observed in intermediates _fac_-$W(CO)_3$-(dppe)(alkyne) rearranged to neutral alkylidenes _e.g._ in _mer_-$W(CO)_3(dppe)(=C=CHR)$[R=Ph, CO_2Me]. The alkylidenes were protonated at the β-carbon yielding cationic carbyne complexes which suffer Cl^- for CO exchange to give species such as $W(CO)_2(dppe)(\equiv C-CH_2Ph)$.[83] Alkylidene $Cp[P(OMe)_3]_2Mo(=C=CHPh)Br$ resulted from addition of $K[BH(CHMeEt)_3$ to $[Cp[P(OMe)_3]_2Mo$-$(BrC\equiv CPh)]$ BF_4 in THF. Secondary reaction of the alkylidene with H^- gave alkylidyne $Cp[P(OMe)_3]_2Mo(\equiv C-CH_2Ph)$.[84]

Formation of μ-alkylidyne ligands via cleavage of the triple bond of diphenylacetylene occurred in reaction of $W_2(OCMe_3)_6$ with the alkyne. The complex formed, $W_2(OCMe_3)_4(\mu-CPh)_2$ was characterized by _X_-ray crystallography.[85] Other interesting high oxidation state species which have been reported include the oxo complexes $Mo(O)_2(2,2'\text{-bipyridyl})R_2$, R=benzyl[86] and neopentyl.[87]

Chisholm has reviewed important chemistry of metal-carbon bonds of Mo and W alkoxide compounds.[88] Reaction pathways, homolytic and free radical, of organochromium complexes have been reviewed by Espenson.[89]

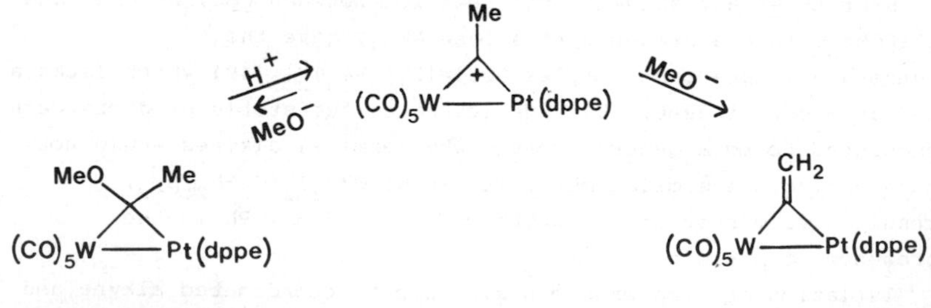

Scheme (7)

The chemistry of the carbyne complex $CpW(CO)_2(\equiv C$-p-tolyl) as a reagent for synthesizing compounds with bonds between W and other metals[90] was reviewed. The prolific chemistry asociated with interconversions of alkylidyne, vinylidene, vinyl and related groups bridging Pt-W bonds has been further explored by Stone et al.[91-96] Representative are the acid-base reactions of Scheme (7).[93]

An annual survey covering Cr, Mo, W in year 1981 was published.[97] Cooper reviewed electron transfer reactions of tungstenocene dialkyls.[98]

6. Group VII

Activation of aromatic C-H bonds was demonstrated in the reaction of CpRe(NO)(CO)H with $Ph_3C^+PF_6^-$ giving the η^2-arene complex $[CpRe(NO)(CO)(\eta^2\text{-Ph-CHPh})_2]^+PF_6^-$ which was deprotonated with NEt_3 to yield the σ-aryl species.[99] Complexes analogous to the last, CpRe(NO)(CO)R (R=Ph, tolyl or $C_6H_4CF_3$), were prepared independently using aryl copper reagents.[100] Protonation of such materials again gave dihapto arene rhenium cations.[101] The molecular structures of two dirhenium species containing ortho-metalled aromatic rings were reported. These are $[(\mu\text{-H})Re_2(CO)_8(\mu\text{-}C_5H_4)]$[102] (the substitution reactivity of which was also discussed) and $[Re_2Cl_3(Ph_2P\text{-}o\text{-py})_2[Ph(C_6H_4)P\text{-}o\text{-py}]]$.[103]

Reaction of CH_2N_2 with sulfur- or telluro-containing manganese complexes gave binuclear bridging thio- or telluro-formaldehyde ligands e.g. (50).[104, 105] Gladyz et al.[106] have also prepared related mononuclear complexes (51) characterized by X-ray.

Matrix isolation studies of the coordinately unsaturated acyl species $Mn(CO)_4(MeC{=}O)$ and FeCp(CO)(MeC=O) were reported.[107]

Bimetallic formyl anions, $(CO)_5M\text{-}M'(CHO)(CO)_4^-$, were prepared by Gladyz et al.[108] (M,M'=Mn, Re) and their reactions with electrophiles such as BzH, $Fe(CO)_5$ and $C_8H_{17}I$ examined. Hydride transfer can occur, e.g. reaction with CH_3I (M,M'=Re) gave 50% CH_4.

Kochi[109] also discovered an unusual radical-chain process which converts the formyl $[(CO)_5ReRe(CHO)(CO)_4]NBu_4^+$ to its corresponding metal hydride using as initiator AIBN, visible light or electrocatalysis. Decarboxylation of the formate complex $CpRe(NO)(PPh_3)OCHO$ at 70-130°C in toluene gave the analogous hydride. Labelling studies indicated the CO_2 extrusion to be

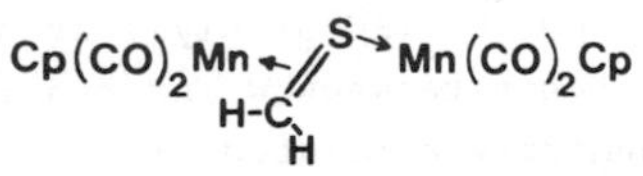

(50)

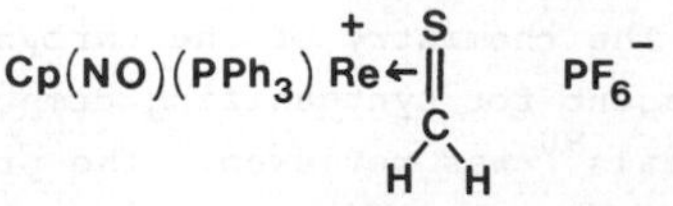

(51)

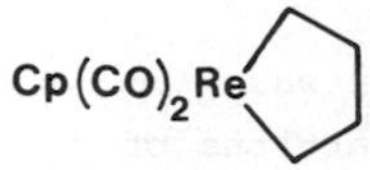

(52)

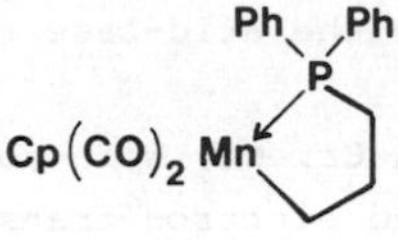

(53)

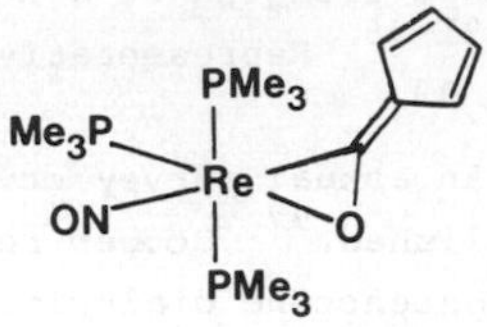

(54)

both intramolecular and without PPh_3 dissociation.[110]

Complexes $[Re(CO)_4(R'NCRNR')]$ (R=Me, Ph; R'=Ph or p-tolyl) containing delocalized bidentate NN'-chelated amidino groups were made by several different routes, including direct reaction of $Re_2(CO)_{10}$ with R'NH-CR=NR'. With R=Ph, isomeric o-metallated complexes were also observed.[111]

Synthesis of a rhenacyclopentane (52) from $CpRe(CO)_2H_2$, $I(CH_2)_4I$ and the amine DBU was reported.[112] Thermolysis (100°) of (52) via ring contraction and reductive elimination gave methylcyclopropane and $CpRe(CO)_3$. Further preparations and reactions of manganese metallocycles containing hetero atoms, such as (53), have appeared.[113,114]

Gladyz reported more of the diverse chemistry of substituents (alkyl, alkylidene, vinyl) attached to $CpRe(NO)(PPh_3)^+$. The stereo- and regiospecificity of α- or β- hydride abstractions from alkyls[115] (Et, Pr, pentyl, CH_2CHMe_2, CH_2CMe_3, $CHMe_2$), coupling of methylidene ligands in $[CpRe(NO)PPh_3(=CH_2)]^+PF_6^-$ [116] to form ethylene (as $[CpRe(NO)(PPh_3)(CH_2=CH_2)]^+PF_6^-$, 50% yield)[117], and reactions of electrophiles at the β-vinyl position of vinyl complexes $CpRe(NO)(PPh_3)(CH=CHR')$ to give alkylidenes[118] were mechanistically delineated. An unusual PMe_3-induced transformation of CpRe(NO)(CO)Me (toluene, 70°C) to the ketene complex (54) (28% yield; with loss of CH_4) observed by Casey[119] has important implications for the ease of $\eta^5 \rightarrow \eta^3 \rightarrow \eta^1$ C_5H_5 isomerization.

Reductive elimination of 1-butene was presumed in the reaction of 3,3-dimethyl cyclopropene with $Re_2(\mu\text{-}H)(\mu\text{-}CH=CHEt)(CO)_8$ forming the μ-alkylidene $Re_2[\mu\text{-}(\eta^1\text{-}\eta^3\text{-}CH\text{-}CH\text{-}CMe_2)](CO)_8$. Further reaction of the last molecule with CO and photolysis finally gave $Re_2(\mu\text{-}H)(\mu\text{-}CH{=}CHCMe{=}CH_2)(CO)_8$.[120] Co-condensation of Re atoms with substituted arenes gave interesting red binuclear μ-arylidene complexes in 15% yield. For example, the reaction of toluene with Re atoms proceeded _via_ C-H activation steps to give (55).[121]

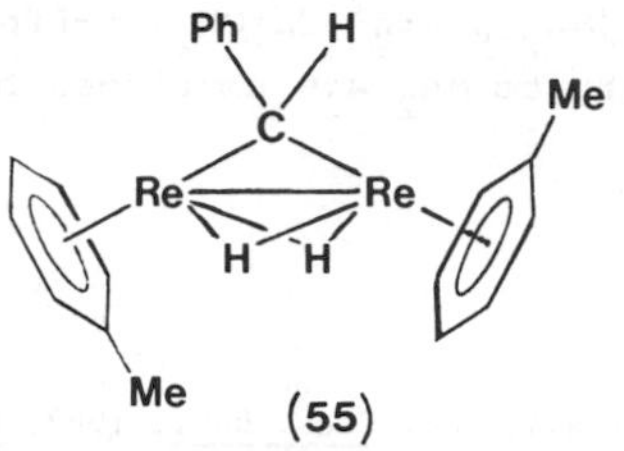

(55)

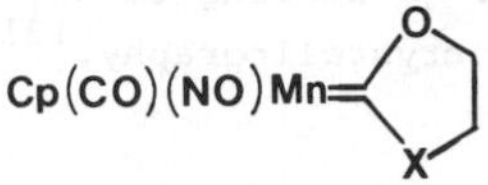

(56)

The manganese alkenylidene complex $CpMn(CO)_2(=C=CHPh)$ afforded olefinic complexes $CpMn(CO)_2(\eta^2\text{-}PhCH{=}CHP(O)(OR)_2)$ on reaction with $P(OR)_3$ (R=Et, Ph). An _X_-ray crystal structure (R=Et) showed the substituents of the π-bound olefin to be mutually _trans_.[122] Fluorocarbene complex $CpMn(CO)_2(=CFPh)$[123] was shown to undergo ready nucleophilic substitution with both BuLi and EtOH to give the corresponding $CpMn(CO)_2(C{=}XPh)$ complexes (X=Bu, OEt).[124] Cyclic alkylidene complexes (56) were formed when three-membered heterocycles (CH_2CH_2X, X=O, S, NH) reacted with the carbonyl cation $CpMn(NO)(CO)_2^+$.[125]

Dimeric Mn(II) species $Mn_2R_4(PR'_3)_2$ ($R=CH_2SiMe_3$, CH_2CMe_3) were conveniently prepared from $MnCl_2$, MgR_2 and PR'_3 (a variety of aryl and alkyl phosphines and phosphites were used) in diethyl ether. Magnetic moments of the dimers were considerably lower than expected for high-spin Mn(II), attributed to antiferromagnetic interactions through the bridging alkyl carbon atoms.[126] Related Mn(II) monomers, $MnR_2(PR'_3)_2$, were formed in the presence of excess PR'_3. The _X_-ray determined structure of $Mn(CH_2CMe_2Ph)_2(PMe_3)_2$ showed severely distorted tetrahedral geometry (P-Mn-P=96.2°, C-Mn-C=137.9°), reflecting the relative ligand sizes.[127] $Mn(III)(acac)_3$ with MeLi and dmpe in diethyl ether gave the disproportionation products, dark orange $Mn(II)Me_2(dmpe)_2$ and yellow $Mn(IV)Me_4(dmpe)$. Green tetrahedral $Mn(IV)R_4$ ($R=CH_2SiMe_3$, CH_2CMe_3) were identified by ESR.[128] Reaction of _trans_-$MnBr(dmpe)_2$ with $MgMe_2$ gave the low-spin _trans_-

$MnMe_2(dmpe)_2$.[129] The structure of a linear homoleptic Mn(II) trimer, $Mn_3(mes)_6$ (mes=2,4,6-$C_6H_2Me_3$) was determined by X-ray crystallography.[130]

High-spin trans- $MnBr_2(dmpe)_2$ was converted to the Mn(l) hydrides $[Mn(AlH_4)(dmpe)_2]_2$ (with $LiAlH_4$) and trans-$MnH(CH_2{=}CH_2)$-$(dmpe)_2$ (with $MgEt_2$).[129] Binuclear Mn(I) hydride $[Mn_2(\mu\text{-}H)$-$(\mu\text{-}PPh_2)(CO)_8]$, prepared from $Mn_2(CO)_{10}$ and $PHPh_2$, reacted with alkynes to give the vinyl complexes $[Mn_2(\mu\text{-}\sigma{:}\eta^2\text{-}CH{=}CH_2)(\mu\text{-}PPh_2)$-$(CO)_7$. π-Bonding to Mn_1 and σ-bonding to Mn_2 was confirmed by X-ray crystallography.[131]

References

1 J.Holton,M.F.Lappert, R.Pearce, and P.I.W.Yarrow, Chem. Rev., 1983, 83, 135.
2 W.A.Hermann, J. Organomet. Chem., 1983, 250, 319.
3 M.Brookhart and M.L.H.Green, J. Organomet. Chem., 1983, 250, 395.
4 R.T.Edidin, B.Longato, B.D.Martin, S.A.Matchett, and J.R.Norton, Organomet. Compd: Synth., Struct., Theory, Proc. Symp. Ind.-Univ. Coop. Chem. Program Dept. Chem., Tex., A&M Univ., Edited by: B.L.Shapiro, Tex. A&M Univ. Press: College Station, Tex, 1983, 260.
5 F.G.A.Stone, Ibid., 1.
6 F.A.Cotton, Ibid. 205.
7 R.F.Fenske, Ibid., 305.
8 H.Schumann, Comments Inorg. Chem., 1983, 2, 247.
9 W.J.Evans, J. Organomet. Chem., 1983, 250, 217.
10 G.B.Deacon, P.I.Mackinnon, and T.D.Tuong, Aust. J. Chem., 1983, 36, 43.
11 G.B.Deacon, P.I.Mackinnon, T.W.Hambley, and J.C.Taylor, J. Organomet. Chem., 1983, 259, 91.
12 A.B.Sigalov, L.F.Rybakova, and I.P.Beletskaya, Izv. Akad. Nauk SSSR, Ser. Khim., 1983, 1808.
13 A.B.Sigalov, L.F.Rybakova, and I.P.Beletskaya, Izv. Akad. Nauk SSSR, Ser. Khim., 1983, 1692.
14 W.J.Evans, I.Bloom, W.E.Hunter, and J.L.Atwood, J. Am. Chem. Soc., 1983, 105, 1401.
15 W.J.Evans, J.H.Meadows, W.E.Hunter, and J.L.Atwood, Organometallics, 1983, 2, 1252.
16 W.J.Evans, I.Bloom, W.E.Hunter, and J.L.Atwood, Organometallics, 1983, 2, 709.
17 P.L.Watson, J. Chem., Soc., Chem. Commun., 1983, 276.
18 P.L.Watson, J. Am. Chem. Soc., 1983, 105, 6491.
19 A.B.Sigalov, L.F.Rybakova, O.P.Syutkina, R.R.Shifrina, Y.S.Bogachev, I.L.Zhuravleva, and I.P.Beletskaya, Izv. Akad, Nauk SSSR, Ser. Khim., 1983, 918.
20 G.Z.Suleimanov, V.I.Bregadze, N.A.Koval'chuk, KH.S.Khalilov, and I.P.Beletskaya, J. Organomet. Chem., 1983, 255, C5.
21 C.Qian, C.Ye, H.Lu, Y.Li, J.Zhou, Y.Ge, and M.Tsutsui, J. Organomet. Chem., 1983, 247, 161.
22 A.L.Wayda, Organometallics, 1983, 2, 565.
23 O.P.Syutkina, L.F.Rybakova, E.N.Egorova, A.B.Sigalov, and I.P.Beletskaya, Izv. Akad, Nauk SSSR, Ser. Khim., 1983, 648.

24 L.Arnaudet, G.Folcher, H.Marquet-Ellis, E.Klaehne, K.Yuenlue, and R.D.Fischer, Organometallics, 1983, 2, 344.
25 L.Arnaudet, F.Brunet-Billiau, G.Folcher, and E.Saito, C. R. Seances Acad. Sci., Ser. 2, 1983, 296, 431.
26 K.G.Moloy, T.J.Marks, and V.W.Day, J. Am. Chem. Soc., 1983, 105, 5696.
27 M.Y.He, R.L.Burwell,Jr., and T.J.Marks, Organometallics, 1983, 2, 566.
28 J.W.Bruno, T.J.Marks, and L.R.Morss, J. Am. Chem. Soc., 1983, 105, 6824.
29 B.E.Bursten and A.Fang, J. Am. Chem. Soc., 1983, 105, 6495.
30 I.L.Fragala, J.Goffart, G.Granozzi, and E.Ciliberto, Inorg. Chem., 1983, 22, 216.
31 D.L.Reger, M.E.Tarquini, and L.Lebioda, Organometallics, 1983, 2, 1763.
32 M.Schiess and D.Seebach, Helv. Chem. Acta, 1983, 66, 1618.
33 E.M.Meyer and A.Jacot-Guillarmod, Helv. Chim. Acta, 1983, 66, 898.
34 V.B.Shur, S.Z.Bernadyuk, V.V.Burlakov, V.G.Andrianov, A.I.Yanovskii, Y.T.Struchkov, and M.E.Vol'pin, J. Organomet. Chem., 1983 243, 157.
35 R.P.Planalp and R.A.Andersen, J. Am. Chem. Soc., 1983, 105, 7774.
36 R.P.Planalp, R.A.Andersen, and A.Zalkin, Organometallics, 1983, 2, 16.
37 R.P.Planalp and R.A.Andersen, Organometallics, 1983, 2, 1675.
38 E.J.Moore, D.A.Straus, J.Armantrout, B.D.Santarsiero, R.H.Grubbs, and J.E.Bercaw, J. Am. Chem. Soc., 1983, 105, 2068.
39 S.Gamabarotta, C.Floriani, A.Chiesi-Villa, C.Guastini, J. Am. Chem. Soc., 1983, 105, 1690.
40 G.Erker, K.Kropp, J.L.Atwood, and W.E.Hunter, Organometallics, 1983, 2, 1555.
41 P.Czisch and G.Erker, J. Organomet. Chem., 1983, 253, C9.
42 D.G.Bickley, N.Hao, P.Bougeard, B.G.Sayer, R.C.Burns, and M.J.McGlinchey, J. Organomet. Chem., 1983, 246, 257.
43 H.W.F.L.Seetz, G.Schat, O.S.Akkerman, and F.Bickelhaupt, Angew. Chem., 1983, 95, 242.
44 G.Erker and U.Dorf, Angew. Chem., 1983, 95, 800.
45 V.Skibbe and G.Erker, J. Organomet. Chem., 1983, 241, 15.
46 S.A.Cohen, P.R.Auburn, and J.E.Bercaw, J. Am. Chem. Soc., 1983, 105, 1136.
47 E.Nakamura and I.Kuwajima, J. Am. Chem. Soc., 1983, 105, 651.
48 V.B.Shur, V.V.Burlakov, A.I.Yanovskii, Y.T.Struchkov, and M.E.Vol'pin, Izv. Akad. Nauk SSSR, Ser. Khim., 1983, 1212.
49 H.Takaya, M.Yamakawa, and K.Mashima, J. Chem. Soc., Chem. Commun., 1983, 1283.
50 M.L.H.Green, N.J.Hazel, P.D.Grebenik, V.S.B.Mtetwa, and K.Prout, J. Chem. Soc., Chem. Commun., 1983, 356.
51 A.Hudson, M.F.Lappert, and R.Pichon, J. Chem. Soc., Chem. Commun., 1983, 374.
52 N.E.Schore, S.J.Young, M.Olmstead, and P.Hofmann, Organometallics, 1983, 2, 1769.
53 B.A.Dolgoplosk, E.I.Tinyakova, I.Sh.Gusman, and L.L.Afinogenova, J. Organomet. Chem., 1983, 244, 137.
54 S.Gambarotta, C.Floriani, A.Chiesi-Villa, and C.Guastini, Inorg. Chem., 1983, 22, 2029.
55 B.Weidmann and D.Seebach, Angew. Chem., 1983, 95, 12.
56 K.Takai and H.Nozaki, Kagaku (Kyoto), 1983, 38, 596.
57 J.R.Stille and R.H.Grubbs, J. Am. Chem. Soc., 1983, 105, 1664.
58 S.L.Buchwald and R.H.Grubbs, J. Am. Chem. Soc., 1983, 105, 5490.
59 L.M.Engelhardt, W.P.Leung, C.L.Raston, and A.H.White, J. Chem. Soc., Chem. Commun., 1983, 386.
60 A.T.Casey and J.B.Raynor, J. Chem. Soc. Dalton Trans., 1983, 2057.
61 R.Mercier, J.Douglade, J.Amaudrut, J.Sala-Pala, and J.E.Guerchais, J. Organomet. Chem., 1983, 244, 145.
62 F.H.Kohler and W.A.Geike, J. Organomet. Chem., 1983, 256, C27.
63 J.C.Green, M.P.Payne and J.Teuben, Organometallics, 1983, 2, 203.
64 C.Moise, J. Organomet. Chem., 1983, 247, 27.

65 P.A.Belmonte, F.Cloke, N.Geoffrey, and R.R. Schrock, J. Am. Chem. Soc., 1983, 105, 2643.
66 J.Schiemann and E.Weiss, J. Organomet. Chem., 1983, 255, 179.
67 H.W.Turner, R.R.Schrock, J.D.Fellmann, and S.J.Holmes, J. Am. Chem. Soc., 1983, 105, 4942.
68 V.C.Gibson, P.D.Grebenik, and M.L.H.Green, J. Chem. Soc., Chem. Commun., 1983, 1101.
69 D.L.Reger, C.A.Swift, and L.Lebioda, J. Am. Chem. Soc., 1983, 105, 5343.
70 M.Luksza, S.Himmel, and W.Malisch, Angew. Chem., 1983, 95, 418.
71 S.C.H.Su and A.Wojcicki, Organometallics, 1983, 2, 1296.
72 C.Lewis and M.S.Wrighton, J. Am. Chem. Soc., 1983, 105, 7768.
73 F.Y.Petillon, J.L.LeQuere, F.LeFloch-Perennou, J.E.Guerchais, M.B.Gomes deLima, L.J.Manojlovic-Muir, K.W.Muir, and D.W.A.Sharp, J. Organomet. Chem., 1983, 255, 231.
74 H.Adams, N.A.Bailey, P.Cahill, D.Rogers, and M.J.Winter, J.Chem. Soc., Chem. Commun., 1983, 831.
75 N.A.Bailey, P.L.Chell, C.P.Manuel, and A.Mukhopadhyay, J.Chem. Soc. Dalton Trans, 1983, 2397.
76 F.R.Kreissl, M.Wolfgruber, W.Sieber, and K.Ackerman, J. Organomet. Chem., 1983, 252, C39.
77 J.J.D'Errico and M.D.Curtis, J. Am. Chem. Soc., 1983, 105, 4479.
78 F.A.Cotton, W.Schwotzer, and E.S.Shamshoum, Organometallics, 1983, 2, 1340.
79 T.W.Coffindaffer, I.P.Rothwell, and J.C.Huffmann, J. Chem. Soc., Chem. Commun., 1983, 1249.
80 L.G.McCullough, M.L.Listemann, R.R.Schrock, M.R.Churchill, and J.W.Ziller, J. Am. Chem. Soc., 1983, 105, 6729.
81 H.C.Foley, L.M.Strubinger, T.Targos, and G.L.Geoffroy, J. Am. Chem. Soc., 1983, 105, 3064.
82 K.A.Mahmoud, A.J.Rest, and H.G.Alt, J. Chem. Soc., Chem Commun., 1983, 1011.
83 K.R.Birdwhistell, S.J.N.Burgmayer, and J.L.Templeton, J. Am. Chem. Soc., 1983 105, 7789.
84 R.G.Beevor, M.Green, A.G.Orpen, I.D.Williams, J. Chem. Soc., Chem Commun., 1983, 673.
85 F.A,Cotton, W.Schwotzer, and E.S.Shamshoum, Organometallics, 1983, 2, 1167.
86 G.N.Schrauzer, L.A.Hughes, M.J.Therien, E.O.Schlemper, F.Ross, and D.Ross, Organometallics, 1983, 2, 1163.
87 F.Ross, D.Ross, and E.O.Schlemper, Organometallics, 1983, 2, 481.
88 M.H.Chisholm, Polyhedron, 1983, 2, 681.
89 J.H.Espenson, Prog. Inorg. Chem., 1983, 30, 189.
90 F.G.A.Stone, ACS Symp. Ser., 1983, 211, 383.
91 J.A.Abad, L.W.Bateman, J.C.Jeffery, K.A.Mead, H.Razay, F.G.A.Stone, and P.Woodward, J. Chem. Soc., Dalton Trans., 1983, 2075.
92 K.A.Mead, I.Moore, F.G.A.Stone, and P.Woodward, J. Chem. Soc., Dalton Trans., 1983, 2083.
93 M.R.Awang, J.C.Jeffery, and F.G.A. Stone, J. Chem. Soc., Dalton Trans., 1983, 2091.
94 M.R.Awang, G.A.Carriedo, J.A.K.Howard, K.A.Mead, I.Moore, C.M.Nunn, and F.G.A.Stone, J. Chem. Soc., Chem. Commun., 1983, 964.
95 R.D.Barr, M.Green, J.A.K.Howard, T.B.Marder, I.Moore, and F.G.A.Stone, J. Chem. Soc., Chem. Commun., 1983, 746.
96 R.R.Awang, J.C.Jeffrey, and F.G.A.Stone, J. Chem. Soc., Chem. Commun., 1983, 1426.
97 J.D.Atwood, J. Organomet. Chem., 1983, 257, 105.
98 J.C.Hayes and N.J.Cooper, Organomet. Compd: Synth., Struct., Theory, Proc. Symp. Ind.-Univ. Coop. Chem. Program Dep. Chem., Tex. A&M Univ., 1983, 353.
99 J.R.Sweet and W.A.G.Graham, Organometallics, 1983, 2, 135.
100 J.R.Sweet and W.A.G.Graham, J. Organomet. Chem., 1983, 241, 45.

101 J.R.Sweet and W.A.G.Graham, J. Am. Chem. Soc., 1983, 105, 305.
102 P.O.Nubel, S.R.Wilson, and T.L.Brown, Organometallics, 1983, 2, 515.
103 T.J.Barder, S.M.Tetrick, R.A.Walton, F.A.Cotton, and G.L.Powell, J. Am. Chem. Soc., 1983 105, 4090.
104 M.Herberhold, W.Ehrenreich, and W.Buehlmeyer, Angew. Chem., 1983 95, 332.
105 W.A.Herrmann, J.Weichmann, R.Serrano, K.Blechschmitt, H.Pfisterer, and M.L.Ziegler, Angew. Chem., 1983, 95, 331.
106 W.E.Buhro, A.T.Patton, C.E.Strouse, J.A.Gladysz, F.B.McCormick, and M.C.Etter, J. Am. Chem. Soc., 1983, 105, 1056.
107 R.B.Hitam, R.Narayanaswamy, and A.J.Rest, J. Chem. Soc., Dalton Trans., 1983, 615.
108 W.Tam, M.Marsi, and J.A.Gladysz, Inorg. Chem., 1983, 22, 1413.
109 B.A.Narayanan, C.Amatore, C.P.Casey, and J.K.Kochi, J. Am. Chem. Soc., 1983, 105, 6351.
110 J.H.Merrifield and J.A.Gladysz, Organometallics, 1983, 782.
111 J.A.Clark and M.Kilner, J. Chem. Soc. Dalton Trans, 1983, 2613.
112 G.K.Yang and R.G.Bergman, J. Am. Chem. Soc., 1983, 105, 6500.
113 E.Lindner, K.A.Starz, H.J.Eberle, and W.Hiller, Chem. Ber., 1983, 116, 1209.
114 E.Lindner, K.A.Starz, N.Pauls, and W.Winter, Chem. Ber., 1983, 116, 1070.
115 W.A.Kiel, G.Y.Lin, G.S.Bodner, and J.A.Gladysz, J. Am. Chem. Soc., 1983, 105, 4958.
116 A.T.Patton, C.E.Strouse,, C.B.Knobler, and J.A.Gladysz, J. Am. Chem. Soc., 1983, 105, 5804.
117 J.H.Merrifield, G.Y.Lin, W.A.Kiel, and J.A.Gladysz, J. Am. Chem. Soc., 1983, 105, 5811.
118 W.G.Hatton and J.A.Gladysz, J. Am. Chem. Soc., 1983, 105, 6157.
119 C.P.Casey and J.M.O'Connor, J. Am. Chem. Soc., 1983, 105, 2919.
120 M.Green, A.G.Orpen, C.J.Schaverien, and I.D.Williams, J. Chem. Soc., Chem. Commun., 1983, 1399.
121 F.G.N.Cloke, A.E.Derome, M.L.H.Green, and D.O'Hare, J. Chem. Soc., Chem. Commun., 1983, 1312.
122 A.B.Antonova, S.V.Kovalenko, E.D.Korniets, A.A.Ioganson, Y.T.Struchkov, A.I.Akhmedov, and A.I.Yanovskii, J. Organomet. Chem., 1983, 244, 35.
123 E.O.Fischer, J.Chen, and K.Scherzer, J. Organomet. Chem., 1983, 253, 231.
124 E.O.Fischer and J.Chen, Z. Naturforsch., B: Anorg. Chem., Org. Chem., 1983, 38B, 580.
125 M.M.Singh and R.J.Angelici, Angew. Chem., 1983, 95, 160.
126 C.G.Howard, G.Wilkinson, M.Thornton-Pett, and M.B.Hursthouse, J. Chem. Soc., Dalton Trans., 1983, 2025.
127 C.G.Howard, G.S.Girolami, and G.Wilkinson, J. Chem. Soc. Dalton Trans., 1983, 2631.
128 C.G.Howard, G.S.Girolami, G.Wilkinson, M.Thornton-Pett, and M.B.Hursthouse, J. Chem. Soc., Chem. Commun., 1983, 1163.
129 G.S.Girolami, G.Wilkinson, M.Thornton-Pett, and M.B.Hursthouse, J. Am. Chem. Soc., 1983, 105, 6752.
130 S.Gambarotta, C.Floriani, A.Chiesi-Villa, and C. Guastini, J. Chem. Soc., Chem. Commun., 1983 1128.
131 J.A.Iggo, M.J.Mays, P.R.Raithby, and K.Hendrick, J. Chem. Soc., Dalton Trans., 1983, 205.

12

Complexes Containing Metal–Carbon σ-Bonds of the Groups Iron, Cobalt, and Nickel

BY A. K. SMITH

1 Introduction

While the general format remains the same as in previous years, the order of the four sections that make up this chapter has been changed. These sections deal with reviews and articles of general interest; metal-carbon σ-bonds of the Group VIII triads in the sequence iron, cobalt, and nickel; carbene and carbyne complexes including complexes containing μ-CH_2 and μ-CH ligands; and a bibliography giving details of papers which have not been included in the main body of the text due to pressure on space.

2 Reviews and Articles of General Interest

Reviews published in 1983 cover the mechanism of decomposition of organometallic compounds,[1] phosphorus ylide complexes,[2] bimolecular homolytic displacement of transition metal complexes from carbon,[3] methylene-bridged complexes,[4] bridged hydrocarbyl or hydrocarbon binuclear transition-metal complexes,[5] and alkyne-substituted carbonyl clusters.[6] Articles of general interest include an account of the mechanistic details of C-H and C-C bond activation by first row Group VIII atomic metal ions,[7] the preparation of highly reactive metal powders which readily lead to a wide range of organometallic complexes,[8] and a study of carbene transfer from aliphatic diazoalkanes to coordinatively unsaturated metal centres as a general synthetic route to organometallic complexes.[9]

3 Metal-Carbon σ-Bonds Involving Group VIII Metals

3.1 The Iron Triad. - The complex $[Fe(C_5Me_5)(CO)(Me)(PMe_3)]$ has been synthesized by the hydride reduction of a carbonyl ligand in the cationic complex $[Fe(C_5Me_5)(CO)_2(PMe_3)]^+$.[10] A similar reduction of $[Fe(C_5Me_5)(CO)_3]^+$ has been shown to be solvent dependent, giving the hydride $[Fe(C_5Me_5)(CO)_2H]$ in water, the hydroxymethyl complex $[Fe(C_5Me_5)(CH_2OH)(CO)_2]$ in CH_2Cl_2, and the methyl derivative $[Fe(C_5Me_5)(CO)_2(Me)]$ together with the hydride in THF.[11] The cyclopentadienyl derivative of the hydroxymethyl complex, $[M(Cp)(CH_2OH)(CO)_2]$ (M=Fe,Ru)[12] and the C_5Me_5Ru derivative $[Ru(C_5Me_5)(CH_2OH)(CO)_2]$[13] have also been prepared by hydride reduction of the corresponding cationic tricarbonyl complexes. The perfluoropropyl-iron complex $[Fe(C_3F_7)(CO)_4I]$ reacts with $[Hg\{N(CF_3)_2\}_2]$ to form $[Fe(C_3F_7)_2(CO)_4]$, and with $[HgF_2]$, $[Hg_2F_2]$, or $[HgF(Ph)]$ to afford $[\{Fe(C_3F_7)(CO)_4\}_2Hg]$ and $[(C_3F_7)Fe(CO)_4Hg(C_3F_7)]$ as major products.[14] The highly reactive complex $[Fe(Cp)(CO)_2(CCl_3)]$ has been prepared by halogen-exchange between $[Fe(Cp)(CO)_2(CF_3)]$ and BCl_3.[15] The chloromethyl complex, $[Fe(Cp)(CO)_2(CH_2Cl)]$, reacts with a range of tertiary phosphines, L, or $AsPh_3$ to give cationic ylide complexes of the type $[Fe(Cp)(CO)_2(CH_2L)]^+$.[16] Treatment of the disilane complex $[(CO)_4\overline{FeSiMe_2CH_2CH_2SiMe_2}]$ with aromatic or aliphatic aldehydes (RCHO) in the presence of PPh_3 or $P(\underline{p}\text{-tol})_3$ affords the ylide complexes $[(CO)_4Fe\{CH(R)PAr_3\}]$.[17] The cationic ylide complex, $[Fe(Cp)(CO)_2\{CH(SiMe_3)PMe_3\}]^+$ is deprotonated by sodium methoxide to give the η^1-ketenyl complex $[Fe(Cp)(CO)\{C(SiMe_3)CO\}(PMe_3)]$.[18] The hydrido-iron complex $[Fe(C_5Me_5)(CO)(H)(PMe_3)]$ reacts with the ylide Me_3PCH_2 to give the methyl complex $[Fe(C_5Me_5)(CO)(Me)(PMe_3)]$ via the phosphonium salt $[Me_4P][Fe(C_5Me_5)(CO)(PMe_3)]$.[19] The η^1-allyl complex $[Fe(Cp)(CO)_2(CH_2CH{=}CH_2)]$ undergoes a [3+3] cyclo-addition reaction with $S\{NS(O)_2Me\}_2$ to afford (1).[20] The preparation of the η^1-2-methoxyallyliron complex $[Fe(Cp)(CO)_2\{P(OCH_2)_3CMe\}\{\eta^1\text{-}CH_2C(OMe){=}CH_2\}]$ has been reported.[21] CO substitution by P donor ligands in $[Fe(\eta^5\text{-}Cp)(\eta^1\text{-}Cp)(CO)_2]$ and $[Fe(Cp)(\eta^1\text{-}CH_2CH{=}CH_2)(CO)_2]$ is suggested to occur by a radical chain mechanism.[22] The treatment of $[Fe(\eta^5\text{-}Cp)(\eta^1\text{-}Cp)(CO)_2]$ with maleic anhydride, dimethylfumarate, dimethyl acetylenedicarboxylate, or 2-chloroacrylonitrile results in the formation of syn-7-bicyclo[2.2.1]hept-5-ene iron derivatives which, on treatment with Ce(IV) in CO-saturated MeOH give the corresponding syn-7-carbomethoxynorbornene compounds[23].

The crystal structure of one of these derivatives, (2), has been determined.[23] The reaction of $[Fe(Cp)(CO)\{P(OPh)_3\}(\eta^2\text{-MeC}\equiv\text{CMe})]$ with the organo-copper reagent $Ph_2Cu(CN)Li_2$ and $[Fe(Cp)(CO)\{$-$P(OPh)_3\}(\eta^2\text{-PhC}\equiv\text{CMe})]$ with $Me_2Cu(CN)Li_2$ yields (3) and (4) respectively, thus demonstrating trans addition of these nucleophiles to the η^2-alkyne complexes.[24] Complexes of the type $[M(Me)(X)(PMe)_{4-2\underline{x}}$-$\{Me_2P(CH_2)_{\underline{n}}PMe_2\}_{\underline{x}}]$ ($\underline{n}$=1-3; X=Cl,Me; M=Fe,Co^+) have been prepared, and their configuration (cis or trans) has been shown to depend upon the number and size of the chelating rings.[25] The structures of two complexes containing Fe-C σ-bonds to aryl ligands, [Fe-$(PhLi)_4(OEt_2)_4$] and $[Fe(C_{10}H_7)_4][LiOEt_2]_2$ ($C_{10}H_7$ = naphthyl), formed by the interaction of $FeCl_3$ and PhLi or $C_{10}H_7Li$ respectively, have been determined by X-ray crystallography.[26,27] Reduction and methylation of $[Fe(Cp)(NO)]_2$ gives $[(Cp)(Me)Fe(\mu\text{-NO})_2Fe(Me)(Cp)]$ which rearranges at 45°C to give $[Fe(Cp)(Me)_2(NO)]$.[28] Addition of PMe_3 to $[Fe(Cp)(Me)_2(NO)]$ leads to migratory insertion of NO to afford $[Fe(Cp)(Me)\{N(O)Me\}(PMe_3)]$.[28]

A number of studies of CO insertion into Fe-C σ-bonds has been reported.[29-34] CO insertion into $M\text{-}CH_3$ bonds of cis-$[MX(Me)(CO)_2$-$(PMe_3)_2]$ (X=I,Me,CN,CN—>BPh_3 ; M=Fe,Ru) has been shown by ^{13}C labelling studies to proceed by cis insertion of the CO group rather than by cis migration of the CH_3 group.[29] $[Fe(Cp)(CO)(Me)$-$(PPh_3)]$ undergoes a rapid redox-catalysed migratory insertion of CO, thus providing the first example of a redox-catalysed carbonylation reaction.[30] The rate of the $P(OMe)_3$ promoted migratory CO insertion in fac-$[Fe(CO)_3(Me)(diars)]^+$ to give $[Fe(CO)_2\{C(O)Me\}$-$(diars)\{P(OMe)_3\}]^+$ has been shown to be first-order in the methyliron complex and zero-order in phosphite.[31] Two reaction pathways are observed when $[Fe(CO)_3(Me)(PMe_3)_2]^+$ is treated with Lewis bases (L) to give $[Fe(CO)_2\{C(O)Me\}(L)(PMe_3)_2]^+$.[32] One pathway involves CO insertion followed by attack of L, and the other pathway involves ligand exchange followed by CO insertion. The relative importance of these two pathways depends upon the nature and concentration of L. The BF_3 promoted carbonylation of the optically active complex, $[Fe(Cp)(CO)(Me)(L)]$ $\{L=PPh_2N(Me)\text{-}(\underline{S})\text{-}CH(Me)Ph\}$ proceeds stereospecifically at temperatures below -20°C.[33] The stereospecificity of the CO insertion reaction in $[Fe(Cp)(CO)(Et)$-$(PPh_3)]$ has been shown to vary widely but reproducibly with reaction conditions.[34] A study of the alkylation of the chiral complexes $[Fe(Cp)(CO)(CH_2Cl)(L)]$ and $[Fe(Cp)(CO)(CH_2{=}CH_2)(L)]^+$ $\{L=PPh_3$, tri(o-biphenyl)phosphite$\}$ by the prochiral nucleophiles

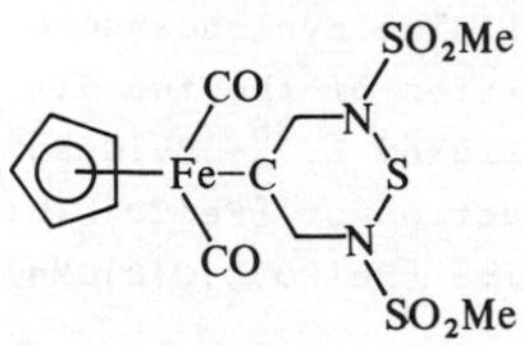

(1)

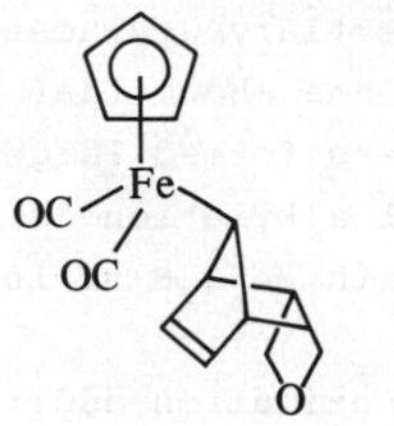

(2)

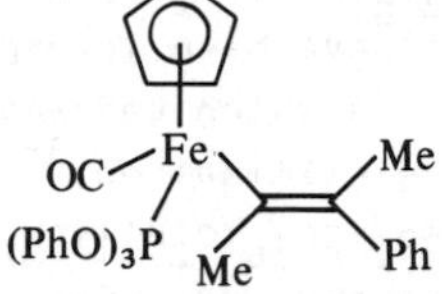

(3)

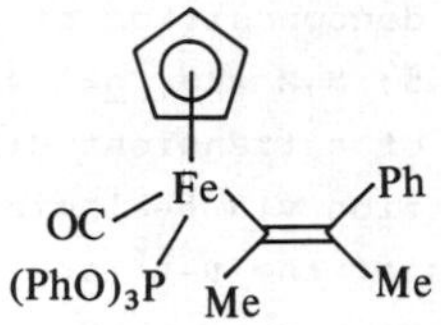

(4)

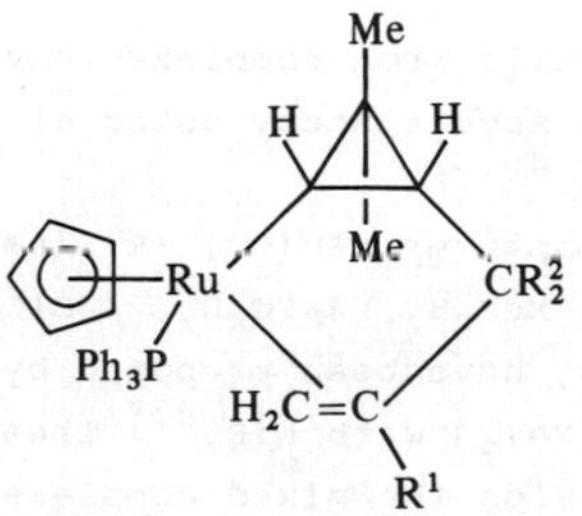

(5)

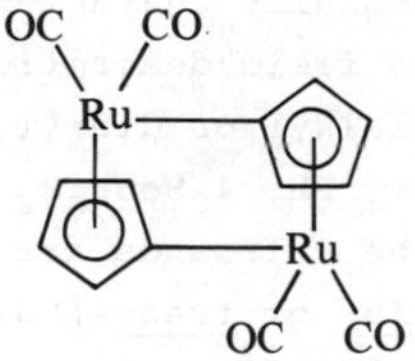

(6)

sodium tertiarybutylacetoacetate and pyrrolidine cyclohexanone enamine, has shown that the relative quantities of the two diastereomers formed largely depends on the ligand L.[35] Evidence for metal alkylation is provided by the reaction of $[Fe(CO)_4\{C(O)OMe\}]^-$ with $MeSO_4F$ at low temperature to give $[Fe(CO)_4\{C(O)OMe\}(Me)]$.[36]

The oxidation addition of CH_4 to photo-excited Fe atoms in a low temperature matrix leads to the formation of [Fe(Me)H], while dinuclear species, Fe_2 have been found to be unreactive toward CH_4.[37] The results of an investigation of the thermal and photochemical decomposition of $[(Cp)(CO)_2M\{\mu-(CH_2)_n\}M'(CO)_2(Cp)]$ (M=M'=Fe, n=3-5; M=M'=Ru, n=3,4; M=Fe,M'=Ru, n=3) have been interpreted in terms of a transient dimetallacyclic species which undergoes decomposition *via* β-elimination and reductive elimination.[38] The reactions of the μ-η^2-acetylide complexes $[Fe_2(CO)_6(C_2R)(PPh_2)]$ (R=Ph, *p*-$MeOC_6H_4$, *p*-BrC_6H_4, Cy, Bu^t) with primary, secondary, and tertiary amines have been studied.[39] These reactions lead to μ-vinylidene, μ-alkylidene, or μ-η^2-alkylidene derivatives, depending upon whether attack occurs at the β- or α-carbon atom of the acetylide.

Various thermally labile metallacyclic iron complexes have been successfully characterized by mass spectroscopy using electron -impact and field desorption techniques.[40,41]

The dialkyl or diaryl ruthenium complexes $[Ru(CO)_2(R)_2(PMe_2Ph)_2]$ (R=Me, Ph, 4-$MeOC_6H_4$, 3-MeC_6H_4, 4-MeC_6H_4, 4-FC_6H_4, 4-ClC_6H_4), in which the R ligands are mutually *cis*, have been prepared by treating *cis*- or *trans*-$[Ru(CO)_2Cl_2(PMe_2Ph)_2]$ with LiR.[42] Treatment of $[Ru(CO)_2R(Cl)(PMe_2Ph)_2]$ with LiR' yields the mixed complexes $[Ru(CO)_2R(R')(PMe_2Ph)_2]$. The dimethyl derivative undergoes reversible carbonylation to form *mono*- and *di*- acetyl complexes but the diaryl derivatives do not react with CO.[42] The structures of two Ru-phenyl complexes $[Ru(CO)(Ph)(Cl)(^tBuNC)(PMe_2Ph)_2]$ [43] and $[Ru(CO)(Ph)\{C(O)Ph\}(^tBuNC)(PMe_2Ph)_2]$[44] have been reported. η^1-Alkenyl complexes of the type $[Ru(CO)_2\{C(CO_2R)=C(CO_2R)Cl\}(Cl)L_2]$ (L=PMe_2Ph, $AsMe_2Ph$; R=Me,Et) have been prepared by the reaction of *trans*-$[Ru(CO)_2Cl_2L_2]$ with $RO_2CC{\equiv}CCO_2R$.[45] The reaction of $[Ru(Cp)Cl(PPh_3)_2]$ with alkylmagnesium halides leads to the formation of $[Ru(Cp)R(PPh_3)_2]$ (R=Et,Pr,^{n}Bu,^{i}Bu) which undergo β-elimination reactions above 50°C to form the corresponding (η^2-alkene)hydridoruthenium complexes, although the iso-butyl derivative may be stabilised by complexation with ethylene as $[Ru(Cp)(^iBu)(CH_2{=}CH_2)$-

$(PPh_3)]$.[46] $[Ru(Cp)Cl(PPh_3)_2]$ also reacts with alkenylmagnesium halides to form complexes of the type $[Ru(Cp)\{(CH_2)_nCR=CH_2\}(PPh_3)_2]$ (n=0,2; R=H; n=3, R=Me).[47] Stable η^1,η^2-4-alkenylruthenium complexes (5) ($R^1=R^2$=H; R^1=Me, R^2=H; R^1=H, R^2=Me) are formed in a similar reaction if isomerization *via* β-elimination is prevented by an *E*-configuration of Ru and the β-H atom in an inflexible cyclopropyl system.[47] The cationic complexes $[M(C_6H_6(C_2H_3R)(H)(PMe_3)]^+$ (M=Ru, R=H; M=Os, R=H, Me) react with NaI to give the alkylmetal complexes $[M(C_6H_6)(C_2H_4R)(I)(PMe_3)]$, and with CH_3I/NH_4PF_6 to give $[M(C_6H_6)(C_2H_3R)(Me)(PMe_3)]^+$ which react in turn with NaI to afford $[M(C_6H_6)(Me)(I)(PMe_3)]$.[48] The addition of $[CPh_3][PF_6]$ to $[Ru(C_6Me_6)(Me)_2(PPh_3)]$ yields $[Ru(C_6Me_6)(C_2H_4)(H)(PPh_3)][PF_6]$ *via* a presumed attack of the trityl cation on a coordinated methyl group.[49] The photolysis of $[Ru(C_6R_6)(H)_2(PR'_3)]$ (R=R'=Me; R=H, R'=^{i}Pr) in arenes C_6H_5R'' (R''=H,Me) leads to the formation of the aryl(hydrido) complexes $[Ru(C_6R_6)(C_6H_4R'')(PR'_3)]$.[50] A similar treatment of $[Ru(C_6Me_6)(H)_2(P^iPr_3)]$ in cyclohexane leads to intramolecular oxidative addition of a methyl C-H bond of the phosphine ligand to give $[Ru(C_6Me_6)(H)\{P(^iPr)_2CH(Me)CH_2\}]$.[50]

Dinuclear oxidative addition takes place when solutions of $[Ru_2(CO)_4(\text{fulvalene})]$ are irradiated leading to the rapid formation of (6).[51] Ruthenacyclopentadiene complexes of the type $[\{(CO)_3RuC_4(R)_4\}Ru(CO)_3]$ (R=CH_2OH, CH_2CH_2OH, Et;[52] R=CO_2Me[53]) have been prepared by treatment of $[Ru_3(CO)_{12}]$ with the appropriate alkyne RC≡CR. A minor product of the pyrolysis of $[Ru_3(CO)_{10}(sp)]$ (sp=2-$CH_2=CHC_6H_4PPh_2$) has been crystallographically characterized as the dinuclear ruthenium complex (7), while the major product is $[Ru_3(CO)_8(H)_2(\mu\text{-}\eta^2,P\text{-}HC_2C_6H_4PPh_2)]$.[54]

The carbido cluster $[Ru_6C(CO)_{16}(CN^tBu)]$ has been synthesized by the pyrolysis of $[Ru_5(CO)_{14}(CN^tBu)(\mu_5\text{-}CN^tBu)]$, and the source of the carbido carbon atom has been established as an isocyanide ligand by labelling experiments.[55] The synthesis of $[Ru_5C(CO)_{15}]$ by carbonylation of $[Ru_6C(CO)_{17}]$ has been described,[56] and a number of derivatives of the pentanuclear carbido cluster have been characterized.[56-59] The hexanuclear carbido clusters $[Ru_6C(CO)_{14}(NO)_2]$,[60] $[Ru_6C(CO)_{15}(NO)(AuPPh_3)]$,[60] and $[Ru_6C(CO)_{15}(H)(NO)]$[61] have been prepared and structurally characterized. A summary of the methods available for the preparation of carbido clusters of Ru and Os has been published.[62]

$[Os(C_6H_6)(CNR)(PMe_3)]$ (R=Me, ^{t}Bu, *p*-tolyl, Ph) reacts with

CH_3I to form $[Os(C_6H_6)(CNR)(Me)(PMe_3)]^+$, which reacts in turn with PMe_3 to form $[Os(CNR)(Me)(PMe_3)_4]^+$.[63] The reduction of *trans*-$[OsCl_2(PMe_3)_4]$ with $NaC_{10}H_8$ leads to Os-C bond formation as shown by the isolation of $[Os(CH_2PMe_2)(H)(PMe_3)_3]$.[64] The irradiation of $[Os_3(CO)_{12}]$ in the presence of methyl acrylate leads to the formation of $[Os(CO)_4(\eta^2-H_2C=C(H)CO_2Me)]$ and the 1,2-diosmacyclobutane derivative $[(CO)_4OsCH_2C(H)(CO_2Me)Os(CO)_4]$.[65] Metallacyclic complexes of the type $[Os\{CH_2OC(Se)Se\}(CO)_2(PPh_3)_2]$, $[Os(CH_2E)(CO)_2-(PPh_3)_2]$, and $[Os(CH_2EMe)(CO)_2(PPh_3)_2]^+$ (E=Se or Te) have been prepared.[66] In the reaction of the cationic derivative with $[BH_4]^-$, the C-E bond is cleaved and $[Os(Me)(EMe)(CO)_2(PPh_3)_2]$ is formed.[66] One of the products of the reaction between $[Os(CS)(CO)-(PPh_3)_3]$ and $PhC\equiv CPh$ is the metallacycle $[Os\{C(S)C(Ph)CPh\}(CO)_2-(PPh_3)_2]$ which is readily alkylated at the S atom to give the metallacyclobutadiene complex $[Os\{C(SMe)C(Ph)CPh\}(CO)_2(PPh_3)_2]^+$.[67] In the metallacyclopentadienyl complexes $[Os_3(CO)_8(C_2R_2)_2]$ (R=Ph, $\underline{p}$-ClC_6H_4) (8), *ortho*-metallation of one of the aryl groups has occurred.[68] Metallation at the position *ortho* to the phenolic group occurs in the reaction between salicylaldimine and $[Os_3-(CO)_{10}(MeCN)_2]$ to give $[Os_3H(CO)_{10}(\mu-OC_6H_3CH=NHCH_2Ph)]$ (9).[69] Two of the three products, (10) and (11), of the thermolysis of (8) contain Os-C σ-bonds.[69] $[Os_3(CO)_{10}(MeCN)_2]$ reacts with benzene to yield the benzyne cluster $[Os_3(CO)_9(H)_2(C_6H_4)]$ (12);[70] similar reactions have been observed with toluene and chlorobenzene.[70]

The synthesis and $\underline{X}$-ray crystal structures of two mixed-metal carbido clusters, $[Os_{10}C(CO)_{24}Cu(NCMe)]^-$ and $[Os_{10}C(CO)_{24}Au(PPh_3)]^-$ have been reported.[71] The first Os_{11} cluster species, $[Os_{11}C-(CO)_{27}]^{2-}$, has been isolated and its $[Cu(NCMe)]^+$ derivative, $[Os_{11}C(CO)_{27}Cu(NCMe)]^-$ has been characterized by $\underline{X}$-ray crystallography.[72]

3.2 The Cobalt Triad.

- The influence of steric factors on the Co-alkyl bond dissociation energies in organocobalt complexes which are models of coenzyme B_{12} have been reported.[73] Treatment of the unstable alkene complexes $[CoCl(alkene)(PMe_3)_3]$ (alkene = C_2H_4, C_3H_6, cyclo-C_5H_8) with MeLi or PhLi gives more stable compounds of the type $[CoR(alkene)(PMe_3)_3]$ (R=Me,Ph).[74] Displacement of the η^6-arene ligand in $[Co(C_6F_5)_2(\eta^6-arene)]$ by ligands L (L=CO, THF, py,tetrahydrothiophene; L_2= bipy) has been used to prepare $[Co-(C_6F_5)_2(CO)_3]$[75] and the tetrahedral compounds $[Co(C_6F_5)_2L_2]$.[76] Similar η^6-arene displacements on $[Ni(C_6F_5)_2(\eta^6-arene)]$ have yielded the square-planar complexes *trans*-$[Ni(C_6F_5)_2L_2]$ and *cis*-

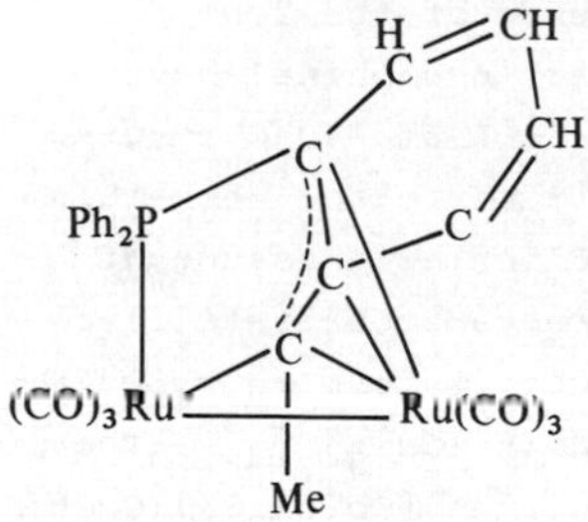
H
CH
C
CH
C
Ph_2P
C
C
C
$(CO)_3Ru$
$Ru(CO)_3$
Me

(7)

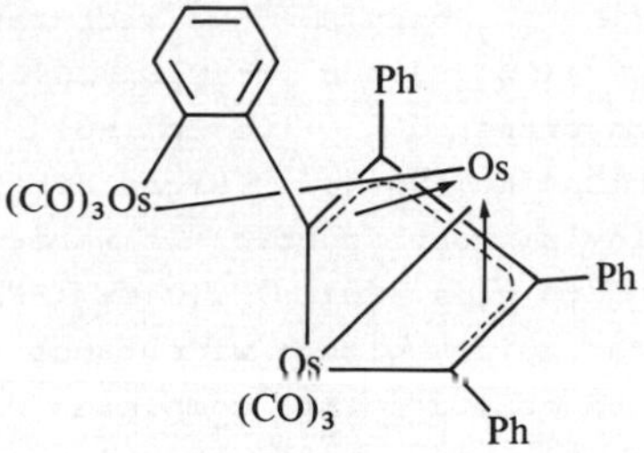
Ph
Os
$(CO)_3Os$
Ph
Os
$(CO)_3$
Ph

(8)

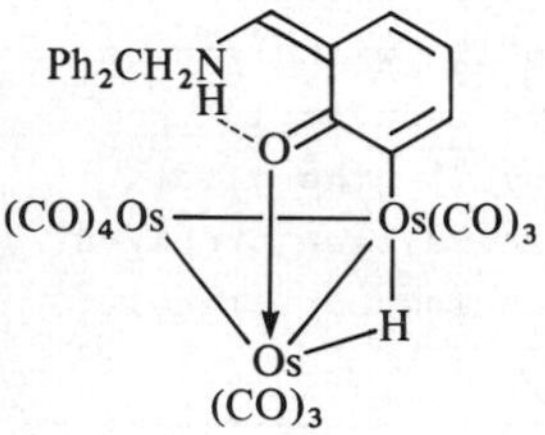
Ph_2CH_2N
H
O
$(CO)_4Os$
$Os(CO)_3$
H
Os
$(CO)_3$

(9)

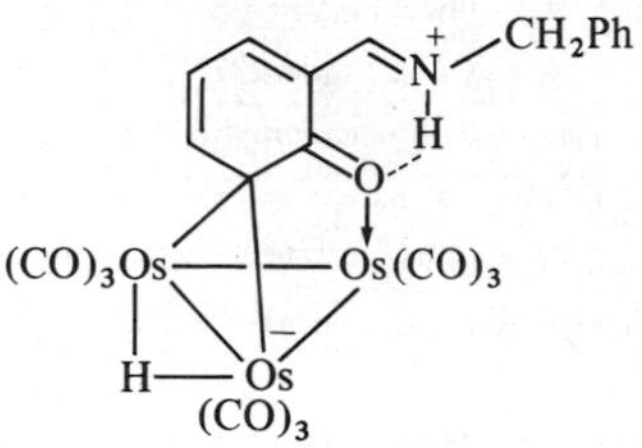
CH_2Ph
N
H
O
$(CO)_3Os$
$Os(CO)_3$
H
Os
$(CO)_3$

(10)

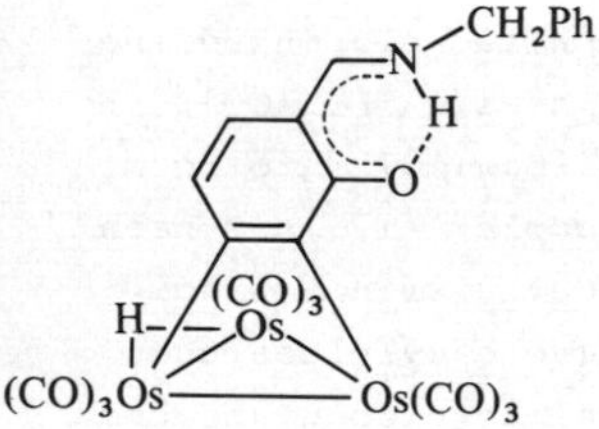
CH_2Ph
N
H
O
$(CO)_3$
H
Os
$(CO)_3Os$
$Os(CO)_3$

(11)

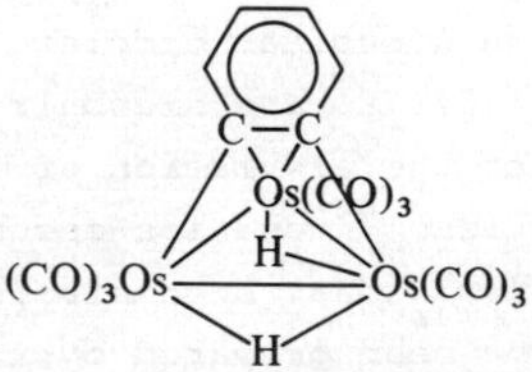
C
C
$Os(CO)_3$
H
$(CO)_3Os$
$Os(CO)_3$
H

(12)

$[Ni(C_6F_5)_2(bipy)]$.[76] A wide range of complexes of the type $[Co(C_6F_5)(CO)_3L]$ and $[Co(C_6F_5)(CO)_2L_2]$ (L=tertiary phosphine) have been prepared.[77] $[Co(Cp)(NO)]^-$ reacts with CH_3I at $-40^{\circ}C$ to give [Co(Cp)(Me)(NO)].[78] Migratory NO insertion takes place when this methylnitrosyl complex is warmed above $-40^{\circ}C$ in the presence of PPh_3 to give $[Co(Cp)\{N(O)Me\}(PPh_3)]$.[78] Tetracyanocobaltate(I) reacts rapidly with a wide range of aryl halides to form σ-arylpentacyanocobaltate(III) complexes of the type $[Co(R)(CN)_5]^{3-}$.[79] The reaction between $[Co(CH_2CN)(CO)_4]$ and NaOMe gives $[Co(CH_2CN)(COOMe)(CO)_3]^-$.[80]

The complexes $[\overline{Co(Cp)(CH_2E})(PMe_3)]$ (E=S,Se) have been prepared by treating $[Co(Cp)(CO)(PMe_3)]$ with CH_2X_2 (X=Br,I) and NaEH.[81] It is thought that the reaction proceeds via an intermediate of the type $[Co(Cp)(CH_2X)(X)(PMe_3)]$. Treatment of $[CoCl_2]$ with lithium bis(dimethylmethylenephosphoranyl) dihydroborate results in the formation of the binary metal ylide complex (13).[82] The ylide complexes $[Co(CH_2PMe_3)(CO)(NO)(L)]$ (L=CO,PMe_3) have been prepared from $[Co(CO)_2(NO)(L)]$ and $CH_2=PMe_3$ via a phosphonium cobaltacylate complex.[83]

Kinetic studies on the formation of cobaltacyclopentadienes by the reaction of alkynes with $[Co(Cp)(PPh_3)(alkyne)]$ indicate that $[Co(Cp)(alkyne)_2]$ is initially formed and this cyclises to a cobaltacyclopentadiene complex by an oxidative coupling reaction. The regioselectivity of this cyclisation is controlled by steric rather than by electronic factors.[84] The cobaltacyclopentadiene complex, $[\overline{Co(CR^1=CR^2\ CR^3=CR^4})(Cp)(PPh_3)]$ reacts with $[Fe_2(CO)_9]$ to produce the dinuclear complex (14), and with $[Co_2(CO)_8]$ to form (15) and (16).[85] The dicobaltacyclohexene complex (17) has been prepared by the alkylation of $[Na][Co_2(Cp)_2(CO)_2]$ with α,α'-dibromo-o-xylene,[86] and its decomposition to give o-xylylene and $[Co_2(Cp)_2(CO)_2]$ has been studied.[87] New cobaltacyclopentane complexes have been prepared by treating the cobalt radical anion $[Co_2(Cp)_2(CO)_2]^{\overline{\cdot}}$ with 1,3-diiodobutane and 2,4-diiodopentane.[88]

The oxidative addition of alkanes to transition metal complexes has been extended to include rhodium complexes. Thus, the irradiation of $[Rh(C_5Me_5)(H)_2(PMe_3)]$ in liquid propane at $-55^{\circ}C$ produces $[Rh(C_5Me_5)(H)(n\text{-propyl})(PMe_3)]$, which can be converted to $[Rh(C_5Me_5)(Br)(n\text{-propyl})(PMe_3)]$ by reaction with $CHBr_3$.[89] $[Rh(C_5Me_5)(H)(Me)(PMe_3)]$ undegoes reductive elimination of methane, and when carried out in C_6D_6, this reaction leads to $[Rh(C_5Me_5)(D)(C_6D_5)(PMe_3)]$.[89] The isocyanide complexes $[Rh(C_5Me_5)(CNCH_2-$

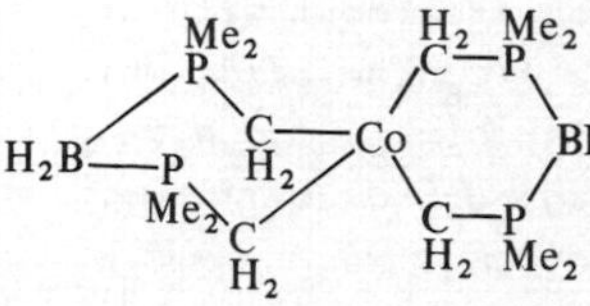

(13)

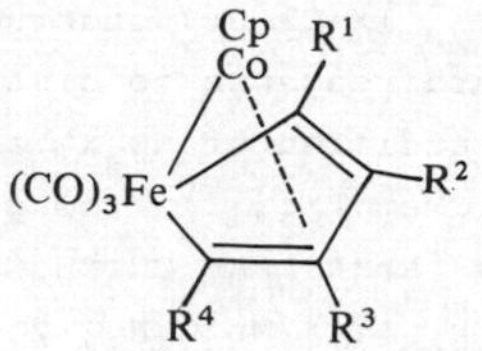

(14)

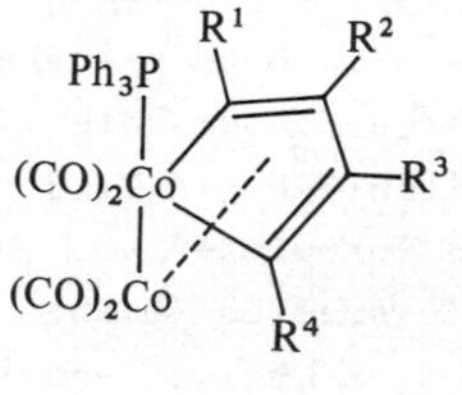

(15)

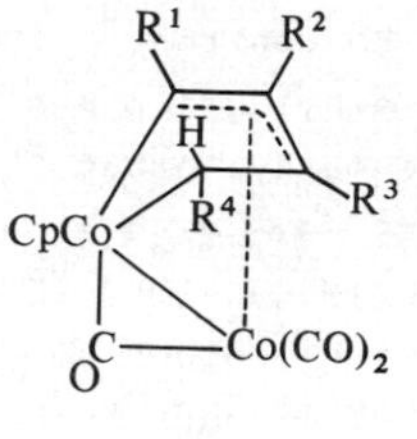

(16)

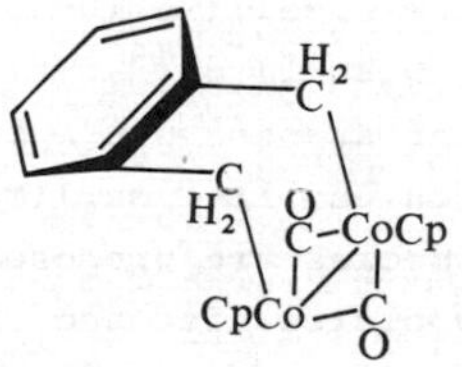

(17)

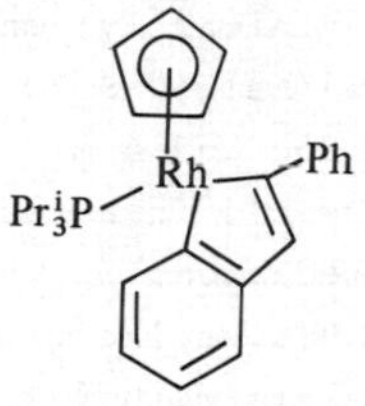

(18)

$CMe_3)X_2]$ (X=Cl,Br) undergo mono-arylation upon treatment with aryl Grignard reagents to give $[Rh(C_5Me_5)(CNCH_2CMe_3)(C_6H_4Me\text{-}\underline{p})X]$, which, after addition of NH_4X, afford the carbenes $[Rh(C_5Me_5)\{C(C_6H_4Me\text{-}\underline{p})\text{-}(NHCH_2CMe_3)\}X_2]$.[90] The preparation of the organorhodium(I) complexes $[Rh(R)\{PhP(CH_2CH_2CH_2PPh_2)_2\}]$ ($R=CH_2CMe_3, CH_2SiMe_3$, 2-MeC_6H_4-4-MeC_6H_4, 2,4-$Me_2C_6H_3$, 2,4,6-$Me_3C_6H_2$, C_2Ph) has been reported.[91] The μ-methylene complex, trans-$[\{Rh(C_5Me_5)Me\}_2(\mu\text{-}CH_2)_2]$reacts with acid in the presence of ligands L (L=MeCN, py, etc.) to give cis- and trans-$[(C_5Me_5)(L)Rh(\mu\text{-}CH_2)_2(MeRhC_5Me_5)]^+$. These cationic complexes equilibrate at temperatures above -40^oC by an intramolecular process which involves methyl group migration between the two Rh centres.[92] The ethyl derivative has also been prepared and the ethyl group has been shown to migrate at a faster rate than the methyl group.[92] $[Al_2Me_6]$ reacts with $[\{M(C_5Me_5)\}_2Cl_4]$ (M=Rh,Ir) to give a heterotrimetallic species formulated as $[\{M(C_5Me_5)Me_3\}_2AlMe]$.[93] This heterotrimetallic species (M=Rh) reacts with acetone to give trans-$[\{Rh(C_5Me_5)Me\}_2(\mu\text{-}CH_2)_2]$, and with other ligands L (L=DMSO, PPh_3, dppm; M=Rh,Ir) to give $[M(C_5Me_5)(Me)_2L]$.[93] The DMSO adducts, $[M(C_5Me_5)(Me)_2(DMSO)]$, react with p-toluenesulphonic acid in acetonitrile to give $[M(C_5Me_5)\text{-}(Me)(DMSO)(MeCN)]^+$, and with CF_3COOH to give $[M(C_5Me_5)(Me)(DMSO)\text{-}(O_2CCF_3)]$ initially, and then $[M(C_5Me_5)(DMSO)(O_2CCF_3)_2]$.[94] $[Ir(C_5Me_5)(Me)_2(DMSO)]$ reacts with a variety of arenes, C_6H_5X, to give methane and a mixture of m- and p-substituted $[Ir(C_5Me_5)(Me)\text{-}(C_6H_4X)(DMSO)]$ in a reaction in which Ir(V) species are proposed as intermediates.[95] The structures of the complexes cis- and trans-$[\{Rh(C_5Me_5)Me\}_2(\mu\text{-}CH_2)_2]$ have been determined.[96] The trans-isomer reacts with $[Al_2Et_6]$ to give the ethyl derivative trans-$[\{Rh(C_5Me_5)Et\}_2(\mu\text{-}CH_2)_2]$.[96] The direct reaction of $[\{M(C_5Me_5)\}_2\text{-}Cl_4]$ (M=Rh,Ir) with $[Al_2Et_6]$ gives $[M(C_5Me_5)(C_2H_4)_2]$ and $[M\text{-}(C_5Me_5)(Cl)(\eta^3\text{-}CH_2CHCHMe)]$, but $[Al_2Et_6]$ reacts with $[Rh(C_5Me_5)\text{-}(Cl)_2(PMe)_3]$ to yield $[Rh(C_5Me_5)(Et)_2(PMe_3)]$.[97]

Oxidative addition of CH_3I to the polar Rh-Mo bond in $[(C_5Me_5)Rh(\mu\text{-}PMe_2)_2Mo(CO)_4]$ occurs regiospecifically to yield $[(C_5Me_5)(Me)Rh(\mu\text{-}PMe_2)_2Mo(CO)_3I]$.[98] Protonation with CF_3COOH of $[Rh(Cp)(C_2Ph_2)(PPr^i{}_3)]$ yields the vinyl complex $[Rh(Cp)\{C(Ph)=CH\text{-}(Ph)\}(O_2CCF_3)(PPr^i{}_3)]$ which reacts with NH_4PF_6 or CF_3COOH in MeOH to form the metallaindene complex (18).[99] Cyclopropyl complexes of the type $[MCl_2(COC_3H_5)(PMePh_2)_2]$ (M=Rh,Ir), $[IrCl_2(C_3H_5)(CO)\text{-}(PMePh_2)_2]$, trans-$[Ir(C_3H_5)(CO)(PMePh_2)_2]$, and $[IrCl(C_3H_5)(CO)\text{-}(PMePh_2)_3][PF_6]$ have been prepared.[100] Cyclometallation of a

3-chloropropylphosphine derivative occurs on treatment of [RhCl-(COD){P(Cy)$_2$CH$_2$CH$_2$CH$_2$Cl}] with P(Cy)$_2$CH$_2$CH$_2$CH$_2$P(Ph)H, to give (19).[101] The direct metallation of arenes occurs on addition of a C_6H_5X (X=H,Me,Cl,OMe) solution of octaethylporphyrinatorhodium(III) chloride to $AgClO_4$ or $AgBF_4$, leading to the formation of the arylrhodium(III) derivative.[102]

Several new heterometallic adducts of $[Rh_6(CO)_{15}C]^{2-}$ have been prepared and characterized.[103] The carbido-cluster anions, $[Rh_{12}C_2(CO)_{24}]^{2-}$,[104] and $[Rh_{12}C_2(CO)_{23}]^{\underline{n}-}$ ($\underline{n}$=3,4)[105] have been reported.

Full details of investigations into the activation of C-H bonds in saturated hydrocarbons upon photolysis of solutions of [Ir(C$_5$Me$_5$)(H)$_2$(PMe$_3$)] have been published.[106] The use of [Ir-(C$_5$Me$_5$)(CO)$_2$] as a precursor for the activation of alkanes has been extended to the activation of methane.[107] Thus, a solution of [Ir(C$_5$Me$_5$)(CO)$_2$] in perfluorohexane under CH_4 gives a 20-25% yield of [Ir(C$_5$Me$_5$)(CO)(H)(Me)] upon irradiation. Activation of methane also occurs when [Ir(C$_5$Me$_5$)(H)(C$_6$H$_{11}$)(PMe$_3$)] is heated to 140°C under CH_4 pressure, giving [Ir(C$_5$Me$_5$)(PMe$_3$)(H)(Me)].[108]

Treatment of [{Ir(C$_5$Me$_5$)}$_2$Cl$_4$] with [Al$_2$Me$_6$] in benzene or toluene gives [{Ir(C$_5$Me$_5$)}$_2$(Me)(Ar)(μ-CH$_2$)$_2$] (Ar=Ph or m- and p-tolyl) as the main product.[109] If CO is introduced into the benzene solution, the products are [Ir(C$_5$Me$_5$)(CO)(R^{1}R^2] (R^1=Me, R^2=Ph; R^1=R^2=Me or Ph).[109] Several papers dealing with various aspects of phosphite-, phosphonite-, or phosphinite-induced CO insertion and C-H addition reactions of iridium-aryl[110-112] and -alkyl[113] complexes have been published. Non-isolable organo-iridium(I) derivatives of the type [Ir(R)(PPh$_3$)$_3$] (R=Ph, 2-MeC$_6$H$_4$, Me, CH$_2$CMe$_3$, CH$_2$SiMe$_3$) are produced when [IrCl(PPh$_3$)$_3$] is treated with LiR.[114] These organoiridium(I) complexes readily lose a PPh$_3$ ligand and undergo ortho-metallation leading to metallacyclic Ir(III) complexes.[114] The multidentate phosphine derivatives [Ir(R){PhP(CH$_2$CH$_2$CH$_2$PPh$_2$)$_2$}] (R=CH$_2$CMe$_3$, CH$_2$SiMe$_3$, 2-MeC$_6$H$_4$, 4-MeC$_6$H$_4$, 2,4,6-Me$_3$C$_6$H$_2$) are more stable, and the X-ray crystal structure of the derivative with R=2-MeC$_6$H$_4$ has been determined.[115] The reaction of trans-[IrCl(N$_2$)(PPh$_3$)$_2$] with dibenzoyldiazomethane yields [IrCl{HC(COPh)$_2$}{P(C$_6$H$_4$)Ph$_2$}(PPh$_3$)], while a similar reaction with [IrCl(PPh$_3$)$_3$] yields [IrCl{C$_6$H$_4$C(O)CC(Ph)OPPh$_2$}(PPh$_3$)$_2$].[116] The first iridium(I) carbon-bonded diketonate complex [Ir(acac-C^3)(cod)(phen)], has been prepared and characterized by

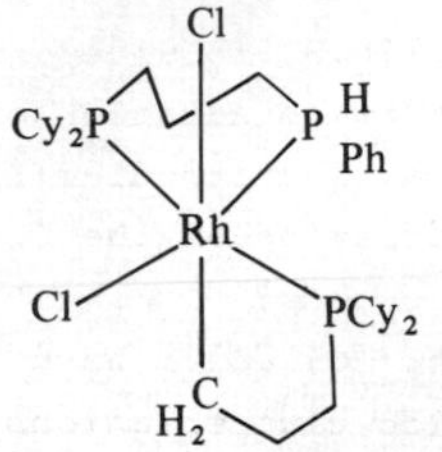
Cl
Cy2P
H
P
Ph
Rh
Cl
PCy2
C
H2

(19)

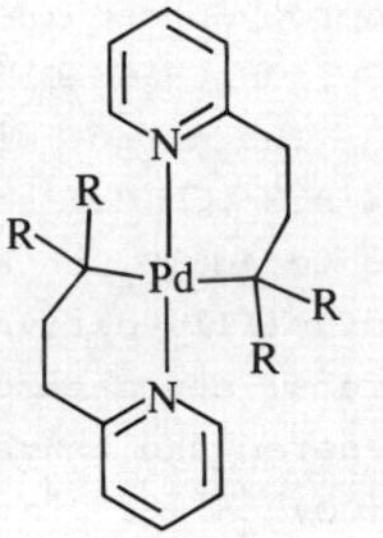
N
R
R
Pd
R
R
N

(20)

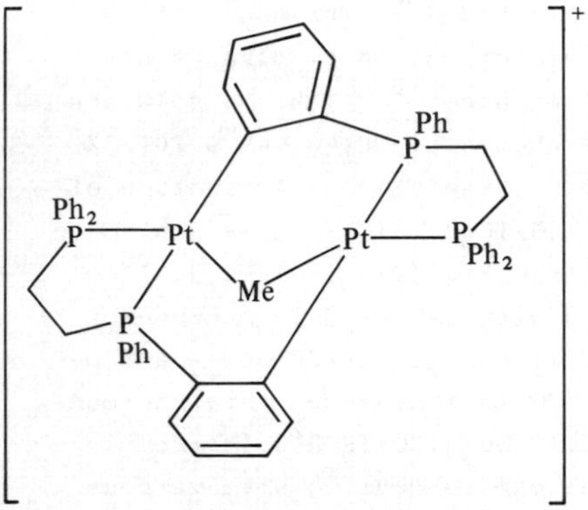
+
Ph
P
Ph2
P
Pt
Pt
P
Ph2
Me
P
Ph

(21)

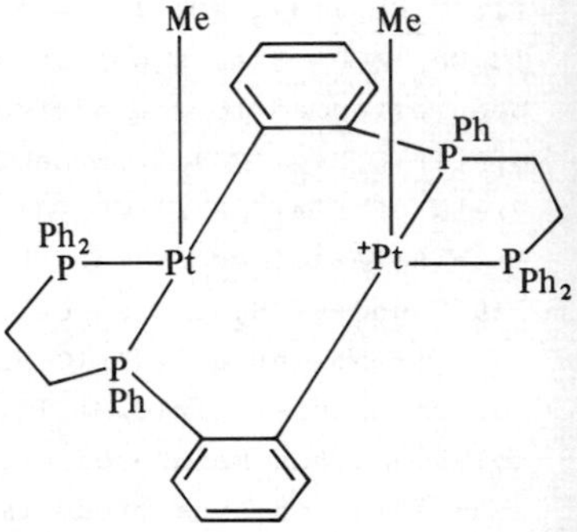
Me
Me
Ph
P
Ph2
P
Pt
+Pt
P
Ph2
P
Ph

(22)

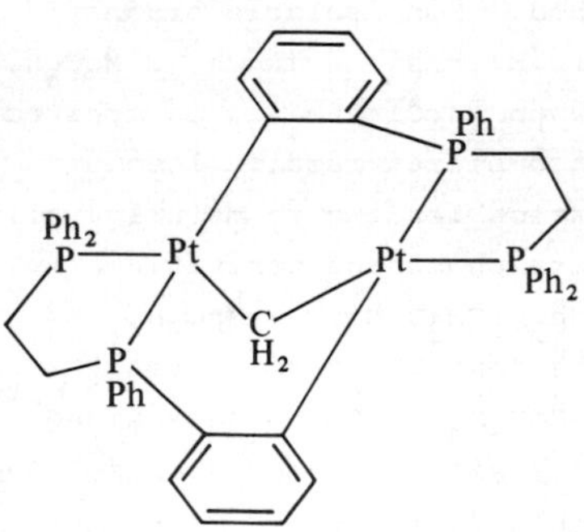
Ph
P
Ph2
P
Pt
Pt
P
Ph2
C
H2
P
Ph

(23)

X-ray crystallography.[117] N.m.r. evidence for a carbon-bonded structure in the complex $[Ir(Hbpy\text{-}C^3,N^1)(bpy\text{-}N,N^1)_2]^{3+}$ has been reported,[118] and the crystal structure of the perchlorate derivative has been determined.[119] The interaction of di-tert-butyl-(cyclopropylmethyl)phosphine with various GroupVIII metal complexes has been described.[120] For example, with $[IrCl(C_8H_{14})_2]_2$, the phosphine reacts to give two diastereomers of the complex $[IrCl(H)\{PBu^t_2CH_2CHCH_2CH\}(PBu^t_2CH_2CHCH_2CH_2)]$.[120] The reaction between $[IrCl(CO)_2(MeC_6H_4NH_2\text{-}p)]$ and $CH(PPh_2)_3$ in the presence of CO and Zn yields the tri-iridium cluster $[Ir_3(CO)_6(Ph)(\mu_3\text{-}PPh)(\mu\text{-}dppm)]$.[121]

3.3 The Nickel Triad. - Mixtures of syn and anti isomers of the complexes trans-$[NiRR'L_2]$ {$R=R'=C_6H_4Me$-2, C_6H_2Br-3-$(OMe)_2$-2,6; R=C-(Cl)CCl_2, $R'=C_6H_4Me$-2, C_6H_2Br-3-$(OMe)_2$-2,6; $L=PMe_2Ph, PMe_3, PMePh_2$} have been obtained and shown to undergo anti- syn isomerization.[122] Steric reasons for the non-reactivity of trans-$[Ni\{C_6H(OMe_2)Br_2\}_2(PMe_3)_2]$ towards CO in contrast to the reactivity of trans-$[Ni\{C_6H_3(OMe)_2\}(PMe_3)_2]$ are clearly indicated by the crystal structures of these two organonickel derivatives.[123] A series of complexes of the type cis-$[Ni(C_6H_4X)(Me)(dmpe)]$ (X=p-OMe, p-Me, H, p-F, o-Me) has been prepared.[124] The addition of tertiary phosphines to these complexes induces the reductive elimination of MeC_6H_4X.[124] A detailed study of the reactions of $[Ni(Me)_2(dipy)]$, $[Ni(Et)_2(dipy)]$, trans-$[Ni(Me)_2(PEt_3)_2]$, $[Ni(Me)_2(dppe)]$, and $[Ni(Me)_2(dppp)]$ with alkyl or aryl halides, RCOX (X=Cl, Br, OPh, OCOPh), and CS_2 has been reported.[125] Treatment of nickelocene with RLi in the presence of alkene yields $[Ni(Cp)(R)(\eta^2\text{-alkene})]$ (R=Et, Bu^n, Bu^i, cyclopropyl; alkene=C_2H_4, C_3H_6).[126] A similar reaction of nickelocene with MeLi and an alkadiene leads to the formation of $[Ni(Cp)(Me)(\eta^2\text{-alkadiene})]$.[127] The complexes $[Ni(Cp)(R)(C_2H_4)]$ (R=Ph, cyclopropyl) react with excess ethylene to give $[Ni(Cp)(CH_2CH_2R)(C_2H_4)]$.[128]

The oxidative addition of $[Hg(Ar)X]$ to $[Ni(PPh_3)_4]$ gives $[Ni(Ar)X(PPh_3)_2]$ (Ar=Ph, X=Cl; Ar=C_6F_5, X=Br), whereas $[Hg(C_6F_5)_2]$ gives $[(Ph_3P)_2Ni(C_6F_5)Hg(C_6F_5)]$.[129] New (polychlorophenyl)nickel(II) complexes, $[NiBr(R)(PPh_3)_2]$ ($R=C_6H_nCl_{5-n}$) have been prepared by the oxidative addition of $C_6H_nBrCl_{5-n}$ to $[Ni(PPh_3)_4]$.[130] New Ni- or Pt- containing cyclic esters such as $[Ni(CH_2CH_2COO)(dppe)]$ and $[Ni(CH_2CH_2CH_2COO)(dipy)]$ have been prepared by the oxidative addition of the appropriate cyclic carboxylic anhydride to Ni(O) or Pt(O) complexes.[131] CO_2 and/or C_2H_4 insertion into the Ni-Ph bond in [Ni(Ph)(PPh_3)(chelate)] {chelate=Ph_2PC(H)C(Ph)O, Ph_2PNC(Ph)O, Ph_2AsC(H)C(Ph)O, OC(Me)C(H)C(Me)-O} occurs to give, after esterification with methanol, hydrocinnamic acid methyl ester, methyl benzoate, styrene , ethylbenzene and butylbenzene.[132]

Nickelacylic complexes, [Ni($CH_2C_6H_4NMe_2$)(PR_3)Cl], are formed on treatment of [$NiCl_2(PR_3)_2$] with o-N,N-dimethylaminobenzyllithium.[133] Insertion of hexafluorobut-2-yne into the Ni-C bonds of these complexes gives [Ni{C(CF_3)=C(CF_3)$CH_2C_6H_4NMe_2$}(PMe_3)Cl] when R=Me, and [Ni{C(CF_3)=C(CF_3)$CH_2C_6H_4NMe_2$}(PEt_3)Cl$]_2$ when R=Et.[133] Depending upon the reaction conditions, 5-, 6-, or7-membered nickelacycloalkanes are obtained from methylenecyclopropane and [Ni(cod)(dipy)].[134] The first structural evidence for the formation of chelating sulphur ylides is provided by the crystal structure of [Ni{$(CH_2)_2$S(O)Me}{MeC(S)C(H)C(S)Me}].[135] Stable, paramagnetic nickel(III) complexes, [Ni{$C_6H_3(CH_2NMe_2)_2$-o,o'}X_2], have been prepared by treating [Ni{$C_6H_3(CH_2NMe_2)_2$-o,o'}X] (X=Cl,Br,I) with $CuCl_2$, $CuBr_2$, or I_2.[136] A spin trapping method, using nitroso-durene, has been used to determine the relative reactivity of Ni-C bonds in complexes containing either σ-or π-bonded groups.[137]

A number of studies of pentahalophenyl-derivatives of palladium have been reported.[138-142] These include the preparation of trans-[Pd$(C_6F_5)_2$(dioxane$)_2$],[138] and its reaction with wet thf to give trans-[Pd$(C_6F_5)_2$(γ-butyrolactone$)_{1.5}$];[139] the preparation of the anionic complexes $[MR_4]^{2-}$, $[M_2(\mu\text{-}X)_2R_4]^{2-}$, cis-$[MClR_2L]^{2-}$ and trans-$[PtCl_2R_2]^{2-}$ (M=Pd or Pt; R=C_6Cl_5; X=Cl or Br; L=PPh_3, PEt_3, py, or $SbPh_3$);[140] the preparation of [LPd(μ-dppm$)_2$Pd(C_6F_5)][BPh_4] { L=PPh_3, P(OPh$)_3$, py, tetrahydrothiophene} and its reactions with isocyanides;[141] and the preparation of [PdR_2L_2] (R=C_6F_5, p-HC_6F_4, m-HC_6F_4; L_2= trans-(py$)_2$, dipy, or phen), cis-[Pd$(C_6F_5)_2$(py$)_2$], and cis- and trans-[Pt$(C_6F_5)_2(PPh_3)_2$].[142]

A review of the mechanisms of nucleophilic attack on Pd(II)-coordinated isocyanides, leading to metal-carbene derivatives and of electrophilic attack on alkyl- and aryl-Pt(II) complexes, has been published.[143] Trans- and cis-[PdR_2L_2] (R=Me,Et,Ph) react with R'Li (R'=Me,Ph) to give alkylpalladate complexes Li[PdR_2R'L] and Li_2[$PdR_2R'_2$] with retention of stereochemistry.[144] Complexes of the type [Pd_2(μ-Cl$)_2R_2L_2$] (R=Ph, C_6H_4Me-p, or Bz; L=PPh_3, $PMePh_2$, PBu_3) are formed by treating [Pd_2(μ-Cl$)_2Cl_2L_2$] with [R_2Hg].[145]

These organopalladium complexes react with L to give trans-[PdCl-(R)L_2], and with CO to give [Pd_2(μ-Cl)$_2$(COR)$_2$$L_2$].[145] The fluxional complexes [Pd(η^5-C_5H_5)(η^1-C_5H_5)(PR_3)] have been prepared from [{PdCl(PR_3)}$_2$(OCOMe)$_2$] and [TlC_5H_5], and the structure of the PPr^i_3 derivative has been determined by X-ray crystallography.[146] Tertiary amines react with [$PdCl_2$(PhCN)$_2$] to give η^1-ylidic complexes of the type [Pd(μ-Cl)Cl{CH(R)C(R')=NR"$_2$}].[147] The allenylpalladium (II) complex, [Pd{CH=C=CMe_2}Cl(PPh_3)$_2$], has been isolated from the reaction of [Pd(PPh_3)$_4$] and prop-2-ynylic chloride or prop-2-ynylic acetate.[148] The oxidative addition of R^1SCH_2Cl (R^1=Me or Ph) to [Pd(PPh_2R)$_4$] (R=Ph or Me) gives the alkylthiomethyl derivatives, [Pd(CH_2SR^1)Cl(PPh_2R)$_2$].[149] The oxidative addition of ortho-bromobenzylamines or ortho-halobenzalimines to [Pd(dba)$_2$] (dba = dibenzylideneacetone) provides a new method for the synthesis of cyclopalladated complexes of primary, secondary, and tertiary amines, and of benzalimines.[150]

As in previous years, numerous cyclopalladation reactions have been reported. Among ligands to undergo such cyclopalladation reactions are hydrazone derivatives of 1-acetylcyclohexene,[151] benzalazines,[152] 2,6-diarylpyridines,[153] the dimethylhydrazone of pinacolone,[154] 2-neopentylpyridine,[155] a benzylgylcinate,[156] benzylamines,[157] 1-methoxynaphthalene,[158] PBu^t_2 ($CH_2\overline{CHCH_2CH_2}$),[159] PPh_2(CH_2Ph),[160] and $PBu^t_2Bu^i$.[161]

Asymmetric cyclopalladation as a tool of enantioselective synthesis has been reviewed.[162] Reactions of cyclopalladated complexes include deprotonation followed by electrophilic attack on the resulting carbanion of complexes of the type [$\overline{M(C_6H_4CH_2PR_2)}_2$] (M=Pd,Pt) to give [$\overline{M(C_6H_4CH(R^1)PR_2)}_2$] {$R^1$=Me, $SiMe_3$, PPh_2, Au-(PPh_3)},[163] insertion of hexafluorobut-2-yne into the M-C bonds in [M(dmba-H)$_2$] (M=Ni, Pd; dmba = N,N-dimethylbenzylamine),[164] and electrophilic attack by Br_2 HgX_2 (X=Br, O_2CMe,) or [$PdCl_2$(SEt_2)$_2$] on complexes of the type [$\overline{M(C_6H_4CH_2PR_2)}_2$] (M=Pd, Pt).[165] [$PdCl_2$] reacts with 2-{1,1-bis(methoxycarbonyl)propyl} pyridine to give the stable organopalladium complex (20).[166]

The crystal structure of [PtH(Ph)(PPr^i_3)$_2$] has been reported.[167] The irradiation of $MeCO_2H$ or CF_3CO_2H/H_2O solutions of [$PtCl_6$]$^{2-}$ and n-hexane or n-heptane gives rise to the formation of the corresponding π-alkene complex [$PtCl_2${CH_2=CH(CH_2)$_n$$CH_3$}]$_2$, presumably via a σ-alkyl complex of Pt(IV).[168] Similarly, σ-acetonyl- and σ-aryl-platinum(IV) complexes are formed by irradiation of [$PtCl_6$]$^{2-}$ solutions with acetone and arenes respectively.[168] The reaction of nucleophiles such as N_3^- and NO_2^- with $[Pt(\eta^2-C_2H_4)(Cl)(tmen)]^+$ yields addition products such as $[PtCl(CH_2CH_2N_3)(tmen)]$ and $[Pt(NO_2)(CH_2CH_2NO_2)(tmen)]$, respectively.[169]

$[Pt(\eta-C_2H_4)(\sigma-C_5H_5)(Cp)]$ is one of the products obtained on treatment of $[Pt_2(\mu-Cl)_2Cl_2(C_2H_4)_2]$ with $[MgCp_2]$.[170] Replacement of Cl^- in cis-$[PtCl_2(CO)(PR_3)]$ by $[C_5H_5]^-$ results exclusively in isomers of $[PtCl(\sigma-C_5H_5)(CO)(PR_3)]$ with C_5H_5 trans to PR_3.[171] $[Pt(\sigma-C_5H_5)_2(CO)(PR_3)]$ may be prepared by using excess $[TlC_5H_5]$. The σ-cyclopentadienyl groups readily exchange in solution with a chloride of cis-$[PtCl_2(CO)(PR_3)]$, again producing exclusively the same isomers of $[PtCl(\sigma-C_5H_5)(CO)(PR_3)]$.[171] The diorgano-platinum complexes $[Pt(\sigma-C_5H_5)(C{\equiv}CR)(CO)L]$ and $[Pt(\sigma-C_5H_5)(Ph)(CO)L]$ react with $[HgCl_2]$ or cis-$[PtCl_2(CO)L]$ to transfer specifically the organic group trans to L (i.e. C_5H_5 in both cases) to Hg or Pt, respectively.[172] Kinetic studies of the methyl for chloride exchange reaction that occurs on treatment of cis-$[Pt(Me)_2(SMe_2)_2]$ with trans-$[PtCl_2(SMe_2)_2]$ suggest that dissociation of SMe_2 to give $[Pt(Me)_2(SMe_2)]$ takes place before the methyl for chloride exchange.[173]

Strong evidence for a free-radical mechanism of oxidation addition in the reaction of Pr^iI with $[PtMe_2(phen)]$ is provided by the isolation of the alkylperoxo complex $[PtMe_2(OOPr^i)(I)(phen)]$, together with $[PtMe_2(Pr^i)(I)(phen)]$ and $[PtMe_2(I)_2(phen)]$, in the product mixture.[174] The alkyl radicals formed in this reaction may be trapped by carrying out the oxidative addition reaction in the presence of alkenes. Thus, $[PtMe_2(phen)]$ reacts with a mixture of RI ($R=Pr^i$ or Bu^t) and $CH_2{=}CHX$ {X=CN, CHO, C(O)Me} to give $[Pt(I)Me_2(CHXCH_2R)(phen)]$.[175]

A theoretical study of CO insertion into the Pt-Me bond in $[Pt(Me)F(CO)(PH_3)]$ has been reported.[176] The 1-cyclohexenyl complex $[Pt(CO_2Me)(C_6H_9)(dppp)]$ {dppp = 1,3-bis(diphenylphosphino)-propane} undergoes reversible CO insertion into the $Pt-C_6H_9$ bond under ambient conditions whereas, under these conditions, the analogous 1-cyclohexenyl complexes containing dppe and cis-1,2-bis(diphenylphosphino)ethylene are unreactive.[177]

The oxidative addition of CH_2Cl_2 to reactive $[PtL_2]$ species formed by photolysis of $[PtL_2(C_2H_4)]$ {L = PPh_3, $P(p\text{-tol})_3$} or $[PtL_2(CH_2)_4]$ (L = PPh_3, PBu^n_3), leads to the formation of trans-$[PtL_2(CH_2Cl)Cl]$, which isomerizes to the cis-complex before bimolecular C_2H_4 elimination occurs.[178] Reductive elimination of disubstituted biphenyls from a series of substituted phenyl complexes of the type cis- [$Pt(PPh_3)_2(phenyl)_2$] has been studied.[179,180] Also, dπ-pπ interactions between Pt and the sp^2-C atom in fluorine-substituted diphenylplatinum complexes have been discussed on the basis of ^{19}F n.m.r. chemical shift data.[181] A kinetic study of the reductive elimination of CH_3CF_3 from cis-[$PtH(CH_2CF_3)(PPh_3)_2$] has been reported.[182] The complexes [$Pt(CH_2CMe_2Ph)_2L_2$] (L_2 = cod, dipy, 2,2'-bipyrimidyl; L = PEt_3, PPh_3) undergo cyclisation reactions to afford [$Pt(2-C_6H_4CMe_2CH_2)L_2$] with the elimination of $C_6H_5CMe_3$ at rates which depend upon L_2.[183] The decomposition of complexes of the type trans-[$Pt(X)(CR_3)(PR'_3)_2$] (R=H,D; R'=Me, CD_3, Et, Ph, Cy ; X=Cl, Br, I, CN) has been studied kinetically[184] and by isotopic labelling.[185] The oxidative addition of $ClCH_2SR$ to [$Pt(PPh_3)_4$] gives trans-[$Pt(PPh_3)_2(CH_2SR)Cl$] (R=Me,Ph, C_6H_4Me-p), which may be reversibly protonated by CF_3COOH to give [$Pt(PPh_3)_2\{CH_2SH(R)\}Cl]^+$.[186] Organoplatinum(IV) compounds of the type [$PtMe_2X_2(\widehat{N\ N})$] (X=I, NO_3; $\widehat{N\ N}$=chelating N-donor ligand) have been prepared.[187]

Functionally substituted platinacyclobutane complexes [$\overline{PtCl_2-(CH_2CR^1CR^2CH_2})(py)_2$] ($R^1$=H, R^2=CH_2OH, CHMeOH, CMe_2OH; R^1=Me, R^2= CH_2OH) have been prepared, and their ring conformations studied using n.m.r. spectroscopy and X-ray crystallography.[188] A detailed examination of the mechanism of the highly stereospecific skeletal rearrangement of platinacyclobutanes, in which the isolobal analogy suggests that [$PtCl_2L(C_3H_6)$] is like the non-classical carbonium ion $C_4H_7^+$ ([$PtCl_2L$]is isolobal to CH^+), leads to a new proposal for the mechanism of the metathesis reaction.[189]

The reaction between [$PtCl_2$(dppe)] and alkynyl stannanes [$Me_3SnC{\equiv}CR$] or [$Me_2Sn(C{\equiv}CR)_2$] gives cis-[$Pt(C{\equiv}CR)_2$(dppe)] (R=H, Me, Ph, $SiMe_3$) quantitatively.[190] Treatment of the acetylide cis-[$Pt(C{\equiv}CH)_2$(dppe)] with R_3B (R=Me, Et, Pr^i) leads to the formation of platinacyclopentadiene complexes of the type [$\overline{Pt\{CH=C(R)C(BR_2)=CH}\}$(dppe)].[191] Benzoplatinacyclopentene complexes [$\overline{Pt(CH_2C_6H_4CH_2})(PR_3)_2$] are formed by δ-hydrogen abstraction from cis-[$Pt(CH_2C_6H_4-Me)_2(PR_3)_2$] (R=Et or Ph).[192]

New methods for the preparation of complexes of general formula [$PtXMe_3\{MeE(CH_2)_nE'Me\}$] (X =Cl, Br, or I ; n=2 or 3; E,E'=S or Se) have been reported, and multinuclear n.m.r. data for these complexes have been obtained and analysed.[193] The complexes [$PtXMe_3(MeSCH_2SCH_2SMe)$] have also been studied by dynamic n.m.r. methods, and the energy barriers associated with pyramidal inversion of the S atoms, Pt-Me scrambling, and chelate ligand rotation have been measured.[194]

Kinetic studies of the oxidative addition of $I(CH_2)_nI$ ($\underline{n}$= 0-5)
to [PtMe$_2$(phen)] show that the platinum atom in the [Pt{$\overline{(CH_2)_n}$I}]
unit activates the C-I bond to further oxidative addition to give
polymethylene-bridged binuclear complexes.[195] A general synthetic
route to complexes of the type [(PR$_3$)$_2$Pt(μ-H)$_2$PtY(PR$'_3$)$_2$]$^+$ (Y=H,Ph)
has been developed.[196] The dinuclear platinum complex [(PPh$_3$)(Ph)-
Pt(μ-H)(μ-PPh$_2$)Pt(PPh$_3$)$_2$]$^+$ has been isolated from the reaction of
[Pt(cod)$_2$], $\underline{trans}$-[Pt(Ph)(acetone)(PPh$_3$)$_2$]$^+$, PPh$_3$, and H$_2$;[197] also,
in a similar reaction, [WH$_2$(Cp)$_2$] reacts with [Pt(Ph)(acetone)(P-
Et$_3$)$_2$]$^+$ to give the bimetallic species [(Cp)$_2$(H)W(μ-H)Pt(Ph)(P -
Et$_3$)$_2$]$^+$ and [(Cp)$_2$W(μ-H)$_2$Pt(Ph)(PEt$_3$)]$^+$.[198] Full details of the
synthesis of $\underline{trans}$-[Pt(C≡CR)$_2$(η^1-dppm)$_2$] and its use in the synthesis of bimetallic complexes have appeared.[199] Studies of other
dppm complexes of platinum include the preparation and crystal
structure of [Pt$_2$(PPh$_3$)$_2$(μ-Ph$_2$PCHPPh$_2$)(μ-dppm)][PF$_6$],[200] the photolysis of [Pt$_2$Me$_3$(μ-dppm)$_2$][PF$_6$] leading to the reductive elimination of ethane,[201] the preparation of [Pt$_2$H(μ-H)Me(μ-dppm)$_2$]
[SbF$_6$] and its reaction with CF$_3$C≡CCF$_3$ in MeCN leading to [Pt$_2$H-
(MeCN)(μ-CF$_3$C≡CCF$_3$)(μ-dppm)$_2$][SbF$_6$] and methane,[202] and the synthesis of the highly reactive complex [Pt$_2$Me$_2$(μ-dppm)$_2$][BF$_4$]$_2$.[203]
A comparison of the chemistry of binuclear platinum complexes containing Me$_2$PCH$_2$PMe$_2$ (dmpm) with those of dppm has also been reported.[204,205]

The oxidative addition of MeI to the cyclometallated dppe
complex [Pt-μ-{$\underline{o}$-C$_6$H$_4$P(Ph)CH$_2$CH$_2$PPh$_2$}]$_2$ gives a cationic complex
which is either the μ-Me complex (21) or a pair of rapidly equil-
ibrating structures containing a terminal Pt-Me bond.[206] (21)
reacts with MeLi to give (22),[206] and with sodium methoxide to give
(23).[207] (23) reacts with [Me$_3$O][PF$_6$] to give the μ-ethyl complex,
[Pt$_2${μ-C$_6$H$_4$P(Ph)CH$_2$CH$_2$PPh$_2$}$_2$(μ-Et)].[207]

4   Carbene and Carbyne Complexes of the
Group VIII Metals

4.1   The Iron Triad. - A general method for the preparation of
compounds of general formula [(CO)$_4$Fe=C(OR)R^1] (R=Et, Li; R^1= Ph,
Me, Bu^n, Bu^t) has been described.[208] The electronic structure and
reactivity of [(CO)$_4$Fe=CH(OH)] has been studied by the $\underline{ab}$ $\underline{initio}$
SCF-MO method.[209] Carbon-carbon coupling reactions take place
when [(CO)$_4$Fe=CR(OEt)] (R=Ph, Bu^n, Bu^t) react with functionalized

alkenes $CH_2 = CHX$ (X = Ph, OEt, CO_2Me, SPh, C_4H_9) with high regioselectivity to give $XCH_2CH{=}CR(OEt)$ and XCH=CHCH(R)(OEt) only.[210] Treatment of [Fe{CH(OMe)R}(Cp)(CO)L] (R=Me, Et, Pr^i; L=CO, PPh_3) with trimethylsilyl triflate generates the corresponding cationic carbene complexes $[Fe(=CHR)(Cp)(CO)L]^+$.[211] The addition of RLi (R=Me, Bu^n, Ph, p-tolyl) to $[(Cp)(CO)_2Fe\{=CH(OMe)\}]^+$ affords the α-methoxyalkyl iron complexes $[(Cp)(CO)_2Fe\{CH(R)OMe\}]$.[212] $[(Cp)(CO)_2Fe{=}CH_2]^+$ reacts with CO to give ketene complex $[(Cp)(CO)_2Fe(CH_2{=}C{=}O)]^+$ which reacts with water or methanol to give $[(Cp)(CO)_2Fe(CH_2CO_2R)]$ (R=H, Me) respectively.[213] Enantioselective cyclopropane synthesis is achieved by ethylidene transfer from the chiral carbene complexes, $(S_{Fe}S_C)$- and $(R_{Fe}S_C)$- $[(Cp)(CO)(PPh_2R^*)Fe{=}CHMe]^+$ (R^*=(S)-2-methylbutyl) to styrene, giving cis- and trans-1-methyl-2-phenylcyclopropanes with high enantiomeric excesses.[214] The carbene complexes $[(Cp)(CO)(L)Fe\{=C(OMe)R\}][SO_3F]$ (R=H, Me; L=PMe_3, PPh_3) are deprotonated by $Me_3P{=}CH_2$ to give η^1-vinyl complexes of the type $[(Cp)(CO)(L)Fe\{C(OMe){=}C(R)R^1\}]$.[215] Treatment of the η^1-vinyl complex $[(Cp)(CO)(L)Fe\{C(OMe){=}CH_2\}]$ with MeX (X=I, OSO_2F) affords a mixture of the carbene complexes [(Cp)(CO)(L)Fe{=C(OMe)-R}][X] (R=Me, Et, Pr^i).[216] The thiocarbene complexes $[(Cp)(CO)(L)Fe\{=CH(SMe)\}]^+$ {L=CO, PPh_3, $P(OPh)_3$} react with nucleophiles L^1 {L^1=PR_3, $P(OR)_3$,[217] py[218]} to form the carbene adducts $[(Cp)(CO)(L)Fe\{CH(SMe)L^1\}]^+$. with NH_3, primary, and secondary amines, $[(Cp)(CO)_2Fe\{=CH(SMe)\}]^+$ react to form the aminocarbene complexes $[(Cp)(CO)_2Fe\{=CH(NR_2)\}]^+$.[218]

The thermolysis and photolysis of binuclear iron complexes $[(Cp)(CO)Fe(\mu\text{-}CH_2)Fe(CO)(Cp)]$, $[(Cp)(CO)_2Fe(CH_2)_nFe(CO)_2(Cp)]$ (n=3,4) and $[(Cp)(CO)_2FeC(O)(CH_2)_nC(O)Fe(CO)_2(Cp)]$ (n=2-4) leading to the formation of hydrocarbons has been investigated.[219] When the μ-pentylidyne complex $[(Cp)_2(CO)_2Fe_2(\mu\text{-}CO)\{\mu\text{-}C(CH_2)_3CH_3\}]^+$ is heated, rearrangement to the μ-1-pentenyl complex $[(Cp)_2(CO)_2Fe_2(\mu\text{-}CO)(\mu\text{-}\eta^1,\eta^2\text{-}(E)\text{-}CH{=}CHCH_2CH_2CH_3)]^+$ occurs.[220] The μ-vinyl cations $[M_2(CO)_2(\mu\text{-}CO)\{\mu\text{-}C(R){=}C(H)R\}(Cp)_2]^+$ (M=Fe or Ru; R=H or Ph) react with $NaBH_4$ to afford the μ-carbene complexes $[M_2(CO)_2(\mu\text{-}CO)\{\mu\text{-}C(R)CH_2R\}(Cp)_2]$.[221] With $[Ru_2(CO)(\mu\text{-}CO)\{\mu\text{-}C(O)C_2Me_2\}(Cp)_2]$, addition of $HBF_4.OEt_2$ produces $[Ru_2(CO)(\mu\text{-}CO)\{\mu\text{-}C(H)(O)C_2Me_2\}(Cp)_2]^+$ which reacts with $NaBH_4$ to give $[Ru_2(CO)(\mu\text{-}CO)\{\mu\text{-}C(Me)C(Me)CH_2\}(Cp)_2]$, thus completing the conversion of a metallacyclic CO to CH_2.[221] The crystal structures of the μ-carbene complex $[Fe_2(CO)_2(\mu\text{-}CO)(\mu\text{-}CHMe)(Cp)_2]$ and the μ-vinyl cation $[Fe_2(CO)_2(\mu\text{-}CO)(\mu\text{-}CHCH_2)(Cp)_2]^+$ have been determined.[222]

A mixture of cis- and trans-$[Fe_2(\mu\text{-}C{=}CH_2)(\mu\text{-}CO)(CO)_2(Cp)_2]$ is obtained by the reaction of $[Fe_2(\mu\text{-}CO)_2(CO)_2(Cp)_2]$ with MeLi followed by treatment with CF_3COOH.[223] Reaction of the trans-isomer with $HBF_4.Et_2O$ gives $[Fe_2(\mu\text{-}CMe)(\mu\text{-}CO)(CO)_2(Cp)_2][BF_4]$. With $[Fe_2(\mu\text{-}CO)(\mu\text{-}dppm)(CO)_6]$, deprotonation occurs on reaction with MeLi to give an anionic species which on protonation gives $[(CO)_3Fe(\mu\text{-}Ph_2PCHPPh_2)FeH(CO)_3]$.[223] $[Fe_2(\mu\text{-}CO)(\mu\text{-}CHCH_2)(CO)_6]^-$ reacts with $[Co_2(CO)_8]$ to give the mixed-metal species $[FeCo(CO)_7(\mu\text{-}CHCH_2)]$ and $[FeCo_2(CO)_9(\mu_3\text{-}CMe)]^-$.[224] $[Fe_2(CO)_9]$ reacts with $[Fe_2(\mu\text{-}C{=}CH_2)(\mu\text{-}CO)(CO)_2(Cp)_2]$ to afford $[Fe_3(\mu_3\text{-}CMe)(\mu\text{-}CO)_2(CO)_6(Cp)]$ which reacts in turn with C_5H_6 to give $[Fe_3(\mu_3\text{-}CMe)(\mu\text{-}CO)_3(Cp)_3]$.[225] $[Fe_3(CO)_9\{\mu_3\text{-}C(Me)O\}]^-$ reacts with HBF_4 to give $[Fe_3(CO)_9\{\mu_3\text{-}C(Me)O\}(\mu\text{-}H)]$, $[Fe_3(CO)_{10}(\mu\text{-}CMe)(\mu\text{-}H)]$, and $[Fe_3(CO)_9(\mu_3\text{-}CMe)(\mu_3\text{-}C(Me)OH]$.[226] On reaction with methyl fluorosulphate, $[Fe_3(CO)_9\{\mu_3\text{-}C(Me)O\}]^-$ affords either $[Fe_3(CO)_9(\mu_3\text{-}CMe)(\mu_3\text{-}OMe)]$ or $[Fe_3(CO)_9(\mu_3\text{-}CMe)(\mu_3\text{-}COMe)]$ depending upon the reaction conditions.[226] The trinuclear complexes $[M_3(\mu\text{-}COMe)(CO)_{10}H]$ (M=Fe, Ru, Os) react with H_2 to give $[M_3(\mu_3\text{-}COMe)(CO)_9H_3]$.[227] The μ_3-COMe ligand undergoes reductive cleavage to give CH_3OMe at 130°C and 3.5 MPa of 1:1 CO/H_2 (for M=Ru).[227] The reduction of $[Fe_3(CO)_9(\mu_3\text{-}CMe)(\mu_3\text{-}COEt)]$ with $[Mn(CO)_5]^-$, or treatment of $[Fe_3(CO)_{10}(\mu_3\text{-}CMe)]^{2-}$ with MeOH, gives the anionic complex $[Fe_3(CO)_9(\mu_3\text{-}C{\equiv}CMe)]^-$ which labelling experiments have shown to arise from the coupling of a carbide (produced by C-O bond cleavage) with the μ_3-CMe ligand of the original cluster.[228]

The difluorocarbene-ruthenium(O) complex, $[Ru(=CF_2)(CO)_2(PPh_3)_2]$ has been obtained from the reaction of $[Ru(CO)_2(PPh_3)_3]$ with $[Cd(CF_3)_2(MeOCH_2CH_2OMe)]$.[229] The CF_2 ligand in this $Ru^{(0)}$ complex reacts with the electrophiles HCl and Ag^+ to give $[Ru(CF_2H)(Cl)(CO)_2(PPh_3)_2]$ and $[Ru(CF_2Ag\{H_2O\})(CO)_2(PPh_3)_2]^+$, respectively.[229] $[Na][RuCp(CO)_2]$ reacts with CH_2Cl_2 to afford the μ-methylene complex $[\{RuCp(CO)_2\}_2(\mu\text{-}CH_2)]$, in which there is no Ru-Ru bond.[230] This complex reacts with CO to produce $[(Cp)(CO)_2RuC(O)CH_2Ru(CO)_2(Cp)]$.[230] The reaction between $[RuCl(Cp)(PPh_3)_2]$ and $HC{\equiv}CC(OH)Me_2$ gives (24), the dimer of the expected dimethylallenylidene complex.[231] (24) is converted to (25) on exposure to moisture or NEt_3, by removal of the β-vinylidene proton.[231] Double insertion of ethyne into the Ru-C bonds of the μ-carbene complex $[Ru_2(CO)_2(\mu\text{-}CO)(\mu\text{-}CMe_2)Cp_2]$ occurs to give (26), whereas a different stereochemistry is obtained from the double insertion of $MeO_2CC{\equiv}CCO_2Me$ into $[Ru_2(CO)_2(\mu\text{-}CO)(\mu\text{-}CH_2)Cp_2]$, when (27) is obtained.[232]

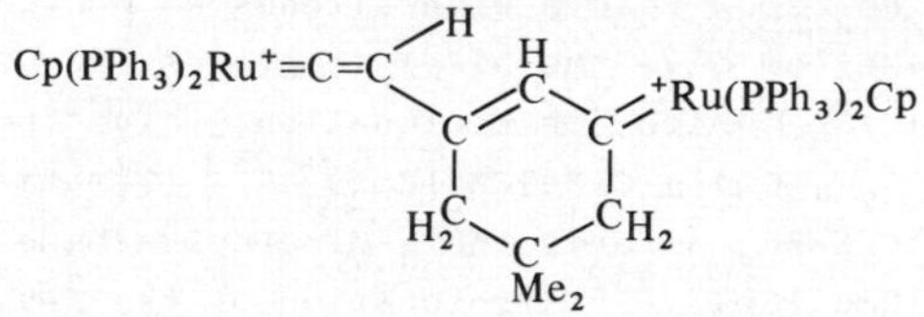
Cp(PPh3)2Ru+=C=C
H
H
C
C
+Ru(PPh3)2Cp
H2C
CH2
C
Me2

(24)

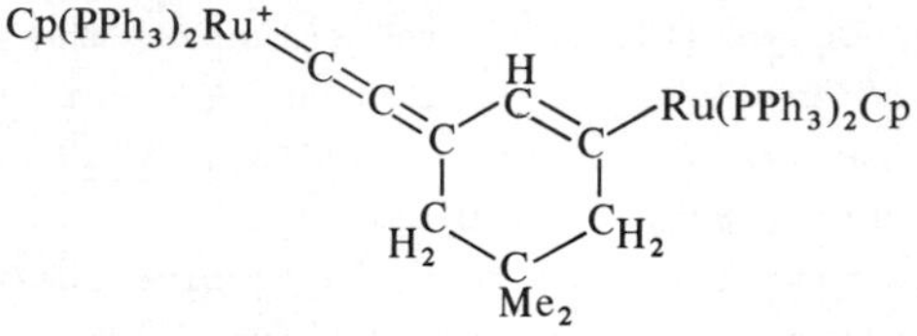
Cp(PPh3)2Ru+
C
C
H
C
Ru(PPh3)2Cp
H2C
CH2
C
Me2

(25)

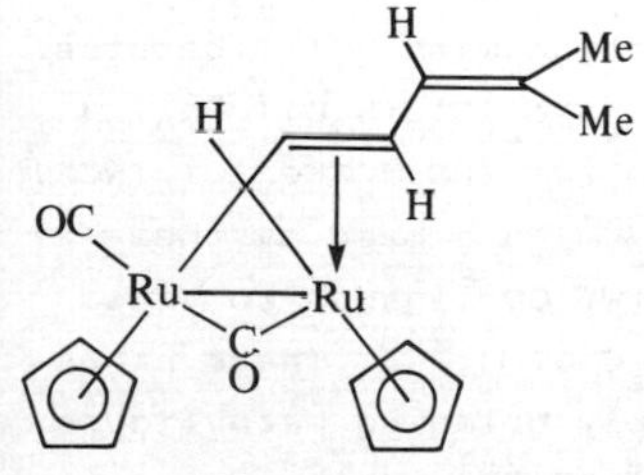
H
Me
Me
H
H
OC
Ru
Ru
C
O

(26)

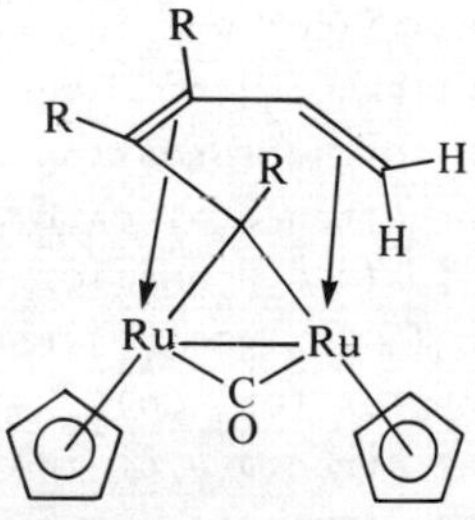
R
R
R
H
H
Ru
Ru
C
O

(27)

$[Ru_2(CO)(\mu\text{-}CO)\{\mu\text{-}C(O)C_2Ph_2\}Cp_2]$ reacts with allenes $R^1CH{=}C{=}CHR^2$ ($R^1{=}R^2{=}H$, Me; $R^1{=}H$, $R^2{=}Me$) to give the allyl complexes $[Ru(CO)\{\eta^3\text{-}C_3H_{4-n}Me_n[2\text{-}Ru(CO)_2(Cp)]\}Cp]$ which, on protonation , give the μ-vinyl species $[Ru_2(CO)_2(\mu\text{-}CO)\{\mu\text{-}C(Me)CH_2\}Cp_2]^+$.[233] Treatment of the μ-vinyl species with $NaBH_4$ affords the μ-dimethylcarbene complex $[Ru_2(CO)_2(\mu\text{-}CO)(\mu\text{-}CMe_2)Cp_2]$.[233] Protonation of the μ-vinylidene complex $[Ru_2(CO)_2(\mu\text{-}CO)(\mu\text{-}C{=}CH_2)Cp_2]$ gives the μ-ethylidyne complex $[Ru_2(CO)_2(\mu\text{-}CO)(\mu\text{-}CMe)Cp_2]^+$.[234] This cationic species reacts with $NaBH_4$ to afford the μ-ethylidene species $[Ru_2(CO)_2(\mu\text{-}CO)(\mu\text{-}CHMe)Cp_2]$.[234] The trinuclear complexes, $[Ru_3H_3(\mu_3\text{-}CX)(CO)_9]$ (X=OMe, Me, Ph) react with RC≡CR to form the corresponding alkene $C_2H_2R_2$ and $[Ru_3H(\mu_3\text{-}\eta^3\text{-}CXCRCR)(CO)_9]$ *via* an alkylidyne-alkyne coupling reaction.[235] This reaction is of interest as a potential model for chain growth in the Fischer-Tropsch reaction. Treatment of $[Ru_3(\mu\text{-}H)\{\mu\text{-}OC(Me)\}(CO)_{10}]$ with MeLi gives the anion $[Ru_3\{\eta^1\text{-}C(O)Me\}(\mu\text{-}H)\{\mu\text{-}OC(Me)\}(CO)_9]^-$, which reacts in turn with $EtOSO_2CF_3$ to give $[Ru_3(\mu\text{-}H)_2\{\eta^2,\mu_3\text{-}C(OEt){=}C(H)\}(CO)_9]$ *via* the carbene complex $[Ru_3\{{=}CMe(OEt)\}(\mu\text{-}H)\{\mu\text{-}OC(Me)\}(CO)_9]$.[236]

Diazomethane reacts with $[OsCl(NO)(PPh_3)_3]$ to give $[Os({=}CH_2)Cl(NO)(PPh_3)_2]$, in which the methylene ligand reacts with electrophilic reagents such as HCl, Cl_2 or MeOH, to give $[Os(CH_2X)Cl_2(NO)(PPh_3)_2]$ (X=H, Cl, or OMe).[237] The carbyne complexes, $[Os(\equiv CPh)(Cl)(CO)(PPh_3)_2]$ and $[Os(\equiv CPh)(CO)_2(PPh_3)_2]^+$ are converted to the carbene complex $[Os\{{=}CH(Ph)\}Cl_2(CO)(PPh_3)_2]$ upon charge transfer excitation in solutions containing HCl.[238] The carbyne complex $[Os(\equiv CC_6H_4NMe_2)Cl(CO)(PPh_3)_2]$ reacts with O_2 to give $[\overline{Os(O_2C}O)(\equiv CC_6H_4NMe_2)Cl(PPh_3)_2]$.[239] Subsequent reaction with HCl liberates CO_2 and leads to the formation of $[Os(\equiv CC_6H_4NMe_2)Cl_2(H_2O)(PPh_3)_2]^+$ from which the octahedral carbyne complexes $[Os(\equiv CC_6H_4NMe_2)Cl_2(NCS)(PPh_3)_2]$ and $[Os(\equiv CC_6H_4NMe_2)Cl_2(CNR)(PPh_3)_2]^+$ have been synthesized.[239] $[Os_3(\mu\text{-}CH_2)(CO)_{11}]$ readily adds two CO ligands to yield the ketene complex $[Os_3(CO)_{12}\{\eta^2(\underline{C},\underline{C})\text{-}\mu\text{-}CH_2C(O)\}]$.[240] The synthesis and crystal structure of the "semi" triply bridging methylidyne complex $[Os_3H(\mu_3\text{-}CH)(CO)_{10}]$ has been described.[241] This complex reacts readily with nucleophiles such as 4-methylpyridine or $LiBEt_3H$ at the methylidyne C atom, providing evidence for the electrophilic nature of the methylidyne ligand.[241]

4.2 The Cobalt Triad. - The Co(I) and Rh(I) carbene complexes (28) and (29) have been prepared from the optically active electron-rich alkenes, (*S*)-$\overline{CN(Me)CH(Me)CH_2N}Me$ and (*S*)-$CN(CH_2CH_2CH_2)CHCH_2NMe$, re-

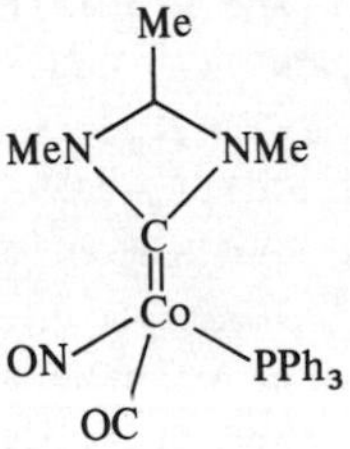
Me
MeN
NMe
C
Co
ON
PPh3
OC

(28)

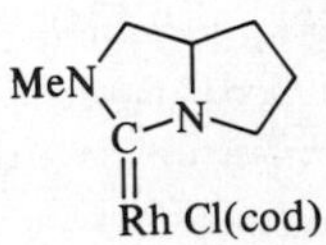
MeN
C
N
Rh Cl(cod)

(29)

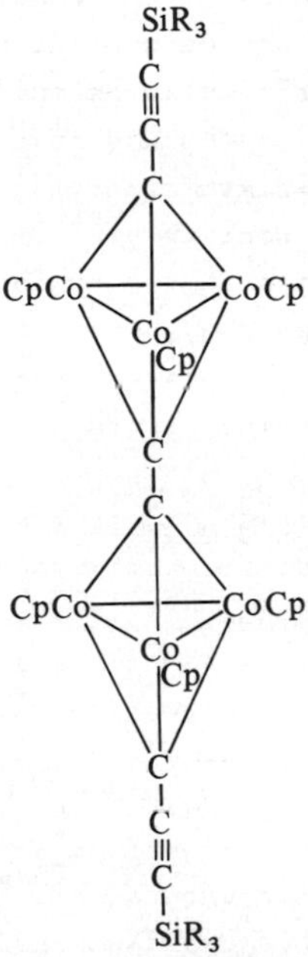
SiR3
C
C
C
CpCo
CoCp
Co
Cp
C
C
CpCo
CoCp
Co
Cp
C
C
C
SiR3

(30)

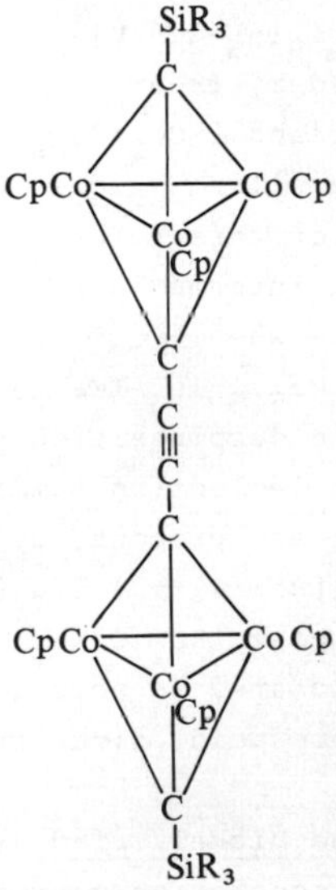
SiR3
C
CpCo
Co Cp
Co
Cp
C
C
C
C
CpCo
Co Cp
Co
Cp
C
SiR3

(31)

spectively.[242] A full account of the synthesis and chemistry of μ-alkylidene complexes of the type $[Co_2(CO)_2(\mu\text{-}CR_2)(\eta^5\text{-}C_5H_4Me)_2]$ ($R=H$, Me, Et, Pr^n) has been published.[243] The μ-alkylidene complexes, $[Co_2(CO)_4(\mu\text{-}CO)(\mu\text{-}CHR)(\mu\text{-}dppm)]$ and $[Co_2(CO)_4(\mu\text{-}CH_2)(\mu\text{-}CHR)\text{-}(\mu\text{-}dppm)]$ ($R=H$ or CO_2Et) have been synthesized, and n.m.r. data indicate that there are rapid μ-CH_2 to terminal CH_2 transformations in these complexes.[244] A wide range of μ-alkylidene complexes of general formula $[Co_2(C_5Me_5)_2(\mu\text{-}CO)_2(\mu\text{-}CRR')]$ have been prepared from $[Co_2(C_5Me_5)_2(\mu\text{-}CO)_2]$ and the appropriate diazoalkane.[245] Alkynes $RC{\equiv}CR^1$ add to the M=M double bonds in $[M_2(C_5Me_5)_2(\mu\text{-}CO)_2]$ (M=Co, Rh) to give metallacyclic complexes of the type $[M_2(C_5Me_5)_2(\mu\text{-}CO)\{\mu\text{-}R^1CC\text{-}(R)C(O)\}]$.[246] In a remarkable reaction, the bis(carbyne)cobalt cluster (30), undergoes simultaneous coupling and decoupling of the carbyne fragments to give (31).[247] Trinuclear alkylidyne complexes $[Co_3(CO)_9(\mu_3\text{-}CCH_2R)]$ and $[(Cp)Mo(CO)_2Co_2(CO)_6(\mu_3\text{-}CCH_2R)]$ have been prepared by treatment of $[Co_2(CO)_6(HC{\equiv}CR)]$ (R=alkyl or aryl group) with H_2 and $[Co_2(CO)_8]$ or $[Mo(CO)_3(Cp)H]$, respectively.[248] Isotope labelling studies and CIDNP evidence suggests that hydrogenolyses of these clusters occurs by a radical pathway, with a vinyl radical intermediate.[248] A μ_3-carbyne ligand is derived from ethylene when $[Co(C_5Me_5)(C_2H_4)_2]$ undergoes thermolysis to afford $[Co_3(C_5Me_5)_3(\mu_3\text{-}CMe)_2]$.[249]

The decomposition of <u>trans</u>-$[(C_5Me_5RhMe)_2(\mu\text{-}CH_2)_2]$ in the presence of 1-electron oxidizers leads to the production of methane, ethene, and propene, by coupling of the C_1 ligands.[250] The reaction of $[Rh_2(C_5Me_5)_2Cl_2(\mu\text{-}CH_2)_2]$ with Na_2CO_3 gives the <u>cis</u>-μ-carbonato complex $[Rh_2(C_5Me_5)_2(CO_3)(\mu\text{-}CH_2)_2]$.[251] This carbonato complex is protonated in acid to give $[Rh_2(C_5Me_5)_2(HCO_3)(\mu\text{-}CH_2)_2]^+$, and with more acid, gives <u>trans</u>-$[Rh_2(C_5Me_5)_2(H_2O)_2(\mu\text{-}CH_2)_2]^{2+}$.[251]

<u>4.3 The Nickel Triad</u>. - Fluorine substituent effects on Ni-carbene bond dissociation energies have been investigated by ion-beam studies.[252]

The complexes $[Pd(C_6F_5)L_2(CNR)][ClO_4]$ (L_2=dipy, tmen, or diphos; R=Cy, <u>p</u>-tol, or Bu^t) react with YH ($Y=Bu^t\bar{N}H_2$, $BzNH_2$, MeOH, or EtOH) to form the corresponding carbene complexes $[Pd(C_6F_5)L_2\{C(NHR)Y\}][ClO_4]$, which are in turn deprotonated with KOH to give the imidoylorganopalladium complexes, $[Pd(C_6F_5)L_2\{C(=NR)Y\}]$.[253] Nucleophilic attack by NHR_2 (R=Me or Et) on $[PdCl(\eta^3\text{-}allyl)(CNC_6H_4X)]$ (η^3-allyl=2-MeC_3H_4; X=Cl, OMe, or Me) leads to the carbene complexes $[PdCl(\eta^3\text{-}allyl)\{C(NHC_6H_4X)(NR_2)\}]$,[254] and a comprehensive mechanistic study of this type of reaction has been reported.[255]

Cleavage of the C-Si bonds in $Me_3SiC{\equiv}CR$ (R=H, Me, or $SiMe_3$) upon reaction with trans-$[PtX(PR^1{}_3)_2(R^{11}OH)][PF_6]$ (X=H or Me; R^{11}= Me or Et) leads to the formation of the alkoxycarbene complexes trans-$[PtX(PR^1{}_3)_2\{C(OR^{11})R\}][PF_6]$.[256] $[Pt(CO)_2Cl_2]$ reacts with $PhC{\equiv}CCO_2Et$ to give $[Pt\{\overline{CC(CO_2Et)=C(Ph)C(CO_2Et)=C(Ph)O}\}(CO)Cl][Pt(CO)Cl_3]$, which contains a 6-membered carbenoid ring bonded to platinum.[257] The aminocarbyne complexes $[Pt_2L_2(\mu\text{-}CNRR^1)(\mu\text{-}dppm)_2]^{n+}$ (L=Cl, Br; R^1=H, Me; n=1; L=CNR; R^1=H,Me; n=3) have been prepared by N-protonation or -alkylation reactions of the corresponding $[Pt_2L_2(\mu\text{-}CNR)(\mu\text{-}dppm)_2]^{n+}$ complexes.[258] A study of the kinetics of the reaction between $[Pt_2X_2(dppm)_2]$ (X=Cl, Br, I, CO, py, NH_3) and CH_2N_2 to give $[Pt_2X_2(\mu\text{-}CH_2)(dppm)_2]$ suggests that CH_2N_2 acts as an electrophile towards the Pt-Pt bond with the rate limiting step being the transfer of an electron pair from the Pt-Pt bond to the CH_2 group of diazomethane.[259] Several Pt-W complexes containing bridging carbene or carbyne ligands have been reported.[260-263] These include $[PtW\{\mu\text{-}C(C_6H_4Me\text{-}4)=CH_2\}(CO)_2(PMe_3)_2(Cp)]$,[260] $[PtW\{\mu\text{-}C(OMe)R\}(\mu\text{-}dppm)(CO)_5]$,[261] $[PtW(\mu\text{-}CR)(\mu\text{-}dppm)(CO)_5][BF_4]$ (R= C_6H_4Me-4),[261] $[PtW\{\mu\text{-}C(OMe)Me\}(\mu\text{-}dppm)(CO)_5]$,[262] $[PtW(\mu\text{-}C{=}CH_2)(dppm)(CO)_5]$,[262] $[Pt_3W_2(\mu_3\text{-}CR)_2(CO)_4Cp_2(cod)_2]$,[263] and $[Pt_2W_3(\mu\text{-}CR)_2(\mu_3\text{-}CR)(CO)_6(Cp)_3]$ (R=C_6H_4Me-4).[263]

5. Bibliography

K.L. Kunze and P.R. Ortiz de Montellano, J. Am. Chem. Soc., 1983, 105, 1380. Formation of a σ-bonded acyliron complex in the reaction of arylhydrazines with hemoglobin and myoglobin.

P. Cocolios, G. Lagrange, and R. Guilard, J. Organomet. Chem., 1983, 253, 65. Synthesis and characterization of alkyl(aryl) ferriporphyrins with M-C σ-bonds.

M.D. Radcliffe and W.M Jones, Organometallics, 1983, 2, 1053. Iron complexes of homocycloheptatrienylidene.

J.-P. Battioni, D. Lexa, D. Mansuy, and J.-M. Savéant, J. Am. Chem. Soc., 1983, 105, 207. Reductive electrochemistry of iron-carbene porphyrins.

S. Abbott, S.G. Davies, and P. Warner, J. Organomet. Chem., 1983, 246, C65. Disubstituted vinylidene complexes of iron and ruthenium: nucleophilic properties of η^1-acetylide ligands.

J. M. Patrick, A.H. White, M.I. Bruce, M.J. Beatson, D.St.C. Black, G.B. Deacon, and N.C. Thomas, J. Chem. Soc., Dalton Trans., 1983, 2121. Crystal structure of bis(benzo[h]quinolin-10-yl-C^{10},N)dicarbonyl-ruthenium(II).

S. Sabo, B. Chaudret, and D. Gervais, Nouv. J. Chim., 1983, 7, 181. Formation of a methylene-bridged Ru-Zr complex.

H.Y. Al-Saigh and T.J. Kemp, J. Chem. Soc., Perkin Trans.II, 1983, 615. Sensitization and quenching processes of alkylcobalt(III) compounds.

A.M. Van den Bergen, D.J. Brockway, and B.O. West, J. Organomet. Chem., 1983, 249, 205. The synthesis and electrochemistry of some perfluoroalkylcobalt complexes.

W.A. Herrmann, C. Bauer, M.L. Ziegler, and H. Pfisterer, J. Organomet. Chem., 1983 243, C54. Ketocarbene addition to a metal carbonyl. A new C-C coupling.

B.D. Murray, M.M. Olmstead, and P.P. Power, Organometallics, 1983, 2, 1700. Synthesis and structure of a rhodium(I) complex containing a bidentate phosphorus alkyl bound ligand.

A. Sebald, B. Wrackmeyer, and W. Beck, Z. Naturforsch., Teil B, 1983, 38, 45. Multinuclear n.m.r. studies of alkynyl derivatives of Pt, Pd, and Ni.

H. Hoberg and B.W. Oster, J. Organomet. Chem., 1983, 252, 359. Nickelacycles as intermediates in the [2+2+2']-cycloaddition of alkynes with isocyanates to give 2-pyridones.

H. Hoberg and K. Sümmermann, J. Organomet. Chem., 1983, 253, 383. Diazanickelacyclopentanone from nickel(O), imines, and isocyanates.

H. Rimml and L.M. Venanzi, J. Organomet. Chem., 1983, 259, C6. Cyclometallation of 1,3-bis{(diphenylphosphino)methyl}benzene.

K. Gehrig, A.J. Klaus, and P. Rys, Helv. Chim. Acta, 1983, 66, 2603. Ortho-cyano-1-arylazonaphthalenes and 3-amino-2-arylbenzo[g] imidazoles from cyclopalladated 1-arylazonaphthalenes.

R. Uson, J. Fornies, P. Espinet,, and E. Lalinde, J. Organomet. Chem., 1983, 254, 371. Insertion of p-MeC_6H_4NC into Pd-C_6F_5 bonds.

K. Hiraki, T. Itoh, K. Eguchi, and M. Onishi, J. Organomet. Chem., 1983, 241, C16. Reactions of $[Pd(Bz)(Cl)(PPh_3)_2]$ with $MeO_2C{\equiv}CCO_2Me$.

H.M. Büch, P. Binger, R. Benn, C. Krüger, and A. Rufińska, Angew. Chem., Int. Ed. Engl., 1983, 22, 774. Phosphine induced rearrangement of an octadienylpalladium complex to 7-and 9-membered metallacycles.

A. Mantovani, J. Organomet Chem., 1983, 255, 385. Synthesis and characterization of $[MCl\{C_5H_3N(6\text{-}Cl)\text{-}\underline{C}2\}(PR_3)_2]$ (M=Pd, Pt).

J. Browning, G.W. Bushnell, K.R. Dixon, and A. Pidcock, Inorg. Chem., 1983, 22, 2226. $[PtCl(PEt_3)\{CH(PPh_2S)_2\}]$, a novel C,S-bonded chelate with dynamic stereochemistry controlled by a metal-ligand pivot.

R.J. Puddephatt, K.A. Azam, R.H. Hill, M.P. Brown, C.D. Nelson, R.P. Moulding, K.R. Seddon, and M.C. Grossel, J. Am. Chem. Soc., 1983, 105, 5642. Mechanisms of A-frame inversion for binuclear platinum(II) complexes with dppm ligands.

R.D.W. Kemmitt, P. McKenna, D.R. Russell, and L.J.S. Sherry, J. Organomet. Chem., 1983, 253, C59. New syntheses of metallacyclobutan-3-one complexes of platinum (II).

A.J. Canty, N.J. Minchin, J.M. Pa ick, and A.H. White, J. Chem. Soc., Dalton Trans., 1983, 1253. Synthetic and X-ray structural studies of complexes formed by metallation of tri(1-pyrazolyl)methane by dimethylplatinum complexes.

L. Chassot and A. von elewsky, Helv. Chim. Acta, 1983, 66, 2443. Cis-bis(2-phenylphyridine)platinum(II).

T.G. Appleton, J.R. Hall, N.S. Ham, F.W. Hess, and M.A. Williams, Aust. J. Chem., 1983, 36, 673. Kinetics of site-exchange reactions in di- and tri-methylplatinum (IV) glycinate complexes from n.m.r. line-shape analysis.

M. Calligaris, G. Carturan, G. Nardin, A. Scrivanti, and A. Wojcicki, Organometallics, 1983, 2, 865. Cycloaddition reactions of η^1-allylplatinum(II) complexes with tetracyanoethylene.

A. Sebald and B. Wrackmeyer, J. Chem. Soc., Chem. Commun., 1983, 309. Novel synthesis of platinum(II) alkenyl complexes via organoboration of platinum(II) acetylides.

M.L. Illingsworth, J.A. Teagle, J.L. Burmeister, W.C. Fultz, and A.L. Rheingold, Organometallics, 1983, 2, 1364. The crystal structure of a dinuclear ortho-metallated platinum-ylide complex.

S. Dholakia, R.D. Gillard, and F.L. Wimmer, Inorg. Chim. Acta., 1983, 69, 179. N-methyl-2,2'-dipyridylium complexes : synthesis and cyclometallation of [M(dipyMe)X_3] (M=Pt, Pd; X=Cl, Br).

A. Avshu, R.D. O'Sullivan, A.W. Parkins, N.W. Alcock, and R.M. Countryman, J. Chem. Soc., Dalton Trans., 1983, 1619. The reaction of Ag^+ with [ML_2X_2] (M=Pd or Pt; X=I, Cl, or SCN; L=amine or phosphine) leading to ortho metallation and catalytic activity.

A.B. Goel and S. Goel, Inorg. Chim. Acta., 1983, 69, 233. The synthesis of hydrido-platinum(II) complexes containing the ligand $Bu_2PC(Me)_2CH_2$.

A.B. Goel and S. Goel, Inorg. Chim. Acta, 1983, 77, L53. A study of CO reactions with platinum(II) complexes containing $SnCl_3$ and $Bu_2PC(Me)_2CH_2$ ligands.

S. Lanza, Inorg. Chim. Acta, 1983, 75, 131. The preparation of complexes of the type [Pt(Ph)$_2$(bipym)] and [Ph_2Pt(μ-bipym)MCl_2] (bipym=bipyrimidine; M=Hg, Mn).

J. Terheijden, G. Van Koten, J.L. de Booys, H.J.C. Ubbels, and C.H. Stam, Organometallics, 1983, 2, 1882. Preparation of stable platinum(IV) complexes of the type [PtX_3{$C_6H_3(CH_2NMe_2)_2$-o,o'}].

D. Carmona, R. Thouvenot, L.M. Venanzi, F. Bachechi, and L. Zambonelli, J. Organomet. Chem., 1983, 250, 589. The synthesis of mono-hydrido-bridged complexes [$(PEt_3)_2$(Ar)Pt(μ_2-H)Pt(Ar)$(PEt_3)_2$][BPh_4] (Ar=Ph, 4-MeC_6H_4, or 2,4-$Me_2C_6H_3$).

G. Read, M. Urgelles, A.M.R. Galas, and M.B. Hursthouse, J. Chem. Soc., Dalton Trans., 1983, 911. Structural and spectroscopic studies of peroxymetallacyclic complexes [Pt{C(CN)$_2$CR_2OO}$(PPh_3)_2$] (R=CN or Me).

U. Baltensperger, J.R. Günter, S. Kägi, G. Kahr, and W. Marty, Organometallics, 1983, 2, 571. Multistep cyclometallation of solid trans-dichloro[3,3'-oxybis-{(diphenylphosphino)methyl}benzene]platinum(II).

G. Lopez, G. Garcia, J. Galvez, and N. Cutillas, J. Organomet. Chem., 1983, 258, 123. [Pt$(C_6F_5)_2$] and its adducts with ketones, arenes, and water.

L.I. Elding, B. Kellenberger, and L.M. Venanzi, Helv. Chim. Acta., 1983, 66, 1676. Ligand substitution reactions on hydrido- or methyl- platinum complexes containing trans-spanning bidentate ligands.

W.J. Youngs and J.A. Ibers, Organometallics, 1983, 2, 979. The intra-molecular activation of C–C and C–H bonds of a cyclopropylphosphine complex of platinum(II).

S. Sostero, O. Traverso, R. Ros, and R.A. Michelin, J. Organomet. Chem., 1983, 246, 325. Photoinduced reductive elimination of cyanoalkanes from hydrido-cyanoalkyl complexes of platinum(II).

J. Geisenberger, U. Nagel, A. Sebald and W. Beck, Chem. Ber., 1983, 116, 911. The structure of [Pt(C≡CPh)$_2$$(PEt_3)_2${1,4-(4-$NO_2C_6H_4$)$_2N_4$}].

A.T. Hutton, D.M. McEwan, B.L. Shaw, and S.W. Wilkinson, J. Chem. Soc., Dalton Trans., 1983, 2011. Formation of [PtCl(PMe_2Ph){ON(=CMe_2)CH=CMe}] by attack of acetoxime on an allene-platinum(II) complex.

References

1. B.A. Dolgoplosk, Russ. Chem. Rev., 1983, 52, 613.
2. H. Schmidbaur, Angew. Chem., Int. Ed. Engl., 1983, 22, 907.
3. M.D. Johnson, Acc. Chem. Res., 1983, 16, 343.
4. W.A. Herrmann, J. Organomet. Chem., 1983, 250, 319.
5. J. Holton, M.F. Lappert, R. Pearce, and P.I.W. Yarrow, Chem. Rev., 1983, 83, 135.
6. E. Sappa, A. Tiripicchio, and P. Braunstein, Chem. Rev., 1983, 83, 203.
7. R. Houriet, L.F. Halle, and J.L. Beauchamp, Organometallics, 1983, 2, 1818.
8. A.V. Kavaliunas, A. Taylor, and R.d. Rieke, Organometallics, 1983, 2, 377.
9. W.A. Herrmann, J. Plank, J.L. Hubbard, G.W. Kriechbaum, W. Kalcher, B. Koumbouris, G. Ihl, A. Schäfer, M.L. Ziegler, H. Pfisterer, C. Pahl, J.L. Atwood, and R.D. Rogers, Z. Naturforsch., Teil B, 1983, 38, 1392.
10. S.G. Davies, S.J. Simpson, and S.E. Thomas, J. Organomet. Chem., 1983, 254, C29.
11. C. Lapinte and D. Astruc, J. Chem. Soc., Chem. Commun., 1983, 430.
12. Y.C. Lin, D. Milstein, and S.S. Wreford, Organometallics, 1983, 2, 1461.
13. G.O. Nelson, Organometallics, 1983, 2, 1474.
14. I.I. Guerus and Y.L. Yagupolskii, J. Organomet. Chem., 1983, 247, 81.
15. T.G. Richmond and D.F. Shriver, Organometallics, 1983, 2, 1061.
16. S. Pelling, C. Botha, and J.R. Moss, J. Chem. Soc., Dalton Trans., 1983, 1495.
17. H. Nakazawa, D.L. Johnson, and J.A. Gladysz, Organometallics, 1983, 2, 1846.
18. S. Voran and W. Malisch, Angew. Chem., Int. Ed. Engl., 1983, 22, 151.
19. W. Angerer, K. Fiederling, G. Grötsch, and W. Malisch, Chem. Ber., 1983, 116, 3947.
20. T.W. Leung, G.G. Christoph, and A. Wojcicki, Inorg. Chim. Acta, 1983, 76, L281.
21. R. Baker, V.B. Rao, and E. Erdik, J. Organomet. Chem., 1983, 243, 451.
22. B.D. Fabian and J. A. Labinger, Organometallics, 1983, 2, 659.
23. M.E. Wright, Organometallics, 1983, 2, 558.
24. D.L. Reger, K.A. Belmore, E. Mintz, N.G. Charles, E.A.H. Griffith, and E.L. Amma, Organometallics, 1983, 2, 101.
25. H.H. Karsch, Chem. Ber., 1983, 116, 1656.
26. T.A. Bazhenova, R.M. Lobkovskaya, R.P. Shibaeva, A.E. Shilov, A.K. Shilova, M. Gruselle, G. Leny, and B. Tchoubar, J. Organomet. Chem., 1983, 244, 265.
27. T.A. Bazhenova, R.M. Lobkovskaya, R.P. Shibaeva, A.K. Shilova, M. Gruselle, G. Leny, and E. Deschamps, J. Organomet. Chem., 1983, 244, 375.
28. M.D. Seidler and R. G. Bergman, Organometallics, 1983, 2, 1897.
29. M. Pańkowski and M. Bigorgne, J. Organomet. Chem., 1983, 251, 333.
30. R.H. Magnuson, R. Meirowitz, S.J. Zulu, and W.P. Giering, Organometallics, 1983, 2, 460.
31. C.R. Jablonski and Y.P. Wang, Inorg. Chim. Acta, 1983, 69, 147.
32. G. Bellachioma, G. Cardaci, and G. Reichenbach, J. Chem. Soc., Dalton Trans., 1983, 2593.
33. H. Brunner, B. Hammer, I. Bernal, and M. Draux, Organometallics, 1983, 2, 1595.
34. T.C. Flood, K.D. Campbell, H.H. Downs, and S. Nakanishi, Organometallics, 1983, 2, 1590
35. J.E. Jensen, L.L. Campbell, S. Nakanishi, and T.C. Flood, J. Organomet. Chem., 1983, 244, 61.
36. W. Petz, Organometallics, 1983, 2, 1044.
37. G.A. Ozin and J.G. McCaffrey, Inorg. Chem., 1983, 22, 1397.
38. M. Cooke, N.J. Farrow, and S.A.R. Knox, J. Chem. Soc., Dalton Trans., 1983, 2435.
39. G.N. Mott and A.J. Carty, Inorg. Chem., 1983, 22, 2727.
40. N. Bild, E.R.F. Gesing, C. Quiquerez, and A. Wehrli, J. Chem. Soc., Chem. Commun., 1983, 172.
41. N. Bild, E.R.F. Gesing, C. Quiquerez, and A. Wehrli, J. Organomet. Chem., 1983, 248, 85.

42. D.R. Saunders, M. Stephenson, and R.J. Mawby, J. Chem. Soc., Dalton Trans., 1983, 2473.
43. Z. Dauter, R.J. Mawby, C.D. Reynolds, and D.R. Saunders, Acta. Cryst., Sect. C, 1983, 39, 1194.
44. S.A. Chawdhury, Z. Dauter, R.J. Mawby, C.D. Reynolds, D.R. Saunders, and M. Stephenson, Acta Cryst., Sect. C., 1983, 39, 985.
45. P.R. Holland, B. Howard, and R.J. Mawby, J. Chem. Soc., Dalton Trans., 1983, 231.
46. H. Lehmkuhl, J. Grundke, and R. Mynott, Chem. Ber., 1983, 116, 159.
47. H. Lehmkuhl, J. Grundke, and R. Mynott, Chem. Ber., 1983 116, 176.
48. R. Werner and H. Werner, Chem. Ber., 1983, 116, 2074
49. H. Kletzin, H. Werner, O. Serhadli, and M.L. Ziegler, Angew. Chem., Int. Ed. Engl., 1983, 22, 46.
50. H. Kletzin and H. Werner, Angew. Chem., Int. Ed. Engl., 1983, 22, 873.
51. K.P.C. Vollhardt and T.W. Weidman, J. Am. Chem. Soc., 1983, 105, 1676.
52. A. Astier, J-C. Daran, Y. Jeannin, and C. Rigault, J. Organomet. Chem., 1983, 241, 53.
53. M.I. Bruce, J.G. Matisons, B.W. Skelton, and A.H. White, J. Organomet. Chem., 1983, 251, 249.
54. M.I. Bruce, M.L. Williams, and B.K. Nicholson, J. Organomet. Chem., 1983, 258, 63.
55. R.D. Adams, P. Mathur, and B.E. Segmüller, Organometallics, 1983, 2, 1258.
56. B.F.G. Johnson, J. Lewis, J.N. Nicholls, J. Puga, P.R. Raithby, M.J. Rosales, M. McPartlin, and W. Clegg, J. Chem. Soc., Dalton Trans., 1983, 277.
57. B.F.G. Johnson, J. Lewis, J.N. Nicholls, J. Puga, and K.H. Whitmire, J. Chem. Soc., Dalton Trans., 1983, 787.
58. A.G. Cowie, B.F.G. Johnson, J. Lewis, J.N. Nicholls, P.R. Raithby, and M.J. Rosales, J. Chem. Soc., Dalton Trans., 1983, 2311.
59. B.F.G. Johnson, J. Lewis, W.J.H. Nelson, J.N. Nicholls, J. Puga, P.R. Raithby, M.J. Rosales, M. Schröder, and M.D. Vargas, J. Chem. Soc., Dalton Trans., 1983, 2447.
60. B.F.G. Johnson, J. Lewis, W.J.H. Nelson, J. Puga, P.R. Raithby, D. Braga, M. McPartlin, and W. Clegg, J. Organomet. Chem., 1983, 243, C13.
61. B.F.G. Johnson, J. Lewis, W.J.H. Nelson, J. Puga, M. McPartlin, and A. Sironi, J. Organomet. Chem., 1983, 253, C5.
62. B.F.G. Johnson, J. Lewis, W.J.H. Nelson, J.N. Nicholls, and M.D. Vargas, J. Organomet. Chem., 1983, 249, 255.
63. H. Werner and R. Weinand, Z. Naturforsch., Teil B, 1983, 38, 1518.
64. H. Werner and J. Gotzig, Organometallics, 1983, 2, 547.
65. M.R. Burke, J. Takats, F.-W. Grevels, and J.G.A. Reuvers, J. Am. Chem. Soc., 1983, 105, 4092.
66. C.E.L. Headford and W. R. Roper, J. Organomet. Chem., 1983, 244, C53.
67. G.P. Elliott and W.R. Roper, J. Organomet. Chem., 1983, 250, C5.
68. R.P. Ferrari and G.A. Vaglio, Transition Met. Chem., 1983, 8, 155.
69. A.J. Arce, A.J. Deeming, and R. Shaunak, J. Chem. Soc., Dalton Trans., 1983, 1023.
70. R.J. Goudsmit, B.F.G. Johnson, J. Lewis, P.R. Raithby, and M.J. Rosales, J. Chem. Soc., Dalton Trans., 1983, 2257.
71. B.F.G. Johnson, J. Lewis, W.J.H. Nelson, M.D. Vargas, D. Braga, and M.McPartlin, J. Organomet. Chem., 1983, 246, C69.
72. D. Braga, K. Henrick, B.F.G. Johnson, J. Lewis, M. McPartlin, W.J.H. Nelson, A. Sironi, and M.D. Vargas, J. Chem. Soc., Chem. Commun., 1983, 1131.
73. F.T.T.Ng, G.L. Rempel, and J. Halpern, Inorg. Chim. Acta, 1983, 77 L165.
74. H.-F. Klein, J. Gross, R. Hammer, and U. Schubert, Chem. Ber., 1983, 116, 1441.
75. T.J. Groshens and K.J. Klabunde, J. Organomet. Chem., 1983, 259, 337.
76. M.M. Brezinski and K.J. Klabunde, Organometallics, 1983, 2, 1116.
77. N. España, P. Gomez, P. Royo, and A. Vazquez de Miguel, J. Organomet. Chem., 1983, 256, 141.

78. W.P. Weiner and R.G. Bergman, J. Am. Chem. Soc., 1983, 105, 3922.
79. T. Funabiki, H. Nakamura, and S. Yoshida, J. Organomet. Chem., 1983, 243, 95.
80. F. Francalanci, A. Gardano, L. Abis, and M. Foà, J. Organomet. Chem., 1983, 251, C5.
81. L. Hofmann and H. Werner, J. Organomet. Chem., 1983, 255, C41.
82. G. Müller, D. Neugebauer, W. Gelke, F.H. Köhler, J. Pebler, and H. Schmidbaur, Organometallics, 1983, 2, 257.
83. W. Malisch, H. Blau, P. Weickert, and K.-H. Griessmann, Z. Naturforsch., Teil B, 1983, 38, 711.
84. Y. Wakatsuki, O. Nomura, K. Kitaura, K. Morokuma, and H. Yamazaki, J. Am. Chem. Soc., 1983, 105, 1907.
85. H. Yamazaki, K. Yasufuku, and Y. Wakatsuki, Organometallics, 1983, 2, 726.
86. W.H. Hersh, F.J. Hollander, and R.G. Bergman, J. Am. Chem. Soc., 1983, 105, 5834.
87. W.H. Hersh, and R.G. Bergman, J. Am. Chem. Soc., 1983, 105, 5846.
88. G.K. Yang and R.G. Bergman, J. Am. Chem. Soc., 1983, 105, 6045.
89. W.D. Jones, and F.J. Feher, Organometallics, 1983, 2, 562.
90. W.D. Jones, and F.J. Feher, Organometallics, 1983, 2, 686.
91. E. Arpac and L. Dahlenburg, J. Organomet. Chem., 1983, 241, 27.
92. S. Okeya, B.F. Taylor, and P.M. Maitlis, J. Chem. Soc., Chem. Commun., 1983, 971.
93. A. Vázquez de Miguel, M. Gómez, K. Isobe, B.F. Taylor, B.E. Mann, and P.M. Maitlis, Organometallics, 1983, 2, 1724.
94. M. Gómez, P.I.W. Yarrow, A. Vázquez de Miguel, and P.M. Maitlis, J. Organomet. Chem., 1983, 259, 237.
95. M. Gómez, D.J. Robinson, and P.M. Maitlis, J. Chem. Soc., Chem. Commun., 1983, 825.
96. K. Isobe, A. Vázquez de Miquel, P.M. Bailey, S. Okeya, and P.M. Maitlis, J. Chem. Soc., Dalton Trans., 1983, 1441.
97. A. Vázquez de Miquel and P.M. Maitlis, J. Organomet. Chem., 1983, 244, C35.
98. R.G. Finke, G. Gaughan, C. Pierpont, and J.H. Noordik, Organometallics, 1983, 2, 1481.
99. H. Werner, J. Wolf, U. Schubert, and K. Ackermann, J. Organomet. Chem., 1983, 243, C63.
100. N.L. Jones, and J.A. Ibers, Organometallics, 1983, 2, 490.
101. R.D. Waid and D.W. Meek, Organometallics, 1983, 2, 932.
102. Y. Aoyama, T. Yoshida, K. Sakurai, and H. Ogoshi, J. Chem. Soc., Chem. Commun., 1983, 478.
103. B.T. Heaton, L. Strona, S. Martinengo, D. Strumolo, V.G. Albano, and D. Braga, J. Chem. Soc., Dalton Trans., 1983, 2175.
104. V.G. Albano, D. Braga, P. Chini, D. Strumolo and S. Martinengo, J. Chem. Soc., Dalton Trans., 1983, 249.
105. D. Strumolo, C. Seregni, S. Martinengo, V.G. Albano, and D. Braga, J. Organomet. Chem., 1983, 252, C93.
106. A.J. Janowicz and R. G. Bergman, J. Am. Chem. Soc., 1983, 105, 3929.
107. J.K. Hoyano, A.D. McMaster, and W.A.G. Graham, J. Am. Chem. Soc., 1983, 105, 7190.
108. Chem. Eng. News, 1983, 61, 33.
109. K. Isobe, A. Vázquez de Miguel, and P.M. Maitlis, J. Organomet. Chem., 1983, 250, C25.
110. L. Dahlenburg and F. Mirzaei, J. Organomet. Chem., 1983, 251, 103.
111. L. Dahlenburg and F. Mirzaei. J. Organomet. Chem., 1983, 251, 113.
112. L. Dahlenburg, J. Organomet. Chem., 1983, 251, 215.
113. L. Dahlenburg and F. Mirzaei, J. Organomet. Chem., 1983, 251, 123.
114. L. Dahlenburg, J. Organomet. Chem., 1983, 251, 347.
115. E. Arpac and L. Dahlenburg, J. Organomet. Chem., 1983, 251, 361.
116. M. Cowie, M.D. Gauthier, S.J. Loeb, and I.R. McKeer, Organometallics, 1983, 2, 1057.
117. L.A. Oro, D. Carmona, M.A. Esteruelas, C. Foces-Foces, and F.H. Cano, J. Organomet. Chem., 1983, 258, 357.

118. P.J. Spellane, R.J. Watts, and C.J. Curtis, Inorg. Chem., 1983, 22, 4060.
119. G. Nord, A.C. Hazell, R.G. Hazell, and O. Farver, Inorg. Chem., 1983, 22, 3429.
120. W.J. Youngs, and J.A. Ibers, J. Am. Chem. Soc., 1983, 105, 639.
121. M.M. Harding, B.S. Nicholls, and A.K. smith, J. Chem. Soc., Dalton Trans., 1983, 1479.
122. M. Wada and M. Kumazoe, J. Organomet. Chem., 1983, 259, 245.
123. D. Xu, Y. Kai, K. Miki, N. Kasai, K. Nishiwaki, and M. Wada, Chem. Lett., 1983, 591.
124. S. Komiya, Y. Abe, A. Yamamoto, and T. Yamamoto, Organometallics, 1983, 2, 1466.
125. T. Yamamoto, T. Kohara, K. Osakada, and A. Yamamoto, Bull. Chem. Soc. Japan, 1983, 56, 2147.
126. H. Lehmkuhl, C. Naydowski, and M. Bellenbaum, J. Organomet. Chem., 1983, 246, C5.
127. H. Lehmkuhl, F. Danowski, R. Benn, A. Rufińska, G. Schroth, and R. Mynott, J. Organomet. Chem., 1983, 254, C11.
128. H. Lehmkuhl, C. Naydowski, R. Benn, A. Rufińska, and G. Schroth, J. Organomet. Chem., 1983, 246, C9.
129. L.S. Isaeva , L.N. Morozova, V.V. Bashilov, P.V. Petrovskii, V.I. Sokolov, and O.A. Reutov, J. Organomet. Chem., 1983, 243, 253.
130. M. Antón, G. Muller, and J. Sales, Transition Met. Chem., 1983, 8, 79.
131. K. Sano, T. Yamamoto, and A. Yamamoto, Chem. Lett., 1983, 115.
132. A. Behr, W. Keim, and G. Thelen, J. Organomet. Chem., 1983, 249, C38.
133. C. Arlen, M. Pfeffer, J. Fischer, and A. Mitschler, J. Chem. Soc., Chem. Commun., 1983, 928.
134. P. Binger, M.J. Doyle, and R. Benn, Chem. Ber., 1983, 116, 1.
135. D.S. Dudis and J.P. Fackler, J. Organomet. Chem., 1983, 249, 289.
136. D.M. Grove, G. VanKoten, R. Zoet, N.W. Murrall, and A.J. Welch, J. Am. Chem. Soc., 1983, 105, 1379.
137. E. Dinjus, D. Walther, R. Kirmse, and J. Stach, Z. Anorg. Allg. Chem., 1983, 501, 121.
138. G. López, G. Garcia, N. Cutillas, and J. Ruiz, J. Organomet. Chem., 1983, 241, 269.
139. G. López, G. Garcia, J. Ruiz, and N. Cutillas, J. Organomet. Chem., 1983, 246, C83.
140. R. Usón, J. Forniés, F. Martínez, M. Tomás, and I. Reoyo, Organometallics, 1983, 2, 1386.
141. R. Usón, J. Forniés, P. Espinet, F. Martínez, C. Fortuño, and B. Menjón, J. Organomet. Chem., 1983, 256, 365.
142. G.B. Deacon, and I.L. Grayson, Transition Met. Chem., 1983, 8, 131.
143. U. Belluco, R.A. Michelin, P. Uguagliati, and B. Crociani, J. Organomet. Chem., 1983, 250, 565.
144. H. Nakazawa, F. Ozawa, and A. Yamamoto, Organometallics, 1983, 2, 241.
145. G.K. Anderson, Organometallics, 1983, 2, 665.
146. H. Werner, H.-J. Kraus, U. Schubert, K. Ackermann, and P. Hofmann, J. Organomet. Chem., 1983, 250, 517.
147. R. McCrindle, G. Ferguson, G.J. Arsenalt, and A.J. McAlees, J. Chem. Soc., Chem. Commun., 1983, 571.
148. C.J. Elsevier, H. Kleijn, K. Ruitenberg, and P. Vermeer, J. Chem. Soc., Chem. Commun., 1983, 1529.
149. H.M. McPherson and J.L. Wardell, Inorg. Chim. Acta, 1983, 75, 37.
150. P.W. Clark and S.F. Dyke, J. Organomet. Chem., 1983, 259, C17.
151. M. Nonoyama, Transition Met. Chem., 1983, 8, 121.
152. J. Granell, J. Sales, J. Vilarrasa, J.P. Declercq, G. Germain, C. Miravitlles, and X. Solans, J. Chem. Soc., Dalton Trans., 1983, 2441.
153. J. Selbin and M.A. Gutierrez, J. Organomet. Chem., 1983, 246, 95.
154. B. Galli, F. Gasparrini, L. Maresca, G. Natile, and G. Palmieri, J. Chem. Soc., Dalton Trans., 1983, 1483.

155 Y. Fuchita, K. Hiraki, and T. Uchiyama, J. Chem. Soc., Dalton Trans., 1983, 897.
156 N. Barr, S.F. Dyke, and S.N. Quessy, J. Organomet. Chem., 1983, 253, 391.
157 P.W. Clark, H.J. Dyke, S.F. Dyke, and G. Perry, J. Organomet. Chem., 1983, 253, 399.
158 J. Dehand, A. Mauro, H. Ossor, M. Pfeffer, R.H. DeA. Santos, and J.R. Lechat, J. Organomet. Chem., 1983, 250, 537.
159 W.J. Youngs, J. Mahood, B.L. Simms, P.N. Swepston, J.A. Ibers, S. Maoyu, H. Jinling, and L. Jiaxi, Organometallics, 1983, 2, 917.
160 K. Hiraki, Y. Fuchita, and T. Uchiyama, Inorg. Chim. Acta, 1983, 69, 187.
161 A.R.H. Bottomley, C. Crocker, and B.L. Shaw, J. Organomet. Chem., 1983, 250, 617.
162 V.I. Sokolov, Pure Appl. Chem., 1983, 55, 1837.
163 H.-P. Abicht, P. Lehniger, and K. Issleib, J. Organomet. Chem., 1983, 250, 609.
164 C. Arlen, M. Pfeffer, O. Bars, and D. Grandjean, J. Chem. Soc., Dalton Trans, 1983, 164.
165 H.-P. Abicht and K. Issleib, Z. Anorg. Allg. Chem., 1983, 500, 31.
166 G.R. Newkome, W.E. Puckett, V.K. Gupta, and F.R. Fronczek, Organometallics, 1983, 2, 1247.
167 G.B. Robertson and P.A. Tucker, Acta Cryst., Sect. A, 1983, 39, 1354.
168 G.B. Shul'pin, G.V. Nizova, and A.E. Shilov, J. Chem. Soc., Chem, Commun., 1983, 671.
169 L. Maresca and G. Natile, J. Chem. Soc., Chem. Commun., 1983, 40.
170 N.M. Boag, R.J. Goodfellow, M. Green, B. Hessner, J.A.K. Howard, and F.G.A. Stone, J. Chem. Soc., Dalton. Trans., 1983, 2585.
171 R.J. Cross and A.J. McLennan, J. Chem. Soc., Dalton Trans., 1983, 359.
172 R.J. Cross and A.J. McLennan, J. Organomet Chem., 1983, 255, 113.
173 J.D. Scott and R.J. Puddephatt, Organometallics, 1983, 2, 1643.
174 G. Ferguson, M. Parvez, P.K. Monaghan, and R.J. Puddephatt, J. Chem. Soc., Chem Commun., 1983, 267.
175 P.K. Monaghan and R.J. Puddephatt, Organometallics, 1983, 2, 1698.
176 S. Sakaki, K. Kitaura, K. Morokuma, and K. Ohkubo, J. Am. Chem. Soc., 1983, 105, 2280.
177 M.A. Bennett and A. Rokicki, J. Organomet. Chem., 1983, 244, C31.
178 C. Bartocci, A. Maldotti, S. Sostero, and O. Traverso, J. Organomet. Chem., 1983, 253, 253.
179 U. Bayer and H.A. Brune, Z. Naturforsch., Teil B, 1983, 38, 226.
180 U. Bayer and H.A. Brune, Z. Naturforsch., Teil B, 1983, 38, 621.
181 U. Bayer and H.A. Brune, Z. Naturforsch., Teil B, 1983, 38, 632.
182 R.A. Michelin, S. Faglia, and P. Uguagliati, Inorg. Chem., 1983, 22, 1831.
183 D.C. Griffiths and G.B. Young, Polyhedron , 1983, 2, 1095.
184 A. Morvillo, G. Favero, and A. Turco, J. Organomet. Chem., 1983, 243, 111.
185 A. Morvillo and A. Turco, J. Organomet. Chem., 1983, 258, 383.
186 H.D. McPherson and J.L. Wardell, J. Organomet. Chem., 1983, 254, 261.
187 H.C. Clark, G. Ferguson, V.K. Jain, and M. Parvez, Organometallics, 1983, 2, 806.
188 J.T. Burton, R.J. Puddephatt, N.L. Jones, and J.A. Ibers, Organometallics, 1983, 2, 1487.
189 C.N. Wilker and R. Hoffmann, J. Am. Chem. Soc., 1983, 105, 5285.
190 A. Sebald and B. Wrackmeyer, Z. Naturforsch., Teil B, 1983, 38, 1156.
191 A. Sebald and B. Wrackmayer, J. Chem. Soc., Chem. Commun., 1983, 1293.
192 S.D. Chappell and D.J. Cole-Hamilton, J. Chem. Soc., Dalton Trans., 1983, 1051.
193 E.W. Abel, K.G. Orrell, and A.W.G. Platt, J. Chem. Soc., Dalton Trans., 1983, 2345.
194 E.W. Abel, M.Z.A. Chowdhury, K.G. Orrell, and V. Šik, J. Organomet. Chem., 1983, 258, 109.
195 P.K. Monaghan, and R.J. Puddephatt, Inorg. Chim. Acta, 1983, 76, L237.
196 F. Bachechi, G. Bracher, D.M. Grove, B. Kellenberger, P.S. Pregosin, L.M. Venanzi, and L. Zambonelli, Inorg. Chem., 1983, 22, 1031.

197 J. Jans, R. Naegeli, L.M. Venanzi, and A. Albinati, J. Organomet. Chem., 1983, 247, C37.
198 A. Albinati, R. Naegeli, A. Togni, and L.M. Venanzi, Organometallics, 1983 2, 926.
199 C.R. Langrick, D.M. McEwan, P.G. Pringle, and B.L. Shaw, J. Chem. Soc., Dalton Trans., 1983, 2487.
200 M.P. Brown, A. Yavari, L. Manojlović-Muir, and K.W. Muir, J. Organomet. Chem., 1983, 256, C19.
201 R.H. Hill and R.J. Puddephatt, Organometallics, 1983, 2, 1472.
202 K.A. Azam and R.J. Puddephatt, Organometallics, 1983, 2, 1396.
203 A.T. Hutton, B. Shabanzadeh, and B.L. Shaw, J. Chem. Soc., Chem. Commun., 1983, 1053.
204 S.S.M. Ling, R.J. Puddephatt, L. Manojlović-Muir, and K.W. Muir, Inorg. Chim. Acta., 1983, 77, L95.
205 S.S.M. Ling, R.J. Puddephatt, L. Manojlović-Muir, and K.W. Muir, J. Crganomet. Chem., 1983, 255, C11.
206 D.P. Arnold, M.A. Bennett, G.M. McLaughlin, G.B. Robertson, and M.J. Whittaker, J. Chem. Soc., Chem. Commun., 1983, 32.
207 D.P. Arnold, M.A. Bennett, G.M. McLaughlin, and G.B. Robertson, J. Chem. Soc., Chem. Commun., 1983, 34.
208 M.F. Semmelhack and R. Tamura, J. Am. Chem. Soc., 1983, 105, 4099.
209 H. Nakatsuji, J. Ushio, S. Han, and T. Yonezawa, J. Am. Chem. Soc., 1983, 105, 426.
210 M.F. Semmelhack and R. Tamura, J. Am. Chem. Soc., 1983,105, 6750.
211 M. Brookhart, J.R. Tucker, and G.R. Husk, J. Am. Chem. Soc., 1983, 105, 258.
212 C.P. Casey and W. H. Miles, J. Organomet. Chem., 1983, 254, 333.
213 T.W. Bodnar and A.R. Cutler, J. Am Chem. Soc., 1983, 105, 5926.
214 M. Brookhart, D. Timmers, J.R. Tucker, and G.D. Williams, J. Am. Chem. Soc., 1983, 105, 6721.
215 G. Grötsch and W. Malisch, J. Organomet. Chem., 1983, 246, C42.
216 G. Grötsch and W. Malisch, J. Organomet. Chem., 1983, 246, C49.
217 Y.S. Yu and R.J. Angelici, Organometallics, 1983, 2, 1018.
218 Y.S. Yu and R.J. Angelici, Organometallics, 1983, 2, 1583.
219 S.C. Kao, C.H. Thiel, and R. Pettit, Organometallics, 1983, 2, 914.
220 C.P. Casey, S.R. Marder, and P.J. Fagan, J. Am. Chem. Soc., 1983, 105, 7197.
221 A.F. Dyke, S.A.R. Knox, M.J. Morris, and P.J. Naish, J. Chem. Soc., Dalton Trans., 1983, 1417.
222 A.G. Orpen, J. Chem. Soc., Dalton Trans., 1983, 1427.
223 G.M. Dawkins, M. Green, J.C. Jeffery, C. Sambale, and F.G.A. Stone, J. Chem. Soc., Dalton Trans., 1983, 499.
224 J. Ros and R. Mathieu, Organometallics, 1983, 2, 771.
225 P. Brun, G.M. Dawkins, M. Green, R.M. Mills, J.-Y. Salaün, F.G.A. Stone, and P. Woodward, J. Chem. Soc., Dalton Trans., 1983, 1357.
226 W.-K. Wong, K.W. Chiu, G. Wilkinson, A.M.R. Galas, M. Thornton-Pett, and M.B. Hursthouse, J. Chem. Soc., Dalton Trans., 1983, 1557.
227 J.B. Keister, M.W. Payne, and M.J. Muscatella, Organometallics, 1983, 2, 219.
228 D. deMontauzon and R. Mathieu, J. Organomet. Chem., 1983, 252, C83.
229 G.R. Clark, S.V. Hoskins, T.C. Jones, and W.R. Roper, J. Chem. Soc., Chem. Commun., 1983, 719.
230 Y.C. Lin, J.C. Calabrese, and S.S. Wreford, J. Am. Chem. Soc., 1983, 105, 1679.
231 J.P. Selegue, J. Am. Chem. Soc., 1983, 105, 5921.
232 P.Q. Adams, D.L. Davies, A.F. Dyke, S.A.R. Knox, K.A. Mead, and P. Woodward, J. Chem. Soc., Chem. Commun., 1983, 222.
233 R.E. Colborn, A.F. Dyke, S.A.R. Knox, K.A. Mead, and P. Woodward, J. Chem. Soc., Dalton Trans., 1983, 2099.
234 R.E. Colborn, D.L. Davies, A.F. Dyke, A. Endesfelder, S.A.R. Knox, A.G. Orpen, and D. Plaas, J. Chem. Soc., Dalton Trans., 1983, 2661.

235 L.R. Beanan, Z.A. Rahman, and J.B. Keister, Organometallics, 1983, 2, 1062.
236 C.M. Jensen and H.D. Kaesz, J. Am. Chem. Soc., 1983, 105, 6969.
237 A.F. Hill, W.R. Roper, J.M. Waters, and A.H. Wright, J. Am. Chem. Soc., 1983, 105, 5939.
238 A. Vogler, J. Kisslinger, and W.R. Roper, Z. Naturforsch., Teil B., 1983, 38, 1506.
239 G.R. Clark, N.R. Edmonds, R.A. Pauptit, W.R. Roper, J.M. Waters, and A.H. Wright, J. Organomet. Chem., 1983, 244, C57.
240 E.D. Morrison, G.R. Steinmetz, G.L. Geoffroy, W.C. Fultz, and A.L. Rheingold, J. Am. Chem. Soc., 1983, 105, 4104.
241 J.R. Shapley, M.E. Cree-Uchiyania, G.M. St. George, M.R. Churchill, and C. Bueno, J. Am. Chem. Soc., 1983, 105, 140.
242 A.W. Coleman, P.B. Hitchcock, M.F. Lappert, R.K. Maskell, and J.H. Müller, J. Organomet. Chem., 1983, 250, C9.
243 K.H. Theopold, and R.G. Bergman, J. Am. Chem. Soc., 1983, 105, 464.
244 W.J. Laws and R.J. Puddephatt, J. Chem. Soc., Chem. Commun., 1983, 1020.
245 W.A. Herrmann, C. Bauer, J. M. Huggins, H. Pfisterer, and M.L. Zeigler, J. Organomet. Chem., 1983, 258, 81.
246 W.A. Herrmann, C. Bauer, and A. Schäfer, J. Organomet. Chem., 1983, 256, 147.
247 N.T. Allison, J.R. Fritch, K.P.C. Vollhardt, and E.C. Walborsky, J. Am. Chem. Soc., 1983, 105, 1384.
248 P.F. Seidler, H.E. Bryndza, J.E. Frommer, L.S. Stuhl, and R.G. Bergman, Organometallics, 1983, 2, 1701.
249 R.B.A. Pardy, G.W. Smith, and M.E. Vickers, J. Organomet. Chem., 1983, 252, 341.
250 A. Nutton, A. Vázquez de Miguel, K. Isobe,and P.M. Maitlis, J. Chem. Soc., Chem. Commun., 1983, 166.
251 N.J. Meanwell, A.J. Smith, H. Adams, S. Okeya and P,M. Maitlis, Organometallics, 1983, 2, 1705.
252 L.F. Halle, P.B. Armentrout, and J.L. Beauchamp, Organometallics, 1983, 2, 1829.
253 R. Usón, J. Forniés, P. Espinet, A. García, and A. Sanáu, Transition Met. Chem., 1983, 8, 11.
254 A. Scrivanti, G. Carturan, and B. Crociani, Organometallics, 1983, 2, 1612.
255 U. Belluco, B. Crociani, R. Michelin, and P. Uguagliati, Pure Appl. Chem., 1983, 55, 47.
256 H.C. Clark, V.K. Jain, and G.S. Rao, J. Organomet. Chem., 1983, 259, 275.
257 F. Canziani, F. Galimberti, L. Garlaschelli, M.C. Malatesta, A. Albinati, and F. Ganazzoli, J. Chem. Soc., Dalton Trans., 1983, 827.
258 K.R. Grundy and K.N. Robertson, Organometallics, 1983, 2, 1736.
259 S. Muralidharan and J.H. Espenson, Inorg. Chem., 1983, 22, 2786.
260 R.D. Barr, M. Green, J.A.K. Howard, T.B. Marder, I. Moore, and F.G.A. Stone, J. Chem. Soc., Chem, Commun., 1983, 746.
261 K.A. Mead, I. Moore, F.G.A. Stone, and P. Woodward, J. Chem. Soc., Dalton Trans., 1983, 2083.
262 M.R. Awang, J.C. Jeffery, and F.G.A. Stone, J. Chem. Soc., Dalton Trans., 1983, 2091.
263 M.R. Awang, G.A. Carriedo, J.A.K. Howard, K.A. Mead, I. Moore, C.M. Nunn, and F.G.A. Stone, J. Chem. Soc., Chem. Commun., 1983, 964.

13

Metal–Hydrocarbon π-Complexes, other than π-Cyclopentadienyl and π-Arene Complexes

BY J. A. S. HOWELL

A. Reviews

Reviews have been published on the use of organometallic complexes in organic synthesis[1], on π-complexes of cobalt[2], on the reactivity of coordinated hydrocarbons[3], on alkyne substituted homo and heterometallic carbonyl clusters[4], and on bridged hydrocarbyl or hydrocarbon binuclear clusters[5]. In addition, reviews on diazabutadiene complexes[6], on bridged methylene complexes[7], on three centre M---H---C bonds[8] and on the mechanisms of substitution and exchange at low valent transition metal centres[9-11] contain material of relevance to this chapter.

B. Allyl Complexes and Complexes Derived from Monoolefins

1. Ni, Pd and Pt

Molecular orbital calculations show an increasing order of $(PH_3)_2Ni$-L back donation (L = CO < C_2H_4 < C_2H_2) which correlates with the decreasing order of their π^* orbital energies; electrostatic interaction becomes larger in the order CO < C_2H_4 $\approx$ C_2H_2.[12] Reaction of C_2H_4 and its derivatives with Ni atoms yields $Ni(ol)_3$ complexes which were not isolated, but yield stable $(PF_3)_2Ni(ol)$ complexes on treatment with PF_3.[13] C_2H_4 may be displaced from $(PPh_3)_2Ni(C_2H_4)$ by methylenecyclopropane and vinyl silanes such as CH_2=$CHSiR_3$ to give $(PPh_3)_2Ni(ol)$ complexes;[14,15] these may alternatively be prepared by reduction of $Ni(acac)_2$ with Et_2AlOEt in the presence of olefin and PPh_3. C_2H_4 displacement of $(PPh_3)_2Pt(C_2H_4)$ by butatrienes yields $(PPh_3)_2Pt(ol)$ complexes in which the butatriene is bound via a terminal C=C bond[16]. Equilibrium constants for a variety of equilibria

$$L_4Ni + ol \rightleftharpoons L_2Ni(ol) + 2L$$

have been reported; K values increase by 10^{10} between cyclohexene and maleic anhydride and by 10^8 between $P(OC_6H_4Me$-$\underline{p})_3$ and $P(OC_6H_4Me$-$\underline{o})_3$. Cyano-olefins exhibit both olefin and nitrile coordination.[17,18] The complexes (triphos)Pt(mesityl P = CPh_2)[19] and (bipy)Ni(xylyl P = CPH_2)[20] contain η^2-phosphaalkene ligands; a crystal structure determination of the latter shows a typical P-C single bond (1.83 Å) and the structure may be regarded as a nickelaphosphacyclopropane.

(Pentamethyldiethylenetriamine)LiOC(Me_2N)Ni(C_2H_4)$_2$ has been prepared by reaction of (cdt)NiCO with $LiNMe_2$ in pentamethyldiethylenetriamine, followed by displacement with C_2H_4; a planar Ni(C_2H_4)$_2$ fragment is σ-bonded to a likewise planar carbamoyl moiety[21].

$[PtCl(tmen)(C_2H_4)]^+$ reacts with cyanate _via_ addition to C_2H_4 to give (1), and similar products of addition are observed in reaction with NO_2^- and N_3^- [22,23]. Treatment of _cis_-$PtCl_2(PMe_2Ph)$-(η^2-allene) with Me_2C=NOH results in nucleophilic attack, followed by 1,3-hydrogen shift and ring closure, to give (2)[24]. A crystal structure determination of the chelated amido complex (3) reveals that the olefin is not quite perpendicular to the square planar Pt, a consequence of steric strain[25].

Treatment of $[PtCl_2(C_2H_4)]_2$ with Cp_2Mg yields CpPt(η^2-C_2H_4)-(η^1-Cp) and the dimer (4) as a mixture of _exo_ and _endo_ isomers; protonation of the former gives $[CpPt(\eta^2-C_2H_4)(\eta^2-C_5H_6)]^+$.[26] A series of CpPd($PPh_3$)($CH_2$=$CHC_6H_4$-_p_-Y) complexes have been characterized by crystallography; the Pd bond length to the internal carbon shortens with increasing electron withdrawing power of Y.[27] Treatment of Cp_2Ni with C_2H_4 in the presence of LiR (R = alkyl) at -78 °C yields CpNi(C_2H_4)R complexes; where R is cyclopropyl or phenyl, rearrangement occurs on warming in the presence of excess C_2H_4 to give CpNi(C_2H_4)CH_2CH_2R. Use of LiMe and butadiene gives the CpNi(η^2-butadiene)Me complex in which the diene adopts an s-_trans_ formation.[28-30] Decomposition of the intramolecularly chelated complexes (5) proceeds differently depending on the value of _n_. Where _n_ = 2, CpNi(allyl) complexes are formed by β-hydrogen elimination, whereas for _n_ = 3, formal elimination of CpNiH occurs to give a series of hydrocarbons; where _n_ = 1, rearrangement to either CpNi(cyclopentyl) or CpNi(vinylallyl) complexes is observed[31]. Treatment of (6) with dmpe yields (7), which spontaneously rearranges at room temperature to give (8); these reactions are of interest with respect to postulated intermediates in the Pd catalysed dimerization of dienes.[32] Coupling of C_2H_4 and CO_2 by reaction with (cdt)Ni yields (9)[33], and an analogous coupling using 2,3-dimethylbutadiene and $(cod)_2Ni$ gives (10).[34,35]

Photoelectron spectra (M = Ni, Pt)[36] and molecular orbital calculations (M = Ni)[37] have been reported for $(allyl)_2M$. Treatment

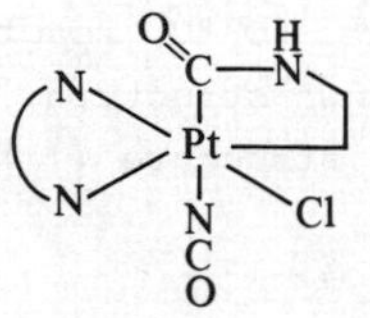

(1)

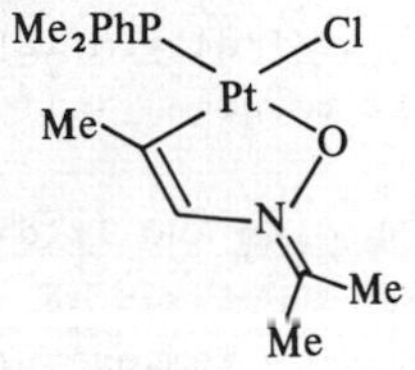

(2)

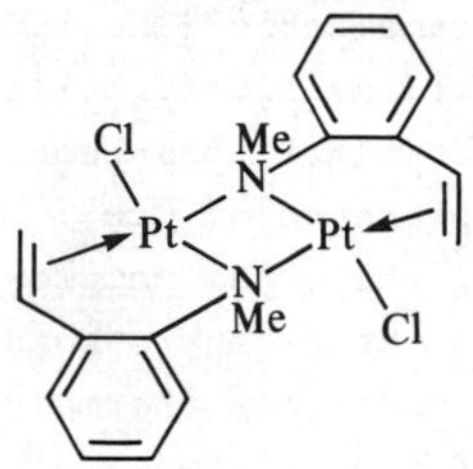

(3)

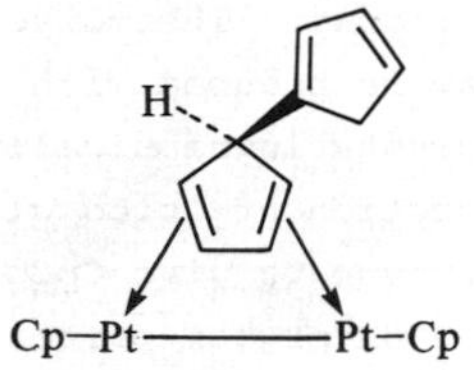

(4) *exo*

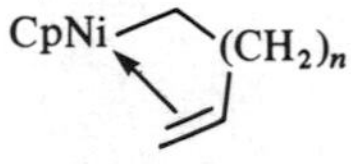

(5)

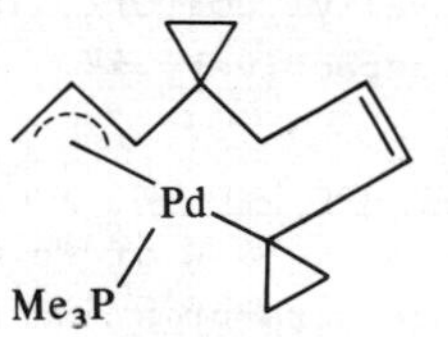

(6)

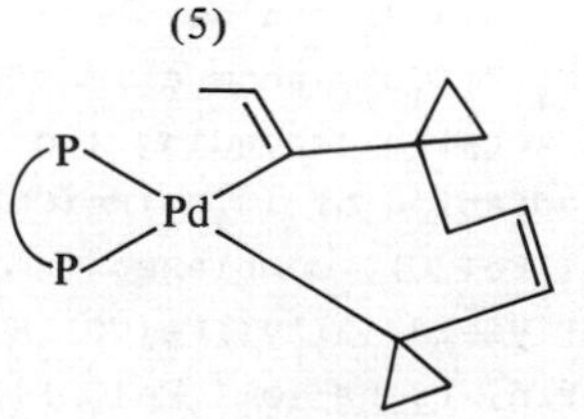

(7)

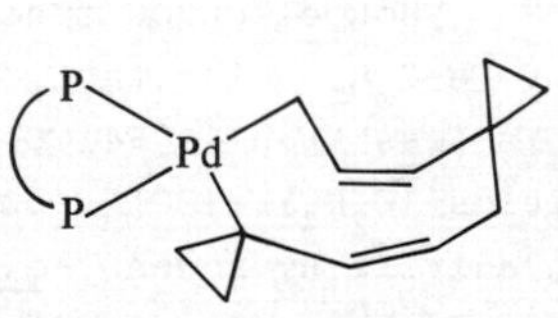

(8)

of CpM(allyl) (M = Ni, Pd, Pt) with H_2S at -78 °C yields (allyl)M-(μ-SH)$_2$M(allyl) dimers which eliminate propene at -60 °C, react with (allyl)$_2$M to give [(allylM)$_2$S]$_x$ polymers of unknown structure, and react with BuLi to give [(allylNi)$_3$S$_2$]$^+$ salts of structure (11).[38-40]

2. Cobalt, Rhodium and Iridium

(PMe$_3$)$_3$CoX compounds (X = Cl, Br, I) form unstable complexes with olefins at low temperature; stable (PMe$_3$)$_3$Co(C_2H_4)R complexes can be isolated after treatment with LiR (R = Me, Ph). A crystal structure determination (R = Ph) shows a trigonal bipyramidal geometry with both C_2H_4 (in the trigonal plane) and Ph in equatorial positions. Thermolysis yields paramagnetic (PMe$_3$)$_3$Co-(C_2H_4) which may be reduced with Li to give Li[(PMe$_3$)$_3$Co(C_2H_4)], characterized crystallographically.[41,42] CpCo(C_2H_4)$_2$ has been prepared by reduction of Cp$_2$Co with K in the presence of C_2H_4; it reacts with 2-butyne to give CpCo(η^6-C$_6$Me$_6$),[43] while the MeCpCo-(C_2H_4)$_2$ analogue undergoes protonation to give the 3-centre bonded complex (12).[44] The Me$_5$CpCo(C_2H_4)$_2$ analogue undergoes thermolysis to give (Me$_5$Cp)$_3$Co$_3$ (μ_3-CMe)$_2$.[45] Reaction of the related Me$_5$CpRh(C_2H_4)$_2$ with PMe$_3$ gives Me$_5$CpRh(C_2H_4)PMe$_3$ which undergoes oxidative addition with MeI to give Me$_5$CpRhCH$_3$(PMe$_3$)I.[46]

Photolyses of (η^3-allyl)Co(CO)$_3$ in argon, nitrogen and CO matrices yields (η^3-allyl)Co(CO)$_2$, (η^3-allyl)Co(CO)$_2$N$_2$ and (η^1-allyl)Co(CO)$_4$ respectively.[47]

3. Iron, Ruthenium and Osmium

Cocondensation at < 18 K of Fe with C_2H_4 yields Fe(C_2H_4) although at higher temperatures and concentrations of Fe, the species Fe$_2$(C_2H_4)$_{\underline{x}}$ ($\underline{x}$ = 1,2) and Fe$_{\underline{p}}$(C_2H_4)$_{\underline{p}}$ ($\underline{p}$ > 2) can be observed.[48] Photoelectron spectra of (C_2H_4)Fe(CO)$_4$ show a significant Fe-C_2H_4 σ-interaction, but the total backbonding (to dxz, dyz) is less than in Fe(CO)$_5$.[49] Irradiation of (C_2H_4)Fe(CO)$_4$ at 77 K yields (C_2H_4)Fe(CO)$_3$, but other (ol)Fe(CO)$_4$ complexes possessing allylic hydrogens (e.g. propene) yield (allyl)Fe(CO)$_3$H complexes. In the presence of excess olefin, trans-(ol)$_2$Fe(CO)$_3$ complexes are formed which on warming to 195 K yield catalytically active (with respect to isomerization) (ol)$_2$Fe$_2$(CO)$_6$ dimers.[50] Limiting low temperature ^{13}C spectra have been obtained for several (ol)Fe(CO)$_4$ complexes (ol = cyclobutene, cycloheptene, cyclooctene); the exchange mechanism appears to be non-Berry and involves separate exchange of the two diastereotopic axial carbonyls with the two

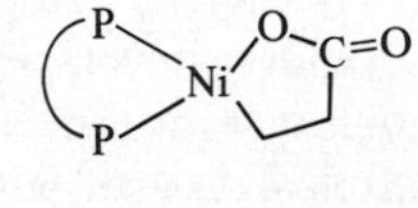

(9)

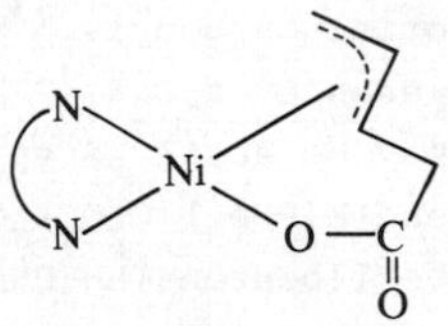

(10)

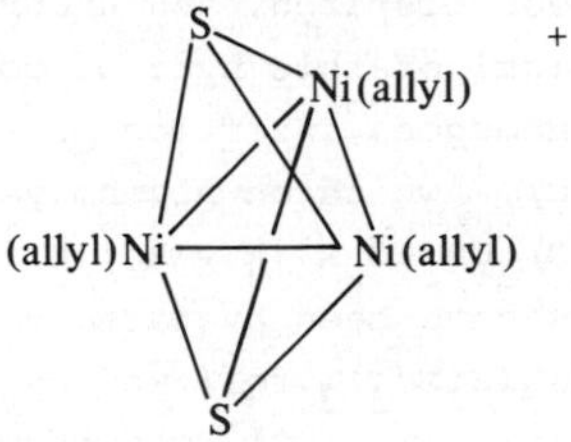

(11)

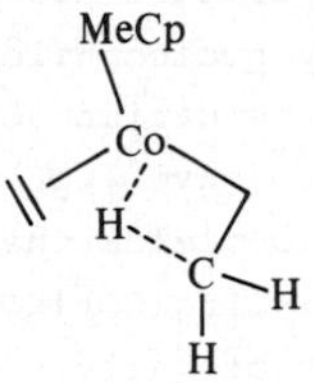

(12)

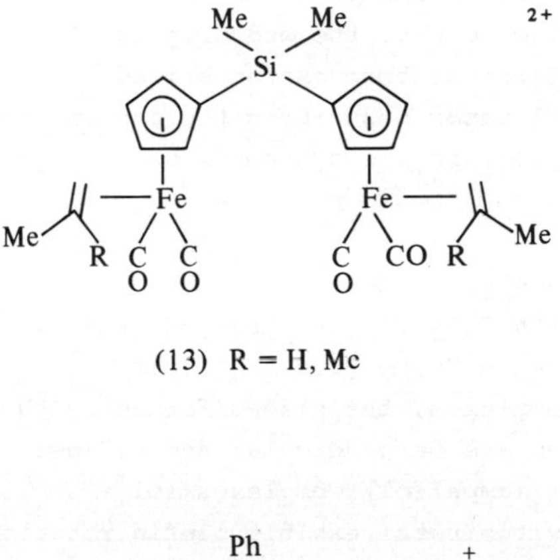

(13) R = H, Me

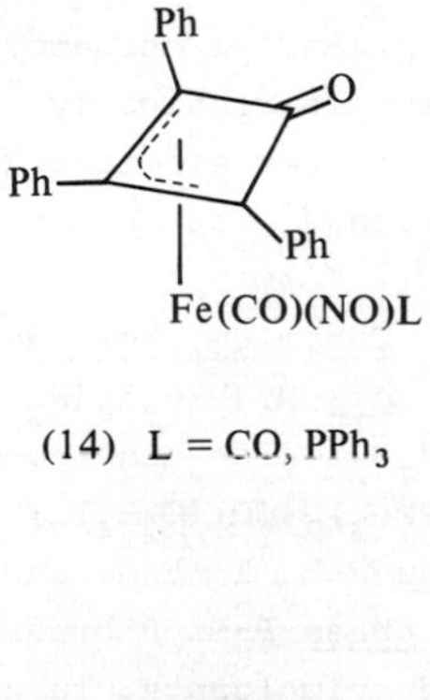

(14) L = CO, PPh_3

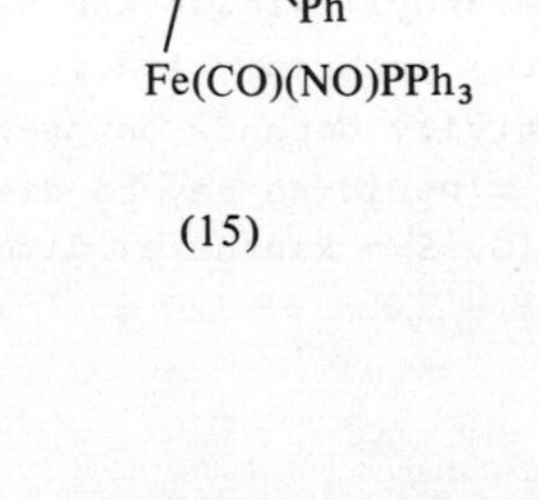

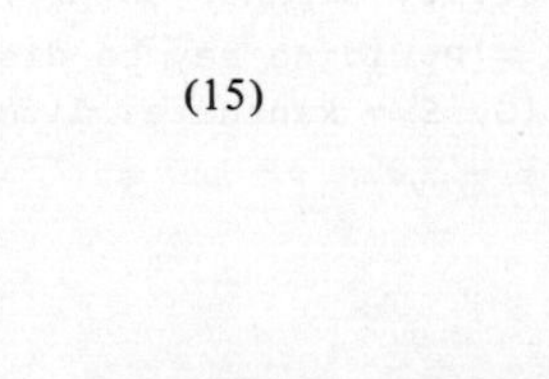

(15)

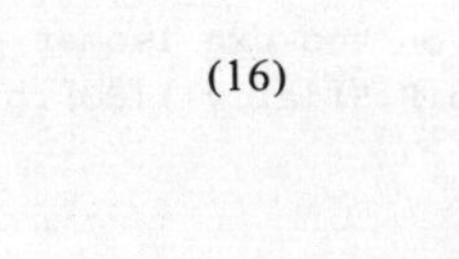

(16)

equatorial carbonyls.[51] Reduction of $Fe(acac)_3$ with Et_2AlOEt in the presence of C_2H_4/$PPhMe_2$ gives $(C_2H_4)_2Fe(PPhMe_2)_3$ from which C_2H_4 may be easily displaced to give (butadiene)$Fe(PPhMe_2)_3$.[52] (Benzene)$M(PMe_3)(C_2H_4)$ complexes (M = Ru, Os) undergo protonation to give $[(benzene)MH(PMe_3)(C_2H_4)]^+$ salts which exist in solution in equilibrium with the alkyl $[(benzene)MC_2H_5(PMe_3)]^+$ cation.[53] Interestingly, the analogous $[(C_6Me_6)RuH(PMePh_2)(C_2H_4)]^+$ cation can also be formed by treatment of $(C_6Me_6)RuMe_2$ with Ph_3C^+.[54]

The bridged dication (13) has been prepared, and undergoes attack by nucleophiles which are typical of this type of complex.[55] The carbene cation $[CpFe(CO)_2CH_2]^+$ undergoes insertion on treatment with CO to give $[CpFe(CO)_2(\eta^2\text{-}CH_2{=}C{=}O)]^+$ which on alcoholysis or hydrolysis yields the neutral $CpFe(CO)_2CH_2CO_2R$ (R = Me, H).[56]

(Allyl)Fe(CO)(NO)(CNMe) complexes have been prepared by treatment of (allyl)$Fe(CO)_2NO$ with $NaN(SiMe_3)_2$ followed by alkylation.[57] Borohydride reduction of (allyl)Fe(CO)(NO)L complexes (L = PPh_3, $P(OR)_3$) results in propene elimination and formation of $FeL_2(NO)_2$, $[Fe(CO)_2(NO)L]^-$ and $[Fe(CO)_3NO]^-$.[58] Reaction of $[Fe(CO)_3NO]^-$ with $Ph_3C_3^+$ yields $(\eta^3\text{-}C_3Ph_3)Fe(CO)_2NO$ and complex (14), both characterized crystallographically. The PPh_3 derivative of (14) may be alkylated to give the cbd complex (15).[59] The σ,π-allyl carbamoyl complex (16) has been characterized crystallographically,[60] and a full paper describing the use of the related ferrelactone derivatives (17) in the synthesis of β-lactams has appeared.[61]

4. Chromium, Molybdenum and Tungsten

cis-$Mo(PMe_3)_4(N_2)_2$ reacts with C_2H_4 to give trans-$Mo(PMe_3)_4$-$(C_2H_4)_2$ which undergoes exchange with CO to give trans,mer-$Mo(PMe_3)_3(CO)(C_2H_4)_2$; in both complexes, the planes formed by the trans-C_2H_4 ligands and the Mo atom are perpendicular and eclipse the trans-P-Mo-P bonds.[62] $W(CO)_3$(dmpe)(ol) complexes (ol = dimethylmalonate, fumarate, methylfumarate) exhibit olefin rotation which may be ascribed unambiguously to a rotation about the M-ol bond.[63]

Nucleophilic attack on pure diastereoisomers of [RCpMo(CO)-(NO)(η^3-1,3-dimethylallyl)]$^+$ (R = neomenthyl) yields the neutral RCpMo(CO)(NO)(ol) complex from which the enantiomerically pure olefin can be liberated. The selectivity depends on preferential attack on the exo isomer cis to NO.[64] Pyridine may be displaced from Mo(S,S)(allyl)$(CO)_2$py complexes (S, S = xanthate, dithio-

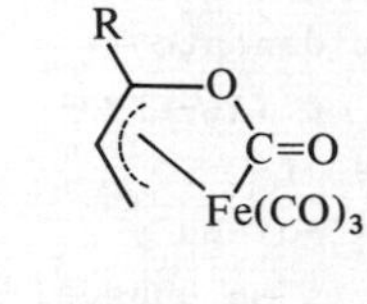

(17)

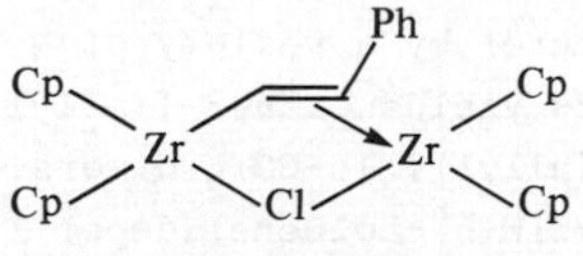

(18)

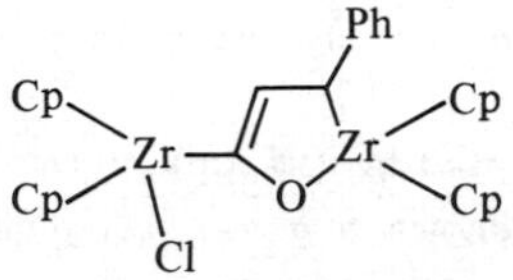

(19)

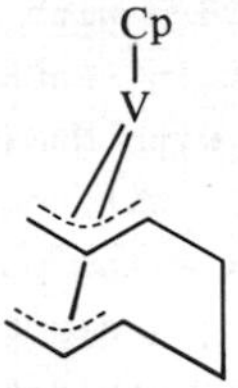

(20)

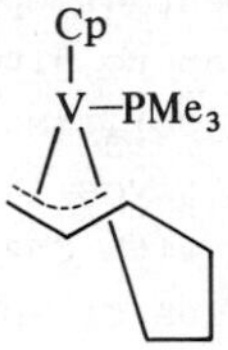

(21)

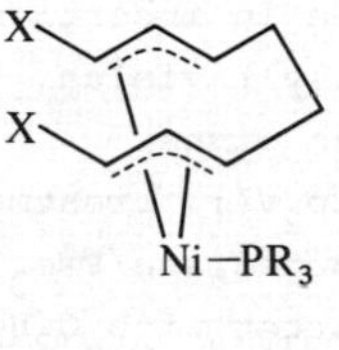

(22) $X = CO_2Me$

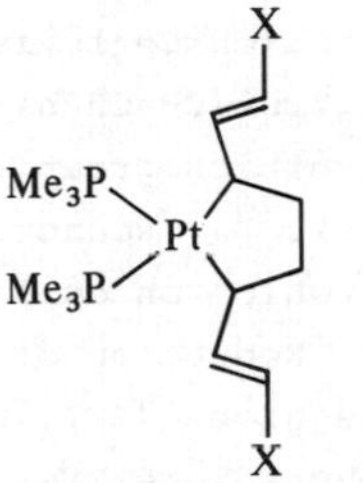

(23) $X = CO_2Me$

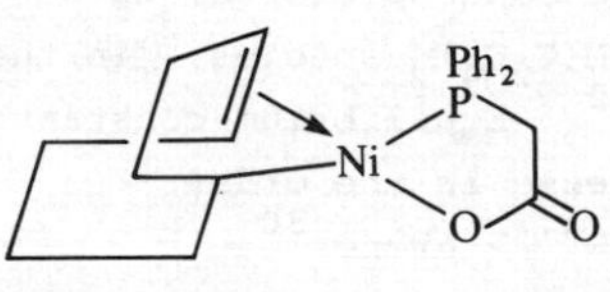

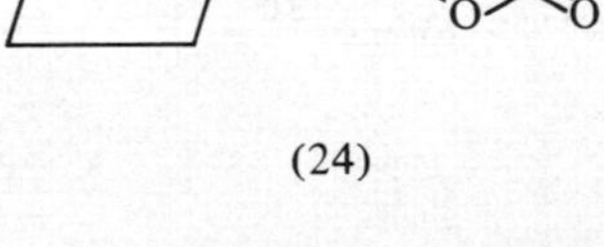

(24)

carbamate) by a variety of N and P donors; use of bidentate ligands yields either [(allyl)Mo(CO)$_2$(S,S)]$_2$(μ-L-L) or [(L-L)Mo-(S,S)(allyl)]$_2$(μ-CO)$_2$ dimers.[65] Crystal structures of [(allyl)W(η^6-toluene)(dmpe)]$^+$ [66] and (allyl)MoBr(N,N'-di-Bu^t-ethanediimine(CO)$_2$ [67] have been reported, but exhibit no unusual features in the M-allyl bonding.

5. Other Metals

Treatment of CpMn(CO)$_2$C=CHPh with P(OEt)$_3$ results in rearrangement to give CpMn(CO)$_2$[η^2-PhCH=CHP(O)(OEt)$_2$].[68] Reaction of (allyl)M(CO)$_3$ (M = Mn, Re) with NaN(SiMe$_3$)$_2$, followed by alkylation, gives (allyl)Mn(CO)$_2$CNR complexes which are trigonal bipyramidal with CNR and CO in axial positions.[69-71]

Reduction of (Me$_5$Cp)$_2$TiCl$_2$ under C$_2$H$_4$ yields (Me$_5$Cp)$_2$Ti(η^2-C$_2$H$_4$); a crystal structure determination shows a great lengthening of the C-C bond (1.438 Å). The C$_2$H$_4$ may be displaced by donor ligands (CO, CNR) and reacts reversibly with C$_2$H$_4$ to give the metallocycle (Me$_5$Cp)$_2$TiCH$_2$(CH$_2$)$_2$CH$_2$.[72] Alkenyl complexes of the type Cp$_2$TiCH$_2$(CH$_2$)$_n$CH=CH$_2$ (n = 0-11) isomerize through a β-hydrogen elimination to give Cp$_2$Ti(η^3-allyl) complexes[73,74]. Hydrozirconation of Cp$_2$ZrCl(CH=CH$_2$) yields (μ-C$_2$H$_4$)[Cp$_2$ZrCl]$_2$; in contrast, use of the Cp$_2$ZrCl(CH=CHPh) derivative gives (18) in which the Zr-C σ-bond exhibits great reactivity, for example in insertion of CO to give (19).[75] Reduction of Cp$_2$V with Li(allyl) yields [CpV(allyl)]$^-$ which on oxidation with allylbromide gives CpV(allyl)$_2$. Reduction of Cp$_2$V with K gives K[Cp$_2$V]; treatment with butadiene gives (20), while treatment with butadiene/PMe$_3$ gives CpV(η^4-butadiene)PMe$_3$. Reaction of the latter with C$_2$H$_4$ gives (21).[76]

Crystal structure determinations of (C$_2$H$_4$)Cu[hydrotris(1-pyrazolyl)borate] and (C$_2$H$_4$)Cu[hydrotris(3,5-dimethyl-1-pyrazoyl)-borate]·CuCl show C=C distances (1.34, 1.35 Å) which are essentially unchanged from free C$_2$H$_4$.[77] Cocondensation of Au or Ag atoms with C$_2$H$_4$ gives both M(C$_2$H$_4$) and M(C$_2$H$_4$)$_2$ species; in the latter, the C$_2$H$_4$ ligands are parallel.[78,79] Equilibrium constants in chloroform for 1:1 Ag(ol)$^+$ adducts decrease in the order RR'C=CH$_2$ > RHC=CR'H > RR'C=CR"H > RR'C=CR"R"' > arene.[80]

C. Complexes Derived from Unconjugated Dienes

1. Nickel, Palladium and Platinum

$(Cod)_2Ni$ reacts with $CH_2=CH-CH=CHCO_2Me$ to give the (cod)Ni-(η^4-diene) complex from which the cod may be displaced by $Cy_2PC_2H_4PCy_2$ to give $(Cy_2PC_2H_4PCy_2)Ni(\eta^4$-diene), although a crystal structure determination of this complex shows a very weak coordination of the carbomethoxy substituted C-C bond. In the presence of PCy_3, $(cod)_2Ni$ reacts with the above diene to give (22), while $(cod)_2Pt$ reacts in the presence of PMe_3 to give (23).[81,82] $(Cod)_2Ni$ undergoes oxidative addition on reaction with $Ph_2PCH_2CO_2H$ to give a 60-40 mixture of (24) and (25).[83]

2. Cobalt, Rhodium and Iridium

Treatment of $[(cod)RhCl]_2$ with R_2S gives $[(cod)RhL_2]^+$ salts, whereas treatment of $[(nbd)_2Rh]^+$ gives five coordinate $[(nbd)_2Rh(SR_2)]^+$ cations; a crystal structure reveals a square pyramidal structure with apical R_2S.[84] A similar square pyramidal structure is found for Ir(acac-C^3)(cod)(phen),[85] while salts of five-coordinate anions of the type $[M(SnCl_3)_2(diene)(PPh_3)]^-$ (M = Rh, diene = cod, nbd; M = Ir, diene = cod) have been prepared and characterised spectroscopically.[86] Dibenzocot acts as a η^4-1,5-diene in its complexes of the type $[M(diene)L_2]^+$ (M = Rh, Ir), and appears to be more electron withdrawing than cod.[87] Crystal structure determinations of 8-hydroxyquinolinato(cod)Rh[88] and (cod)Rh(3,5-dimethylpyrazole)Cl[89] have been reported. $[(Cod)RhCl]_2$ reacts with indole in the presence of AgX to give the $[(cod)Rh(\eta^6$-indole$)]^+$ cation which on treatment with methanolic KOH gives $[(cod)RhOMe]_2$.[90] Reaction of $[(cod)MCl]_2$ (M = Rh, Ir) with $LiPPh_2$ gives $[(cod)M(\mu-PPh_2)]_2$ from which cod may be displaced by both monodentate and bidentate phosphines.[91] Complexes (26), (27) and (28) contain (cod)Rh moieties bridged by 1,6-bis(2'-benzimidazolyl)-2,5-dithiahexane,[92] glyoxalbis(isopropylimine),[93] and 1,3-di-t-butyl-2,4-difluorocyclodiphosphazene.[94]

Reaction of $[(cod)RhCl]_2$ with LiMe in the presence of chd or cod yields (29) and (30) respectively; the methyl group is <u>endo</u>, indicating a transfer of methyl from Rh to diene.[95] Cocondensation of Rh with cod, followed by addition of PPh_3 yields the related complex (31).[96]

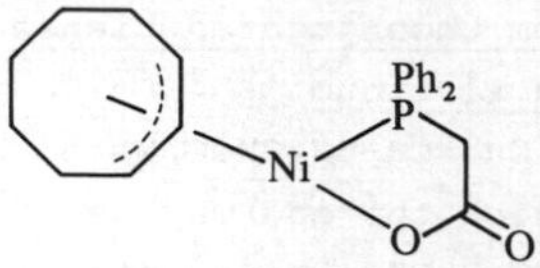

(25)

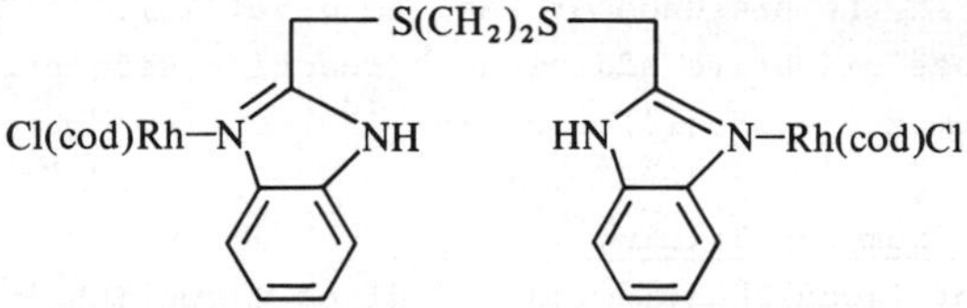

(26)

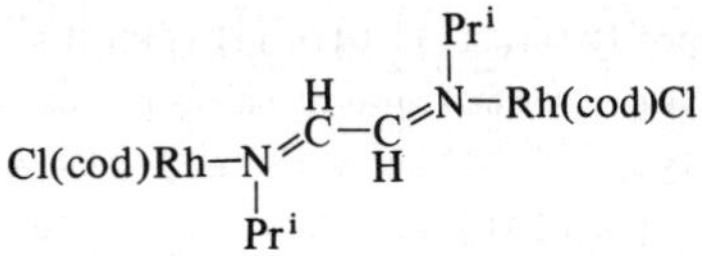

(27)

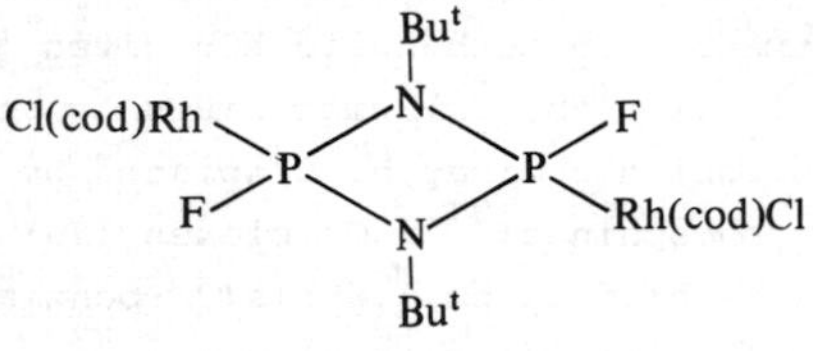

(28)

3. Other Metals

A molecular orbital description of (nbd)$Fe(CO)_3$ has appeared[97] while [CpRu(nbd)PPh_3]ClO_4 has been prepared by reaction of $CpRu(PPh_3)_2Cl$ with nbd in the presence of $NaClO_4$.[98] Crystal structure determinations of the 1:1, 2:1 and 2:3 adducts of cot with $AgNO_3$ all show a 1,5-bonding of the cot ligand.[99-101] In (benzocot)$AgClO_4$, silver has a distorted trigonal bipyramidal configuration consisting of a 1,5-cot ligand, two perchlorate oxygens, and one aromatic C=C bond of a neighbouring benzocot ligand;[102] [(dibenzocot)CuCl]$_4$ contains only 1,5-bonded dibenzocot.[103]

D. Complexes Derived from Conjugated Dienes

1. Iron, Ruthenium and Osmium

i. Acyclic Dienes (Pentadienyl)$_2$M complexes (M = Fe, Ru) adopt a *gauche*-eclipsed ligand conformation in the solid state.[104,105] *In situ* generation of the ylid (32), followed by reaction with $(CO)_3Fe(\eta^4$-CH_3CH=CH-CH=CHCHO) yields the highly conjugated (33) as a mixture of (*E*)- and (*Z*)-isomers,[106] and $Fe(CO)_3$ complexation to natural products of similar structure may be used in the assignment of (*E*)- and (*Z*)-stereochemistry.[107]

The solid state structures of (η^4-benzylideneacetone)$Fe(CO)_2L$ (L = $PPhMe_2$, PEt_3) can be described as square pyramidal with axial phosphine.[108] Cocondensation of Ru with butadiene, followed by reaction with L, yields (butadiene)$_2$RuL complexes (L = CO, PF_3, $CNBu^t$) in reasonable yield.[109]

ii. Cyclic Dienes Molecular orbital calculations on (cbd)$Fe(CO)_3$, including configuration interaction, suggest a strong polarization of the six cbd-M bonding electrons towards the cbd ring.[110] A liquid crystal nmr study of (cbd)$Fe(CO)_3$ has also been reported.[111] [(cbd)$Fe(CO)_2NO]^+$ undergoes nucleophilic addition by $PhNMe_2$ to give (34); the *exo* addition was confirmed by crystallography.[112] Trapping experiments following Me_3NO oxidation of (35) show that free triplet trimethylenemethane is liberated directly.[113]

Low temperature reaction of [CpFe(CO)(diphos)]$^+$ with H^- gives CpFe(CO)(diphos)H which on heating in toluene undergoes exclusive endo transfer of hydrogen to give (η^4-C_5H_6)Fe(CO)(diphos),[114] and nmr methods for the assignment of stereochemistry in 5-substituted cyclopentadiene complexes have been described.[115] (1,1-dimethylsilole)$Fe(CO)_3$ and (1,1-dimethylgermole)$Fe(CO)_3$ have been prepared.[116]

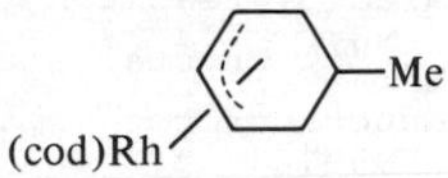

(29)

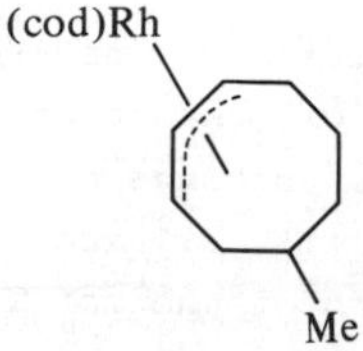

(30)

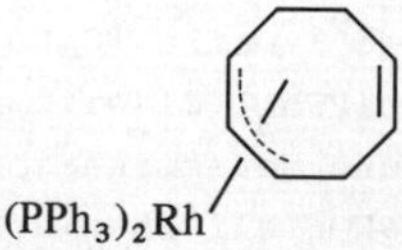

(31)

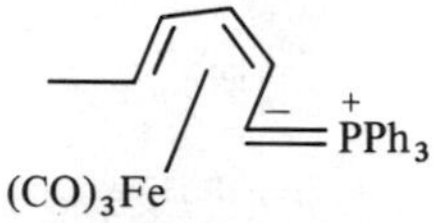

(32)

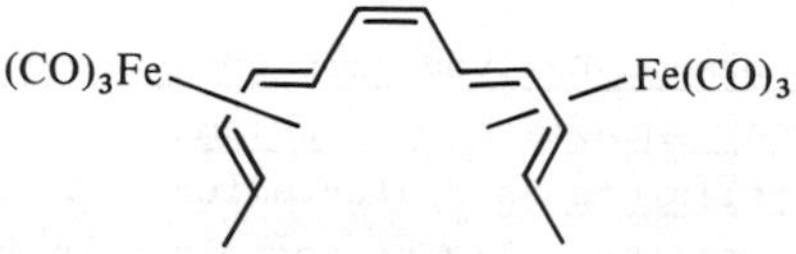

(33)

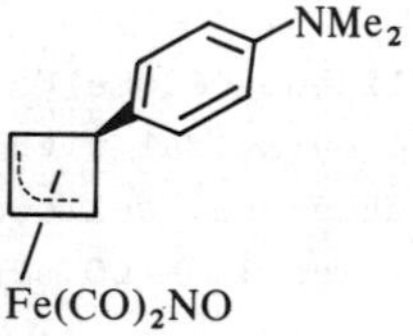

(34)

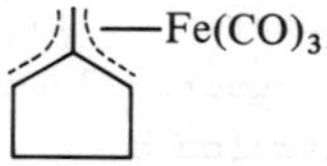

(35)

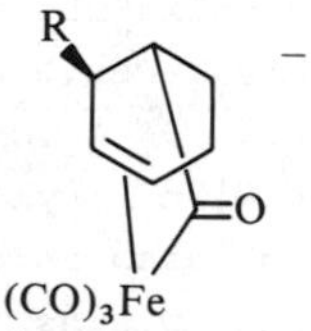

(36)

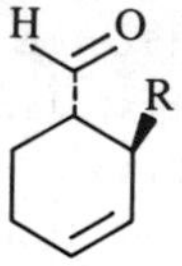

(37)

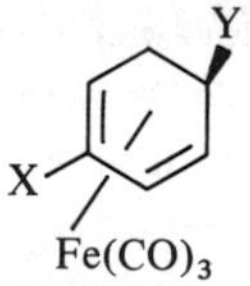

(38) X = H, Y = aryl
(39) X = H, OMe, Y = CH_2CO_2Me
(41) X = H, Y = OH

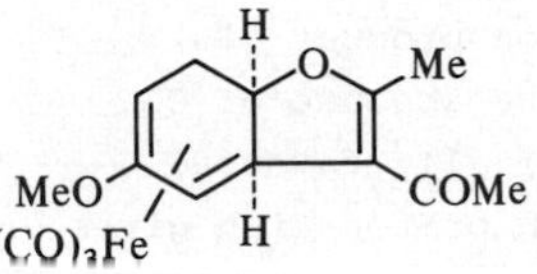

(40)

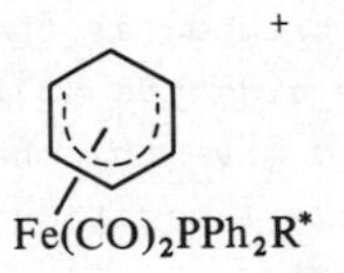

(42) R* = menthyl

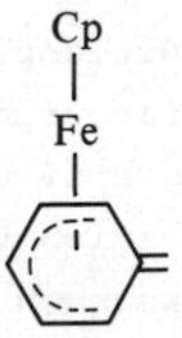

(43)

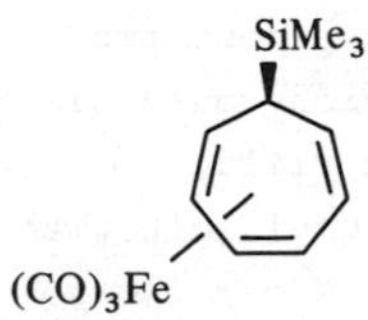

(44)

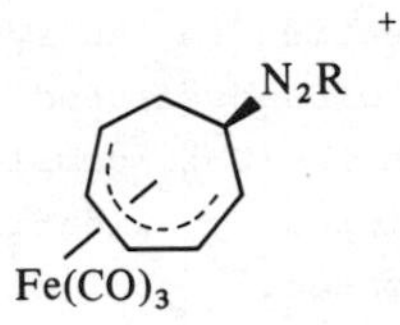

(45)

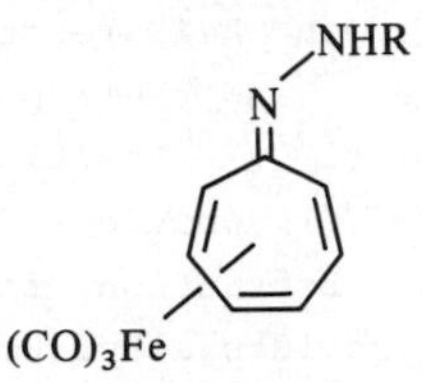

(46)

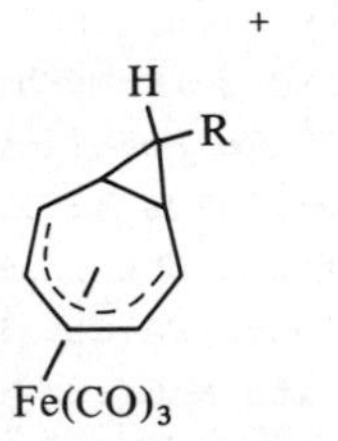

(47) R = p—$NO_2C_6H_4$
(51) R = H

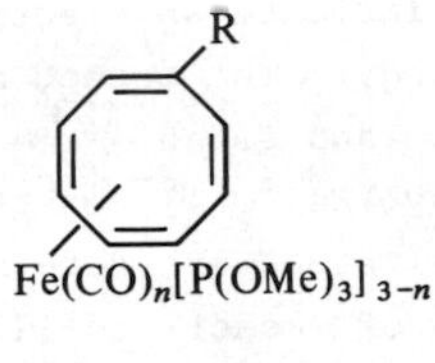

(48) R = p—$NO_2C_6H_4$; n = 3
(50) R = N = NR; n = 1, 2

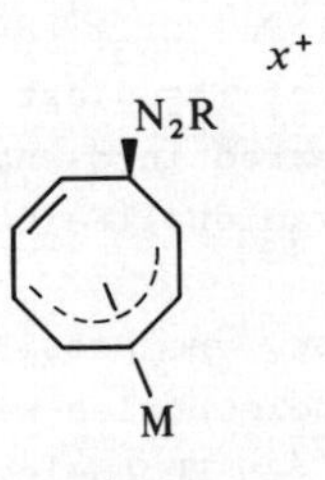

(49a) M = $Fe(CO)_n[P(OMe)_3]_{3-n}$; n = 1, 2; x = 1
(49b) M = $Fe(CO)[P(OMe)_3]P(O)(OMe)_2$; x = O

Nucleophilic attack by R^- on (chd)$Fe(CO)_3$ yields the kinetically favoured product of attack at the 2-position which under a pressure of CO may be trapped as the insertion product (36); protonation yields the aldehyde (37). In the absence of CO, rearrangement occurs to give the thermodynamically favoured product of attack at the 1-position, which on protonation gives substituted cyclohexenes. Similar results are obtained with (butadiene)$Fe(CO)_3$ and its derivatives.[117,118] Applications of [(cyclohexadienyl)$Fe(CO)_3]^+$ salts continue, and full papers have appeared on their reactions with O-silylated enolates and allyl silanes[119] and their potential in the synthesis of aspidosperma alkaloids.[120] Reaction of substituted tertiary cyclohexen-2-ols with $Fe(CO)_5$ provides (chd)$Fe(CO)_3$ complexes in good yield; hydride abstraction, followed by oxidation of the dienyl salt with Ce^{4+}, provides alkyl benzenes.[121] [(1,2-disubstituted-4-alkoxy-cyclohexadienyl)$Fe(CO)_3]^+$ salts have been prepared, but in general show a low regioselectivity in their reactions with nucleophiles.[122] [(Cyclohexadienyl)$Fe(CO)_3]^+$ salts may be arylated using aryltrimethylsilanes and stannanes to give (38); kinetic studies indicate an electrophilic attack by the dienyl salt on the arene ring. The method may also be extended to 2-trimethylsilylthiophene and furan.[123] Reaction of $BrZnCH_2CO_2Me$ and [(cyclohexadienyl)$Fe(CO)_3]^+$ or its 2-methoxy analogue yields (39).[124] A crystal structure determination of (40) confirms the initial exo addition of $(acac)^-$ to [(2-methoxylcyclohexadienyl)$Fe(CO)_3]^+$, followed by endo proton removal on treatment with MnO_2.[125] Reversible addition of OH^- to [(cyclohexadienyl)$Fe(CO)_3]^+$ yields (η^5-cyclohexadienyl)$Fe(CO)_2COOH$ in dead-end equilibrium. In polar solvents, this complex rearranges to give (41).[126] [(2-methoxy-cyclohexadienyl)$Fe(CO)_3]^+$ has been resolved into pure enantiomers via separation of its diastereoisomeric menthyl ethers,[127,128] and it may be converted into enantiomerically pure [(2-methoxy-5--methylcyclohexadienyl)$Fe(CO)_3]^+$ via a Wittig reaction followed by protonation.[129]

In addition, the possibility of asymmetric induction in reactions of nucleophiles with complexes of type (42) has been demonstrated.[127] Photoelectron spectra of derivatives of type (43) show them to be normal d^6 complexes,[130] and a general review of their chemistry has appeared.[131]

A crystal structure determination of (44) (formed from [$(C_7H_7)Fe(CO)_3]^-$) confirms the exo nature of the substituent; this complex shows a much lower barrier to 1,3-shift compared to (chpt)Fe$(CO)_3$, and a photoelectron study indicates that this may be the result of stronger Fe-diene bonding.[132,133] Arene diazonium salts undergo nucleophilic attack on (chpt)Fe$(CO)_3$ to give (45) which on deprotonation gives the hydrazone (46) as a syn/anti mixture. In contrast, (cot)Fe$(CO)_3$ undergoes arylation to give (47) in which the R group is endo to the metal, implying an initial metal attack; deprotonation yields (48). In contrast, (cot)Fe$(CO)_{\underline{n}}L_{3-\underline{n}}$ (L = P$(OMe)_3$, n = 1,2) react to give salts of structure (49a) which on deprotonation yields (50).[134,135] Reaction of (51) with I^- results in ring opening to give (52), whereas (49a) (n = 1) reacts with I^- with elimination of MeI to give the phosphonate (49b).[136]

A variety of complexes of structure (53) have been prepared; in general, thermolysis provides low yields of the bis-Fe$(CO)_3$ complex (54) *via* metal transfer, except in the case where X = Ph where only (55), the product of metal hydride shift isomerization, is observed.[137] (3,7,7-trimethylchpt)Fe$(CO)_3$ has been prepared, and undergoes 1,3- and 1,6-cycloaddition with TCNE in a 4:1 ratio; only the 1,3- adduct is found with (chpt)Fe$(CO)_3$.[138] Dichlorocarbene adds to the uncoordinated double bond of (chpt)Fe$(CO)_3$ and [(N-ethoxycarbonyl)azepine]Fe$(CO)_3$.[139]

Matrix photolysis of (η^4-cot)Fe$(CO)_3$ results in tub ⇌ chair interconversion of the cot ligand *via* the intermediacy of (η^2-cot)-Fe$(CO)_3$; smaller amounts of (η^4-cot)Fe$(CO)_2$ are also formed.[140] Magnetization transfer experiments on (cot)Fe$(CO)_2$CNR complexes have been used to demonstrate the correctness of the Woodward-Hoffmann mechanism for ring whizzing in these complexes.[141] Reaction of [(1-3:5,6-η-cyclooctadienyl)M$(CO)_3]^+$ (M = Fe, Ru) with PPh_3 yields initially (56) which isomerizes on standing to (57). Reaction with I^- yields the neutral complex (58), while other anionic nucleophiles (OMe^-, CN^-) yield neutral analogues of (56).[142,143] (cod)Ru(η^6-cotr) undergoes reaction with P$(OMe)_3$ to give (cod)Ru(η^4-cotr)P$(OMe)_3$ in which the cotr is 1,5-bound.[144] Acetylenes (RC≡CH) undergo cycloaddition with (cod)Ru(cotr) to give complexes of structure (59),[145] and a similar cycloaddition occurs on reaction of (nbd)Ru(chpt) with acetylene to give (60).[146]

(52)

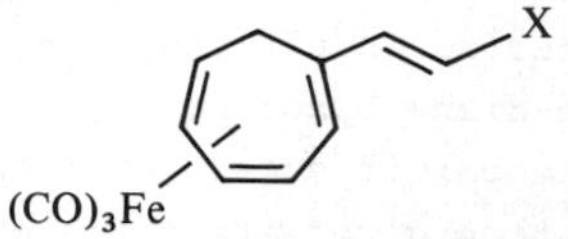

(53) X = Ph, CO_2R, COR

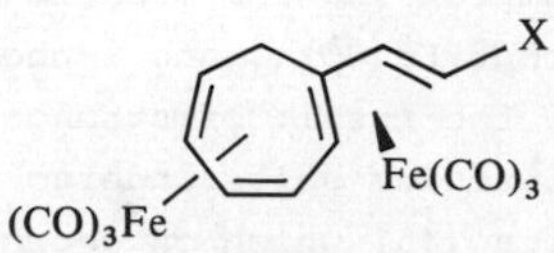

(54)

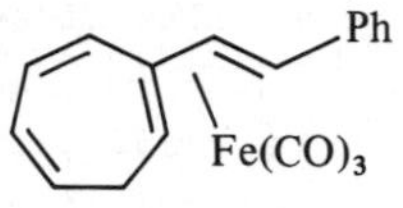

(55)

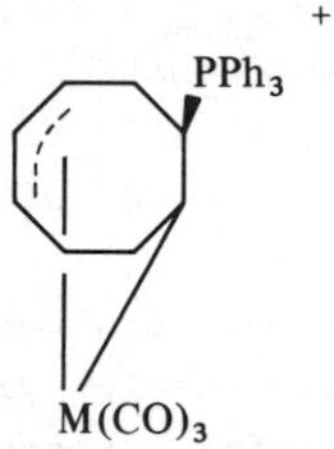

(56) M = Fe, Ru

(57) L = PPh_3, n = 1
(58) L = I, n = 0

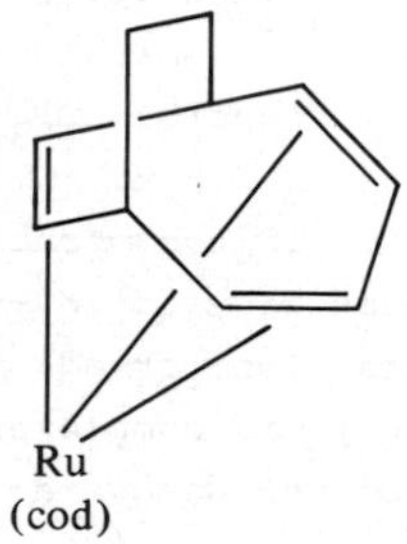

(59)

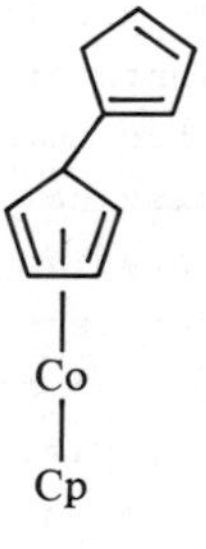

(61)

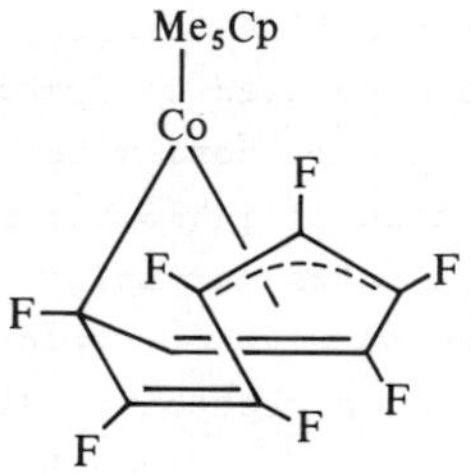

(62)

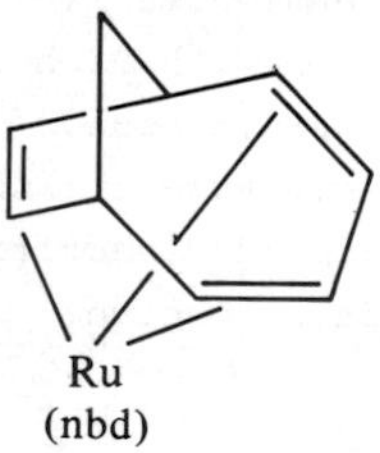

(60)

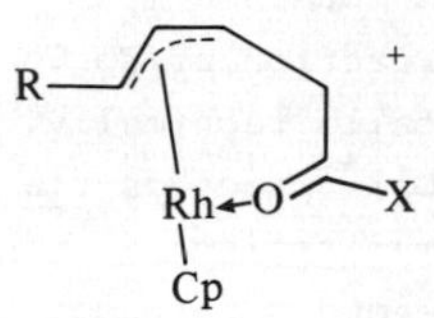

(63) R = Me, Ph; X = Me, OMe
(64) R = PhCH=CH; X = Ph

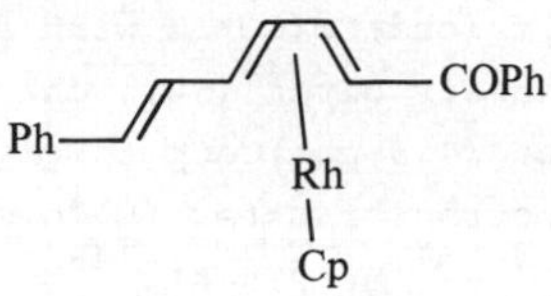

(65)

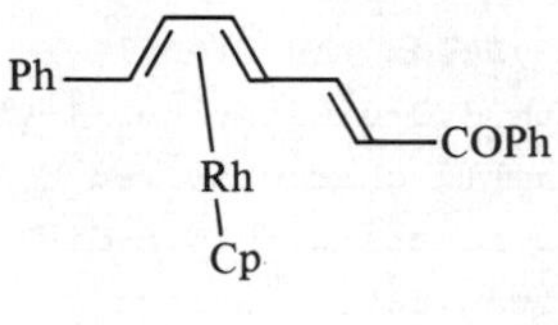

(66)

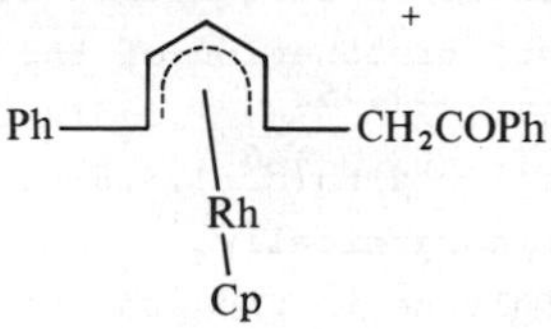

(67)

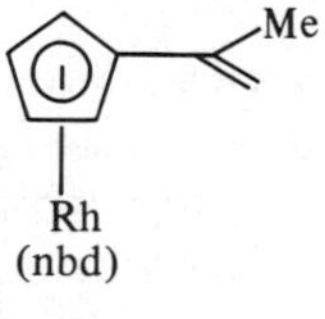

(68)

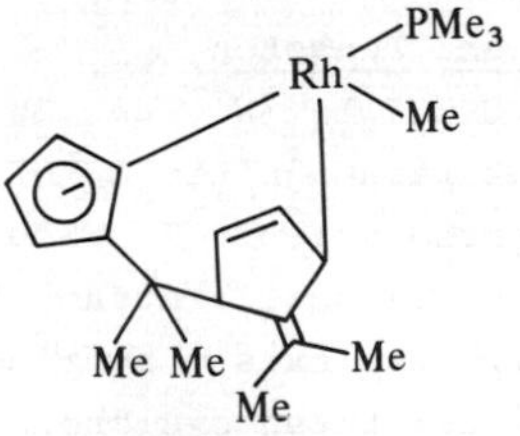

(69)

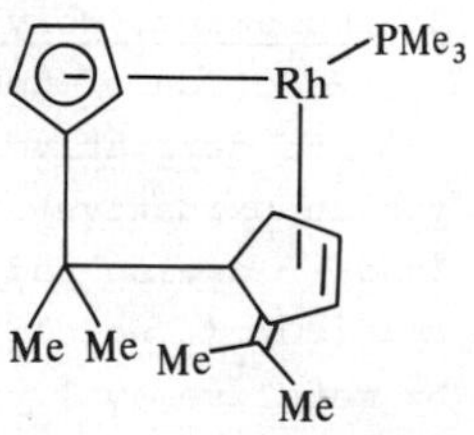

(70)

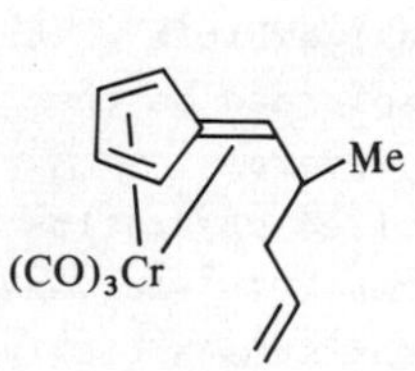

(71)

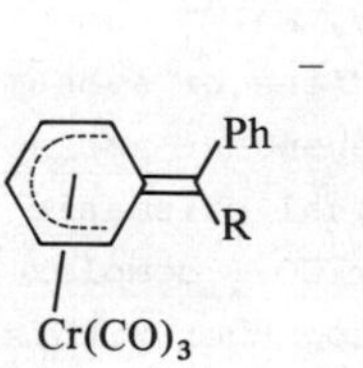

(72) R = H, Ph

2. Cobalt, Rhodium and Iridium

Oxidation of (cbd)$Co_2(CO)_6$ with Ph_3C^+ yields $[(cbd)Co(CO)_3]^+$ and the PPh_3 analogue $[(cbd)Co(CO)_2PPh_3]^+$ can be obtained by treatment of (cbd)$Co(CO)_2I$ with Ag^+/PPh_3.[147] Reaction of Cp_2Co with C_5H_6 in air gives (61), characterized crystallographically.[148] Reduction of (1,5-cot)CoCp to give $[(1,3\text{-cot})CoCp]^-$ proceeds via initial electron transfer followed by rapid isomerization ($k = 2 \times 10^3\ s^{-1}$ at 298 K).[149] $Me_5CpCo(CO)_2$ reacts with F_8cot to give (1,5-η^4-cot)CoCp and the σ,π-allyl (62).[150]

Protonation of CpRh(η^4-RCH=CHCH=CHCOX) (R = Me, Ph; X = OMe, Me) yields salts of structure (63), and the related complex (64) is obtained on protonation of the triene complex (65). In contrast, protonation of the isomeric (66) yields the dienyl salt (67).[151,152] $[(cod)RhCl]_2$ reacts with 3,3-dimethylbut-1-yne to give $[(cod)Rh(\eta^6\text{-}1,3,6\text{-tri-t-butylfulvene})]^+$ characterized crystallographically,[153] while $[(nbd)RhCl]_2$ reacts with 6,6-dimethylfulvene in the presence of MeLi to give (68). Bridge cleavage to give (cod)RhCl(PMe_3), followed by reaction with 6,6-dimethylfulvene and either MeLi or Pr^iMgBr gives (69) and (70) respectively.[154]

3. Chromium, Molybdenum and Tungsten

Reaction of $CpMoX(CF_3C{\equiv}CCF_3)_2$ (X = Cl, Br, I) with butadiene and its derivatives yields paramagnetic $CpMoX_2(\eta^4\text{-diene})$ complexes via an oxidative disproportionation.[155] The related $[CpMo(CO)_2(chd)]^+$ cation undergoes nucleophilic attack to give substituted $CpMo(CO)_2(\eta^3\text{-cyclohexenyl})$ complexes; hydride abstraction followed by metal removal yields substituted cyclohexadienes.[156]

Octahedral (η^4-diene)$M(CO)_3L$ (M = Cr, L = PMe_3, $P(OMe)_3$),[157] (η^4-diene)$M(CO)_2L_2$ (M = Mo, W; L = $P(OMe)_3$),[158] and $(\eta^4\text{-diene})_2M(CO)_2$ (M = Cr, Mo)[159] complexes have been prepared photochemically (diene = butadiene or substituted derivative); all exhibit temperature dependent nmr spectra which may be explained on the basis of trigonal prismatic transition states. Several (η^6-fulvene)$Cr(CO)_3$ complexes have been characterized crystallographically, together with several related (η^6-arene)M(η^6-fulvene) derivatives (M = Mo, W). In all cases, C_6 is bent substantially towards the metal.[160,161] Reaction of $NaCpCr(CO)_3$ with allylchloride provides the fulvene complex (71) in low yield.[162]

α-Hydrogen abstraction from $Cr(CO)_3$ complexes of di- and triphenylmethanes yields anions which can best be described as (72).[163]

Reduction of (naphthalene)$Cr(CO)_3$ with Na yields a dianion; monoprotonation occurs in an endo fashion to give (73).[164] Anions of similar structure (74) are formed by nucleophilic attack by $[\overline{C(Me)S(CH_2)_3S}]^-$ and on reaction with MeI undergo carbonyl insertion and endo methyl transfer to give (75).[165] An example of endo-H^- attack on a cationic complex has been reported in the reaction of (76) to give (77).[166]

$W(CO)_3(NCR)_3$ complexes (R = Et, Pr) give better yields than the MeCN derivative in the preparation of (chpt)$W(CO)_3$ and (chd)$_2$-$W(CO)_2$.[167] Room temperature iodination studies place the Cr-chpt bond strength in the order mesitylene > chpt > toluene > benzene, an order different from that obtained from high temperature studies.[168] Complexation of tropone to $M(CO)_3$ greatly enhances the basicity of the ketonic oxygen, although little dependence on metal is found (Mo ≈ W > Cr).[169,170] In contrast, a larger metal dependence is observed on transfer of MeO^- to $[(C_7H_7)M(CO)_3]^+$ (Mo > W > Cr), although all complexes show much slower rates than free $C_7H_7^+$.[171,172]

Photochemical reaction of $W(CO)_6$ with cot yields (η^6-cot)W-$(CO)_3$ via the intermediacy of (η^4-1,5-cot)$W(CO)_4$.[173] (Cot)$Cr(CO)_3$ reacts photochemically with 6-mono and 6,6-disubstituted fulvenes to give products of structure (78).[174] The identical product (R = R' = Me) can be obtained from (6,6-dimethylfulvene)$Cr(CO)_3$ and cot. Photolysis of (8,8-dimethylheptafulvene)$Cr(CO)_3$ and 2,3-dimethylbutadiene yields (79) which on treatment with CO yields (80), the product of formal (4+6) cycloaddition.[175]

4. Other Metals

$PhMn(CO)_5$ reacts photochemically with 6,6-dimethylfulvene to give the $PhMn(CO)_4(\eta^2$-ol) derivative which undergoes insertion on heating to give cyclopentenyl complexes of structure (81 a,b).[176] Similar products (82) are formed from reaction of butadiene or 1,3-pentadiene with $PhMn(CO)_5$ (R = CH_2Ph)[176] or $Mn_2(CO)_{10}$ (R = $CH{=}CH_2$).[178] $Re_2(CO)_{10}$ reacts with a variety of olefins and polyolefins to give mainly products of structure (83).[177] Cocondensation of Mn with chd followed by treatment with CO gives (chd)$_2$-MnCO having a square pyramidal geometry with apical CO; use of 1,3-cod gives a mixture of (η^5–C_8H_{11})$Mn(CO)_3$ and (η^3-C_8H_{13})$Mn(CO)_4$ while with chpt, (η^5-C_7H_9)$Mn(CO)_3$ and (chpt)$_2$MnCO are formed.[179] Reaction of $MnBr_2$ with the 2,4-dimethylpentadienyl anion in the presence of PMe_3 gives (84) of related structure.[180]

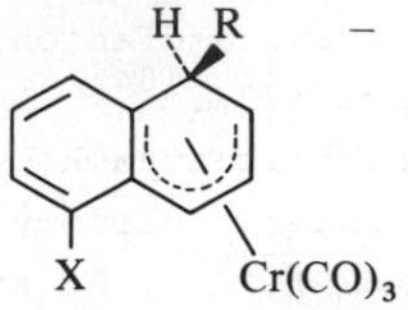

(73) R = X = H
(74) R = $\overline{C(Me)S(CH_2)_3S}$
X = OMe

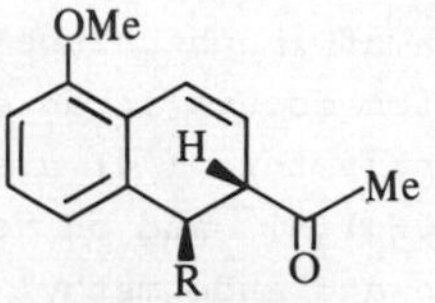

(75)

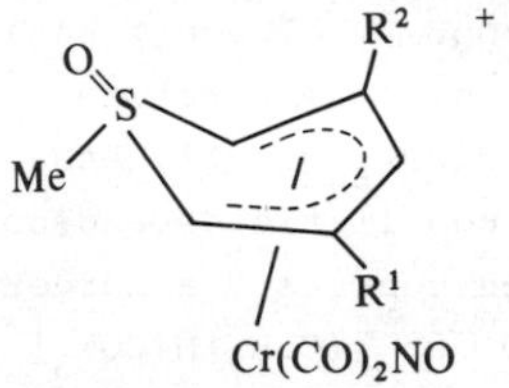

(76) $R^1 = Bu^t$; $R^2 = Ph$

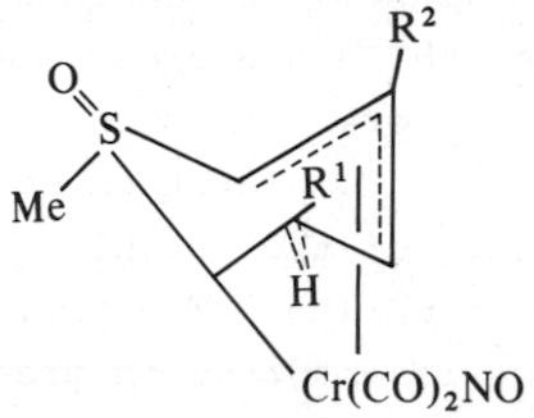

(77)

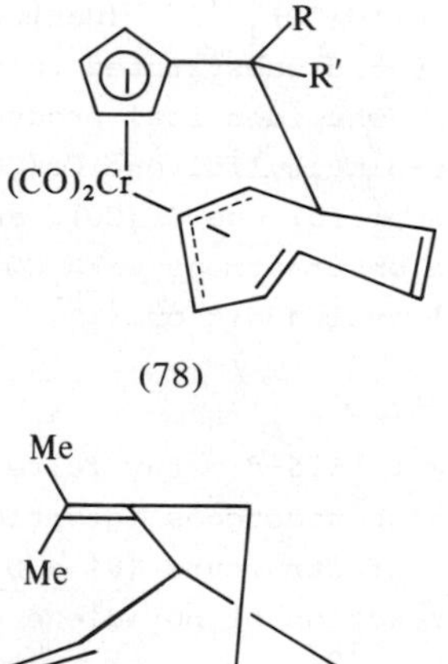

(78)

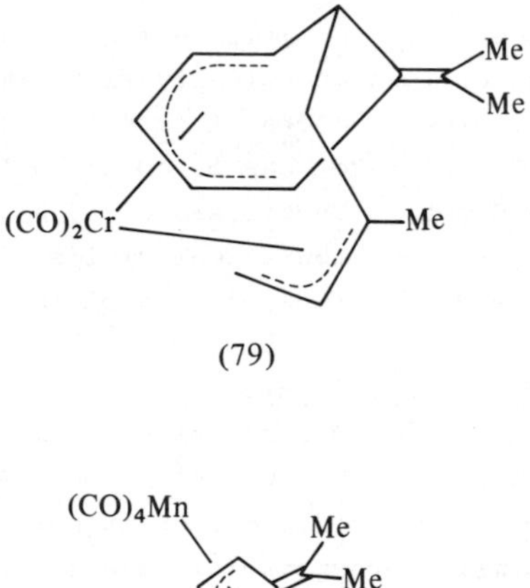

(79)

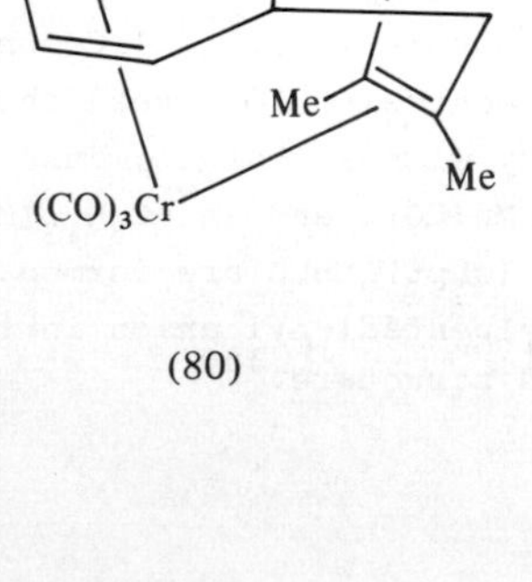

(80)

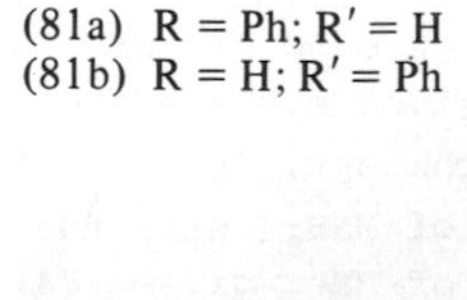

(81a) R = Ph; R′ = H
(81b) R = H; R′ = Ph

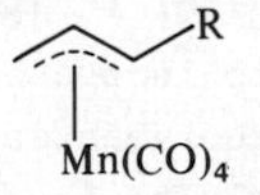

(82) R = CH_2Ph, Me, $CH{=}CH_2$

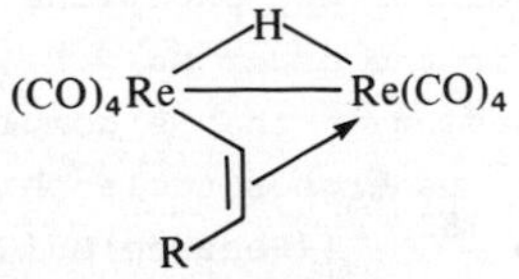

(83)

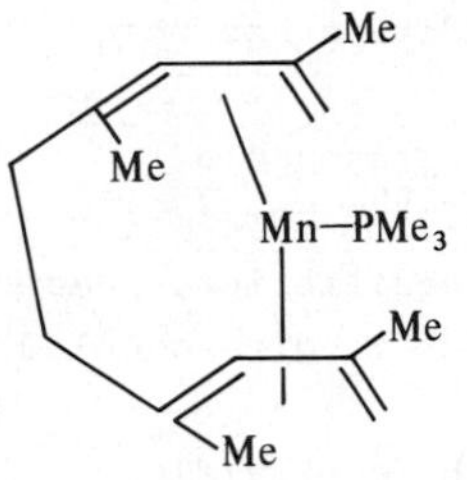

(84)

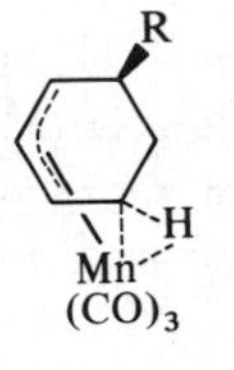

(85) R = H, Me

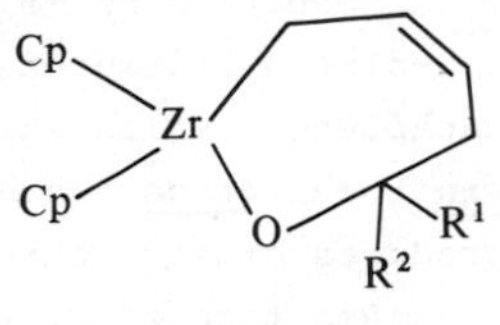

(86)

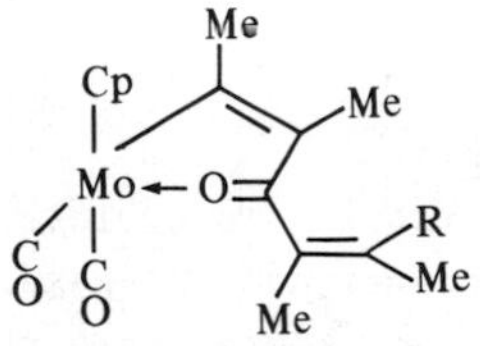

(87) R = H
(88) R = Me

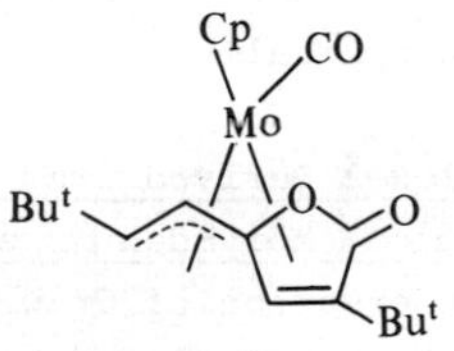

(89)

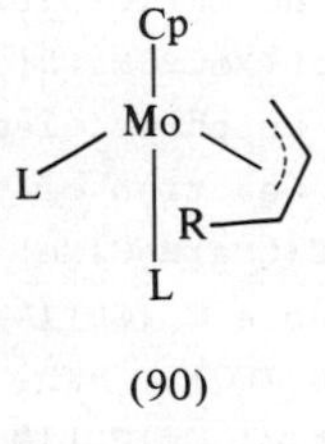

(90)

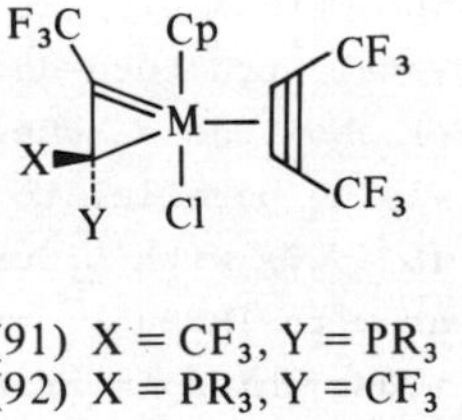

(91) X = CF_3, Y = PR_3
(92) X = PR_3, Y = CF_3

Rates for _exo_-phosphine addition to [arene)$M(CO)_3]^+$ (M = Mn, Re) are in the order Mn > Re; this is attributed to increased π-backbonding in the Re complex.[181] [(1,2-dimethoxybenzene)-$Mn(CO)_3]^+$ undergoes nucleophilic attack exclusively at the _ortho_ position.[182] [(Benzene)$Mn(CO)_3]^+$ undergoes _exo_ addition of two moles of H^- to generate [(chd)$Mn(CO)_3]^-$ which may be protonated or alkylated to give (85); a similar reduction of [(toluene)$Mn(CO)_3]^+$ is not regiospecific.[183] [(η^5-cyclohexadienyl)$Mn(CO)_2NO]^+$ salts also undergo reaction with H^-, but here, the addition has been demonstrated to be _endo_.[184]

Reduction of Me_5CpMCl_3 (M = Zr, Hf) in the presence of 2,3-dimethylbutadiene yields $Me_5CpM(\eta^4$-diene)Cl.[185] $Cp_2Zr(\eta^4$-butadiene), which exists in an s-_trans_ ⇌ s-_cis_ equilibrium, reacts _via_ the s-_trans_ isomer with ketones and aldehydes to give coupled products of structure (86).[186]

Reaction of $Ti(toluene)_2$ with $AlEtCl_2$ in thf/chpt yields $[Ti(\eta^7-C_7H_7)(thf)(\mu-Cl)]_2$. Bridge cleavage with dmpe, followed by reaction with RMgX, gives $(\eta^7-C_7H_7)$Ti(dmpe)R, characterized crystallographically.[187]

E. Complexes Derived from Acetylenes

1. Chromium, Molybdenum and Tungsten

Cocondensation of Cr with PhC≡CPh results in cyclotrimerization to give (η^6-PhC≡CPh)Cr(η^6-hexaphenylbenzene) and (η^6-hexaphenyl-benzene)$_2$Cr as two of the products.[188] Low temperature photolysis of $W(CO)_5$[C(OMe)Ph] in the presence of acetylenes yields $W(CO)_4$(R'C≡CR)[C(OMe)Ph] complexes which at room temperature decompose to polyacetylenes in the case of terminal alkynes or to substituted indenes (R = R' = Ph or R = Me, R' = Ph).[189] Acetylene--carbene complexes have also been isolated from the alkylation of CpW(CO)(HC≡CH)COR with Et_3O^+ to give $\{CpW(CO)(HC{\equiv}CH)[C(OEt)R]\}^+$.[190]

Crystal structure determinations of $[CpMoL(MeC{\equiv}CMe)_2]^+$ (L = CO, NCMe) show the alkynes (formally regarded as three electron donors) lying parallel to the Mo-L axis.[191] Reaction of this cation (L = CO) with $LiCuMe_2$ yields Cp(CO)(MeC≡CMe)MoC(Me)=CMe_2; treatment with $[HBBu^s_3]^-$ similarly yields unstable Cp(CO)(MeC≡CMe)-MoC(Me)=CHMe which on treatment with CO yields (87). The methyl derivative (88) can be obtained from treatment of Cp(CO)(MeC≡CMe)-MoC(Me)=CMe_2 with CO. In contrast, reaction of $[CpMo(CO)(Bu^tC{\equiv}CH)_2]^+$ with $[HBBu^s_3]^-$ followed by treatment with CO yields (89).[192] Displacement by PR_3 and $P(OR)_3$ of CO and/or acetylene from

$[CpMo(CO)(MeC{\equiv}CR)_2]^+$ salts yields $[CpMoLL'(MeC{\equiv}CR)]^+$ complexes ($L = CO$, $L' = PEt_3$; $L = L' = P(OMe)_3$; $R = Me, Et, Pr^i$) which on reduction with $[HBBu^s_3]^-$ yield allyl complexes of structure (90). The reaction proceeds via formation of the $16e^-$ $CpLL'MoC(Me){=}CHR$ complex, and indeed, reduction of $[CpMoL_2(HC{\equiv}CBu^t)]^+$ ($L = P(OMe)_3$) in the presence of excess L allows isolation of $CpL_3Mo(\sigma\text{-}(E)\text{-}CH{=}\text{-}CHBu^t)$.[193] Reaction of $[CpMoL_2(BrC{\equiv}CPh)]^+$ with $[HBBu^s_3]^-$ yields $CpBrL_2Mo{=}C{=}CHPh$ which may be further reduced to the carbyne complex $CpL_2Mo{\equiv}CCH_2Ph$.[194] Protonation or ethylation of the ketene complexes $CpW(CO)(PMe_3)(\eta^2\text{-}O{=}C{=}CR)$ ($R = Me$, p-tolyl) yields related $[CpW(CO)(PMe_3)(R'OC{\equiv}CR)]^+$ salts ($R' = H, Et$).[195,196]

Phosphines undergo nucleophilic attack on $CpMCl(CF_3C{\equiv}CCF_3)_2$ complexes ($M = Mo, W$) to give the η^2-vinyl derivative (91) as the kinetic product; isomerization, most probably via a η^1-vinyl intermediate, yields the thermodynamic product (92).[197] Nucleophilic attack by SR^- (SR^- = pyridine-2-thiolate and related ions) on $CpMoCl(CF_3C{\equiv}CCF_3)_2$ yields products of related structure (93).[198] A full report on the reaction of $CpMCl(CF_3C{\equiv}CCF_3)_2$ ($M = Mo, W$) with $Co_2(CO)_8$ to give $CpM(CO)(\mu\text{-}CF_3C{\equiv}CCF_3)_2Co(CO)_2$ has appeared.[199]

Reaction of $[W(CO)_4Br_2]_n$ with alkynes yields $[WBr_2(CO)(RC{\equiv}CR)_2]_2$ dimers which may be cleaved with $L(CNR, PR_3)$ to give $WBr_2(CO)(RC{\equiv}CR)L_2$ complexes.[200] Reaction of $Mo(CO)_2(PEt_3)_2Br_2$ with $PhC{\equiv}CH$ yields the analogous $MoBr_2(CO)(PEt_3)_2(PhC{\equiv}CH)$.[201] $[W(CO)_4Br_2]_n$ also reacts with dienes (cod, nbd, cot) to give $[WBr_2(CO)(diene)]_2$ dimers which may be cleaved on the addition of Lewis bases to give $WBr_2(CO)(nbd)L$ and $WBr_2(CO)(diene)L_2$ complexes.[202] The complex $Mo(SBu^t)_2(Bu^tNC)_2(RC{\equiv}CR')$ ($R = R' = H, Ph$), prepared from cis-$Mo(SBu^t)_2(Bu^tNC)_4$ and alkyne, has a trigonal bipyramidal geometry with axial $CNBu^t$; the acetylene is parallel to the RNC-Mo-CNR axis, but no rotation is evident up to 100 °C.[203]

A kinetic study of the reaction of $Mo(CO)(RC{\equiv}CR')(S_2CNMe_2)_2$ with alkynes to give $Mo(RC{\equiv}CR')_2(S_2CNMe_2)_2$ shows that the rate is independent of the concentration of entering alkyne, and that electron withdrawing substituents on the bound alkyne increase lability.[204] A crystal structure of a pyrrole-N-carbodithiato derivative shows an octahedral configuration with two cis- parallel acetylenes; nmr studies reveal an acetylene rotation with a barrier of 13.7 kcal mol^{-1}.[205] Reaction of $Mo(RC{\equiv}CR)_2(S_2CNR)_2$ complexes with PEt_3 yields bridged dimers of structure (94).[206]

Bridged thiolato complexes of the type $Cp_2Mo_2(\mu\text{-}S)_2(\mu\text{-}S_2CR_2)$ undergo adduct formation with acetylenes to give (95); similar, but less stable adducts are formed with alkenes.[207] $CO_2MeC{\equiv}CCO_2Me$ undergoes insertion into the Mo-S bond of $[Mo_2O_2(\mu\text{-}S)_2(S_2)_2]^{2-}$ to give (96).[208]

2. Cobalt, Rhodium and Iridium

$CoBr(PMe_3)_3$ reacts with PhC≡CPh in toluene to give (PhC≡CPh)-$CoBr(PMe_3)_3$; in acetone, dissociation to $[(PhC{\equiv}CPh)Co(PMe_3)_3]Br$ is observed, which on addition of MeCN gives $[(PhC{\equiv}CPh)Co(PMe_3)_3$-(NCMe)]Br. The two cations have been characterized crystallographically as distorted tetrahedral and trigonal bipyramidal, and contain $4e^-$ and $2e^-$ donor acetylenes respectively.[209] The related $16e^-$ complex $ClRh(PPr^i_3)_2(PhC{\equiv}CH)$ exists in equilibrium in solution with $ClRh(H)(C{\equiv}CPh)(PPr^i_3)_2$, and addition of pyridine allows isolation of $ClRh(H)(C{\equiv}CPh)(PPr^i_3)_2py$. This reacts with NaCp to give $CpRh(PPr_3)C{=}CHPh$, whereas $ClRh(PhC{\equiv}CH)(PPr^i_3)_2$ yields CpRh-$(PhC{\equiv}CH)(PPr^i_3)$.[210] Disubstituted derivatives $CpRh(RC{\equiv}CR)PPr^i_3$ have also been prepared, and undergo protonation (R = Me) to give $[CpRh(1\text{-methylallyl})(PPr^i_3)]^+$.[211]

Reaction of $(tmhd)Rh(C_2H_4)(CF_3C{\equiv}CCF_3)$ (tmhd = 2,2,6,6-tetramethylheptane-3,5-dionate) with ER_3 (E = P, As, Sb) yields $(tmhd)Rh(CF_3C{\equiv}CCF_3)(ER_3)_n$ (n = 1,2) complexes; the mono-$AsPh_3$ complex reacts with further $CF_3C{\equiv}CCF_3$ to give the rhodacyclopentadiene complex (97). In contrast, reaction of $(tmhd)Rh(C_2H_4)$-$(CF_3C{\equiv}CCF_3)$ with nitrogen donors yields rhodacyclopentene complexes of structure (98).[212] Treatment of $(tmhd)Rh(C_2H_4)(CF_3C{\equiv}CCF_3)$ with cyclopropane or propene yields the 1,4-diene complex (99), and complex (100) of related structure is isolated from reaction with tetramethylallene.[213]

Kinetic studies of the reaction of $CpCo(PPh_3)(RC{\equiv}CR)$ with alkynes to give complexes of structure (101) indicate the intermediacy of $CpCo(RC{\equiv}CR)_2$ which cyclizes <u>via</u> a spontaneous oxidative coupling. Regioselectivity is governed by the size of R.[214]

3. Other Metals

$[(HC{\equiv}CH)Cu(2,2'\text{-dipyridylamine})]^+$ has been characterized crystallographically; like the ethylene complexes mentioned previously, little change in the C≡C bond length relative to free acetylene is observed.[215] Unstable $R_5CpCu(Me_3SiC{\equiv}CSiMe_3)$ complexes (R = H, Me) have also been prepared, but undergo elimination of free acetylene at room temperature.[216]

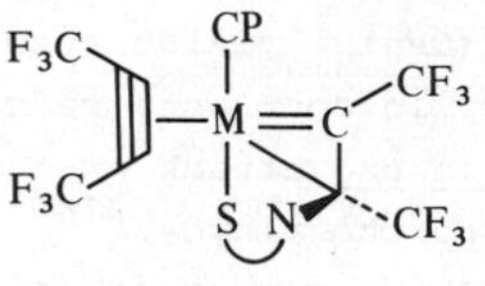

(93)

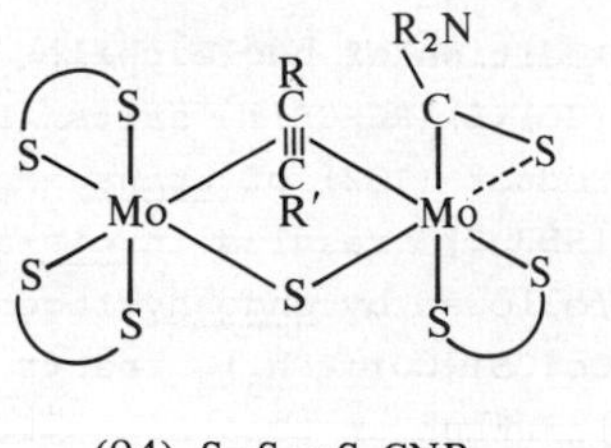

(94) S—S = S_2CNR_2

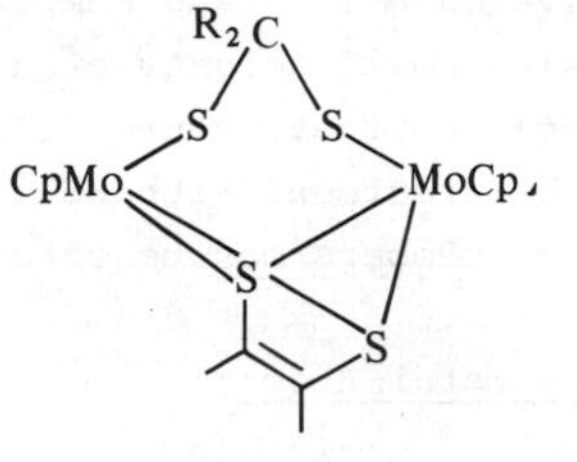

(95)

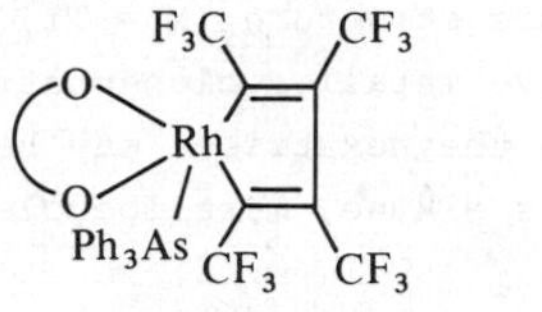

(97)

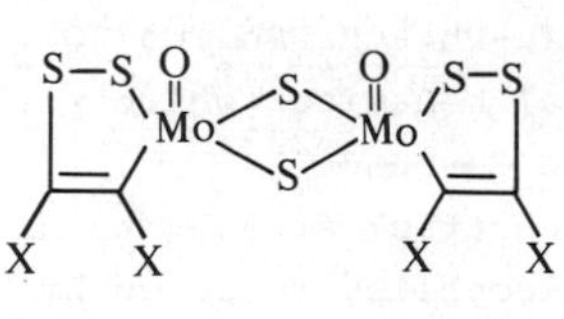

(96) X = CO_2Me

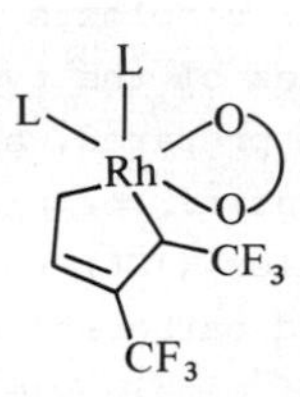

(98)

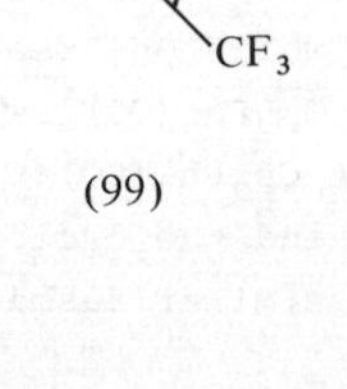

(99)

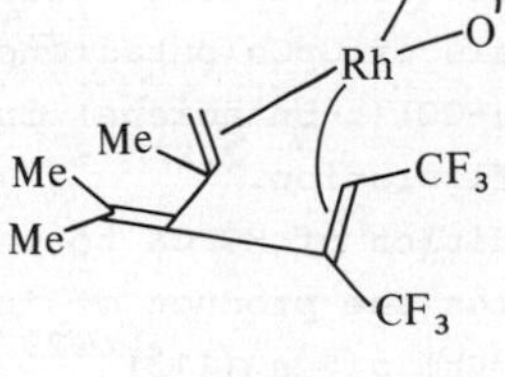

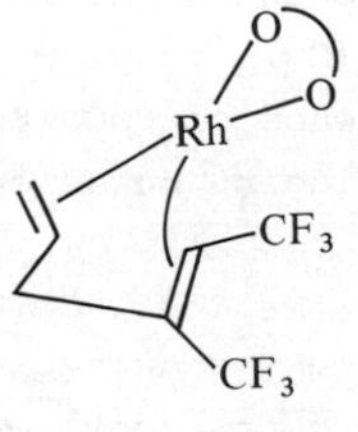

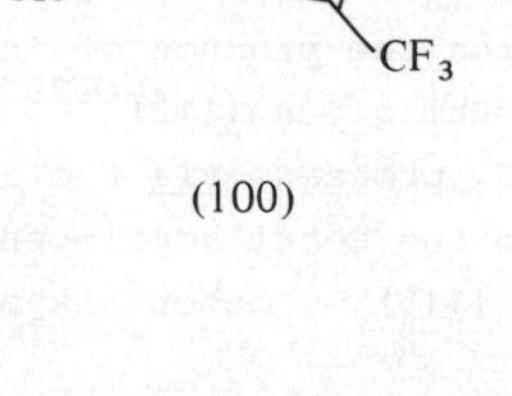

(100)

Addition of nucleophiles (X^-) derived from organocuprates to $[CpFe(CO)(L)(RC{\equiv}CR)]^+$ salts (L = PPh_3, $P(OPh)_3$) yields specifically the product (102) of trans-addition.[217] In contrast, reduction with $[HBBu^s_3]^-$ results in cis-addition, via exo attack on the Cp ring followed by endo hydrogen transfer to the alkyne.[218]

$Os(CS)(CO)(PPh_3)_3$ reacts with PhC≡CPh to give Os(PhC≡CPh)(CS)-$(PPh_3)_2$ which on treatment with CO gives the metallocycle (103). HCl addition to the acetylene complex gives $Os(CS)Cl(PPh_3)_2$-(η^1-CPh=CHPh) which undergoes insertion of CS to give (104).[219]

Cp_2ZrH_2 reacts with PhC≡CPh to give (105a),[220] and complexes of this structure (M = Ti, Zr) react with small amounts of alkynes to give metallocycloheptatriene derivatives of structure (105b) from which the hexatriene may be liberated by treatment with HCl. With excess alkyne, metallocycles of up to 17 members may be obtained.[221]

F. Complexes Containing More Than One Metal Atom

1. Binuclear Complexes

Electron transfer catalysis via the labile $[Co_2(CO)_6(\mu\text{-}RC{\equiv}CR')]^-$ radical ion has been used to prepare a variety of $Co_2(CO)_5L$-(μ-RC≡CR') complexes (L = MeCN, PR_3, $P(OR)_3$).[222] Substituted derivatives of the type $(CO)_2Co(\mu\text{-}RC{\equiv}CR')(\mu\text{-}Ph_2PCH_2PPh_2)Co(CO)_2$ have been prepared, and exhibit a fluxional behaviour which yields an apparent mirror plane containing the Co-Co bond.[223]

Complex (106) undergoes electrophilic attack at $=CH_2$ to give stabilized carbocations; addition of nucleophile, followed by removal of metal, yields the organic product of Ad_E addition.[224] Complex (108), obtained from reaction of (107) with $Fe(CO)_5$, contains a bridging ligand of similar structure.[225] α-Halo-acetylenes react with (109) via a 1,2- shift of halogen to give the vinylidene complex (110).[226] Cobaltacylopentadiene complexes of structure (101) undergo reaction with $Fe_2(CO)_9$ to give (111).[227] Photolysis of CpCo(butadiene) in the presence of $CpCo(CO)_2$ yields $Cp_2Co_2(\mu\text{-}CO)(\mu\text{-}butadiene)$ in which the bridging butadiene adopts a syn configuration.[228]

Addition of HC≡CH to the M=M bond of $[Me_5CpM(\mu\text{-}CO)]_2$ (M = Co, Rh) yields the product of insertion (112) which on treatment with SO_2 (M =Rh) gives (113).[229] Reaction of $Cp_2Rh_2(CO)_3$ with $CF_3C{\equiv}CCF_3$ proceeds via the intermediacy of $Cp_2Rh_2(CO)(CF_3C{\equiv}CCF_3)_2$ to yield the metallacyclopentadiene (114) and the cyclopentadienone complex (115). Other alkynes react in a similar fashion.[230]

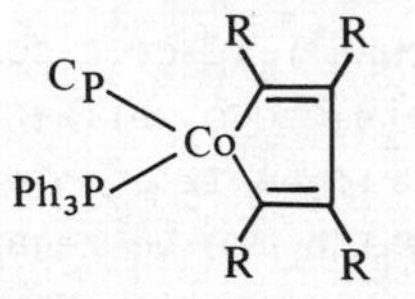

(101)

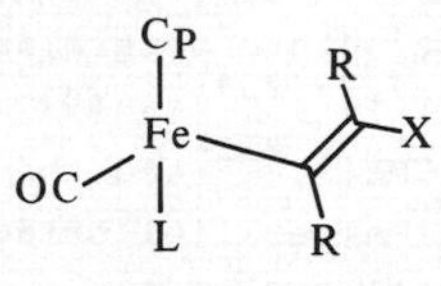

(102)

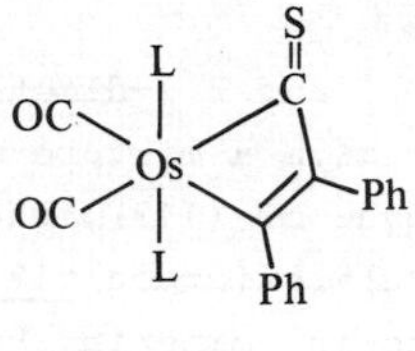

(103)

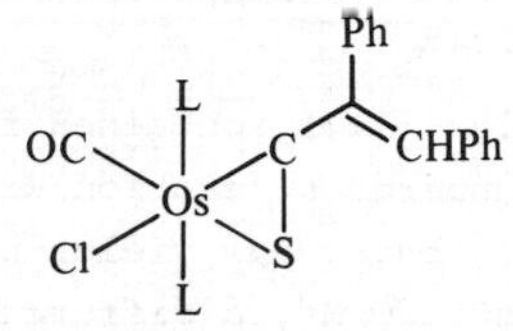

(104)

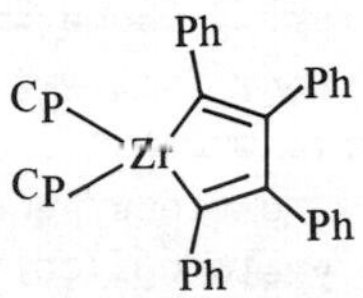

(105a)

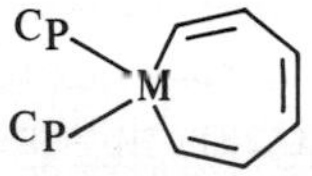

(105b) M = Ti, Zr

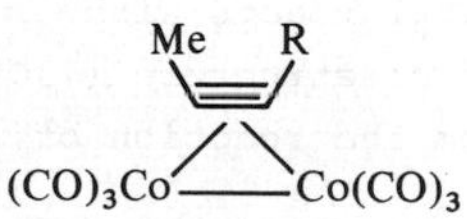

(106) R = C(Me)=CH_2
(107) R = CH_2OH

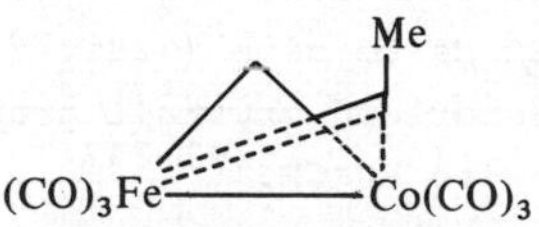

(108)

Crystal structure determinations of $[Rh_2(CNBu^t)_4(\mu-CF_3C{\equiv}CCF_3)-(\mu-Ph_2PCH_2PPh_2)_2]^{2+}$,[231] $[Rh(P(OMe)_3)(\mu-CO)(\mu-CO_2MeC{\equiv}CCO_2Me)(\mu-Ph_2-PCH_2PPh_2)_2RhO_2CMe]^+$,[232] and the mixed valence bridged Ir complex (116)[233] reveal a parallel bridging acetylene which can be regarded as a cis-dimetallated olefin. The product of the reaction between $Co(CN)_5^{3-}$ and $CO_2MeC{\equiv}CCO_2Me$ has been characterized crystallographically as the trans-ethylene complex $[(CN)_5CoC(CO_2Me)=C(CO_2Me)Co-(CN)_5]^{6-}$.[234a]

Complex (117), obtained from $Cp_2Mo_2(CO)_4$ and 3,3-dimethylcyclopropene, undergoes reaction with $Bu^tC{\equiv}CH$ to give a mixture of (118) and (119); but-2-yne yields only the analogue of (119), whereas with $CO_2MeC{\equiv}CCO_2Me$, a different complex (120) is formed via insertion.[234b] Thermolysis of (117) results in competing hydrogen shifts to give $Cp_2Mo_2(CO)_4(\mu-HC{\equiv}CPr^i)$ and complex (121), together with smaller amounts of a cluster complex.[235]

Protonation of $Cp_2Mo_2(CO)_4(\mu-RC{\equiv}CR)$ complexes by coordinating acids yields bridged vinyl complexes of structure (122).[236] Thermolysis of the related Ru derivative (123) results in isomerization to $Cp_2(CO)_2Ru_2(\mu-CO)(\mu-C=CH_2)$ which on protonation, followed by reaction with hydride, yields $Cp_2(CO)_2Ru_2(\mu-CO)(\mu-CHMe)$.[237] Protonation of (123) (M = Fe, Ru) yields μ-vinyl cations such as (124) which on treatment with borohydride also yield $Cp_2(CO)_2M_2-(\mu-CO)(\mu-CHMe)$ via hydride additon to the β-carbon.[238,239] The substituted derivative (127) may be obtained by protonation of (126), the product of the reaction of (125) with allene; borohydride reduction yields $Cp_2(CO)_2Ru_2(\mu-CO)(\mu-CMe_2)$.[240] This complex undergoes a double insertion on reaction with HC≡CH to give (128). In contrast, $Cp_2(CO)_2Ru_2(\mu-CO)(\mu-CH_2)$ reacts with $CO_2Me_2C{\equiv}CCO_2Me$ to give (129).[241] Products of structure (130) have been characterized crystallographically from the reaction of $Ru_3(CO)_{12}$ with RC≡CR.[242,243]

Nucleophilic attack by primary and secondary amines on the bridged acetylide (131) yields either (132), the product of attack at the β-carbon followed by hydrogen transfer, or (133), the product of α-attack and hydrogen transfer; amines of large bulk favour β-attack, while the reaction pathway is also sensitive to the size of R'.[244]

Reaction of $Os_2(CO)_8^{2-}$ with $X(CH_2)_nX$ yields bridged $(CO)_4Os-[\mu-(CH_2)_n]Os(CO)_4$ complexes (n = 1-3); the n = 3 complex may also be prepared by insertion of C_2H_4 into the n = 1 derivative.[245] An analogous complex $(CO)_4Os(\mu-CH_2CHCO_2Me)Os(CO)_4$ is formed, in

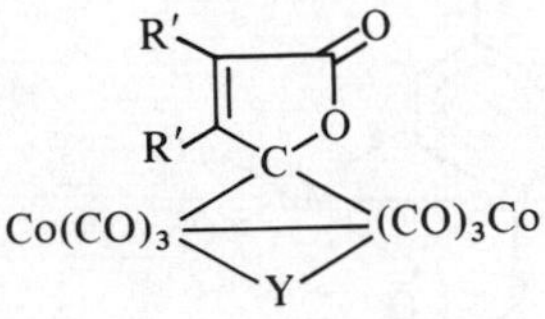

(109) Y = CO

(110) Y = C=C<X, R

X = halogen,
R = alkyl, halogen
R′ = alkyl

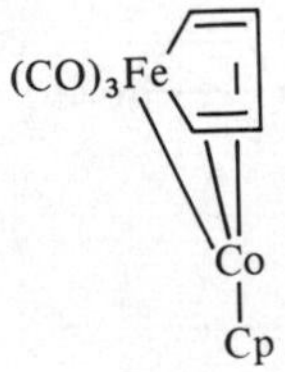

(111)

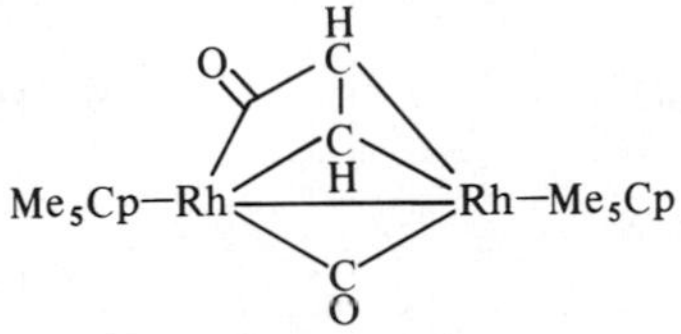

(112)

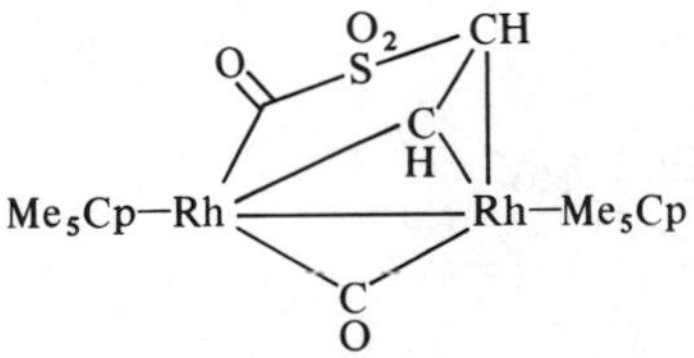

(113)

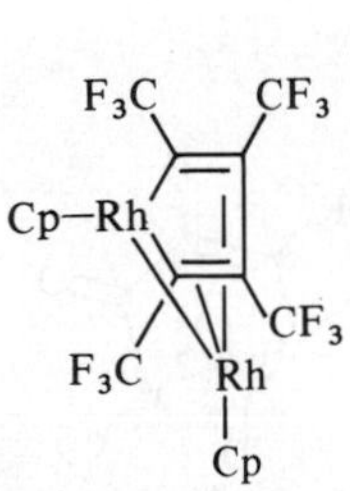

(114)

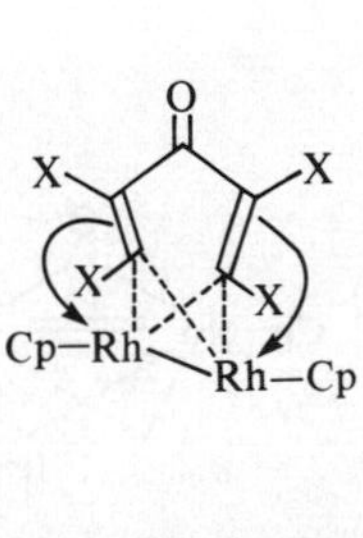

(115) X = CF$_3$

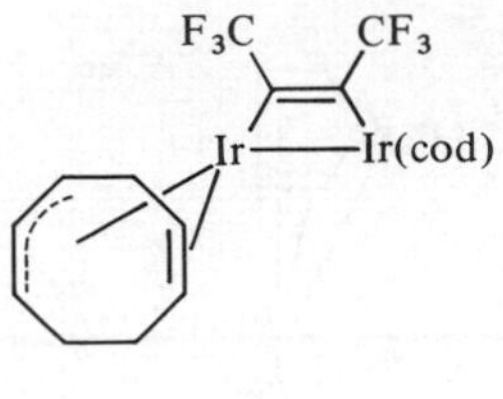

(116)

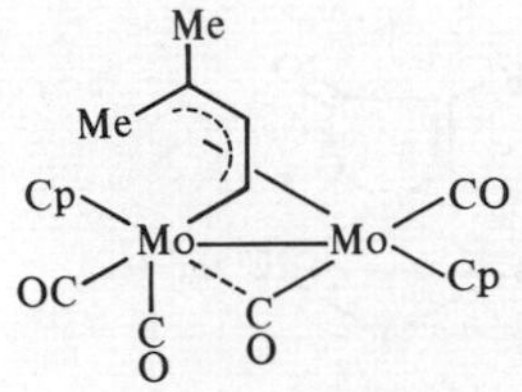

(117)

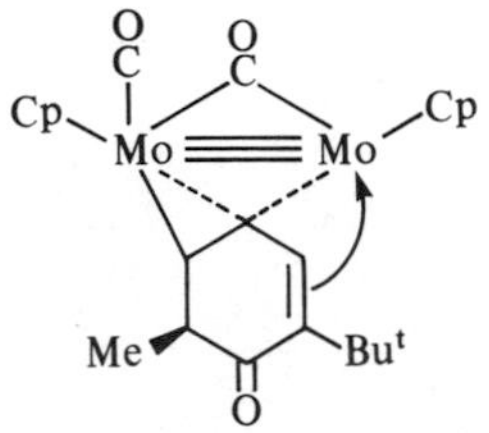

(118)

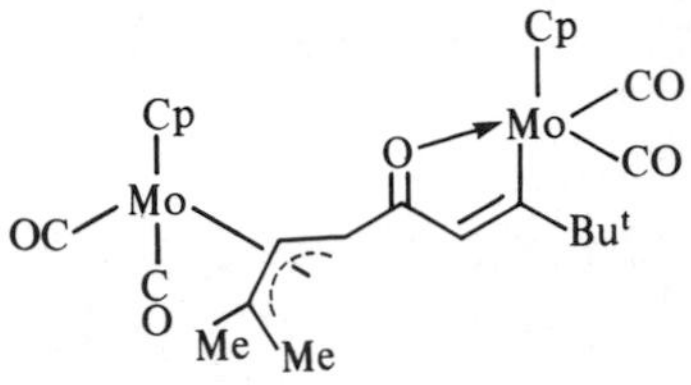

(119)

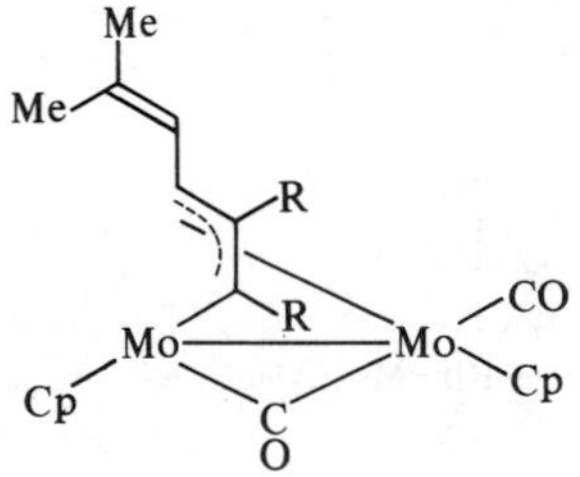

(120) R = CO_2Me

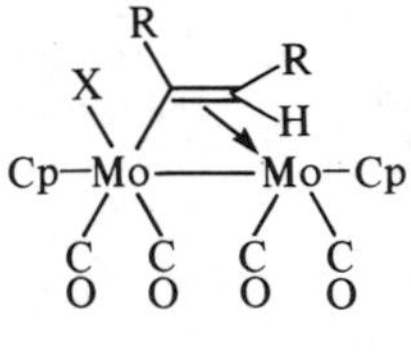

(122) X = CF_3CO_2

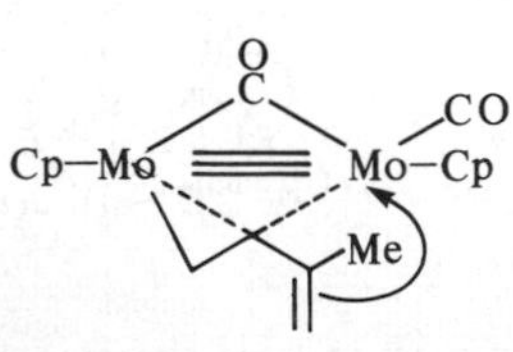

(121)

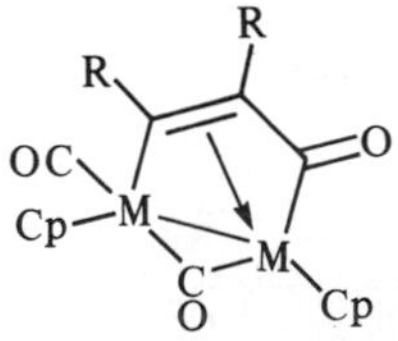

(123) M = Fe, Ru; R = H
(125) M = Ru; R = Ph

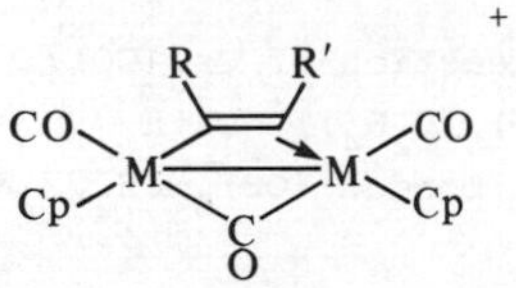

(124) M = Fe, Ru; R = R′ = H
(127) M = Ru; R = Me; R′ = H

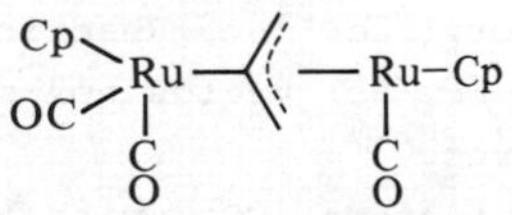

(126)

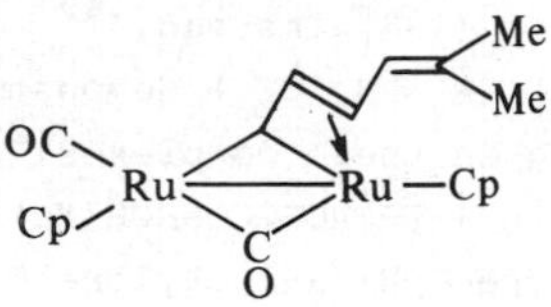

(128)

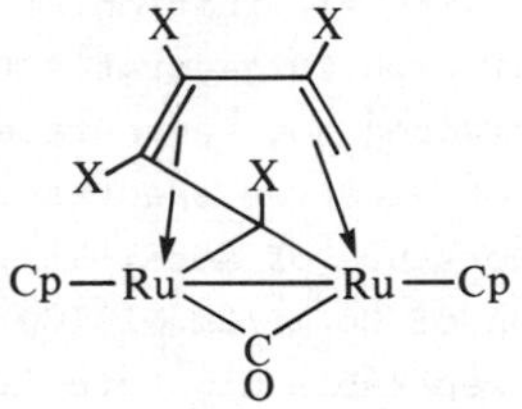

(129) $X = CO_2Me$

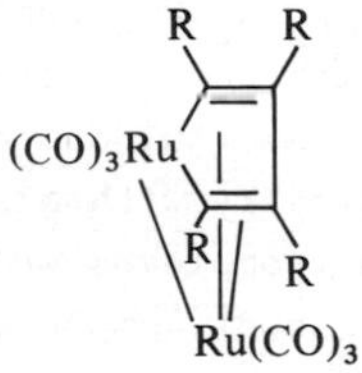

(130) $R = CO_2Me, CH_2OH$

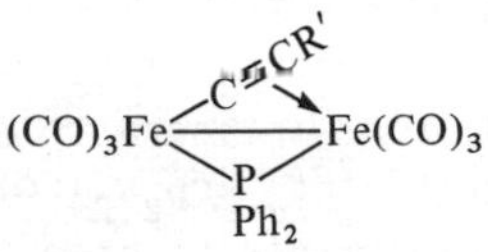

(131)

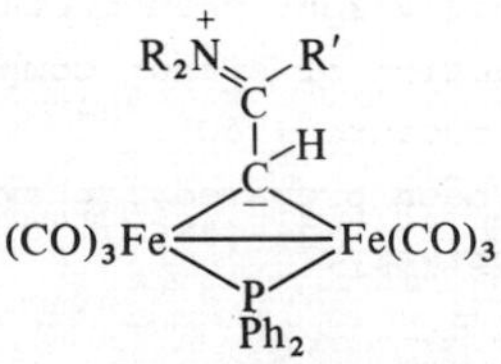

(132)

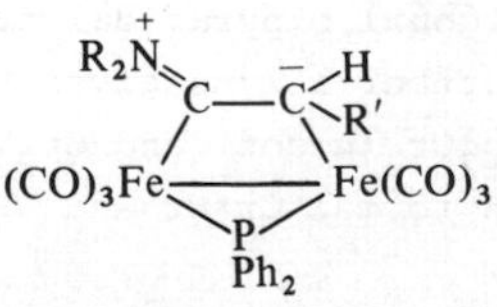

(133)

addition to (ol)$Os(CO)_4$, on photochemical reaction of $Os_3(CO)_{12}$ with methylacrylate.[246] Reaction of $[M(CO)_5(C_2H_4)]^+$ with $M(CO)_5^-$ (M = Mn, Re) yields non-metal-metal bonded $(CO)_5MCH_2CH_2M(CO)_5$ complexes.[247]

Treatment of $Mn_2(CO)_{10}$ with $HPPh_2$ gives $Mn_2(\mu\text{-}H)(\mu\text{-}PPh_2)(CO)_8$ which undergoes insertion with HC≡CH to give (134).[248] The related Re complex (135) undergoes reaction with 3,3-dimethylpropene with elimination of 1-butene to give (136), and treatment of (136) with CO generates the μ-carbene complex (137). Photolysis of (137) does not regenerate (136) but gives (138) instead.[249]

Photoelectron spectra of Cp_2Ni_2(RC≡CR) (R = H, CF_3) demonstrate the importance of metal-alkyne back bonding in these complexes.[250] Reaction of Cp_2Ni/Cp^1_2Ni (Cp^1 = MeO_2CCp) with (-)-PhC≡CCONHCH(Me)Ph yields separable diastereoisomers of the type CpNi(μ-acetylene)$NiCp^1$ which are configurationally stable with respect to change of cluster chirality.[251] Reaction of $Cp_2Pt_2(CO)_2$ with $Bu^tC{\equiv}CBu^t$ in the presence of Me_3NO yields $Cp_2Pt_2(\mu\text{-}Bu^tC{\equiv}CBu^t)$; using PhC≡CPh, the coupled product (139) is obtained instead.[26] One electron chemical or electrochemical oxidation of $(Ph_5Cp)_2Pd_2$(PhC≡CPh) yields paramagnetic $[(Ph_5Cp)_2Pd_2(PhC{\equiv}CPh)]^+$ which undergoes reaction with cod or Lewis bases to regenerate starting material and give $[Ph_5CpPdL_2]^+$.[252] $Cp_2V_2(CO)_5$ reacts photochemically with alkynes at low temperature to give $Cp_2V_2(CO)_4$(μ-alkyne) complexes which on warming give either $CpV(CO)_2$(RC≡CR) or $CpV(CO)_2$(cbd) complexes depending on the basicity of the acetylene.[253]

While reaction of $W_2(OPr^i)_6(py)_2$ with HC≡CH yields symmetrically bridged $(py)(OPr^i)_2W(\mu\text{-}HC{\equiv}CH)(\mu\text{-}OPr^i)_2W(OPr^i)_2(py)$, MeC≡CMe reacts with $W_2(OCH_2Bu^t)_6(py)_2$ to give $(py)(OCH_2Bu^t)_3W(\mu\text{-}MeC{\equiv}CMe)(\mu\text{-}OCH_2Bu^t)W(OCH_2Bu^t)_2(py)$ containing inequivalent pseudooctahedral and trigonal bipyramidal tungstens. Reaction of either complex with further alkyne gives complexes of structure (140).[254]

$Cp_2Cr_2(\mu\text{-cot})$ and $Cp_2V_2(\mu\text{-cot})$ have been prepared; they contain formal Cr=Cr and V≡V bonds respectively.[255,256]

2. Polynuclear Complexes

PhC≡CCl reacts with $Ru_3(CO)_{12}$ to give the bridged acetylide (141).[257] The well known hydride analogues (142) show photoelectron spectra which confirm the (σ + 2π) interaction of the acetylide, but show little overlap population for the hydride bridged M-M bond.[258] Derivatives of (142) (M = Ru) in which hydride has been replaced by HgX or Hg have also been prepared.[259]

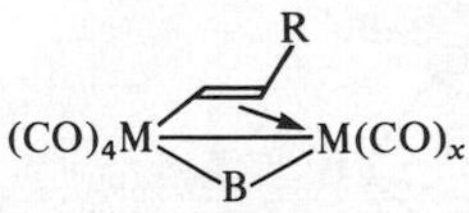

(134) B = PPh_2; x = 3; R = H; M = Mn
(135) B = H; x = 4; R = Et; M = Re

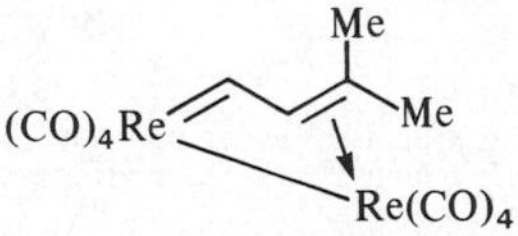

(136)

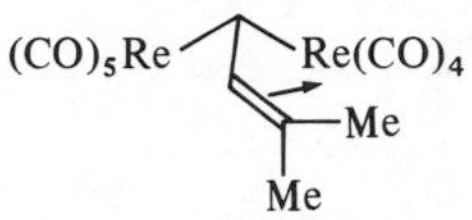

(137)

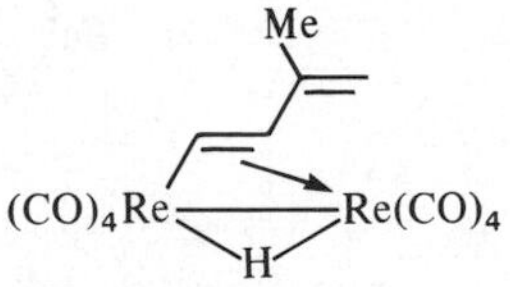

(138)

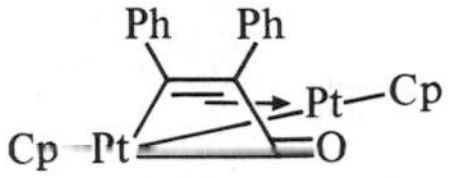

(139)

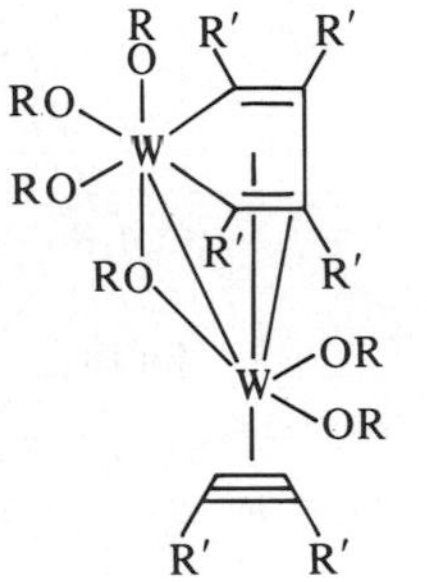

(140) OR = OPr^i; R′ = H
OR = OCH_2Bu^t; R′ = Me

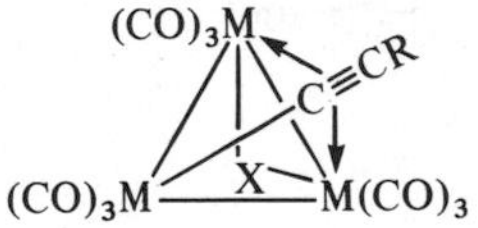

(141) M = Ru; X = Cl; R = Ph
(142) M = Ru, Os; X = H; R = Bu^t
(147) M = Os; X = H; R = CH_2CH_2OH

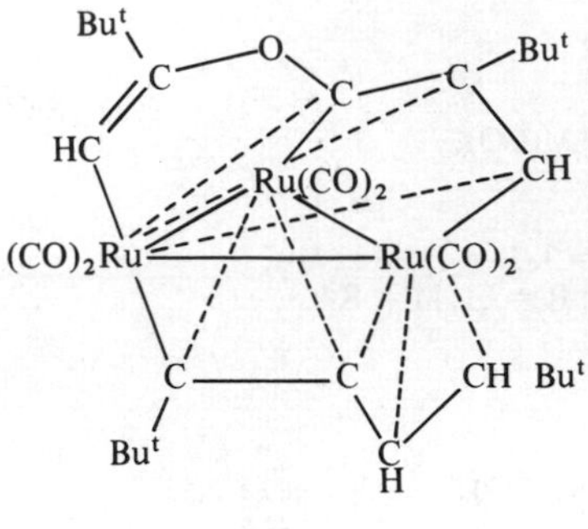

(143)

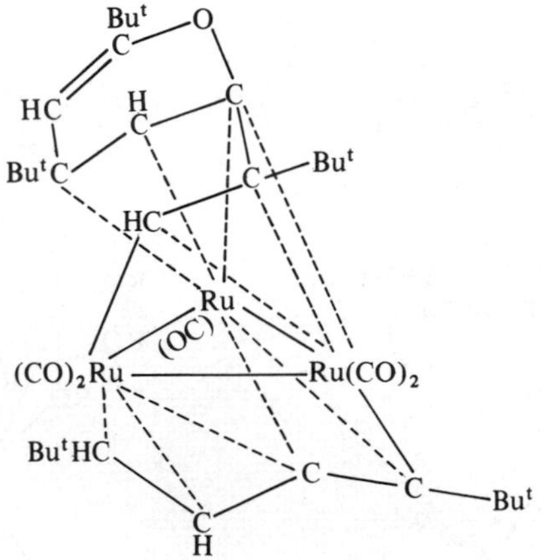

(144)

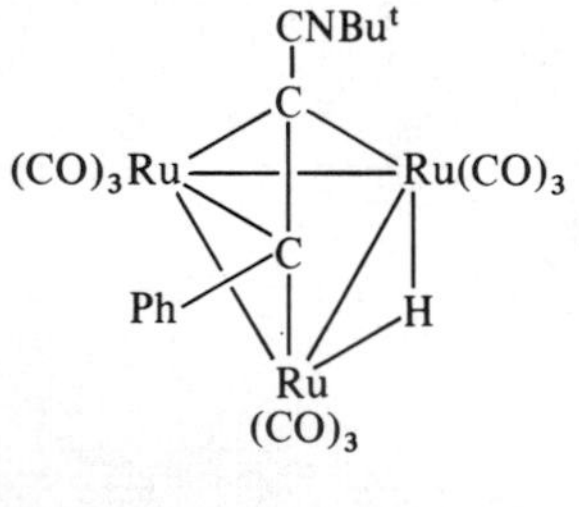

(145)

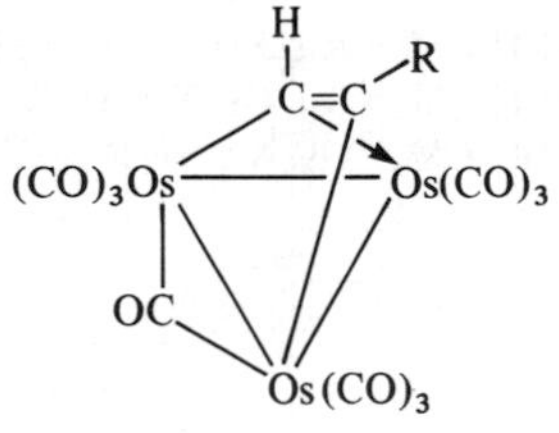

(146) R = CH_2CH_2OH
(149) R = H

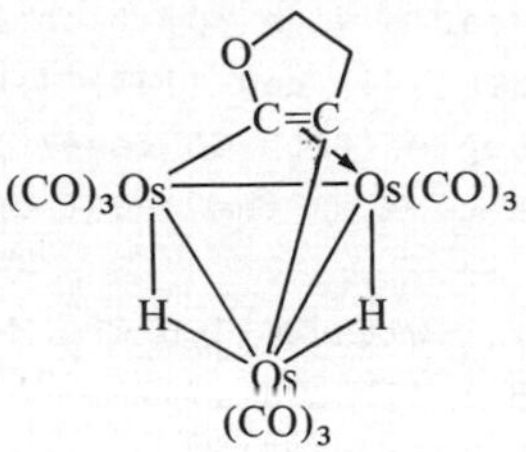

(148)

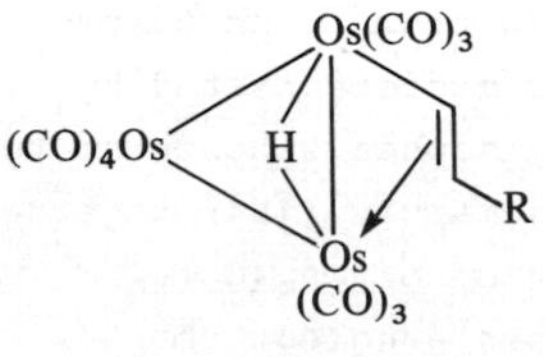

(150) R = Bu^t
(151) R = H

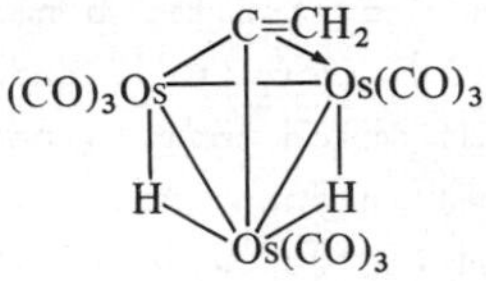

(152)

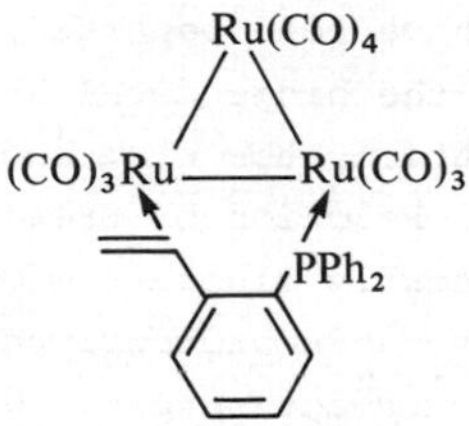

(153)

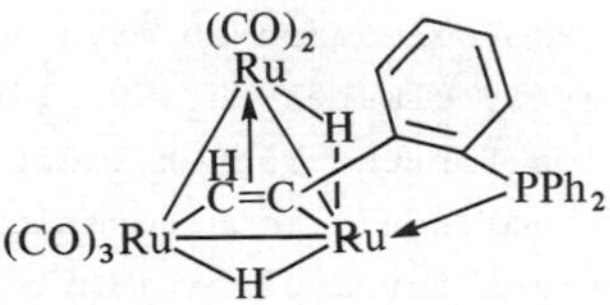

(154)

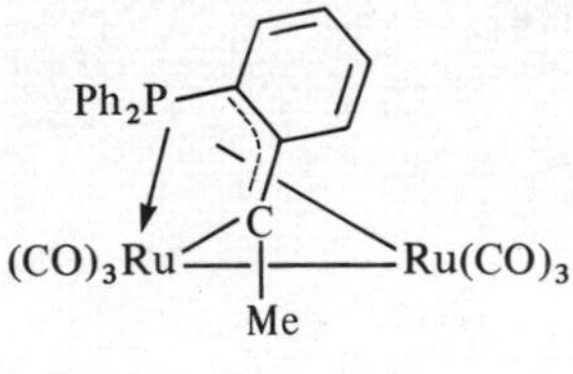

(155)

Complex (142) (M = Ru) reacts with additional $HC{\equiv}CBu^t$ to give sequentially the clusters (143) and (144) (and isomers) via oligomerization of alkyne, coupled with CO insertion.[260] Isonitrile undergoes nucleophilic attack at the β-carbon of (142) to give (145).[261]

Reaction of $Os_3(CO)_{12}$ with $HC{\equiv}CCH_2CH_2OH$ yields (143) via the probable intermediacy of (146) and (147).[262] A detailed vibrational analysis of (149) coupled with electron energy loss data shows that HC≡CH adsorbed on M(111) faces (M = Pt, Pd) has the same bonding mode as in (149).[263] $H_2Os_3(CO)_{10}$ reacts with $HC{\equiv}CBu^t$ to give (150);[264] a detailed vibrational analysis for the parent (151) and the vinylidene (152) derived from it on thermolysis show that the two bonding modes can be differentiated by infrared.[265] $Ru_3(CO)_{12}$ reacts with 2-styryldiphenylphosphine to yield (153) which on mild thermolysis yields mainly (154) together with a small amount of (155).[266,267] Reaction of $Os_5(CO)_{19}$ with HC≡CH yields $Os_4(CO)_{12}(HC{\equiv}CH)$ and the pentanuclear complex (156).[268] Me_3NO assisted substitution of $CpWOs_3(CO)_{12}H$ by alkynes yields clusters of structure (157); thermolysis results in C-C cleavage to give the bis(carbyne) complex (158).[269]

Photoelectron spectra for $Fe_3(CO)_9(\mu_3\text{-}EtC{\equiv}CEt)$[270] and the butterfly cluster $Ru_4(CO)_{12}(\mu_4\text{-}PhC{\equiv}CH)$[271] have been reported; results for the latter indicate that while the hinge metal atoms are involved in both donation and back donation, the wing metal atoms are involved primarily in donation. Reaction of $HFe_3(CO)_{12}$ with alkynes in apolar solvents yields butterfly clusters of structure (159); in polar solvents, the products are derived from $[FeCo_3(CO)_{12}]^-$ formed by deprotonation.[272] Reaction of $[RuCo_3(CO)_{12}]^-$ with PhC≡CPh yields the butterfly cluster (160) which on treatment with HCl gives (161).[273] Treatment of either $HOs_3(CO)_9(C_2Bu^t)$ or $HOs_3(CO)_{10}(HC{=}CHBu^t)$ with $Cp_2Ni_2(CO)_2$ yields the butterfly cluster (162).[274]

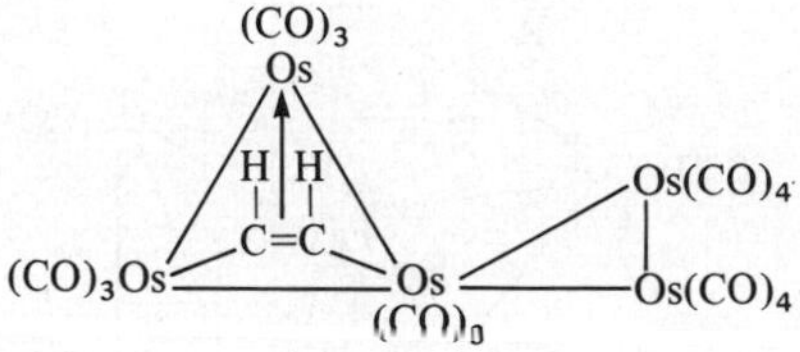

(156)

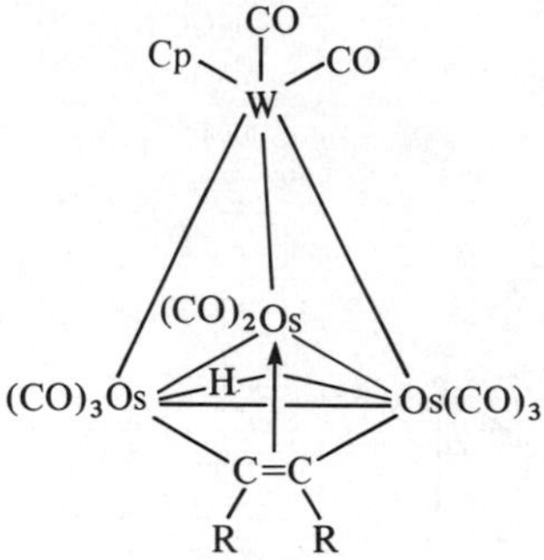

(157)

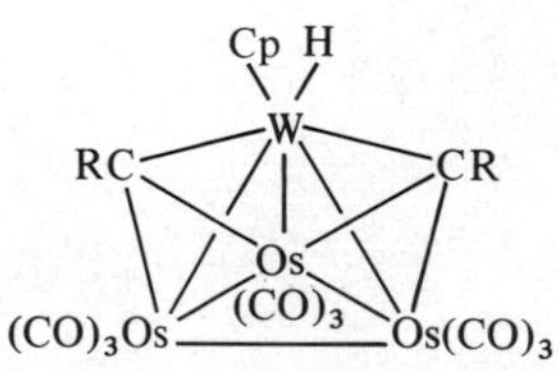

(158)

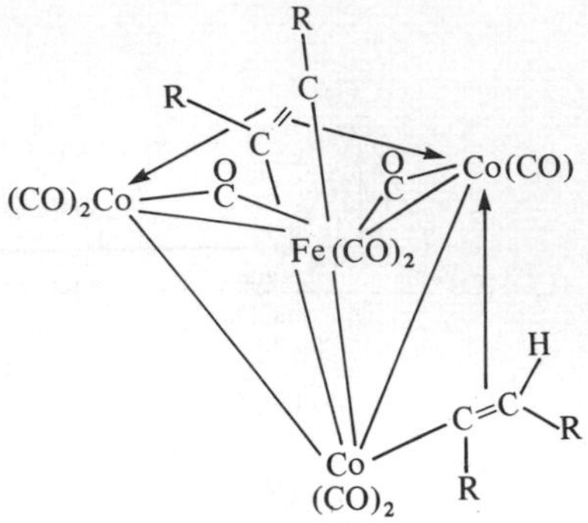

(159)

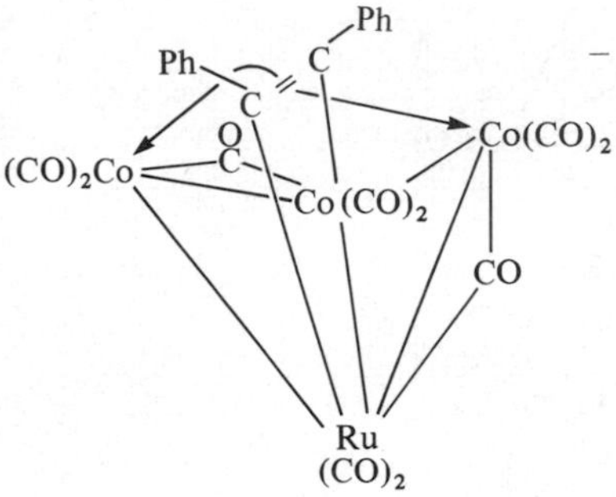

(160)

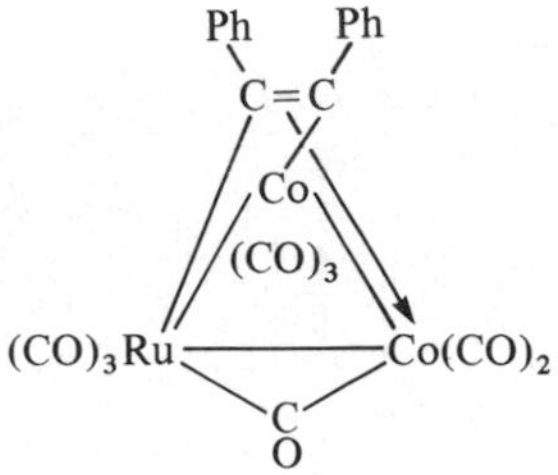

(161)

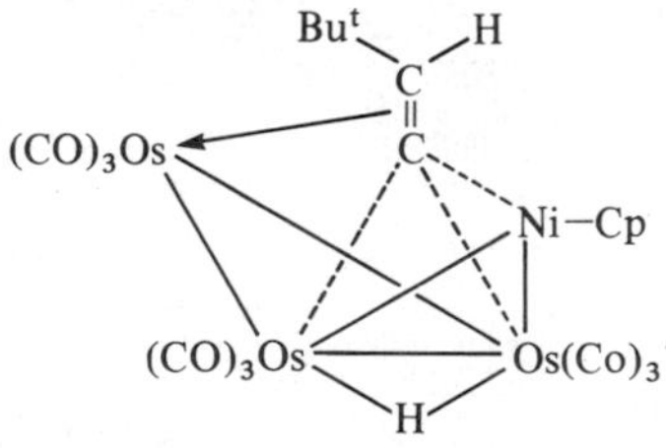

(162)

References

1. S.V. Ley, R.A. Porter, P.F. Gordon and A.J. Nelson, Gen. Synth. Methods, 1983, 6, 218.
2. T. Funabiki, Rev. Inorg. Chem., 1982, 4, 329.
3. L.A.P. Kane-Maguire, Mech. Inorg. Organomet. React., 1983, 273 and 346.
4. E. Sappa, A. Tiripicchio and P. Braunstein, Chem. Rev., 1983, 83, 203.
5. J. Holt, M.F. Lappert and P.I.W. Yarrow, Chem. Rev., 1983, 83, 135.
6. G. van Koten and K. Vrieze, Adv. Organometallic Chem., 1982, 21, 152.
7. W.A. Hermann, J. Organometal. Chem., 1983, 250, 319.
8. M. Brookhart and M.L.H. Green, J. Organometal. Chem., 1983, 250, 395.
9. J.A.S. Howell and P.M. Burkinshaw, Chem. Rev., 1983, 83, 557.
10. J.D. Atwood, M.J. Wovkulich and D.C. Sonnenberger, Accts. Chem. Res., 1983, 16, 350.
11. D.J. Darensbourg, Adv. Organometallic Chem., 1982, 21, 113.
12. S. Sakaki, K. Kutaura, K. Morokuma and K. Ohkubo, Inorg. Chem., 1983, 22, 104.
13. A.D. Berry, Organometallics, 1983, 2, 895.
14. L.S. Isaeva, T.A. Pegamova, P.V. Petrovskii, P.F. Kayumov, F.G. Yusupova and V.P. Yureb, J. Organometal. Chem., 1983, 248, 375.
15. L.S. Isaeva, T.A. Pegamova, P.V. Petrovskii, D.B. Furman, S.V. Zotova, A.V. Kudryashev and O.V. Bragin, J. Organometal. Chem., 1983, 258, 367.
16. M.R. White and P.J. Stang, Organometallics, 1983, 2, 1654.
17. C.A. Tolman, W.C. Seidel and L.W. Gosser, Organometallics, 1983, 2, 1391.
18. C.A. Tolman, Organometallics, 1983, 2, 614.
19. S.I. Al Resayes, S.I. Klein, H.W. Kroto, M.F. Meidine and J.F. Nixon, J. Chem. Soc., Chem. Comm., 1983, 930.
20. T.A. vanderKamp, L.W. Jenneskens, H. Meeuwissen, F. Bickelhaust, D. Walther, E. Dinjus, E. Uhlig and A.L. Spek, J. Organometal. Chem., 1983, 254, C33.
21. K. Porschke, G. Wilke and C. Kruger, Angew. Chem. Int. Ed., 1983, 22, 546.
22. L. Maresca, G. Natile, A.M.M. Lanfredi and A. Tiripicchio, J. Am. Chem.Soc. 1982, 104, 7661.
23. L. Maresca and G. Natile, J. Chem. Soc., Chem. Comm., 1983, 40.
24. A.T. Hutton, D.M. McEwan, B.L. Shaw and S.W. Wilkinson, J. Chem. Soc., Dalton Trans., 1983, 2011.
25. M.K. Cooper, P.V. Stevens and M. McPartlin, J. Chem. Soc., Dalton Trans., 1983, 553.
26. N.M. Boag, R.J. Goodfellow, M. Green, B. Hessner, J.A.K. Howard and F.G.A. Stone, J. Chem. Soc., Dalton Trans., 1983, 2585.
27. K. Miki, O. Shiotami, Y. Kai, V. Kasai, H. Kamatami and H. Kurosawa, Organometallics, 1983, 2, 585.
28. H. Lehmkuhl, F. Danowski, R. Benn, A. Rufinska, G. Schroth and R. Mynott, J. Organometal. Chem., 1983, 254, C11.
29. H. Lehmkuhl, C. Naydowski and M. Bellenbaum, J. Organometal. Chem., 1983, 246, C5.
30. H. Lehmkuhl, C. Naydowski, R. Benn, A. Rufinska and G. Schroth, J. Organometal. Chem., 1983, 246, C9.
31. H. Lehmkuhl, C. Naydowski, R. Benn, A. Rufinska, G. Schroth, R. Mynott and C. Kruger, Chem. Ber., 1983, 116, 2447.
32. H.M. Buch, P. Bruger, R. Benn, C. Kruger and A. Rufinska, Angew. Chem. Int. Ed., 1983, 22, 774.
33. H. Hoberg and D. Schafter, J. Organometal. Chem., 1983, 251, C51.
34. D. Walther and E. Dinjus, Z. Chem., 1982, 22, 228.
35. D. Walther, E. Dinjus, J. Sieler, N.N. Thanh, W. Schade and I. Leban, Z. Naturforsch., 1983, 38B, 835.
36. M.C. Bohm and R. Gleiter, Angew. Chem. Int. Ed., 1983, 22, 329.
37. G.C. Hancock, N.M. Kostic and R.F. Fenske, Organometallics, 1983, 2, 1089.
38. B. Bogdanovic, M. Rubach and K. Seevogel, Z. Naturforsch., 1983, 38B, 592.
39. B. Bogdanovic, P. Gotsch and M. Rubach, Z. Naturforsch., 1983, 38B, 599.

40. R. Benn, B. Bogdanovic, P. Gotsch and M. Rubach, Z. Naturforsch., 1983, 38B, 602.
41. H.F. Klein, J. Gross, P. Hammer and U. Schubert, Chem. Ber., 1983, 116, 1441.
42. H.F. Klein, H. Witty and U. Schubert, J. Chem. Soc., Chem. Comm., 1983, 231.
43. K. Jonas, E. Deffense and D. Habermann, Angew. Chem. Int. Ed., 1983, 22, 716.
44. M. Brookhart, M.L.H. Green and R.B.A. Pardy, J. Chem. Soc., Chem. Comm., 1983, 691.
45. R.B.A. Pardy, G.W. Smith and M.E. Vickers, J. Organometal. Chem., 1983, 252, 341.
46. R. Klingert and H. Werner, Chem. Ber., 1983, 116, 1450.
47. A.J. Rest and D.J. Taylor, J. Chem. Soc., Dalton Trans., 1983, 1291.
48. S.F. Parker, C.H.F. Peden, P.H. Barrett and R.G. Pearson, Inorg. Chem., 1983, 22, 2813.
49. D.B. Beach and W.L. Jolly, Inorg. Chem., 1983, 22, 2137.
50. J.C. Mitchener and M.S. Wrighton, J. Am. Chem. Soc., 1983, 105, 1065.
51. M. Cosandey, M. van Buren and H.J. Hansen, Helv. Chim. Acta, 1983, 66, 1.
52. S. Komiya, H. Minato, T. Ikariya, T. Yamamoto and A. Yamamoto, J. Organometal. Chem., 1983, 254, 83.
53. R. Werner and H. Werner, Chem. Ber., 1983, 116, 2074.
54. H. Kletzin, H. Werner, O. Serhadli and M.L. Zeigler, Angew. Chem. Int. Ed., 1983, 22, 46.
55. W.W. McConnell, G.O. Nelson and M.F. Wright, Inorg. Chem., 1983, 22, 1689.
56. T.W. Bodner and A.R. Cutler, J. Am. Chem. Soc., 1983, 105, 5926.
57. M. Moll, H. Behrens, H.J. Siebold and P. Merbach, Z. Naturforsch., 1983, 38B, 409.
58. G. Cardaci, J. Organometal. Chem., 1983, 244, 153.
59. K.J. Jens, T. Valeri and E. Weiss, Chem. Ber., 1983, 116, 2872.
60. A.S. Batsanov and Y.T. Struchkov, J. Organometal. Chem., 1983, 248, 101.
61. G.D. Annis, E.M. Hebblethwaite, S.T. Hodgson, D.M. Hollinshead and S.V. Ley, J. Chem. Soc., Perkin Trans. I, 1983, 2851.
62. E. Carmona, J.M. Marin, M.L. Poveda and J.L. Atwood, J. Am. Chem. Soc., 1983, 105, 3014.
63. C.G. Kreiter and U. Koemm, Z. Naturforsch., 1983, 38B, 943.
64. J.W. Faller and K.H. Chao, J. Am. Chem. Soc., 1983, 105, 3893.
65. M.F. Perpignon, L. Ballester and A. Santos, J. Organometal. Chem., 1983, 241, 215.
66. A. Gourdon and K. Prout, Acta Cryst., 1983, 39C, 865.
67. A.J. Graham, D.A. Krigg and B. Sheldrick, Acta Cryst., 1983, 39C, 192.
68. A.B. Antonova, S.V. Kovalenko, E.D. Korniyets, A.A. Johansson, Y.T. Struchkov, A.I. Ahmedov, A.I. Yanovsky, J. Organometal. Chem., 1983, 244, 35.
69. M. Moll, H. Behrens, H.T. Seibold and P. Merbach, J. Organometal. Chem., 1983, 248, 329.
70. M. Moll, H.J. Seibold, J. Organometal. Chem., 1983, 248, 343.
71. G. Liehr, H.S. Seibold and H. Behrens, J. Organometal. Chem., 1983, 248, 351.
72. S.A. Cohen, P. Auburn and J.E. Bercaw, J. Am. Chem. Soc., 1983, 105, 1136.
73. H. Lehmkuhl, Y.L. Tsien, E. Janssen and R. Mynott, Chem. Ber., 1983, 116, 2426.
74. H. Lehmkuh, E. Janssen and R. Schwickardi, J. Organometal. Chem., 1983, 258, 171.
75. G. Erker, K. Krupp, J.L.Atwood and W.E. Hunter, Organometallics, 1983, 2, 1555.
76. K. Jonas and V. Wiskamp, Z. Naturforsch., 1983, 38B, 1113.
77. I.S. Thompson, R.L. Harlow and J.F. Whitney, J. Am. Chem. Soc., 1983, 105, 3522.
78. D. Cohen and H. Basch, J. Am. Chem. Soc., 1983, 105, 6980.
79. P.H. Kasai, J. Am. Chem. Soc., 1983, 105, 6704.
80. W. Offermann and U. Fritzche, Inorg. Chim. Acta, 1983, 73, 113.

81. H.M. Buch, P. Binger, R. Goddard and C. Kruger, J. Chem. Soc., Chem. Comm., 1983, 648.
82. H.M. Buch, G. Schroth, R. Mynott and P. Binger, J. Organometal. Chem., 1983, 247, C63.
83. M. Penckert and W. Keim, Organometallics, 1983, 2, 594.
84. M. Hiramatsu, K. Shiozaki, T. Fujinami and S. Sakai, J. Organometal. Chem., 1983, 246, 203.
85. A. Tiripicchio, M.T. Camellini, C. Claver, A. Ruiz and L.A. Oro, J. Organometal. Chem., 1983, 241, 77.
86. L.A. Oro, M.A. Esteruelas, C. Foces-Foces and F.H. Cano, J. Organometal. Chem., 1983, 258, 357.
87. M. Kretschmer, P.S. Pregosin and M. Garralda, J. Organometal. Chem., 1983, 244, 175.
88. D.R. Anton and R.H. Crabtree, Organometallics, 1983, 2, 621.
89. J.G. Liepoldt and E.C. Grobler, Inorg. Chim. Acta, 1983, 72, 17.
90. M.J.Decker, D.O.K. Fjeldsted, S.R. Stobart and M.J. Zaworotko, J. Chem. Soc., Chem. Comm., 1983, 1525.
91. R. Uson, L.A. Oro, J.A. Cabeza, C. Foces-Foces, F.H. Cano and S. Garcia-Blanco, J. Organometal. Chem., 1983, 246, 73.
92. P.E. Kreter and D.W. Meek, Inorg. Chem., 1983, 22, 319.
93. L.A. Oro, D. Carmona and J. Reedijk, Inorg. Chim. Acta, 1983, 71, 115.
94. H. Tom Dieck and J. Klaus, J. Organometal. Chem., 1983, 246, 301.
95. J.C.T.R. Burckett-St. Laurent, P.B. Hitchcock and J.F. Nixon, J. Organometal. Chem., 1983, 249, 243.
96. J. Muller and B. Passon, J. Organometal. Chem., 1983, 247, 13.
97. G. Vitulli, A. Raffaeli, P.A. Costanino, C. Barberino, F. Marchetti, S. Merlino and P.S. Skell, J. Chem. Soc., Chem. Comm., 1983, 232.
98. M. Boehm, Theochem., 1982, 6, 165.
99. R. Uson, L.A. Oro, M.A. Ciriano, M. Naval, M.C. Apreda, C. Foces-Foces, F.H. Cane and S. Garcia-Blanco, J. Organometal. Chem., 1983, 256, 331.
100. W.C. Ho and T.C.W. Mak, J. Organometal. Chem., 1983, 241, 131.
101. T.C.W. Mak and W.C. Ho, J. Organometal. Chem., 1983, 243, 233.
102. T.C.W. Mak, J. Organometal. Chem., 1983, 246, 331.
103. T.C.W. Mak, W.C. Ho and N.Z. Huang, J. Organometal. Chem., 1983, 255, 123.
104. R.D. Ernst and T.H. Cymbaluk, Organometallics, 1983, 2, 1220.
105. L. Stahl and R.D. Ernst, Organometallics, 1983, 2, 1229.
106. A. Hofner, J.H. Bieri, R. Prewo, W. von Phillipsborn and A. Salzer, Angew. Chem. Int. Ed., 1983, 22, 713.
107. J. Clelland and G.R. Knox, J. Chem. Soc., Chem. Comm., 1983, 1219.
108. E.J.S. Vichi, P.R. Raithby and M. McPartlin, J. Organometal. Chem., 1983, 256, 111.
109. D. Minniti and P.L. Timms, J. Organometal. Chem., 1983, 258, C12.
110. J.W. Chinn and M.B. Hall, Inorg. Chem., 1983, 22, 2759.
111. P. Diehl, F. Moia, H. Boesinger and J. Wirz, J. Mol. Structure, 1983, 98, 297.
112. J.C. Calabrese, S.D. Ittel, D.A. Sweigart, H.S. Choi and S.G. Davis, Organometallics, 1983, 2, 226.
113. J.A. Mondo and J.A. Berson, J. Am. Chem. Soc., 1983, 105, 3340.
114. S.G. Davies, J. Hibberd and S.J. Simpson, J. Organometal. Chem., 1983, 246, C16.
115. S.G. Davies, S.D. Moon, S.J. Simpson and S.E. Thomas, J. Chem. Soc., Dalton Trans., 1983, 1805.
116. A. Laporterie, H. Iloughmane and J. Dubac, J. Organometal. Chem., 1983, 244, C12.
117. M.F. Semmelhack, J.W. Herndon and J.P. Springer, J. Am. Chem. Soc., 1983, 105, 2497.
118. M.F. Semmelhack and J.W. Herndon, Organometallics, 1983, 2, 363.
119. A.J. Birch, L.F. Kelly and A.S. Narula, Tetrahedron, 1982, 38, 1813.
120. A.J. Pearson, D.C. Rees and C.W. Thornber, J. Chem. Soc., Perkin Trans. I, 1983, 619.
121. D. Farcasiu and G. Marino, J. Organometal. Chem., 1983, 253, 243.

122. A.J. Pearson, T.R. Perrior and D.A. Griffin, J. Chem. Soc., Perkin Trans.I, 1983, 625.
123. G.R. John, L.A.P. Kane-Maguire, T.I. Odiaka and C. Eaborn, J. Chem. Soc., Dalton Trans., 1983, 1721.
124. A.J. Pearson and I.C. Richards, Tet. Letters, 1983, 24, 2465.
125. B.F. Anderson and G.B. Robertson, Acta Cryst., 1983, 39C, 428.
126. J.G. Atton and L.A.P. Kane-Maguire, J. Organometal. Chem., 1983, 246, C23.
127. J.A.S. Howell and M.J. Thomas, J. Chem. Soc., Dalton Trans., 1983, 1401.
128. B.M.R. Bandara, A.J. Birch, L.F. Kelly and T.C. Khor, Tet. Letters, 1983, 24, 2491.
129. J.A.S. Howell and M.J. Thomas, J. Organometal. Chem., 1983, 247, C21.
130. J.C. Green, M.R. Kelly, M.P. Payne, F.A. Seddon, D. Astruc, J.R. Hamon and P. Michaud, Organometallics, 1983, 2, 211.
131. D. Astruc, Tetrahedron, 1983, 39, 4027.
132. L.L.K. Li Shing Man, J.G.A. Reuvers and J. Takats, Organometallics, 1983, 2, 28.
133. I.L. Fragala, J. Takats and M.A. Zerbo, Organometallics, 1983, 2, 1502.
134. N.G. Connelly, A.R. Lucy and J.B. Sheridan, J. Chem. Soc., Dalton Trans., 1983, 1465.
135. N.G. Connelly, A.R. Lucy and M.W. Whitely, J. Chem. Soc., Dalton Trans., 1983, 111.
136. N.G. Connelly, A.R. Lucy and M.W. Whitely, J. Chem. Soc., Dalton Trans., 1983, 117.
137. Z. Goldschmidt, Y. Bakal, D. Hoffer, A. Eisenstadt and P.F. Lindley, J. Organometal. Chem., 1983, 258, 53.
138. Z. Goldschmidt and S. Antebi, J. Organometal. Chem., 1983, 259, 119.
139. T. Ishitu, K. Haramo, N. Hori, M. Yasuda and K. Kanematsu, Tetrahedron, 1983, 39, 1281.
140. R.B. Hitam, R. Narayamaswamy and A.J. Rest, J. Chem. Soc., Dalton Trans., 1983, 1351.
141. M.J. Hails, B.E. Mann and K.M. Spencer, J. Chem. Soc., Chem. Comm., 1983, 120.
142. C. Paradisi and G. Schiavon, J. Organometal. Chem. , 1983, 146, 197.
143. G. Schiavon and C. Paradisi, J. Organometal. Chem., 1983, 243, 351.
144. P. Pertici, G. Vitulli, W. Porzio, M. Zocchi, P.L. Barili and G. Deganello, J. Chem. Soc., Dalton Trans., 1983, 1553.
145. K. Itoh, K. Mukai and H. Nagashima, Chem. Lett., 1983, 499.
146. H. Nagashima, H. Matsuda and K. Itoh, J. Organometal. Chem., 1983, 258, C15.
147. P. Harter, P. Pauson and S.S. Ullah, J. Organometal. Chem., 1983, 247, C27.
148. H. Boennemann, M. Radermacher, C. Krueger and H.J. Kraus, Helv. Chim. Acta, 1983, 66, 185.
149. M. Grzezczuk, D.E. Smith and W.E. Geiger, J. Am. Chem. Soc., 1983, 105, 1772.
150. R.P. Hughes, D.E. Samkoff, R.E. Davis and B.B. Laird, Organometallics, 1983, 2, 195.
151. P. Powell, J. Organometal. Chem., 1983, 244, 393.
152. P. Powell, J. Organometal. Chem., 1983, 243, 205.
153. G. Moran, M. Green and A.G. Orpen, J. Organometal. Chem., 1983, 250, C15.
154. J. Muller, R. Stock and J. Pickardt, Z. Naturforsch., 1983, 38B, 1454.
155. J.L. Davidson, K. Davidson and E.W. Lindsell, J. Chem. Soc., Chem. Comm., 1983, 452.
156. J.W. Faller, H.H. Murray, D.L. White and K.H. Chao, Organometallics, 1983, 2, 400.
157. M. Kotzian, C.G. Kreiter, G. Michael and S. Ozkar, Chem. Ber., 1983, 116, 3637.
158. S. Ozkar and C.G. Kreiter, J. Organometal. Chem., 1983, 256, 57.
159. C.G. Kreiter and S. Oskar, Z. Naturforsch., 1983, 38B, 1424.
160. B. Lubke, F. Edelmann and U. Behrens, Chem. Ber., 1983, 116, 11.
161. M.L.H. Green, A. Izquerido, J.J. Martin-Polo, V.S.B. Mtetwa and K. Prout, J. Chem. Soc., Chem. Comm., 1983, 538.
162. P. Legzdins and A.D. Hunter, Organometallics, 1982, 2, 525.

163. A. Ceccon, A. Gambaro, A.M. Romanin and A. Venzo, J. Organometal. Chem., 1983, 254, 199.
164. W.P. Henry and R.D. Rieke, J. Am. Chem. Soc., 1983, 105, 6314.
165. E.P. Kundig and D.P. Simmons, J. Chem. Soc., Chem. Comm., 1983, 1320.
166. L. Weber and R. Boese, Angew. Chem. Int. Ed., 1983, 22, 498.
167. G.J. Kubas, Inorg. Chem., 1983, 22, 692.
168. C.D. Hoff, J. Organometal. Chem., 1983, 246, C53.
169. K. Lal, N.T. Leckey and W.E. Watts, J. Organometal. Chem., 1983, 254, 193.
170. K. Lal, N.T. Leckey and W.E. Watts, J. Organometal. Chem., 1983, 258, 205.
171. C.A. Bunton, M.M. Whole, J.R. Moffatt and W.E. Watts, J. Organometal.Chem., 1983, 253, C33.
172. C.A. Bunton, K. Lal and W.E. Watts, J. Organometal. Chem., 1983, 247, C14.
173. S. Ozkar and C.G. Kreiter, Z. Anorg. Chem., 1983, 502, 215.
174. C.G. Kreiter and H. Kurz, Z. Naturforsch., 1983, 38B, 841.
175. E. Michels and C.G. Kreiter, J. Organometal. Chem., 1983, 252, Cl.
176. C.G. Kreiter and W. Lipps, J. Organometal. Chem., 1983, 253, 339; 241,185
177. K.Franzreb and C.G. Kreiter, J. Organometal. Chem., 1983, 246, 189.
178. M. Layendecker and C.G. Kreiter, J. Organometal. Chem., 1983, 249, C31.
179. E.P. Kundig, P.L.Timms, B.A. Kelly and P. Woodward, J. Chem. Soc., Dalton Trans., 1983, 901.
180. J.R. Bleeke and J.J. Kotyk, Organometallics, 1983, 2, 1263.
181. Y.K. Chung, E.D. Honig and D.A. Sweigart, J. Organometal. Chem., 1983, 256, 277.
182. A.J. Pearson and I.C. Richards, J. Organometal. Chem., 1983, 258, C41.
183. M. Brookhart, W. Lamanna and A.R. Pinhas, Organometallics, 1983, 2, 638.
184. Y.K. Chung, E.D. Honig, W.T. Robinson, D.A. Sweigart and N.C. Connelly, Organometallics, 1983, 2, 1479.
185. G. Erker, H.J. de Liefde Meijer and J.H. Feuben, Organometallics, 1983, 2, 1173,
186. G. Erker, K. Engel, J.L. Atwood and W.E. Hunter, Angew. Chem. Int. Ed., 1983, 22, 494.
187. M.L.H. Green, N.J. Hazel, P.D. Orchenik, V.S.B. Mtetwa and K. Prout, J. Chem. Soc., Chem. Comm., 1983, 356.
188. L.P. Yureva, N.N. Zaitseva, N.V. Zakurin, A.Y. Vasilkov and N.I. Vasyukova, J. Organometal. Chem., 1983, 247, 287.
189. H.C. Foley, L.M. Strubinger, T.S. Targos and G.L. Geoffroy, J. Am. Chem.Soc. 1983, 105, 3064.
190. H.G. Alt, J. Organometal. Chem., 1983, 256, C12.
191. K.A. Mead, H. Morgan and P. Woodward, J. Chem. Soc., Dalton Trans., 1983, 271.
192. S.R. Allen, M. Green, N.C. Norman, K.E. Paddick and A.G. Orpen, J. Chem. Soc., Dalton Trans., 1983, 1625.
193. S.R. Allen, P.K. Baker, S.G. Barnes, M. Bottrill, M. Green, A.G. Orpen, I.D. Williams and A.J. Welch, J. Chem. Soc., Dalton Trans., 1983, 927.
194. R.G. Beevor, M. Green, A.G. Orpen and I.D. Williams, J. Chem. Soc., Dalton Trans., 1983, 673.
195. J.C. Jeffrey, J.C.V. Laurie, I. Moore and F.G.A. Stone, J.Organometal. Chem., 1983, 258, C37.
196. F.R. Kreissl, W.J. Sieber and M. Wolfgruber, Z. Naturforsch., 1983, 38B, 1419.
197. J.L. Davidson, W.F. Wilson, L. Manojlovic-Muir and K.W. Muir, J. Organometal. Chem., 1983, 254, C6.
198. J.L. Davidson, I.E.P. Murray, P.N. Preston and M.V. Russo, J. Chem. Soc., Dalton Trans., 1983, 1783.
199. J.L. Davidson, J. Chem. Soc., Dalton Trans., 1983, 1667.
200. J.L. Davidson and G. Vasapollo, Polyhedron, 1983, 2, 205.
201. P.W. Winston, S.J.N. Burgmeyer and J.L. Templeton, Organometallics, 1983, 2, 167.
202. J.L. Davidson and G. Vasapollo, J. Organometal. Chem., 1983, 241, C24.
203. K. Hirotsu, T. Higuchi, M. Kido, K. Tatsumi, T. Yoshida and S. Otsuka, Inorg. Chem., 1983, 22, 2416.

204. R.S. Herrick, D.M. Leazer and J.C. Templeton, Organometallics, 1983, 2, 834.
205. R.S. Herrick, S.J.N. Burgmeyer and J.L. Templeton, Inorg. Chem., 1983, 22, 3275.
206. R.S. Herrick, S.J. Neiter-Burgmeyer and J.L. Templeton, J. Am. Chem. Soc., 1983, 105, 2599.
207. M. McKenna, L.L. Wright, D.J. Miller, L. Tanner, R.C. Haltiwanger and M.R. Dubois, J. Am. Chem. Soc., 1983, 105, 5329.
208. T.R. Halbert, W.H. Pan and E.I. Steifel, J. Am. Chem. Soc., 1983, 105, 5476.
209. B. Capelle, M. Dartiguenave, Y. Dartiguenave and A.L. Beauchamp, J. Am. Chem. Soc., 1983, 105, 4662.
210. J. Wolf, H. Werner, O. Serhadli and M.C. Zeigler, Angew. Chem. Int. Ed., 1983, 22, 414.
211. H. Werner, J. Wolf, U. Schubert and K. Ackermann, J. Organometal. Chem., 1983, 243, C63.
212. M.E. Howdon, R.D.W. Kemmitt and M.D. Schilling, J. Chem. Soc., Dalton Trans., 1983, 2459.
213. R.D.W. Kemmitt and M.D. Schilling, J. Chem. Soc., Dalton Trans., 1983, 1887.
214. Y. Wakatsuki, O. Nomura, K. Kitaura, K. Morokuma and H. Yamazaki, J. Am. Chem. Soc., 1983, 105, 1907.
215. J.S. Thompson and J.F. Whitney, J. Am. Chem. Soc., 1983, 105, 5488.
216. D.W. Macomber and M.D. Rausch, J. Am. Chem. Soc., 1983, 105, 5325.
217. D.L. Reger, K.L. Belmore, E. Mintz, N.G. Charles, E.A.H. Griffith and E.L. Amma, Organometallics, 1983, 2, 101.
218. D.L. Reger, K.A. Belmore, J.L. Atwood and W.E. Hunter, J. Am. Chem. Soc., 1983, 105, 5710.
219. G.P. Elliott and W.R. Roper, J. Organometal. Chem., 1983, 250, C5.
220. D.G. Bickley, N. Hao, P.Bougeard, B.G. Sayer, R.C. Burns and M.J. McGlinchey, J. Organometal. Chem., 1983, 246, 257.
221. A. Famili, M.F. Farona and S. Thanedar, J. Chem. Soc., Chem. Comm., 1983, 435.
222. M. Arewgoda, B.H. Robinson and J. Simpson, J. Am. Chem. Soc., 1983, 105, 1893.
223. B.E. Hanson and J.S. Mancini, Organometallics, 1983, 2, 126.
224. A. Schegolov, W.A. Smit, Y. Kalyan, M.Z. Krimer and R. Caple, Tet. Letters, 1982, 23, 4419.
225. S. Aime, D. Osella, L. Milone and A. Tiripicchio, Polyhedron, 1983, 2, 77.
226. I.T. Horvath, G. Palyi, L. Marko and G.D. Andreeti, Inorg. Chem., 1983, 22, 1049.
227. H. Yamazaki, K. Yasufuku and Y. Wakatsuki, Organometallics, 1983, 2, 726.
228. K.P.C. Vollhardt and J.A. King, Organometallics, 1983, 2, 684.
229. W.A. Herrmann, C. Bauer and J. Weichmann, J. Organometal. Chem., 1983, 243, C21.
230. P.A. Corrigan, R.S. Dickson, S.H. Johnson, G.N. Pain and M. Yeoh, Austral. J. Chem., 1982, 35, 2203.
231. J.T. Mague, Inorg. Chem., 1983, 22, 1159.
232. J.T. Mague, Inorg. Chem., 1983, 22, 45.
233. G.W. Bushnell, D.O.K. Fjeldstad, S.R. Stobart and M.J. Zaworotko, J. Chem. Soc., Chem. Comm., 1983, 580.
234a. K.P. Grande, A.J. Kumin, L.S. Stuhl and B.M. Foxman, Inorg. Chem., 1983, 22, 1791.
234b. M. Green, A.G. Orpen, C.J. Schaverien and I.D. Williams, J. Chem. Soc., Chem. Comm., 1983, 181.
235. M. Green, A.G. Orpen, C.J. Schaverien and I.D. Williams, J. Chem. Soc., Chem. Comm., 1983, 583.
236. R.F. Gerlach, D.N. Duffy and M.D. Curtis, Organometallics, 1983, 2, 1172.
237. R.E. Colborn, D.L. Davies, A.F. Dyke, A. Endesfelder, S.A.R. Knox, A.G. Orpen and D. Plaas, J. Chem. Soc., Dalton Trans., 1983, 2661.
238. A.F. Dyke, S.A.R. Knox, M.J. Morris and P.J. Naish, J. Chem. Soc., Dalton Trans., 1983, 1417.
239. A.G. Orpen, J. Chem. Soc., Dalton Trans., 1983, 1427.

240. R.E. Colborn, A.F. Dyke, S.A.R. Knox, K.A. Mead and P. Woodward, J. Chem. Soc., Dalton Trans., 1983, 2099.
241. P.Q. Adams, D.L. Davies, A.F. Dyke, S.A.R. Knox, K.A. Mead and P. Woodward, J. Chem. Soc., Chem., Comm., 1983, 222.
242. M.I. Bruce, J.G. Matisons, B.W. Skelton and A.H. White, J. Organometal.Chem. 1983, 251, 249.
243. A. Astier, J.C. Daran, Y. Jeannin and C. Rigault, J. Organometal. Chem., 1983, 241, 53.
244. G.N. Mott and A.J. Carty, Inorg. Chem., 1983, 22, 2726.
245. K.M. Motyl, J.R. Norton, C.K. Schauer and O.P. Anderson, J. Am. Chem. Soc., 1982, 104, 7325.
246. M. Burke, J. Takats, F.W. Grevels and J.G.A. Reuvers, J. Am. Chem. Soc., 1983, 105, 4092.
247. K. Raab, U. Nagel and W. Beck, Z. Naturforsch., 1983, 38B, 1466.
248. J.A. Iggo, M.J. Mays, P.R. Raithby and K. Henrick, J. Chem. Soc., Dalton Trans., 1983, 205.
249. M. Green, A.G. Orpen, C.J. Schaverien and I.P. Williams, J. Chem. Soc., Chem. Comm., 1983, 1399.
250. M. Casarin, D. Ajo, G. Granozzi, E. Tondello and S. Aime, J. Chem. Soc., Dalton Trans., 1983, 869.
251. H. Brunner and M. Muschiol, J. Organometal. Chem., 1983, 248, 233.
252. K. Broadley, G.A. Lane, N.G. Connelly and W.E. Geiger, J. Am. Chem. Soc., 1983, 105, 2486.
253. L.N. Lewis and K.G. Caulton, J. Organometal. Chem., 1983, 258, 57.
254. M.H. Chisholm, K. Folting, J.C. Huffmann, J. Leonelli and D.M. Hoffmann, J. Chem. Soc., Chem. Comm., 1983, 589.
255. E. Elsenbroich, J. Heck, W. Massa, E. Nun and R. Schmidt, J. Am. Chem. Soc., 1983, 105, 2905.
256. E. Elsenbroich, J. Heck, W. Massa and R. Schmidt, Angew. Chem. Int. Ed., 1983, 22, 330.
257. S. Aime, D. Osella, A.J. Deeming, A.M.M. Lanfredi and A. Tiripicchio, J. Organometal. Chem., 1983, 244, C47.
258. G. Granozzi, E. Tondello, R. Bertoncello, S. Aime and D. Osella, Inorg. Chem., 1983, 22, 744.
259. S. Ermer, K. King, K.I. Hardcastle, E. Rosenberg, A.M.M. Lanfredi, M.T. Camellini and A. Tiripicchio, Inorg. Chem., 1983, 22, 1339.
260. G. Gervasio, E. Sappa, A.M.M. Lanfredi and A. Tiripicchio, Inorg. Chim.Acta 1983, 68, 71.
261. S.A. MacLaughlin, J.P. Johnson, N.J. Taylor, A.J. Carty and E. Sappa, Organometallics, 1983, 2, 352.
262. S. Aime and A.J. Deeming, J. Chem. Soc., Dalton Trans., 1983, 1807.
263. C.E. Anson, B.T. Keiller, I.A. Oxton, D.B. Powell and N. Sheppard, J. Chem. Soc., Chem. Comm., 1983, 470.
264. E. Sappa, A. Tiripicchio and A.M.M. Lanfredi, J. Organometal. Chem., 1983, 249, 391.
265. J. Evans and G.S. McNulty, J. Chem. Soc., Dalton Trans., 1983, 639.
266. M.I. Bruce, B.K. Nicholson and M.L. Williams, J. Organometal. Chem., 1983, 243, 69.
267. M.I. Bruce, M.L. Williams and B.K. Nicholson, J. Organometal. Chem., 1983, 258, 63.
268. B.F.G. Johnson, J. Lewis, P.R. Raithby and M.J. Rosales, J. Chem. Soc., Dalton Trans., 1983, 2645.
269. J.T. Park, J.R. Shapley, M.R. Churchill and C. Bueno, J. Am. Chem. Soc., 1983, 105, 6182.
270. G. Granozzi, E. Tondello, M. Casarin, S. Aime and D. Osella, Organometallics, 1983, 2, 430.
271. G. Granozzi, R. Bertoncello, M. Acampara, D. Ajo, D. Osella and S. Aime, J. Organometal. Chem., 1983, 244, 383.
272. S. Aime, D. Osella, L. Milone, A.M.M. Lanfredi and A. Tiripicchio, Inorg. Chim. Acta, 1983, 71, 141.

273. P. Braunstein, J. Rose and O. Bars, J. Organometal. Chem., 1983, 252, C101.
274. E. Sappa, A. Tiripicchio and M.T. Camellini, J. Organometal. Chem., 1983, 246, 287.

14

π-Cyclopentadienyl, π-Arene, and Related Complexes†

BY W. E. WATTS

1 Introduction

The organisation of this Report is similar to that of Volume 12. Cyclopentadienyl and arene complexes that contain metal-metal bonds, carbene, carbyne, or hydrocarbyl ligands, and reactions that involve replacement or elaboration of other ligands present, generally are not included (see Chapters 9-13). Only those metalla- -borane and -carbaborane complexes that incorporate a Cp or arene ligand are described, and crystal-structure determinations are included only where results of particular significance or solutions to structural problems are provided.

2 Studies of General Interest

The organic chemistry of metal-complexed Cp and arene ligands has been reviewed.[1] Other reviews have appeared dealing with complexes containing pentafulvene,[2] cyclopentadienylide,[2] and functionally substituted Cp ligands,[3] organo-lanthanide[4] and -actinide[5] complexes, chelate complexes containing Cp and arene ligands,[6] and with metal- -vapour syntheses.[7] The role of the metal atom in borane and carbaborane complexes (including those incorporating CpM residues) has been discussed,[8] and there have been calculations of the electronic structures of a variety of Cp and arene complexes,[9] and of the energetics of $\eta^5 \rightleftarrows \eta^6$ ligand-slip rearrangements of indenyl and fluorenyl complexes.[10] Stabilisation of carbanions by organo-metallic groups has been discussed,[11] and there has been a review of e.s.r. and n.m.r. spectroscopic studies of paramagnetic Cp_2M and and $(arene)_2M$ complexes.[12] Reorientational motions in $CpM(CO)_{\underline{n}}$ (M = V, Mn, Re) and benchrotrene have been investigated by neutron- -scattering[13] and spin-lattice relaxation[14] methods. The mass spectra of various Cp and arene complexes of V, Cr, and Mn have been analysed.[15]

† Throughout this review, the abbreviations Cp, Cp', Cp*, and hmb explicitly denote $(\eta^5\text{-}C_5H_5)$, $(\eta^5\text{-}C_5H_4Me)$, $(\eta^5\text{-}C_5Me_5)$, and $(\eta^6\text{-}C_6Me_6)$ ligands respectively.

3 Mono-(π-cyclopentadienyl)metal Complexes

Studies of General Interest.- The striking differences often found in the properties of analogous Cp and Cp* complexes have been related[16] to the large increase in the basicity of $(C_5H_5)^-$ caused by pentamethylation. The ease of thermal displacement of CO from $(\eta^5\text{-L})M(CO)_n$ by PPh_3 increases markedly through the series L = Cp <<< indenyl < fluorenyl.[17] Photolytic reactions of CpM complexes have been reviewed,[18] and there have been studies of the ^{19}F n.m.r. spectra of p-$FC_6H_4(\eta\text{-}C_5H_4)M(L)_n$,[19] and the e.s.r. spectra of radical anions formed by interaction of $CpM(CO)_n(L)$ with ^{60}Co γ-rays[20] and by spin-trapping of $CpM(CO)_n$ (M = Mo, W, Fe) with di-t--butyl-o-quinone.[21] The ligand $[C_5Me_4CH(Ph)Et]^-$ has been synthesised in an optically-active form and incorporated into complexes of Ti and Mo.[22]

Vanadium, Niobium, and Tantalum.- Displacement of halide from $(R_3P)_2VX_3$ (X = Cl, Br) by $(C_5H_5)^-$ to give $CpV(PR_3)_2X_2$ is successful only when the phosphine ligand is not too bulky.[23] There have been n.m.r. spectroscopic studies (^1H, ^{51}V, ^{93}Nb) of $CpM(CO)_4$ (M = V, Nb),[24] $Cp^*M(L)_2(X)H_3$ (M = Nb, Ta; L = phosphine, phosphite; X = H, Cl),[25] and $[CpV(CO)_3X]^-$ (X = H, halide, CN).[26]

Chromium, Molybdenum, and Tungsten.- A variety of complexes of the types $(\eta\text{-}C_5H_4R)M(CO)_2(NO)$ and $(\eta\text{-}C_5H_4R)M(CO)_3Me$ (M = Cr, Mo, W) has been synthesised from $(C_5H_4R)^-Na^+$ (R = CHO, COMe, CO_2Me).[3,27] Photochemical reactions of $(\eta^6\text{-cot})Cr(CO)_3$ with pentafulvenes and of (η-pentafulvene)$Cr(CO)_3$ with cot have given products incorporating both an $(\eta^5\text{-}C_5H_4)$ and an $(\eta^3\text{-}C_8H_8)$ ligand bonded to a $Cr(CO)_2$ residue.[28] Evidence has been presented for the generation of HCO and $CpM(CO)_3$ radicals upon low-temperature photolysis of $CpM(CO)_3H$ (M = Mo, W) in a CO matrix.[29] There have been multinuclear n.m.r. spectroscopic studies of $CpMo(CO)_3CH_2Ar$,[30] $CpMo(CO)_2(NO)$,[31] and $CpM(CO)_2(L)(FBF_3)$ (M = Mo, W)[32] and related complexes.

Manganese and Rhenium.- Co-condensation of Mn vapour with cyclopentadiene gives a highly reactive product which reacts with CO to give $CpMn(CO)_3$ and $CpMn(CO)_2(\eta^2\text{-}C_{10}H_{12})$.[33] Whereas cymantrene reacts with an excess of $(CF_3CO_2)_2Hg$ to give a ring-pentamercuriated product, the complex $CpMn(CO)_2(PPh_3)$ reacts at the metal atom to give $[CpMn(CO)_2(PPh_3)]_2[\mu\text{-}Hg(OCOCF_3)_2]_2$.[34] Friedel-Crafts acetylation of $(\eta^5\text{-indenyl})M(CO)_3$ (M = Mn, Re) occurs in both ligand rings,[35a] but lithiation occurs exclusively in the five-membered ring.[35b]

The equilibrium mixture of $CpRe(CO)_3$ and $(\eta^1-C_5H_5)Re(CO)_3(PMe_3)_2$, formed by reaction of the former with an excess of PMe_3, is slowly converted into the substitution product $CpRe(CO)_2(PMe_3)$.[36a] Under similar conditions, CpRe(CO)(NO)Me gives the η^2-ketene complex (1) which undergoes reversible insertion of acetone to give the cyclic ketal (2).[36b]

The mass spectra of a large number of ring-substituted cymantrenes have been analysed,[37] and oxidation potentials have been measured for complexes of the type $Cp'Mn(CO)_2(L)$ (L = phosphine, amine, ene, etc.).[38]

Iron and Ruthenium.- There have been spectroscopic studies of CpFe(CO)COMe (i.r., in CH_4 matrix at 12K),[39] FpX (X = halide) (^{13}C and ^{57}Fe n.m.r.),[40] $[CpFe(CO)(L^1)(L^2)]^+$ (i.r., Mössbauer),[41] $Fp(B_2H_5)$ (u.v. photoelectron),[42] and $(FpEPh_3)^-$ (E = Si, Ge, Sn) (e.s.r.).[43] The chemistry of the dianions $[(OC)_2Fe(\eta-C_5H_4)X(\eta-C_5H_4)Fe(CO)_2]^{2-}$ (X = $SiMe_2$, $SiMe_2CH_2CH_2SiMe_2$) has been explored.[44] Reaction of $CpRu(PPh_3)_2Cl$ with $(C_5X_5)Tl$ (X = CO_2Me) in air gives the ruthenocene derivative $CpRu(\eta-C_5X_5)$; the substituted ligand in this product is readily displaced by phosphines.[45] Lithiation of the Cp rings in FpR (R = Ph, CH_2Ph) and Fp_2 can be effected, and the resulting lithio-compounds have been used to prepare other ring-substituted derivatives.[46]

Cations of the type $[CpFe(CO)(L)(\eta-R^1C{\equiv}CR^2)]^+$ (L = PPh_3, phosphite) undergo exo-addition of deuteride from $(Et_3BD)^-$ to the Cp ligand; spontaneous transfer of the endo-hydrogen atom of the resulting η^4-cyclopentadiene complex to the alkyne ligand affords the ring-deuteriated σ-alkenyl product $(\eta-C_5H_4D)Fe(CO)(L)-C(R^1)=CHR^2$.[47] In a detailed study[48] of hydride (deuteride) additions to cations of the type $[CpM(L)_n]^+$ (M = Fe, Ru; L = CO and/or bi- or tri-dentate polyphosphines), various reaction modes have been discovered. For example, $[CpFe(CO)(\eta^2-dppe)]^+$ reacts with $LiAlH_4$ in THF at -78°C to give $CpFe(CO)(\eta^1-dppe)H$; when the reaction is conducted at 70°C in the same solvent, a mixture of $CpFe(\eta^2-dppe)Me$ and $(\eta^4-C_5H_6)Fe(CO)(\eta^2-dppe)$ results. Furthermore, whereas $[CpFe(L_3)]^+$ (L_3 = a tridentate triphosphine) adds hydride at the metal atom to give a product in which the triphosphine is bidentate, the corresponding ruthenium complex undergoes exo-addition of hydride to the Cp ligand to give $(\eta^4-C_5H_6)Ru(L_3)$.[48]

Cobalt and Rhodium.- The electronic structures of a series of 16-electron complexes of Co and Rh have been calculated,[49] e.g. $CpM(PH_3)$, $[CpM(Me)]^-$, $[CpM(NO)]^+$. Chemical and electrochemical one-

-electron oxidations of $CpM(CO)(PR_3)$ (M = Co, Rh) to the corresponding radical cations have been reported.[50]

A new synthesis of CpCo(η-cod) has been achieved (79% yield) by treatment of a mixture of cyclopentadiene and cod with $Co(acac)_3$ and an Mg-anthracene "solvate".[51] Many new complexes of the types $CpRh(\eta^3$-allyl)X (X = Br, CN, OAc, SCN, etc.)[52] and $Cp^*Rh(L^1)(L^2)$ (L^1,L^2 = CO, PR_3, ene, CS_2, _etc._)[53] have been synthesised, and the complexes (3) and (4) have been obtained from the reaction of (η-cod)$Rh(PMe_3)Cl$ with MeLi and 6,6-dimethylpentafulvene.[54]

Treatment of $CpCo(\eta\text{-}C_2H_4)_2$, synthesised in 85% yield by reaction of Cp_2Co with K and C_2H_4 in ether at low temperature, with but-2-yne affords CpCo(hmb) which is an effective catalyst for the cyclotrimerisation of alkenes and alkynes at room temperature.[55] Construction of the tetrahydroprotoberberine skeleton has been achieved by a reaction sequence in which the key step is an alkyne cyclocondensation catalysed by $CpCo(CO)_2$.[56]

Nickel, Palladium, and Platinum.- Reactions in which one of the Cp ligands of Cp_2Ni is replaced have afforded series of complexes of the type $[CpNi(PMe_3)_2]^+$ and then $CpNi(PMe_3)X$ (X = Me, CN),[57] and CpNi(η-ene)R (R = alkyl).[58] The fluxional behaviour (_i.e._ "ring-whizzing" and $\eta^1 \rightleftarrows \eta^5$ Cp site-exchange) of $CpPd(\eta^1\text{-}C_5H_5)(PR_3)$ has been studied by n.m.r. spectroscopy.[59] Additions of phosphines and phosphites (L) to $Cp^*Pd(\eta^3$-allyl) afford $Cp^*Pd(\eta^1$-allyl)(L) which are precursors of complexes in which a $Pd(L)_2$ residue is attached to the exocyclic double bond of 1,2,3,4-tetramethylpentafulvene.[60] The crystal structure of $Cp^*Pd(PPr^i{}_3)Cl$ shows that the Cp* ligand is attached asymmetrically to the metal atom.[61]

Reaction of $(C_5H_5)_2Mg$ with $[(\eta\text{-}C_2H_4)PtCl_2]_2$ affords $(CpPt\text{-}PtCp)(\mu\text{-}C_{10}H_{12})$ and $CpPt(\eta^1\text{-}C_5H_5)(\eta\text{-}C_2H_4)$; protonation of the latter product affords $[CpPt(\eta^2\text{-}C_5H_6)(\eta\text{-}C_2H_4)]^+$; the corresponding reaction of $[Cl_4Pt_2(CO)_2]^{2-}$ gives $[CpPt(CO)]_2$ and $CpPt(\eta^1\text{-}C_5H_5)(CO)$.[62] There has been a reinvestigation,[63] with contrary results, of the synthesis of 7-substituted nbd derivatives from Cp_2Ni (see Volume 12, ref.139).

Copper.- Complexes of the type Cp*Cu(L) (L = CO, PR_3, yne) have been synthesised.[64]

Lanthanide and Actinide Elements.- Reactions of Cp_2Hg with MCl_3 in THF have afforded $CpMCl_2(THF)_4$ (M = La, Sm, Eu, Tm, Yb).[65] Both $(\eta^5\text{-}C_5Ph_5)LuCl_2(THF)$ and $(\eta\text{-}C_5Ph_5)_2LuCl$ have been obtained from reactions of $(C_5Ph_5)Na$ with $LuCl_3$ in THF.[66] Catalysis by Cp* complexes of actinide elements has been reviewed.[67]

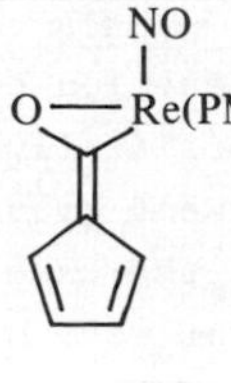

(1)

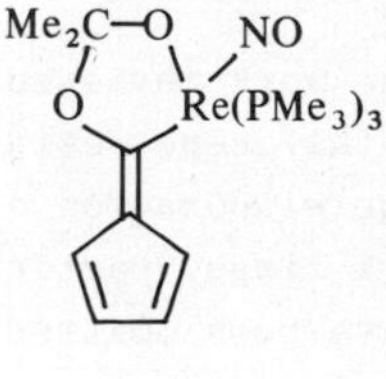

(2)

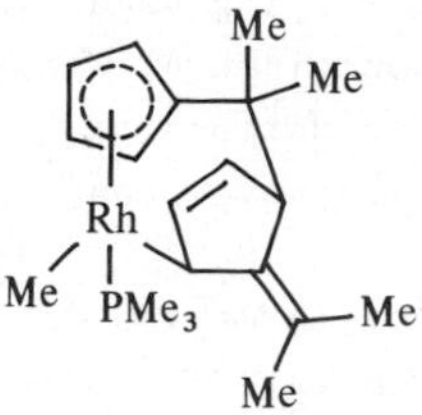

(3)

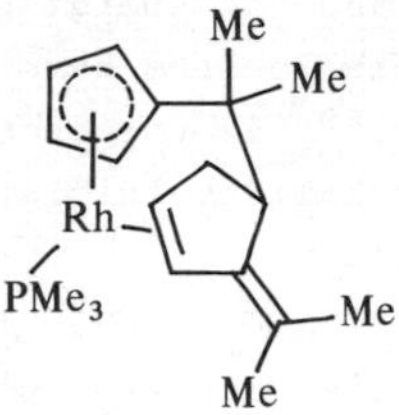

(4)

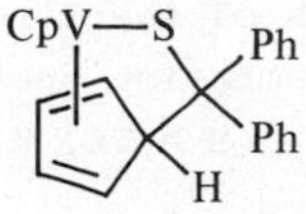

(5)

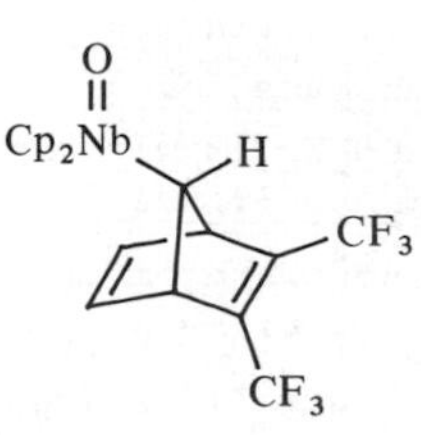

(6)

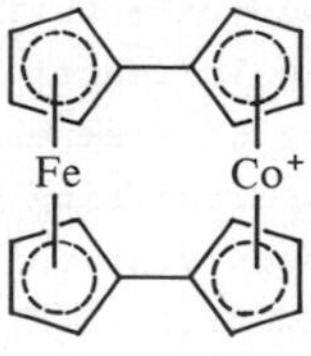

(7)

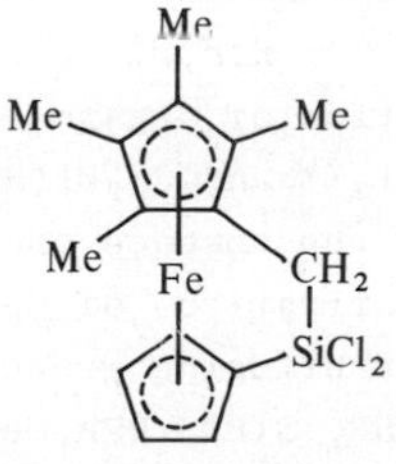

(8)

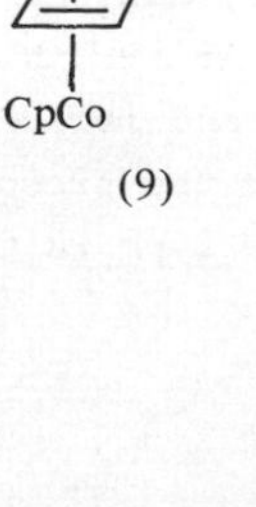

(9)

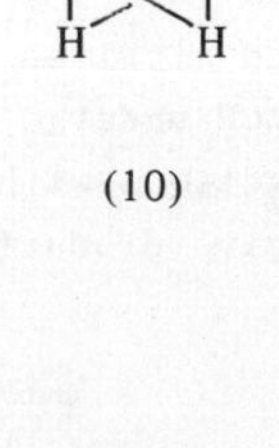

(10)

4 Bis-(π-cyclopentadienyl)metal Complexes

In the main, the work reviewed in this section deals with chemistry in which the metallocene residue is specifically involved. Papers dealing with the elaboration of substituent groups attached to the cyclopentadienyl rings, particularly in the case of ferrocene derivatives, have been omitted except for work of particular interest.

Studies of General Interest.- The metal-ligand charge separation in various metallocenes Cp_2M has been assessed by analysis of i.r. absorption intensities,[68] and the energies of low-lying negative-ion states of the first-row metallocenes have been obtained by electron--transmission spectroscopy.[69] Electrochemical oxidations of Cp_2MX_2 (M = Ti, Mo, W; X = halide, thiolate, etc.) have been studied.[70]

Titanium, Zirconium, and Hafnium.- Enthalpies of formation of Cp_2TiAr_2 have been determined by solution calorimetry.[71] A ^{13}C n.m.r. spectroscopic study of complexes of the type $Cp_2M(OAr)_2$ (M = Ti, Zr, Hf) has been reported;[72] ^1H and ^{13}C n.m.r. spectroscopy has been used in a study of the reduction of N_2 by Cp_2TiCl_2-Mg in THF.[73] E.s.r. spectroscopy has been used to investigate the generation of organometallic radicals during photolyses of Cp_2TiCl_2[74] and $Cp_2Zr(R)X$,[75] and in reactions of Cp_2ZrCl_2 with Mg in THF in the absence and presence of alkenes, alkynes, etc.[76]

Derivatives of Cp_2MCl_2 incorporating one- and two-atom inter-annular bridging groups have been synthesised, e.g. $X(\eta\text{-}C_5H_4)_2TiCl_2$ (X = CHR, SiHR, PPh, etc.),[77] $Me_2Si(\eta\text{-}C_5H_4)_2MCl_2$ (M = Ti, Zr, Hf),[77] and $[Me_2C(\eta\text{-}C_5H_4)]_2MCl_2$ (M = Ti, Zr).[78] Treatment of $CpTiCl_3$ with $(Ph_2PC_5H_4)Tl$ has given $CpTi(\eta\text{-}C_5H_4PPh_2)Cl_2$ which has been converted into a range of other complexes by reactions involving either replacement of chloride or coordination of the phosphine group to another transition metal atom.[79]

The use of $Pr^iMgBr\text{-}Cp_2TiCl_n$ (n = 1, 2) systems as catalysts for the isomerisation of hexa-1,5-diene has been investigated.[80] The complex $Cp_2Ti(CH_2)ZnX_2$ (X = halide), obtained by the reaction of Cp_2TiCl_2 with ICH_2ZnI, acts as a methylene-transfer reagent in reactions with ketones ($R_2C{=}O \rightarrow R_2C{=}CH_2$), alkynes [$RCH_2C{\equiv}CSiMe_3 \rightarrow RCH{=}C{=}C(Me)SiMe_3$], and nitriles [$ArCN \rightarrow ArC(Me){=}NH \rightarrow ArCOMe$].[81] Terminal alkynes $R^1C{\equiv}CH$ undergo stereo- and regio-selective carbozincation in reactions with R_2Zn in the presence of Cp_2ZrX_2 (X = halogen) which give products of the type $RR^1C{=}CHZnX$.[82] The chemistry of complexes, such as $Cp_2\overline{Ti(CH_2)_3O}Me$, which incorporate an intramolecularly metal-coordinated alkoxy group has been reviewed.[83]

Vanadium, Niobium, and Tantalum.- There have been e.s.r. spectroscopic studies of $[Cp_2M(L)_2]^{n+}$ (M = V, Nb; n = 0, 1),[84] $Cp_2V(OCHRCH_2Cl)_2$ (R = H, Me),[85] $Cp_2Ta(SR)_2$ (R = Me, Ph),[86] and and Cp'_2NbCl_2 doped in crystalline Cp'_2ZrCl_2.[87] Photoelectron spectra of a series of complexes of the type Cp_2VX (X = halide, alkyl, aryl) have been recorded.[88] The electrochemistry of vanadocene and its derivatives has been investigated; one-electron reductions of Cp_2VCl_n (n = 1, 2) and $Cp^*_2VCl_2$ give anions which readily eliminate Cl^-; one-electron oxidation of Cp_2VCl gives an unstable cation which disproportionates to $(Cp_2V)^{2+}$ and Cp_2VCl_2; Cp_2V itself undergoes reversible one-electron reduction.[89] Chemical reduction of Cp_2V with metallic K in THF affords $(Cp_2V)^-K^+$, which can be isolated as an etherate and which is a useful starting material for the synthesis of complexes of the type $CpV(L)_n$.[90]

Reactions of $Cp^*_2VCl_2$ with $(NH_4)_2S_5$[91] and of Cp^*_2V with S_8 or COS[92] afford $Cp^*_2V(\eta\text{-}S_2)$. In an unusual reaction, thermolysis of $Cp_2V(\eta^2\text{-}SCPh_2)$, obtained by addition of Ph_2CS to Cp_2V, leads to the formation of $(CpVS)_n$ and 6,6-diphenylpentafulvene; the reaction probably occurs via the radical $Cp_2V\text{-}S\dot{C}Ph_2$ and then the η^4-cyclopentadiene complex (5).[93] The properties of $Cp_2V(\eta^1\text{-}C_5H_5)$ in solution have been investigated by ^1H and ^{13}C n.m.r. spectroscopy; the η^1-ring is fluxional and intermolecular ligand-exchange occurs.[94] The related complex $Cp_2Nb(\eta^1\text{-}C_5H_5)_2$ reacts with hfb to give a product whose structure (6) has been established by crystallography; the source of the oxygen atom was not elucidated.[95] Complexes of the types Cp_2NbBr_n (n = 1, 2), $Cp_2Nb(L)Br$ (L = CO, phosphines, yne, etc.), and $(Cp_2NbBr)_2(\mu\text{-}Br)$ have been synthesised.[96]

Chromium, Molybdenum, and Tungsten.- The chemistry of Cp_2Cr has been reviewed.[97] The electronic structures of Cp_2M (M = Mo, W), generated by photolysis of Cp_2MH_2 in argon matrices, have been discussed in the light of u.v. and magnetic circular dichroism spectra.[98]

Co-condensation of Mo vapour with spiro[2.4]hepta-4,6-diene affords $[(\eta\text{-}C_5H_4)CH_2CH_2]_2Mo$ which has been converted into a variety of complexes of the type $(L)_nMo[(\eta\text{-}C_5H_4)CH_2CH_2X]_2$ (X = halide, SPh, etc.).[99] The interannularly bridged chromocene carbonyl $[Me_2C(\eta\text{-}C_5H_4)]_2Cr(CO)$ has been synthesised,[100] and complexes of the type $CpMo(\eta\text{-}C_5H_4Ar)(NO)I$ have been obtained by reactions of

$[CpMo(NO)I]_2$ with $(C_5H_4Ar)Tl$.[101]

Manganese and Rhenium.- The electronic structures of $(C_5H_4Pr^i)_2Mn$[102] and $(L)Mn(C_5H_4)_2(CH_2)_3$ (L = N-bonded 3,5-dimethylpyridine)[103] have been investigated. Low-temperature photolyses of Cp_2ReH in CO and N_2 matrices have afforded Cp_2Re and $CpRe(\eta^3-C_5H_5)(L)H$ (L = CO, N_2).[104]

Iron and Ruthenium.- The chemistry of ferrocene derivatives containing another transition metal atom has been reviewed.[105] A variable-temperature X-ray study of crystalline Cp_2Fe has shed light upon the structures, molecular packings, and lattice energies of the different low-temperature phases.[106] Spectroscopic investigations include studies of the carbonyl stretching absorptions in the i.r. spectra of acylferrocenes,[107] the ^{57}Fe and ^{99}Ru n.m.r. spectra of derivatives of Cp_2Fe[108] and Cp_2Ru,[109] and the mass[110] and Mössbauer[111] spectra of substituted ferrocenes.

Appearance potentials of $(Cp_2Fe)^+$, $(CpFe)^+$, and Fe^+ in the photo-fragmentation of ferrocene vapour have been measured.[112] In liquid SO_2, Cp_2Fe can be oxidised electrochemically to $(Cp_2Fe)^{2+}$ which is stable on a coulometric time-scale.[113] The electrochemistries of $FcCH_2\overset{+}{N}Me_2R\ Br^-$ (R = Bu^n, CH_2Ph)[114] and substrates containing two Fc groups[115] have also been studied. The electrical conductivities of charge-transfer adducts of poly(1,1'-ferrocenylene) with electron acceptors have been measured,[116] and there has been a study of the effective polarisability anisotropies, the electric quadrupole moments, and the magnetic anisotropies of Cp_2Fe and Cp_2Ru.[117] The interaction of HO radicals with carboxylates of the type $Fc(CH_2)_nCO_2^-$ (n = 0, 1, 2, 4) has been studied.[118]

Two intriguing syntheses of ferrocene derivatives have been discovered. Thus, co-condensations of Fe vapour with RC≡CR (R = Me, Et) afford the decaalkylferrocenes $(\eta-C_5R_5)_2Fe$ together with peralkylated benzenes and cyclooctatetraenes,[119] and the reaction of FpI with PhC(Li)=CHCH=C(Li)Ph followed by treatment of the product with $(Me_3O)^+(BF_4)^-$ gives 1-methoxy-2,5-diphenylferrocene.[120] Ethylation of Cp_2Fe can be achieved by reaction with Et radicals (ex $HgEt_2$) in the vapour phase,[121] and by treatment with $ClCH_2CH_2OH$ or $BrCH_2CH_2Br$ in the presence of $AlCl_3$-$NaBH_4$.[122] A variety of ferrocenyl sulphides and polysulphides has been isolated from reactions of Cp_2Fe with S_8-$Fe_3(CO)_{12}$ in refluxing benzene.[123] Full details of the syntheses of $FcNH_2$ Fc_2NH, and Fc_3N (see Volume 12, ref.219) have been published.[124] The crystal structures of the 2,1'-dilithio derivative of $FcCH(Me)NMe_2$ (as TMED adduct)[125] and

$Fc\overset{+}{P}(CH_2Ph)Ph_2Cl^-$[126] have been discussed in relation to their chemical reactivities.

A kinetic study of the oxidation of FcCH(OH)Ar to FcCOAr by $(Ph_3SiO)_2CrO_2$ has been reported.[127] The use of derivatives of Cp_2Fe in asymmetric synthesis has been reviewed,[128] and there have been further such studies of cyclopalladation reactions[129] and cross-coupling of ArCH(R)ZnX and CH_2=CHBr.[130] There have been reports of kinetic studies of alkene hydrogenation catalysed by complexes incorporating ferrocenylphosphine residues,[131] and of trans-acylation of β-cyclodextrin by p-nitrophenyl ferrocenyl- and ruthenocenyl-acrylates.[132]

The properties of ferrocenylalkyl cations $Fc\overset{+}{C}(R^1)R^2$ have been discussed in the light of calculations[133] and ^{13}C and ^{57}Fe n.m.r. spectra.[134] Differences in the electrophilic reactivities of $Fc_2\overset{+}{C}H$ and the related [1.1]ferrocenophan-1-yl cation have been rationalised in terms of structural and conformational effects.[135] The transition states in S_N1 hydrolyses of substrates of the type $FcCR_2X$ (R = H, D; X = OCOR, $\overset{+}{N}$-quinolinium) have been investigated by measurements of the secondary deuterium kinetic isotope effects (k^H/k^D 1.25-1.5).[136] The exceptional stability of cations of the type $Fc\overset{+}{C}R_2$ enables "one-pot" acid-catalysed conversions of $FcC(OH)R_2$ into $FcC(X)R_2$ (X = py^+, Me_2S^+, Ph_3P^+, CNS, HO_2CCH_2NH, etc.).[137] Several Fc-stabilised phosphenium cations $Fc\overset{+}{P}X$ (X = Cl, NMe_2, Fc) have been isolated as $(AlCl_4)^-$ salts.[138] Rates of methoxy-thallation of $FcCH=CH_2$ and $PhCH=CH_2$, by reaction with $Tl(OAc)_3$ in MeOH, have been compared.[139] Unexpectedly, $FcCH=CH_2$ reacts with $(3\text{-iodopentane-2,4-dionato})_3Co$ to give 2- and 3-ferrocenyl-4-acetyl-5-methyl-2,3-dihydrofurans.[140]

The preparation and properties of bridged ferrocenes continue to attract attention. There have been reports of syntheses of [m]ferrocenophanes with interannular P(R),[141] As(Ph),[141] $SCH_2C(CH_2Br)_2CH_2S$,[142] CH_2SeCH_2,[143] and linked bis(biphenylyl)[144] bridges, of [m.m]ferrocenophanes with two $Sn(Bu^n)_2$,[145] CH=CR (R = H, CN),[146] and $CO_2C_6H_4OCO$[147] bridges linking two ferrocenylene residues, and of [m](1,3)ferrocenophanes incorporating m- and p-$(CH_2CH_2)_2C_6H_4$ bridges.[148] A range of [m]- and [m.m]-ferrocenophanes incorporating interannular poly(ethyleneoxy) bridges, and thia-analogues, has been synthesised,[149-151] and the efficiency of such compounds and unbridged analogues in the extraction of alkali metal cations has been investigated.[149] The amazing conformational mobility of [1.1]ferrocenophane has been established.[152]

The inter-ring bridge in the phosphine $PhP(\eta-C_5H_4)_2Fe$ is cleaved by PhLi and the resulting product $(\eta-C_5H_4Li)Fe(\eta-C_5H_4)PPh_2$ has been converted into a variety of 1'-substituted derivatives of $FcPPh_2$.[153] Completely stereospecific addition of Grignard reagents to the carbonyl groups of 3-substituted [5]ferrocenophane-1,5-diones has been found; in the diol products, both OH groups are attached to the bridge in configurations trans to the substituent at C(3).[154] Further studies of the properties of cations resulting from one--electron oxidation of substrates containing two or more Fc groups have been made[115,155] (see Volume 11, ref.199; Volume 12, ref.230); most substrates in which Fc/Fc^+ groups are either directly connected or linked by a conjugating chain, and those in which the Fe/Fe^+ atoms are held in close proximity, show evidence of intervalence electron-transfer but this is lacking with cations resulting from oxidation of compounds such as Fc_2CH_2, Fc_2PPh, and Fc_3B.[155] The polymetallic complex $FcC[Co(CO)_3]_3$ undergoes both one-electron oxidation and one-electron reduction, reversibly; the former process involves the Fc Group and the latter the $Co_3(CO)_9$ cluster.[156]

The paramagnetic properties of salts of the cation (7) have been taken to indicate a metal-metal interaction;[157] cf., neutral ferrocenes and cobalticenium cations are diamagnetic 18-electron species. The soluble polymer $(CH=CHC_6H_4CH=CHX)_n$ (X = 1,1'-ferrocenylene), doped with I_2, behaves as an air-stable, photo-active semiconductor.[158] The ferrocene derivative (8) has been synthesised and used as a surface-derivatising reagent for n-type Si and Pt electrodes; such treated electrodes show an improved response compared to that of the corresponding "naked" electrodes.[159]

Cobalt, Rhodium, and Iridium.- Cobalticenium salts can be synthesised at a Co anode by electrolysis of solutions of cyclopentadienes in DMSO containing LiCl.[160] The products of reaction of Cp_2Co with cyclopentadiene in air have been shown to be the complex (9) and the corresponding 2-substituted cyclopentadiene isomer.[161] Several new 1,1'-disubstituted rhodicenium salts have been synthesised, including the dicarboxylic acid and diamine.[162] The reaction of $CpIr(\eta^3\text{-allyl})Cl$ with PhC≡CPh in the presence of $AgBF_4$ affords the 1,2-diphenyliridicenium salt.[163]

Nickel.- Displacements of one Cp ligand from Cp_2Ni occur in reactions with $R_2P(H)S$,[164] dienes in the presence of MeLi,[165] PhC≡CCONHCH(Me)Ph,[166] and $RCCo_3(CO)_9$.[167]

Lanthanide and Actinide Elements.- Syntheses of Cp_2M and Cp_3M complexes of the lanthanide elements, and ring-alkylated derivatives, have been described.[168] The mode of bonding of the BH_4 ligand in $(\eta\text{-}C_5H_3R_2)_2M(BH_4)$ [R_2 = 1,3-$(SiMe_3)_2$] is dependent on the metal, i.e. bidentate for M = Sc, Y, and Yb but tridentate for M = La, Pr, Nd, and Sm;[169a] addition of Cl^- to $[(\eta\text{-}C_5H_3R_2)_2MCl]_2$ affords the anions $[(\eta\text{-}C_5H_3R_2)_2MCl_2]^-$ (M = Y, La, Pr, Nd, Dy, Tm), isolable in salts.[169b]

Calculations of the valence electronic structures of Cp_2UR_2 (R = Me, Cl) have been carried out.[170] Metal-ligand disruption energies have been determined by calorimetric measurements on $Cp^*_2ThR_2$ (R = H, alkyl, aryl, amino, etc.),[171] and the nature of Cp-Np bonding in complexes of the type $Cp_2Np(L)_2$ has been discussed in the light of ^{237}Np Mössbauer spectra.[172] Reaction of Cp_3UCl with MeCN and buta-1,3-diene in the presence of O_2 has given an unusual product in which the ions $[Cp_3U(NCMe)_2]^+$ and $(Cl_4UO_2)^{2-}$ are present.[173] Complexes of the type $Cp^*U(L)_nCl_{3-n}$ (L = Cp or substituted Cp; n = 1 or 2) have been synthesised; the derivative $Cp^*U(\eta\text{-}C_5H_4CHMePh)_2Cl$ has been obtained as a mixture of one racemic and two meso compounds.[174]

5 Mono-(π-arene)metal Complexes

Studies of General Interest.- The applications of arene-iron complexes in organic synthesis have been the subject of a comprehensive review.[175] New preparative routes to salts of $[Cp^*M(\eta\text{-arene})]^{2+}$ (M = Rh, Ir) and $[(\eta\text{-arene})_2Ru]^{2+}$ have been described.[176] A comparative kinetic study of additions of phosphorus nucleophiles to the arene ligands of $[(\eta\text{-arene})M(CO)_3]^+$ (M = Mn, Re) and $[(\eta\text{-arene})_2M]^{2+}$ (M = Fe, Ru, Os) have been reported; with the former, the Mn complex is slightly more reactive towards PBu_3 than is the Re analogue, while for the latter sandwich dications, reactivity towards $P(OBu)_3$ and PPh_3 decreases through the series Fe ≫ Ru > Os.[177]

Complexes have been synthesised in which a folded (crystal structures) arene ligand (C_6H_6, C_6Me_6, C_6Et_6, and p-$C_6H_4Me_2$) bridges the metal atoms of CpFe-FeCp, (L)Co-Co(L) (L = bidentate $Cy_2PCH_2CH_2PCy_2$), and $CpV\text{-}VCp(\mu\text{-}H)_2$ residues;[178] the ^{1}H n.m.r. spectroscopic properties of the last example (10) show that the bridging benzene ligand is rotating relative to the two V atoms. Displacement of the toluene ligands of $(\eta\text{-}PhMe)Co(C_6F_5)_2$ and $(\eta\text{-}PhMe)Ni(SiX_3)_2$ (X = F, Cl) by CO occurs readily.[179]

Vanadium.- In the low-temperature reaction of V vapour with benzene to form $(\eta\text{-}PhH)_2V$, evidence has been obtained from e.s.r.

spectroscopy for the initial formation of (η-PhH)V and there have been MO calculations of the electronic structure of this "half--sandwich" species.[180] Several methods of synthesis of (η-mesitylene)$_2$V by reduction of (η-mesitylene)$_2$VI have been found,[181] and triple-decker complexes of the type CpV(arene)VCp have been characterised.[182]

Chromium, Molybdenum, and Tungsten.- New mono- and bis-[Cr(CO)$_3$] complexes of arenes have been synthesised by reactions of Cr(CO)$_6$ or (L)$_3$Cr(CO)$_3$ with (PhCH$_2$)$_2$Hg,[183] ArP=PAr (Ar = 2,4,6-tri-t-butyl-phenyl),[184] a disilane in which two SiMe$_2$ bridges link two o-phenylene groups,[185] and the biphenyl derivatives (11) and (12).[186] Only the exo-stereoisomers (13) of the mono-[Cr(CO)$_3$] complexes of (11) and (12) are formed when R = H, but when R = Me the exo- and the endo-stereoisomer (14) are formed in equal proportions; these are separable but can interconvert.[186] Reactions of 9-phenyl- and 9,10-diphenyl-anthracene with Cr(CO)$_6$ give two mono-[Cr(CO)$_3$] complexes, in which the metal atom is attached either to a Ph ring or to a terminal anthracene ring, and the corresponding bis-[Cr(CO)$_3$] complexes; the former isomers interconvert on heating in solution, the Ph complex being the less stable.[187] Further examples have been reported (see Volume 12, ref.275, and earlier Volumes) of the synthesis of derivatives of (η^6-naphthalene)Cr(CO)$_3$ from (OC)$_5$Cr=C(OMe)R and alkynes,[188a] and the reaction has been used as the key step in a synthesis of α-tocopherol.[188b]

The conformational mobilities of (η-PhH)M(CO)$_3$ (M = Cr, Mo, W)[189] and (η^6-C_6Et_6)Cr(CO)$_2$(CS)[190] have been studied spectroscopically, and the negative-ion mass spectra of a range of ring--substituted benchrotrenes have been analysed.[191] Electrochemical oxidations[192] and reductions[193] of various benchrotrene derivatives have been studied. Taft substituent constants (σ_I, σ_R^o) for Ph groups complexed with Cr(CO)$_2$(L) residues (L = CO, PPh$_3$, AsPh$_3$) have been calculated.[194] The mechanism of photochemical conversion of (η-arene)Cr(CO)$_3$ into Cr(CO)$_6$ has been studied by a number of spectroscopic methods.[195] Relative Cr-arene bond strengths [arene = mesitylene (> chpt) > PhMe > PhH] have been assessed from measurements of heats of iodination (I$_2$/THF) of (η-arene)Cr(CO)$_3$.[196]

The activation of aryl halides towards nucleophilic displacement of halide through complexation with a Cr(CO)$_3$ residue has been exploited in syntheses of chroman,[197] aryl sulphides,[198] and complexed arylcarbaboranes.[199] A synthetic route to m-substituted phenols and anilines has been developed which utilises the

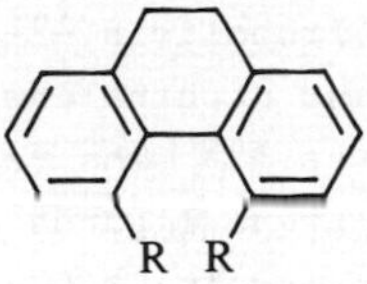

(11)

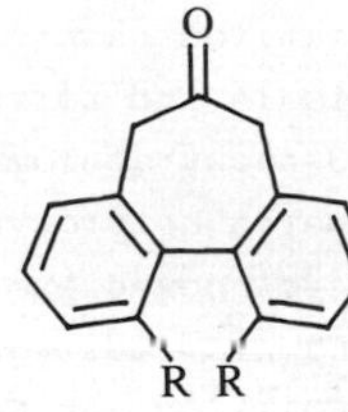

(12)

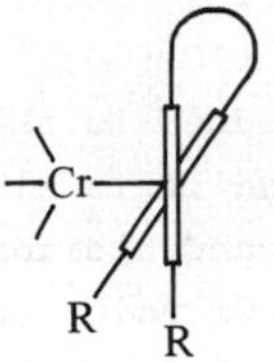

(13)

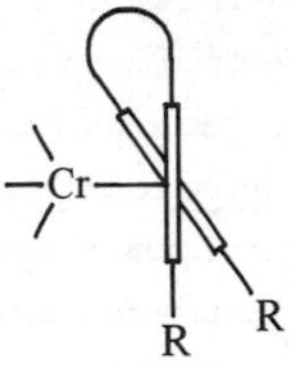

(14)

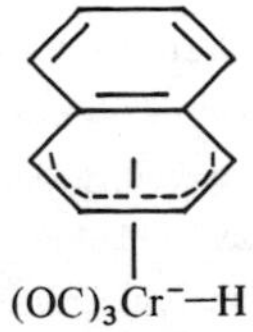

(15)

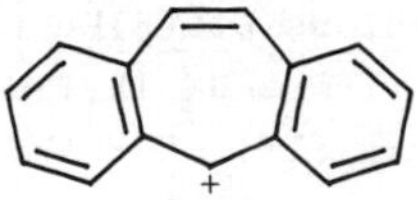

(16)

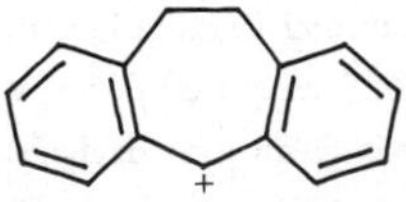

(17)

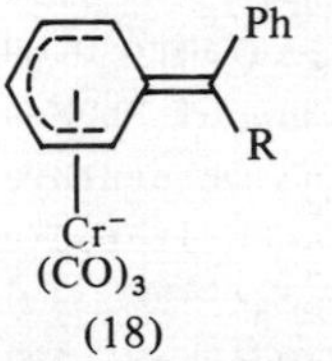

(18)

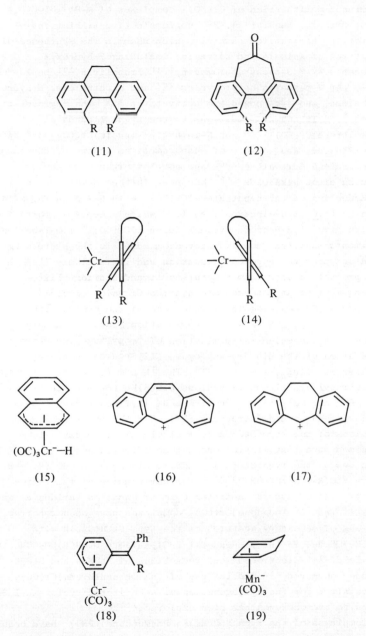

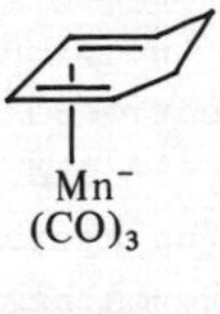

(19)

predominant *m*-lithiation of $Cr(CO)_3$ complexes of *m*-$Pr^i{}_3SiOC_6H_4X$ (X = H, $OSiPr^i{}_3$, $NMeSiPr^i{}_3$).[200] Regioselective lithiations of benchrotrene derivatives have also been used in the syntheses of derivatives of anisole and tetralin, including 7-hydroxy-calamenenes,[201a] 3-deoxyrabelomycin,[201b] frenolicin,[202] benzo-fused 5-, 6-, and 7-membered heterocycles,[203] and naphthalene, dihydro-naphthalene, and cyclohexadiene derivatives.[204] The regioselectivities of carbanionic additions to the arene ligands of (η-alkylarene)$Cr(CO)_3$[205a] and *N*-methyl(η^6-indole)$Cr(CO)_3$[206a] have been discussed, and the use of such reactions in the synthesis of cyclohexadiene derivatives[206b] and complex organic molecules, including ergot alkaloids,[205b] has been further demonstrated.

Reduction of (η^6-naphthalene)$Cr(CO)_3$ with Na-Hg in THF affords a dianion (*i.e.* two-electron transfer) which accepts a proton from water to give [1,2-benzo(η-cyclohexadienyl)$Cr(CO)_3]^-$; D-labelling experiments show that the hydrogen atom added to the hydrocarbon ligand occupies the *endo*-configuration and it is likely that the final product results from a metal-to-ligand hydride-shift rearrangement of an intermediate hydrido-Cr anion (15).[207]

Whereas $Cr(CO)_3$-complexation of one of the benzene rings destabilises (ΔpK_{R^+} = -2.4) the carbocation (16), presumably because of a consequential diminution of tropylium-like character, complexation of the dihydro-analogue (17) causes substantial stabilisation (ΔpK_{R^+} = +3.5).[208a] The ^1H n.m.r. spectra of simple benchrotrenyl-stabilised carbanions $Bt\bar{C}(R)Ph$ (Bt = benchrotrenyl; R = H, Ph) are more in accord with a 6-alkylidene(η^5-cyclohexa-dienyl) ligand-bonding structure (18).[208b] Wittig rearrangement of carbanions of the type $Bt\bar{C}HOR$ (R = alkyl, allyl, benzyl) is suppressed such that alkylations can be effected normally.[209] Carbanions of the type $Bt\bar{C}(R^1)R^2$ add to aldehydes R^3CHO (R^3 = H, Ph) to give $BtC(R^1)(R^2)CH(OH)R^3$.[210] Further studies of the regio-selectivities of hydride addition (*exo* or *endo*) to various α- and β-substituted (η^6-indanone)$Cr(CO)_3$ complexes have been reported.[211] Reactions of secondary acetates of the type BtCH(OAc)R with CH_2=C(Me)$OSiMe_3$ in the presence of $ZnCl_2$ afford BtCH(R)CH_2COMe in high yield.[212] Electrochemical reductions of $Ph\overset{+}{S}Me_2$ and $Bt\overset{+}{S}Me_2$ have been compared,[213] and the use of (η^6-phenanthrene)$Cr(CO)_3$ as a catalyst for the hydrogenation of substituted cyclohexa-1,3--dienes to cyclohexenes has been explored.[214]

Complexes of the type PhP(Me)(η-Ph)Mo(dppe)(PR_3) have been synthesised and incorporated as phosphine ligands (L) in complexes such as (η-cod)Rh(L)Cl.[215] A variety of complexes of the type (η-PhMe)W$(L)_n$ have been obtained by ligand-displacement reactions of $(η\text{-PhMe})_2$W.[216]

Manganese and Rhenium.- The salts $[(η\text{-arene})Mn(CO)_3]^+(BF_4)^-$ (arene = anisole, 1,2-dimethoxybenzene, 4-bromo-1,2-dimethoxybenzene) have been obtained by reactions of $[(OC)_5Mn]^+(BF_4)^-$ with the arenes.[217] The mechanism of reduction of $[(η\text{-PhH})Mn(CO)_3]^+$ by $R_3\bar{B}H$ (R = Et, Pr^iO), to give the highly reactive anionic (η-cyclohexa-1,3-diene) complex (19), involves stepwise exo-addition of two hydride ions to adjacent ring carbon atoms; with ring-methylated analogues, e.g. $[(η\text{-mesitylene})Mn(CO)_3]^+$, however, hydride addition to the endo-face of the arene ligand cannot be ruled out.[218] In a related study, evidence has been obtained to show that $\bar{B}H_4$ transfers hydride stereospecifically to the endo-face of the hydrocarbon ligands of $[(η\text{-cyclohexadienyl})Mn(CO)_2(NO)]^+$ and related cations to give the corresponding neutral ($η^4$-cyclohexa-1,3-diene) products.[219]

Co-condensations of Re vapour with methylarenes (ArMe) have given products of the type $[(η\text{-arene})Re]_2(μ\text{-H})_2(μ\text{-CHAr})$; with PhEt, isomers incorporating μ-C(Me)Ph and μ-$CHCH_2$Ph bridges were both formed.[220]

Iron and Ruthenium.- Co-condensation of Fe vapour with PhMe and $(EtC)_2B_4H_6$ has afforded (η-arene)Fe$[(EtC)_2B_4H_4]$.[221] The chemistry of complexes of the type (hmb)Ru$(PR_3)H_2$ has been investigated,[222] and a variety of new (hmb)Ru$(L)_n$ complexes has been synthesised including (hmb)Ru(Ar)H[222] and (hmb)Ru$(η\text{-}C_2H_4)_2$.[223] Complexes of the type (η-arene)Ru(O_2CR)X (R = Me, CF_3; X = Cl, Br) incorporating bidentate carboxylato ligands have been prepared.[224]

The photoelectron spectra of the 19-electron complexes CpFe(η-arene) and Cp*Fe(hmb) have been reported,[225] and the chemistry of these and derived complexes has been further developed[226] (see Volume 12, refs.309, 310). Liberation of the arene ligand from $[CpFe(η\text{-arene})]^+$ can be achieved by electrochemical reduction in the presence of $P(OMe)_3$[227] and by photolysis in MeCN;[228] the latter reaction affords $[CpFe(NCMe)_3]^+$ which is stable at low temperatures in MeCN but whose nitrile ligands are readily displaced by phosphines, phosphites, isonitriles, etc.[228]

Ligand-exchange reactions of Cp_2Fe with dibenzodioxin, phenoxathiin, thianthrene, and phenoxazine have afforded the corresponding mono- and bis-$(CpFe)^+$ sandwich complexes.[229a] Oxidations of $(CpFe)^+$ complexes of Ph_2CH_2, fluorene, and 9,10-dihydroanthracene with $KMnO_4$ have given the corresponding acylarene complexes (i.e. $CH_2 \rightarrow C{=}O$) which cannot be obtained directly by ligand-exchange reactions.[229b] Reactions of anthracene or 9,10-dihydroanthracene with $[Cp^*Fe(CO)_2]_2$ in the presence of Al_2Cl_6 give the same product mixture of the $(Cp^*Fe)^+$ complexes of 9,10-dihydroanthracene and 1,2,3,4-tetrahydroanthracene (2:3, respectively), suggesting that these cationic complexes interconvert under the reaction conditions.[230]

Cobalt, Nickel, Rhodium, and Iridium.- The arene ligands of complexes of the type $(\eta\text{-arene})M(C_6F_5)_2$ (M = Co, Ni) are labile and can be exchanged for other arenes; arene-metal bond strengths decrease through the series: mesitylene > PhMe ≥ PhOMe > PhH.[179,231] Salts of the type $[(\eta\text{-arene})M(L)_2]^+X^-$ [M = Rh, L = CO; M = Rh and Ir, $(L)_2$ = an η^4-tetrafluorobenzobarrelene] have been synthesised,[232] and the related salts $[(\eta\text{-arene})Rh(diphos)]^+X^-$ have been found to catalyse the hydrogenation of anthracenes to 1,2,3,4-tetrahydroanthracenes under mild conditions.[233] Cluster complexes $(\eta\text{-PhR})Rh[Os(CO)_3]_3(\mu\text{-}H)_3$[234] and several cationic (η^6-indole) complexes of Rh and Ir[235] have been synthesised.

6 Bis-(π-arene)metal Complexes

Studies of General Interest.- The metal-vapour method has been used widely for the synthesis of bis(η-arene) complexes of Ti,[236,237] V,[236] Cr,[237,238] Co,[239] Nb,[236] Mo,[236,240] Ta,[236] and W.[236] In a variant of this method, reduction of metal chlorides in THF by K atoms in the presence of an arene has given bis(η-arene) complexes of Ti, V, Cr, and Mo.[241]

Chromium.- The e.s.r. spectrum of $[(\eta\text{-PhCN})_2Cr]^-$ suggests that the unpaired electron resides in an orbital of predominant ligand character.[242] Nucleophilic displacements of chloride from $[(\eta\text{-PhCl})_2Cr]^+$ by alkoxides and secondary amines have been achieved.[243]

Iron and Ruthenium.- Dicationic complexes in which an Fe[244] or Ru[245] atom is sandwiched between two η^6-[2.2]paracyclophane ligands have been isolated in salts. Electrochemical reductions of $[(hmb)_2Ru]^{2+}$ and $[(hmb)Ru(\eta^6\text{-cyclophane})]^{2+}$ have been carried out.[246] See also refs.175-177.

7 Complexes with Related Ligands

Bora-benzene and -cyclopentadienyl Complexes.- The electronic structures of various ring-rotational conformers of bis(η-borabenzene) complexes of Fe and Co have been calculated.[247a] Several π-bonded B-methylborabenzene (≡ L) complexes have been synthesised including $[(L)Cr(CO)_3]^-Na^+$, $(L)Cr(CO)_3H$, and the sandwich complex (20).[248a] This last complex is more reactive than is ferrocene towards electrophilic substitution, with the ring positions adjacent to the B atom the most reactive, e.g. in H^+/D^+ exchange (CF_3CO_2D) and Vilsmeier formylation; Friedel-Crafts acetylation at the 2- and 6-positions can be effected with MeCOCl-$AsCl_3$, but with $AlCl_3$ as catalyst extrusion of the BMe group occurs and the cations $[(\eta\text{-arene})Co(\eta\text{-}C_4Me_4)]^+$ (arene = PhMe and 2-acetyltoluene) are formed.[248a]

Thermal reaction of Fp_2 with B-phenyl-4,5-dihydroborepin affords a mixture of products with ring-contracted ligands, viz. CpFe(η-2-methyl-B-phenylborabenzene) and the triple-decker complex CpFe(L)FeCp (L = 2-ethyl-B-phenylboracyclopentadienyl); a related product $(OC)_3Mn(L)Mn(CO)_3$ was obtained from reaction with $Mn_2(CO)_{10}$.[248b] A Rh complex of B-(diisopropylamino)boracyclopentadienyl has been synthesised.[248c]

Pyrrolyl Complexes.- The phase structure of azaferrocene has been studied by measurements of proton spin-lattice relaxation times and by differential thermal analysis.[249] Lithiation (Bu^nLi) then methylation (MeI) of this complex affords a mixture of the 2-Me, 1'-Me, and 2,1'-Me_2 derivatives.[250] Halide-replacement reactions of CpTi(η-pyrrolyl)Cl_2 and the (η^5-indenyl) analogue have been reported.[251]

Phosphacyclopentadienyl Complexes.- The electronic structures of phospha- and 1,1'-diphospha-ferrocene and their reactivities towards electrophiles and nucleophiles have been discussed in the light of MO calculations; electrophilic substitutions are highly regioselective because such reactions are charge-controlled, but orbital and charge effects are important in nucleophilic substitutions.[247b] Several new 2,5-diphenyl and 3,4-dimethyl derivatives of phospha- and 1,1'-diphospha-ferrocene have been synthesised.[252]

Thiophene Complexes.- Isomeric complexes in which a $Cr(CO)_3$ group is attached to the 5- and the 7-membered rings of thiophenocycloheptatrienes have been obtained from $Cr(CO)_6$ by direct synthesis.[253] Lithiation (Bu^nLi-THF) of (η-thiophene)$Cr(CO)_3$ occurs extremely readily at the 2- and 5-positions of the ring.[254]

Dibora-benzene and -cyclopentadienyl Complexes.- Complexes in which a 1,4-diferrocenyl-1,4-diborabenzene ligand (a 4-electron donor) is

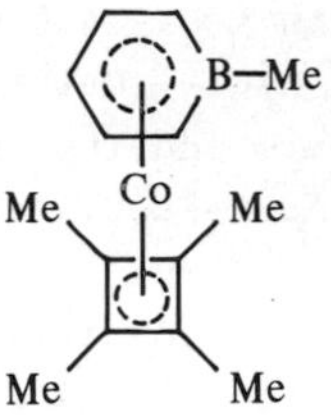

(20)

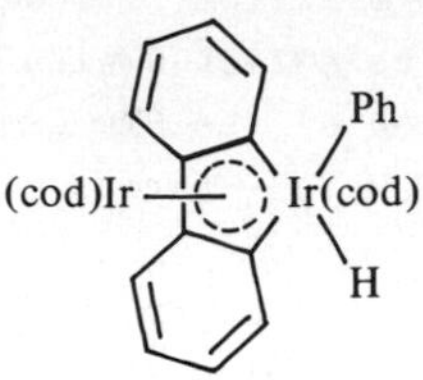

(21)

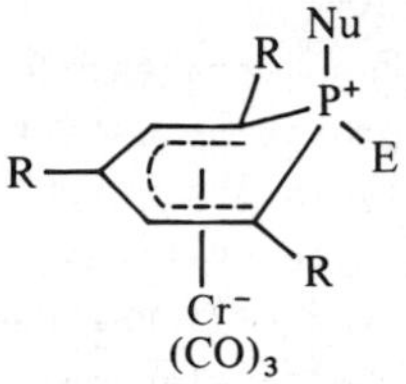

(22)

π-complexed with $M(CO)_3$ (M = Fe, Ru, Os) and $M(CO)_4$ (M = Cr, Mo, W) residues have been synthesised.[248d] Sandwich, triple-decker, and quadruple-decker complexes incorporating 1,3-diboracyclopentadienyl ligands (≡ L) have been prepared, including CpCo(L), $CpM^1(L)M^2Cp$ ($M^1 = M^2$ = Co, Ni; M^1 = Co, M^2 = Fe, Ni), and CpCo(L)Ni(L)CoCp.[255]

Azaboracyclopentadienyl Complexes.- The syntheses, structures, and chemical reactivities of complexes of the type CpCo(L) and $(L)_2Co$ (L = substituted 1-aza-2-boracyclopentadienyl) have been studied.[256]

Iridacyclopentadienyl Complex.- The synthesis and crystal structure of the unusual bimetallic iridacyclopentadienyl complex (21) have been described.[257]

Pyridine Complexes.- Displacements of the CO ligands of $Cr(CO)_3$ π-complexes of 2,6-dimethyl- and 2,4,6-trimethyl-pyridine by PPh_3 and $P(OMe)_3$ have been reported,[258] and the syntheses of $(\eta\text{-L})Mo(Ph_2PMe)_3$ (L = pyridine and 4-picoline) have been achieved.[259]

Phosphabenzene Complexes.- Nucleophiles (Nu^-) add exo to the P atoms of the $Cr(CO)_3$ complexes of 2,4,6-trisubstituted (Ph_3, $Bu^t{}_3$) phosphabenzenes giving the corresponding anions which add electrophiles (E^+) endo, with formation of zwitterionic products (22; Nu = alkyl, Ph, OMe; E = alkyl, H).[260]

Thiabenzene Complexes.- There have been further studies (see Volume 12, ref.348) of the chemistry of $M(CO)_3$ complexes (M = Cr, Mo, W) of 1-alkyl- and 1-oxo-3,5-diphenylthiabenzenes.[261]

Pentafulvene Complexes.- The extent of ligand-folding, at C(1), in $Cr(CO)_3$ complexes of various 6-substituted pentafulvenes has been investigated by crystal-structure determinations.[262] Similar ligand deformations have also been found in the structure of pentafulvenes π-complexed with $(\eta\text{-cod})Rh^+$[263] and (η-arene)M (M = Mo, W)[264] residues. Rotational barriers (C===N) have been measured for $M(CO)_3$ complexes (M = Cr, Mo, W) of 6-(dimethylamino)-pentafulvene and -2,3-dibenzopentafulvene.[265] Reactions of $[CpCr(CO)_3]^-$ and $[Cp'Cr(CO)_3]^-$ with allylic chlorides have afforded as by-products (6-allylpentafulvene)$Cr(CO)_3$ complexes, and the mechanism of their formation has been discussed.[266] Photochemical displacements of one CO ligand from $PhMn(CO)_5$ by a ring double bond of 6,6-disubstituted pentafulvenes have been reported;[267] see also ref.60.

References

1 W.E.Watts, in 'Comprehensive Organometallic Chemistry', eds. G.Wilkinson, F.G.A.Stone, and E.W.Abel, Pergamon, Oxford, 1982, Vol.8, ch.59.

2 V.I.Zdanovich, A.Zh.Seitembetova, and V.N.Setkina, Russ. Chem. Rev. (Engl. Transl.), 1982, 51, 659.

3 D.W.Macomber, W.P.Hart, and M.D.Rausch, Adv. Organomet. Chem., 1982, 21, 1; see also D.W.Macomber, W.C.Spink, and M.D.Rausch, J. Organomet. Chem., 1983, 250, 311.

4 W.J.Evans, J. Organomet. Chem., 1983, 250, 217.

5 T.J.Marks, Science (Washington, D.C., 1883-), 1982, 217(4564), 989; Chem. Abs., 1982, 97, 163038.

6 M.I.Rybinskaya and V.V.Krivykh, Koord. Khim., 1983, 9, 435.

7 P.L.Timms, Transition Met. Chem., Proc. Workshop 1980, publ. 1981, p. 23-34; Chem. Abs., 1983, 98, 89409; G.A.Ozin, M.P.Andrews, L.F.Nazar, H.X. Huber, and C.G.Francis, Coord. Chem. Rev., 1983, 48, 203.

8 R.N.Grimes, Acc. Chem. Res., 1983, 16, 22.

9 M.C.Boehm, Z. Naturforsch., 1982, 37A, 1193; THEOCHEM, 1983, 9, 73; Chem. Abs., 1983, 98, 198387; J.W.Chinn and M.B.Hall, J. Am. Chem. Soc., 1983, 105, 4930.

10 T.A.Albright, P.Hofmann, R.Hoffmann, C.P.Lillya, and P.A.Dobosh, J. Am. Chem. Soc., 1983, 105, 3396.

11 R.C.Kerber, J. Organomet. Chem., 1983, 254, 131.

12 S.P.Solodovnikov, Russ. Chem. Rev. (Engl. Transl.), 1982, 51, 961.

13 G.Lucazeau, K.Chhor, C.Sourisseau, and A.J.Dianoux, Chem. Phys., 1983, 76, 307.

14 D.F.R.Gilson, G.Gomez, I.S.Butler, and P.J.Fitzpatrick, Can. J. Chem., 1983, 61, 737.

15 J.Mueller, E.Baumgartner, and C.Haensch, Int. J. Mass Spectrom. Ion Phys., 1983, 47, 523; Chem. Abs., 1983, 98, 143591.

16 F.G.Bordwell and M.J.Bausch, J. Am. Chem. Soc., 1983, 105, 6188; see also P.G.Gassmann, D.W.Macomber, and J.W.Hershberger, Organometallics, 1983, 2, 1470.

17 M.E.Rerek, L.-N.Ji, and F.Basolo, J. Chem. Soc., Chem. Commun., 1983, 1208.

18 A.G.Davies, Pure Appl. Chem., 1982, 54, 23.

19 N.A.Ogorodnikova and A.A.Koridze, Polyhedron, 1983, 2, 941.

20 M.C.R.Symons, S.W.Bratt, and J.L.Wyatt, J. Chem. Soc., Dalton Trans., 1983, 1377.

21 K.Sarbasov, S.P.Solodovnikov, B.L.Tumanskii, N.N.Bubnov, and A.I.Prokof'ev, Izv. Akad. Nauk SSSR, Ser. Khim., 1982, 1509.

22 A.Dormond, A.El Bouadili, and C. Moise, Tetrahedron Lett., 1983, 24, 3087.

23 J.Nieman, J.H.Teuben, J.C.Huffman, and K.G.Caulton, J. Organomet. Chem., 1983, 255, 193.

24 D.Rehder, K.Paulsen, and W.Basler, J. Magn. Reson., 1983, 53, 500.

25 J.M.Mayer, P.T.Wolczanski, B.D.Santarsiero, W.A.Olson, and J.E.Bercaw, Inorg. Chem., 1983, 22, 1149.

26 R.Talay and D.Rehder, Inorg. Chim. Acta, 1983, 77, L175.

27 D.W.Macomber and M.D.Rausch, Organometallics, 1983, 2, 1523; J. Organomet. Chem., 1983, 258, 331.

28 C.G.Kreiter and H.Kurz, Z. Naturforsch., 1983, 38B, 841.

29 K.A.Mahmoud, A.J.Rest, and H.G.Alt, J. Organomet. Chem., 1983, 246, C37.

30 R.T.C.Brownlee, A.F.Masters, M.J.O'Connor, A.G.Wedd, H.A.Kimlin, and J.D. Cotton, Org. Magn. Reson., 1982, 20, 73.

31 M.Minelli, J.L.Hubbard, K.A.Christensen, and J.H.Enemark, Inorg. Chem., 1983, 22, 2652.

32 K.Sünkel, G.Urban, and W.Beck, J. Organomet. Chem., 1983, 252, 187.

33 E.P.Kündig, P.L.Timms, B.A.Kelly, and P.Woodward, J. Chem. Soc., Dalton Trans., 1983, 901.

34 L.G.Kuzmina, A.G.Ginzburg, Yu.T.Struchkov, and D.N.Kursanov, J. Organomet. Chem., 1983, 253, 329.

35 (a) I.A.Lobanova, V.I.Zdanovich, and N.E.Kolobova, Izv. Akad. Nauk SSSR, Ser. Khim., 1982, 2606; (b) I.A.Lobanova, V.I.Zdanovich, N.E.Kolobova, and P.V.Petrovskii, ibid., 1983, 684.

36 (a) C.P.Casey, J.M.O'Connor, W.D.Jones, and K.J.Haller, Organometallics, 1983, 2, 535; (b) C.P.Casey and J.M.O'Connor, J. Am. Chem. Soc., 1983, 105, 2919.

37 V.F.Sizoi and Yu.S.Nekrasov, Izv. Akad. Nauk SSSR, Ser. Khim., 1982, 285; Yu.S.Nekrasov, V.K.Mavrodiev, V.F.Sizoi, and V.I.Khvostenko, ibid., p. 1299.

38 J.W.Hershberger and J.K.Kochi, Polyhedron, 1983, 2, 929.

39 R.B.Hitam, R.Narayanaswamy, and A.J.Rest, J. Chem. Soc., Dalton Trans., 1983, 615.

40 A.A.Koridze, N.M.Astakhova, and P.V.Petrovskii, Izv. Akad. Nauk SSSR, Ser. Khim., 1982, 956.

41 E.Roman and R.Latorre, Bol. Soc. Chil. Quim., 1982, 27, 5; Chem. Abs., 1982, 97, 144979.

42 R.L.DeKock, P.Deshmukh, T.P.Fehlner, C.E.Housecroft, J.S.Plotkin, and S.G. Shore, J. Am. Chem. Soc., 1983, 105, 815.

43 D.Milholová and A.A.Vlček, Inorg. Chim. Acta, 1983, 73, 249.

44 W.W.McConnell, G.O.Nelson, and M.E.Wright, Inorg. Chem., 1983, 22, 1689; K.D.Janda, W.W.McConnell, G.O.Nelson, and M.E.Wright, J. Organomet. Chem., 1983, 259, 139.

45 M.I.Bruce, R.C.Wallis, M.L.Williams, B.W.Skelton, and A.H.White, J. Chem. Soc., Dalton Trans., 1983, 2183.

46 T.Yu.Orlova, V.N.Setkina, V.F.Sizoi, and D.N.Kursanov, J. Organomet. Chem., 1983, 252, 201.

47 D.L.Reger, K.A.Belmore, J.L.Atwood, and W.E.Hunter, J. Am. Chem. Soc., 1983, 105, 5710.

48 S.G.Davies, J.Hibberd, and S.J.Simpson, J. Organomet. Chem., 1983, 246, C16; S.G.Davies, S.J.Simpson, and S.E.Thomas, ibid., 1983, 254, C29; S.G. Davies, S.J.Simpson, H.Felkin, F.Tadj, and O.Watts, J. Chem. Soc., Dalton Trans., 1983, 981; S.G.Davies, S.J.Simpson, H.Felkin, and T.Fillebeen-Khan, Organometallics, 1983, 2, 539; see also C.Lapinte and D.Astruc, J. Chem. Soc., Chem. Commun., 1983, 430.

49 P.Hofmann and M.Padmanabhan, Organometallics, 1983, 2, 1273.

50 K.Broadley, N.G.Connelly, and W.E.Geiger, J. Chem. Soc., Dalton Trans., 1983, 121; N.G.Connelly, A.R.Lucy, J.D.Payne, A.M.R.Galas, and W.E.Geiger, ibid., p. 1879.

51 H.Bönnemann, B.Bogdanović, R.Brinkmann, D.-W.He, and B.Spliethoff, Angew. Chem. Int. Ed. Engl., 1983, 22, 728.

52 Z.L.Lutsenko, M.G.Kuznetsova, A.V.Kisin, and A.Z.Rubezhov, Izv. Akad. Nauk SSSR, Ser. Khim., 1983, 178.

53 B.Klingert and H.Werner, Chem. Ber., 1983, 116, 1450.

54 J.Müller, R.Stock, and J.Pickardt, Z. Naturforsch., 1983, 38B, 1454.

55 K.Jonas, E.Deffense, and D.Habermann, Angew. Chem. Int. Ed. Engl., 1983, 22, 716.

56 R.L.Hillard, C.A.Parnell, and K.P.C.Vollhardt, Tetrahedron, 1983, 39, 905.

57 N.Kuhn and M.Winter, Chem.-Ztg., 1982, 106, 438; Chem. Abs., 1983, 98, 143616.

58 H.Lehmkuhl, C.Naydowski, and M.Bellenbaum, J. Organomet. Chem., 1983, 246, C5; see also H.Lehmkuhl, C.Naydowski, R.Benn, A.Rufińska, and G.Schroth, ibid., p. C9.

59 H.Werner, H.-J.Kraus, U.Schubert, K.Ackermann, and P.Hofmann, J. Organomet. Chem., 1983, 250, 517; see also R.J.Cross and A.J.McLennan, J. Chem. Soc., Dalton Trans., 1983, 359.

60 H.Werner, G.T.Crisp, P.W.Jolly, H.-J.Kraus, and C.Krüger, Organometallics, 1983, 2, 1369.

61 H.-J.Kraus, H.Werner, and C.Krüger, Z. Naturforsch., 1983, 38B, 733.

62 N.M.Boag, R.J.Goodfellow, M.Green, B.Hessner, J.A.K.Howard, and F.G.A. Stone, J. Chem. Soc., Dalton Trans., 1983, 2585.

63 C.Bleasdale and D.W.Jones, J. Chem. Soc., Chem. Commun., 1983, 214.

64 D.W.Macomber and M.D.Rausch, J. Am. Chem. Soc., 1983, 105, 5325.

65 G.Z.Suleimanov, Ya.A.Nuriev, O.P.Syutkina, T.Kh.Kurbanov, and I.P. Beletskaya, Izv. Akad. Nauk SSSR, Ser. Khim., 1982, 1671.
66 R.Zhang and M.Tsutsui, Youji Huaxue, 1982, 435; Chem. Abs., 1983, 98, 198374.
67 P.J.Fagan, E.A.Maata, A.M.Seyam, and T.J.Marks, Rare Earths Mod. Sci. Technol., 1982, 3, 77; Chem. Abs., 1982, 97, 198237.
68 V.T.Aleksanyan and I.I.Grinval'd, THEOCHEM, 1982, 7, 35; Chem Abs., 1983, 98, 89420.
69 J.C.Giordan, J.H.Moore, J.A.Tossell, and J.Weber, J. Am. Chem. Soc., 1983, 105, 3431.
70 J.C.Kotz, W.Vining, W.Coco, R.Rosen, A.R.Dias, and M.H.Garcia, Organometallics, 1983, 2, 68.
71 M.J.Calhorda, A.R.Dias, M.S.Salema, and J.A.Martinho Simões, J. Organomet. Chem., 1983, 255, 81.
72 G.Li and D.Zhang, Huaxue Xuebao, 1982, 40, 1177; Chem. Abs., 1983, 98, 126294.
73 P.Sobota and Z.Janas, J. Organomet. Chem., 1983, 243, 35.
74 P.R.Brindley, A.G.Davies, and J.A.-A.Hawari, J. Organomet. Chem., 1983, 250, 247.
75 A.Hudson, M.F.Lappert, and R.Pichon, J. Chem. Soc., Chem. Commun., 1983, 374.
76 E.Samuel, Inorg. Chem., 1983, 22, 2967.
77 H.Köpf and N.Klouras, Z. Naturforsch., 1983, 38B, 321; Monatsh. Chem., 1983, 114, 243.
78 H.Schwemlein and H.H.Brintzinger, J. Organomet. Chem., 1983, 254, 69.
79 M.D.Rausch, B.H.Edwards, R.D.Rogers, and J.L.Atwood, J. Am. Chem. Soc., 1983, 105, 3882.
80 H.Lehmkuhl and Y.-L.Tsien, Chem. Ber., 1983, 116, 2437.
81 J.J.Eisch and A.Piotrowski, Tetrahedron Lett., 1983, 24, 2043.
82 E.-i.Negishi, D.E.Van Horn, T.Yoshida, and C.L.Rand, Organometallics, 1983, 2, 563.
83 I.Omae, Kagaku Kogyo, 1982, 33, 989; Chem. Abs., 1983, 98, 160750.
84 A.T.Casey and J.B.Raynor, J. Chem. Soc., Dalton Trans., 1983, 2057.
85 S.Sabo, R.Choukroun, and D.Gervais, J. Chem. Soc., Dalton Trans., 1983, 2695.
86 A.H.Al-Mowali, Inorg. Chim. Acta, 1983, 77, L51.
87 J.L.Petersen and J.W.Egan, Inorg. Chem., 1983, 22, 3571.
88 J.C.Green, M.P.Payne, and J.H.Teuben, Organometallics, 1983, 2, 203.
89 Y.Mugnier, C.Moise, and E.Laviron, Nouv. J. Chim., 1982, 6, 197; see also Y.Mugnier, A.Fakhr, M.Fauconet, C.Moise, and E.Laviron, Acta Chem. Scand., 1983, 37B, 423.
90 K.Jonas and V.Wiskamp, Z. Naturforsch., 1983, 38B, 1113.
91 S.A.Koch and V.Chebolu, Organometallics, 1983, 2, 350.
92 S.Gambarotta, C.Floriani, A.Chiesi-Villa, and C.Guastini, J. Chem. Soc., Chem. Commun., 1983, 184.
93 M.Pasquali, P.Leoni, C.Floriani, A.Chiesi-Villa, and C.Guastini, Inorg. Chem., 1983, 22, 841.
94 F.H.Köhler and W.A.Geike, J. Organomet. Chem., 1983, 256, C27.
95 R.Mercier, J.Douglade, J.Amaudrut, J.Sala-Pala, and J.E.Guerchais, J. Organomet. Chem., 1983, 244, 145.
96 L.Acedo, A.Otero, and P.Royo, J. Organomet. Chem., 1983, 258, 181.
97 J.Kalousová, J.Holeček, J.Votinský, and L.Beneš, Z. Chem., 1983, 23, 327.
98 P.A.Cox, P.Grebenik, R.N.Perutz, M.D.Robinson, R.Grinter, and D.R.Stern, Inorg. Chem., 1983, 22, 3614.
99 A.Baretta, K.S.Chong, F.G.N.Cloke, A.Feigenbaum, and M.L.H.Green, J. Chem. Soc., Dalton Trans., 1983, 861.
100 H.Schwemlein, L.Zsolnai, G.Huttner, and H.H.Brintzinger, J. Organomet. Chem., 1983, 256, 285.
101 A.A.Koridze, A.I.Yanovskii, Yu.I.Slovokhotov, V.G.Andrianov, and Yu.T. Struchkov, Koord. Khim., 1982, 8, 541.
102 F.H.Köhler and N.Hebendanz, Chem. Ber., 1983, 116, 1261.

103 J.T.Weed, M.F.Rettig, and R.M.Wing, J. Am. Chem. Soc., 1983, 105, 6510.
104 J.Chetwynd-Talbot, P.Grebenik, R.N.Perutz, and M.H.A.Powell, Inorg. Chem., 1983, 22, 1675.
105 P.N.Gaponik, A.I.Lesnikovich, and Yu.G.Orlik, Russ. Chem. Rev. (Engl. Transl.), 1983, 52, 168.
106 G.Calvarin, G.Clec'h, and J.F.Berar, J. Phys. Chem. Solids, 1982, 43, 785; F.Calvarin, J.F.Berar, and G.Clec'h, ibid., p. 791; Chem. Abs., 1983, 98, 107484, 107485.
107 A.Perjéssy and Š.Toma, Collect. Czech. Chem. Commun., 1983, 48, 1635.
108 E.Haslinger, K.Koci, W.Robien, and K.Schlögl, Monatsh. Chem., 1983, 114, 495.
109 C.Brevard and P.Granger, Inorg. Chem., 1983, 22, 532.
110 G.Innorta, F.Scagnolari, A.Modelli, S.Torroni, A.Foffani, and S.Sorriso, J. Organomet. Chem., 1983, 241, 375; G.Fu, Y.Xu, and Y.Zhang, Youji Huaxue, 1982, 438; Chem Abs., 1983, 99, 5742; Y.Zhang, J.Lu, P.Lu, and W. Xu, Huaxue Tongbao, 1983, 19; Chem. Abs., 1983, 99, 22615.
111 M.Clemance, R.M.G.Roberts, and J.Silver, J. Organomet. Chem., 1983, 243, 461; 1983, 247, 219; M.Katada, Y.Uchida, H.Sano, H.H.Wei, H.Sakai, and Y. Maeda, Radiochem. Radioanal. Lett., 1982, 54, 55.
112 R.Baer, T.Heinis, C.Nager, and M.Jungen, Chem. Phys. Lett., 1982, 91, 440.
113 P.R.Sharp and A.J.Bard, Inorg. Chem., 1983, 22, 2689.
114 R.J.Gale, K.M.Motyl, and R.Job, Inorg. Chem., 1983, 22, 130.
115 T.Solomun, I.Tabakovic, and V.Rapic, Glas. Hem. Drus. Beograd, 1982, 47, 399; Chem. Abs., 1983, 99, 5773.
116 T.Yamamoto, K.-I.Sanechika, A.Yamamoto, M.Katada, I.Motoyama, and H.Sano, Inorg. Chim. Acta, 1983, 73, 75.
117 G.L.D.Ritchie, M.K.Cooper, R.L.Calvert, G.R.Dennis, L.Phillips, and J. Vrbancich, J. Am. Chem. Soc., 1983, 105, 5215.
118 S.R.Logan and G.A.Salmon, J. Chem. Soc., Perkin Trans. 1, 1983, 1781.
119 L.H.Simons and J.J.Lagowski, J. Organomet. Chem., 1983, 249, 195.
120 R.Ferede and N.T.Allison, Organometallics, 1983, 2, 463.
121 G.A.Razuvaev, S.G.Yudenich, O.N.Druzhkov, and V.A.Dodonov, J. Gen. Chem. USSR (Engl. Transl.), 1982, 52, 2108.
122 V.P.Tverdokhlebov, B.V.Polyakov, I.V.Tselinskii, and L.I.Golubeva, J. Gen. Chem. USSR (Engl. Transl.), 1982, 52, 1807.
123 N.S.Nametkin, V.D.Tyurin, S.A.Sleptsova, A.M.Krapivin, and A.Ya.Sideridu, Izv. Akad. Nauk SSSR, Ser. Khim., 1982, 955.
124 M.Herberhold, M.Ellinger, and W.Kremnitz, J. Organomet. Chem., 1983, 241, 227.
125 I.R.Butler, W.R.Cullen, J.Reglinski, and S.J.Rettig, J. Organomet. Chem., 1983, 249, 183.
126 W.E.McEwen, C.E.Sullivan, and R.O.Day, Organometallics, 1983, 2, 420.
127 J.Holeček, K.Handlíř, M.Nádvorník, and M.Vlček, Collect. Czech. Chem. Commun., 1982, 47, 3375.
128 T.Hayashi and M.Kumada, Acc. Chem. Res., 1982, 15, 395; see also W.S. Knowles, ibid., 1983, 16, 106.
129 V.I.Sokolov, L.L.Troitskaya, and N.S.Khrushchova, J. Organomet. Chem., 1983, 250, 439.
130 T.Hayashi, T.Hagihara, Y.Katsuro, and M.Kumada, Bull. Chem. Soc. Jpn., 1983, 56, 363.
131 G.E.Rodgers, W.R.Cullen, and B.R.James, Can. J. Chem., 1983, 61, 1314; W.R.Cullen, T.-J.Kim, F.W.B.Einstein, and T.Jones, Organometallics, 1983, 2, 714.
132 R.Breslow, G.Trainor, and A.Ueno, J. Am. Chem. Soc., 1983, 105, 2739; see also W.J.le Noble, S.Srivastava, R.Breslow, and G.Trainor, ibid., p. 2745.
133 W.C.Herndon and I.Agranat, Isr. J. Chem., 1983, 23, 66.
134 A.A.Koridze, N.M.Astakhova, and P.V.Petrovskii, Izv. Akad. Nauk SSSR, Ser. Khim., 1982, 957; J. Organomet. Chem., 1983, 254, 345.
135 C.A.Bunton and W.E.Watts, J. Chem. Soc., Perkin Trans. 2, 1983, 1591.
136 D.Šutič, S.Ašperger, and S.Borčič, J. Org. Chem., 1982, 47, 5120; see also S.Ursic, Croat. Chem. Acta, 1982, 55, 305; Chem. Abs., 1983, 98, 89596.

137 V.I.Boev and A.V.Dombrovskii, J. Gen. Chem. USSR (Engl. Transl.), 1982, 52, 1497; see also A.Ceccon, A.Gambaro, D.Paolucci, and A.Venzo, Polyhedron, 1983, 2, 183.

138 S.G.Baxter, R.L.Collins, A.H.Cowley, and S.F.Sena, Inorg. Chem., 1983, 22, 3475.

139 B.Floris, Gazz. Chim. Ital., 1982, 112, 489; see also M.Vlček, K.Handlíř, and J.Holeček, Z. Chem., 1983, 23, 294.

140 N.A.Lewis, B.Patel, and P.S.White, J. Chem. Soc., Dalton Trans., 1983, 1367.

141 I.R.Butler, W.R.Cullen, F.W.B.Einstein, S.J.Rettig, and A.J.Willis, Organometallics, 1983, 2, 128.

142 B.Czech, A.Piorkó, and R.Annunziata, J. Organomet. Chem., 1983, 255, 365.

143 A.Ratajczak and A.Piorkó, Pol. J. Chem., 1981, 55, 935; Chem. Abs., 1982, 97, 72526.

144 I.Shimizu, Y.Kamei, T.Tezuka, T.Izumi, and A.Kasahara, Bull. Chem. Soc. Jpn., 1983, 56, 192; A.Kasahara, T.Izumi, I.Shimizu, T.Oikawa, H.Umezawa, and I.Hoshino, ibid., p. 1143.

145 A.Clearfield, C.J.Simmons, H.P.Withers, and D.Seyferth, Inorg. Chim. Acta, 1983, 75, 139.

146 T.Izumi, I.Shimizu, and A.Kasahara, Kenkyu Hokoku - Asahi Garasu Kogyo Gijutsu Shoreikai, 1981, 38, 223; Chem. Abs., 1982, 97, 216400.

147 A.Kasahara, T.Izumi, and K.Saito, Bull. Chem. Soc. Jpn., 1983, 56, 2865.

148 M.Hisatome, M.Yoshihashi, K.Yamakawa, and Y.Iitaka, Tetrahedron Lett., 1983, 24, 5757; see also S.El-Tamany, F.-W.Raulfs, and H.Hopf, Angew. Chem. Int. Ed. Engl., 1983, 22, 633.

149 S.Akabori, H.Fukuda, Y.Habata, M.Sato, and S.Ebine, Chem. Lett., 1982, 1393; M.Sato, H.Watanabe, S.Ebine, and A.Akabori, ibid., p. 1753; S. Akabori, Y.Habata, Y.Sakamoto, M.Sato, and S.Ebine, Bull. Chem. Soc. Jpn., 1983, 56, 537; S.Akabori, M.Ohtomi, M.Sato, and S.Ebine, ibid., p. 1455; S.Akabori, Y.Habata, M.Sato, and S.Ebine, ibid., p. 1459.

150 B.Czech, A.Ratajczak, and K.Nagraba, Monatsh. Chem., 1982, 113, 965; see also A.Ratajczak and A.Piorkó, Pol. J. Chem., 1981, 55, 1953; Chem. Abs., 1983, 99, 53920.

151 P.J.Hammond, A.P.Bell, and C.D.Hall, J. Chem. Soc., Perkin Trans. 1, 1983, 707; P.J.Hammond, P.D.Beer, and C.D.Hall, J. Chem. Soc., Chem. Commun., 1983, 1161.

152 V.K.Kansal, W.E.Watts, U.T.Mueller-Westerhoff, and A.Nazzal, J. Organomet. Chem., 1983, 243, 443.

153 I.R.Butler and W.R.Cullen, Can. J. Chem., 1983, 61, 147; see also J.D. Fellmann, P.E.Garrou, H.P.Withers, D.Seyferth, and D.D.Traficante, Organometallics, 1983, 2, 818.

154 J.Mirek, S.Rachwał, and B.Kawałek, J. Organomet. Chem., 1983, 248, 107.

155 F.Delgado-Pena, D.R.Talham, and D.O.Cowan, J. Organomet. Chem., 1983, 253, C43; J.Kotz, G.Neyhart, W.J.Vining, and M.D.Rausch, Organometallics, 1983, 2, 79; H.H.Wei, C.Y.Lin, and S.J.Chang, Proc. Natl. Sci. Counc., Repub. China, 1983, 7B, 35; Chem. Abs., 1983, 99, 38606.

156 S.B.Colbran, B.H.Robinson, and J.Simpson, Organometallics, 1983, 2, 943, 952.

157 K.E.Schwarzhans and H.Schottenberger, Z. Naturforsch., 1983, 38B, 1493.

158 R.Gooding, C.P.Lillya, and C.W.Chien, J. Chem. Soc., Chem. Commun., 1983, 151.

159 S.Chao, J.L.Robbins, and M.S.Wrighton, J. Am. Chem. Soc., 1983, 105, 181.

160 Yu.A.Ol'dekop, N.A.Maier, V.L.Shirokii, V.F.Romanova, E.A.Chernyshev, A.V. Bukhtiarov, and A.P.Tomilov, J. Gen. Chem. USSR (Engl. Transl.), 1982, 52, 1701.

161 H.Bönnemann, M.Radermacher, C.Krüger, and H.-J.Kraus, Helv. Chim. Acta, 1983, 66, 185.

162 J.E.Sheats, W.C.Spink, R.A.Nabinger, D.Nicol, and G.Hlatky, J. Organomet. Chem., 1983, 251, 93.

163 Z.L.Lutsenko and A.Z.Rubezhov, Izv. Akad. Nauk SSSR, Ser. Khim., 1982, 1412.

164 E.Lindner, F.Bouachir, and S.Hoehne, Chem. Ber., 1983, 116, 46.

165 H.Lehmkuhl, F.Danowski, R.Benn, A.Rufińska, G.Schroth, and R.Mynott, J. Organomet. Chem., 1983, 254, C11.

166 H.Brunner and M.Muschiol, J. Organomet. Chem., 1983, 248, 233.

167 M.Mlekuz, P.Bougeard, M.J.McGlinchey, and G.Jaouen, J. Organomet. Chem., 1983, 253, 117.

168 N.P.Chernyaev, Yu.B.Zverev, E.M.Gavrishchuk, I.V.Runovskaya, S.G. Chesnokova, and V.F.Kutsepin, Russ. J. Inorg. Chem. (Engl. Transl.), 1982, 27, 1259; G.Z.Suleimanov, T.Kh.Kurbanov, Ya.A.Nuriev, L.F.Rybakova, and I.P.Beletskaya, Dokl. Akad. Nauk SSSR, 1982, 265, 896; G.B.Deacon, T.D. Tuong, and D.G.Vince, Polyhedron, 1983, 2, 969.

169 (a) M.F.Lappert, A.Singh, J.L.Atwood, and W.E.Hunter, J. Chem. Soc., Chem. Commun., 1983, 206; (b) M.F.Lappert, A.Singh, J.L.Atwood, W.E.Hunter, and H.-M.Zhang, ibid., p. 69.

170 B.E.Bursten and A.Fang, J. Am. Chem. Soc., 1983, 105, 6495.

171 J.W.Bruno, T.J.Marks, and L.R.Morss, J. Am. Chem. Soc., 1983, 105, 6824.

172 D.G.Karraker, Inorg. Chem., 1983, 22, 503.

173 G.Bombieri, F.Benetollo, E.Klähne, and R.D.Fischer, J. Chem. Soc., Dalton Trans., 1983, 1115.

174 A.Dormond, J. Organomet. Chem., 1983, 256, 47.

175 D.Astruc, Tetrahedron, 1983, 39, 4027.

176 M.I.Rybinskaya, A.R.Kudinov, and V.S.Kaganovich, J. Organomet. Chem., 1983, 246, 279.

177 Y.K.Chung, E.D.Honig, and D.A.Sweigart, J. Organomet. Chem., 1983, 256, 277.

178 K.Jonas, G.Koepe, L.Schieferstein, R.Mynott, C.Krüger, and Y.-H.Tsay, Angew. Chem. Int. Ed. Engl., 1983, 22, 620; see also K.Jonas, V.Wiskamp, Y.-H.Tsay, and C.Krüger, J. Am. Chem. Soc., 1983, 105, 5480.

179 T.J.Groshens and K.J.Klabunde, J. Organomet. Chem., 1983, 259, 337.

180 M.P.Andrews, H.X.Huber, S.M.Mattar, D.F.McIntosh, and G.A.Ozin, J. Am. Chem. Soc., 1983, 105, 6170.

181 T.Avilés and J.H.Teuben, J. Organomet. Chem., 1983, 253, 39.

182 A.W.Duff, K.Jonas, R.Goddard, H.-J.Kraus, and C.Krüger, J. Am. Chem. Soc., 1983, 105, 5479.

183 G.A.Razuvaev, A.N.Artemov, T.G.Kasatkina, and E.N.Poselenova, Izv. Akad. Nauk SSSR, Ser. Khim., 1982, 1131.

184 M.Yoshifuji and N.Inamoto, Tetrahedron Lett., 1983, 24, 4855.

185 O.B.Afanasova, N.G.Komalenkova, Yu.E.Zubarev, V.M.Nosova, A.V.Kisin, V.A. Sharapov, A.I.Gusev, N.V.Alekseev, N.I.Kirillova, and E.A.Chernyshev, J. Gen. Chem. USSR (Engl. Transl.), 1982, 52, 1903.

186 H.Kalchhauser, K.Schlögl, W.Weissensteiner, and A.Werner, J. Chem. Soc., Perkin Trans. 1, 1983, 1723; K.Schlögl, A.Werner, and M.Widhalm, ibid., p. 1731.

187 S.D.Cunningham, K.Öfele, and B.R.Willeford, J. Am. Chem. Soc., 1983, 105, 3724.

188 (a) K.H.Dötz, J.Mühlemeier, U.Schubert, and O.Orama, J. Organomet. Chem., 1983, 247, 187; K.H.Dötz and W.Kuhn, ibid., 1983, 252, C78; (b) Angew. Chem. Int. Ed. Engl., 1983, 22, 732.

189 K.Chhor, C.Sourisseau, and G.Lucazeau, J. Mol. Struct., 1982, 80, 485.

190 M.J.McGlinchey, J.L.Fletcher, B.G.Sayer, P.Bougeard, R.Faggiani, C.J.L. Lock, A.D.Bain, C.Rodger, E.P.Kündig, D.Astruc, J.-R.Hamon, P.Le Maux, S. Top, and G.Jaouen, J. Chem. Soc., Chem. Commun., 1983, 634.

191 G.A.Vaglio, P.Volpe, and L.Operti, Org. Mass Spectrom., 1982, 17, 617.

192 S.N.Milligan, I.Tucker, and R.D.Rieke, Inorg. Chem., 1983, 22, 987; M.G. Peterleitner, M.V.Tolstaya, V.V.Krivykh, L.I.Denisovitch, and M.I. Rybinskaya, J. Organomet. Chem., 1983, 254, 313.

193 S.N.Milligan and R.D.Rieke, Organometallics, 1983, 2, 171.

194 T.E.Bitterwolf, Polyhedron, 1983, 2, 675.

195 E.A.Domogatskaya, V.N.Setkina, N.K.Baranetskaya, V.N.Trembovler, B.M. Yavorskii, A.Ya.Shteinshneider, and P.V.Petrovskii, J. Organomet. Chem., 1983, 248, 161.

196 C.D.Hoff, J. Organomet. Chem., 1983, 246, C53.
197 R.P.Houghton, M.Voyle, and R.Price, J. Organomet. Chem., 1983, 259, 183.
198 A.Alemagna, P.Del Buttero, C.Gorini, D.Landini, E.Licandro, and S. Maiorana, J. Org. Chem., 1983, 48, 605; A.Alemagna, P.Cremonesi, P.Del Buttero, E.Licandro, and S.Maiorana, ibid., p. 3114.
199 L.I.Zakharkin and G.G.Zhigareva, J. Gen. Chem. USSR (Engl. Transl.), 1983, 53, 841.
200 N.F.Masters and D.A.Widdowson, J. Chem. Soc., Chem. Commun., 1983, 955.
201 (a) M.Uemura, N.Nishikawa, K.Take, M.Ohnishi, K.Hirotsu, T.Higuchi, and Y.Hayashi, J. Org. Chem., 1983, 48, 2349; M.Uemura, K.Isobe, K.Take, and Y.Hayashi, ibid., p. 3855; (b) M.Uemura, K.Take, and Y.Hayashi, J. Chem. Soc., Chem. Commun., 1983, 858.
202 M.F.Semmelhack and A.Zask, J. Am. Chem. Soc., 1983, 105, 2034.
203 M.Ghavshou and D.A.Widdowson, J. Chem. Soc., Perkin Trans. 1, 1983, 3065.
204 E.P.Kündig, V.Desobry, and D.P.Simmons, J. Am. Chem. Soc., 1983, 105, 6962; E.P.Kündig and D.P.Simmons, J. Chem. Soc., Chem. Commun., 1983, 1320.
205 (a) M.F.Semmelhack, J.L.Garcia, D.Cortes, R.Farina, R.Hong, and B.K. Carpenter, Organometallics, 1983, 2, 467; (b) M.F.Semmelhack, Pure Appl. Chem., 1981, 53, 2379.
206 (a) J.-C.Boutonnet, J.Levisalles, E.Rose, G.Precigoux, C.Courseille, and N.Platzer, J. Organomet. Chem., 1983, 255, 317; (b) J.-C.Boutonnet, J. Levisalles, J.-M.Normant, and E.Rose, ibid., p. C21.
207 W.P.Henry and R.D.Rieke, J. Am. Chem. Soc., 1983, 105, 6314.
208 (a) A.Ceccon, A.Gambaro, A.M.Romanin, and A.Venzo, Angew. Chem. Int. Ed. Engl., 1983, 22, 559; (b) J. Organomet. Chem., 1983, 254, 199; see also S.Top, G.Jaouen, B.G.Sayer, and M.J.McGlinchey, J. Am. Chem. Soc., 1983, 105, 6426.
209 S.G.Davies, N.J.Holman, C.A.Laughton, and B.E.Mobbs, J. Chem. Soc., Chem. Commun., 1983, 1316.
210 J.Lebibi, J.Brocard, and D.Couturier, Bull. Soc. Chim. Fr. II, 1982, 357.
211 B.Caro and G.Jaouen, Tetrahedron, 1983, 39, 83, 93.
212 M.T.Reetz and M.Sauerwald, Tetrahedron Lett., 1983, 24, 2837.
213 A.Ceccon, A.Gambaro, A.M.Romanin, and A.Venzo, J. Organomet. Chem., 1983, 254, 207.
214 P.Le Maux, R.Dabard, and G.Simonneaux, J. Organomet. Chem., 1983, 241, 37.
215 R.Luck and R.H.Morris, J. Organomet. Chem., 1983, 255, 221.
216 P.R.Brown, F.G.N.Cloke, M.L.H.Green, and N.J.Hazel, J. Chem. Soc., Dalton Trans., 1983, 1075.
217 A.J.Pearson and I.C.Richards, J. Organomet. Chem., 1983, 258, C41.
218 M.Brookhart, W.Lamanna, and A.R.Pinhas, Organometallics, 1983, 2, 638; M.Brookhart and A.Lukacs, ibid., p. 649.
219 Y.K.Chung, E.D.Honig, W.T.Robinson, D.A.Sweigart, N.G.Connelly, and S.D. Ittel, Organometallics, 1983, 2, 1479.
220 F.G.N.Cloke, A.E.Derome, M.L.H.Green, and D.O'Hare, J. Chem. Soc., Chem. Commun., 1983, 1312.
221 R.P.Micciche and L.G.Sneddon, Organometallics, 1983, 2, 674; see also R.G. Swisher, E.Sinn, and R.N.Grimes, ibid., p. 506.
222 H.Werner and H.Kletzin, J. Organomet. Chem., 1983, 243, C59; H.Kletzin and H.Werner, Angew. Chem. Int. Ed. Engl., 1983, 22, 873.
223 M.A.Bennett, T.N.Huang, T.W.Matheson, and A.K.Smith, Inorg. Synth., 1982, 74.
224 D.A.Tocher, R.O.Gould, T.A.Stephenson, M.A.Bennett, J.P.Ennett, T.W. Matheson, L.Sawyer, and V.K.Shah, J. Chem. Soc., Dalton Trans., 1983, 1571.
225 J.C.Green, M.R.Kelly, M.P.Payne, E.A.Seddon, D.Astruc, J.-R.Hamon, and P. Michaud, Organometallics, 1983, 2, 211.
226 P.Michaud and D.Astruc, Angew. Chem. Int. Ed. Engl., 1982, 21, 918; J.-R. Hamon, J.-Y.Saillard, A.Le Beuze, M.J.McGlinchey, and D.Astruc, J. Am. Chem. Soc., 1982, 104, 7549; J.-R.Hamon and D.Astruc, ibid., 1983, 105, 5951.
227 A.Darchen, J. Chem. Soc., Chem. Commun., 1983, 768.
228 T.P.Gill and K.R.Mann, Inorg. Chem., 1983, 22, 1986.

229 (a) C.C.Lee, A.Piorkó, and R.G.Sutherland, J. Organomet. Chem., 1983, 248, 357; see also C.C.Lee, A.Piorkó, B.R.Steele, U.S.Gill, and R.G.Sutherland, ibid., 1983, 256, 303; (b) C.C.Lee, K.J.Demchuk, U.S.Gill, and R.G. Sutherland, ibid., 1983, 247, 71.
230 V.Guerchais and D.Astruc, J. Chem. Soc., Chem. Commun., 1983, 1115.
231 M.M.Brezinski and K.J.Klabunde, Organometallics, 1983, 2, 1116.
232 R.M.Scotti and G.M.Valderrama, Bol. Soc. Chil. Quim., 1982, 27, 3; G.M. Valderrama and S.R.Ganz, ibid., p. 19; Chem. Abs., 1982, 97, 144989, 144990; R.Usón, L.A.Oro, D.Carmona, M.A.Esteruelas, C.Foces-Foces, F.H. Cano, and S.Garcia-Blanco, J. Organomet. Chem., 1983, 254, 249.
233 C.R.Landis and J.Halpern, Organometallics, 1983, 2, 840.
234 S.G.Shore, W.-L.Hsu, M.R.Churchill, and C.Bueno, J. Am. Chem. Soc., 1983, 105, 655.
235 A.Tiripicchio, M.Tiripicchio Camellini, R.Usón, L.A.Oro, and J.A.Cabeza, J. Organomet. Chem., 1983, 244, 165; R.Usón, L.A.Oro, J.A.Cabeza, C. Foces-Foces, F.H.Cano, and S.Garcia-Blanco, ibid., 1983, 246, 73.
236 F.G.N.Cloke, A.N.Dix, J.C.Green, R.N.Perutz, and E.A.Seddon, Organometallics, 1983, 2, 1150; see also M.L.H.Green, N.J.Hazel, P.D.Grebenik, V.S. B.Mtetwa, and K.Prout, J. Chem. Soc., Chem. Commun., 1983, 356.
237 S.P.Kolesnikov, S.L.Povarov, and A.Ya.Shteinshneider, Izv. Akad. Nauk SSSR, Ser. Khim., 1982, 415.
238 A.Yu.Vasil'kov, N.V.Zakurin, I.A.Uralets, N.N.Zaitseva, N.I.Vasyukova, A.G.Makarovskaya, and L.P.Yur'eva, Izv. Akad. Nauk SSSR, Ser. Khim., 1982, 2144; V.V.Litvak, P.P.Kun, and V.D.Shteingarts, Zh. Org. Khim., 1982, 18, 2153; L.P.Yur'eva, N.N.Zaitseva, N.V.Zakurin, A.Yu.Vasil'kov, and N.I. Vasyukova, J. Organomet. Chem., 1983, 247, 287; P.Lumme, P.von Bagh, J. Kahima, and H.Karrus, Inorg. Chim. Acta, 1983, 71, 209; R.Faggiani, N. Hao, C.J.L.Lock, B.G.Sayer, and M.J.McGlinchey, Organometallics, 1983, 2, 96; C.Elschenbroich, E.Bilger, and R.Möckel, Z. Naturforsch., 1983, 38B, 1357.
239 T.S.Kurtikyan, Arm. Khim. Zh., 1983, 36, 142; Chem. Abs., 1983, 99, 53947.
240 B.E.Wilburn and P.S.Skell, J. Am. Chem. Soc., 1982, 104, 6989.
241 P.N.Hawker and P.L.Timms, J. Chem. Soc., Dalton Trans., 1983, 1123.
242 S.P.Solodovnikov, L.P.Yur'eva, and N.N.Zaitseva, Izv. Akad. Nauk SSSR, Ser. Khim., 1982, 2392.
243 V.V.Litvak, P.P.Kun, and V.D.Shteingarts, Zh. Org. Khim., 1982, 18, 1966.
244 J.Elzinga and M.Rosenblum, Organometallics, 1983, 2, 1214.
245 W.D.Rohrbach and V.Boekelheide, J. Org. Chem., 1983, 48, 3673.
246 R.G.Finke, R.H.Voegeli, E.D.Laganis, and V.Boekelheide, Organometallics, 1983, 2, 347.
247 (a) N.M.Kostić and R.F.Fenske, Organometallics, 1983, 2, 1319; (b) ibid., p. 1008
248 (a) G.E.Herberich and A.K.Naithani, J. Organomet. Chem., 1983, 241, 1; G.E.Herberich and D.Söhnen, ibid., 1983, 254, 143; (b) G.E.Herberich, J. Hengesbach, G.Huttner, A.Frank, and U.Schubert, ibid., 1983, 246, 141; (c) G.E.Herberich, B.Hessner, and D.Söhnen, ibid., 1983, 256, C23; (d) G.E.Herberich and M.M.Kucharska-Jansen, ibid., 1983, 243, 45.
249 A.Kubo, R.Ikeda, and D.Nakamura, Chem. Lett., 1982, 1487.
250 N.I.Pyshnograeva, V.N.Setkina, and D.N.Kursanov, J. Organomet. Chem., 1983, 251, C41.
251 G.S.Sodhi, A.K.Sharma, and N.K.Kashik, Synth. React. Inorg. Metal-Org. Chem., 1982, 12, 947.
252 B.Lukas, R.M.G.Roberts, J.Silver, and A.S.Wells, J. Organomet. Chem., 1983, 256, 103.
253 M.El Borai, A.Akelah, and M.A.Hassen, Egypt J. Chem., 1981, 24, 91; Chem. Abs., 1983, 99, 53912.
254 M.N.Nefedova, V.N.Setkina, and D.N.Kursanov, J. Organomet. Chem., 1983, 244, C21.
255 J.Edwin, M.C.Böhm, N.Chester, D.M.Hoffman, R.Hoffmann, H.Pritzkow, W. Siebert, K.Stumpf, and H.Wadepohl, Organometallics, 1983, 2, 1666; J. Edwin, M.Bochmann, M.C.Böhm, D.E.Brennan, W.E.Geiger, C.Krüger, J.Pebler,

H.Pritzkow, W.Siebert, W.Swiridoff, H.Wadepohl, J.Weiss, and U.Zenneck, J. Am. Chem. Soc., 1983, 105, 2582.
256 G.Schmid, U.Höhner, D.Kampmann, D.Zaika, and R.Boese, Chem. Ber., 1983, 116, 951; J. Organomet. Chem., 1983, 256, 225; G.Schmid and R.Boese, Z. Naturforsch., 1983, 38B, 485; G.Schmid, U.Höhner, and D.Kampmann, ibid., p. 1094.
257 J.Müller, C.Hänsch, and J.Pickardt, J. Organomet. Chem., 1983, 259, C21.
258 H.W.Choi and M.S.Sollenberger, J. Organomet. Chem., 1983, 243, C39.
259 R.H.Morris and J.M.Ressner, J. Chem. Soc., Chem. Commun., 1983, 909.
260 K.Dimroth and H.Kaletsch, J. Organomet. Chem., 1983, 247, 271.
261 L.Weber and R.Boese, Chem. Ber., 1983, 116, 197, 514; L.Weber, ibid., p. 2022.
262 B.Lubke, F.Edelmann, and U.Behrens, Chem. Ber., 1983, 116, 11.
263 G.Moran, M.Green, and A.G.Orpen, J. Organomet. Chem., 1983, 250, C15.
264 M.L.H.Green, A.Izquierdo, J.J.Martin-Polo, V.S.B.Mtetwa, and K.Prout, J. Chem. Soc., Chem. Commun., 1983, 538.
265 B.N.Strunin, V.I.Bakhmutov, N.I.Vasyukova, V.N.Trembovler, V.N.Setkina, and D.N.Kursanov, J. Organomet. Chem., 1983, 246, 169.
266 A.D.Hunter and P.Legzdins, Organometallics, 1983, 2, 525.
267 C.G.Kreiter and W.Lipps, J. Organomet. Chem., 1983, 253, 339.

15
Homogeneous Catalysis by Transition-metal Complexes

BY M. E. FAKLEY

1 Introduction and General Reviews

A Book describes recent advances in organometallic chemistry[1a] and another catalysis with metal phosphine complexes[1b]. Catalysis via metal-carbon double and triple bonds has been reviewed[2]. The chemistry of bridged hydrocarbyl and hydrocarbon binuclear transition metal complexes has been discussed with reference to olefin polymerisation, oligomerisation and metathesis[3]. The preparation of supported catalysts from organometallics is reviewed[4a] and surface organometallic chemistry relevant to heterogeneous catalysis has been described[4b]. The role of metal clusters in catalysis has again been discussed[5].

A review of homogeneous metal catalysis in organic photochemistry has appeared[6]. Catalytic rearrangements of epoxides are reviewed[7]. The termination step in palladium-catalysed insertion reactions has been discussed[8].

Dibenzo[*a*,*e*]cyclooctatetraene (1) is a selective poison for homogeneous catalysts and can be used to distinguish homogeneous from heterogeneous processes[9].

Conformational interconversions of Wilkinson's catalyst and related compounds $[MX(PR_3)_2]$ determined from analysis of solid-state structural data are discussed with reference to mechanisms of asymmetric induction[10a]. Reviews on asymmetric hydrogenation[10b], rhodium-catalysed enantioselective hydrogenation[10c] and hydrogenation of carbon monoxide to ethanol and ethylene glycol[10d] have appeared. Point-group selection rules for ΔS = 2 multiplicity change in d^4, d^5 and d^6 configurations and are relevant to catalysis by, and reactivity of, organotransition metal complexes[11].

The versatility of $[OsHBr(CO)(PPh_3)_3]$ as a catalyst for C=C bond migration, hydrogenation and hydroformylation under moderate conditions has been demonstrated[12].

The papers presented at the Fourth International Symposium on Relations between Homogeneous and Heterogeneous Catalysis have been published[13].

2 Hydrogenation

Hydrides of the early transition metals are catalysts for the hydrogenation of aromatics. For example, $[ZrH_2(\eta-C_5H_5)_2]_n$ catalyses the reduction of naphthalene to tetralin and dihydronaphthalene[14]. Solid, reduced materials isolated from the reaction of $ZrC\ell_4$ or $HfC\ell_4$ with BuLi and treated with H_2/Et_3N or H_2/Et_2O catalyse the hydrogenation of benzene with up to ~90% initial selectivity to cyclohexene[14]. Hydrogenation of α-terpinene or 2,3-dihydroanisole catalysed by $[Cr(CO)_3(phen)]$ is accelerated by small amounts of a ketone[15]. Aldehydes and ketones can be hydrogenated by $[Fe(CO)_5]$ in tertiary amine solvents, $[HFe(CO)_4]^-$ being the active species[16]. The complex, dichloro (triphenylphosphine)[1-*N,N*-α-dimethylaminoethyl-2-diphenylphosphinoferrocene]ruthenium(II), catalyses the hydrogenation of terminal olefins under mild conditions[17]. Kinetic and spectroscopic data suggest catalysis *via* a Ru(I) complex that dissociates a phosphine ligand to generate the active species. Several complexes of 1-substituted 3,4-dimethylphospholes (RDMP), $[RuC\ell_2(CO)_x(RDMP)_{(3-x)}]$ (R = Me, Bu, Me_3C, Ph, $PhCH_2$), catalyse the hydrogenation of 1-heptene at ambient temperature and pressure at rates comparable to $[RuC\ell_2(PPh_3)_3]$[18]. The hydrogenation of allyl bromide by $[RuC\ell_2(PPh_3)_3]$ is a zero order with respect to substrate and H_2 but 0.85 order in catalyst; an activation energy of 11.13 kcal mol^{-1} was obtained. Cyclooctadiene isomers (1,3- and 1,5-cod) are selectively hydrogenated to cyclooctene in THF at 20°C and 1 atmosphere H_2 pressure by [Ru(1,5-cod)(1,3,5-cot)][20]. Isomerisation of 1,5-cod to 1,3-cod is reported[20]. The co-ordination chemistry and catalytic properties of hydrido(phosphine)ruthenate complexes have been further investigated[21]; in particular the hydrogenation of anthracene which is thought to proceed as in Scheme 1.

Arenes react with the cations $[Rh(MeOH)_2(diphos)]^+$ to form 1:1 adducts $[Rh(arene)(diphos)]^+$. The adducts of anthracene and its derivatives catalyse the hydrogenation of the arene to the

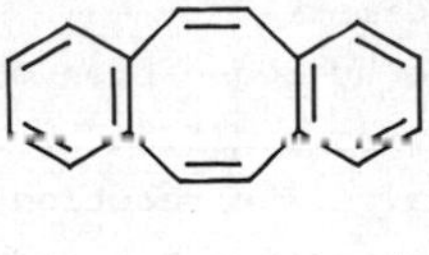

$[RuH(anthracene)(PPh_3)_2]^- \rightleftharpoons [RuH_5(PPh_3)_2]^-$

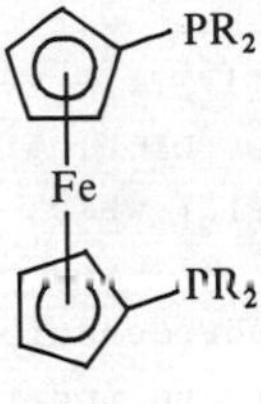

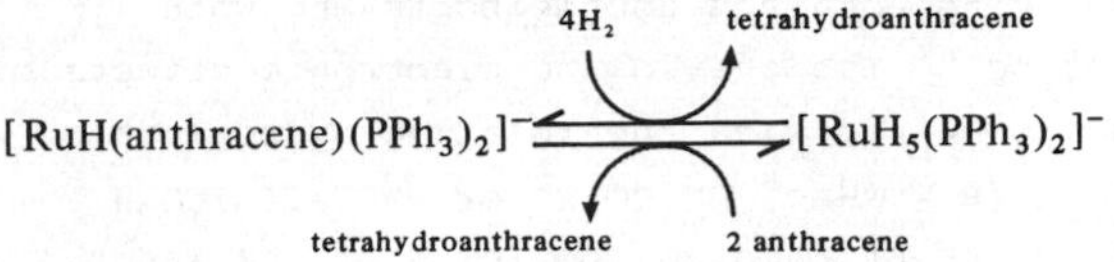

Scheme 1

$\frac{1}{2}[Rh_2Cl_2(nbd)_2] + (amphos)NO_3 \longrightarrow [RhCl(nbd)(amphos)]NO_3$
(3)

$\downarrow$ (amphos)NO$_3$

$[Rh(nbd)(amphos)_2]^{3+}$
(4)

Scheme 2

$[Sm(\mu^5 - C_5Me_5)_2(THF)_2] \xrightarrow[\text{pentane}]{PhC\equiv CPh} [Sm_2(\eta - PhC=CPh)(\mu^5 - C_5Me_5)_4]$ (1)
(5)   (6)

$(6) + H_2 \longrightarrow [Sm_2H_2(\mu^5 - C_5Me_5)_4]$ (2)
(7)

$Ph_2C=CH_2 \xrightarrow[\text{aq, HBF}_4, 4\text{-Me(CH}_2)_{11}C_6H_4SO_3Na]{Co_2(CO)_6(PBu_3)_2, C_6H_6, 55\,^\circ C} Ph_2CHMe$ (3)

corresponding 1,2,3,4-tetrahydroanthracenes under mild conditions (50-75°C, 1 atm H_2)[22].

The activation energy for the hydrogenation of cinnamate ion in aqueous NaCℓ solution with $[CoH(CN)_5]^{3-}$ ions is reported to be 14.5 ± 0.5 kcal mol^{-1} [23]. Reactions of CO with $[RhH(P\text{-}P)_2]$ (P-P = dppe, dppp, or DIOP) give the new dimers $[Rh_2(\mu\text{-}CO)_2(CO)_2(P\text{-}P)_2]$ which catalyse the hydrogenation of terminal olefins[24]. Wilkinson-type complexes modified by Et_3N are catalysts for the hydrogenation of ketones. The reaction is first order in substrate and rhodium concentration but independent of the partial pressure of hydrogen.

The structure of the hydrogenation precursor $[Rh(nbd)L]C\ell O_4$(1) (L = 2, R = Me_3C) shows the rhodium atom in a distorted square planar environment involving both nbd double bonds and both phosphorus atoms chelating. The steric and electronic effects of (1; L = 2, R = Ph) are compared with the relative hydrogenation kinetics of $CH_2=C(R)CO_2H$ (R = NHAc, CH_2CO_2H) or $PhCH=C(R)CO_2H$ (R = NHAc, Me)[26]. A series of rhodium complexes of the water soluble phosphine $Ph_2PCH_2CH_2NMe_3^+$ (amphos)are reported (Scheme 2). Aqueous solutions of (4) catalyse the hydrogenation or hydroformylation of water-immiscible olefins in a 2-phase system as well as hydrogenation of water soluble olefins[27].

Catalytic hydrogenation of methyl linoleate and methyl linolenate by the complexes $[MC\ell_2(PR_3)_2]$ (M = Pd, Pt; R = n-alkyl, *p*-n-alkylphenyl) in the presence of $SnC\ell_2$ proceeds faster for M = Pt when R changes to *p*-n-alkyl but slower for M = Pd under similar conditions[28]. Dichlorobis(2,4,6-trimethylpyridine) platinum(II) is a good catalyst for the hydrogenation of olefins and α,β-unsaturated aldehydes and ketones under mild conditions[29]. The half hydrogenation of $CH_2=C(R)CH=CH_2$ (R = H,Me) and $CH_2=CMeCMe=CH_2$ to the mono-olefins is catalysed by $[PtH(\eta^3\text{-}C_3H_5)(P\{CMe_3\}_3)]$ with 100% selectivity in each case[30]. The heteronuclear clusters $[Pt_2Co_2(CO)_8(PPh_3)_2]$ and $[Pt(C_6H_{11}NC)_2\{Mo(\eta^5\text{-}C_5H_5)(CO)_3\}_2]$ are poor catalysts for the reduction of alkenes, but reasonably good for the hydrogenation of terminal acetylenes[31]. The latter catalyst can be recovered unaltered from the reaction mixture, unlike the former which undergoes complex rearrangements.

A well characterised series of low valent samarium complexes (5-7) initiate the catalytic hydrogenation of alkynes[32].

Lanthanide metal vapours (Er, Lu, Sm, Yb, Nd) cocondense with RC≡CR (R = Et, Me_3Si) or PPh_3. The products hydrogenate 3-hexyne at room temperature and 1 atm H_2 giving 75-99% selectivity to cis-3-hexene at 35-100% conversion[33]. Activated olefins are hydrogenated in 95-99% yield on treatment with $[Co_2(CO)_8]$ or $[Co_2(CO)_6(PBu_3)_2]$ under acidic phase transfer conditions (Reaction 3)[34a].

Under buffered phase-transfer conditions, stereospecific reduction of naphthalene and p-methylanisole are catalysed by $[Rh_2Cl_2(1,5\text{-hexadiene})_2]$[34b]. $[RuCl_2(PPh_3)_3]$ catalyses the phase-transfer reduction of RNO_2 (R = p-MeC_6H_4, m-MeC_6H_4, p-$PhCH{=}CH.C_6H_4$, $C_{10}H_7$, p-$PhOC_6H_4$, p-$MeOC_6H_4$, p-$C\ell C_6H_4$) by CO/H_2 (1:1) at room temperature and 1 atm to give 20-71% RNH_2[35]. The reaction did not occur with H_2 alone.

2.1 Asymmetric Hydrogenation.- The synthesis of α-amino acids by asymmetric hydrogenation has been discussed[36]. The high optical yields in asymmetric hydrogenation of prochiral alkenes catalysed by chiral metal complexes is explained by higher reactivity of the minor diastereoisomer of the catalyst-substrate adduct as is the temperature dependence of enantioselection[37].

Hydrogenation of terpinen-4-ol is catalysed by $[Ir(cod)(P\{Cy\}_3)_3(py)]PF_6$(8) with about 100:1 preference for H_2 addition to the face bearing the hydroxyl group. Chelation of the alcohol occurs as shown since a catalyst-substrate complex (9) can be detected by nmr which is related to the proposed intermediate[38].

Similarly, hydrogenation of indenones (10) (R = CO, β-hydroxyl) catalysed by (8) gives exclusively cis-indanone (11) but for (R = OH), a 96% yield of trans-indanone is produced[39]. Hydroxyl-directed hydrogenation may be a reaction of considerable generality[39].

The hydrogenation of cycloalkenones (12) (n = 0,1; R = H, Me) with $[Co_2(CO)_6(PPh_2R)_2]$ (R = neomenthyl) as catalyst gives optically active cyclic ketones with optical yields of 1-16%[40].

A new route to chiral bis-tertiary phosphine ligands has appeared[41], which facilitates synthesis of (-)-(13). $[Rh_2C\ell_2L_2]$ (L = 1,4-cod) with (-)-(13) catalyses hydrogenation of methyl(N-benzoyl)- and (N-benzoyl)dehydrophenylalanine to products of 100% optical purity[41]. X-ray structural and kinetic studies of the

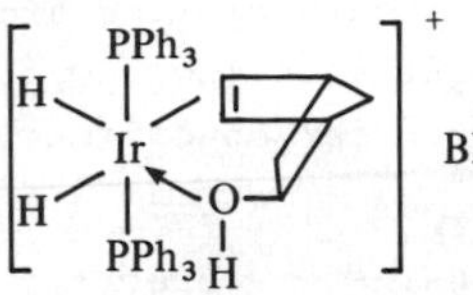

(9)

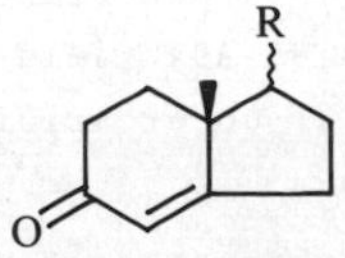

(10)

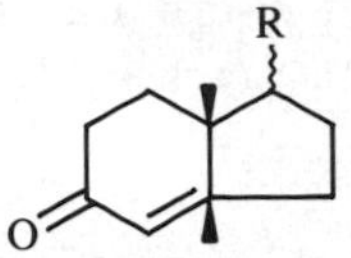

(11)

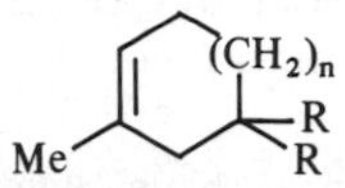

(12)

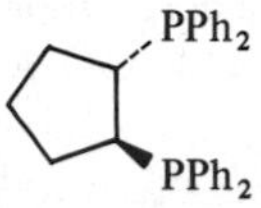

(13)

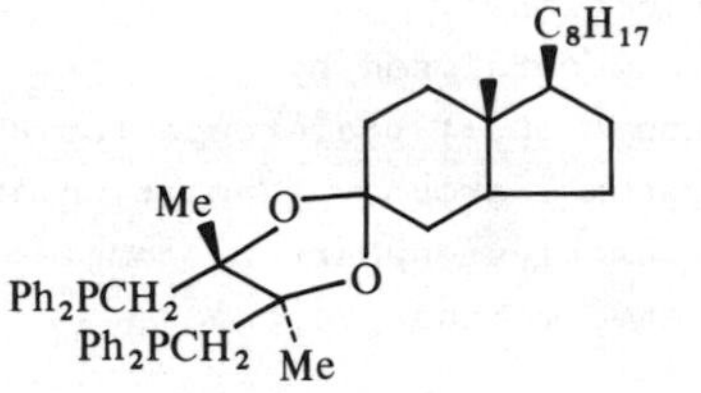

(14)

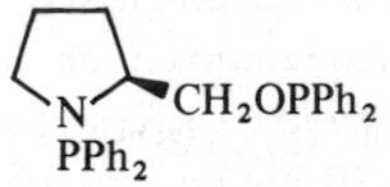

(15)

(S)–$Ph_2PNEtCHEtCH_2OPPh_2$

(16)

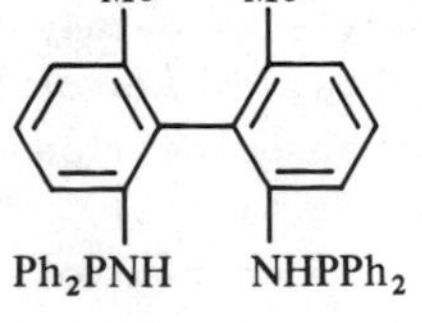

(17)

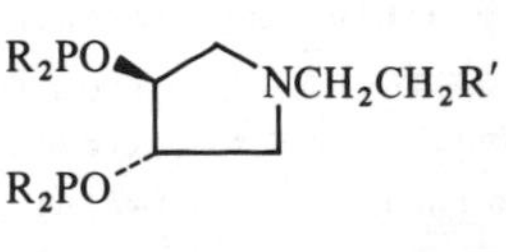

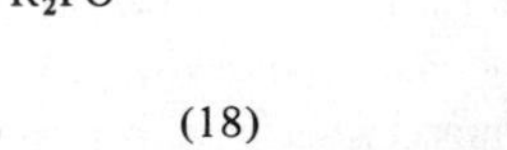

(18)

rhodium (I) complex with (R)-eycpos[(R)-1,2-bis(diphenylphosphino-1-cyclohexylethane)] reveal high optical yields can be achieved at good reaction rates[42]. Optical yields of up to 93% are obtained in hydrogenations catalysed by Rh complexes of the steroidal phosphine(14)[43].

Hydrogenation of $RR'C=C(NHR^2)CO_2H$ (R,R' = H, Me, Ph; R^2 = Ac, PhCO) in the presence of $[RhL(cod)]ClO_4$ (17; L = 15, 16) gives (S)-$RR'C=C(NHR^2)CO_2H$ with 45-96% and 23-64% optical yields respectively[44]. The rhodium complex of (17), a new atropisomeric chiral bis-phosphine, is highly effective for the asymmetric hydrogenation of 2-acetamidoacrylic acid under mild conditions[45] Rh(I) complexes of the new chiral diphosphinites (18; R = Ph, p-anisyl; R' = Me_2CH, Me_2N, Me_2NCH_2) are very effective for the hydrogenation of dehydrodipeptides[46]; an electrostatic effect may contribute to the asymmetric induction.

Synthesis of structural analogues of DIOXOP and the use of their rhodium complexes for various hydrogenations have been described[47]. Chelate ligands with lateral asymmetry at N atoms form complexes such as (19) where the asymmetric centres interact with the co-ordination sites and convert prochiral substrates into optically active products[40].

The hydrogenation kinetics of 1-hexene and (*Z*)-$PhCH=C(CO_2Me)NHAc$ using catalysts derived from $[Rh(nbd)LL']^+$ (L,L' = PPh_3; L,L' = $Ph_2P(CH_2)_nPPh_2$, {4R,5R-bis(diphenylphosphinomethyl)-2,2-dimethyldioxolane; *n* = 3-5} have been determined[44]. The rate of hydrogenation increases as the diphosphine chelate ring size increases. The dihydride $[RhH_2(O_2COH)(P\{i\text{-}Pr\}_3)_2]$ catalyses the stereoselective half-hydrogenation of alkynes to *trans*-alkenes[50], as does the dinuclear hydride $[Rh_2(\mu\text{-}H)_2(P\{OCHMe_2\}_3)_4]$[51]. Various prochiral olefins are reduced by $[Rh_6(CO)_{10}((-)DIOP)_3]$ with optical yields up to 47%, although the active species may be mono-nuclear[52].

An nmr study suggests that asymmetric hydrogenation of dehydro-amino acids involves rhodium enamide complexes which interconvert by an intra-molecular mechanism requiring an intermediate from which the alkene has dissociated[53].

2.2 Heterogenised Catalysts.- Well-dispersed, highly active catalysts for alkene hydrogenation can be prepared by impregnation of $[M(CO)_6]$ (M = Cr, Mo or W) on suitable supports followed by careful activation[54]. A highly dispersed rhodium catalyst active for toluene hydrogenation is obtained by treatment of silica-supported diallylrhodium with H_2[55]. Intercalated clay catalysts, formed from Na^+-hectorite with $[Rh(nbd)LL']^+$ (L = L' = PPh_2; LL' = dppe), exhibit a lower tendency to isomerise 1-hexene under hydrogenation conditions when compared with homogeneous analogues[56].

Polymer supported $[RhL(CO)_2]$ (L = polymer functionalised with pentane-2,4-dionate or ethyl dithiocarbamate) and $[RhCl(CO)_2L]$ (L = polymeric resin functionalised with ethylamine or diphenylphosphine) respond differently to thermal or photochemical treatments as well as in their reactions with NO and H_2. Catalytic activities of these complexes are determined by the ease of CO loss from the supported species - cyclohexene hydrogenation occurs rapidly[57]. Ethylene hydrogenation is catalysed by materials obtained from dispersed rhodium/polyphosphine complexes[58], a tetraruthenium cluster anchored to silica via a phosphine ligand[59] and by $[H_2RuOs_3(CO)_{13}]$ adsorbed on γ-Al_2O_3[60].

3 Hydrogen-Transfer and Dehydrogenation Reactions

The reduction of trichloromethyl compounds by hydrogen donors catalysed by transition metal carbonyls, their complexes, or their salts has been reviewed[61]. Allylic alcohols undergo hydrogen transfer reactions in the presence of catalytic amounts of trans-$[Mo(N_2)_2(dppe)_2]$[62]. Cyclohexane, -heptane and -octane are selectively dehydrogenated to the corresponding cycloalkenes by $[ReH_7(PR_3)_2]$ (R = C_6H_4F-p, Ph, C_6H_4Me-p) catalysts and $Me_3CCH{=}CH_2$ as acceptor[63]. A mechanistic study of hydrogen transfer from 1-phenylethanol to cyclohexanone catalysed by $[RuCl_2(PPh_3)_3]$ with a rate which depends upon the order of addition of the reactants[64]. $[Co_2(CO)_8]$ and $[Co_4(CO)_{12}]$ catalyse hydrogen transfer from 9,10-dihydroanthracene or 9-fluorenol to $Ph_2C{=}CH_2$ possibly via a radical pathway[65]. A comprehensive examination of transfer hydrogenation of α,β-unsaturated ketones by alcohols in the presence of $[RhH(PPh_3)_4]$ shows that cleavage of the O-H bond is the rate determining step and the order of reagent addition is critical[66]. $[Rh_2Cl_2(cot)_4]$ and $[M_2Cl_2(cod)_2]$ (M = Rh, Ir) in the presence of tertiary phosphines, catalyse hydrogen transfer from propan-2-ol to 4-t-butylcyclohexanone[67].

A very active catalyst for dehydrogenation of propan-2-ol is formed from $[Rh_2(OAc)_4]/PPh_3$ although $[Rh_2(OAc)_4]$ itself is inactive[68]. Aromatic nitro compounds can be reduced to amines by selective hydrogen transfer from propan-2-ol catalysed by $[IrC\ell(CH_2=CH_2)(LL)]$ and $[Ir(1,5\text{-hexadiene})]C\ell O_4$ (LL = bipy, phen, etc) at rates greater than those reported for this reaction when catalysed by Ru and Rh phosphine complexes[69].

The system formed in situ from $[Rh_2C\ell_2L_2]$ (L = 1,5-hexadiene) and (-)-2-pyridinylphenylethylimine catalyses the asymmetric hydrogen transfer from propan-2-ol to several ketones but rate and enantiomeric excess (<23%) are decreased by steric hindrance[70]. A mass spectrometric characterisation of the active sites of polymer-bound iridium complexes in hydrogen transfer reactions reveals a possible reaction scheme involving both π-alkene and σ-alkyl complexes[71].

4 Hydroformylation, Hydrocarbonylation and Carbonylation Reactions

4.1 Hydroformylation.- Infra-red evidence is presented that photolysis of $[MH(\eta^5\text{-}C_5H_5)(CO)_3]$ (M = Mo, W) complexes at high dilution in 5% C_2H_4-doped CH_4 matrices at 12 K gives the 16 electron species $[MH(\eta^5\text{-}C_5H_5)(CO)_2]$ and the 18 electron species cis- and trans-$[MH(\eta^5\text{-}C_5H_5)(C_2H_4)(CO)_2]$. Further photolysis produces the 16 electron insertion product $[M(Et)(\eta^5\text{-}C_5H_5)(CO)_2]$[72]. The rhodium catalyst for the hydroformylation of formaldehyde to glycolaldehyde is stabilised by excess phosphine and its activity is enhanced by small amounts of amine[73]. Reaction of $BuCH=CH_2$ with paraformaldehyde in THF at 120°C in the presence of $[RhH_2(O_2COH)(P\{CHMe_2\}_3)_2]$ gives $C_6H_{13}CHO$, $C_6H_{13}CH_2OH$, $C_6H_{13}CO_2Me$ and C_6H_{14}, predominantly as the linear isomers[74]. Hydroformylation of pent-1-ene-5-d_3 in the presence of $[Ru_3(CO)_{12}]$ or $[H_4Ru_4(CO)_8\{(-)\text{-DIOP}\}_2]$ under 7-50 atmospheres pressure of CO gives a high percentage of the n-isomer with complete deuterium retention only under high pressures[75]. The oligomeric complex $[Co(\mu\text{-}PPh_2)(CO)_3]_x$ is a catalyst precursor for the hydroformylation of 1-hexene, but is transformed irreversibly into $[Co_2(CO)_6(PPh_2\{C_6H_{13}\text{-}n\})_2]$[76]. Hydroformylation of acrylonitrile using $[Co_2(CO)_8]/L$ (L = $HN(CH_2CN)_2$, PCy_3, NMP, $Me_2N(CH_2)_2NHMe$, PPh_3) in methanol follows four reaction pathways depending upon L; the more basic amines producing ~50% yields of $NCCHMe(CH_2)CN$ / $NC(CH_2)_4CN$ in 10:1 ratio and a 30% yield of CH_3CH_2CN[77]. Partial decomposition of polymerised alkylidyne tricobalt nonacarbonyl complexes (20) occurs under hydroformylation conditions; mono- and/or dinuclear species are thought to be the catalysts[78], contrary to the previous interpretation.

Reactions of $[Co_2(CO)_8]$ with ferrocenylphosphine oligomers (21a; n = 1-4) and (21b; n = 1,2) produce catalysts for 1-hexene hydroformylation which are similar in performance to Co/PPh_3 systems[79]. The polymers 21c (mol wt 8900-161000 amu) have a mol wt dependent activity and a lower selectivity to alcohols than the above[79]. Resins containing the units $\{Co(CO)_3(PPh_2C_6H_4)\}_2$ and $\{Co(CO)_3(PPh_2CH_2C_6H_4)\}_2$ give similar selectivities in 2-pentene hydroformylations as their homogeneous analogues unlike resins containing a diphosphine unit which are more selective for n-hexanal[80]. A comparison of the hydroformylation of 1,1-diphenylethene catalysed by $[Co_2(CO)_8]$ and $[Rh_2C\ell_2(CO)_4]$ concludes that the former is very likely a free radial reaction whilst the latter follows a conventional pathway[81]. C_2H_4 hydroformylation with equimolar H_2/D_2 mixtures and $[Rh_4(CO)_{12}]$ or $[Co_2(CO)_8]$ as catalyst precursor gives ~50% EtCDO having d_0/d_2 ratios of ~3 and ~2.6 respectively[82]. Rhodium/$P(aryl)_3$ complexes catalyse the low pressure production of α- and β-carbonyl-propionaldehydes (predominantly branched) from vinyl esters[83]. $H_2C{=}CHCH_2CR_2CH_2CH{=}CHR'$ (R = H,R' = Me, CH_2CO_2Et; R = COMe, R' = Me) are hydroformylated specifically at the terminal double bond with good selectivity to the n-aldehyde using $[RhH(CO)(PPh_3)_3]$ (22) as catalyst and CO/H_2 = 1:2[84]. It is claimed that more thermally stable rhodium catalysts result from the addition of phosphinous acids to conventional Rh/PPh_3 systems via formation of the new species $[Rh_2(CO)_2(PPh_3)_2(R_2PO)_2]$[85]. A comparative study of 1-hexene hydroformylation at 80°C and 5 atm CO/H_2 reveals that $[Rh_2(\mu\text{-}SCMe_3)_2(CO)_2(P\{OMe\}_3)_2]$ is a more efficient catalyst (after a short induction period) than (22) or $[Rh_2(\mu\text{-}SCMe_3)_2(P\{OMe\}_3)_4]$[86]. The photochemical hydroformylation reaction of 1-octene catalysed by $RhC\ell_3$/norbornadiene proceeds at room temperature[87a], and by Co/phosphine complexes in MeOH at 80°C to give >90% selectivity for n-aldehydes[87b].

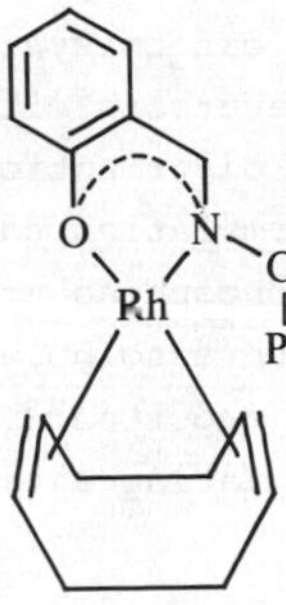

(19)

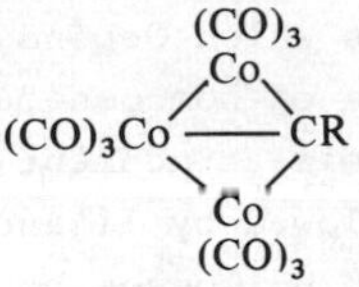

(20) R = $CO_2CH_2CH_2O_2CCMe{=}CH_2$, $C_6H_4COCMe{=}CH_2$-*p*, $C_6H_4CH{=}CH_2$-*p*

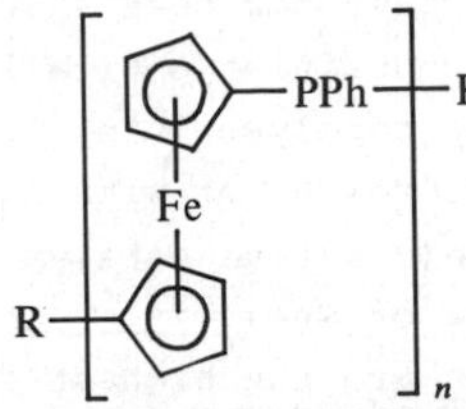

(21) a; R = H, R′ = Ph
b; R = Ph_2P, R′ = Ph
c; R = H, R′ = OH

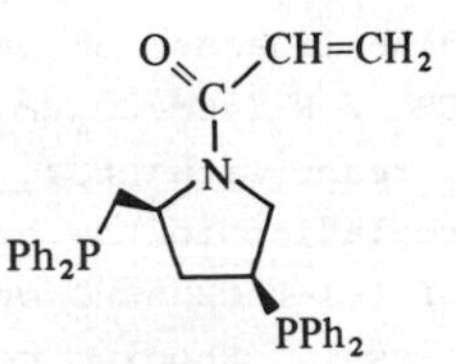

(24)

$$RCH_2X + CO(1\ atm) \xrightarrow[1M\ aq \cdot NaOH—Bu_4\overset{+}{N} \cdot \overset{-}{H}SO_4\ +\ Fe(CO)_5]{CH_2Cl_2,\ C_6H_6\ or\ PhMe} RCH_2CO_2H \qquad (4)$$

(R = Ph, 3-MeC_6H_4, 2-MeC_6H_4, 3-NCC_6H_4, 4-BrC_6H_4, *β*-naphthyl, 3-$MeOC_6H_4$, 2,4,6-$Me_3C_6H_2$; X = Cl, Br)

$$RC{\equiv}CH + CO(1\ atm) + MeI \xrightarrow[5M\ aq \cdot NaOH—Me(CH_2)_{11}\overset{+}{N}MeCl^-]{C_6H_6,\ Co_2(CO)_8,\ Ru_3(CO)_{12}}$$

$$RCH(CO_2H)CH_2COMe \qquad (5)$$

(R = Ph, 4-MeC_6H_4, 4-$MeOC_6H_4$, 4-$EtOC_6H_4$, cyclohexyl)

The silylated diphosphines $\underline{p}$-(EtO)$_{\underline{n}}$SiMe$_{3-\underline{n}}$C$_6$H$_4$OCH$_2$CH(PPh$_2$)CH$_2$PPh$_2$ (23) can be attached to a silica gel surface in a random fashion. Rhodium complexes of (23) both unsupported and supported on silica can be synthesised, the latter being hydroformylation catalyst precursors with specific activities which depend upon rhodium distribution[88]. Insoluble analogues of homogeneous Rh hydroformylation catalysts are obtained by ionic attachment of an amino-phosphine to anion exchange resin followed by ligand exchange with a soluble metal complex. Low rhodium losses by leaching are reported[89]. Further studies of the Pt(II)/SnCℓ_2 hydroformylation catalyst system are reported[90,91].

4.1.1 Asymmetric Hydroformylation. Previously estimated optical yields in the asymmetric hydroformylation of styrene over chiral Pt catalysts reported to be up to 95%[92] have now been shown to be considerably lower (73%)[93]. Polymers containing optically active chelating phosphine ligands, prepared by copolymerising (24) with HOCH$_2$CH$_2$O$_2$CCMe=CH$_2$ / H$_2$C=CMeCO$_2$CH$_2$CH$_2$O$_2$CCMe=CH$_2$ or with H$_2$C=CHPh / H$_2$C=CHC$_6$H$_4$CH=CH$_2$ react with PtCℓ_2 to yield active catalysts for styrene hydroformylation in the presence of SnCℓ_2 [94]. The 20% optical yields of (+)-PhCHMeCHO obtained are the highest claimed with such a catalyst. Rhodium complex catalysts of the steroidal phosphine (14) produce optical yields of <34%[43].

4.2 Hydrocarbonylation.- The catalytic hydrocarbonylation of alcohols (and in particular methanol) has been reviewed[95]. Stoichiometric hydrogenation of [Co(MeCO)(CO)$_3$(PMePh)$_2$] in methanol yields similar products to those observed during methanol hydrocarbonylation catalysed by [Co(CO)$_3$(PMePh$_2$)]$_2$, suggesting the intermediacy of an acetyl group[96]. Further studies of the cobalt catalysed homologation reaction of methanol have appeared covering such aspects as H/D exchange[97], intermediacy of MeI[97,98], reaction kinetics[99], effect of reaction variables on product distribution[100] and added ruthenium[101]. The Mn, Rh and Ru systems which catalyse methanol hydrocarbonylation in the presence of amines at 300 atm CO/H$_2$ and 200°C are described in greater detail[102].

Homologation of dimethyl ether and methyl esters to ethyl esters is more effectively catalysed by mixtures of Co(OAc)$_2$ and Ru(acac)$_3$ than by the individual metals[103a, b]. Iodide promotes the Ru catalysed reaction and a cation effect is observed[104].

4.3 Carbonylation.- Acetic anhydride synthesis via methyl acetate carbonylation with Group VIII compounds has been surveyed and possible co-catalysts and promoters examined[105]. Phase transfer systems are efficacious for the carbonylation of benzylic halides catalysed by $[Fe(CO)_5]$, reaction (4)[106], $[Pd(dba)_2]$[107] and $[Co_2(CO)_8]$[107].

The kinetics and mechanism of phase-transfer catalysed carbonylation of benzyl bromide using $[Co(CO)_4]^-$ are reported and cleavage of the acylcobalt intermediate by $Bu_4N^+OH^-$ at the liquid-liquid interface is rate determining[108]. New syntheses of benzolactams, benzolactones[109a] and carboxylic acids[109b] are achieved via 'sunlamp' irradiated, cobalt catalysed, carbonylations of aryl or vinyl halides. Cobalt catalysed carbonylation of optically active α-phenylethyl halides under phase transfer conditions gives hydratropic acid of the opposite absolute configuration[110]. A new heterogenised catalyst comprising $[Co(CO)_4]^-$ supported on anionic exchange resins is active for the carbonylation of organic halides[111].

Propargyl amines, $HC{\equiv}CCH_2NR_2$, (NR_2 = NEt_2, morpholino) are carbonylated in the presence of $Co_2(CO)_8$ to the corresponding citraconic acid diamides $R_2NCOCMe{=}CHCONR_2$[112]. Mono-substituted acetylenes undergo a bimetallic phase transfer catalysed carbonylation, reaction (5)[113a]. A related reaction yields allylic amides from azadienes[113b].

Hydro-esterification of acrylonitrile with CO and MeOH in the presence of H_2 is catalysed by novel base-promoted Co catalysts which give high yields of $NCCH_2CH_2CO_2Me$ and $NCCHMeCO_2Me$[114]. The reaction of $RCONH_2$ (R = C_{1-17} alkyl, Ph) with HCHO, CO and $[Co_2(CO)_8]$ produces $RCON(CH_2CO_2H)_2$ and some $RCONH(CH_2CO_2H)$[115].

Aromatic azides, p-$RC_6H_4N_3$ (R = H, NO_2) are carbonylated at 1 atm to the corresponding isocyanates in the presence of $[Rh(dppe)_2]Cl$, $[Rh(dpp)_2]Cl$, $[RhCl(CO)(PPh_3)_2]$ and $[Rh_2Cl_2(CO)_2(dpp)_2]$[116]. $RhCl_3.3H_2O$ catalyses both carbonylation of azobenzenes and nitro compounds to give isocyanates; pyridine, compounds of V, Mo and Fe inhibit the former reaction but promote the latter[117]. Carbonylation of diphenyl acetylene catalysed by $[Rh_4(CO)_{12}]/Na_2CO_3$ gives 5-ethoxy-3,4-diphenyl-2(5H)-furanone in 87% yield[118].

Treating C_2H_2 with $[NiBr_2(Et_2NH)_2]$ and CO gives products which retain the triple bond, viz, $HC{\equiv}CCONEt_2$ and $Et_2NCOC{\equiv}CCONEt_2$[119].

Unsymmetrical ketones RCOR' (R = Ph, substituted Ph; R' = alkyl, cyclohexyl, CH_2Ph, $4\text{-}C\ell C_6H_4CH_2$) are prepared selectively by the reaction of RI with R'I or $R^2CH_2C\ell$ in the presence of a stoichiometric amount of Zn-Cu and a catalytic amount of $[Pd(PPh_3)_4]$ under 1 atm of CO[120]. Insertion of CO into vinyl halides bearing a secondary amine or alcohol catalysed by $Pd(OAc)_2/PPh_3$ produces five-, six- and seven membered lactams and lactones with α-methylene groups[121]. Pd-catalysed carbonylation of $H_2C{=}C(CF_3)Br$ (25) with RNHCONHR' (26; R = R' = H, Me) gives yields of 26% and 70% respectively of the dihydrouracils (27) whereas (25) with (26) (R = Me, R' = H) gives 51% (27) (R = Me, R' = H) and 9% (27) (R = H, R' = Me)[122].

Vinylic dibromides react with CO, Pd(O) compounds and a phase-transfer agent in a 2-phase system (C_6H_6, 5N NaOH) to give diynes in reasonable yields[123]. Under similar conditions, vinylic dibromides derived from aliphatic aldehydes or ketones, produced α, β-unsaturated monoacids, whilst in tert-amyl alcohol, dicarbonylation of vinyl dibromides to diacids occurred[123]. Reaction of tetralkyl-substituted dipropargyl amines with CO and H_2O or alcohols under ambient conditions with $PdX_2/(H_2N)_2C{=}S$ (X = $C\ell$, Br, I) as catalyst leads to pyrrolidone derivatives (28)[124]. Aliphatic and aromatic allylic carbonates undergo facile decarboxylation-carbonylation on treatment with 10 atm CO using $Pd(OAc)_2/2PPh_3$ as catalyst and yield β,γ-unsaturated esters in reasonable yields[125]. Treatment of cis-2-decene with O_2 and CO in THF containing $PdC\ell_2/CuC\ell_2$/aq $HC\ell$ gives the hydrocarboxylated product $Me(CH_2)_7CHMeCO_2H$ in 59% yield[126]. The catalytic synthesis of oxalate esters by the treatment of alcohols with O_2 and CO in the presence of palladium catalysts is significantly improved by the inclusion of small amounts of 1,4-benzoquinone[127]. The halides $MC\ell_3.3H_2O$ (M = Rh, Ir, Ru) promote the Co catalysed carbonylation of methanol to acetaldehyde[128].

$RhC\ell_3.3H_2O$ stoichiometrically liberates HI when reacted in the presence of LiI and CO which in turn generates MeI from MeOH. A novel carboxymethylation of styrene is catalysed with marked asymmetric induction by Pd/CF_3CO_2H[129a]. Hydroesterification and

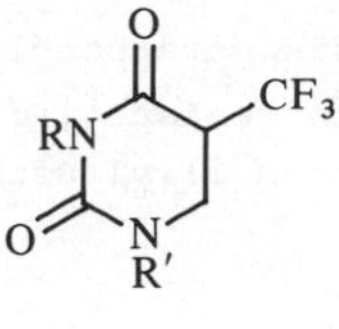

(27)

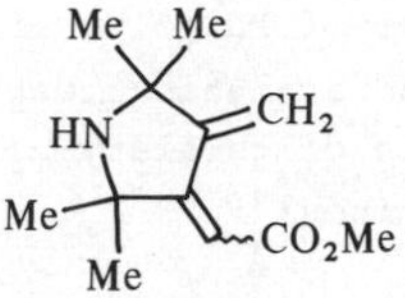

(28)

$$\mathrm{MeOH + CO_2 + H_2 \xrightarrow[\text{[HRu}_3\text{(CO)}_{11}]^-]{\text{33 atm, 125°C}} HCO_2Me} \qquad (6)$$

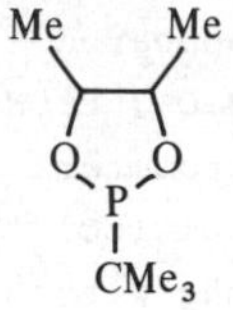

(29)

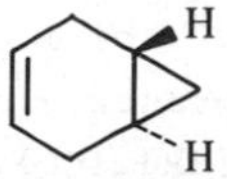

(31)

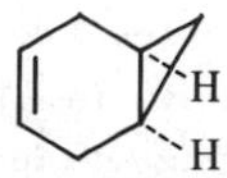

(32)

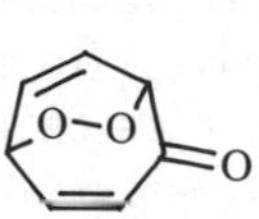

(34)

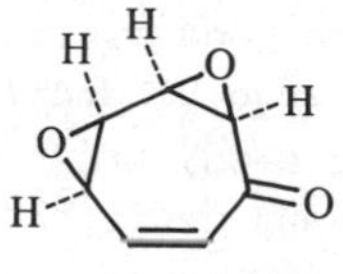

(35)

(36)

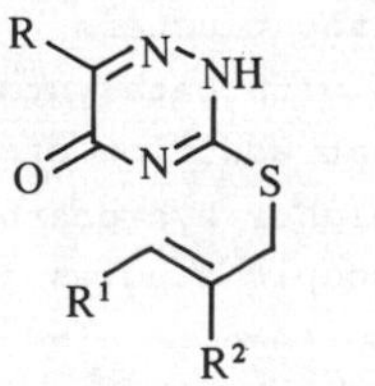

(37)

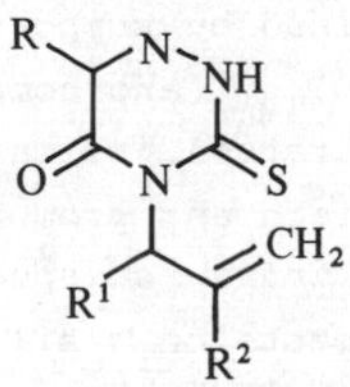

(38)

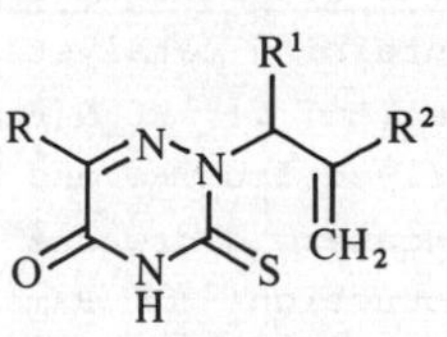

(39)

hydrocarboxylation of $CF_3CH=CH_2$ occur with high regiospecificity using $SnCl_2$/[$PdCl_2$(dppf)] (dppf = 1,1'-bis(diphenylphosphino) ferrocene)[129b].

A series of Pd/polystyrene catalysts decarbonylate simple aliphatic and aromatic aldehydes and can replace [$Pd(PPh_3)_4$] for the formation of tertiary allylamines from allyl esters and secondary amines[130].

5 Carbon Monoxide Reduction, The Water Gas Shift Reaction and Reactions of Carbon Dioxide

5.1 Carbon Monoxide Reduction.- A book based on 270 patents issued between 1977 and 1981 in the C_1 chemistry area has been published. Two further volumes in a series describing catalysis by metal complexes cover catalytic reactions of synthesis gas[132a,b]. A review of the factors which determine selectivity in homogeneous hydrogenation of CO to oxygenated products has appeared[133].

A detailed account of ruthenium catalysed MeOH, EtOH and $HOCH_2CH_2OH$ synthesis in the presence of iodide promoters presents spectroscopic evidence that $[HRu_3(CO)_{11}]^-$ and $[RuI_3(CO)_3]^-$ are essential for optimum activity[134]. EtO_2CEt along with byproducts RO_2CEt (R = Me, Pr, Bu) are produced by reaction of CO/H_2 with $EtCO_2H$ in the presence of RuO_2, [$Ru_3(CO)_{12}$] or [$Ru(acac)_3$] and quaternary ammonium or preferably phosphonium salts[135]. (Pyrolysis of EtO_2CEt gives good yields of ethylene.)[135] Equimolar mixtures of [$Ru(acac)_3$] and [$Rh(acac)_3$] in molten Bu_4PBr catalyse the production of MeOH, EtOH and $HOCH_2CH_2OR$ (R = H, Me, Et) from CO/H_2 at 430 atm and 220°C[136a]. C_2-oxygenate esters ROAc (R=Me, Et), $AcOCH_2CH_2OH$ and in particular $(AcOCH_2)_2$ are obtained in good selectivity using a mixture of Ru and Rh catalysts (10:1) in AcOH with Et_3N or Ca^{2+} as promoters[136b].

5.1.1 Heterogenised Catalysts. Studies of CO hydrogenation by Fe containing catalysts obtained by supporting the clusters [$Fe_3(CO)_{12}$], $Et_4N[Fe_2Mn(CO)_{12}]$ (and compared with catalysts derived from Fe and Mn nitrates) indicate that addition of manganese decreases CH_4 yield and promotes higher hydrocarbon production[137]. Raman and infra-red spectroscopic studies indicate that [$H_2FeOs_3(CO)_{13}$] fragments on a silica surface to give Os hydrido-carbonyl species[138]. The resulting catalyst exhibits Fischer-Tropsch selectivity and activity intermediate between those of $[Os_3(CO)_{12}]$ and $[Fe_3(CO)_{12}]$[128].

Treatment of $R(CH_2)_3NH_2$ (R = silica gel) with $[HFeCo_3(CO)_{12}]$ in Me_2CO reproducibly gives $R(CH_2)_3NH_3[FeCo_3(CO)_{12}]$ which after decarbonylation-activation is a highly active Fischer-Tropsch catalyst[139]. $[Ru_3(CO)_{12}]$ and $[H_4Ru_4(CO)_{12}]$ supported on Al_2O_3 and MgO show differences in catalytic behaviour which are attributed to surface organometallic chemistry; in particular $[H_3Ru_4(CO)_{12}]^-$ on the support surface reacts with synthesis gas to give $[Ru_6C(CO)_{16}]^{2-}$[140].

5.2 The Water Gas Shift Reaction and its Applications.- The cationic compounds $[Rh(diene)L_2]^+$ (diene = cod or nbd; L_2 = mono- or bidentate P ligand) and Rh or Ir carbonyls catalyse the WGS reaction and the selective reduction of benzalacetone to $PhCH_2CH_2COMe$ with CO/H_2O[141]. A direct route to $HO(CH_2)_4OH$ from allyl alcohol is catalysed by $[Rh_6(CO)_{16}]/(Me_2NCH_2)_2CH_2$ under WGS conditions[142]. Replacing the diamine by aminopyridine gives good yields of butyrolacetone[142]. $[Rh_6(CO)_{16}]$ with 3,4,7,8-Me_4phen is a catalyst precursor for the reduction of nitrobenzene[143]. Pyridine undergoes hydrodenitrification with CO/H_2O to give $CH_2(CH_2CH_2R)_2$ (R = piperidino) as the main product[144].

5.3 Reactions of Carbon Dioxide.- A book describes the organic and bio-organic chemistry of CO_2 with some reference to organometallic systems[145]. A review of metal carbonato and carbon dioxide complexes covers references from 1969 to 1982 with particular mention of fixation and insertion reactions[146].

Anionic ruthenium clusters, $[HRu_3(CO)_{11}]^-$ and its CO_2 insertion product $[HCO_2Ru_3(CO)_{10}]^-$ are catalyst precursors for hydrogenation of CO_2 to methyl formate (reaction 6); $[H_3Ru_4(CO)_{12}]^-$ being recovered at the end of the reaction[147].

Pd(O) - catalysed dimerisation reaction of isoprene with H_2O to give dimethyloctadienols is enhanced by the presence of CO_2 and a base[148]. The selective synthesis of the δ-lactone 2-ethylidene-6-hepten-5-olide and of the γ-lactone 2-ethyl-2, 4-heptadien-4-olide occur via Pd-catalysed reaction of CO_2 and butadiene[149]. Visible light irradiation of CO_2 solutions in

$(HOCH_2CH_2)_3N$/DMF containing respectively $[Ru(bipy)_3]^{2+}/[Co(bipy)_3]^{2+}$ or the single component $[ReX(bipy)(CO)_3]$ (X = Cℓ, Br) produce a mixture of CO + H_2 simultaneously or CO[150].

6 Alkene Isomerisation and Skeletal Rearrangements

Isomerisation of *cis*-1,2-divinylcyclobutane to 4-vinylcyclohexene takes place in the presence of Ni catalysts modified with optically active 1,3,2-dioxaphospholanes, *eg* (29) with maximum asymmetric induction of 22%[151], cf cyclodimerisation of butadiene over the same catalyst. Cationic Pt(II) complexes $[Pt(\eta^3\text{-}CH_2CMeCH_2)(CH_2{=}CHR)(PPh_3)]C\ell O_4$ (30, R = CH_2Ph, Et, Me) show high catalytic activity in double bond migration reactions under mild conditions[152]. The unprecedented isomerisation of *trans*-cyclopropane (31) to *cis*-cyclopropane (32) is catalysed by Rh and Pt compounds[153].

trans-$RC_6H_4CH{=}CHCO_2H$ (33, R = H, 4-Cℓ, 3-Br, 4-Me, 4-CF_3, 2-Me, 4-MeO) are isomerised to *cis*-(33) in 28-47% yield in a Pd(O) phase transfer catalysed reaction under carbon monoxide[154]. The phosphinidene stabilised cluster $[(\mu\text{-}H)_2Ru_3(CO)_9\{\mu_3\text{-}PCH_2CH_2Si(OEt)_3\}]$ supported on SiO_2, Al_2O_3, TiO_2, MgO and ZnO catalyses 1-pentene isomerisation[155]. The reactivity of α-olefins toward isomerisation by a $[Ni(CO)_2(PPh_3)_2]/Al_2O_3$ catalyst decreases in the order 1-hexene $\sim$4-methyl-1-pentene 1-heptene 2-methyl-1-pentene[156]. $[Os_3(CO)_{12}]$ supported on SiO_2 remains intact whilst catalysing 1-butene isomerisation[157], but on γ-Al_2O_3, break-up to Os(II) carbonyl complexes occurs during 1-hexene isomerisation[158].

Co *meso*-tetraphenylprophyrin catalyses the rearrangement of cycloheptene 1,4-endoperoxides to diepoxides in moderate to quantitative yields[159], *eg* tropone *endo*peroxide (34) gave 40% diepoxide (35).

New evidence is presented which suggests that rearrangement of *endo*-tricyclo[3.2.1.0^{2,4}]oct-6-ene to (36) catalysed by Ir(II) complexes proceeds via a carbocyclic mechanism[160]. The Cope rearrangment of acyclic 1,5-dienes having an elctron-withdrawing group at C-3 and an alkyl group at C-2 is catalysed by $PdC\ell_2$ and can be used for γ-allylation of vinyl esters or acids[161]. Pd(II) salts also catalyse the S→N allylic transposition of 3-(allylthio)-1,2,4-triazin-5(2H)-ones (37; R = H, Me, Ph; R' = H, Me; R^2 = H, Me) to (38) and (39) (same R, R^1, R^2)[162].

The rearrangements of diademane to triquinacene and of diademane to snoutene (39) are catalysed by Cu or Ag and Au or Rh compounds respectively[163].

The cyclic allene (40) rearranges in the presence of various catalysts (eg, AgOAc, $HgSO_4$, $ZnBF_4$) in AcOH to give (41-43)[164]. $[TiC\ell(\eta^5-C_5H_5)_2]$ and $[TiC\ell_2(\eta^5-C_5H_5)_2]$ in conjunction with $Me_2CHMgBr$ catalyse the cyclisation of hexa-1,5-diene to (44) and (45) and isomerisation to several linear hexadienes[165].

Cyclization of the substituted cyclohexenes (46; R = H, Bu) to the spiro compounds (47; R = H, Bu) in 80 and 60% yield respectively is catalysed by $[Pd(PPh_3)_4]$ in refluxing MeCN[166].

7 Alkene and Alkyne Metathesis

Metathesis of cis- and trans-cyclooctene in the presence of $[MoC\ell_2(NO)_2(PPh_3)_2]/EtAlC\ell_2$ catalysts gives predominantly cis- and trans-polyoctene respectively; retention of configuration is a function of ring strain[167]. Functionalised acetylenes are metathesised by $[Mo(CO)_6]$/p-$C\ell C_6H_4OH$ in refluxing octane, eg. $PhC{\equiv}CCH_2CO_2Me$ gives $PhC{\equiv}CPh$ and $MeO_2CCH_2C{\equiv}CCH_2CO_2Me$ in 98% yield[168]. Similarly, $BuC{\equiv}CPr$ is metathesised by $[MoO_2(acac)_2]/AlEt_3$ in high selectivity[169]. Careful studies of the reaction of $[WBr_2(\eta^2-CHCMe_3)(OCH_2CMe_3)_2]$ with Ga_2Br_6 at various Ga/W ratios show that catalytic activity in this system is related to the presence of a chiral complex $[WBr(\eta^2-CHCMe_3)(OCH_2CMe_3)_2]^+$ in solution[170]. Electrochemical reduction of $WC\ell_6$ or $MoC\ell_5$ in chlorinated solvents under controlled potential at a Pt cathode, with an Al anode gives in situ formation of the catalytically active species[171]. Evidence has appeared which shows that the initial carbene for C_2H_4 or C_3H_6 metathesis over well defined, heterogenised Mo catalysts, is formed by a 1,2-hydrogen shift via a molybdenum-n-alkenyl intermediate[172].

8 Oligomerisation and Poymerisation of Alkenes and Alkynes

8.1 Alkenes.- The synthesis and X-ray structure of

(40)

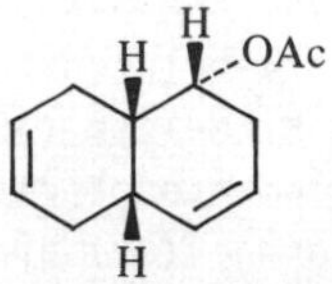

(41)

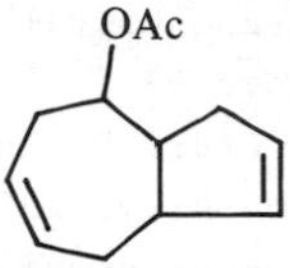

(42)

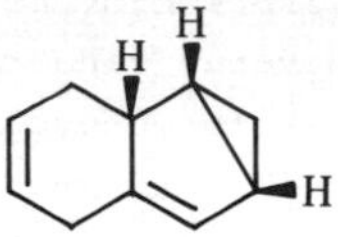

(43)

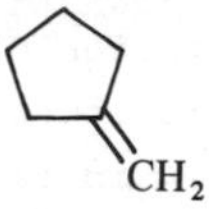

(44)

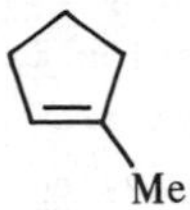

(45)

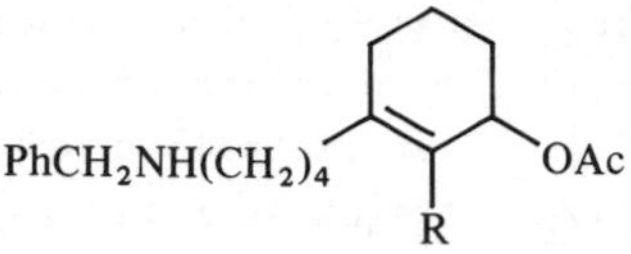

(46)

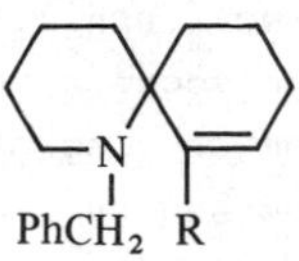

(47)

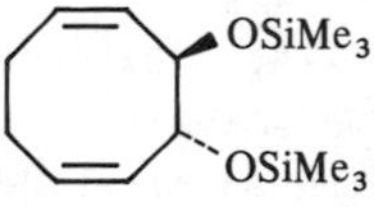

(48)

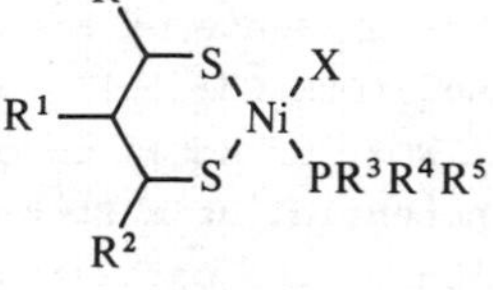

(49)

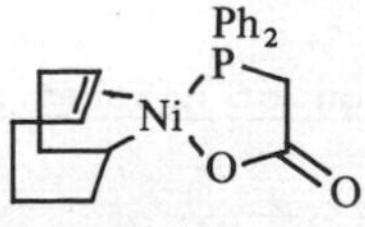

(50)

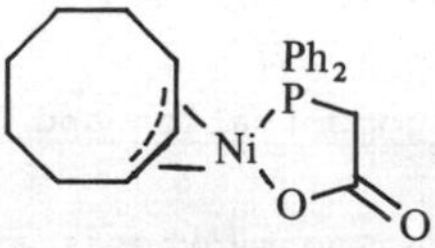

(51)

$[Ti(\eta\text{-}C_2H_4)(\eta^5\text{-}C_5Me_5)_2]$ have been published and the complex is found to catalyse the conversion of C_2H_4 to butadiene and ethane specifically even though it is a poor catalyst for C_2H_4 polymerisation[173]. The $TiC\ell_4/Et_2AlC\ell$ system catalyses the [$\pi 6_s$ + $\pi 2_s$] cycloaddition of cycloheptatriene to buta-1,3-diene, norbornadiene and alkynes[174]. The dimerisation of propene to give methylpentenes is catalysed by two active Co species[175]: the system $Co(acac)_3/HAlEt_2/L$ (L = substituted phosphine) in cycloocta-1,5-diene produces '$HCoL_3$' which is thirty-fold more active and selective for 2-methyl-1-pentene than $[HCo(cod)_2]$. A different isomerisation mechanism than that generally accepted is invoked to explain the structure of propene trimers obtained from the Ni(II)/$EtAlC\ell_2$ catalyst system in the presence of nitrogen bases[176]. Cyclodimerisation of substituted dienes in benzene containing Ni(O) catalyst gives trans-1,2-disubstituted cycloocta-3,7-dienes regio-and sterioselectively, eg. reaction of $H_2C{=}CHCH{=}CHOSiMe_3$ gives 90% of (48) exclusively[177]. Linear unsaturated esters are produced in the reaction of 1,3-dienes with Me acrylates catalysed by $[Pd(MeCH{=}CH_2)L_2]Y$ (L_2 = cod, Y = BF_4; L_2 = 1,5-hexamethyl-phosphoramide, Y = BF_4; L = PBu_3, Y = BF_4; L = PPh_3, Y = PF_6; L = $CH_2{=}CHCO_2Me$, Y = BF_4)[178]. Diels-Alder by-product formation is reduced by the addition of basic phosphines.

Cyclooctenylnickel complexes containing cyclic 1,2-diketones, α-acylcycloalkanones, or substituted 1,3-propanediones as chelating ligands are active catalysts for the linear oligomerisation of but-1-ene[179]. A linear correlation between ligand acidity and activity is noted[179]. Dithio nickel (II) chelates (49) are precursors to highly active olefin oligomerisation catalysts, eg. with (49; $R\text{-}R^2$ = H, $R^3\text{-}R^4$ = Ph, X = $C\ell$) a 10:74:16 mixture of dimethylbutenes, methylpentenes and hexanes respectively are obtained from propylene with a turnover of 22,000 mol $MeCH{=}CH_2$/mol Ni/h[180].

New nickel complex precursors for the oligomerisation of ethylene include a 60:40 mixture of (50)-(51) which gives 99% α-olefins in 93-99% yield[181] and (52) which gives 68-80% linear products in 99.9% yield[182].

A ^{31}P NMR study of ethylene dimerisation in the $[NiX(\sigma\text{-}C_6Y_5)(PPh_3)_2]/AgC\ell O_4$ (X = Br, Y = F; X = Y = $C\ell$) system indicates that [Ni($C\ell O_4$)(σ-C_6X_5)(PPh_3)] (same X) is the catalytically active species[183].

The allyl group is apparently retained on the catalyst in the dimerisation of ethylene to butenes (and of styrene to PhCH=CHCHMePh) catalysed by [Pd(η^3-C_3H_5)(CD_3NO_2)$_2$]BF_4 [184]. Heterogenised Ni/Al_2O_3 catalysts for ethylene dimerisation[185] and ethylene/propylene co-dimerisation[186] require activation by $Et_3Al_2C\ell_3$.

8.2 Alkynes.- ($HC{\equiv}CCR_2$)$_2$NH (R=Me, R_2=(CH_2)$_5$) react with $R'C{\equiv}CR^2$ (R'=CMe_2NH_2, CMe_2OH, Ph; R^2=H, Ph) and catalytic amounts of [NiL_4] (L = P($OCHMe_2$)$_3$, PPh($OCHMe_2$)$_2$) or [Ni(cod)$_2$]/P($OCHMe_2$)$_3$ to give 20-82% isoindoles (53)[187].

$RC{\equiv}CH$ (R=Me, Pr, Bu, n-C_5H_{11} n-$C_{10}H_{21}$) are dimerised by Pd(acac)$_2$/L/Et_3Al systems (L = Ph_3P, Bu_3P, (PhO)$_2$P(O)NMe_2, depe, py, Et_3N) to the corresponding $RC{\equiv}CCR{=}CH_2$ and some higher oligomers, in yields which decreased in the stated order of the ligand L [188].

9 Catalysed Additions to Alkenes, Alkynes and Telomerisation Reactions

A new model for characterisation of nucleophiles is based on the regioselectivity of their reaction with trans-PhCH=CHCH(OAc)CN catalysed by [Pd(PPh_3)$_4$][189]. trans-[Mo(N_2)$_2$(dppe)$_2$] catalyses reactions of allylic compounds[190], eg. allylation of $MeCOCH_2CO_2Me$ by allylic ethers and amines, and [W(MeCN)$_3$(CO)$_3$] (with bipy or 4,5-phenanthroline) the alkylation of allylic substrates[191], eg, (E)-$PhCH{=}CHCH_2O_2COMe$ reacts with NaCH(CO_2Me)$_2$ to give 91% (MeO_2C)$_2$CHCHPhCH=CH_2. The intramolecular Diels-Alder reaction of (54; R=Me, R'=R^2=H; R=H, R'=R^2=Me) to give (55; same R,R',R^2) is catalysed by [$MC\ell$(η^3-C_3H_5)(CO)$_2$(NCMe)$_2$] (M=Mo, W)[192].

Hydrogenation of acetylenes is catalysed regioselectively by cyanocobaltate prepared at CN:Co 5:1[193] and hydrocyanation of α, ω-diynes catalysed by tetracyanonickelate produces novel acyclic dinitriles[194]. The reaction of $PhC{\equiv}CPh$ with PhNCO catalysed by Ni(O)/TMEDA (TMEDA = $Me_2NCH_2CH_2NMe_2$) produces pyrimidinediones via 5-and 7-membered azanickel heterocycles[195].

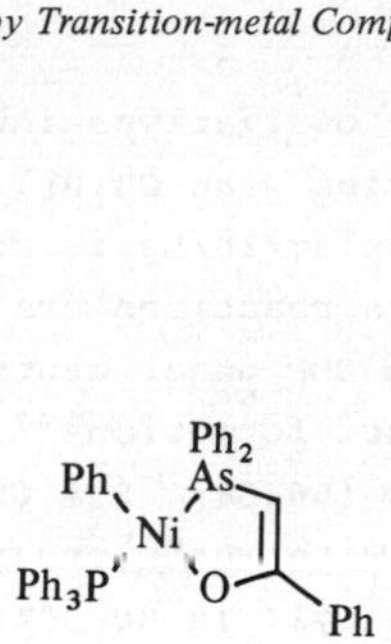

(52)

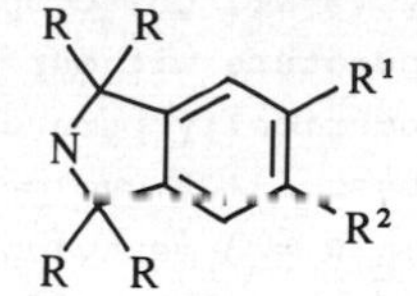

(53)

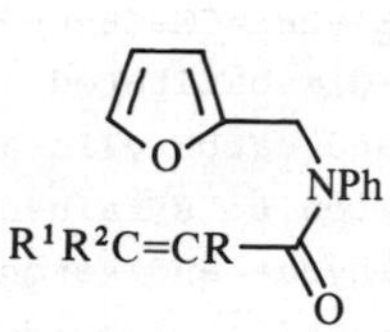

(54)

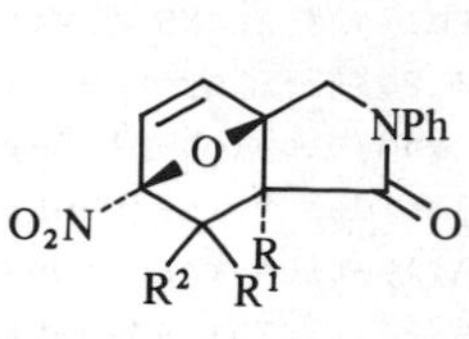

(55)

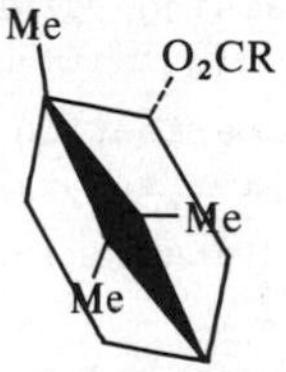

(56)

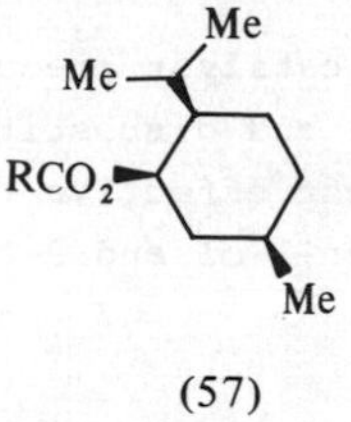

(57)

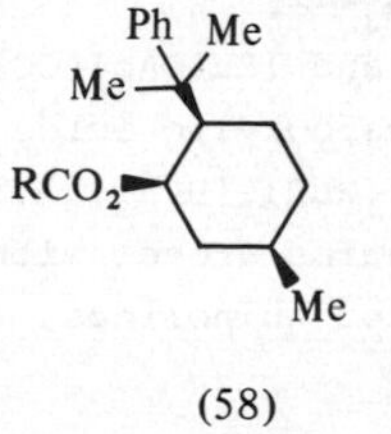

(58)

A new synthesis of cyclopropane derivatives involves the treatment of electron-deficient olefins with CH_2Br_2 in the presence of an appropriate Ni(O) complex/Zn/Lewis acid (or alkali halide) system[196]. Michael-addition reactions are catalysed by $M(OAc)_2$/bipy (M=Ni, Co) complexes in DMF under neutral conditions at room temperature without by-product formation[197].

Enantiomerically pure acrylates (56-58; R = CH_2=CH) undergo Ni(O)-catalysed [3+2] cycloaddition with methylenecyclopropane to give (56-58; R = 3-methylenecyclopentyl) in 80-90% yields with 3-64% asymmetric induction[198].

Interest in the palladium catalysed allylation and arylation of alkenes and alkynes continues, eg. ethylene plus benzoyl chlorides giving styrene and trans-stilbene derivatives[199]; isoprene plus PhI and Et_2NH giving 17% (E)-CH_2=CMeCH=CHPh and 35% Et_2NCH_2CMe=$CHCH_2Ph$[200]; synthesis of 2,3-disubstituted bicyloheptanes and bicylohept-5-enes[201] and carboxylic acids, esters and anhydrides by sequential addition to strained olefins[202]. Allyation reactions of aldehydes and ketones produce α-allyl ketones[203]. The Pd-catalysed reaction of azidoformate with allylic ethers has been reinvestigated[204]. Decarboxylative allylation of active methylene compounds proceeds under mild conditions using allyl carbonates[205], eg. $EtCOCHMeCO_2Me$ with H_2C=$CMeCH_2OCO_2Me$ in THF containing Pd/PPh_3 gives 92% H_2C=$MeCH_2CMe(CO_2Me)COEt$.

Palladium catalysed telomerisation reactions of butadiene with amino alcohols[206], water[148,207], and sulphones[208] and of isoprene with sulphones[208], diethylamine[209a, b] and alcohols[210] are described.

Allyl sulphones can be prepared by the reaction of RSO_3H (R=Ph, Bu) with allyl ethers, esters, amines and sulphides catalysed by the $Pd(acac)_2/PPh_3/AlEt_3$ system[211]. The addition of $Me_3SnSnMe_3$ to HC≡CR (R=H, Bu, Ph, CH_2Ph, CH_2OMe, CH_2OPh) is catalysed by $[Pd(PPh_3)_4]$ and yields 29-80% (Z)Me_3SnRC=$CHSnMe_3$[212].

$[Ru_3(CO)_{12}]$ and $[Ru(OAc)(CO)_2]_n$ are catalyst precursors for the addition of carboxylic acids to mono- and disubstituted acetylenes[213]. $[RuC\ell_2(PPh_3)_3]$ is the most effective catalyst for the reaction of amino arenes with 2-propen-1-ol and 2-buten-1-ol to give substituted quinolines, (59)[214].

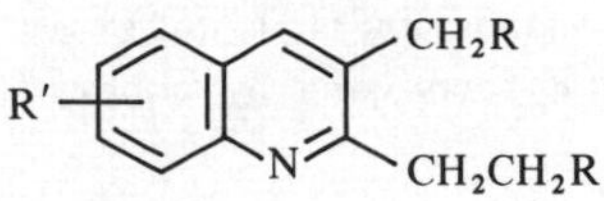

(59) R = H, Me; R′ = H, MeO, Me, Cl

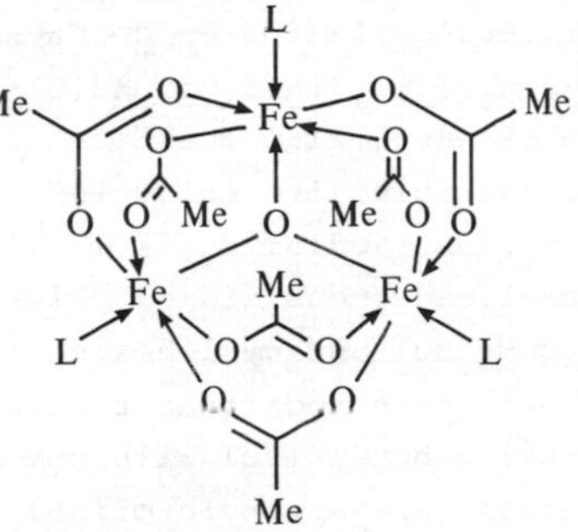

(60)

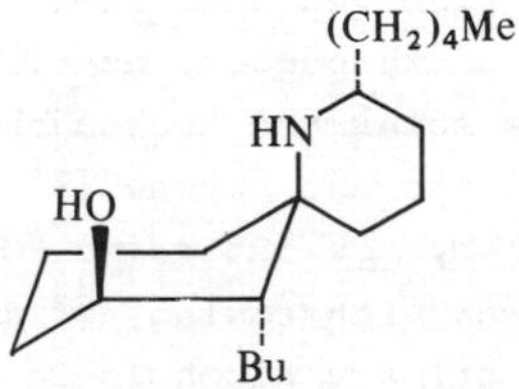

(61)

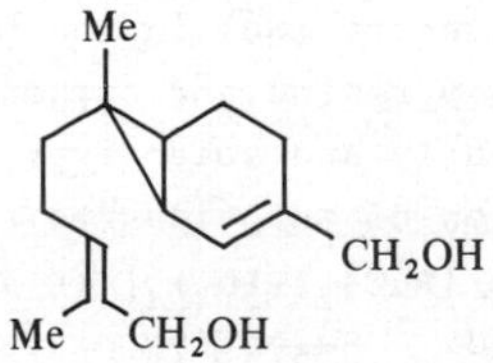

(62)

The ruthenium catalysed hydroacylation of ethylene and propylene with RCHO gives RCOR' (R=R'=Et, Pr; R=Me, R'=Et)[215]; reactions of ethylene with HCO_2Me produce $EtCO_2Me$ probably via an initial fragmentation of the formate[216].

Kinetic evidence indicates that $CuC\ell^+$ is the reactive species in the chlorination of cylcohexene by copper (II) chloride in MeCN[217].

10 Oxidation Reactions

Reviews of catalytic methods for selective oxidation[218], ruthenium catalysed oxidation reactions[219] and selective oxidation via palladium catalysis[220] have appeared. The synthesis and oxidative ability of metalloflavins as coenzyme models have been investigated[221]. $[(Ph_3P)_2N]_5[(NC)_5CoOOMo(O)(OH_2)(CN)_5]$ heterogenised by ion exchange with a silica gel having $(CH_2)_3NH_3^+$ $C\ell^-$ groups catalyses the selective oxidation of cyclohexene at 170°C in the presence of molecular O_2 [222]. $[Mo_2O_3(S_2NEt_2)_4]$ supported on functionalised cross-linked polystyrene catalyses oxidation of Me_2SO to Me_2SO_2 and cyclohexene to cyclohexene oxide with Bu^tOOH better than the homogeneous counterpart[223]. Epoxidation of 3-methyl-2-buten-1-ol with cumeme hydroperoxide is catalysed by $[MoO_2(acac)_2]$/L-N-methylprolinol yielding chiral products until the ligand is oxidatively destroyed[224]. A two phase H_2O/C_6H_6 system, incorporating $[Mn(TPP)]C\ell$ + phase transfer agent and ascorbate, activates dioxygen for selective epoxidation of olefins and hydroxylation of alkanes under mild conditions (20°C, pH 8.5)[225]. Epoxidations and hydroxylations using iodosyl compounds and iron porphyrins have received attention[226]. Tri-iron clusters (60) in the presence of Zn powder, acetic acid, aqueous pyridine and oxygen oxidise saturated hydrocarbons efficiently and selectively to alcohols and ketones[227]. The oxidation of sulphides to sulphones by cis-$[RuC\ell_2(Me_2SO)_4]$[228] and $RC_3H_4IO/[RuC\ell_2(PPh_3)_3]$ (R is electron withdrawing)[229] are reported. Selective oxidation of primary alcohols in the presence of secondary alcohols is catalysed by $[RuC\ell_2(PPh_3)_3]/Me_3SiOOSiMe_3$ [330].

The oxidation of 2,6-di-tert-butylphenol by O_2 giving both oxidative coupling and partial oxygen insertion products is catalysed by $[Co(bipy)_2]^{2+}$ via intermediate $[Co_2(\mu\text{-}OOH)(\mu\text{-}OH)(bipy)_4]$[231] to 2-$MeCOC_6H_4NHCHO$). The [Co(II)(salen)] - catalysed oxygenolysis of 3-methylindole to 2-$MeCOC_6H_4NHCHO$ proceeds via [Co(III)(salen)][2-$MeCOC_6H_4N$=CHO] ie, activation of the substrate rather than the molecular oxygen[232]. Oxygenation kinetics of cycloocta-1,5-diene by $[RhC\ell(O_2)(PPh_3)_2]$ at room temperature leads to cyclo-octane-1,4-dione suggesting reaction at two isolated olefin centres in one molecule[233]. 2,3-dimethylbut-2-ene, butadiene and isoprene undergo rhodium catalysed oxidative double bond cleavage to acetone, acrolein and related compounds[234]. $[Rh(Ph_2PCH_2CH_2SR)_2]^+$ (R = Me, Et, Ph) catalyse air oxidation of terminal olefins, primary alcohols, and secondary alcohols to the corresponding methyl ketones, acetals, and ketones, but at incredibly low rates[235]. Oxidation of $BuCH=CH_2$ to BuCOMe by a Rh(III)/Cu(II) couple proceeds via the co-ordination of in situ generated hydrogen peroxide[236]. A study of palladium catalysed ethylene acetoxylation in acetic acid/dioxane/nitric acid proposes formation of an intermediate Pd nitrito complex[237]. Pd/Cu systems for the oxidation of polyunsaturated ethers and esters[238] and amino akenes[239] are described. Treating terminal olefins containing electron withdrawing substituents with (R,R)-$(HOCHMe)_2CH_2$ in $MeOCH_2CH_2OMe$ containing $PdC\ell_2/CuC\ell$ and O_2 gives cyclic acetals[240]. Copper(II)-complex catalysed oxidative coupling of 2,6-diaklyphenols produces dimers in the presence of N-substituted ethylene diamines[241a] but low mol wt polymers in the presence of KOH/pyridine[241b].

11 Synthesis

Reviews of control in metal-catalysed organic synthesis by discontinuous titration[242], catalytic synthesis by vitamin B_{12} and related cobalt complexes[243], palladium-catalysed synthesis of heterocycles[244] and palladium- or nickel-catalysed cross coupling reactions[245] have appeared.

$RuC\ell_3/PBu_3$ generates a much more active and stable catalyst than $[RuC\ell_2(PPh_3)_3]$ for the conversion of primary to secondary amines[246]. $[RuC\ell_2(PPh_3)_3]$ is the most active catalyst for the N-alkylation of amides with alcohols[247]. Polymer supported $[HFe(CO)_4]^-$ produced by ion-exchange with an anionic resin selectively reduces nitroarenes to amines and reacts with 1,2-dibromoalkanes to give alkenes in good yield[248]. This new reduction technique is claimed to be generally applicable. Cyclopropanation of dimethyl fumarate and maleate with gem-dihalides is catalysed by Co(O) or Ni(O) complexes and zinc[249]. In a synthesis of ketones from carboxylic acids by the Grignard reaction in the presence of $[NiC\ell_2(dppe)]$, formation of alcohols is almost negligible[250]. Large numbers of palladium catalysed syntheses continue to appear, for example: cyclopentanoids via cycloaddition[251], cytochalasins via macrocyclization[252], total synthesis of perhydrohistrionicotoxin (61)[253], preparation of α,β-unsaturated ketones via allyl enol carbonates[254], facile synthesis of 4-substituted 3-formylindoles[255], a new total synthesis dl-sirenin (62)[256].

Regiospecific C-alkyenylation of phenols with $Me_2C{=}C{=}CH_2$ in $CHC\ell_3$ and $Pr_4N[PtC\ell_3(C_2H_4)]$ gives o-isopentenylphenols and 2,2-dimethylchromans[257]. Dehydration of RCH=NOH (R=n-heptyl, n-nonyl, $PhCH_2CH_2$, E-styryl, Ph, p-tolyl, p-anisyl, m-$O_2NC_6H_4$) in the presence of $Cu(OAc)_2$ gives 80-98% RCN[258].

Cross-coupling reactions catalysed by cobalt[259,260] nickel[259, 261] palladium[259,260,262] and copper[263] receive considerable attention.

11.1 Asymmetric Synthesis.- Enantioselective routes to cyanohydrins[264a], their derivatives[264a] and propargylic alcohols[264b] use acetal templates and $TiC\ell_4$ catalysts. Regio-and stereoselective formation of δ-lactones from 1,3-diene mono-epoxides is catalysed by Pd(O) complexes[265]. New stereospecific synthesis of the pheromone bombykol and its three geometrical isomers are achieved by $[Pd(PPh_3)_4]$ catalysed cross-coupling. For example (E)-$HO(CH_2)_9CH{=}CHB(OH)_2$ with (Z)-PrCH=CHBr in refluxing C_6H_6/NaOEt gives [10E, 12Z-$HO(CH_2)_9CH{=}CHCH{=}CHPr$] in 82% yield[266].

References

1a Advances in Organometallic Chemistry, Vol. 21, ed. F. G. A. Stone and R. West, Academic Press, New York, 1982.

1b Modern Inorganic Chemistry: Homogeneous Catalysis with Metal Phosphine Complexes, ed. L. H. Pignolet, Plenum, New York, 1983.

2 R. R. Schrock, Science, 1983, 219 (4580), 13; ACS Symp. Ser., 1983, 211, 369.

3 J. Holton, M. F. Lappert, R. Pearce, and P. I. W. Yarrow, Chem. Rev., 1983, 83, 135.

4a Y. I. Yermakov, J. Mol. Catal., 1983, 21, 35.

4b J. M. Basset, and A. Choplin, J. Mol. Catal., 1983, 21, 95.

5a R. D. Adams, Accounts Chem. Res., 1983, 16, 67.

5b E. L. Muetterties, and M. J. Krausse, Angew. Chem. Internat. Edn., 1983, 22, 135.

6 R. G. Salmon, Tetrahedron, 1983, 39, 485.

7 K. Arata, and K. Tanabe, Catal. Rev. - Sci. Eng., 1983, 25, 365.

8 M. Catellini, and G. P. Chiusoli, J. Organomet. Chem., 1983, 250, 509.

9 D. R. Anton, and R. H. Crabtree, Organometallics, 1983, 2, 855.

10a K. Chandrasekhar, and H. B. Buergi, J. Am. Chem. Soc., 1983, 105, 7081.

10b W. S. Kennedy, Accounts Chem. Res., 1983, 16, 106.

10c H. Brunner, Angew. Chem. Internat. Edn., 1983, 22, 897.

10d D. B. Dombek, Adv. in Catalysis, 1983, 32, 326.

11 Y. N. Chun, and L. Y. C. Chun, J. Chem. Phys., 1983, 78, 1905.

12 R. A. Sanchez-Delgado, A. Andriollo, and N. Valencia, J. Chem. Soc., Chem. Commun., 1983, 444.

13 Papers presented at the Fourth International Symposium on Relations between Homogeneous and Heterogeneous Catalysis, ed. B. C. Gates, J. Mol. Catal., 1983, Vol 21.

14 C. P. Pez, and R. K. Crissey, J. Mol. Catal., 1983, 21, 393.

15 P. Le Maux, R. Dabard, and G. Simmoneaux, J. Organomet. Chem., 1983, 241, 37.

16 L. Marko, and J. Palagyi, Transition Met. Chem., 1983 8, 207.

17 G. E. Rodgers, W. R. Cullen, and B. R. James, Can. J. Chem., 1983, 61, 1314.

18 L. M. Wilkes, J. H. Nelson, L. B. McCusker, K. Seff, and F. Mathey, Inorg. Chim. Acta, 1983, 68, 179.

19 S. R. Patil, D. N. Sen, and R. V. Chaudhari, J. Mol. Catal., 1983, 19, 233.

20 M. Airoldi, G. Deganello, G. Dia, and G. Gennaro, Inorg. Chim. Acta, 1983, 68, 179.

21 R. Wilczynski, W. A. Fordyce, and J. Halpern, J. Am. Chem. Soc., 1983, 105, 2066.

22 C. R. Landis, and J. Halpern, Organometallics, 1983, 2, 840.

23 R. Lykvist, and R. Larsson, J. Mol. Catal., 1983, 19, 1.

24 B. R. James, D. Mahajan, and S. J. Rettig, Organometallics, 1983, 2, 1452.

25 S. Toros, L. Kollar, and B. Heil, J. Organomet. Chem., 1983, 253, 375.

26 W. R. Cullen, T. J. Kim, F. W. B. Einstein, and T. Jones, Organometallics, 1983, 2, 714.

27 R. T. Smith, R. K Ungar, L J Sanderson, and M. C. Baird, Organometallics, 1983, 2, 1138.

28 D. H. Goldsworthy, F. R. Hartley, and S. G. Murray, J. Mol. Catal., 1983, 19 257; ibid, 1983, 19, 269.

29 R. Rumin, J. Organomet. Chem., 1983, 247, 351.

30 R. Bertani, G. Carturan, and A. Scrivanti, Angew. Chem., 1983, 95, 241.

31 A. Fusi, R. Ugo, R. Psara, P. Braunstein, and J. Dehand, Phil. Trans. R. Soc. London, Ser. A, 1982, 308, 125.

32 W. J. Evans, I. Bloom, W. E. Hunter, and J. L. Atwood, J. Am. Chem. Soc., 1983, 105, 1401.

33 W. J. Evans, I. Bloom, and S. C. Engerer, J. Catal., 1983, 84, 468.

34a H. Alper, and J. Heveling. J. Chem. Soc., Chem. Commun., 1983, 365;

34b H. Alper, and K. R. Januszkiewicz, Organometallics, 1983, 2, 1055.
35 H. Alper, and K. R. Januszkiewicz, J. Mol. Catal., 1983, 19, 139.
36 B. D. Vineyard, W. S. Knowles, and M. J. Sabacky, J. Mol. Catal., 1983, 19, 159.
37 J. Halpern, Pure Appl. Chem., 1983, 55, 99.
38 R. H. Crabtree, and M. W. Davis, Organometallics, 1983, 2, 681.
39 G. Stark, and D. E. Kahne, J. Am. Chem. Soc., 1983, 105, 1072.
40 P. Le Maux, and G. Simmoneaux, J. Organomet. Chem., 1983, 252, C60.
41 D. L. Allen, V. C. Gibson, M. L. H. Green, J. F. Skinner, J. Bashkin, and P. D. Grebenik, J. Chem Soc., Chem. Commun., 1983, 895.
42 J. D. Oliver, and D. P. Riley, Organometallics, 1983, 2, 1032.
43 S. Gladiali, G. Faedda, M. Marchetti, and C. Botteghi, J. Organomet. Chem., 1983, 244, 289.
44 E. Cesarotti, A. Chiesa, G. Ciani, and A. Sironi, J. Organomet. Chem., 1983, 251, 79.
45 A. Uehara, T. Kubota, and R. Tsuchiya, Chem. Lett., 1983, 441.
46 M. Yatagai, M. Zama, T. Yamagishi, and M. Hida, Chem. Lett., 1983, 1203.
47a D. Lafont, D. Sinou, and G. Descotes, Nouv. J. Chim., 1983, 7, 283;
47b D. Sinou, D. Lafont, G. Descotes, and T. Dayrit, Nouv. J. Chim., 1983, 7, 291.
48 H. Brunner. A. F. M. M. Rahman, Z. Naturforsch, 1983, 38B, 1332.
49 C. R. Landis, and J. Halpern, J. Organomet. Chem., 1983, 250, 485.
50 T. Yoshida, W. J. Youngs, T. Sakaeda, T. Ueda, S. Otsuka and J. A. Ibers, J. Am. Chem. Soc., 1983, 105, 6273.
51 R. R. Burch, A. J. Shusterman, E. L. Muetterties, R. G. Teller, and J. M. Williams, J. Am. Chem. Soc., 1983, 105, 3546.
52 R. Mutin, W. Arbound, J. M. Basset, and D. Sinou, Polyhedron, 1983, 2, 539.
53 J. M. Brown, P. A. Chaloner, and G. A. Morris, J. Chem. Soc., Chem. Commun., 1983, 664.
54 T. J. Thomas, and A. Brenner, J. Mol. Catal., 1983, 18, 197.
55 H. C. Foley, S. J. DeCanio, K. D. Tau, K. J. Chao, J. H. Onuferko, C. Dybowski, and B. C. Gates, J. Am. Chem. Soc., 1983, 105, 3074.
56 R. Raythatha, and T. J. Pinnavaia, J. Catal., 1983, 80, 47.
57 S. Bhadari, and H. Khwaja, J. Chem. Soc., Dalton Trans., 1983, 419.
58 F. Pinna, M. Gonizzi, G. Strukul, G. Cocco, and S. Enzo, J. Catal., 1983, 82, 171.
59 Y. Doi, K. Soga, Y. Yano, and Y. Ono, J. Mol. Catal., 1983, 19, 359.
60 J. R. Budge, J. P. Scott, and B. C. Gates, J. Chem. Soc., Chem. Commun., 1983, 342.
61 E. C. Chukovskaya, R. Kh. Freidlina, and N. A. Kuz'mina, Synthesis, 1983, 773.
62 Y. Lin, and X. Lu, J. Organomet. Chem., 1983, 251, 321.
63 D. Baudry, M. Ephritikhine, H. Felkin, R. Holmes-Smith, J. Chem. Soc., Chem. Commun., 1983, 788.
64 S. M. Pillai, S. Vancheesan, J. Rejaram, and J. C. Kuriacose, J. Mol. Catal., 1983, 20, 169.
65 Y. Matsui, T. E. Nalesnik, and M. Orchin, J. Mol. Catal., 1983, 19, 303.
66a D. Beaupere, L. Nadjo, R. Uzan, and P. Bauer, J. Mol. Catal., 1983, 18, 73;
66b ibid., 1983, 20, 185;
66c ibid., 1983, 20, 195.
67 R. Spogliarich, A. Tencich, J. Kaspar, and M. Graziani, J. Organomet. Chem., 1982, 240, 453.
68 S. Shinoda, T. Kojima, and Y. Saito, J. Mol. Catal., 1983, 18, 99.
69 G. Maestroni, G. Zassinovich, C. Del Bianco, and A. Camus, J. Mol. Catal., 1983, 18, 33.
70 G. Zassinovich, and F. Grisoni, J. Organomet, Chem., 1983, 247, C24.
71 J. Azran, O. Buchman, G. Hoehne, H. Schwarz, and J. Blum, J. Mol. Catal., 1983, 18, 105.
72 H. G. Alt, K. A. Mahmoud, and A. J. Rest, J. Organomet. Chem., 1983, 243, C5.

73 A. S. C. Chan, W. E. Carroll, and D. E. Willis, J. Mol. Catal., 1983, 19, 377.
74 T Okano, T. Kobayashi, H. Konishi, and J. Kiji, Tetrahedron Lett., 1982, 23, 4967.
75 M. Bianchi, G. Menchi, P. Frediani, U. Matteoli, and F. Piacenti, J. Organomet. Chem., 1983, 247, 89.
76 A. D. Harley, G. J. Guskey, and G. L. Geoffroy, Organometallics. 1983, 2, 53.
77 R. A. Dubois, and P. E. Garrou, J. Organomet. Chem., 1983, 241, 69.
78 D. Seyferth, and H. P. Withers, Inorg. Chem., 1983, 22, 2931.
79 J. D. Fellman, P. E. Garrou, H. P. Withers, and D. Seyferth, Organometallics, 1983, 2, 818.
80 D. Chen, and C. U. Pittman Jr., J. Mol. Catal., 1983, 21, 405.
81 Y. Matsui, and M. Orchin, J. Organomet. Chem., 1983, 246, 57.
82 P. Pino, F. Oldani, and G. Consiglio, J. Organomet. Chem., 1983, 250, 491.
83 A. G. Abatjoglou, D. R. Bryant, and L. C. D'Esposito, J. Mol. Catal., 1983, 18, 381.
84 R. Grigg, G. J. Reimer, and A. R. Wade, J. Chem. Soc., Perkin Trans. 1, 1983, 1929.
85 M. Matsumoto, and M. Tamura, J. Mol. Catal, 1983, 19, 365.
86 P. Kalck, J. M. Frances, P. M. Ffister, T. G. Southern, and A. Thorez, J. Chem. Soc., Chem Commun., 1983, 510.
87a A. Saus, N. P. Tuyet, M. J. Mirbach, and M. F. Mirbach, J. Mol. Catal., 1983, 18, 117.
87b M. J. Mirbach, N Topalsavoglou, N. P. Tuyet, M. F. Mirbach, and A. Saus, Chem. Ber., 1983, 116, 1422.
88 J. P. Collman, J. A. Belmont, and J. I. Brauman, J. Amer. Chem. Soc., 1983, 105, 7288.
89 M. E. Ford, and J. E. Premecz, J. Mol. Catal., 1983, 19, 99.
90 G. Cavinato, and L. Toniolo, J. Organomet. Chem., 1983, 241, 275.
91 A. B. Goel, and S. Goel, Inorg. Chim. Acta, 1983, 77, L53.
92 C. U. Pittmann Jr., Y. Kawabata, and L. I. Flowers, J. Chem. Soc., Chem. Comm., 1982, 473.
93 G. Consiglio, P. Pino, L. I. Flowers, and C. U. Pittman Jr., J. Chem. Soc., Chem Commun., 1983, 612.
94 J. K. Stille, and G. Parrinello, J. Mol. Catal., 1983, 21, 203.
95 M. E. Fakley, and R. A. Head, Appl. Catal., 1983, 5, 3.
96 J. T. Martin, and M. C. Baird, Organometallics, 1983, 2, 1073.
97 M. Roeper, and H. Loevenich, J. Organomet. Chem., 1983, 255, 95.
98 M. Roeper, H. Loevenich, and J. Korff, J. Mol. Catal., 1983, 17, 315.
99 P. B. Francoisse, and F. C. Thyrion, Ind. Eng. Chem. Prod. Res. Dev., 1983, 22, 542.
100 P. Andrianary, G. Jenner, and A. Kienneman, J. Organomet. Chem, 1983, 252, 209.
101 G. Doyle, J. Mol. Catal., 1983, 18, 251.
102 M. J. Chen, H. M. Feder, and J. W. Rathke, J. Mol. Catal., 1983, 17, 331.
103a H. Kheradmand, A. Kiennemann, and G. Jenner, J. Organomet. Chem., 1983, 251, 339.
103b H. Kheradmand, A. Kiennemann, and A. Deluzarche, J. Mol. Catal., 1983, 18, 61.
104 G. Braca, G. Sbrana, G. Valentini, and M. Cini, J. Mol. Catal., 1982, 17, 323.
105 G. Luft, and M. Schrod, J. Mol. Catal., 1983, 20, 175.
106 G. Tanguy, B. Weinberger, and H. Des Abbayes, Tetrahedron Lett., 1983, 24, 4005.
107 V. Galamb, and H. Alper, J. Chem Soc., Chem Commun., 1983, 88.
108 H. Des Abbayes, A. Buloup, and G. Tanguy, Organometallics, 1983, 2, 1730.
109a J. J. Brunet, C. Sidot, and P. Caubere, J. Org. Chem., 1983, 48, 1166.
109b J. J. Brunet, C. Sidot, and P. Caubere, J. Org. Chem., 1983, 48, 1919.

110 F. Francalanci, A. Gardano, L. Abis, T. Fiorani, and M. Foa, J. Organomet. Chem., 1983, 243, 87.
111 M. Foa, F. Francalanci, A. Gardano, and G. Cainelli, J. Organomet. Chem., 1983, 248, 225.
112 I. T. Horvath, I. Pelczer, G. Szabo, and G. Palyi, J. Mol. Catal., 1983, 20, 153.
113a H. Alper, and J. F. Petrignani, J. Chem. Soc., Chem. Commun., 1983, 1154.
113b H. Alper, and S. Amaratunga, Can. J. Chem., 1983, 61, 1309.
114 F. Pesa, and T. Haase, J. Mol. Catal., 1983, 18, 237.
115 R. Stern, D. Reffet, A Hirschauer, D. Commereuc, and Y. Chauvin, Synth. Commun., 1982, 12, 1111.
116 G. La Monica, C. Monti, and S. Cenini, J. Mol Catal., 1983 18, 93.
117 V. I. Manov-Yuvenskii, K. B. Petrovskii, and A. L. Lapidus, Izv. Akad. Nauk SSSR, Ser. Khim., 1983, 606.
118 T. Mise, P. Hong, and H. Yamazaki, J. Org. Chem., 1983, 48, 238.
119 H. Hoberg, and H. J. Riegel, J. Organomet. Chem., 1983, 241, 245.
120 Y. Tamaru, H. Ochiai, Y. Yamada, and Z. Yoshida, Tetrahedron Lett., 1983, 24, 3869.
121 M. Mori, Y. Washioka, T. Urgayama, K. Yoshiura, K. Chiba, and Y. Ban, J. Org. Chem., 1983, 48, 4058.
122 T. Fuchikami, and I. Ojima, Tetrahedron Lett., 1982, 23, 4099.
123 V. Galamb, M. Gopal, and H. Alper, Organometallics, 1983, 2, 801.
124 G. P. Chiusoli, M. Costa, E. Masarati, and G. Salerno, J. Organomet. Chem., 1983, 255, C35.
125 J. Tsuji, K. Sato, and H. Okumoto, Tetrahedron Lett., 1982, 23, 5189.
126 H. Alper, J. B. Woell, B. Despeyroux, and D. J. H. Smith, J. Chem. Soc., Chem. Commun., 1983, 1270.
127 S. P. Current, J. Org. Chem., 1983, 48, 1779.
128 G. R. Steinmetz, and T. H. Larkins, Organometallics, 1983, 2, 1879.
129a G. Comitti, and G. P. Chiusoli, J. Organomet. Chem., 1982, 236, C31.
129b T. Fuchikawa, K. Ohishi, and I. Ojima, J. Org. Chem., 1983, 48, 3803.
130 D. E. Bergbreiter, B. Chen, and T. J. Lynch, J. Org. Chem., 1983, 48, 4179.
131 M. T. Trillies, ed., "Chemical Technology Review No. 209", Noyes Data Corporation, Park Ridge, New Jersey, 1982.
132a R. A. Sheldon, ed., "Catalysis by Metal Complexes, Vol. 3", D. Reidel Publishing Corp., Dordrecht, The Netherlands, 1982;
132b W. Keim, ed., "Catalysis by Metal Complexes, Vol. 4", D. Reidel Publishing Corp., Dordecht, The Netherlands, 1983.
133 L. C. Costa, Cat. Rev.-Sci. Eng., 1983, 25, 3253.
134 B. D. Dombek, J. Organomet. Chem., 1983, 250, 467.
135 J. F. Knifton, J. Catal., 1983, 79, 147.
136a J. F. Knifton, J. Chem. Soc., Chem Commun., 1983, 729.
136b R. Whyman, J. Chem. Soc., Chem. Commun., 1983, 1439
137 V. L. Kuznetsov, A. F. Danilyuk, I. E. Kolosova, and Y. I Ermakov, React. Kinet. Catal. Lett., 1982, 21, 249.
138 A. Choplin, M. Leconte, J. M. Basset, S. G. Shore, and W. L. Hsu, J. Mol. Catal., 1983, 21, 389.
139 R. Hemmerich, W. Keim, and M. Roeper, J. Chem. Soc., Chem Commun., 1983, 428.
140 R. Pierantozzi, E. G. Valagene, A. F. Nordquist, and P. N. Dyer, J. Mol. Catal., 1983, 21, 189.
141 J. Kaspar, R. Spogliarich, A. Cernogoraz, and M. Graziani, J. Organomet. Chem., 1983, 255, 371.
142 K. Kaneda, T. Imanaka, and S. Teranishi, Chem. Lett., 1983, 1465.
143 E. Alessio, G. Zassinovich, and G. Mestroni, J. Mol. Catal., 1983, 18, 113.
144 R. M. Laine, J. Mol. Catal., 1983, 21, 119.
145 S. Inoue, ed., "Organic and Bio-organic Chemistry of Carbon Dioxide", Wiley, New York, 1982.
146 D. A. Palmer, and R. Van Eldik, Chem. Rev., 1983, 83, 651.

147 D. J. Darensbourg, C. Ovalles, and M. Pala, J. Am. Chem. Soc., 1983, 105, 5937.
148 Y. Inoue, M. Sato, M. Satake, and H. Hashimoto, Bull. Chem. Soc. Jap., 1983, 56, 637.
149 A. Behr, and K. D. Jusnak, J. Organomet. Chem., 1983, 255, 263.
150 J. Hawecker, J-M. Lehn, and R. Ziesel, J. Chem. Soc., Chem. Commun., 1983, 536.
151 W. J. Richter, J. Mol. Catal., 1983, 18, 145.
152 H. Kurosawa, and N. Asada, Organometallics, 1983, 2, 251.
153 P. G. Gassman, and S. M. Bonser, Tetrahedron Lett., 1983, 24, 3431.
154 V. Galamb, and H. Alper, Tetrahedron Lett., 1983, 24, 2965.
155 S. L. Cook, and J. Evans, J. Chem. Soc., Chem. Commun., 1983, 713.
156 D. B. Furman, N. V. Volchkov, L. A. Makhlis, S. A. Chernov, V. E. Vasserberg, and O. V. Bragin, Izv. Akad. Nauk SSSR, Ser. Khim., 1983, 568.
157 R. Barth, B. C. Gates, Y. Zhao, H. Knoezinger, and J. Holse, J. Catal., 1983, 82, 147.
158 X. J. Li, and B. C. Gates, J. Catal., 1983, 84, 55.
159 M. Balci, and Y. Sutbeyaz, Tetrahedron Lett., 1983, 24, 311.
160 H. Campbell, and P. W. Jennings, Organometallics, 1983, 2, 1460.
161 L. E. Overman, and A. F. Renaldo, Tetrahedron Lett., 1983, 24, 3757.
162 M. Mizutani, Y. Sanemitsu, Y. Tamaru, and Z. Yoshida, J. Org. Chem., 1983 48, 4585.
163 D. Kaufmann. O. Schallner, L. U. Meyer, H. H. Fick, and A. De Meijere, Chem. Ber., 1983, 116, 1377.
164 R. W. Thies, J. L. Boop, M. Schiedler, D. C. Zimmerman, and T. H. LaPage, J. Org. Chem., 1983, 48, 2021.
165 H. Lehmkuhl, and Y. L. Tsien, Chem. Ber., 1983, 116, 2437.
166 W. Carruthers, and S. A. Cumming, J. Chem. Soc., Chem Commun., 1983, 360.
167 C. Larroche, J. P. Laval, A. Lattes, M. Leconte, F. Quignard, and J. M. Basset, J. Chem. Soc., Chem. Commun., 1983, 220.
168 D. Villemin, and P. Cadiot, Tetrahedron Lett., 1982, 23, 5139.
169 M. Petit, A. Mortreux, and F. Petit, J. Chem. Soc., Chem Commun., 1982, 1385.
170 J. Kress, and J. A. Osborn, J. Am. Chem. Soc., 1983, 105, 6346.
171 M. Gilet, A. Mortreux, J. C. Folest, and F. Petit, J. Am. Chem. Soc., 1983, 105, 3876.
172 Y. Iwasawa, and H. Hamamura, J. Chem. Soc., Chem Commun., 1983, 130.
173 S. A. Cohen, P. R. Auburn, and J. E. Bercaw, J. Am. Chem. Soc., 1983, 105, 1136.
174 K. Mach, H. Antropiusova, P. Sedmera, V. Hanus, and F. Turecek, J. Chem. Soc. Chem. Commun., 1983, 805.
175 F. Petit. H. Masotti, G. Peiffer, and G. Buono, J. Organomet. Chem., 1983, 244, 273.
176 A. Pruvot, D. Commereuc, and Y. Chauvin, J. Mol. Catal., 1983, 22, 179.
177 P. Brun. A. Tenaglia, and B. Waegell, Tetrahedron Lett., 1983, 24, 385.
178 P. Grenouillet, D. Neibecker, and I. Tkatchenko, J. Chem. Soc., Chem. Commun., 1983, 542.
179 W. Keim, A. Behr, and G. Kraus, J. Organomet. Chem., 1983, 251, 377.
180 K. J. Cavell, and A. F. Masters, J. Chem. Res., Synop., 1983, 72.
181 M. Peukert, and W. Keim, Organometallics, 1983, 2, 594.
182 W. Keim, A. Behr, B. Limbaeker, and C. Krueger, Angew. Chem., 1983, 95, 505.
183 Y. Ishimura, K. Maruya, Y. Nakamura, T. Mizoroki, and A. Ozaki, Bull. Chem. Soc. Jap., 1983, 56, 818.
184 A. Sen, and T. W. Lai, Organometallics, 1983 2, 1059.
185 D. B. Furman, N. V. Volchok. L. A. Makhlis, P. E. Matkovskii, G. P. Belov, V. E. Vasserberg, and O. V. Bragin, Izv. Akad. Nauk SSR, Ser. Khim., 1983, 573.
186 N. V. Volchkov, D. B. Furman, L. I. Lafer, S. A. Chernov, A. V. Kudryashev, V. I. Yakerson, and O. V. Bragin, Izv. Akad. Nauk SSSR, Ser. Khim., 1983, 573.

187 G. P. Chiusoli, L. Pallini, and G. Terenghi, Transition Met. Chem., 1983, 8, 189.
188 A. F. Selimov, O. G. Rutman, and U. M. Dzhemilev, Zh. Org. Khim., 1983, 19, 1853.
189 E Keinan, and Z. Roth, J. Org. Chem., 1983, 48, 1769.
190 T. Tatsumi, K. Hashimoto, H. Tominaga, Y. Mizuta, K. Hata, M. Hidai, and Y. Uchida, J. Organomet. Chem., 1983, 252, 105.
191 B. M. Trost, and M. H. Hung, J. Am. Chem. Soc., 1983, 105, 7757.
192 M. S. Bailey, B. J. Brisdon, D. W. Brown, and K. M. Stark, Tetrahedron Lett., 1983, 24, 3037.
193 T. Funabiki, Y. Yamazaki, Y. Sato, and S. Yoshida, J. Chem. Soc., Perkin Trans., 2, 1983, 1915.
194 T. Funabiki, Y. Yoshihiro, and S. Yoshida, Bull. Chem. Soc. Jap., 1983, 56, 2863.
195 H. Hoberg, and B. W. Oster, J. Organomet. Chem., 1983, 252, 359.
196 H. Kanai, N. Hiraki, and S. Iida, Bull. Chem. Soc. Jpn., 1983, 56, 1025.
197 K. Watanabe, K. Miyazu, K. Irie, Bull. Chem. Soc. Jpn., 1982, 55, 3212.
198 P. Binger, A. Brinkmann, and W. J. Richter, Tetrahedron Lett., 1983, 24, 3599.
199a A. Spencer, J. Organomet. Chem., 1983, 247, 117;
199b ibid, 1983, 258, 101.
200 J. M. O'Conner, B. J. Stallman, W. G. Clark, A. Y. L. Shu, R. E. Spada, T. M. Stevenson, and H. A. Dieck, J. Org. Chem., 1983, 48, 807.
201 M. Catellani, and G. P. Chiusoli, Tetrahedron Lett., 1982, 23, 4517.
202 M. Catellani, and G. P. Chiusoli, Tetrahedron Lett., 1983, 24, 813.
203a J. Tsuji, I. Minami, and I. Shimizu, Tetrahedron Lett., 1983, 24, 1793;
203b Chem. Lett., 1983, 1325.
204 T. Migita, M. Chiba, K. Takahashi, N. Saitoh, S. Nakaido, and M. Kosugi, Bull. Chem. Soc. Jpn., 1982, 55, 3943.
205 J. Tsuji, I Shimizu, and I. Minami, Tetrahedron Lett., 1982, 23, 4809.
206 A. Groult, and A. Guy, Tetrahedron, 1983, 39, 1543.
207 U. M. Dzheimilev, V. V. Sidorova, and R. V. Kunakova, Izv. Akad. Nauk SSSR, Ser. Khim., 1983, 584.
208 G. A. Tolstikov, O. A. Rozentsvet, R. V. Kunakova, and N. N. Novitskaya, Izv. Akad. Nauk SSSR, Ser. Khim., 1983, 589.
209a W. Keim, M. Roeper, and M. Schieren, J. Mol. Catal., 1983, 20, 139;
209b W. Keim, K. R. Kurtz, and M. Roeper, J. Mol. Catal., 1983, 20, 129.
210 A. Behr, and W. Keim, Chem. Ber., 1983, 116, 862.
211 R. V. Kunakova, R. Gaisin, M. M. Sirazova, and U. M. Dzhemilev, Izv. Akad. Nauk SSSR, Ser. Khim., 1983, 157.
212 T. N. Mitchell, A. Amamria, H. Killing, and D. Rutschow, J. Organomet. Chem., 1983, 241, C45.
213 M. Rotem, and Y. Shro, Organometallics, 1983, 2, 1689.
214 Y. Watanabe, Y. Tsuji, Y. Ohsugi, and J. Shida, Bull. Chem. Soc. Jap., 1983, 56, 2452.
215 P. Isnard, B. Denise, R. P. A. Sneeden, J. M. Cognion, and P. Durual, J. Organomet. Chem., 1982, 240, 285.
216 P. Isnard, B. Denise, R. P. A. Sneeden, J. M. Cognion and P. Durual, J. Organomet. Chem., 1983, 256, 135.
217 P. H. Gamlen, M. S. Henty, and H. L. Roberts, J. Chem. Soc., Dalton Trans., 1983, 1373.
218 R. A. Sheldon, J. Mol. Catal., 1983, 20, 1.
219 E. S. Gore, Platinum Met. Rev., 1983, 27, 111.
220 J. E. Baeckvall, Pure Appl. Chem., 1983, 55, 1669.
221 S. Shinkai, I. Yuichi, and O. Manabe, Bull. Chem. Soc. Jap., 1983, 56, 1694.
222 O. Leal, M. R. Goldwasser, H. Martinez, M. Garmendia, and R. Lopez, J. Mol. Catal., 1983, 22, 117.
223 S. Bhaduri, and H. Khwaja, J. Chem. Soc., Dalton Trans., 1983, 415.

224 S. Coleman-Kammula, and E. T. Duim-Koolstra, J. Organomet. Chem., 1983, 246, 53.
225 D. Mansuy, M. Fontecave, and J-F. Bartoli, J. Chem. Soc. Chem. Commun., 1983, 253.
226a J. T. Groves, and T. E. Nemo, J. Am. Chem. Soc., 1983, 105, 5786.
226b J. T. Groves, and R. S. Myers, J. Am. Chem. Soc., 1983, 105, 5791.
226c J. T. Groves, and T. E. Nemo, J. Am. Chem. Soc., 1983, 105, 6243.
227a D. H. R. Barton, M. J. Gastiger, and W. B. Motherwell, J. Chem. Soc., Chem Commun., 1983, 731;
227b ibid, 1983, 41.
228 D. P. Riley, Inorg. Chem., 1983, 22, 1965.
229 P. Mueller, and J. Godoy, Helv. Chim. Acta, 1983, 66, 1790.
230 S. Kanemoto, K. Oshima, S. Matsubara, K. Takai, and H. Nozaki, Tetrahedron Lett., 1983, 24, 2185.
231 S. A. Bedell, and A. E. Martell, Inorg. Chem., 1983, 22, 364.
232 A. Nishinaga, H. Ohara, H. Tomita, and T. Matsuura, Tetrahedron Lett., 1983, 24, 213.
233 L. Carlton, G. Read, and M. Urgelles, J. Chem. Soc., Chem. Commun., 1983, 586.
234 H. Boennemann, W. Nunez, and D. M. M. Rohe, Helv. Chim. Acta, 1983, 66, 177.
235 M. Bressan, F. Morandini, and P. Rigo, J. Organomet. Chem., 1983, 247, C8.
236 E. D. Nyberg, D. C. Fribich, and R. S. Drago, J. Am. Chem. Soc., 1983, 105, 3538.
237 M. G. Volkhonskii, V. A. Likholobov, Y. I. Ermakov, Kinet. Katal., 1983, 24, 578.
238 G. A. Dzhemileva, V. N. Odinokov, U. M. Dzhemilev, and G. A. Tolstikov, Izv. Akad. Nauk SSSR, Ser. Khim., 1983, 343.
239 B. Pugin, and L. M. Venanzi, J. Am. Chem. Soc., 1983, 105, 6877.
240 T. Hosokawa, T. Ohta, and S. Murahashi, J. Chem. Soc., Chem Commun., 1983, 848.
241a K. Kushioka, J. Org. Chem., 1983, 48, 4948.
241b S. Tsuruya, K. Kinumi, K. Hagi, and M. Masai, J. Mol. Catal., 1983, 22, 47.
242a P. Heimbach, J. Kluth, and H. Schenkluhn, Kontakte, 1982, 33.
242b T. Bartik, P. Heimbach, nd H. Schenkluhn, ibid, 1983, 16.
243 R. Scheffold. G. Rytz, and L. Walder, Mod. Synth. Methods, 1983, 3, 355.
244 L. S. Hegedus, J. Mol. Catal., 1983, 19, 201.
245 E. Negishi, Curr. Trends Org. Synth., Proc. 4th Int. Conf., 1982, p 269, Pergamon, Oxford, 1983.
246 C. W. Jung, J. D. Fellmann, and P. E. Garrou, Organometallics, 1983, 2, 1042.
247 Y. Watanabe, T. Ohta, and Y. Tsuji, Bull. Chem. Soc. Jap., 1983, 56, 2647.
248 G. P. Boldrini, G. Cainelli, and A. Umani-Ronchi, J. Organomet. Chem., 1983, 243, 195.
249 H. Kanai, Y. Nishiguchi, and H. Matsuda, Bull. Chem. Soc. Jap., 1983, 56, 1592.
250 V. Fiandanese, G. Marchese, and L. Ronzini, Tetrahedron Lett., 1983, 24, 3677.
251a B. M. Trost, and D. M. T. Chan, J. Am. Chem. Soc., 1983, 105, 2315.
251b ibid, 1983, 105, 2326.
252 B. M. Trost, and S. J. Brickner, J. Am. Chem. Soc., 1983, 105, 568.
253 S. A. Godleski, D. J. Heacock, J. D. Meinhart, and S. Van Wallendael, J. Org. Chem., 1983, 48, 2101.
254 I. Shimizu, I. Minami, and J. Tsuji, Tetrahedron Lett, 1983, 24, 1797.
255 M. Somei, T. Hasegawa, and C. Kaneko. Heterocycles, 1983, 20, 1983.
256 T. Mandai, K. Hara, M. Kawada, and J. Nokami, Tetrahedron Lett., 1983, 24, 1517.
257 A. De Renzi, A. Panunzi, A. Saporito, and A. Vitagliano, J. Chem. Soc., Perkin Trans. 2, 1983, 993.
258 O. Atlanasi, P. Palma, and F. Serra-Zanetti, Synthesis, 1983, 741.
259 S. Uemura, S. Fukuzawa, and S. R. Patil, J. Organomet. Chem., 1983, 243, 9.

260 N. Oguni, T. Omi. Y. Yamamoto, and A. Nakamura, Chem. Lett., 1983, 841.
261a Y. Rollin, G. Meyer, M. Troupel, J-F. Fauvarque, and J. Perichon, J. Chem. Soc., Chem Commun., 1983, 793.
261b K. Tamao, S. Kodama, I. Nakajima, M. Kumada, A. Minato, and K. Suzuki, Tetrahedron, 1982, 38, 3347;
261c Y. Ishii, M. Kawahara, T. Noda, H. Ishigaki, and M. Ogawa, Bull. Chem. Soc. Jap., 1983, 56, 2181;
261d G. Consiglio, F. Morandini, and O. Piccolo, J. Chem. Soc., Chem. Commun., 1983, 112;
261e T. Hayashi, M. Konishi, M. Fukushima, K Kanehira, T. Hioki, and M. Kumada, J. Org. Chem., 1983, 48, 2195.
262a E. Negishi, H. Matsushita, M. Kobayashi, and C. L. Rand, Tetrahedron Lett., 1983, 24, 3823;
262b E. Negishi, and F. Luo, J. Org. Chem., 1983, 48, 1560;
262c K. Kikukawa, A. Abe, F. Wada, and T. Matsuda, Bull. Chem. Soc. Jpn., 1983, 56, 961;
262d T. Hayashi, Y. Okamoto, and M. Kumada, Tetrahedron Lett., 1983, 24, 807;
262e S. Koyama, Z. Kumazawa, and N. Kashimura, Nucleic Acids Symp. Ser., 1982, 11, 41;
262f T. Jeffery-Luong, and G. Linstrumelle, Synthesis, 1983, 32;
262g F. K. Sheffy, and J. K. Stille, J. Am. Chem. Soc., 1983, 105, 7173;
262h I. Pri-Bar, P. S. Pearlman, and J. K. Stille, J. Org. Chem., 1983, 48, 4629;
262i J. W. Labadie, D. Tueting, and J. K. Stille, J. Org. Chem., 1983, 48, 4634;
262j J. W. Labadie, and J. K. Stille, Tetrahedron Lett., 1983, 24, 4283;
262k J. W. Labadie, and J. K. Stille, J. Am. Chem. Soc., 1983, 105, 669;
262l A. K. Yatsimirskii, S. A. Deiko, and A. D. Ryabov, Tetrahedron, 1983, 39, 2381.
263a K. J. Shea, and P. Q. Pham, Tetrahedron Lett., 1983, 24, 1003;
263b D. E. Bergstrom, M. W. Ng, and J. J. Wong, J. Org. Chem., 1983, 48, 1902;
263c J. E. Arrowsmith, C. W. Greengrass, M. J. Newmann, Tetrahedron, 1983, 39, 2469.
264a J. D. Elliott, V. M. F. Choi, and W. S. Johnson, J. Org. Chem., 1983, 48, 2294;
264b W. S Johnson, R. Elliott, and J. D. Elliott, J. Am. Chem. Soc., 1983, 105, 2904.
265 T. Takahashi, H. Kataoka, and J. Tsuji, J. Am. Chem. Soc., 1983, 105, 147.
266 N. Miyaura, H. Suginome, and A. Suzuka, Tetrahedron Lett., 1983, 24, 1527.

16

Organometallic Compounds in Biological Chemistry

BY B. RIDGE

1 Introduction

This Chapter surveys aspects of the chemistry of organometallic compounds which relate to biological chemistry and follows the pattern laid down in previous volumes[1] in this Series. Emphasis has been placed on those model compounds and reactions which mimic some important aspects of the behaviour of cobalamin-dependent enzyme systems. The central theme concerns the various modes of cleavage of Co-C bonds, and includes discussion of Co-C bond dissociation energies, the electronic and steric factors responsible for labilising such bonds, and the mechanistic fate of the organyl moiety. The literature is surveyed for the period January 1982 to December 1983.

2 Cobaloxime Model Systems

2.1 Formation of Co-C Bonds.- Alkylcobaloximes are used as simple models for the reactions of cobalamin-dependent enzymes. Reductive arylation has been used to secure the substituted arylcobaloximes (1;R=*p*-$CF_3\cdot C_6H_4$, *m*- and *p*-$MeO_2C\cdot C_6H_4$ or *p*-Ac$\cdot C_6H_4$,L=py)[2]. Reductive alkylation using the halides (2a or 3a) yields the corresponding alkyl(pyridine)cobaloximes (2b or 3b), whereas the halides (4a or 5a) yield a mixture in which the products (2b or 3b) predominate,[3] presumably due to electron transfer and radical equilibration. Improved reductive methods of preparation involve the use of bromo(pyridine)cobaloxime and hydroborate,[4] or of an aprotic medium[5] (although the use of the precursor cyanoethylcobaloxime makes the method cumbersome). Reaction of (bisdimethylglyoximato)-dipyridinecobalt(II) with alkyl iodides or α-bromoesters in the presence of zinc under aprotic anaerobic conditions[6] (Scheme 1) is generally applicable, whereas the reaction of the cobalt(II) species with alkyl halides under basic conditions has restricted application.[4]

The following organylcobaloximes (1;R=Me, $BrCH_2$, Et or Pr^i, L=3-BzlA or 3-EcmA) have been prepared[7] from the preformed organyl derivatives by base replacement. Optically pure organylcobaloximes which contain an α-chiral centre in the alkyl moiety, have been prepared by diastereomeric resolution[8] using a chiral base (Scheme 2). A series of organylcobaloxime analogues (6;R=Me, Et, Pr^i, L=py;R=Me, L=$P(C_6H_{11})_3$, PPh_3, or $P(OMe)_3$) containing a sterically minimal equatorial ligand system has been prepared by reductive alkylation.[9]

<u>2.2 Structural Studies</u>.- *X*-ray crystal diffraction analysis of a range of organylcobaloximes has revealed the following types of structural distortion: variation in Co-C bond length[10-13] [*e.g.* (1;R=Pr^i, L=$P(OCH_2)_3C{\cdot}Me$)[10]], variation in Co-L bond length[11,14] [*e.g.* (1;R=Pr^i,L=$2NH_2py$)[14]], variation in Co-C^α-X^β bond angle[13] [*e.g.* (1;R=$BrCH_2$, L=PPh_3)[13]], variation in C^α-Co-L bond angle[13] [*e.g.* (1;R=CF_3CH_2, L=PPh_3)[15]], deformation of the equatorial ligand system towards the axial alkyl group[11,13] [*e.g.* (1;R=$NCCH_2$, L=PPh_3)[13]], deformation of the equatorial ligand system away from the axial alkyl group[12,16] [*e.g.* (1;R=Me_3CCH_2, L=PMe_3)[12]], and displacement of the cobalt atom from the equatorial plane[11] [*e.g.* (1;R=Me, L=$P(C_6H_{11})_3$)[11]]. The factors influencing these distortions are the electronic *trans* influence,[11,14,16,17] the steric *trans* influence of the axial ligand[10-13,16] (including consideration of the cone angle[10,11,18] subtended by the axial base), and the steric *cis* influence.[11,14,18] These electronic and steric effects have important implications for the reactivity of the organylcobalamins.

In(1;R=XCH_2, L=PPh_3) the steric bulk of X increases across the series thus: $NO_2 \lessdot CN \sim CO_2Me < Me \sim CF_3 < Br < SiMe_3 \lessdot CMe_3$.[13] Differences in structural, spectroscopic, and kinetic parameters of the compounds (6;R=Me, L=PPh_3 or $P(OMe)_3$), which contain a sterically minimal equatorial ligand system, compared to the corresponding alkylcobaloximes are a consequence of electronic effects[9] rather than steric effects.

<u>2.3 Cleavage of Co-C Bonds</u>.- A recently devised thermodynamic method for the determination of metal-alkyl group bond dissociation energies (for a summary see ref. 1b, Vol.11, p.390), has been applied to two series of α-phenylethylcobaloximes, one containing 4-substituted pyridines as the axial base,[19] and the other con-

$$RCo(dmgH)_2L$$

(1)

$$CH_2 = CHCH_2CH(X)CO_2Et$$

(2) a; X = Br
b; X = Co(dmgH)$_2$py

$$CH_2 = CHCH(CH_2X)CO_2Et$$

(3) a; X = I
b; X = Co(dmgH)$_2$py

(4) a; X = I
b; X = Co(dmgH)$_2$py

(5) a; X = I
b; X = Co(dmgH)$_2$py

$$Co^{II}(dmgH)_2py_2 \xrightarrow[\text{ii}]{\text{i}} BrCo^{III}(dmgH)_2py \mid R\cdot + py$$

$$R\cdot + Co^{II}(dmgH)_2py_2 \longrightarrow RCo(dmgH)_2py + py$$

Reagents: i, RBr; ii, Zn, py

Scheme 1

$$\begin{array}{c} Co(dmgH)_2 \\ + \\ CH_2{=}CHR' \end{array} \xrightarrow{\text{i}} R_{\underline{RS}}Co(dmgH)_2B_{\underline{S}} \xrightarrow{\text{ii}} \begin{array}{c} R_{\underline{R}}Co(dmgH)_2B_{\underline{S}} \\ + \\ R_{\underline{S}}Co(dmgH)_2B_{\underline{S}} \end{array}$$

Reagents and conditions: i, B$_{\underline{S}}$; ii, fractional crystallisation
[R' = CO$_2$Me or CN, R = CH$_3$CHR', and B = PhCH(CH$_3$)NH$_2$,
subscripts indicate the configuration of chiral centres]

Scheme 2

taining substituted phosphines of varying cone angle.[20] In the former case the Co–C bond dissociation energies increase systematically with the increase of the basicity of the ligand[19] (the steric factor remains constant), and in the latter case the bond dissociation energies decrease dramatically as the cone angle of the phosphine increases (*i.e.* bulky axial ligands significantly weaken the Co–C bond).[20]

The thermodynamic method has been augmented by a kinetic method[21] which involves determination of the enthalpy of activation of the thermally induced homolytic process shown in Scheme 3. Based on the valid assumption that the enthalpy of activation of the reverse (or radical combination) process is small, the bond dissociation energy can be estimated. Radical traps [XH=$nC_8H_{17}SH$[21] or (7)[22]] are used for efficient scavenging of the alkyl radical, thus preventing secondary reactions from interfering. The method has been applied to (8;R=Pr^n, Pr^i, Me_3CCH_2,Bzl),[21,22] and (9;R=Me_3CCH_2 and Bzl, L=I).[22]

These methods, which promise to have a wide application, have been reviewed:[23,24] the factors influencing Co–C bond dissociation energies are of considerable importance for coenzyme B_{12}-dependent enzymic processes.

Electron capture by methyl or benzyl(pyridine)cobaloxime, in a rigid matrix at 77 K gave, in the former case, a σ^* anion[25,26] (the electron being confined to a modified $3d_{z^2}$ orbital of the metal) and in the later case, a benzyl anion (the strongly perturbed reducing electron occupies the $3d_{z^2}$ orbital). In the former case, thermolysis yields a methyl radical and a Co(I) species[26] (this is also observed by photolysis[25]) while in the latter case, a benzyl anion and a Co(II) species is produced.[26] X-ray induced racemisation of (10) in the crystalline state occurs *via* a selective inversion of the chiral centres in only one of the two types of crystallographically independent molecules present.[27] Ethyl(aquo) cobaloxime (1;R=Et, L=H_2O) is a potent quencher[28] of a wide variety of excited states, generated by laser flash photolysis. The photodecomposition of (1) is sensitised by a number of dye quenchers.[28]

The cyclic voltammetry of alkylcobaloximes under various conditions of solvent, added base, low temperature, and scan rate, shows total electrochemical irreversibility due to fast irreversible competing reactions of the reduced species.[29] The Co–C bond is broken during the two step controlled-potential electrochemical

$$\text{Pr}^i\text{Co(saloph)py} \underset{k_{-1}}{\overset{k_1}{\rightleftharpoons}} \text{Co(saloph)py} + \text{Pr}^i\cdot$$

$$\downarrow k_0 \qquad\qquad\qquad\qquad k_2 \downarrow \text{XH}$$

$$\text{Co(saloph)py} + \bigwedge\!\!\!\diagdown \qquad\qquad \bigwedge\!\!\!\diagdown + \text{X}\cdot(\rightarrow \tfrac{1}{2}\text{X}_2)$$

[where $k_1 \gg k_0$]

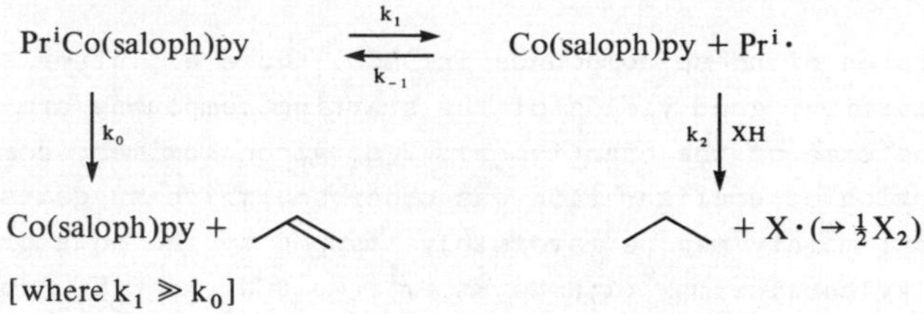

Scheme 3

RCo(GH)$_2$L

(6)

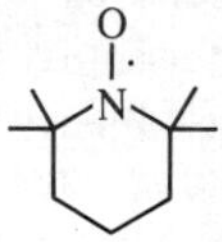

(7)

RCo(saloph)py

(8)

RCo(tn)L

(9)

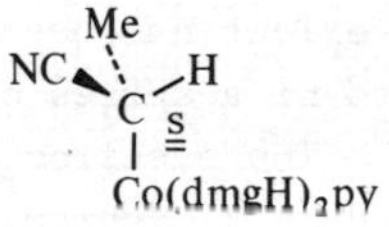

$$\begin{array}{c} \text{Me} \\ \text{NC} \diagdown \underset{\overset{|}{C}}{|} \diagup \text{H} \\ \underset{\text{Co(dmgH)}_2\text{py}}{} \end{array}$$

(10)

Reagents: i, BrĊHCN; ii, Co(dmgH)$_2^-$; iii, Cl$_3$CSO$_2$Cl

Scheme 4

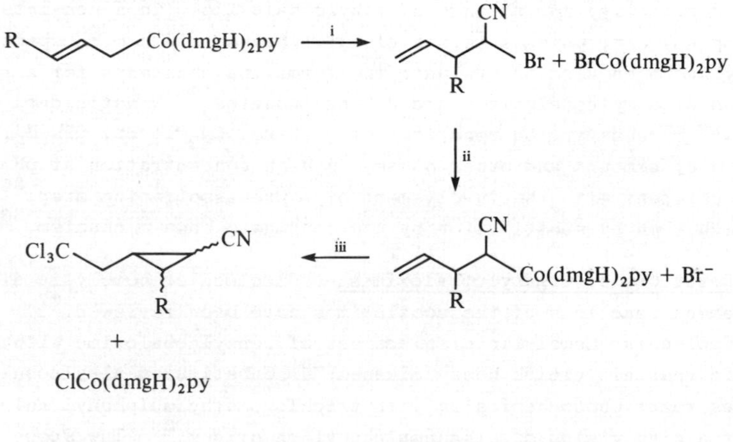

reduction of these compounds in DMSO. However, after subsequent reoxidation, good yields of the starting compounds are recovered.[30] In the case of the enantiomeric 2-octylcobaloximes, complete retention of configuration was observed, which suggests that the organyl moiety may be reversibly trapped by the equatorial ligand.[30]

Alkylcobaloximes (1;R=Me,Et,Pr^i, Me_3CCH_2, c-C_6H_{11}) undergo slow cleavage with anhydrous hydrogen chloride in chlorinated hydrocarbon solvents to give the corresponding alkanes, presumably *via* electrophilic attack by protons on the Co-C bond of protonated cobaloxime.[31] However in the case of (1;R=Bzl) the predominant reaction is homolysis yielding bibenzyl and a Co(II) species.[31] Compounds (1;R=Me, CH_2Cl, $CHCl_2$, Et, Pr^i, Bu^n, $CH_2CH_2CH{=}CH_2$, $CH_2CHMeCH{=}CH_2$) in the presence of trifluoroacetic acid in chloroform, undergo rapid monoprotonation followed by subsequent slow release of RH and formation of a red high spin octahedral Co(II) complex.[32]

The β-effect has been quantitatively examined by studying the pK_a values of a series of carboxymethyl and of 1-carboxyethylcobaloximes.[33] The electron donation by the β-effect appears to be mediated by σ-π hyperconjugation. Ground state *cis* and *trans* effects have been quantitatively evaluated by studying the temperature dependence of the formation constants[34] (determined spectrophotometrically) for a range of alkylcobaloximes, in a non-interacting solvent, using a series of axial bases. ^{1}H n.m.r. spectroscopy has been used to evaluate the formation constants for a series of alkylcobaloximes and 3-benzyladenine.[7] Kinetic data on the water substitution reactions of (1;R=Me, CH_2Cl, Et, CF_3CH_2, L=H_2O) by ammonia and other bases, in high concentration at pH=9.0, is consistent with the involvement of a pre-association step,[35] and at higher pH by substitution by the conjugate base mechanism.[36]

2.4 Reactions of Organylcobaloximes.- Bimolecular homolytic displacement reactions of the cobaloximes have been reviewed.[37]

Bimolecular homolytic displacement of benzylcobaloxime with aliphatic radicals yields benzylalkanes.[38] Substituted alkylcobaloximes react photochemically with trichloromethylsulphonyl chloride to give high yields of alkanesulphonyl chlorides.[39] The scope of the regiospecific allyl-transfer reactions from substituted allyl cobaloximes to a variety of bromodiesters, has been further explored.[40]

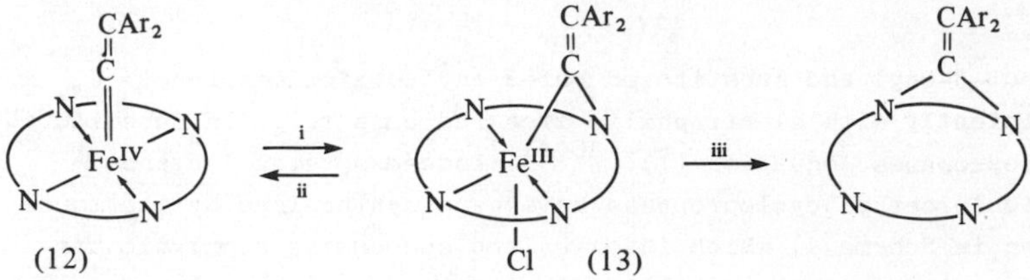

(12) (13)

Reagents: i, $Fe^{III}Cl_3$ or $Cu^{II}Cl_2$; ii, $Na_2S_2O_4$;
iii, excess $Fe^{III}Cl_3$

[where Ar = p-Cl·C_6H_4,  = TPP, p-TTP, T(p-MeO)PP]

Scheme 7

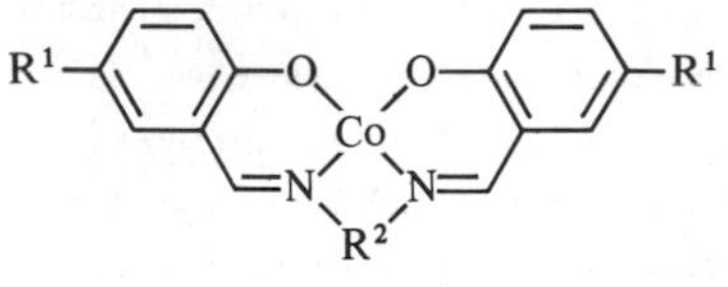

(16)

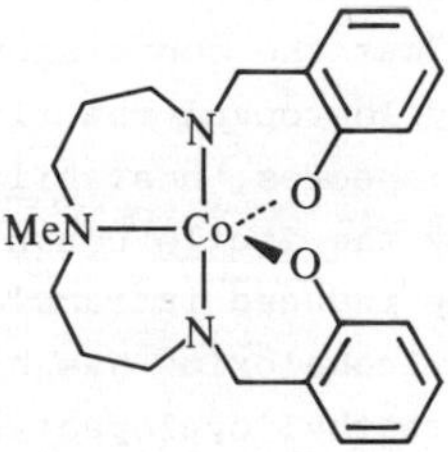

(17)

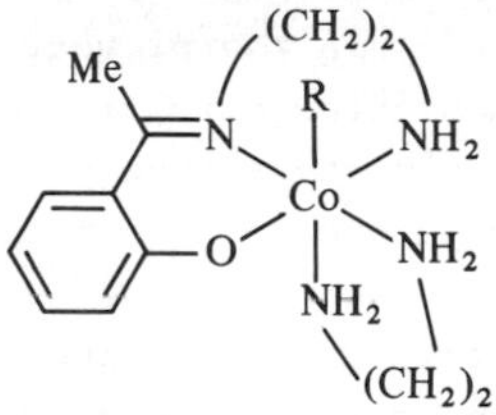

(18)

RCo(TPP)Cl
(19)

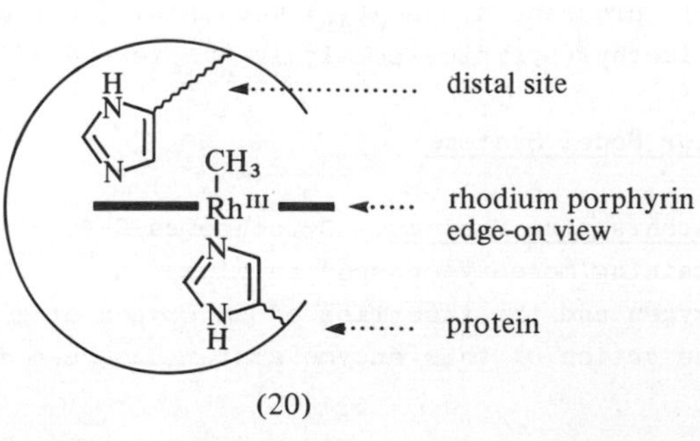

(20)

But-3-enyl and substituted but-3-enylcobaloximes react efficiently with electrophilic free radicals to yield substituted cyclopropanes [equation (1)].[40,41] Indeed several 1-cyano-2-(trichloroethyl)cyclopropanes have been synthesised by the route shown in Scheme 4, which involves two successive homolytic displacements of cobaloxime(II), first from an allylcobaloxime and secondly from the derived but-3-enylcobaloxime.[41]. But-3-enyl (pyridine)cobaloximes react with diethyl bromomalonate to give cyclopropylmethylmalonic esters,[40] [equation (1), where $X\cdot = \cdot CH(CO_2Et)_2$].

Under acidic conditions 1 and 2-methyl-substituted but-3-enyl (pyridine)cobaloximes undergo a fast equilibration, such that the 2-methyl isomer predominates. Conversely, under similar conditions the isomeric carboxyethylcobaloximes (2b and 3b) undergo a slow equilibration in which the 1-carboxyethyl isomer (2b) predominates, presumably due to the stabilising metal-carbonyl interaction present.[3] That the conversion of (3b) to (2b) probably proceeds through a cyclopropane reaction intermediate (probably *via* a η^3-homoallylic species, destabilised by the carbethoxy group) is indicated by the facile conversion of (4b) to (2b).[3]

Thermally induced intramolecular homolysis of the organyl chain of an organylcobaloxime has been used for the ingenious synthesis of (trichloroethyl)cyclopentanes (Scheme 5).[42] In the presence of cobaloxime(I), 1,1-dichloro-2,2-diarylethanes undergo rearrangement to *trans* stilbenes[43] in good yield, presumably *via* a cobalt-carbenoid complex. Methylcobaloxime is formed when cobaloxime(II) and tertiary butyl hydroperoxide react in aqueous solution.[44]

The non-equivalence of the equatorial methyl groups in the ^1H n.m.r. spectrum of α-cyanoethyl(pyridine)cobaloxime has been re-examined.[45] ^{17}O n.m.r. spectroscopy has been used to demonstrate that the hydrolysis of 2-acetyl-[^{17}O]-oxyethyl(pyridine)cobaloxime in dioxan-water, proceeds by the $B_{AL}1$ mechanism.[46] The hydration of [^{17}O]-formylmethyl(pyridine)cobaloxime is very difficult.[46]

3 Other Model Systems

3.1 Manganese-containing Systems.- Cytochromes P-450 (a class of haem-iron containing monooxygenases) catalyse the reductive activation of dioxygen and the insertion of an oxygen atom into organic compounds. The action of this enzyme system has been mimicked by a variety of synthetic metalloporphyrins. Manganese(III) porphyrins catalyse alkane (also alkene) functionalisation *via* high-valent manganese-oxo (or nitrido) species. Table 1 lists the systems or compounds which have been investigated together with the type of alkane oxidation observed.

Table 1 Manganese porphyrins which catalyse alkane functionalisation

System/Compound	Conversion	Ref.
Mn(III)(TPP)(X) + PhIO (System 1, X = Cl, Br, I, or N_3)	RH → RX RH → ROH alkene → oxirane	47,48
Mn(IV)(TPP)$(OMe)_2$[a]	-	49[b]
[Mn(IV)(X)(TPP)]$_2$O[a] (X = N_3 or OCN)	RH → ROH RH → RX alkene → allylic products	50[b],51
[Mn(IV)(X)(TPP)(PhIO)]$_2$O[a] (X = Cl or Br)	RH → ROH RX → RX alkene → oxirane alkene → allylic products	51,52[b]
Mn(IV)(TPP)[PhI(OAc)O]$_2$[a]	RH → ROH alkene → oxirane	53
Mn(III)(TMP)(OAc) + NaOCl + phase transfer catalyst + pyridine (System 2)	alkene → oxirane	54-56
[Mn(V)(TMP)O]Cl[c]	alkene → oxirane	57
[Mn(III)(T(*p*-poly)PP)(OAc)][d] + NaOCl + phase transfer catalyst + pyridine	alkene → oxirane	58
Mn(III)(TPP)(Cl) + O_2 + ascorbate + phase transfer catalyst	RH → ROH R'R"CHOH → R'R"CO alkene → oxirane	59
Mn(V)(TMP)(O)(X)	alkene → oxirane	60
Mn(III)(TPP) + air + $NaBH_4$	alkene → ROH	61
Mn(III)(TPP)(Cl) + O_2 + $[NBu_4][BH_4]$	terminal alkene → methyl ketones, plus ROH by further reduction	62
Mn(III)(TPP)(Cl) + O_2 + H_2/colloidal Pt + N-MeIm	alkene → oxirane RH → ROH	63
Mn(III)(TMP)(X) + PhIO + NH_3		64
Mn(III)(TPP)(X) + NaOCl + NH_3 (System 3)	alkene → aziridine	65
Mn(V)(TMP)N	alkene → aziridine	64
Mn(V)(TPP)N[e]		65[b]
Mn(V)(T(*p*-MeO)PP)N		66[b]

a Compound derived from System 1.
b Reference deals with structure only.
c Compound derived from System 2.
d Catalyst is supported on a polymer matrix, the isolation of the sites prevents formation of less reactive dimers, thereby slightly enhancing the rate of epoxidation in this triphasic system.
e Compound derived from System 3.

Free alkyl radicals have been implicated in a number of these alkane modification reactions.[47,48,51,53] The role, if any of transient alkylmanganese species has not been elucidated but conflicting views have been expressed.[48,51]

3.2 Iron-containing Systems.- Iron(III) porphyrins [*e.g.* Fe(III)(TPP)(Cl) and Fe(III)(*o*-TTP)(Cl)] in the presence of iodosylbenzene, catalyse the epoxidation of alkenes[67,68] and the hydroxylation of alkanes.[69,70] The asymmetric induction of the epoxidation of prochiral olefins has been explored using chiral metalloporphyrin atropisomers[71] [*e.g.* chloro[5α,10β,15α,20β-tetrakis (*o*-[(*R*)-2-phenylpropionamido]phenyl)porphyrinato]iron(III) and the corresponding chloro[5α,10β,15α,20β-tetrakis(*o*-[(*S*)-2'-carboxymethyl-1,1'-binaphthyl-2-carboxamido]phenyl)porphyrinato]iron(III)]. Alkyl radicals are involved in the alkane hydroxylation reaction[69,70] but the role, if any, of transient organometallic species is unknown.

The biphasic haem system [Fe(III)(TPP)(Cl), ascorbate, phase transfer catalyst] is able to mimic a range of microsomal cytochrome P-450-dependent reductions of substrates.[72]

Haem proteins such as cytochrome P-450[73] and haemoglobin[74] react with arylhydrazines and suffer inactivation.[73,74] It has been shown that phenyldiazene[73] (produced by oxidation of the reagent), and also methyldiazene,[75] react with the haem protein to form an unstable Fe(II)diazene,[75,76] which decomposes to an Fe(III) σ-aryl derivative,[76] and in the presence of acid and oxygen the aryl group migrates to form an *N*-arylmetalloporphyrin.[77] Model studies have elegantly confirmed this reaction sequence[78-80] which is shown in Scheme 6.

Iron(III) porphyrins containing a σ-alkyl or aryl group can be prepared from a Grignard reagent and iron(III)porphyrin,[79,81-83] an alkyl halide and iron(II)porphyrin in the presence of a reductant,[84] or organyl radicals and iron(I)porphyrins[85] (both species being generated electrochemically). Pulse radiolysis experiments show that methyl radicals react with both iron(II) and iron(III) deuteroporphyrin.[86]

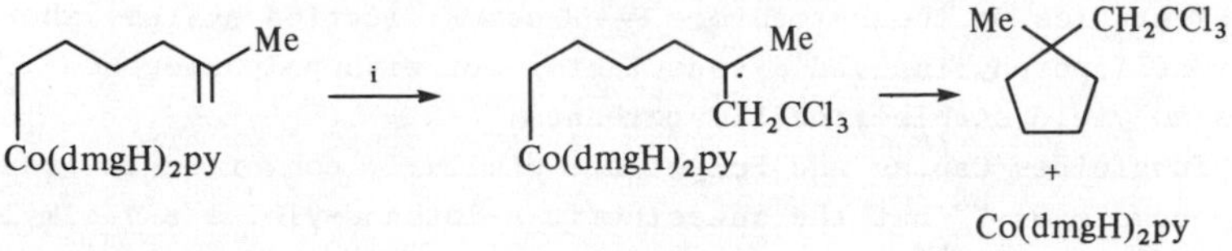

Reagent: i, $Cl_3C\cdot$

Scheme 5

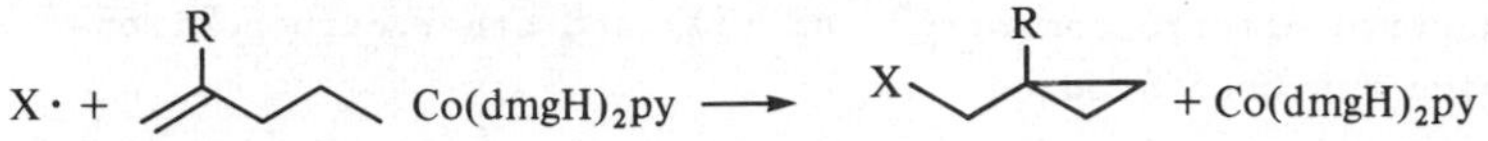 (1)

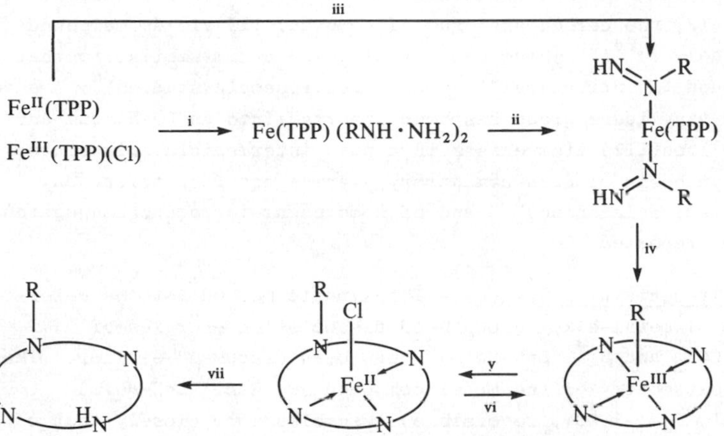

Reagents: i, $RNH\cdot NH_2$; ii, O_2; iii, RN:NH; iv, O_2 or $Fe^{III}Cl_3$; v, $HCl/MeOH/O_2$; vi, $Na_2S_2O_4$; vii, HCl

Scheme 6

$(TPP)Fe^{IV}{=}C{=}Fe^{IV}(TPP)$

(11)

$Me_2Co(pn)$

(14)

$[Me_2Co(tim)]^+$

(15)

Model studies of the cytochrome P-450 detoxification system, show that iron(II)porphyrins and a reductant react with polyhalogenoalkanes to yield stable iron(II) carbenes.[87]

The fungicides Captan and Folpet are similarly converted to iron-carbene complexes,[88] but the anaesthetic halothane yields a σ-alkyl iron(III) porphyrin.[89] Theoretical calculations[90] of the structure of (11) concur with the subsequently determined X-ray crystal structure,[91] and the Mössbauer spectroscopic properties.[92] The reductive electrochemistry[93] of (11) and other carbene-iron-porphyrins has been studied.

The catalytic cycle of the haemoproteins, catalase and peroxidase, involves two porphyrin-iron-oxene species known as compound I (formally iron(V)) and compound II (formally iron (IV)). The reversible one electron oxidation of the iron-porphyrin-carbene complex (12; the carbon analogue of compound II) yields compound (13) (Scheme 7)[94,95] which has a visible spectrum similar to that of compound I. Structural[96,97] and spectroscopic studies[98,99] show that the vinylidene group has been inserted into an Fe-N bond and that the iron(III) atom exists in a pure intermediate spin state.[98]

The synthesis of *DL*-β-cymantrenylalanine and *DL*-β-tricarbonyl (cyclobutadienylalanine)[100] and of some sugar-ferrocene conjugates[101] have been reported.

3.3 Cobalt-containing Systems.- The kinetic method for the determination of metal-alkyl group bond dissociation energies of (9; R=Me_3CCH_2 and Bzl, L=N-MeIm)[22] has been discussed earlier. The cyclic voltammetry of the model compound (9; R=Me, L=N-MeIm) (CoIII/II, $E_{\frac{1}{2}}$= -1.20V, reversible) resembles more closely that of methylcobalamin and methylcobinamide (CoIII/II, $E_{\frac{1}{2}}$= -1.47V, reversible) than that of methylcobaloxime (CoIII/-, two waves E_p= -2.3 or -2.5V, irreversible).[29] The oxidative cleavage of some dimethylcobalt(III) macrocycles (14 or 15) both by chemical and electrochemical methods leads to the selective cleavage of one methyl group as a radical[102] (as evident by trapping experiments) with the formation of ethane *via* a highly labile dimethylcobalt(IV) cation formed by an irreversible outer sphere process.

The kinetic determination of the Co-C bond dissociation energy of compounds (8; R=Pr^n, Pr^i, Me_3CCH_2,Bzl) has been discussed earlier.[21,22] The electrochemistry[103] of some perfluoroalkyl derivatives (8; R=CF_3,C_2F_5, or C_3F_7) has been studied. The kinetics for the formation of organyl derivatives from electrogenerated diastereoisomeric cobalt(I) chelates (derived from 16; R^1 = H and R^2 = $(CH_2)_2$,$(CH_2)_4$,$CH_2CH(CH_3)$ ((+) or (±) forms), $CH_2CH(Ph)$((-) form), $CH(CH_3)CH(CH_3)$ ((+) or meso forms), CH(Ph)CH(Ph)((+) or meso forms) and alkyl halides have been evaluated in terms of stereochemical perturbation.[104] In the case of t-butyl bromide, the cobalt(II) chelate, isobutene and hydrogen are produced. The water-soluble cobalt(II) compounds (16; R^1 = SO_3^-, R^2 = $(CH_2)_2$,$(CH_2)_3$, $CH_2CH(CH_3)$,*o*-C_6H_4) can be converted to water soluble organometallic derivatives (Me,Et,Pr^n) by reductive alkylation.[105] The compound (17) selectively catalyzes the oxidation of terminal olefins by dioxygen to the corresponding 2-ketones and 2-alcohols.[106] The organocobalt(III) complex (18) undergoes homolysis in the presence of protons.[107,108]

A further study of the transient species involved in the reaction of $\cdot CH_2CMe_2OH$ radicals with cobalt(II) tetrasulphophthalocyanine, using pulse radiolysis in aqueous solution, shows a dependence of reaction mechanism on pH.[109] The results (see Scheme 8) suggest that the pK_a of a water molecule coordinated *trans* to the organyl group has a controlling influence on the mechanism of hydrolysis of the Co-C bond. This subtle control may be relevant to reactions catalysed by coenzyme B_{12}.

The organyl group of (19; R=Et,EtO_2C or Ph) suffers reversible migration to a porphyrin *N* atom upon electrochemical oxidation or acid treatment[111] (Co → N), or hydroborate reduction[110,111] (N → Co).

3.4 Rhodium-containing System.- The reconstitution of sperm whale myoglobin with the organometallic porphyrin, methylrhodium(III)-mesoporphorin IX, gives rise to a coordination structure formulated as (20).[112] The organyl group is a potential probe of the distal site in the haem proteins.

4 Studies of Cobalamins

4.1 Formation of Co-C Bonds.- Reductive alkylation has been used for the synthesis of stereoisomerically pure samples of (2*R*)- and (2*S*)-2-hydroxypropyl-, (2*R*)- and (2*S*)-2,3-dihydroxypropyl-,(3*S*)-3,4-dihydroxybutyl-, and (4*S*)-4,5-dihydroxypentyl-cobalamin,[113] and analogues of neopentylcobalamin in which one or two β-methyl groups have been replaced by ethoxycarbonyl[114] (in the latter series the use of ammonium iodide allows the isolation of product in the protonated 'base off' form). 'Base on', 2,2-bis(ethoxycarbonyl)-ethylcobalamin has also been prepared.[114]

4.2 Cleavage of Co-C Bonds.- The Co-C bond dissociation energies, measured for several organylcobalt model compounds,[19-24] are sufficiently low to be consistent with the proposed bond homolysis[24] in adenosylcobalamin-promoted reactions. The factors which may help to labilise this bond include : steric influences[14,21,115] (*e.g.* sterically hindered axial bases weaken the Co-C bond[19,20,23]), electronic influences[14] (*e.g.* the Co-C bond dissociation energy increases with the basicity of the axial ligand[19,23]), the mobility of the axial ligand[109] (*i.e.* base substitution[24] or dissociation[109]), the oxidation state of the metal[24], and induced conformational changes upon protein-coenzyme-substrate binding.[14,22,116,117]

A pulse radiolysis study of the reaction of B_{12r} with the radicals $\cdot CH_2CMe_2OH$, $\cdot CMe_2OH$, $\cdot CH_2CHO$, or $\cdot CH(OH)CH_2OH$ reveals that transient alkylcobalt(III) species are formed initially.[118] At 77 K, methyl[119], ethyl, and adenosylcobalamin undergo electron capture[120] into a π^* corrin orbital. On warming, the methyl compound gave a species with an unpaired electron in the $Co(d_{x^2-y^2})$ orbital[119], whereas the other two compounds gave B_{12r} (with an electron in the d_{z^2} orbital). Solid state photolysis[121] of the above mentioned compounds results in homolysis of the Co-C bond with formation of B_{12r} (in most cases *via* a Co(II)R· radical pair).[119]

Organylcobalamins of the type (21) exist in the 'base on' form in neutral aqueous solution and undergo spontaneous decomposition by sterically-induced Co-C bond homolysis[114] (under aerobic conditions some demethylated products are produced). In contrast the 'base on' derivative (22) decomposes spontaneously by *syn* β-elimination.[114]

4.3 Electrochemistry.- The electrochemistry of the cobalamins has been reviewed[122] and comparisons with model compounds[29] have already been mentioned. The one electron electrochemical reduction of adenosylcobalamin or methylcobalamin yields B_{12s} (in the latter case ethane is produced quantitatively),[123] whereas the two electron oxidation yields B_{12a} (with methanol the major product in the latter case).[123] The spectroelectrochemical determination of relative free energies of formation and free energies of interconversion of various cobalamins, as a function of pH, shows that

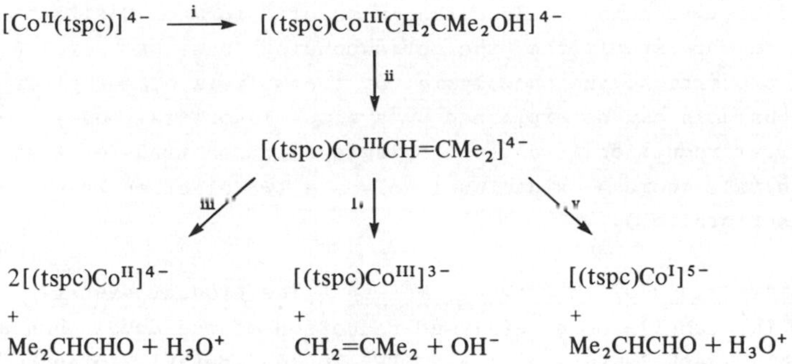

$$[\text{Co}^{\text{II}}(\text{tspc})]^{4-} \xrightarrow{\quad i \quad} [(\text{tspc})\text{Co}^{\text{III}}\text{CH}_2\text{CMe}_2\text{OH}]^{4-}$$

$$\downarrow ii$$

$$[(\text{tspc})\text{Co}^{\text{III}}\text{CH}=\text{CMe}_2]^{4-}$$

iii ↙	i↓	↘ v

$$2[(\text{tspc})\text{Co}^{\text{II}}]^{4-} \qquad [(\text{tspc})\text{Co}^{\text{III}}]^{3-} \qquad [(\text{tspc})\text{Co}^{\text{I}}]^{5-}$$

$$+ \qquad\qquad + \qquad\qquad +$$

$$\text{Me}_2\text{CHCHO} + \text{H}_3\text{O}^+ \qquad \text{CH}_2=\text{CMe}_2 + \text{OH}^- \qquad \text{Me}_2\text{CHCHO} + \text{H}_3\text{O}^+$$

Reagents: i, $\cdot\text{CH}_2\text{CMe}_2\text{OH}$; ii, $-\text{H}_2\text{O}$; iii, $[\text{Co}^{\text{III}}(\text{tspc})]^{3-}$, H_2O (pH $\sim$ 3);

iv, H_2O (pH $\sim$ 6); v, H_2O (pH $>$ 9)

Scheme 8

$$\begin{array}{c} \text{Me} \diagdown \diagup \text{CO}_2\text{Et} \\ \diagup \diagdown \text{CO}_2\text{Et} \\ | \\ [\text{Co}] \end{array}$$

(21)

$$\begin{array}{c} \text{H} \diagdown \diagup \text{CO}_2\text{Et} \\ \diagup \diagdown \text{CO}_2\text{Et} \\ | \\ [\text{Co}] \end{array}$$

(22)

(23)

$$(\text{R}^1\text{CO})_2\text{O} + \text{R}^2\text{CH}=\text{CR}^3\text{Z} \xrightarrow[\text{DMF, B}_{12}]{\text{e}^-,\, h\nu} (\text{R}^1\text{CO})\text{R}^2\text{CH}-\text{R}^3\text{CHZ} + \text{R}_1\text{CO}_2^-$$

$$[\text{where } Z = \text{CHO}, \text{CR}^4\text{O}, \text{CO}_2\text{R}^5, \text{CN}] \tag{2}$$

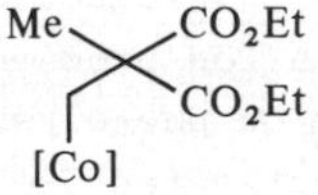

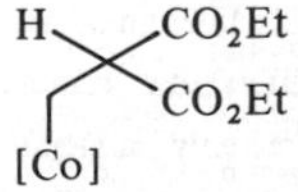

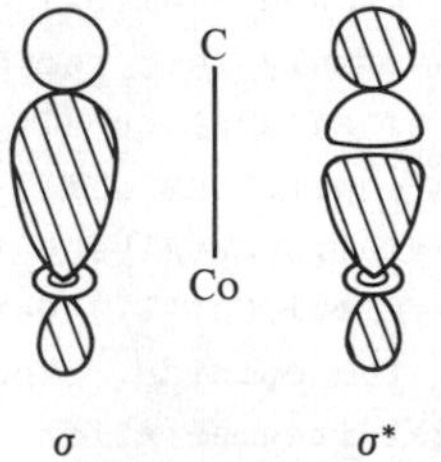

Co(II) cobalamin species are more stable than the corresponding Co(III) or Co(I) species (indeed the 'base off' form of Co(II) cobalamin is more stable than the corresponding 'base on' form).[124] The redox properties, the photolysis, or thermolysis of methyl or adenosylcobalamin can be explained by a simple σ-orbital model (23)[124] (*i.e.* reduction involves a single electron transfer into the σ^* orbital, whereas oxidation involves a two electron transfer out of the σ orbital).

4.4 Transformations of the Organyl Moiety.- The product distribution in the cob(I)alamin catalysed reduction of the double bonds in 3β-methoxy-4-cholesten-6α-ol and (1*R*)-10,10-dimethyl-2-pinene-10-carbonitrile has been rationalised.[125] In the former case, normal Markownikoff attack by the catalyst on **either face** of the Δ^4 double-bond is followed by reductive cleavage with retention of configuration. In the latter case, nucleofugal fragmentation of the intermediate tertiary alkylcobalamin occurs.

4-Oxo aldehydes, ketones, esters, or nitriles, can be formed from the corresponding carboxylic anhydride and activated olefin, using catalytic amounts of vitamin B_{12a} under irradiation with visible light and chemical or electrochemical reduction [equation (2)]. Presumably this regiospecific acylation proceeds *via* initial formation of an acyl cobalt species which on photolysis releases an acyl anion equivalent which adds to the activated olefin.[126] In the cob(I)alamin catalysed transformation shown in equation (3),[127] a β-hydroxyalkyl Co(III) derivative is formed by nucleophilic ring opening of the epoxide. Upon reduction with zinc/acetic acid, electrofugal fragmentation yields (24), further attack by the catalyst leads to the formation of a secondary alkyl derivative which undergoes reductive cleavage under protic conditions to yield (25).

Anaerobic reductive cleavage of the Co-C bond in the protonated 'base-off' derivative (26) yields mainly the rearranged product [equation (4)].[114]

The cyanide-induced formation of methyl acetate from the compound (27), yields kinetic data consistent with the mechanism shown [equation (5)].[128] Kinetic isotope effects and studies with chirally labelled substrate indicate, in part, that protonation of the enolate occurs preferentially on one face in an intermediate π-complex.

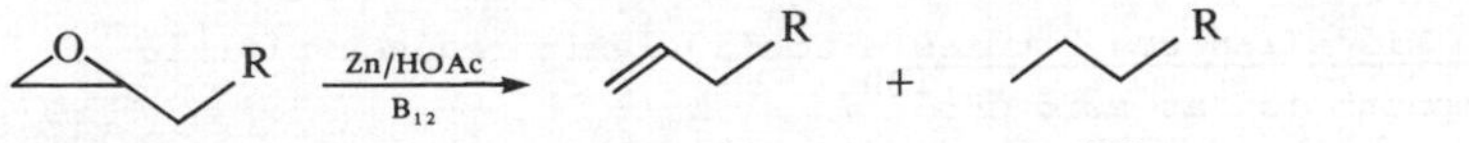

(24)   (25)

[where R = $(CH_2)_8O_2CMe$]

$$\text{(4)}$$

(26)   25 : 1

$$\text{(5)}$$

(27)

(28)   (29)   (31)

$$Co(tn)Cl \xrightarrow[\text{MeOH}]{\text{MeO}^-} Co^{II}(tn)Cl + CH_3CHO + CH_3O\overset{O}{\overset{\|}{C}}OCH_3 + CH_3O\overset{O}{\overset{\|}{C}}O^-$$

$$\text{(6)}$$

(30)   equivalents: 1   0.95   0.5   0.5

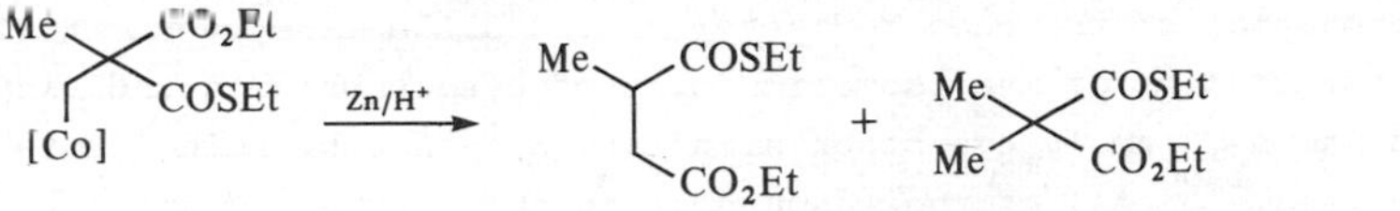

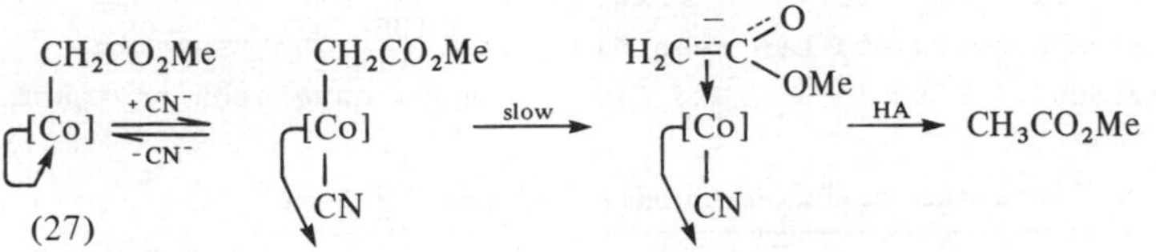

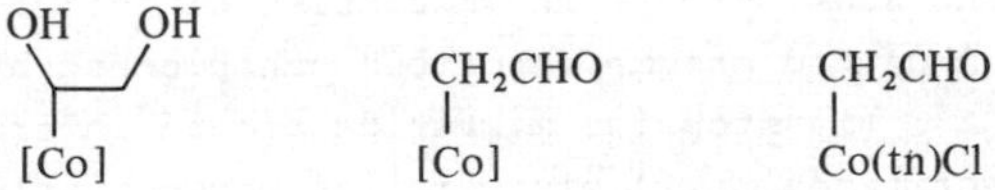

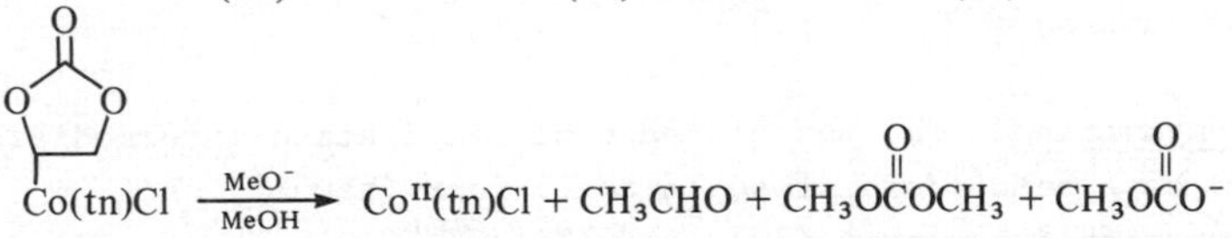

(32)   (33)

4.5 Miscellaneous Studies.- Cob(I)alamin reduces nitrate to ammonium ion at acid pH.[129]

The ^{31}P nmr spectrum of methylcobalamin, in contradistinction to that of cyanocobalamin, does not show an upfield shift in the phosphorus resonance on conversion to the 'base off' form.[130,131] However, below pH = 2.17 a substantial upfield shift occurs due to protonation of the phosphodiester of the nucleotide loop. Further ^{1}H nmr studies of the isomeric forms of cyanocobalamin and dicyanocobinamide, namely cyanocobalamin' and dicyanocobinamide' have been reported.[132] All of the resonances in the ^{1}H and ^{13}C nmr spectra of heptamethyl dicyanocobyrinate have been assigned,[133] and the X-ray structure analysis of the anhydrous form has been reported.[134] The mass spectrometry of methylcobalamin and its analogues using fast atom bombardment has been discussed.[135] The magnetic properties of vitamin B_{12} and its analogues have been reviewed.[136]

5 Studies of B_{12}-dependent Enzyme Systems

5.1 General.- According to theoretical calculations, the 1,2 radical migration within the substrate in the mechanism of action of adenosylcobalamin-mediated enzyme reactions can proceed by three distinct pathways[137]: (i) stepwise migration *via* an intermediate complex (diol dehydratase), (ii) dissociation-recombination (glutamate mutase), and (iii) concerted migration or (ii) (methylmalonyl-CoA-mutase).

5.2 Diol Dehydratase.- Two postulated diol dehydratase intermediates (28 and 29) have been synthesised in model form (namely, the carbonate protected compound (30), and (31)).[138] Substance (30) undergoes catalytic deprotection with methoxide to yield the products shown [equation (6)]. However, the compound (31) is stable and is not an intermediate in the deprotection of (30).[139] These observations argue against cobalt-participation in the enzymic rearrangement step, and suggest that it is the protein and not the cofactor which is involved in the 'bound radical mechanism'.[139] 2,2-Diethoxyethyl(pyridine)cobaloxime (32), in the presence of [^{17}O]-water and HCl, yields [^{17}O]-formylmethyl(pyridine)cobaloxime (33), which at pH = 6.8, slowly loses the label, presumably *via* [hydroxy-^{17}O]-dihydroxyethyl(pyridine)cobaloxime.[46] These compounds are also possible intermediates for the diol dehydratase system.

Reductive alkylation of compound (34) with the diastereoisomeric 6,7-dihydroxycycloundecyl iodides (35) gives cycloundecanone [equation (7)], *via* a transannular 1,6-radical migration, followed by loss of a proton and elimination of OH.[140] This reaction mimics the diol dehydratase-catalysed reaction.

5.3 Ethanolamine Ammonia-lyase.- Studies of the mechanism of ethanolamine ammonia-lyase using the series of adenosylcobalamin analogues (36;x = 1-6), indicate that Co-C bond cleavage is induced by conformational changes in the corrin ring, mediated by the protein, but triggered by cofactor binding.[141] Chirally labelled ethanolamine[142] substrate has been used to show that the enzyme-catalysed rearrangement is fully stereospecific, and proceeds with migration of the (1*S*)-hydrogen atom.[143]

5.4 Methylmalonyl-CoA Mutase.- The anaerobic reductive rearrangement of the organyl moiety in the derivative (26), depicted in [equation (4)], is related to the methylmalonyl-CoA mutase catalysed reaction[114] for which a carbanion mechanism has been proposed (Scheme 9).

5.5 α-Methyleneglutarate Mutase.- The organylcobaloximes (2b-5b) undergo slow acid-catalysed equilibration (Section 2.4). Accordingly the corresponding organylcobalamins would be insufficiently reactive to fulfill their postulated role as intermediates in the cobalamin-dependent α-methyleneglutarate mutase reaction.[3] However, the enzymic reaction may proceed *via* intermediate carboxy-substituted but-3-enyl and cyclopropylmethyl radicals.

5.6 Glutamate Mutase.- Stereospecifically labelled glutamic acid has been prepared for studies of the stereochemistry of the co-enzyme B_{12}-mediated glutamate mutase-catalysed rearrangement of glutamic acid to β-methylaspartic acid.[144]

6 Alkyl-transfer Reactions and Alkyl Metal Derivatives

6.1 General.- The biological transmethylation of metals and metalloids by methylcobalamin, methylsulphonium salts, and methylmetals has been reviewed.[145] Methylcobalamin undergoes methyl-transfer to

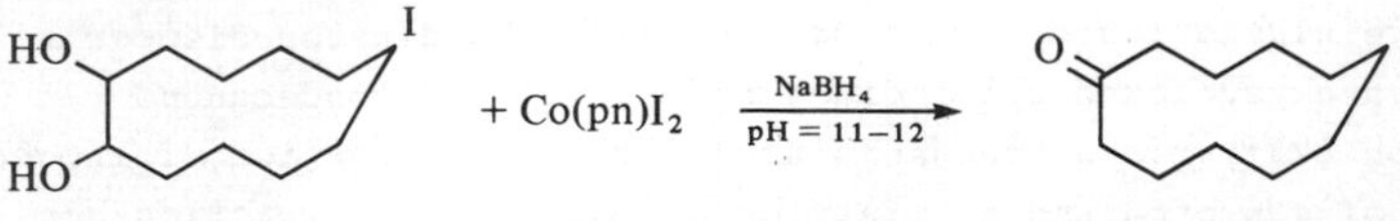

(7)

(35) (34)

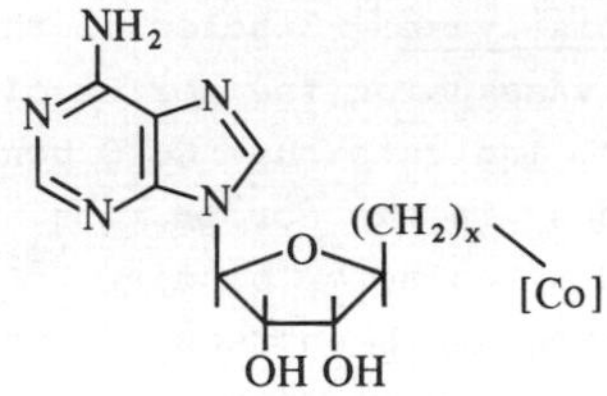

(36)

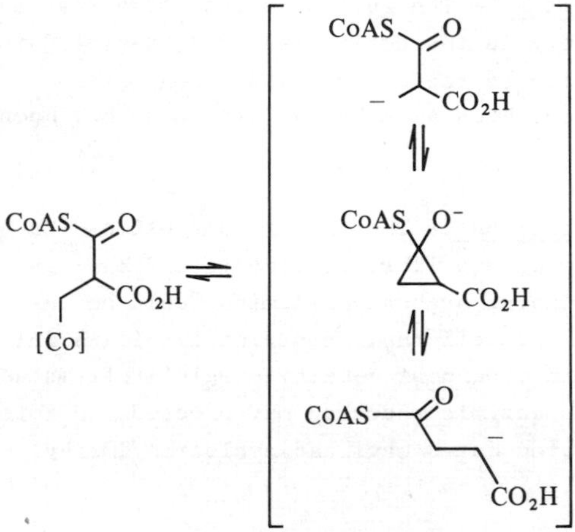

Scheme 9

$$M^{2+} \xrightarrow{(14)\text{ or }(15)} MeM^{+} \xrightarrow{(14)\text{ or }(15)} Me_2M \qquad (8)$$

[where M = Zn or Cd]

$$Pb^{2+} \xrightarrow{(14)} [MePb^{+}] \xrightarrow{(14)} Me_2Pb \xrightarrow{(14)} Me_3Pb^{+} \xrightarrow{(14)} Me_4Pb \qquad (9)$$

tetracyanoethylene.[146]

6.2 Gold.- The one-electron oxidative demethylation of methylcobalamin by tetrahaloaurates in acidic solution, under anaerobic conditions, yields aquocobalamin with an oxidised corrin ring, methyl halide, and metallic gold.[147] Ethylcobalamin is similarly dealkylated by tetrahaloaurates and the kinetic data supports the electron transfer mechanism.[148]

6.3 Zinc and Cadmium.- The *trans*-dimethylcobalt(III) complexes (14 and 15) rapidly transfer methyl groups to Zn^{2+} and Cd^{2+} in acetonitrile solution [equation (8)].[149] The $MeZn^{+}$ or $MeCd^{+}$ produced, evolve methane slowly,whereas the transient products Me_2Zn and Me_2Cd rapidly evolve methane by solvolysis reactions.

6.4 Mercury.- The highly toxic species (37) accumulates in certain higher marine organisms by a vitamin B_{12}-mediated process. Interaction of thiolated models of B_{12} and the appropriate mercury(II) salts does not yield the compound (37).[150] The structure of substance (38) has been determined.[150]

6.5 Tin.- The formation of complexes between $MeSnCl_3$ or Me_2SnCl_2 and a series of methylcobalt macrocycles has been studied in solution by ^{1}H n.m.r. spectroscopy.[151]

6.6 Lead.- The compound (14) rapidly methylates Pb^{2+} in acetonitrile solution, ultimately forming Me_4Pb, [equation (9)], and may relate to the corresponding biotransformation.[152]

6.7 Arsenic.- Cobalt(II) cobaloximes react with organohaloarsines to yield compounds such as (39) which may have some relevance for the biological methylation of arsenic.[153]

7 Books and Reviews

The period under review has seen the publication of several major reference works. The two volumes modestly titled 'B_{12}' give an up-to-date, critical, and comprehensive review of all the major chemical, biochemical, and medical aspects of vitamin B_{12}.[154] The preparations, properties and structures, and reactions of organyl-

cobalt model systems related to vitamin B_{12} have been reviewed
comprehensively[155-158] and in one case experimental detail is
included.[156] Specific reviews have appeared on cobaloximes,[159]
the determination of Co-C bond dissociation energies,[23,24]
bimolecular homolytic displacement of transition-metal complexes
from carbon,[37] mechanistic studies of bio-inorganic systems,[160] and
the electrochemistry[122] and magnetic properties[136] of vitamin B_{12}.
The mechanism of adenosylcobalamin-dependent enzymic reactions[161]
and the biological methylation of metals and metalloids[145] has been
reviewed. The abstracts of the First International Conference on
Bio-inorganic Chemistry contain several papers of relevance to
organo-cobalt chemistry.[162] A monograph[163] on 'Metals in Biochemistry' has appeared.

$$\begin{array}{c} I \diagdown \diagup I \\ Hg \\ MeS \diagup \diagdown \end{array} \Bigg\}_2$$

MeHgSMe   Co(dmgH)$_2$py

(37)   (38)

$$\begin{array}{c} Me \diagdown \\ Ph-AsCo(dmgH)_2Cl \\ HO \diagup \end{array}$$

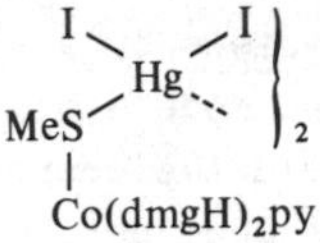

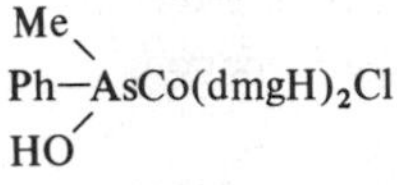

(39)

References

1a D.J.Cardin, in 'Organometallic Chemistry', ed. E.W.Abel and F.G.A.Stone (Specialist Periodical Reports), The Chemical Society, London, 1975, Vol.5, p.435; 1976, Vol.6, p.397; 1977, Vol.7, p.400; 1980, Vol.8, p.434.

1b B.Ridge, in 'Organometallic Chemistry', ed. E.W.Abel and F.G.A.Stone (Specialist Periodical Reports), The Royal Society of Chemistry, London, 1981, Vol.9, p.408; 1983, Vol.11, p.387.

2 K.L.Brown and R.Legates, J.Organomet. Chem., 1982, 233, 259.

3 B.T.Golding, S.Muresigye-Kibende, J. Chem. Soc., Chem. Commun., 1983, 1103.

4 M.R.Ashcroft, M.P.Atkins, B.T. Golding, M.D.Johnson, and P.J.Sellars, J. Chem. Res.(S), 1982, 216.

5 D.G.H.Livermore and D.A.Widdowson, J. Chem. Soc., Perkin Trans. 1, 1982, 1019.

6 P.F.Roussi and D.A.Widdowson, J. Chem. Soc., Perkin Trans. 1, 1982, 1025.

7 P.J.Toscano, C.C.Chiang, T.J.Kistenmacher, and L.G.Marzilli, Inorg. Chem., 1981, 20, 1513.

8 Y.Ohgo, S.Takeuchi, Y.Natori, J.Yoshimura, Y.Ohashi, and Y.Sasada, Bull. Chem. Soc. Jpn., 1981, 54, 3095.

9 P.J. Toscano, T.F.Swider, L.G.Marzilli, N.Bresciani-Pahor, and L.Randaccio, Inorg. Chem., 1983, 22, 3416.

10 N.Bresciani-Pahor, G.Nardin, L.Randaccio, and E.Zangrando, Inorg. Chim. Acta, 1982, 65, L143.

11 N.Bresciani-Pahor, L.Randaccio, P.G.Toscano, A.C.Sandercock, and L.G.Marzilli, J. Chem. Soc., Dalton Trans., 1982, 129.

12 N.Bresciani-Pahor, M.Calligaris, G. Nardin, and L.Randaccio, J. Chem. Soc., Dalton Trans., 1982, 2549.

13 N.Bresciani-Pahor, L.Randaccio, M.Summers, and P.J.Toscano, Inorg. Chim. Acta, 1983, 68, 69.

14 M.F.Summers, P.J.Toscano, N.Bresciani-Pahor, G.Nardin, L.Randaccio, and L.G.Marzilli, J. Am. Chem. Soc., 1983, 105, 6259.

15 Quoted in Ref. 13 and 14.

16 N.Bresciani-Pahor, L.Randaccio, P.J.Toscano, and L.G.Marzilli, J. Chem. Soc., Dalton Trans., 1982, 567.

17 N.Bresciani-Pahor, M.Calligaris, L.Randaccio, and P.J.Toscano, J. Chem. Soc., Dalton Trans., 1982, 1009.

18 N.Bresciani-Pahor, L.Randaccio, and P.J.Toscano, J. Chem. Soc., Dalton Trans., 1982, 1559.

19 F.T.T.Ng, G.L.Rempel, and J.Halpern, J. Am. Chem. Soc., 1982, 104, 621.

20 F.T.T.Ng, G.L.Rempel, and J.Halpern, Inorg. Chim. Acta, 1983, 77, L165.

21 T-T.Tsou, M.Loots, and J.Halpern, J. Am. Chem. Soc., 1982, 104. 623.

22 R.G.Finke, B.L.Smith, B.J.Mayer, and A.A.Molinero, Inorg. Chem., 1983, 22, 3677.

23 J.Halpern, Acc. Chem. Res., 1982, 15, 238.

24 J.Halpern, Pure Appl. Chem., 1983, 55, 1059.

25 D.N.Ramakrishna Rao and M.C.R.Symons, J. Organomet. Chem., 1983, 244, C43

26 M.Hoshino, S.Konishi, Y.Terai, and M.Imamura, Inorg. Chem., 1982, 21, 89.

27 Y.Ohashi, K.Yanagi, T.Kurihara, Y.Sasada, and Y.Ohgo, J. Am. Chem. Soc., 1982, 104, 6353.

28 H.Y.Al-Saigh and T.J.Kemp, J. Chem. Soc., Perkin Trans. 2, 1983, 615.

29 C.M.Elliot, E.Hershenhart, R.G.Finke, and B.L.Smith, J. Am. Chem. Soc., 1981, 103, 5558.

30 M.D.LeHoang, Y.Robin, J.Devynck, C.Bied-Charreton, and A.Gaudemer, J. Organomet. Chem., 1981, 222, 311.

31 S.B.Fergusson and M.C.Baird, Inorg. Chim. Acta, 1982, 63, 41.

32 N.W.Alcock, M.P.Atkins, B.T.Golding, and P.J.Sellars, J. Chem. Soc., Dalton Trans., 1982, 337.

33 K.L.Brown and E.Zahonyi-Budo, J. Am. Chem. Soc., 1982, 104, 4117.

34 D.G.Brown and R.B.Flay, Inorg. Chim. Acta, 1982, 57, 63.

35 R.Dreos Garlatti, G.Tauzer, and G.Costa, Inorg. Chim. Acta, 1983, 70, 83.

36 R.Dreos Garlatti, G.Tauzher, and G.Costa, Inorg. Chim. Acta, 1983, 71, 9.
37 M.D.Johnson, Acc. Chem. Res., 1983, 16, 343.
38 R.C.McHatton, J.H.Espenson, and A.Bakac, J. Am. Chem. Soc., 1982, 104, 3531.
39 P.Bougeard, M.D.Johnson, and G.M.Lampman, J. Chem. Soc., Perkin Trans. 1, 1982, 849.
40 M.Veber, K.N.-V.-Duong, A.Gaudemer, and M.D.Johnson, J. Organomet. Chem. 1981, 209, 393..
41 A.Bury, S.T.Corker, and M.D.Johnson, J. Chem. Soc., Perkin Trans. 1,1982, 645.
42 P.Bougeard, A.Bury, C.J.Cooksey, M.D.Johnson, J.M.Hungerford, and G.M.Lampman, J. Am. Chem. Soc., 1982, 104, 5230.
43 F.Nome, M.C.Rezende, and N.Sergio de Sonza, J. Org. Chem., 1983, 48, 5357.
44 J.H.Espenson and J.D.Melton, Inorg. Chem., 1983, 22, 2779.
45 B.Clifford and W.R.Cullen, J. Chem. Educ., 1983, 554.
46 E.H.Curzon, B.T.Golding, and A.K.Wong, J. Chem. Soc., Chem. Commun., 1982, 63.
47 C.L.Hill and B.C.Schardt, J. Am. Chem. Soc., 1980, 102, 6374.
48 J.T.Groves, W.J.Kruper, Jr., R.C.Haushalter, J. Am. Chem. Soc., 1980, 102, 6374.
49 M.J.Camenzind, F.J.Hollander, and C.L.Hill, Inorg. Chem., 1982, 21, 4301.
50 B.C.Schardt, F.J.Hollander, and C.L.Hill, J. Am. Chem. Soc., 1982, 104, 3964.
51 J.A.Smegal and C.L.Hill, J. Am. Chem. Soc., 1983, 105, 3515.
52 J.A.Smegal, B.C.Schardt, and C.L.Hill, J. Am. Chem. Soc., 1983, 105, 3510.
53 J.A.Smegal and C.L.Hill, J. Am. Chem. Soc., 1983, 105, 2920.
54 E.Guilmet and B.Meunier, Nouv. J. Chim., 1982, 6, 511.
55 E.Guilmet and B.Meunier, Tetrahedron Lett., 1982, 23, 2449.
56 M.E.Carvalho and B.Meunier, Tetrahedron Lett., 1983, 24, 3621.
57 O.Bortolini and B.Meunier, J. Chem. Soc., Chem. Commun., 1983, 1364.
58 A.W. van der Made, J.W.H.Smeets, R.J.M.Nolte, and W.Drenth, J. Chem. Soc., Chem. Commun., 1983, 1204.
59 D.Mansuy, M.Fontecave, and J.-F.Bartoli, J. Chem. Soc., Chem. Commun., 1983, 253.
60 J.T.Groves, Y.Watanabe, and T.J.McMurry, J. Am. Chem. Soc., 1983, 105, 4489.
61 I.Tabushi and N.J.Koga, J. Am. Chem. Soc., 1979, 101, 6456.
62 M.Perrée-Fauvet and A.Gaudemer, J. Chem. Soc., Chem. Commun., 1981, 874.
63 I.Tabushi and A.Yazaki, J. Am. Chem. Soc., 1981, 103, 7371.
64 J.T.Groves and T.Takahashi, J. Am. Chem. Soc., 1983, 105, 2073.
65 J.W.Buchler, C.Dreher, and K.L.Lay, Z. Naturforsch., Teil B, 1982, 37, 1155.
66 C.L.Hill and F.J.Hollander, J. Am. Chem. Soc., 1982, 104, 7318.
67 J.T.Groves and T.E.Nemo, J. Am. Chem. Soc., 1983, 105, 5786.
68 J.R.Lindsay Smith and P.R.Sleath, J. Chem. Soc., Perkin Trans. 2, 1982, 1009.
69 J.T.Groves and T.E.Nemo, J. Am. Chem. Soc., 1983, 105, 6243.
70 J.R.Lindsay Smith and P.R.Sleath, J. Chem. Soc., Perkin Trans. 2, 1983, 1165.
71 J.T.Groves and R.S.Myers, J. Am. Chem. Soc., 1983, 105, 5791.
72 D.Mansuy and M.Fontecave, Biochem. Biophys. Res. Commun., 1982, 104, 1651.
73 H.G.Jonen, J.Werringloer, R.A.Prough, and R.W.Estabrook, J. Biol. Chem., 1982, 257, 4404.
74 H.A.Itano and J.L.Matteson, Biochemistry, 1982, 21, 2421.
75 D.Mansuy, P.Battioni, J.-P.Mahy, and G.Gillett, Biochem.Biophys.Res.Commun., 1982, 106, 30.
76 K.L.Kunze and P.R.Ortiz de Montellano, J. Am. Chem. Soc., 1983, 105, 1380.
77 O.Augusto, K.L.Kunze, P.R.Ortiz de Montellano, J. Biol. Chem., 1982, 257, 6231.
78 P.Battioni, J.P.Mahy, G.Gillet, and D.Mansuy, J. Am. Chem. Soc., 1983, 105, 1399.
79 P.R.Ortiz de Montellano, K.L.Kunze, and O.Augusto, J. Am. Chem. Soc., 1982, 104, 3545.
80 D.Mansuy, J.-P.Battioni, D.Dupré, E.Sartori, and G.Chottard, J. Am. Chem. Soc., 1982, 104, 6159.
81 P.Cocolios, E.Laviron, and R.Guilard, J. Organomet. Chem., 1982, 228, C39.

82 H.Ogoshi, H.Sugimoto, Z.-I.Yoshida, H.Kobayashi, H.Saki, and Y.Maeda, J. Organomet. Chem., 1982, 234, 185.
83 P.Cocolios, G.Lagrange, and R.Guilard, J. Organomet. Chem., 1983, 253, 65.
84 D.Mansuy, M.Fontecave, and J.-P.Battioni, J. Chem. Soc., Chem. Commun., 1982, 317.
85 D.Lexa and J.-M.Savéant, J. Am. Chem. Soc., 1982, 104, 3503.
86 D.Brault and P.Neta, J. Am. Chem. Soc., 1981, 103, 2705.
87 P.Guerin, J-P.Battioni, J-C.Chottard, and D.Mansuy, J. Organomet. Chem., 1981, 218, 201.
88 J-P.Battioni, J-C.Chottard, and D.Mansuy, Inorg. Chem., 1982, 21, 2056.
89 D.Mansuy and J-P.Battioni, J. Chem. Soc., Chem. Commun., 1982, 638.
90 K.Tatsumi and R. Hoffmann, J. Am. Chem. Soc., 1981, 103, 3328.
91 V.L.Goedken, M.R.Deakin, and L.A.Bottomley, J. Chem. Soc., Chem. Commun., 1982, 607.
92 D.R.English, D.N.Hendrickson, and K.S.Suslick, Inorg. Chem., 1983, 22, 367.
93 J-P.Battioni, D.Lexa, D.Mansuy, and J.M.Savéant, J. Am. Chem. Soc., 1983, 105, 207.
94 L.Latos-Grazynski, R-J.Cheng, G. N.LaMar, and A.L.Balch, J. Am. Chem. Soc., 1981, 103, 4270.
95 T.J.Wisnieff, A.Gold, S.A.Evans, Jr., J. Am. Chem. Soc., 1981, 103, 5616.
96 B.Chevrier, R.Weiss, M.Lange, J-C.Chottard, and D.Mansuy, J. Am. Chem. Soc., 1981, 103, 2899.
97 M.M.Olmstead, R-J.Cheng, and A.L.Balch, Inorg. Chem., 1982, 21, 4143.
98 D.Mansuy, I.Morgenstern-Badarau, M.Lange, and P.Gans, Inorg. Chem., 1982, 21, 1427.
99 G.Chottard, D.Mansuy, and H.J.Callot, Inorg. Chem., 1983, 22, 362.
100 J.C.Brunet, E.Cuinget, H.Gras, P.Marcincal, A.Mocz, C.Sergheraert, and A.Tartar, J. Organomet. Chem., 1981, 216, 73.
101 M.Schneider and M.Wenzel, J. Labelled Comp. Radiopharm., 1981, 18, 293.
102 W.H.Tamblyn, R.J.Klingler, W.S.Hwang, and J.K.Kochi, J. Am. Chem. Soc., 1981, 103, 3161
103 A.M.van den Bergen, D.J.Brockway, and B.O.West, J. Organomet. Chem., 1983, 249, 205.
104 A.Puxeddu and G.Costa, J. Chem. Soc., Dalton Trans., 1982, 1285.
105 K.J.Berry, F.Maya, K.S.Murray, A.M.B. van den Bergen, and B.O.West, J. Chem. Soc., Dalton Trans., 1982, 109.
106 A.Zombeek, D.E.Hamilton, and R.S.Drago, J. Am. Chem. Soc., 1982, 104, 6782.
107 I.Ya Levitin, A.L.Sigan, R.M.Bodnor, R.G.Gasanov, and M.E.Vol'pin, Inorg. Chim. Acta, 1983, 76, L169.
108 I.Ya Levitin and M.E.Vol'pin, Inorg. Chim. Acta, 1983, 79, 9.
109 Y.Sorek, H.Cohen, W.A.Mulac, K.H.Schmidt, and D.Meyerstein, Inorg. Chem., 1983, 22, 3040.
110 D.Dolphin, D.J.Halko, and E.Johnson, Inorg. Chem., 1981, 20, 4348.
111 H.J.Callot and F.Metz, J. Chem. Soc., Chem. Commun., 1982, 947.
112 Y.Aoyama, K.Aoyagi, H.Toi, and H.Ogoshi, Inorg. Chem., 1983, 22, 3046.
113 R.M.Dixon, B.T.Golding, O.W.Howarth, and J.L.Murphy, J. Chem. Soc., Chem. Commun., 1983, 243.
114 J.H.Grate, J.W.Grate, and G.N.Schrauzer, J. Am. Chem. Soc., 1982, 104, 1588.
115 H.Halpern, Inorg. Chim. Acta, 1983, 79, 1.
116 J.M.Pratt, Inorg. Chim. Acta, 1983, 79, 27.
117 L.G.Marzilli, P.J.Toscano, M.Summers, L.Randaccio, N.Bresciani-Pahor, J.P.Glusker, and M.Rossi, Inorg. Chim. Acta, 1983, 79, 29.
118 W.A.Mulac and D.Meyerstein, J. Am. Chem. Soc., 1982, 104, 4124.
119 D.N.Ramakrishna Rao, and M.C.R.Symons, J. Chem. Soc., Chem. Commun., 1982, 954.
120 D.N.Ramakrishna Rao, and M.C.R.Symons, J. Chem. Soc., Faraday Trans. 1, 1983, 79, 269.
121 D.N.Ramakrishna Rao, and M.C.R.Symons, J. Chem. Soc., Perkin Trans. 2, 1983, 187.
122 D.Lexa and J-M.Savéant, Acc. Chem. Res., 1983, 16, 235.

123 K.A.Rubinson, E.Itabashi, and H.B.Mark, Jr., Inorg. Chem., 1982, 21, 3571.
124 K.A.Rubinson, H.V.Parekh, E.Itabashi, H.B.Mark, Jr., Inorg. Chem., 1983, 22, 458.
125 A.Fischli, Helv. Chim. Acta, 1982, 65, 2697
126 R.Scheffold and R.Orlinski, J. Am. Chem. Soc., 1983, 105, 7200.
127 A.Fischi, Helv. Chim. Acta, 1982, 65, 1167.
128 W.W.Reenstra, R.H.Abeles, and W.P.Jencks, J. Am. Chem. Soc., 1982, 104, 1016.
129 P.N.Balasubramanian and E.S.Gould, Inorg. Chem., 1983, 22, 2635.
130 K.L.Brown and J.M.Hakimi, Inorg. Chim. Acta, 1982, 67, L29.
131 K.L.Brown, J.M.Hakimi, and D.S.Marynick, Inorg. Chim. Acta, 1983, 79, 120.
132 P.K.Mishra, R.K.Gupta, P.C.Goswami, P.N.Venkatasubramanian, and A.Nath, Polyhedron, 1982, 1, 321.
133 A.R.Battersby, C.Edington, C.J.R.Fookes, and J.M.Hook, J. Chem. Soc., Perkin Trans. 1, 1982, 2265.
134 K.Kamiya and O.Kennard, J. Chem. Soc., Perkin Trans. 1, 1982, 2279.
135 J.M.Miller, J. Organomet. Chem., 1983, 249, 299.
136 S.C.Agarwal, Bull. Soc. Chim. Fr., 1979, I-100.
137 J.J.Russell, H.S.Rezepa, and D.A.Widdowson, J. Chem. Soc., Chem. Commun., 1983, 625.
138 R.G.Finke, W.P.McKenna, D.A.Schiraldi, B.L.Smith, and C.Pierpoint, J. Am. Chem. Soc., 1983, 105, 7592.
139 R.G.Finke and D.A.Schiraldi, J. Am. Chem. Soc., 1983, 105, 7605.
140 P.Müller and J.Rétey, J. Chem. Soc., Chem. Commun., 1983, 1342.
141 J.S.Krouwer, B.Holmquist, R.S.Kipner, and B.M.Babior, Biochim. Biophys. Acta, 1980, 612, 153.
142 D.Gani and D.W.Young, J. Chem. Soc., Chem. Commun., 1982, 867.
143 D.Gani, O.C.Wallis, and D.W.Young, J. Chem. Soc., Chem. Commun., 1982, 898.
144 S.J.Field and D.W.Young, J. Chem. Soc., Perkin Trans.1, 1983, 2387.
145 J.S.Thayer and F.E.Brinckman, Adv. Organomet. Chem., 1982, 20, 314.
146 Y-T.Fanchiang, J. Chem. Soc., Chem. Commun., 1982, 1369.
147 Y-T.Fanchiang, Inorg. Chem., 1982, 21, 2344.
148 Y-T.Fanchiang, Inorg. Chem., 1983, 22, 1693.
149 J.H.Dimmit and J.H.Weber, Inorg. Chem., 1982, 21, 700.
150 R.Kergoat, M.M.Kubicki, J.E.Guerchais, and J.A.K.Howard, J. Chem. Res. (S), 1982, 340.
151 M.H.Darbieu and G.Cros, J. Organomet. Chem., 1983, 252, 327.
152 J.H.Dimmit and J.H.Weber, Inorg. Chem., 1982, 21, 1554.
153 L.M.Mihichuk, M.J.Mombourquette, F.W.B.Einstein, and A.C.Willis, Inorg. Chim. Acta, 1982, 63, 189.
154 'B_{12}', ed. D. Dolphin, John-Wiley, New York, 1982, volumes 1 and 2.
155 R.D.W.Kemmitt and D.R.Russell, in 'Comprehensive Organometallic Chemistry', ed. Sir Geoffrey Wilkinson, F.G.A.Stone, and E.W.Abel, Pergamon, Oxford 1982 chapter 34, p.80.
156 E.Langer, in 'Houben Weyl-Methoden der Organischen Chemie', ed. A.Segnitz, George Thieme, Stuttgart, 1984, volume 13, part 9b, p.1.
157 P. J.Toscano and L.G.Marzilli, Prog. Inorg. Chem., 1984, 31, 105.
158 B.W.Rockett and G.Mar, J. Organomet. Chem., 1982, 237, 257.
159 L.G.Marzilli, P.J.Toscano, J.H.Ramsden, L.Randaccio, and N.Bresciani-Pahor, Adv. Chem. Ser., 1982, 196, 86.
160 A.G.Lappin, in 'Inorganic Reaction Mechanisms', ed. A.G. Sykes (Specialist Periodical Reports), The Royal Society of Chemistry, London, 1981, Vol. 7, p.303.
161 B.Zagalak, Naturwissenschaften, 1982, 69, 63.
162 Inorg. Chim. Acta, 1983, 79, 1.
163 P.M.Harrison and R.J.Hoare, 'Metals in Biochemistry', Chapman and Hall, London, 1980.

17
Structures of Organometallic Compounds determined by Diffraction Methods

BY D. R. RUSSELL

1 Introduction

This Chapter contains a comprehensive list of organometallic compounds whose structures have been determined by X-ray, neutron or electron diffraction methods and reported during 1983. Some reports not covered in the previous volume 12 have also been included. Compounds containing metal-to-carbon bonds (except to cyanide if this is the only such bond) are included. Metals are defined as all elements except C,H,N,P,O,S, the halogens, and the inert gases. Some compounds containing silicon which are regarded by this reviewer as "purely organic" are not included.

Unfortunately it has not been possible to give as much information as in previous volumes, due both to the continuing growth in numbers of determinations, and the restrictions of space imposed by the new format. In fact, there are a growing number of reports with no published details. Accordingly, structural information has been condensed into a Main Table, with compounds (or ions) ordered alphabetically by element symbol, subdivided by numbers of each element in a similar manner to that used in the Structural Index of Comprehensive Organometallic Chemistry (Vol. 9, Pergamon Press, 1982). The five columns of the Main Table consist of:

1 An identifying number for each compound
2 The molecular formula, using the modified Hill system (all metal atoms, C, H, then other elements alphabetically)
3 A line formula which attempts to describe the structural identity of the compound. For compounds containing bridging polyfunctional groups, an indication of which atoms are bonded to each metal is given (see for example compounds 26 and 27 below). The symbol η is reserved (but not exclusively) for delocalized multiply-bonded systems (see for example compounds 437 and 451 below)
4 Under "Details" appear the letters N (for neutron), E (for

electron), otherwise X-ray diffraction (X) is to be assumed. Numbers refer to the temperature(K) of the determination if below ambient.

5 Reference numbers refer to the listing given at the end of the Chapter.

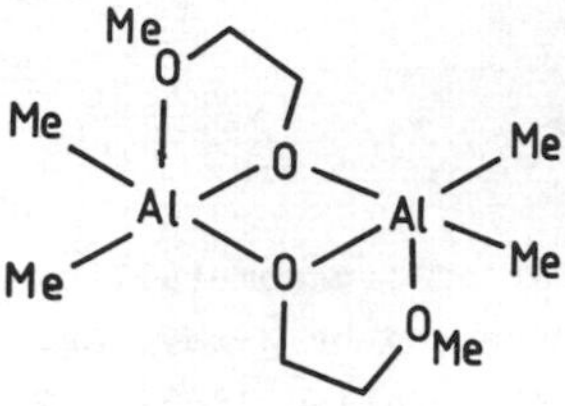

26 $(AlMe_2)_2\{\mu\text{-}(O,\mu\text{-}O)\text{-}OCH_2CH_2OMe\}_2$

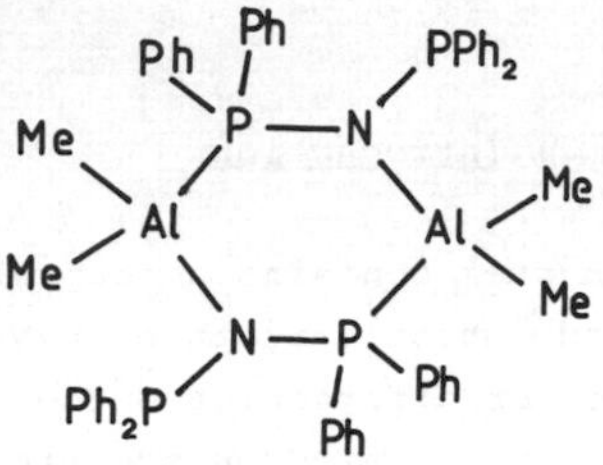

27 $[AlMe_2\{\mu\text{-}(P,N)\text{-}PPh_2NPPh_2\}]_2$

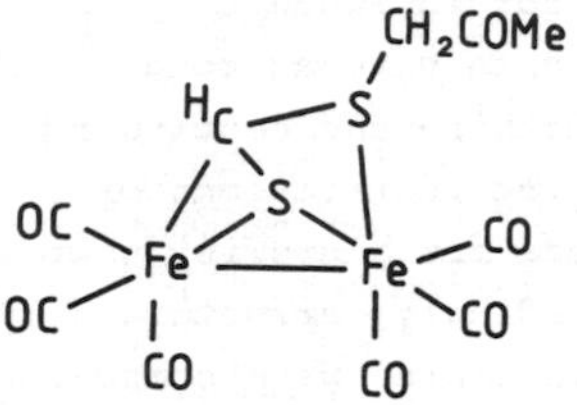

437 $(OC)_3Fe\{\mu\text{-}(CS\eta^2,S,S)\text{-}SCHSCH_2COMe\}Fe(CO)_3$

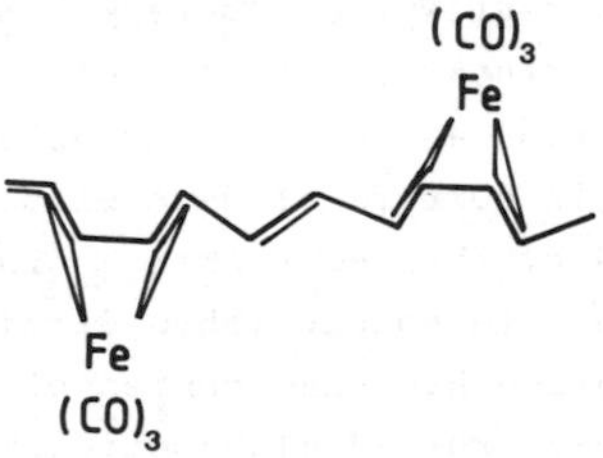

451 <u>trans</u>-$\{Fe(CO)_3\}_2\{\mu\text{-}(2\text{-}5\eta^4,8\text{-}11\eta^4)\text{-}$dodecapentene

Mixed metal compounds will appear only once in the Main Table, found under the earliest metal alphabetically. A Metals Cross Reference Table is provided to locate mixed metal compounds which appear alphabetically under another metal in the Main Table.

The Main Table contains 1409 unique compounds (numbers 19, 635, and 1165 are not allocated). There are on average 1.3 structures per cited reference, one paper describes 12 separate structures. The most popular metal is Fe, which occurs in 183 compounds, with Mo(134), Ru(123), Si(122), Os(104) and Co(103) all appearing in over 100 compounds. The largest total number of atoms (235) in one molecule is found in compound 962, compounds 78 and 79 both contain a total of 24 metal atoms. The largest number of different metals in one molecule (five) is found in compound 47, which also shares with compound 87 in having the record number of nine different elements.

Abbreviations used in Main Table

bipy	2,2′-bipyridyl	Bz	benzyl
cod	η-cycloocta-1,5-diene	cot	cyclooctatetraene
Cp	η-cyclopentadienyl	Cy	cyclohexyl
dcpe	$Cy_2PCH_2CH_2PCy_2$	detc	diethyldithiocarbamate
diars	o-phenylenebis(dimethyl)arsine	dmgH	dimethylglyoximate
dmpe	$Me_2PCH_2CH_2PMe_2$	dmpm	$Me_2PCH_2PMe_2$
dmtc	dimethyldithiocarbamate	doppe	$Ph_2P(O)CH_2CH_2P(O)Ph_2$
dpam	$Ph_2AsCH_2AsPh_2$	dppe	$Ph_2PCH_2CH_2PPh_2$
dppm	$Ph_2PCH_2PPh_2$	dppp	$Ph_2PCH_2CH_2CH_2PPh_2$
en	ethylenediamine	hfacac	hexafluoroacetylacetonate
ind	indenyl	Me_2pz	3,5-dimethylpyrazolyl
Me_5dien	$\{Me_2NCH_2CH_2N(Me)CH_2-\}_2$	mTol	m-tolyl
nbd	norbornadiene	oTol	o-tolyl
oxine	8-oxoquinolate	phen	1,10-phenanthroline
[PPN]	$[(Ph_3P)_2N]^+$	pTol	p-tolyl
py	pyridine	pyz	1,4-pyrazine
pz	pyrazolyl	SacSac	dithioacetylacetonate
TCNQ	7,7,8,8-tetracyanoquinodimethane	THF	tetrahydrofuran
tmeda	$Me_2NCH_2CH_2NMe_2$	tmpda	$Me_2NCH_2CH_2CH_2NMe_2$
triphos	$(Ph_2PCH_2)_3CMe$		

2 Main Table

No.	Formula	Structure	Details	Ref.
1	$AgC_8H_8NO_3$	(cot).$AgNO_3$		1
2	$[AgC_8H_{14}O_8S_4]^-$	$K[Ag\{CHSO_2(CH_2)_3SO_2\}_2].H_2O$		2
3	$AgC_{12}H_{10}ClO_4$	(benzocyclooctatetraene).$AgClO_4$		3
4	$AgC_{15}H_{17}O_{11}$	$[Ag\{\mu\text{-}(\eta^2,2O)\text{-}C_5(CO_2Me)_5\}(OH_2)]_n.1.5nH_2O$		4
5	$AgC_{16}H_{16}NO_3$	$Ag(NO_3)(\eta^4\text{-cot})_2$		5
6	$[AgC_{24}H_{24}]^+$	$[Ag\{\eta^2\text{-}\underline{p}\text{-}CH_2C_6H_4CH_2\text{-}\}_3][ClO_4]$	153	6
7	$[AgC_{38}H_{58}N_2]^+$	$[Ag\{CNC_6H_2(Bu^t{}_3\text{-}2,4,6)\}_2][PF_6]$		7
8	$AgCuRu_4C_{48}H_{32}O_{12}P_2$	$(Ph_3P)_2AgCuRu_4(CO)_{12}(\mu_3\text{-}H)_2.CH_2Cl_2$		8
9	$[AgFe_2C_{27}H_{20}NO_6P]^+$	$[AgFe_2(CO)_6(\mu\text{-}PPh_2)\{\mu\text{-}CHC(Ph){=}NHMe\}]$- $[ClO_4].PhH.PhMe$		9
10	$[AgRh_2C_{48}H_{40}O_2P_2]^+$	$[Ag\{Rh(CO)(PPh_3)(Cp)\}_2][PF_6].PhMe$		10
11	$[AgRh_{12}C_{32}O_{30}]^{3-}$	$[PPh_4]_3[Ag\{Rh_6(CO)_{15}C\}_2]$		11
12	$[AgSe_6C_6H_{12}]^+$	$[Ag\{(\text{-}SeCH_2\text{-})_3\}_2][AsF_6]$		12
13	$[Ag_2C_{30}H_{48}N_6]^{2+}$	$[Ag_2\{CNC(Me)_2CH_2CH_2C(Me)_2NC\}_3][ClO_4]_2$		13
14	$Ag_2C_{66}H_{60}O_{20}P_2$	$[Ag\{\mu\text{-}(\eta^2,2O)\text{-}\{C_5(CO_2Me)_5\}\}(PPh_3)]_2$		4
15	$Ag_2Ru_4C_{48}H_{32}O_{12}P_2$	$(Ph_3P)_2Ag_2Ru_4(\mu_3\text{-}H)_2(CO)_{12}.CH_2Cl_2$		8
16	$Ag_3C_{16}H_{16}NO_3$	2cot.3$AgNO_3$		14
17	$Ag_4C_{36}H_{44}$	$[Ag(\mu\text{-Mes})]_4$		15
18	$AlCrSi_2C_{28}H_{41}NO_5P$	$(OC)_5CrPPh_2\{Al(NMe_3)(CH_2SiMe_3)_2\}$		16,17
20	$AlFeC_{28}H_{37}NO_2$	$(Cp)(OC)Fe\{\eta^2\text{-}OC(Me)P(Ph)_2N(Bu^t)AlEt_2\}$		18
21	$AlNaSi_4C_{12}H_{36}$	$NaAl(SiMe_3)_4$		19
22	$AlNaSi_4C_{26}H_{52}$	$NaAl(SiMe_3)_4.2PhMe$		19
23	$AlSi_2C_{24}H_{54}ClF_2N_2$	$AlCl\{(\mu\text{-}F)(\mu\text{-}NBu^t)SiBu^t{}_2\}_2$		20
24	$AlSi_3C_{13}H_{37}O$	$Al(SiMe_3)_3(OEt_2)$		19
25	$AlWC_{12}H_{28}Cl_4O_2P_3$	$WCl(CO)(PMe_3)_3(\eta^2\text{-}HC_2OAlCl_3)$		21
26	$Al_2C_{10}H_{26}O_4$	$(AlMe_2)_2\{\mu\text{-}(O,\mu\text{-}O)\text{-}OCH_2CH_2OMe\}_2$		22
27	$Al_2C_{52}H_{52}N_2P_4$	$[AlMe_2\{\mu\text{-}(P,N)\text{-}P(Ph)_2NPPh_2\}]_2.0.5PhH$		23
28	$Al_2Co_2C_{22}H_{66}N_4P_6$	$[(Me_3P)_3CoNN(\mu\text{-}AlMe_2)]_2$		24
29	$Al_2Cr_2Si_4C_{58}H_{80}O_{12}P_2$	$[(OC)_5CrPPh_2(CH_2)_4O\{\mu\text{-}Al(CH_2SiMe_3)_2\}]_2$		25

No.	Formula	Structure	Details	Ref.
30	$\mathrm{Al_2MnC_{40}H_{51}N_2O_4P_2}$	$\mathrm{(OC)_3Mn(CHPPh_2NBu^tAlMe_2O)(PPh_2NBu^tAlMeHCH_2)}$		26
31	$\mathrm{Al_2Si_4C_{14}H_{42}Cl_2N_6}$	$\mathrm{[ClAlN(Me)Si(Me)_2N(Me)Si(Me)_2NMe]_2}$		27
32	$\mathrm{Al_2Si_6C_{24}H_{70}N_2}$	$\mathrm{\{(Me_3Si)_3Al\}_2(\mu\text{-}tmeda)}$		28
33	$\mathrm{Al_2Y_2C_{32}H_{56}Cl_2N_2}$	$\mathrm{[(Cp)_2Y(\mu\text{-}Cl).H_3AlNEt_3]_2}$		29
34	$\mathrm{[Al_4C_{10}H_{30}O_2]^{2-}}$	$\mathrm{[AsMe_4]_2[\{Me_2Al(\mu\text{-}OAlMe_3)\}_2].PhH}$		30
35	$\mathrm{[Al_7C_{16}H_{48}O_6]^-}$	$\mathrm{K[\{Me_2AlO(Me)Al(Me)_2O\}_3AlMe].PhH}$		31
36	$\mathrm{[Al_7C_{16}H_{48}O_6]^-}$	$\mathrm{Cs[\{Me_2AlO(Me)Al(Me)_2O\}_3AlMe].3PhMe}$		31
37	$\mathrm{AsC_3H_9}$	$\mathrm{AsMe_3}$	E	32
38	$\mathrm{AsC_4H_{10}NO_5}$	$\mathrm{HO_3AsCH_2CH_2CH(NH_3)CO_2H.H_2O}$		33
39	$\mathrm{[AsC_4H_{12}]^+}$	$\mathrm{[AsMe_4]_2[\{Me_2Al(\mu\text{-}OAlMe_3)\}_2].PhH}$		30
40	$\mathrm{[AsC_6H_9NO_4]^+}$	$\mathrm{[C_6H_3\{AsO(OH)_2)\}(3\text{-}NH_3)(4\text{-}OH)]Cl.H_2O}$		34
41	$\mathrm{AsC_7H_6N}$	$\mathrm{C_6H_4\{1,2\text{-}(As{=}CHNH)\}}$		35
42	$\mathrm{AsC_{12}H_8ClS}$	$\mathrm{S(C_6H_4)_2(2,2'\text{-}AsCl)}$		36
43	$\mathrm{AsC_{18}H_{15}}$	$\mathrm{AsPh_3}$		37
44	$\mathrm{[AsC_{20}H_{30}]^+}$	$\mathrm{[As(\eta\text{-}C_5Me_5)_2][BF_4]}$		38
45	$\mathrm{[AsC_{24}H_{20}]^+}$	$\mathrm{[AsPh_4]^+}$		39-54
46	$\mathrm{AsC_{29}H_{20}F_5O_2}$	$\mathrm{Ph_3AsC(COPh)C(O)C_2F_5}$		55
47	$\mathrm{AsCoFeMoWC_{19}H_{16}O_7S}$	$\mathrm{(Cp)_2MoWCoFe(CO)_4(\mu\text{-}CO)_3(\mu\text{-}AsMe_2)(\mu_3\text{-}S)}$		56
48	$\mathrm{AsCo_2FeMoC_{15}H_{11}O_8S}$	$\mathrm{(Cp)MoCo_2Fe(CO)_5(\mu\text{-}CO)_3(\mu\text{-}AsMe_2)(\mu_3\text{-}S)}$		56
49	$\mathrm{AsCo_2MoC_{17}H_{20}O_6P}$	$\mathrm{(Cp)MoCo_2(CO)_6(\mu\text{-}AsMe_2)(\mu_3\text{-}PBu^t)}$		57
50	$\mathrm{AsCo_2MoRuC_{15}H_{11}O_8S}$	$\mathrm{(Cp)MoCo_2Ru(CO)_5(\mu\text{-}CO)_3(\mu\text{-}AsMe_3)(\mu_3\text{-}S)}$		56
51	$\mathrm{AsCr_2Si_2C_{16}H_{18}NO_{10}}$	$\mathrm{As\{Cr(CO)_5\}_2\{N(SiMe_3)_2\}}$		58
52	$\mathrm{AsFeC_{20}H_{22}N}$	$\mathrm{Fe(\eta\text{-}C_5H_4AsPh)\{\eta\text{-}C_5H_3CH(Me)(NMe_2)\}}$		59
53	$\mathrm{AsFe_3C_{15}H_{15}O_9S}$	$\mathrm{Fe_3(CO)_9(\mu_3\text{-}SBu^t)(\mu\text{-}AsMe_2)}$	221	60
54	$\mathrm{AsHgMoC_{16}H_{16}NO_2}$	$\mathrm{(NC)HgMo(CO)_2(AsMe_2Ph)(Cp)}$		61
55	$\mathrm{AsPdC_{26}H_{21}Cl_2N_2}$	*cis*-$\mathrm{PdCl_2\{(As,N)\eta^2\text{-}(6\text{-}Mequinolin\text{-}8\text{-}yl)_2AsPh\}}$		62
56	$\mathrm{AsPdC_{26}H_{21}I_2N_2}$	*cis*-$\mathrm{PdI_2\{(As,N)\eta^2\text{-}(6\text{-}Mequinolin\text{-}8\text{-}yl)_2AsPh\}}$		62
57	$\mathrm{AsRuC_{22}H_{15}O_4}$	$\mathrm{Ru(CO)_4(AsPh_3)}$		63
58	$\mathrm{AsSi_2C_{25}H_{48}P}$	$\mathrm{\{(2,4,6\text{-}Bu^t{}_3)C_6H_2\}P{=}AsCH(SiMe_3)_2}$		64
59	$\mathrm{As_2C_{12}H_{16}}$	$\mathrm{\{AsC(Me)CHCHCMe\}_2\text{-}(\underline{AsAs})}$		65

No.	*Formula*	*Structure*	*Details*	*Ref.*
60	$As_2C_{26}H_{20}$	PhAsC(Ph)=C(Ph)AsPh		66
61	$As_2Co_2W_2C_{20}H_{18}O_{20}P_2$	$\{(OC)_5W\}_2As_2Co_2(CO)_4\{P(OMe)_3\}_2$		67
62	$As_2CrSi_2C_{30}H_{48}O_5$	$(2,4,6\text{-}Bu^t{}_3C_6H_2)As{=}As\{CH(SiMe_3)_2\}Cr(CO)_5$		68
63	$As_2Hg_2C_{54}H_{66}N_4O_{12}$	$[Hg(\mu\text{-}NO_3)_2As(Mes)_3]_2$		69
64	$As_2Mn_2C_{33}H_{22}O_8$	$Mn_2(CO)_8(\mu\text{-dpam})$		70
65	$[As_2MoC_{20}H_{36}N_3OS_4]^+$	$[Mo(detc)_2(NO)(diars)][BF_4]$		71
66	$As_2RuC_{22}H_{20}Cl_2O_2$	$RuCl_2(CO)_2\{$<u>o</u>$\text{-}C_6H_4(AsPhMe)_2\}$		72
67	$As_2Si_2C_{25}H_{48}$	$(2,4,6\text{-}Bu^t{}_3C_6H_2)As{=}AsCH(SiMe_3)_2$		73
68	$As_2Sn_2C_{62}H_{50}N_2O_{12}$	$\{SnPh_2(\eta^2\text{-}NO_3)(OAsPh_3)\}_2(\mu\text{-}C_2O_4)$		74
69	$As_2Sn_2C_{62}H_{54}Cl_2O_2$	$\{ClPh_3SnOAs(Ph)_2CH_2\text{-}\}_2$		75
70	$[As_3NiC_{60}H_{57}NP]^+$	$[Ni\{N(CH_2CH_2AsPh_2)_3\}(PPh_3)][ClO_4]$		76
71	$As_4C_8F_{24}N_4$	$[(CF_3)_2AsN]_4$		77
72	$[As_4PdC_{20}H_{32}Cl_2]^{2+}$	<u>trans</u>-$[Pd(diars)_2Cl_2][ClO_4]_2$		78
73	$As_4RhC_{21}H_{32}ClO_2$	<u>trans</u>-$RhCl(diars)_2(\eta^1\text{-}(C)\text{-}CO_2)$		79
74	$As_4Ru_3C_{51}H_{40}O_7$	$Ru_3(CO)_7(\mu\text{-}H)\{\mu_3\text{-}(As,\mu\text{-}As)\text{-}AsPh_2CH_2AsPh\}\text{-}$ (dpam)$.CH_2Cl_2$	X,N	80
75	$As_4UC_{52}H_{48}Br_4O_2$	$UBr_4\{OAs(Ph)_2CH_2CH_2AsPh_2\}_2$		81
76	$As_5Mo_2C_{19}H_{25}O_4$	$\{(Cp)(OC)_2Mo\}_2\{\mu$-<u>catena</u>-$(AsMe)_5\}$		82
77	$As_6C_{36}H_{30}$	<u>catena</u>-$(AsPh)_6$		83
78	$As_8Co_{16}C_{56}H_{20}O_{32}$	$[Co_8(\mu_6\text{-}As)(\mu_4\text{-}As)(\mu_4\text{-}AsPh)_2(CO)_{16}]_2$		84
79	$As_{24}C_{48}H_{96}N_8O_{24}$	$\{N(CH_2CH_2)_3\}_8(As_4O_4)_6$		85
80	$[AuC_{18}ClF_{15}]^-$	$[Au_3Cl_2(dppm)_2][AuCl(C_6F_5)_3]$		86
81	$AuC_{33}H_{30}O_{10}P$	$Au\{\eta^1\text{-}C_5(CO_2Me)_5\}(PPh_3)$		87
82	$[AuC_{37}H_{22}F_{10}P_2]^+$	$[Au(C_6F_5)_2(dppm)][ClO_4].Me_2CO$		86
83	$AuC_{40}H_{28}ClN_2$	$AuCl\{C(Ph){=}C(Ph)C(Ph){=}C(Ph)\}(phen)$		88
84	$AuBC_{11}H_{15}N_6$	$AuMe_2\{\eta^2\text{-}(pz)_3BH\}$		89
85	$AuB_8NiC_{25}H_{30}P$	<u>nido</u>-$[9\text{-}(Cp)\text{-}\mu_{10,11}\text{-}(AuPPh_3)\text{-}7,8,9\text{-}C_2NiB_8H_{10}]$		90
86	$AuFe_3C_{32}H_{25}NO_9P$	$Fe_3(CO)_9(\mu\text{-}AuPPh_3)\{\mu_3\text{-}(C,N,\eta^2)\text{-}HC{=}NBu^t\}$		91
87	$AuOsC_{37}H_{32}ClINOP_2$	$IAu(\mu\text{-}CH_2)OsCl(NO)(PPh_3)_2$		92
88	$AuOs_3C_{49}H_{35}N_2O_8P_2$	$Os_3(CO)_8(PPh_3)(\mu\text{-}AuPPh_3)\{\mu\text{-}(N,\mu\text{-}N)\text{-}NC_5H_4(NH\text{-}2)\}$		93

No.	Formula	Structure	Details	Ref.
89	$[AuOs_{10}C_{43}H_{15}O_{24}P]^-$	$[PPh_3Me][Os_{10}C(CO)_{24}AuPPh_3]$		94
90	$[AuRhC_{36}H_{36}NP]^{2+}$	$[(\eta\text{-}C_5Me_5)Rh\{\mu\text{-}(\eta^6,N)\text{-indolyl}\}AuPPh_3][ClO_4].CH_2Cl_2$		95
91	$AuRu_3C_{29}H_{20}O_{10}P$	$Ru_3(CO)_9(\mu\text{-}H)_2(\mu\text{-}AuPPh_3)(\mu_3\text{-}COMe)$		96
92	$AuRu_3C_{30}H_{18}O_{11}P$	$Ru_3(CO)_{10}(\mu\text{-}AuPPh_3)(\mu\text{-}COMe)$	220	96
93	$AuRu_5C_{33}H_{15}BrO_{14}$	$Ru_5C(CO)_{14}(\mu\text{-}AuPPh_3)(\mu\text{-}Br)$		97
94	$AuRu_5C_{34}H_{15}ClO_{15}$	$Ru_5C(CO)_{15}(\mu\text{-}AuPPh_3)Cl$		97
95	$AuRu_5C_{50}H_{30}IO_{13}P_2$	$Ru_5C(CO)_{13}(PPh_3)(\mu\text{-}AuPPh_3)(\mu\text{-}I)$		98
96	$AuRu_6C_{34}H_{15}NO_{16}P$	$Ru_6C(CO)_{15}(NO)(\mu_3\text{-}AuPPh_3)$		99
97	$[AuW_2C_{30}H_{24}O_4]^+$	$[\{(Cp)(OC)_2W\{\mu\text{-}C(pTol)\}\}_2Au][PF_6]$		100
98	$Au_2Co_2Ru_2C_{48}H_{30}O_{12}P_2$	$(Ph_3P)_2Au_2Ru_2Co_2(CO)_{10}(\mu\text{-}CO)_2$		101
99	$Au_2Fe_3C_{45}H_{30}O_9P_2S$	$(Ph_3P)_2Au_2Fe_3(CO)_9(\mu_3\text{-}S)$		101
100	$Au_2Os_3C_{22}H_{30}O_{10}$	$Os_3(CO)_{10}(\mu\text{-}AuPEt_3)_2$		102
101	$Au_2Os_5C_{51}H_{30}O_{14}P_2$	$Os_5C(CO)_{14}(\mu\text{-}AuPPh_3)_2$		103
102	$Au_2Os_8C_{58}H_{30}O_{22}P_2$	$Os_8(CO)_{22}(AuPPh_3)_2$		104
103	$[Au_2RhC_{58}H_{50}N_4P_2]^+$	$[(Ph_3P)_2Au_2\{\mu\text{-}(2,2'\text{-dibenzimidazole})\}\text{-}Rh(cod)][ClO_4].CHCl_3$		105
104	$Au_2Ru_3C_{47}H_{34}O_{10}P_2$	$(Ph_3P)_2Au_2Ru_3(\mu\text{-}H)(\mu\text{-}COMe)(CO)_9$		106
105	$Au_2Ru_3C_{62}H_{45}O_8P_3S$	$(Ph_3P)_2Au_2Ru_3(\mu_3\text{-}S)(CO)_8(PPh_3).0.5CH_2Cl_2$		106
106	$Au_3Ru_3C_{65}H_{48}O_{10}P_3$	$(Ph_3P)_3Au_3Ru_3(\mu_3\text{-}COMe)(CO)_9$		96
107	$Au_3Ru_4C_{66}H_{46}O_{12}P_3$	$(Ph_3P)_3Au_3Ru_4(CO)_{12}(\mu\text{-}H)$		107
108	$Au_4Fe_2C_{58}H_{44}O_8P_4$	$[(dppe)Au_2Fe(CO)_4]_2.THF$		108
109	$Au_4Fe_2C_{60}H_{48}O_8P_4$	$[(dppm)Au_2Fe(CO)_4]_2.Pr^iOH$		108
110	$Au_5C_{45}H_{55}$	$[Au(\mu\text{-}Mes)]_5.2THF$		109
111	$Au_6Co_2C_{80}H_{60}O_8P_4$	$[Au_6(PPh_3)_4\{Co(CO)_4\}_2$		110
112	$BC_{12}H_{11}O$	$Ph_2BOH.Ph_2BOCH_2CH_2NMe_2$		111
113	$BC_{14}H_{20}NS_2$	$(2,6\text{-}Me_2C_6H_3)NHBC(Et)=C(Et)SS$		112
114	$BC_{15}H_{18}NO$	$Ph_2BOCH_2CH_2CH_2NH_2$		111
115	$BC_{16}H_{20}NO$	$Ph_2BOCH_2CH_2NMe_2$		111
116	$BC_{16}H_{20}NO$	$Ph_2BOCH_2CH_2NMe_2.Ph_2BOH$		111
117	$BC_{17}H_{14}NO_2$	$C_5H_4N(1,2\text{-}O_2BPh_2)$		113

No.	*Formula*	*Structure*	*Details*	*Ref.*
118	$BC_{18}H_{20}NS_2$	$Et_2NBC(Ph)=C(Ph)SS$		112
119	$[BC_{24}H_{20}]^-$	$[BPh_4]^-$		114-123
120	$BC_{24}H_{26}NO_2$	$C_6H_2N(2\text{-}Cy)(4\text{-}Me)(1,6\text{-}O_2BPh_2)$		113
121	$BC_{24}H_{42}N_4S$	$\{Bu^n{}_2BNCNC(NHCy)N(Cy)\}(2,3\text{-}CH=CHS)$		124
122	$BCuC_{17}H_{26}N_6$	$Cu\{(Me_2pz)_3BH\}(\eta\text{-}C_2H_4)$	173	125
123	$BCu_2C_{11}H_{14}ClN_6$	$(\eta\text{-}C_2H_4)Cu\{\mu\text{-}(2N,N)\text{-}pz_3BH\}CuCl$	173	125
124	$BFe_2C_{22}H_{23}$	$(Cp)Fe\{\mu\text{-}\eta^5\text{-}B(Ph)C(Et)CHCHCH\}Fe(Cp)$		126
125	$BFe_4C_{12}H_3O_{12}$	$B(\mu\text{-}H)_2Fe_4(\mu\text{-}H)(CO)_{12}$		127
126	$BMn_2C_{18}H_{13}O_6$	$(OC)_3Mn\{\mu\text{-}\eta^5\text{-}B(Ph)C(Et)CHCHCH\}Mn(CO)_3$		126
127	$BMoC_{12}H_7N_6O_3$	$Mo(CO)_3(pz_3BH)$		128
128	$BMoC_{12}H_{10}BrO_3$	$MoBr(pz_3BH)(CO)_3$		129
129	$BMoC_{13}H_{12}N_6O_3$	$Mo(pz_3BH)(CO)_2(\eta\text{-}OCMe)$		129
130	$BMoC_{15}H_{12}ClN_8O_2$	$Mo(CCl)(CO)_2\{(Me_2pz)_3BH\}$		130
131	$BMoC_{18}H_{15}N_6O_3$	$Mo(\eta^2\text{-}OCPh)(CO)_2(pz_3BH)$		129
132	$BMoC_{24}H_{33}N_6O_3$	$Mo(\eta^2\text{-}OCCy)(CO)_2\{(Me_2pz)_3BH\}$		131
133	$BMoC_{25}H_{29}N_6O_3$	$Mo\{\eta^2\text{-}OC(pTol)\}(CO)_2\{(Me_2pz)_3BH\}$		131
134	$BOs_3C_{10}H_3O_{10}$	$Os_3(CO)_9(\mu\text{-}H)_3(\mu_3\text{-}BCO)$		132
135	$BReC_9H_9BrO_6$	<u>fac</u>-$(OC)_3Re\{(CMeO)_3BBr\}$		133
136	$BReC_9H_9ClO_6$	<u>fac</u>-$(OC)_3Re\{(CMeO)_3BCl\}$		133
137	$BScSi_4C_{22}H_{46}$	$Sc\{\eta\text{-}C_5H_3(1,3\text{-}SiMe_3)_2\}_2(\mu\text{-}H)_2BH_2$		134
138	$BSi_8C_{16}H_{50}F_3N_6$	$FB\{NSi(Me)_2N(SiMe_2F)Si(Me)_2NHSiMe_2\}_2$		135
139	$BTaC_{13}H_{25}ClN_4$	$TaClMe_3\{(Me_2pz)_3BH\}$		136
140	$BZrC_{26}H_{46}N_7O$	$ZrMe(OBu^t)\{\eta^2\text{-}C(Me)NBu^t\}\{(Me_2pz)_3BH\}$		137
141	$B_2C_{18}H_{12}N_2$	$(PhCH=NBMe_2)_2$		138
142	$B_2C_{21}H_{21}NO_2$	$Me_2C=NOB(Ph)OBPh_2$		139
143	$B_2C_{24}H_{22}N_8$	$\{B(Ph)(pz)(\mu\text{-}pz)\}_2$		140
144	$B_2C_{26}H_{20}N_2O_4$	$[B(Ph)\{\mu\text{-}OC_6H_4(2\text{-}CHNO)\}]_2$		141
145	$B_2C_{36}H_{44}O$	$(Mes_2B)_2O$		142
146	$B_2CoC_{10}H_{18}N_2$	$Co\{\eta^5\text{-}B(Me)N(Me)CHCHCH\}_2$		143
147	$B_2CoC_{16}H_{30}N_2$	$Co\{\eta^5\text{-}B(Me)N(Bu^t)CHCHCH\}_2$		144

No.	Formula	Structure	Details	Ref.
148	$B_2CoC_{17}H_{29}$	$Co(Cp)(\eta^5\text{-}BEtCEtCEtBEtCHMe)$		145
149	$B_2CoFeC_{19}H_{27}$	$(Cp)Co\{\mu\text{-}\eta^5\text{-}(BMeCEtCEtBMeCH)\}Fe(Cp)$		146
150	$B_2CoNiC_{19}H_{27}$	$(Cp)Co\{\mu\text{-}\eta^5\text{-}(BMeCEtCEtBMeCH)\}Ni(Cp)$		146
151	$B_2CrC_{20}H_{36}N_2O_4$	$Cr(CO)_4\{\eta^4\text{-}(Bu^nBNBu^t)_2\}$		147
152	$B_2FeC_8H_{14}N_2$	$Fe(\eta^5\text{-}BMeNHCHCHCH)_2$		148
153	$B_2MnC_{12}H_{36}P_4$	$Mn\{CH_2P(Me)_2BH_2P(Me)_2CH_2\}_2$		149
154	$B_2Mo_2C_{22}H_{14}N_{12}O_4$	$Mo_2(CO)_4(pz_3BH)_2$		128
155	$B_2Ni_2C_{19}H_{27}$	$(Cp)Ni\{\mu\text{-}\eta^5\text{-}(BMeCEtCEtBMeCH)\}Ni(Cp)$		146
156	$B_4Co_2SnC_{28}H_{44}$	$[(Cp)Co\{\mu\text{-}\eta^5\text{-}(BMeCEtCEtBMeCH)\}]_2Sn$		150
157	$B_4FeC_{12}H_{20}$	$Fe(\eta^5\text{-}C_2B_4H_4Me_2)(\eta^6\text{-}C_8H_{10})$		151
158	$B_4FeC_{12}H_{20}$	$Fe(\eta^5\text{-}C_2B_4H_4Et_2)(\eta\text{-}C_6H_6)$		152
159	$B_4FeC_{13}H_{22}$	$Fe(\eta^5\text{-}C_2B_4H_4Et_2)(\eta\text{-}C_6H_5Me)$		153
160	$B_4FeC_{15}H_{26}$	$Fe(\eta^5\text{-}C_2B_4H_4Et_2)(\eta\text{-}Mes)$		152
161	$B_4FeC_{18}H_{32}$	$Fe(\eta^5\text{-}C_2B_4H_4Et_2)(\eta\text{-}C_6Me_6)$		152
162	$B_4Na_4C_{16}H_{50}O$	$[NaBMe_3(\mu\text{-}H)]_4.Et_2O$		154
163	$B_4NpC_4H_{24}$	$Np(BH_3Me)_4$		155
164	$B_4ThC_4H_{24}$	$Th(BH_3Me)_4$		155
165	$B_4UC_4H_{24}$	$U(BH_3Me)_4$		155
166	$B_4VC_{14}H_{22}$	$V(\eta^5\text{-}C_2B_4H_4Et_2)(\eta\text{-}C_8H_8)$		156
167	$B_4ZrC_4H_{24}$	$Zr(BH_3Me)_4$		155
168	$B_5CoC_9H_{16}$	$1,2\text{-}Me_2\text{-}3,1,2\text{-}(Cp)CoC_2B_5H_5$		157
169	$B_5OsC_{45}H_{49}ClOP_3$	$(Ph_3P)_2(CO)Os(\mu\text{-}H)(\mu\text{-}B_5H_7)PtClP(Me_2Ph)$		158
170	$B_6C_{42}H_{38}N_2O_6$	$\{(PhBO)_3\}_2\{\mu\text{-}C_6H_4(NH_2)_2\text{-}1,4\}.2C_6H_4(NH_2)_2$		159
171	$B_6C_{42}H_{42}N_2O_6$	$\{(PhBO)_3\}_2\{\mu\text{-}N(CH_2CH_2)_3N\}.3PhH$	193	159
172	$B_7CrC_{17}H_{32}$	$(Cp)Cr\{\eta^6\text{-}(3C,3B)\text{-}C_4(Et)_4B_7H_7\}$		160
173	$[B_8CoC_6H_{14}]^-$	$[NMe_4]$[<u>closo</u>-$\{1\text{-}(Cp)\text{-}1,10\text{-}CoCB_8CH_9\}$]		161
174	$B_8CrC_{17}H_{33}$	$(Cp)Cr\{\eta^6\text{-}(2C,4B)\text{-}C_4(Et)_4B_8H_8\}$		160
175	$B_8Cr_2C_{12}H_{20}$	<u>closo</u>-$[2,3\text{-}(Cp)_2\text{-}2,3,1,7\text{-}Cr_2C_2B_8H_{10}]$		90
176	$B_8IrC_{39}H_{41}O_2P_2$	<u>closo</u>-$\mu\text{-}1,2\text{-}(MeCOO)\text{-}2\text{-}H\text{-}2,10\text{-}(PPh_3)_2\text{-}1,2\text{-}CIrB_8H_7$		162
177	$B_8IrC_{55}H_{54}P_3$	<u>closo</u>-$1,H\text{-}1,1,2\text{-}(PPh_3)_3\text{-}1,9\text{-}IrCB_8H_8$		163

No.	Formula	Structure	Details	Ref.
178	$B_8NiC_7H_{16}$	nido-[9-(Cp)-7,8,9-$C_2NiB_8H_{11}$]		90
179	$B_8PtC_{14}H_{40}P_2$	nido-[9-H-9,10-$(Et_3P)_2$-7,8,9-$C_2PtB_8H_9$]		164
180	$B_8PtC_{14}H_{42}P_2$	nido-[9-H-9,9-$(Et_3P)_2$-$\mu_{10,11}$-H-7,8,9-$C_2PtB_8H_{10}$]		164
181	$[B_9RhC_{20}H_{26}Br_2P]^-$	[$PHPh_3$][closo-3-Ph_3P-3,3-Br_2-3,1,2-$RhC_2B_9H_{11}$]		165
182	$[B_9RhC_{28}H_{30}FNO_2P]^-$	[PPN][closo-2,2-{OCON=C(C_6H_4F)}-2-(Ph_3P)-2,1,7-$RhC_2B_9H_{11}$]		166
183	$[B_9RhC_{38}H_{41}P_2]^-$	[NMe_4][closo-2,2-$(Ph_3P)_2$-2,1,7-$RhC_2B_9H_{11}$]		167
184	$[B_9RhC_{38}H_{41}P_2]^-$	[K(18-crown-6)][closo-3,3-$(Ph_3P)_2$-3,1,2-$RhC_2B_9H_{11}$]		167
185	$B_{10}C_{10}H_{19}P$	closo-3,4-$PhP(CH_2)_2$-3,4-$C_2B_{10}H_{10}$		168
186	$B_{10}Co_2C_{14}H_{22}$	1,7-$(Cp)_2$-1,7,2,6-$Co_2C_2B_5H_6$-3-$C_2B_5H_6$		169
187	$B_{10}Co_2C_{21}H_{60}P_3$	1-Me-4-(Et_3P)-$\mu_{4,7}$-{Co$(PEt_3)_2$-μ-$(H)_2$}-1,2,4-$C_2CoB_{10}H_{10}$}		170
188	$B_{10}ReC_{15}H_{22}NO_4$	$(OC)_4$Re{η^2-(N,B^2)-1-(Me_2NCH_2)-7-Ph-1,7-$C_2B_{10}H_9$}		171
189	$B_{18}Ni_2C_{42}H_{50}O_2P_2$	[closo-3-(μ-CO)-8-PPh_3-3,1,2-$NiC_2B_9H_{10}$$]_2$		172
190	$B_{18}PtC_4H_{22}$	3,3′-Pt(1,2-$C_2B_9H_{11}$$)_2$		173
191	$B_{18}Rh_2C_{16}H_{50}$	[1-(Et_3P)-1,2,3-$RhC_2B_9H_{10}$$]_2$		174
192	$B_{20}C_5H_{24}S_2$	(1,2-$C_2B_{10}H_{11}$-12-S$)_2CH_2$		175
193	$[B_{20}Rh_2C_{38}H_{59}N_2P_2]^-$	[$Bu^n{}_4N$][{1-H-1-(Ph_3P)-2-NH_2-1,2-$RhCB_{10}H_{10}$$\}_2$]		176
194	$B_{20}SnC_{10}H_{36}$	{closo-1,2-Me_2-1,2-$C_2B_{10}H_9$-9-$\}_2SnMe_2$		177
195	$BiC_{21}H_{21}$	Bi$(pTol)_3$		178
196	$BiC_{28}H_{23}ClNO$	BiCl$(Ph)_3$(2-Me-8-quinolato)		179
197	$BiSi_6C_{21}H_{57}$	Bi{CH$(SiMe_3)_2$$\}_3$		180
198	$Bi_2C_{24}H_{20}$	Bi_2Ph_4		181
199	$CoC_8H_{18}N_4O_7P$	Co(Me)$(dmgH)_2$P$(OMe)_3$		182
200	$CoC_{11}H_9S_2$	Co(1,2-$S_2C_6H_4$)(Cp)		183
201	$CoC_{11}H_{11}N_2$	Co{1,2-$(NH)_2C_6H_4$}(Cp)		184
202	$CoC_{13}H_{18}$	Co{η^4-1,2-$(CH_2)_2C_6H_4$}(Cp)		185
203	$CoC_{13}H_{29}O_{12}P_3$	Co{P$(OMe)_3$$\}_3$($\eta^2$-CH=CHC(O)OCO)		186
204	$CoC_{14}H_5F_8O$	Co{F_8bicyclo[3.3.0.]octa-2,7-diene-4,6-diyl}-		

No.	Formula	Structure	Details	Ref.
		(CO)(Cp)		187
205	$[CoC_{14}H_{26}N_4O]^+$	$[(Et)Co\{\underline{o}\text{-}OC_6H_4C(Me){=}N(CH_2)_2NH_2\}(en)]Br$		188
206	$CoC_{16}H_{27}N_6O_4$	$Co(Pr^i)(dmgH)_2(NC_5H_4NH_2)$		189
207	$CoC_{16}H_{31}P_2$	$Co(dmpe)(\eta\text{-}C_5Me_5)$		190
208	$CoC_{17}H_{35}P_2$	$Co(PEt_3)_2(Cp)$	173	191
209	$[CoC_{17}H_{35}P_2]^+$	$[Co(PEt_3)_2)(Cp)][BF_4]$	213	191
210	$CoC_{17}H_{36}P_3$	$Co(Ph)(PMe_3)_3(\eta\text{-}C_2H_4)$		192
211	$CoC_{23}H_{24}N_4O_4P$	$Co(Me)(dmgH)_2(PPh_3)$		182
212	$[CoC_{23}H_{37}P_3]^+$	$[Co(PMe_3)_3(\eta\text{-}C_2Ph_2)][BPh_4]$		119
213	$CoC_{25}H_{27}N_3O_2P$	$Co(CO)(NO)(PPh_3)\{(\underline{S})\text{-}\overline{CN(Me)CH(Me)CH_2NMe}\}$		193
214	$[CoC_{25}H_{40}NP_3]^+$	$[Co(NCMe)(PMe_3)_3(\eta\text{-}C_2Ph_2)][BPh_4]$		119
215	$CoC_{27}H_{31}BrN_4O_4P$	$Co(CH_2Br)(dmgH)_2(PPh_3)$		194
216	$CoC_{28}H_{31}N_5O_4P$	$Co(CH_2CN)(dmgH)_2(PPh_3)$		194
217	$CoC_{42}H_{39}OP_2$	$Co\{(1\text{-}2\eta,5\text{-}7\eta)\text{-}CH_2CHCH_2OCHCHCH_2\}(PPh_3)_2$		195
218	$CoCr_4C_{28}H_{38}S_6$	$Co\{Cr_2(Cp)_2(\mu\text{-}SBu^t)(\mu_3\text{-}S)_2\}_2$		196
219	$CoFeC_{16}H_{11}O_3$	$(Cp)Co[\mu\text{-}(\eta^4,\eta^1,\eta^1)\text{-}\{1,2\text{-}(CH)_2\}C_6H_4]Fe(CO)_3$		197
220	$CoFeC_{27}H_{20}O_6P$	$(Ph_3P)(OC)_2CoFe\{\mu\text{-}(\eta^1,\mu\text{-}\eta^2)\text{-}CH_2C_2CH_2OH\}(CO)_3$		198
221	$CoFeC_{28}H_{21}O_7$	$(Cp)Co\{\mu\text{-}(\eta^4,\eta^1,\eta^1)\text{-}C(Ph)C(CO_2Me)C(CO_2Me)CPh\}\text{-}Fe(CO)_3$		199
222	$CoFeMnC_{14}H_8O_8P$	$(Cp)MoFeCo(CO)_8(\mu_3\text{-}PMe)$		57
223	$CoFeMnC_{14}H_9O_{10}P$	$CoFeMn(CO)_9(\mu\text{-}CO)(\mu_3\text{-}PBu^t)$		200
224	$CoFeNiC_{15}H_{14}O_6P$	$(Cp)NiCoFe(CO)_6(\mu_3\text{-}PBu^t)$		201
225	$CoFe_2C_{16}H_{14}O_7P$	$CoFe_2(CO)_6(\mu\text{-}CO)(\mu_3\text{-}PBu^t)(Cp)$		200
226	$CoFe_2C_{21}H_{11}O_9P_2$	$CoFe_2(CO)_9(\mu\text{-}PHPh)(\mu_3\text{-}PPh)$		202
227	$CoFe_2NiC_{15}H_{11}O_8P_2$	$(Cp)NiCoFe_2(CO)_8(\mu_4\text{-}PMe)_2$		201
228	$CoLiC_{11}H_{31}P_3$	$Li\{Co(PMe_3)_3(\eta\text{-}C_2H_4)\}$	243	203
229	$CoOs_3C_{14}H_8O_9$	$(Cp)CoOs_3(CO)_9(\mu\text{-}H)_3$		204
230	$CoRhC_{22}H_{30}O_2$	$CoRh(\mu\text{-}CO)_2(\eta\text{-}C_5Me_5)_2$	200	205
231	$CoSnC_{27}H_{42}P_3$	$(Me_3P)_3CoSnPh_3$		206
232	$CoWC_{15}H_5O_8$	$(OC)_4CoW(CO)_4(CPh)$		207

No.	Formula	Structure	Details	Ref.
233	$CoWC_{26}H_{27}O_3$	$(\eta\text{-}C_5Me_5)(OC)CoW\{\mu\text{-}C(pTol)(CO)_2(Cp)$		208
234	$Co_2C_8O_8$	$Co_2(CO)_8$	100	209
235	$Co_2C_{11}H_3F_6NO_5$	$\{\mu\text{-}\eta\text{-}C_2(CF_3)_2\}Co_2(CO)_5(MeCN)$		210
236	$Co_2C_{12}H_6O_{10}$	$Co_2(CO)_6\{\mu\text{-}\eta^2\text{-}C_2(CO_2Me)_2\}$		211
237	$Co_2C_{15}H_{16}O$	$(Cp)_2Co_2(\mu\text{-}CO)\{\mu\text{-}(\eta^2,\eta^2)\text{-}C_4H_6\}$		212
238	$Co_2C_{15}H_{16}O_2$	$Co_2(CO)_2(\mu\text{-}CH_2)(\eta\text{-}C_5H_4Me)_2$		213
239	$Co_2C_{15}H_{16}O_2$	$\{Co(Cp)\}_2(\mu\text{-}CO)_2(\mu\text{-}C_3H_6)$		214
240	$[Co_2C_{16}H_6N_{10}O_4]^{6-}$	$K_6[(NC)_5CoC(CO_2Me)\text{=}C(CO_2Me)Co(CN)_5].6H_2O$		215
241	$Co_2C_{16}H_{11}I_2O_8$	$Co_2(CO)_6\{\mu\text{-}CCI_2\}\{\mu\text{-}CCH\text{=}C(C_5H_{11})C(O)O\}$		216
242	$Co_2C_{16}H_{18}$	$(Cp)Co\{\mu\text{-}(1\text{-}3\eta:1,4,5\eta)\text{-}CHCHCHCHCHMe\}Co(Cp)$		217
243	$Co_2C_{16}H_{18}$	$(Cp)Co\{\mu\text{-}(1\text{-}3\eta:1,4,5\eta)\text{-}C(Me)CHCHCHCH_2\}Co(Cp)$		217
244	$Co_2C_{16}H_{18}O_6$	$Co_2(CO)_6(\mu\text{-}\eta^2\text{-}C_2Bu^t{}_2)$	200	211
245	$Co_2C_{20}H_{10}Cl_4S_4$	$(Cp)_2Co_2\{\mu\text{-}(2\mu\text{-}S)\text{-}S_2C_{10}Cl_4S_2\}$		218
246	$Co_2C_{20}H_{10}O_6$	$Co_2(CO)_6(\mu\text{-}\eta^2\text{-}C_2Ph_2)$		211
247	$[Co_2C_{20}H_{29}O_4P_2]^+$	$[(Cp)Co(\mu\text{-}PMe_2)\{\mu\text{-}(O,P,\eta^2)\text{-}Me_2PC(CO_2Me)\text{=}CHCO_2Me\}\text{-}$ $Co(Cp)][BF_4]$		219
248	$Co_2C_{20}H_{36}O_4P_2$	$\{Co(CO)_2(\mu\text{-}PBu^t{}_2)\}_2$		220
249	$Co_2C_{22}H_{18}S_4$	$(Cp)_2Co_2\{\mu\text{-}(S,\mu\text{-}S)\text{-}(1,2\text{-}S_2)C_6H_4\}_2$		183
250	$Co_2C_{22}H_{22}O_2$	$(\eta\text{-}C_5H_4Me)_2Co_2\{\mu\text{-}(1,2\text{-}(CH_2)_2C_6H_4\}$		185
251	$Co_2C_{22}H_{25}O_2P$	$(\eta\text{-}C_5H_4Me)Co(\mu\text{-}CO)_2Co(PMe_2Ph)(\eta\text{-}C_5H_4Me)$		185
252	$Co_2C_{23}H_{32}O_2$	$\{(\eta\text{-}C_5Me_5)(OC)Co\}_2(\mu\text{-}CH_2)$		221
253	$Co_2C_{24}H_{54}O_2P_4$	$[Co(CO)(PMe_3)(\mu\text{-}PBu^t{}_2)]_2$		220
254	$Co_2C_{28}H_{22}O_7$	$(Cp)Co\{\mu\text{-}(\eta^1,\eta^1,\eta^3)\text{-}PhCC(CO_2Me)C(Ph)C(CO_2Me)\}\text{-}$ $(\mu\text{-}CO)Co(CO)_2$		199
255	$Co_2C_{30}H_{20}O_6P_2$	$Co_2(CO)_6(\mu\text{-}PPh_2)_2$		222
256	$Co_2C_{32}H_{24}O_6P_2$	$(OC)_3Co(\mu\text{-}CO)_2Co(CO)(dppe)$		223
257	$Co_2C_{46}H_{50}O_2P_2$	$\{(PhEt_2P)(OC)Co\}_2(\mu\text{-}PPh_2)_2$		224
258	$Co_2C_{60}H_{106}P_4$	$\{Co(dcpe)\}_2\{\mu\text{-}(\eta^3,\eta^3)\text{-}C_6H_4Me_2\}$		225
259	$Co_2Cr_2C_{16}O_{16}P_2$	$\{Co_2(CO)_6\}(\mu_4\text{-}P_2)\{Cr(CO)_5\}_2$		226
260	$Co_2FeC_9HNO_9$	$FeCo_2(CO)_9(\mu_3\text{-}NH)$		227

No.	Formula	Structure	Details	Ref.
261	$Co_2FeC_{27}H_{20}O_3$	$(OC)_3FeCo_2(Cp)_2(\mu_3\text{-}CPh)_2$		228
262	$Co_2FeRuC_{23}H_{10}O_{10}$	$(OC)_3RuCo_2(CO)_5(\mu\text{-}CO)Fe(CO)(Cp)\text{-}$ $\{\mu_4\text{-}(\eta^1,\eta^1,\eta^1,\eta^2)\text{-}C_2Ph\}$		229
263	$Co_2Fe_2C_{23}H_{10}O_{11}P_2$	$(OC)_6Fe_2Co_2(CO)_4(\mu\text{-}CO)(\mu_4\text{-}PPh)_2$		202
264	$Co_2MoC_{18}H_{19}NO_6S$	$(Cp)(OC)_2MoCo_2(CO)_4(\mu_3\text{-}S)\{\mu\text{-}(C,N)\text{-}MeC{=}NCy\}$		230
265	$Co_2Mo_2C_{14}H_{10}O_4S_3$	$(Cp)_2Mo_2(\mu_4\text{-}S)(\mu_3\text{-}S)_2Co_2(CO)_4$		231
266	$Co_2Pd_2C_{57}H_{44}O_7P_4$	$(OC)_3CoPd_2(\mu\text{-}dppm)_2Co(CO)_4.2.5THF$		232
267	$Co_2RuC_{23}H_{10}O_9$	$RuCo_2(CO)_9\{\mu_3\text{-}(\eta^1,\eta^1,\eta^2)\text{-}C_2Ph_2\}$		229,233
268	$Co_2RuWC_{23}H_{12}O_{10}$	$(Cp)(OC)WCo_2Ru(CO_7)(\mu\text{-}CO)_2\{\mu_3\text{-}C(pTol)\}$		229
269	$Co_2Ru_2C_{12}H_2O_{12}$	$(OC)_6(\mu\text{-}H)Ru_2Co_2(\mu_3\text{-}H)(CO)_6$		234
270	$Co_2Ru_2C_{25}H_{10}O_{11}$	$(OC)_3Ru\{(OC)_2Co(\mu\text{-}CO)\}_2Ru(CO)_2\{\mu_4\text{-}C_2Ph_2\}$		234
271	$Co_3C_9H_4NO_7S_2$	$Co_3(CO)_7(\mu_3\text{-}S)\{\mu\text{-}(N,S)\text{-}NHC(Me)S\}$		235
272	$Co_3C_{22}H_{26}NS$	$Co_3(Cp)_3(\mu_3\text{-}CNCy)(\mu_3\text{-}S)$		236
273	$Co_3C_{26}H_{36}S$	$\{(Cp)Co\}_3(\mu\text{-}SMe)(\mu_3\text{-}CBu^n)_2$		237
274	$Co_3C_{34}H_{51}$	$Co_3(CO)_9(\mu_3\text{-}CMe)_2$		238
275	$Co_3C_{60}H_{46}O_6P_3$	$Co_3H(CO)_6(PPh_3)_3$		239
276	$Co_3FeC_{16}H_7O_9$	$(Cp)FeCo_3(\mu_4\text{-}C{=}CH_2)(\mu\text{-}CO)_2(CO)_7$	220	240
277	$Co_3FeC_{37}H_{21}O_9$	$Co_3Fe(CO)_7(\mu\text{-}CO)_2\{\mu_4\text{-}C_2Ph_2\}\{\mu\text{-}(\eta^1,\eta^2)\text{-}CPhCHPh\}$		241
278	$[Co_3RuC_{12}O_{12}]^-$	$[PPN][RuCo_3(CO)_9(\mu\text{-}CO)_3]$		242
279	$[Co_3RuC_{24}H_{10}O_{10}]^-$	$[NEt_4][RuCo_3(CO)_8(\mu\text{-}CO)_2\{\mu_4\text{-}(2\eta^1,2\eta^2)\text{-}C_2Ph_2\}]$		233
280	$Co_4C_{20}H_{16}O_8P_2$	$[(OC)_3Co\{\mu_3\text{-}(\eta^4,\mu\text{-}P)\text{-}\overline{PCHC(Me)C(Me)CH}\}Co(CO)]_2$		243
281	$[Co_6C_{16}O_{16}P]^-$	$[PPh_4][Co_6P(CO)_{16}]$		244
282	$Co_6C_{21}O_{19}S$	$\{Co_3(CO)_9C\}_2SCO$		245
283	CrC_5O_7S	$Cr(CO)_5\{\eta^1\text{-}(S)\text{-}SO_2\}$		246
284	$[CrC_6HO_6S]^-$	$[PPN][Cr(CO)_5SCHO]$		247
285	$CrC_6H_3ClO_4$	<u>trans</u>-$CrCl(CMe)(CO)_4$	100	248
286	$CrC_7H_6N_2O_5S$	$Cr(SNNMe_2)(CO)_5$		249
287	$CrC_9H_8O_6$	$Cr\{{=}C(Me)OEt\}(CO)_5$	X,N 100	248
288	$CrC_{11}H_6O_3S_2$	$Cr(CO)_3\{\eta^6\text{-}C_5H_4\overline{CSCHCHS}\}$		250
289	$CrC_{11}H_8O_3S_2$	$Cr(CO)_3\{\eta^6\text{-}C_5H_4\overline{CSCH_2CH_2S}\}$		250

No.	Formula	Structure	Details	Ref.
290	$CrC_{11}H_{10}O_3S_2$	$Cr(CO)_3\{\eta^6\text{-}C_5H_4C(SMe)_2\}$		250
291	$CrC_{11}H_{10}O_6$	$Cr(CO)_5\{C(OMe)C(Me){=}CHMe\}$	233	251
292	$CrC_{11}H_{12}N_4O_7P_2$	$Cr(CO)_5[P_2\{MeNC(O)NMe\}_2]$		252
293	$CrC_{12}H_{10}NO_3$	$Cr(CO)_3\{\eta^6\text{-}(NMe)indole\}$		253
294	$CrC_{14}H_9N_3O_4$	$Cr(CO)_4\{(\underline{o}\text{-}NC_5H_4)_2NH\}$		254
295	$CrC_{14}H_{16}N_4O_4$	<u>cis</u>-$Cr(CO)_4\{CN(Me)CHCHNMe)_2$	243	255
296	$[CrC_{14}H_{16}N_4O_4]^+$	<u>cis</u>-$[Cr(CO)_4\{CN(Me)CHCHNMe)_2][PF_6]$	243	255
297	$CrC_{14}H_{18}N_2O_4$	<u>cis</u>-$Cr(CO)_4(CNBu^t)_2$		256
298	$CrC_{14}H_{38}P_4$	$CrMe_2(dmpe)_2$		257
299	$CrC_{15}H_{14}O_3$	$Cr(CO)_3\{\eta^6\text{-}CH(CH)_4C{=}(CCH_2CH_2)_2\}$		250
300	$CrC_{15}H_{17}NO_4$	$Cr(CO)_2(CNBu^t)(\eta\text{-}C_6H_5CO_2Me)$		258
301	$CrC_{17}H_{10}N_2O_5S$	$Cr(SNNPh_2)(CO)_5$		249
302	$CrC_{17}H_{16}N_2O_5$	$Cr(CO)_3\{\eta^6\text{-}C_6H_4(1,2\text{-}C_8H_{12}N_2O_2)\}$		259
303	$CrC_{17}H_{20}O$	$Cr(CO)[\{\eta\text{-}C_5H_4C(Me)_2\text{-}\}_2]$		260
304	$CrC_{18}H_{21}NO_4S$	$Cr(CO)_2(NO)\{1\eta^1,3\text{-}5\eta^3\text{-}CHCH(Bu^t)CHC(Ph)PCHSOMe\}$		261
305	$[CrC_{21}H_{18}NO_3S]^+$	$[Cr(CO)_2(NO)\{\eta^5\text{-}CHPhCHCPhCHSEt\}][PF_6]$		262
306	$CrC_{21}H_{30}O_2S$	$Cr(CO)_2(CS)(\eta\text{-}C_6Et_6)$		263
307	$CrC_{24}H_{16}F_5P$	$Cr(\eta\text{-}C_6H_6)(\eta\text{-}C_6F_5PPh_2)$	203	264
308	$CrC_{29}H_{15}O_5$	$Cr(CO)_5P(C_2Ph)_3$		265
309	$CrC_{30}H_{39}O_3$	$Cr(Ph)_3(THF)_3$		266
310	$CrC_{31}H_{30}O_2P_2$	$Cr(CO)_2\{\eta^6\text{-}C_6H_4(Me\text{-}2)P(oTol)CH_2P(oTol)_2\}$		267
311	$CrC_{40}H_{44}$	$Cr(PhC{=}CMe_2)_4$		268
312	$CrFe_3C_{18}H_9O_{14}PS$	$Fe_3(CO)_9\{\mu_3\text{-}SCr(CO)_5\}(\mu_3\text{-}PBu^t)$	233	269
313	$CrRhC_{20}H_{21}O_4$	$(\eta\text{-}C_6H_6)(OC)Cr(\mu\text{-}CO)_2Rh(CO)(\eta\text{-}C_5Me_5)$		270
314	$CrSeC_{17}H_{10}O_5S$	$(OC)_5CrSe(Ph)SPh$		271
315	$CrSi_4C_{19}H_{38}O_5P_2$	$(OC)_5Cr\{PCH(SiMe_3)_2\}_2$		272
316	$CrSnC_{13}H_{17}NO_5S_2$	$(OC)_5CrSn\{(SCH_2CH_2)_2NBu^t\}$		273
317	$CrWC_{13}H_8N_2O_7$	$(Cp)(OC)_2WNN(Me)Cr(CO)_5$	98	274
318	$CrWC_{29}H_{30}O_4$	$(\eta\text{-}C_6Me_6)(OC)_2CrW\{\mu\text{-}C(pTol)\}(CO)_2(Cp).CH_2Cl_2$		208
319	$CrZrC_{28}H_{22}ClO_5P$	$(OC)_5Cr(\mu\text{-}Ph_2PCH_2)ZrCl(Cp)_2$	140	275

No.	Formula	Structure	Details	Ref.
320	$CrZrC_{40}H_{34}O_4P_2$	$(OC)_4Cr(\mu\text{-}Ph_2PCH_2)_2Zr(Cp)_2$.0.5diglyme	140	275
321	$[Cr_2C_{10}HO_{10}]^-$	$K[\{Cr(CO)_5\}_2(\mu\text{-}H)]$		276
322	$Cr_2C_{10}H_{10}Cl_4$	<u>trans</u>-$[CrCl(\mu\text{-}Cl)(Cp)]_2$		277
323	$Cr_2C_{14}H_{10}O_4S$	$\{Cr(CO)_2(Cp)\}_2(\mu\text{-}S)$		278
324	$Cr_2C_{15}H_{10}O_5S_2$	$(Cp)(CO)_3Cr\{\mu\text{-}(\eta^1,\eta^2)\text{-}S_2\}Cr(CO)_2(Cp)$		278
325	$Cr_2C_{18}H_{18}$	$\{(Cp)Cr\}_2\{\eta\text{-}(\eta^5,\eta^5)\text{-}C_8H_8\}$		279
326	$Cr_2C_{20}H_{24}N_8O_{12}P_4$	$\{Cr(CO)_4\}_2[\mu\text{-}P_2\{MeNC(O)NMe\}_2]_2$		252
327	$Cr_2C_{22}H_{10}O_{10}P_2$	$\{(OC)_5Cr\}_2(\mu\text{-}P_2Ph_2)$		280
328	$Cr_2C_{22}H_{14}O_{10}$	$\{(OC)_3Cr\}_2\{\mu\text{-}(\eta^6,\eta^6)\text{-}C_6H_4(2\text{-}CO_2Me)\text{-}1\text{-}\}_2$		281
329	$Cr_2C_{23}H_{16}O_{10}$	$\{(OC)_3Cr\}_2\{\mu\text{-}(\eta^6,\eta^6)\text{-}C_6H_4(2\text{-}C_3H_5O_2)C_6H_4(2\text{-}CO_2Me)\}$		281
330	$Cr_2C_{24}H_{36}N_6O_6$	$Cr_2(CO)_6\{\mu\text{-}(N,N)\text{-}N_2C_2Me_4\}_3$		282
331	$Cr_2FeC_{22}H_{28}O_6$	$(Cp)_2(\mu\text{-}OBu^t)_2Cr_2Fe(CO)_4$		283
332	$Cr_2HgC_{18}H_{10}O_6$	$\{(OC)_3Cr(\eta\text{-}C_6H_5)\}_2Hg$		284
333	$Cr_2Pd_2C_{28}H_{40}O_6P_2$	$(PEt_3)_2Pd_2Cr_2(Cp)_2(\mu\text{-}CO)_4(\mu_3CO)_2$		285
334	$Cr_2Si_2C_{16}H_{18}NO_{10}P$	$\{\mu\text{-}PN(SiMe_3)_2\}\{Cr(CO)_5\}_2$	140	58
335	$Cr_3C_{27}H_{10}O_{15}P_2$	$\{(OC)_5Cr\}_3\{\mu\text{-}(\eta^1,\eta^1,\eta^2)\text{-}P_2Ph_2\}$.0.5$CH_2Cl_2$		280
336	$Cr_3FeC_{20}H_{24}O_2S_3$	$(Cp)_3Cr_3Fe(CO_2Bu^t)(\mu_3\text{-}S)_4$		286
337	$Cr_3Si_4C_{26}H_{36}N_2O_{14}P_2$	$[\{\mu\text{-}PN(SiMe_3)_2\}\{Cr(CO)_5\}]_2Cr(CO)_4$	140	58
338	$[Cr_4C_{17}O_{17}S]^{2-}$	$[PPN]_2[(OC)_5Cr(\mu_4\text{-}S)Cr_3(CO)_9(\mu\text{-}CO)_3]$		287
339	$Cr_4C_{24}H_{28}OS_3$	$Cr_4(\mu_3\text{-}O)(\mu_3\text{-}S)_3(\eta\text{-}C_5H_4Me)_4$.$CuBr_2$		288
340	$Cr_4C_{24}H_{28}S_4$	$\{(\eta\text{-}C_5H_4Me)Cr(\mu_3\text{-}S)\}_4$	173	228,289
341	$CuC_3H_5O_4S$	$[Cu(OSO_2Et)(CO)]_n$		290
342	$[CuC_{12}H_{11}N_3]^+$	$[Cu\{(NC_5H_4)_2NH\}(\eta\text{-}C_2H_2)][BF_4]$	173	291
343	$[CuC_{18}H_{22}]^-$	$[Cu(dppe)_2][Cu(Mes)_2]$		292
344	$CuC_{37}H_{30}ClP_2$	$CuCl\{C(PPh_3)_2\}$		293
345	$CuIrC_{59}H_{49}ClOP_4$	$ClCuIr(C_2Ph)(CO)(\mu\text{-}dppm)_2$		294
346	$CuMoC_{17}H_{18}NO_6$	$(4\text{-}MeC_5H_4N)CuMo(\mu\text{-}CO)_2(\mu\text{-}CO_2Me)_2(Cp)$		295
347	$CuMoC_{28}H_{54}BrN_4S_2$	$(Bu^tNC)_4Mo(\mu\text{-}SBu^t)_2CuBr$		296
348	$[CuOs_{10}C_{27}H_3NO_{24}]^-$	$[PPh_3Me][Os_{10}C(CO)_{24}Cu(NCMe)]$		94
349	$[CuOs_{11}C_{30}H_3NO_{27}]^-$	$[PPh_3Me][Os_{11}C(CO)_{27}Cu(NCMe)]$.$CH_2Cl_2$		297

No.	*Formula*	*Structure*	*Details*	*Ref.*
350	$CuPtWC_{31}H_{45}O_2P_2$	$(\eta\text{-}C_5Me_5)CuPt(PMe_3)_2W(CO)_2(Cp)\{\mu_3\text{-}C(pTol)\}$		298
351	$CuRh_2C_{32}H_{45}O_2$	$(\eta\text{-}C_5Me_5)CuRh_2(Cp)_2(\mu_3\text{-}CO)_2$		298
352	$CuWC_{44}H_{35}O_3P_2$	$(Ph_3P)_2CuW(CO)_3(Cp)$		299
353	$Cu_2C_6H_{12}Cl_2O_2$	$\{Cu(\mu\text{-}Cl)(\eta^2\text{-}CH_2{=}CHCH_2OH)\}_2$		300
354	$Cu_2C_{17}H_{32}O_5$	$\{Cu(tmeda)\}_2(\mu\text{-}CO)(\mu\text{-}C_4O_4)$		301
355	$Cu_2C_{44}H_{38}N_4O_2$	$[Cu\{CN(pTol)\}_2(\mu\text{-}OPh)]_2$		302
356	$Cu_2Ru_4C_{48}H_{32}O_{12}P_2$	$(Ph_3P)_2Cu_2Ru_4(\mu_3\text{-}H)_2(CO)_{12}.CH_2Cl_2$		8
357	$Cu_2Si_4C_{24}H_{44}N_2$	$[2\text{-}(Me_3Si)_2CC_5H_4NCu]_2$		303
358	$Cu_4C_{20}H_{36}O_8$	$[Cu(CO)(\mu_3\text{-}OBu^t)]_4.1.5THF$	111	304
359	$Cu_4C_{44}H_{32}N_4O_8$	$[Cu(2\text{-}Mequinolin\text{-}8\text{-}olate)(CO)]_4$		305
360	$Cu_4C_{44}H_{60}S_2$	$Cu_4(\mu\text{-}Mes)_4(C_4H_8S)_2$		306
361	$Cu_4C_{64}H_{48}Cl_4$	$[CuCl(\eta^4\text{-}dibenzocot)]_4$		307
362	$Cu_5C_{45}H_{55}$	$[Cu(\mu\text{-}Mes)]_5$		306
363	$Cu_5C_{56}H_{60}N_{22}$	$Cu\{\mu\text{-}(1,2\text{-}N_3)C_6H_4\}_6\{Cu(CNBu^t)\}_4$		308
364	$[FeC_3NO_4]^-$	$[PPN][Fe(CO)_3(NO)]$		309
365	$FeC_5H_3NO_3$	$Fe(CO)_3(CNMe)_2$		310
366	$FeC_8H_8Cl_4N_3O_2P_3$	$Fe\{P(Me)NPCl_2NPCl_2N\}(CO)_2(Cp)$		311
367	$[FeC_9H_8N_3O_2]^+$	$[Fe(CO)_2(N_3C_2H_3)(Cp)][HSO_4]$	173	312
368	$FeC_{10}H_{10}$	$[Fe(Cp)_2][(NC)_2C{=}C(CN)O].0.5Fe(Cp)_2$		313
369	$[FeC_{10}H_{10}]^+$	$[Fe(Cp)_2][(NC)_2C{=}C(CN)O].0.5Fe(Cp)_2$		313
370	$[FeC_{10}H_{10}]^+$	$[Fe(Cp)_2][Bi_4Br_{16}]$		314
371	$FeC_{10}H_{22}O_3$	$Fe(CO)_3\{\eta^4\text{-}(C_8H_{11})_2\}$		315
372	$[FeC_{11}O_{10}]^{2-}$	$[AsPh_4]_2[Fe_3(CO)_9(\mu_3\text{-}CCO)]$		50
373	$[FeC_{11}HO_{10}]^-$	$[AsPh_4][Fe_3(CO)_9(\mu\text{-}CO)(\mu_3\text{-}CH)]$		50
374	$FeC_{12}H_9NO_6S$	$Fe_2(CO)_6\{\mu\text{-}(C,N,\mu\text{-}S)\text{-}SC(CH_2)_4C{=}NH\}$		316
375	$FeC_{12}H_{14}O_2S_4$	$Fe(SacSac)_2(CO)_2$		317
376	$FeC_{12}H_{16}N_2O_6S_3$	$Fe\{CHCH_2N(SO_2Me)SN(SO_2Me)CH_2\}(CO)_2(Cp)$		318
377	$FeC_{13}H_{15}F_3O_5S$	$Fe(OSO_2CF_3)(CO)_2(\eta\text{-}C_5Me_5)$		319
378	$FeC_{14}H_{14}N_2O_3$	$Fe(CO)_2(NO)\{\eta^3\text{-}C_4H_4(C_6H_4NMe_2\text{-}4)\}$	173	320
379	$FeC_{14}H_{14}N_2O_7$	$(OC)_3Fe\{C(CO_2Me){=}C(CO_2Me)NNC_5H_8\}$		321

No.	Formula	Structure	Details	Ref.
380	$FeC_{14}H_{17}P$	$Fe\{(\eta-C_5H_4)_2PBu^t\}$		59
381	$[FeC_{14}H_{18}O_5PS]^+$	$[Fe\{CH(SMe)(P(OCH_2)_3CMe)\}(CO)_2(Cp)][PF_6]$		322
382	$[FeC_{14}H_{20}N]^+$	$[Fe(Cp)(\eta-C_5H_4CH_2NMe_3)][B_3H_8]$		323
383	$[FeC_{14}H_{20}N]^+$	$[Fe(Cp)(\eta-C_5H_4CH_2NMe_3)]_2[B_{10}H_{10}]$		324
384	$[FeC_{14}H_{30}NO_2P_2]^+$	$[Fe(CO)_2(PMe_3)_2(\eta^2-MeC{=}NBu^t)][BPh_4]$		121
385	$FeC_{15}H_{12}N_2$	$(Cp)Fe\{\eta^5-C_5H_4C(Me){=}C(CN)_2\}$		325
386	$FeC_{15}H_{14}O_6$	$Fe(CO)_3\{(4-7)\eta-3-MeCO-3a,7a-H_2-6-MeO-$ 2-Mebenzofuran}		326
387	$FeC_{15}H_{18}O_3$	$Fe(CO)_2\{(1-4,1',2')\eta^6-(1-CHCHCOMe)-$ $(2,6,6-Me_3)C_6H_4\}$		327
388	$FeC_{16}H_{12}O_5$	$Fe\{\eta^1$-bicyclo[2.2.1]hept-5-ene-2,3-<u>endo</u>-$C_2O_3\}-$ $(CO)_2(Cp)$		328
389	$FeC_{16}H_{13}P$	$Fe\{(\eta-C_5H_4)_2PPh\}$		59
390	$FeC_{16}H_{14}O_2S_4$	$Fe(CO)_2(MeSC_6H_4S)_2$		329
391	$FeC_{17}H_{18}O_2$	$Fe(Cp)\{\eta-C_5H_4(C_7H_9O_2)\}$ (two isomers)		330
392	$FeC_{18}H_{25}O_3P$	$Fe(CO)_2(PEt_3)\{\eta^4-O{=}C(Me)CH{=}CHPh\}$		331
393	$FeC_{19}H_{16}O_4$	$(OC)_3Fe\{\eta^4$-1,3-epoxy-2,3-$(=CH_2)_2-$ 1,2,3,4,4a,9,9a,10-H_8anthracene}		332
394	$FeC_{19}H_{21}O_3P$	$Fe(CO)_2(PPhMe_2)\{\eta^4-O{=}C(Me)CH{=}CHPh\}$		331
395	$FeC_{20}H_{24}N_2O_7$	$(OC)_3Fe\{C(CO_2Me)CH_2C(Me){=}C(Me)CH_2C(CO_2Me)NNC_5H_8\}$		321
396	$FeC_{21}H_{17}NO_4$	$(OC)_3Fe\{\eta^2,\eta^1-CH(Ph)CHC(Me)N(CH_2Ph)CO\}$	153	333
397	$FeC_{22}H_{26}NP$	$Fe\{(\eta-C_5H_4)P(Ph)(\eta-C_5H_3CH(Pr^i)(NMe_2)\}$		59
398	$FeC_{23}H_{15}NO_3$	$Fe(CO)_2(NO)(\eta^3-C_3Ph_3)$		334
399	$FeC_{24}H_{15}NO_4$	$Fe(CO)_2(NO)(\eta^3$-oxoPh_3cyclobutenyl)		334
400	$[FeC_{25}H_{20}O_2P]^+$	$[Fe(CO)_2(PPh_3)(Cp)]Cl.3H_2O$		335
401	$FeC_{28}H_{16}$	$Fe\{\eta^5,\eta^5$-diindeno(5,4-c:4',5'-g)phenanthrene}		336
402	$FeC_{28}H_{21}O_2P$	$Fe(CO)_2(PPh_3)(\eta^4$-benzocyclobutadiene)		335
403	$FeC_{28}H_{30}NOP$	(<u>RS</u>)-$FeMe(CO)(PPh_2NMeCHMePh)(Cp)$		337
404	$[FeC_{29}H_{26}P]^+$	$[Fe(Cp)(\eta^5-C_5H_4PPh_2CH_2Ph)]Cl$		338
405	$FeC_{29}H_{29}O_2P$	$Fe\{COCH(Me)Et\}(Cp)(CO)PPh_3$		339

No.	***Formula***	***Structure***	***Details***	***Ref.***
406	$FeC_{29}H_{36}N_2O_3$	$Fe(CO)_3[N\{(2,6\text{-}Pr^i)_2C_6H_3\}=CHCH=N\{C_6H_3(Pr^i{}_2\text{-}2,6)\}]$		340
407	$FeC_{30}H_{24}O_4P_2$	$Fe(CO)_4(\eta^1\text{-}dppe)$		341
408	$FeC_{30}H_{29}O_3P$	$Fe\{\eta^1\text{-}\underline{E}\text{-}C(CO_2Et)=CHMe\}(CO)(PPh_3)(Cp)$		342
409	$[FeC_{32}H_{29}OP_2]^+$	$[Fe(CO)(dppe)(Cp)][BF_4]$		335
410	$FeC_{34}H_{31}O_4P$	$Fe\{\eta^1\text{-}\underline{Z}\text{-}C(Me)=CMePh\}(CO)\{P(OPh)_3\}(Cp)$		343
411	$FeC_{37}H_{31}NO_2P_2$	$FeH(CO)(NO)(PPh_3)_2$	115	344
412	$FeC_{38}H_{30}O_{10}S$	$Fe(CO)_2\{P(OPh)_3\}_2\{\eta^1\text{-}(S)\text{-}SO_2\}$		345
413	$FeC_{40}H_{58}O_4P_2$	$Fe(CO)_4\{\eta^1\text{-}P_2(2,4,6\text{-}Bu^t{}_3C_6H_2)_2\}$		346
414	$FeC_{44}H_{42}O_{10}S$	$Fe(CO)_2\{P(OoTol)_3\}_2\{\eta^1\text{-}(S)\text{-}SO_2\}$		345
415	$FeGeC_{28}H_{22}O_3$	$Fe(CO)_3\{(1\text{-}4)\eta\text{-}\underline{exo}\text{-}7\text{-}(Ph_3Ge)C_7H_7\}$		347
416	$FeIrC_{53}H_{40}O_5P_2$	$(Ph_3P)(OC)_3FeIr(\mu\text{-}PPh_2)(CO)_2(PPh_3)$		348
417	$FeLiC_{34}H_{43}O_5S_4$	$Fe(CO)(\underline{o}\text{-}MeSC_6H_4S)_2(PhCO)Li(THF)_3.THF$	233	329
418	$FeLi_2C_{48}H_{48}O_2$	$(Et_2OLi)_2Fe(naphthyl)_4$		349
419	$FeLi_4C_{38}H_{60}O_4$	$(Et_2OLi)_4FePh_4$		350
420	$FeMnC_{12}H_7O_6$	$\{(OC)_4MnFe(CO)(Cp)\}(\mu\text{-}CH_2)(\mu\text{-}CO)$	163	351
421	$FeMnC_{17}H_{11}O_6$	$(Cp)(OC)_2FeCH_2C(O)(\eta\text{-}C_5H_4)Mn(CO)_3$		352
422	$[FeMoC_{17}H_{13}O_6]^+$	$[(Cp)(OC)_2Fe\{\mu\text{-}(C,O)\text{-}COMe\}Mo(CO)_3(Cp)][SbF_6]$		353
423	$FePdC_{28}H_{23}PS_2$	$(Ph_3P)PdFe(\eta\text{-}C_5H_4S)_2.0.5PhMe$		354
424	$FeRhC_{12}H_9O_3$	$(Cp)Rh\{\mu\text{-}(\eta^4:\eta^1,\eta^1)\text{-}C_4H_4\}Fe(CO)_3$		197
425	$[FeRhC_{33}H_{52}]^+$	$[(nbd)Rh\{P(Bu^t)_2(\eta\text{-}C_5H_4)Fe(\eta\text{-}C_5H_4)P(Bu^t)_2\}][ClO_4]$		355
426	$FeRuC_{13}H_{10}F_4N_3P_3$	$FeRu(CO)_2(\mu\text{-}CO)\{\mu\text{-}P=NP(F)_2=NP(F)_2=N\}(Cp)_2$		356
427	$FeRuC_{22}H_{22}O_6$	$(OC)_3Fe\{\mu\text{-}(\eta^4:\eta^1,\eta^1)\text{-}(H_{12}C_8)(C_8H_{10})\}Ru(CO)_3$		315
428	$FeRu_2C_{44}H_{32}O_{10}P_2$	$(OC)_4FeRu_2(CO)_4(PPh_3)_2(\mu\text{-}OH)_2$		357
429	$FeSbC_{12}H_{23}Br_2P_2$	$(Cp)(Me_3P)(OC)FeSbBr_2(PMe_3)$		358
430	$[FeSe_6C_{45}H_{42}N_3]^+$	$[Fe\{Se_2CN(CH_2Ph)_2\}_3][BF_4]$		359
431	$FeSi_2C_{27}H_{25}O_2$	$(Cp)(OC)_2FeSi(Me_2)SiPh_3$		360
432	$FeWC_{22}H_{16}N_2O_2$	$(Cp)Fe\{\mu\text{-}(\eta^5,\eta^3)\text{-}C_5H_4C(CH_2)=C(CN)_2\}W(CO)_2(Cp)$		361
433	$FeWC_{23}H_{16}N_2O_3$	$(Cp)Fe(\eta\text{-}C_5H_4)C\{C(CN)_2\}CH_2W(CO)_3(Cp)$		361
434	$FeW_2C_{31}H_{24}O_6$	$FeW_2(O)\{\mu_3\text{-}(\eta^2,\eta^2,\eta^1)\text{-}C_2(pTol)_2\}(CO)_5(Cp)_2$	200	362
435	$FeW_2C_{32}H_{24}O_6$	$FeW_2\{\mu_3\text{-}(\eta^2,\eta^2,\eta^1)\text{-}C_2(pTol)_2\}(CO)_6(Cp)_2$	200	362

No.	Formula	Structure	Details	Ref.
436	$Fe_2C_9H_6O_6S_2$	$Fe(CO)_6(\mu\text{-}\eta^2\text{-}SCSMe_2)$		363
437	$Fe_2C_{10}H_6O_7S_2$	$(OC)_3Fe\{\mu\text{-}(CS\eta^2,S,S)\text{-}SCHSCH_2COMe\}Fe(CO)_3$		364
438	$Fe_2C_{10}H_6O_7S_2$	$(OC)_6Fe_2\{\mu\text{-}2(\mu\text{-}S)\text{-}S_2CHCH_2COMe\}$		364
439	$Fe_2C_{12}H_5Cl_2O_6P$	$Fe_2(CO)_6(\mu\text{-}Cl)(\mu\text{-}PClPh)$		365
440	$[Fe_2C_{12}H_7O_5]^-$	$[AsPh_4][Fe_2(CO)_4(\mu\text{-}CO)\{\mu\text{-}(\eta^3,\eta^4)\text{-}C_7H_7\}]$		41
441	$Fe_2C_{13}H_{11}O_5P$	$Fe_2(CO)_2(\mu\text{-}CO)(\mu\text{-}POOH)(Cp)_2$		366
442	$Fe_2C_{14}H_{10}O_4S_3$	$\{(Cp)(OC)_2Fe\}_2(\mu\text{-}S_3)$		367
443	$Fe_2C_{14}H_{10}O_4S_4$	$\{(Cp)(OC)_2Fe\}_2(\mu\text{-}S_4)$		367
444	$Fe_2C_{14}H_{16}O_3$	$(Cp)Fe\{\mu\text{-}(\eta^5,\eta^1)\text{-}C_5H_4COCH_2\}Fe(CO)_2(Cp)$		352
445	$Fe_2C_{14}H_{18}O_6P_2$	$Fe_2(CO)_6(\mu\text{-}\eta^2\text{-}P_2Bu^t{}_2)$		368
446	$Fe_2C_{14}H_{22}$	$Fe\{\eta^5\text{-}CH_2C(Me)CHC(Me)CH_2\}_2$		369
447	$[Fe_2C_{15}H_{13}O_3]^+$	$[Fe_2(CO)_2(\mu\text{-}CO)(\mu\text{-}CHCH_2)(Cp)_2][BF_4]$	220	370
448	$Fe_2C_{15}H_{14}O_3$	$Fe_2(CO)_2(\mu\text{-}CO)(\mu\text{-}CHMe)(Cp)_2$	220	370
449	$Fe_2C_{17}H_{18}O_6S$	$Fe_2(CO)_6\{\mu\text{-}\eta^2\text{-}S{=}CC(Me)_2CH_2CH_2CMe_2\}$		371
450	$Fe_2C_{18}H_{10}Cl_2O_6P_2$	$Fe_2(CO)_6(\mu\text{-}PClPh)_2$	243	365
451	$Fe_2C_{18}H_{16}O_6$	<u>trans</u>-$\{Fe(CO)_3\}_2\{\mu\text{-}(2\text{-}5\eta^4,8\text{-}11\eta^4)\text{-}$dodecapentene$\}$	133	372
452	$[Fe_2C_{20}H_{16}]^+$	$[\{Fe(\eta\text{-}C_5H_4)_2\}_2][C_6H_2N_2O_7].0.5C_6H_6O_2$		373
453	$Fe_2C_{20}H_{17}NO_7$	$Fe_2(CO)_6\{\mu\text{-}(O,C,C)\text{-}OC(Ph)CNHCy\}$		374
454	$Fe_2C_{21}H_{14}O_6$	$(OC)_3Fe[\mu\text{-}\{(3\text{-}6)\eta^4,(1,2,1',2')\eta^4\}\text{-}C_7H_7CH{=}CHPh]\text{-}Fe(CO)_3$		375
455	$Fe_2C_{21}H_{16}N_2O$	$(Cp)Fe\{\mu\text{-}(\eta^5,\eta^3)\text{-}C_5H_4C(CH_2){=}C(CN)_2\}Fe(CO)(Cp)$		361
456	$Fe_2C_{21}H_{29}O_9S_2$	$Fe_2(CO)_5\{P(OMe)_3\}(\mu\text{-}SMe)\{\mu\text{-}(S,C)\text{-}SCO(\text{adamantyl})\}$		376
457	$Fe_2C_{22}H_{16}N_2O_2$	$(Cp)Fe(\eta\text{-}C_5H_4)C\{C(CN)_2\}CH_2Fe(CO)_2(Cp)$		361
458	$Fe_2C_{22}H_{28}$	$Fe_2(Cp)_2\{\mu\text{-}(\eta^4,\eta^4)\text{-}C_8Me_8\}$		225
459	$Fe_2C_{24}H_{21}O_6PS$	$Fe_2(CO)_6(\mu\text{-}SCy)(\mu\text{-}PPh_2)$	238	60
460	$Fe_2C_{30}H_{10}Cl_8S_7$	$(Cp)_2Fe_2\{\mu\text{-}(C,\mu\text{-}S)\text{-}C_{10}Cl_4S_3\}\{\mu\text{-}(S,\mu\text{-}S)\text{-}C_{10}Cl_4S_4\}$		218
461	$Fe_2C_{31}H_{22}O_6P_2$	$(OC)_3Fe\{\mu\text{-}(PC\eta^2,P)\text{-}Ph_2PCHPPh_2\}FeH(CO)_3$		377
462	$Fe_2C_{32}H_{26}O_6P_2$	$(OC)_3Fe(\mu\text{-}PPh_2)\{\mu\text{-}(C,O)\text{-}COMe\}Fe(CO)_2\text{-}(PPh_2Me)$	150	378

No.	Formula	Structure	Details	Ref.
463	$Fe_2C_{33}H_{22}NO_7P$	Fe$_2$(CO)$_6$(μ-PPh$_2$){μ-(O,C)-OCCH=C(Ph)NHPh}		379
464	$Fe_2C_{38}H_{34}I_2O_6P_2$	{FeI(CO)(Cp)}$_2$(doppe)		380
465	$Fe_2C_{44}H_{54}O_{10}P_2$	(OC)$_4$Fe[μ-(P,PPη^2)-{P(OC$_6$H$_2$Bu^t$_3$-2,4,6)}$_2$]-Fe(CO)$_4$	140	272
466	$[Fe_2HgC_8O_8]^{2-}$	[Na(THF)$_2$]$_2$[Hg{Fe(CO)$_4$}$_2$]		381
467	$Fe_2Hg_2C_2H_6N_4O_4S_2$	Fe$_2$(NO)$_4$(μ-SHgMe)$_2$		382
468	$Fe_2Hg_2C_8H_6O_6S_2$	Fe$_2$(CO)$_6$(μ-SHgMe)$_2$		382
469	$Fe_2Mn_2C_{20}H_{10}O_{10}P_2$	{Fe$_2$(CO)$_6$}(μ$_3$-P)$_2${Mn(CO)$_2$(Cp)$_2$}$_2$		383
470	$Fe_2MoC_{24}H_{36}O_8S_4$	(OC)$_2$MoFe$_2$(CO)$_6$(μ-SBu^t)$_4$		384
471	$Fe_2Mo_2C_{18}H_{14}O_6S_4$	[(η-C$_5$H$_4$Me)Mo(μ$_3$-S)$_2$Fe(CO)$_3$]$_2$		385
472	$Fe_2Mo_2C_{20}H_{14}O_8S_2$	(η-C$_5$H$_4$Me)$_2$Mo$_2$Fe$_2$(μ$_3$-S)$_2$(CO)$_6$(μ-CO)$_2$		386
473	$Fe_2Rh_2C_{56}H_{40}O_8P_4$	[(OC)$_3$Fe(μ-CO)(μ-PPh$_2$)Rh(μ-PPh$_2$)]$_2$		387
474	$Fe_2SiC_{11}H_{12}O_6$	(OC)$_6$Fe$_2${μ-2(μ-S)-SCH(SiMe$_3$)CH$_2$S}		388
475	$Fe_2SiC_{16}H_{14}I_2O_4$	{FeI(CO)$_2$(η-C$_5$H$_4$)}$_2$SiMe$_2$		380
476	$Fe_2SiC_{39}H_{36}O_2P_2$	Me$_2$Si{η-C$_5$H$_4$)Fe(μ-CO)}$_2$(μ-dppm)		389
477	$Fe_2SiC_{41}H_{40}O_2P_2$	Me$_2$Si{η-C$_5$H$_4$)Fe(μ-CO)}$_2$(μ-dppp)		389
478	$Fe_2Si_4C_{20}H_{36}N_2O_8P_2$	[P{N(SiMe$_3$)$_2$}{Fe(CO)$_4$}]$_2$	140	58
479	$Fe_2Si_4C_{22}H_{37}O_8P_2$	{Fe(CO)$_4$}$_2$[μ-{PCH(SiMe$_3$)$_2$}$_2$]	140	390
480	$Fe_2Sn_2C_{36}H_{52}$	Fe$_2${μ-(η^5,η^5)-C$_5$H$_4$Sn(Bu^n)$_2$C$_5$H$_4$}$_2$		391
481	$Fe_2V_2C_{12}H_{14}N_2O_2S_4$	(η-C$_5$H$_4$Me)$_2$V$_2$Fe$_2$(NO)$_2$(μ$_3$-S)$_4$		392
482	$Fe_2WC_{20}H_{10}O_9S_2$	(OC)$_2$(Cp)FeC{=SW(CO)$_5$}SFe(CO)$_2$(Cp)		393
483	$Fe_2WC_{22}H_{12}O_8$	(OC)$_6$Fe$_2$W(CO)$_2$(Cp){μ$_3$-(η^2,η^2,η^1)-C$_2$(pTol)}		394
484	$Fe_2WC_{22}H_{12}O_9$	(OC)$_6$Fe$_2$W(CO)$_2$(Cp){μ$_3$-C(pTol)}(μ-CO)	200	362
485	$[Fe_3C_9O_{10}]^{2-}$	[NMe$_3$CH$_2$Ph]$_2$[Fe$_3$(CO)$_9$(μ$_3$-O)]		395
486	$[Fe_3C_{11}HO_{11}]^-$	[PPN][Fe$_3$(CO)$_{10}$(μ-H)(μ-CO)]		396
487	$Fe_3C_{11}H_4O_{10}$	Fe$_3$(μ-H)(CO)$_9$(μ$_3$-MeCO)		397
488	$[Fe_3C_{12}H_5O_{10}]^-$	[PPh$_4$][Fe$_3$(CO)$_9${μ$_3$-(μ-C,μ-O)-COEt}]		398
489	$Fe_3C_{12}H_6O_{10}$	Fe$_3$(CO)$_9$(μ$_3$-CMe)(μ$_3$-OMe)		397
490	$Fe_3C_{13}H_6O_{10}$	Fe$_3$(CO)$_9$(μ$_3$-CMe)(μ$_3$-COMe)		397
491	$Fe_3C_{14}H_9NO_9$	Fe$_3$(CO)$_9${μ$_3$-(η^1,η^2,η^2)-CNBu^t)		399

No.	***Formula***	***Structure***	***Details***	***Ref.***
492	$Fe_3C_{15}H_5O_9PS$	$Fe_3(CO)_9(\mu_3\text{-}S)(\mu_3\text{-}PPh)$		400
493	$Fe_3C_{15}H_8O_8$	$Fe_3(\mu_3\text{-}CMe)(\mu\text{-}CO)_2(CO)_6(Cp)$	220	240
494	$Fe_3C_{16}H_7O_{10}PS$	$Fe_3(CO)_9(\mu_3\text{-}S)(\mu_3\text{-}PC_6H_4OMe\text{-}4)$		400
495	$Fe_3C_{16}H_9NO_{11}$	$Fe_3(CO)_9(CNBu^t)(\mu\text{-}CO)_2$		399
496	$Fe_3C_{19}H_8N_4O_7S_2$	$[(OC)_3Fe\{\mu_3\text{-}(\mu\text{-}S,\mu\text{-}N)\text{-}SC_6H_4(NHN\text{-}2)\}]_2Fe(CO)$		401
497	$Fe_3C_{21}H_{11}NO_8$	$Fe_3(CO)_8\{\mu_3\text{-}(\mu\text{-}N,\eta^1,\eta^6)\text{-}C_6H_4(CH_2NPh\text{-}2)\}$	153	402
498	$Fe_3C_{21}H_{16}O_9S_2$	$Fe_3(CO)_9(\mu_3\text{-}SCy)(\mu\text{-}SPh)$	233	60
499	$Fe_3C_{23}H_{16}N_2O_7P_2$	$Fe_3(CO)_7(NCMe)_2(\mu_3\text{-}PPh)_2$		403
500	$Fe_3GeMnC_{20}H_7O_{14}$	$\{(OC)_4Fe\}_2Ge\{(OC)_4FeMn(CO)_2(\eta\text{-}C_5H_4Me)\}$		404
501	$Fe_3SeC_{21}H_{16}O_9S$	$Fe_3(CO)_9(\mu_3\text{-}SCy)(\mu\text{-}SePh)$	243	60
502	$Fe_3WC_{18}H_9O_{14}PS$	$Fe_3(CO)_9\{\mu_3\text{-}SW(CO)_5\}(\mu_3\text{-}PBu^t)$	243	269
503	$[Fe_3ZnC_{21}H_{15}O_6]^-$	$[NBu^n{}_4][Zn\{Fe(CO)_2(Cp)\}_3]$		405
504	$Fe_4C_{25}H_{14}O_{11}P_2$	$Fe_4(CO)_{10}(\mu\text{-}CO)\{\mu_4\text{-}P(pTol)\}_2$		202
505	$Fe_4GeC_{16}O_{16}$	$Ge\{Fe_2(CO)_8\}_2$	295,153	404,406
506	$Fe_4Li_{10}C_{72}H_{110}N_8O_2$	$[(\eta\text{-}C_5H_4Li)Fe\{\eta\text{-}C_5H_3(Li)CH(Me)NMe_2\}]_4\text{-}$ $[LiOEt]_2(tmeda)_2$		407
507	$Fe_4UC_{56}H_{54}O_8P_2$	$[Fe_2(CO)(\mu\text{-}CO)(Cp)_2\{\mu\text{-}(C,\eta^3)\text{-}C(O)C(O)CHPPhMe_2\}]_2\text{-}$ $U(Cp)_2$		408
508	$Fe_4UC_{66}H_{58}O_8P_2$	$[Fe_2(CO)(\mu\text{-}CO)(Cp)_2\{\mu\text{-}(C,\eta^3)\text{-}C(O)C(O)CHPPh_2Me\}]_2\text{-}$ $U(Cp)_2$		408
509	$[Fe_5C_{13}Br_2O_{12}]^{2-}$	$[Et_4N]_2[Fe_5C(CO)_{12}Br_2]$		409
510	$[GaSi_3C_{12}H_{34}]^-$	$K[GaH(CH_2SiMe_3)_3]$		410
511	$Ga_4C_{24}H_{24}Cl_8$	$[(\eta^6\text{-}C_6H_6)_2Ga.GaCl_4]_2$		411
512	$[Ga_4Mo_2C_{18}H_{27}N_8O_6]^-$	$[pzH_2][Mo_2(\mu\text{-}MeCO_2)\{\mu\text{-}OGa(Me)(pz)\}_4].2THF$		412
513	$GeC_8H_{20}O_2$	$Bu^t{}_2Ge(OH)_2$		413
514	$GeC_{10}H_{21}NO_3$	$Bu^tGe(OCH_2CH_2)_3N$		414
515	$GeC_{12}H_{14}$	$Ge(\eta\text{-}C_5H_4Me)_2$	E	415
516	$GeC_{20}H_{15}F_3O_2$	$GePh_3(OCOCF_3)$		416
517	$GeC_{22}H_{22}ClN$	$(Ph)ClGe\{C_6H_3(5\text{-}Me)\text{-}2\text{-}\}_2NEt$		417
518	$GeC_{24}H_{18}O$	$Ph_2Ge\{C_6H_4\text{-}2\text{-}\}_2O$		417

No.	Formula	Structure	Details	Ref.
519	$GeC_{26}H_{21}Br_2N$	$Ph_2Ge\{C_6H_3(5\text{-}Br)\text{-}2\text{-}\}_2NEt$		417
520	$GeC_{26}H_{23}N$	$Ph_2Ge\{C_6H_4\text{-}2\text{-}\}_2NEt$		417
521	$GeMnC_8F_9O_5$	$(F_3C)_3GeMn(CO)_5$		418
522	$GeOs_2C_8Cl_4O_8$	$(OC)_5OsOsCl(GeCl)(CO)_3$		419
523	$Ge_2C_{10}H_{24}N_2S_2$	$[Me_3GeN(Me)C(S)\text{-}]_2$		420
524	$Ge_2C_{13}H_{26}N_2O_6$	$\{N(CH_2CH_2O)_3Ge\}_2CH_2$		421
525	$Ge_2C_{36}H_{30}S_3$	$(Ph_3Ge)_2S_3$		422
526	$Ge_2C_{36}H_{66}S_3$	$(Cy_3Ge)_2S_3$		422
527	$Ge_3C_9H_{27}P_7$	$P_7(GeMe_3)_3$		423
528	$Ge_3C_{24}H_{54}O_3$	$[Bu^t{}_2GeO]_3$	188	413,424
529	$Ge_5C_{60}H_{50}$	$(GePh_2)_5$		425
530	$Ge_6C_{24}H_{54}O_9$	$(Bu^tGe)_6(\mu\text{-}O)_9$	188	426
531	$HfC_{12}H_{16}$	$HfMe_2(Cp)_2$		427
532	$HfC_{14}H_{18}Cl_2$	$HfCl_2(\eta\text{-}C_5H_4Et)_2$		428
533	$HfC_{18}H_{30}P_2$	$Hf(PEt_2)_2(Cp)_2$		429
534	$Hf_2C_{36}H_{30}$	$\{HfH(\eta\text{-}C_5H_4Bu^t)_2(\mu\text{-}H)\}_2$	powder-X	430
535	$HgC_5H_8N_2S$	$HgMe(SC{=}NHCH{=}CHNMe)$		431
536	$[HgC_5H_9N_2S]^+$	$[HgMe(S{=}CNHCH{=}CHNMe)][NO_3]$		431
537	$[HgC_{14}H_{19}N_6O]^+$	$[HgMe\{\eta^1\text{-}(N\text{-}Meimidazol\text{-}2\text{-}yl)_3COH\}][NO_3]$		432
538	$HgC_{16}H_{14}$	Hg(<u>trans</u>-CH=CHPh)$_2$		433
539	$HgC_{18}H_{24}N_2$	$Hg\{C_6H_4(2\text{-}CH_2NMe_2)\}_2$		434
540	$HgC_{28}H_{36}F_6O_6$	$Hg(CF_3)_2[\eta\text{-}(4O)\text{-}O(C_2H_4O)_4\{(6\text{-}CH_2)(2\text{-}Me)C_6H_3\}_2\}]$		435
541	$HgMnC_{12}H_{10}BrO_8$	$Mn\{C(CO_2Et){=}C(HgBr)C(OEt)O\}(CO)_4$		436
542	$HgMoRu_3C_{23}H_{14}O_{12}$	$(Cp)(OC)_3MoHgRu_3(CO)_9(\mu_3\text{-}C_2Bu^t)$		437
543	$HgRu_6C_{30}H_{18}O_{18}$	$Hg[Ru_3(CO)_9\{\mu_3\text{-}(\eta^1,\eta^2,\eta^2)\text{-}C_2Bu^t\}]_2$		437
544	$HgTeC_{12}H_{10}I_2$	$HgI_2(TePh_2)$		438
545	$Hg_2Mn_2C_{58}H_{40}F_{12}O_{12}P_2$	$[(Cp)_2(Ph_3P)(OC)_2MnHg(O_2CCF_3)\{\mu\text{-}(O)\text{-}O_2CCF_3\}]_2$	153	439
546	$Hg_2Pd_4C_{28}H_{60}Br_2O_4P_4$	$Pd_4(PEt_3)_4(\mu\text{-}CO)_4(\mu_3\text{-}HgBr)_2$	153	440
547	$[Hg_3C_2HO_2]^+$	$[(\mu_3\text{-}O)Hg_3CCHO][NO_3].H_2O$		441

No.	Formula	Structure	Details	Ref.
548	$Hg_4C_3H_6Cl_4OS$	$C(HgCl)_4.Me_2SO$		442
549	$Hg_5C_{15}F_{30}$	$\{(F_3C)_2CHg\}_5.2py.2H_2O$		443
550	$[InSi_4C_{16}H_{44}]^-$	$K[In(CH_2SiMe_3)_4]$		410
551	$IrC_{10}H_{19}$	$IrH_4(\eta\text{-}C_5Me_5)$		444
552	$IrC_{13}H_{30}Cl_2F_2O_2P_3$	$IrCl_2(POF_2)(CO)(PEt_3)_2$		445
553	$[IrC_{13}H_{40}OP_4]^+$	$[IrH(CH_2OH)(PMe_3)_4][PF_6]$	173	446
554	$IrC_{14}H_{26}O_3P$	$Ir(\eta\text{-}C_3H_5)(cod)\{P(OMe)_3\}$		447
555	$IrC_{19}H_{38}O_{10}P_3$	$IrH\{C(O)C_6H_3(6\text{-}Et)(2\text{-}CH_2)\}\{P(OMe)_3\}_3$		448
556	$[IrC_{20}H_{16}F_4]^+$	$[Ir(\eta^4\text{-}C_6H_6C_6F_4)\{\eta\text{-}C_6H_4(Me_2\text{-}1,4)\}][BF_4]$		449
557	$IrC_{20}H_{40}O_{10}P_3$	$IrH\{C(O)C_6H_3(6\text{-}Et)(2\text{-}CHMe)\}\{P(OMe)_3\}_3$		448
558	$[IrC_{24}H_{24}F_4]^+$	$[Ir(\eta^4\text{-}C_6H_6C_6F_4)(\eta\text{-}C_6Me_6)][BF_4]$		449
559	$IrC_{25}H_{27}N_2O_2$	$Ir\{\eta^1\text{-}CH(COMe)_2\}(phen)(cod)$		450
560	$[IrC_{30}H_{23}N_6]^{2+}$	$[Ir\{\eta^2\text{-}(C,N)\text{-}C_{10}N_2H_7\}(bipy)_2][ClO_4]_2.0.33H_2O$		451
561	$IrC_{37}H_{30}Cl_3OP_2$	$IrCl_3(CO)(PPh_3)_2$		452
562	$IrC_{39}H_{40}ClO_2P_2$	$IrCl[1,2\text{-}\{5,5'\text{-}(C_6H_4)_2PCH_2\}_2CCOCMe_2O](cod)$		453
563	$[IrC_{43}H_{35}Cl_2N_3O_2P_2]^+$	<u>trans</u>-$[Ir\{C_6H_2(1\text{-}NHNH_2)(2\text{-}NO)(3\text{-}Cl)\}Cl(CO)\text{-}(PPh_3)_2]Cl$		454
564	$IrC_{43}H_{44}P_3$	$Ir(oTol)\{PhP(CH_2CH_2CH_2PPh_2)_2\}$		455
565	$IrC_{51}H_{40}ClO_2P_2$	$IrCl\{(2\text{-}PPh_2)C_6H_4\}(PPh_3)\{PhC(O)CHC(O)Ph\}$		456
566	$IrC_{63}H_{49}ClO_2P_3$	$IrCl\{C_6H_4C(O)CC(Ph)OPPh_2\}(PPh_3)_2$		456
567	$IrOs_3C_{12}H_2ClO_{12}$	$(OC)_3IrOs_3(CO)_9(\mu\text{-}H)_2(\mu\text{-}Cl)$	200	457
568	$IrRhC_{28}H_{28}F_6O_4$	(<u>S</u>)-$Rh(CO)_2L.$(<u>R</u>)-$Ir(CO)_2L$ L=$(3\text{-}CF_3CO)$camphorate		458
569	$IrRuC_{23}H_{44}Cl_4F_2OP_3$	$IrCl_2(CO)(PEt_3)_2(\mu\text{-}PF_2)RuCl_2(\eta\text{-}C_6H_4MePr^i)$		445
570	$IrSiC_{14}H_{38}O_{10}P_3$	$Ir(CH_2SiMe_3)(CO)\{P(OMe)_3\}_3$		459
571	$[IrSnC_{16}H_{49}Cl_2O_{15}P_5]^+$	$[IrH\{P(OMe)_3\}_4\{P(OMe)_2(OSnCl_2Me_2)\}][SnMe_2Cl_3]$		460
572	$IrWC_{47}H_{36}O_5P_3$	$(OC)_4W(\mu\text{-}PPh_2)_2IrH(CO)(PPh_3)$		461
573	$Ir_2C_{22}H_{30}N_4$	$[Ir(cod)(\mu\text{-}pz)]_2$		462
574	$Ir_2C_{23}H_{26}F_6N_2$	$(cod)Ir(\mu\text{-}C_4F_6)(\mu\text{-}pz)Ir(\eta^5\text{-}1,2,3,5,6\text{-}C_8H_{11})$		463
575	$Ir_2C_{26}H_{26}F_{12}N_4$	$[Ir(cod)\{\mu\text{-}N_2C_3H(CF_3)_2\text{-}3,5\}]_2$		462
576	$Ir_2C_{26}H_{32}F_6N_4$	$[Ir(cod)\{\mu\text{-}N_2C_3H(CF_3)Me\text{-}3,5\}]_2$		462

No.	Formula	Structure	Details	Ref.
577	$Ir_2C_{34}H_{38}$	$(cod)Ir\{\mu-(\eta^4,\eta^1,\eta^1)-H_4C_6C_6H_4)IrH(Ph)(cod)$		464
578	$Ir_2C_{38}H_{30}Cl_2O_2P_{10}S_6$	$[IrCl(CO)(PPh_3)(\mu-P_4S_3)]_2$		465
579	$[Ir_2C_{63}H_{50}N_2O_3P_4]^{2+}$	$[Ir_2(CO)_2(\mu-CO)(\mu-NC_5H_4CH(PPh_2)_2\}_2][BF_4]_2$		466
580	$Ir_2PtC_{26}H_{60}Cl_6F_4O_2P_6$	$[IrCl_2(CO)(PEt_3)_2(\mu-PF_2)]_2PtCl_2$		445
581	$Ir_2Rh_2C_{27}H_{30}O_7$	$(OC)_4(\mu-CO)Ir_2Rh_2(\mu_3-CO)_2(\eta-C_5Me_5)_2$		457
582	$Ir_3C_{43}H_{32}O_6P_3$	$Ir_3(Ph)(\mu_3-PPh)(CO)_6(\mu-dppm)$		467
583	$[Ir_6C_{18}H_5O_{16}]^-$	$[PPh_4][Ir_6(CO)_{15}(COEt)]$		468
584	$Ir_6C_{26}H_{45}O_{26}P_5$	$Ir_6(CO)_7(\mu-CO)(\mu_3-CO)_3\{P(OMe)_3\}_5$		469
585	$LiNiC_{16}H_{37}N_4O$	$Li\{N(Me)(CH_2CH_2NMe_2)_2\}OC(NMe_2)Ni(\eta-C_2H_4)_2$		470
586	$LiSi_3C_{20}H_{45}N_2$	$(tmeda)Li\{\eta-C_5H_2(SiMe_3)_3-1,2,4\}$		471
587	$LiSi_3C_{23}H_{52}N_3$	$(Me_5dien)Li\{\eta-C_5H_2(SiMe_3)_3-1,2,4\}$		471
588	$LiSi_3C_{29}H_{41}O$	$Li\{C(SiMe_2Ph)_3\}(THF)$		472
589	$[LiSi_6C_{20}H_{54}]^-$	$[Li(THF)_4][Li\{C(SiMe_3)_3\}_2]$		473
590	$Li_2C_{30}H_{46}N_4$	$Li_2(tmpda)_2(\mu-CCPh)_2$		474
591	$Li_2C_{32}H_{44}N_2O_2$	$[Et_2OLi\{\mu-C_{10}H_6(NMe_2)-8\}]_2$		475
592	$Li_2Mo_2C_{42}H_{56}N_4O_{16}P_2$	<u>cis</u>-$[(OC)_4Mo\{C(Ph)OLi\}\{EtOP(OC_2H_4NMeCH_2)_2\}]_2$		476
593	$Li_2Si_4C_{20}H_{56}N_2$	$[Li\{N(SiMe_3)_2\}(OEt_2)]_2$		477
594	$Li_2Si_4C_{24}H_{44}N_2$	$[\{2-(Me_3Si)_2C\}C_5H_4NLi]_2$		303
595	$Li_4C_{18}H_{50}Br_2O_4$	$(Et_2OLi)_4(\mu_3-Br)_2(\mu_3-C_3H_5)_2$	253	478
596	$Li_4C_{20}H_{44}O_4$	$Li_4(\mu-CH(Me)CH_2CH_2OMe)_4$		479
597	$Li_4C_{30}H_{45}BrO_3$	$(PhLi.OEt_2)_3.LiBr$	135	480
598	$Li_4C_{40}H_{60}O_4$	$[PhLiOEt_2]_4$	135	480
599	$Li_4C_{52}H_{68}N_4$	$(PhCCLi)_4\{Me_2N(CH_2)_6NMe_2\}_2$		481
600	$MgSi_2C_{12}H_{34}N_2$	$(Me_3Si)_2Mg(tmeda)$		28
601	$[MnC_2N_2O_4]^-$	$[PPN]\{Mn(CO)_2(NO)_2]$		482
602	$MnC_6H_3O_5$	$Mn(Me)(CO)_5$		483
603	$MnC_8H_8NO_3$	$Mn(CO)_3(CNMe)(\eta-C_3H_5)$		484
604	$MnC_{10}H_{11}O_3$	$Mn(CO)_3\{\eta^5-(4C,H)-CHCHCHCH_2CH(Me)CH_2\}$	N,25	485
605	$MnC_{10}H_{28}P_2$	$Mn(Me)_4(dmpe)$		486
606	$MnC_{12}H_{14}NO_2$	$Mn(CO)_2(CNBu^t)(Cp)$		487

No.	Formula	Structure	Details	Ref.
607	$MnC_{12}H_{18}NO_4$	$(OC)_3Mn\{\eta^2-(N,O)-ONC(Me)_2CH_2CH_2CH_2CMe_2\}$	218	488
608	$MnC_{13}H_{14}NO_4$	(<u>R</u>,<u>S</u>)-$Mn(CO)_3\{\eta-C_5H_3(1-CHO)(2-CHMeNMe_2)\}$	153	489
609	$MnC_{13}H_{16}O$	$Mn\{(1-4)\eta-C_6H_8\}_2(CO)$		490
610	$MnC_{14}H_{38}P_4$	$Mn(Me)_2(dmpe)_2$		491
611	$MnC_{16}H_{16}NO_4$	$Mn(CO)_2(NO)\{(1-4)\eta^4-C_6H_4(2-OMe)(5-Me)(6-Ph)\}$		492
612	$MnC_{17}H_{12}O_4P$	$Mn(CO)_4\{\eta^2-(P,C)-H_2CPPh_2\}$		493
613	$MnC_{17}H_{31}P$	$Mn(PMe_3)\{\eta^4,\eta^4-CH_2=C(Me)CH=C(Me)(CH_2)_2C(Me)=CH-C(Me)=CH_2\}$		494
614	$MnC_{18}H_{17}Cl_2N$	$Mn(NC_5H_3Cl_2-3,5)\{(\eta-C_5H_4)_2C_3H_6\}$		495
615	$MnC_{19}H_{14}O_6P$	$Mn(CO)_4\{CH_2C(O)CH_2OPPh_2\}$		496
616	$MnC_{19}H_{22}O_5P$	$Mn(CO)_2(Cp)\{\eta^2-PhCH=CHP(O)(OEt)_2\}$		497
617	$MnC_{20}H_{40}P_4$	Mn{<u>o</u>-$(CH_2)_2C_6H_4\}(dmpe)_2$		498
618	$MnC_{22}H_{17}O_2$	$(Cp)(OC)_2MnC\{$<u>o</u>-$C_6H_4CH_2-)_2\}$		499
619	$MnC_{24}H_{19}O_3$	$(\eta-C_5H_4Me)(OC)_2MnC(O)C\{($<u>o</u>-$C_6H_4CH_2-)_2\}$		499
620	$MnC_{25}H_{21}BrO_3PS_2$	$MnBr(CO)_3\{(SS)\eta^2-S_2CCMe_2PPh_3\}$	163	500
621	$MnC_{26}H_{44}P_2$	$Mn(CH_2CMe_2Ph)_2(PMe_3)_2$		498
622	$MnC_{30}H_{20}NO_3$	$Mn(\eta^4-C_4Ph_4)(CO)_2(NO)$		501
623	$MnMoC_{24}H_{14}O_7P$	$(CO)_4MnMo(CO)_3\{\eta^5-C_5H_4PPh_2\}$		502
624	$MnPdC_{53}H_{44}BrO_3P_4$	$(OC)_3Mn(\mu-dppm)_2PdBr$		503
625	$MnSiC_8H_8Cl_3O_2$	$(\eta-C_5H_4Me)(OC)_2Mn(\mu-H)SiCl_3$		504
626	$MnSi_2C_8H_7Cl_6O_2$	$Mn(SiCl_3)_2(CO)_2(\eta-C_5H_4Me)$		504
627	$MnTiC_{29}H_{25}Cl_2O_2P$	$(Cp)(OC)_2MnPPh_2(\eta^5-C_5H_4)TiCl_2(Cp)$		505
628	$Mn_2C_{20}H_{10}O_6S_2$	$Mn_2(CO)_6\{\mu-(2\mu-S,\eta^2)-SC(Ph)=C(Ph)S\}$		506
629	$Mn_2C_{21}H_{13}O_7P$	$Mn_2(CO)_7\{\mu-(\eta^1,\eta^2)-CH=CH_2\}(\mu-PPh_2)$		507
630	$Mn_2C_{24}H_8F_2O_{10}S_2$	<u>trans</u>-$C_2(C_6H_4F-4)_2\{SMn(CO)_5\}_2$		506,508
631	$Mn_2C_{26}H_{62}P_2$	$Mn_2(CH_2CMe_3)_2(PMe_3)_2(\mu-CH_2CMe_3)_2$		509
632	$Mn_2C_{28}H_{29}N_2O_6P$	$Mn_2(\mu-H)(\mu-PPh_2)(CO)_6(CNBu^t)_2$		507
633	$Mn_2C_{34}H_{46}P_2$	$Mn_2(CH_2Ph)_2(PMe_3)_2(\mu-CH_2Ph)_2$		509
634	$Mn_2C_{54}H_{52}F_2O_{16}S_2$	<u>trans</u>-$C_2(C_6H_4F-4)_2\{SC(CO_2Cy)=C(CO_2Cy)-Mn(CO)_4\}_2$		506,508

No.	Formula	Structure	Details	Ref.
636	$Mn_2C_{56}H_{46}N_2O_5P_4$	$Mn_2(CO)_4\{\mu\text{-}(C,\mu\text{-}N)\text{-}C(O)CH_2N_2\}(\mu\text{-}dppm)_2.2CH_2Cl_2$		510
637	$Mn_2Pd_2C_{59}H_{44}O_9P_4$	$(CO)_5MnPd_2Mn(CO)_2(\mu\text{-}CO)_2(\mu\text{-}dppm)_2.PhCl$		511
638	$Mn_2SeC_{15}H_{12}O_4$	$\{Mn(CO)_2(Cp)\}_2\{\mu\text{-}(\eta^2,\mu\text{-}Se)\text{-}SeCH_2\}$		512
639	$Mn_2Se_2C_{18}Br_2O_6$	$Mn_2(CO)_6(\mu\text{-}Br)_2(\mu\text{-}Se_2Ph_2)$		513
640	$Mn_2Si_4C_{42}H_{70}P_2$	$Mn_2(CH_2SiMe_3)_2(PMePh_2)_2(\mu\text{-}CH_2SiMe_3)_2$		509
641	$Mn_3C_{24}H_{36}$	$Mn\{(\eta^5\text{-}CH_2CHCMeCHCH_2)_2Mn\}_2$		514
642	$Mn_3C_{61}H_{74}$	$Mn\{(\mu\text{-}Mes)_2Mn(Mes)\}_2$		515
643	$Mn_3TeC_{21}H_{15}O_6$	$Te\{Mn(CO)_2(Cp)\}_3$		516
644	$MoC_{11}H_{11}IO_3$	$MoI(CO)_2\{CO(CH_2)_2CH_2\}(Cp)$		517
645	$MoC_{12}H_{14}O_3$	$Mo(CO)_2\{\eta^2\text{-}(C,O)\text{-}CHOCH_2CH_2CH_2\}(\eta\text{-}C_5H_4Me)$		518
646	$MoC_{12}H_{16}N_2O_4$	$Mo\{N(Pr^i)CHCHNPr^i\}(CO)_4$		519
647	$MoC_{12}H_{27}NO_2P_2S_2$	$Mo(Ac)(S_2CNMe_2)(CO)(PMe_3)_2$		520
648	$MoC_{13}H_{14}O_2$	$Mo(CO)_2(\eta^3\text{-}CHCHCHCH_2CH_2CH_2)(Cp)$		521
649	$MoC_{14}H_9N_3O_4$	$Mo(CO)_4\{(NC_5H_4\text{-}2)_2NH\}$		254
650	$MoC_{14}H_{16}$	$Mo\{\eta^5\text{-}C_5H_4CMe_2)(\eta\text{-}C_6H_6)$		522
651	$MoC_{14}H_{16}N_4O_4$	cis-$Mo(CO)_4(CNMeCHCHNMe)_2$		523
652	$MoC_{14}H_{16}N_4O_4$	trans-$Mo(CO)_4(CNMeCHCHNMe)_2$		523
653	$[MoC_{14}H_{17}O]^+$	$[Mo(CO)(\eta\text{-}C_2Me_2)_2(Cp)][BF_4]$	220	524
654	$MoC_{14}H_{18}$	$Mo(CH_2CH_2CH_2CH_2)(Cp)_2$		525
655	$MoC_{14}H_{35}P_3$	$Mo(CO)(PMe_3)_3(\eta\text{-}C_2H_4)_2$		526
656	$[MoC_{15}H_{20}N]^+$	$[Mo(NCMe)(\eta\text{-}C_2Me_2)_2(Cp)][BF_4]$	200	524
657	$MoC_{15}H_{25}BrN_2O_2$	$MoBr(CO)_2(Bu^tNCHCHNBu^t)(\eta\text{-}C_3H_5)$		527
658	$MoC_{16}H_{16}N_4$	$Mo(pz)_2(Cp)_2$		528
659	$MoC_{16}H_{18}O_3$	$Mo\{C(Me)=C(Me)C(O)C(Me)=CHMe\}(CO)_2(Cp)$		529
660	$MoC_{16}H_{18}O_4$	$MoH\{C(CO_2Me)=CHCO_2Me\}(Cp)_2$		530
661	$[MoC_{16}H_{19}N_2O]^+$	$[Mo(pzCMe_2O)(Cp)_2][PF_6]$		528
662	$MoC_{16}H_{44}P_4$	trans-$Mo(PMe_3)_4(\eta\text{-}C_2H_4)_2$		526
663	$MoC_{18}H_{20}N_2S_4$	$Mo(S_2CNC_4H_4)_2(\eta\text{-}C_2Me_2)_2$		531
664	$MoC_{19}H_{17}O_4PS$	$Mo(CO)_4(Ph_2PCH_2CH_2SMe)$		532
665	$MoC_{19}H_{20}ClF_{12}P$	$MoCl\{\eta^2\text{-}C(CF_3)C(CF_3)(PEt_3)\}(\eta\text{-}C_2(CF_3)_2)(Cp)$		533

No.	***Formula***	***Structure***	***Details***	***Ref.***
666	$MoC_{19}H_{29}BrO_6P_2$	Mo{C=CHPh}Br{P(OMe)$_3$}$_2$(Cp)		534
667	$MoC_{19}H_{29}OP$	Mo(CO)(PEt$_3$)(η^3-anti-1-MeC$_3$H$_4$)(η^5-ind)		535
668	$[MoC_{20}H_{20}O_4PS]^+$	[Mo(CO)$_4$(Ph$_2$PCH$_2$CH$_2$SMe$_2$)][BF$_4$]		532
669	$MoC_{20}H_{26}O_3$	Mo(CO)(Cp){η^3,η^2-Bu^{t}CHCHCCH=C(Bu^t)C(O)O}	230	529
670	$MoC_{20}H_{30}N_2O_2$	Mo(CH$_2$CMe$_3$)$_2$O$_2$(bipy).0.5Et$_2$O		536
671	$MoC_{20}H_{38}N_2S_2$	Mo(SBu^t)$_2$(CNBu^t)$_2$(η-C$_2$H$_2$).0.5C$_6$H$_{14}$		537
672	$MoC_{20}H_{43}O_9P_3$	Mo{P(OMe)$_3$}$_3$(CH=CHBu^t)(Cp)	195	535
673	$MoC_{21}H_{36}Br_2OP_2$	MoBr$_2$(CO)(PEt$_3$)$_2$(η-HCCPh)		538
674	$MoC_{21}H_{42}O_5S$	Mo(CO)$_3$(PPr$^i_3$)$_2${η^1-(S)-SO$_2$)		539
675	$MoC_{24}H_{20}$	Mo(η^5-C$_5$H$_4$CPh$_2$)(η-C$_6$H$_6$)		522
676	$MoC_{24}H_{22}N_2O_2$	Mo(CH$_2$Ph)$_2$(O)$_2$(bipy)$_2$		540
677	$MoC_{25}H_{39}NO_3$	Mo(CO)(NO)(η^2-CHMe=CHC$_5$H$_{10}$CHO)(η^5-C$_5$H$_4$C$_{10}$H$_{19}$)		541
678	$MoC_{26}H_{21}O_5P$	Mo(CO)$_6$P(pTol)$_3$		542
679	$MoC_{26}H_{24}N_2O_2$	Mo(CO){η^2-(N,C)-HN(Me)C(Ph)(C$_5$H$_4$N)}-{η^2-(O,C)-OCH(Ph)}(Cp)		543
680	$MoC_{29}H_{21}N_2O_2P$	Mo(CO)$_2${η^2-CH(CN)=C(CN)PPh$_3$}(Cp)		544
681	$MoC_{29}H_{26}N_2O_2$	Mo{N(CH$_2$Ph)C(Ph)N(CHMePh)}(CO)$_2$(Cp)		545
682	$MoC_{29}H_{26}N_2O_2$	(Cp)(OC)$_2$Mo{CPh$_2$N(Me)C(Ph)=NMe}		547,546
683	$MoC_{30}H_{28}N_2O_2$	Mo{N(CHMePh)C(Ph)N(CHMePh)}(CO)$_2$(Cp)		545
684	$MoC_{30}H_{36}N_2O_4$	Mo{N(R)=CHCH=NR}(CO)$_4$ [R=C$_6$H$_3$(Pr^i)$_2$-2,6]		519
685	$MoC_{31}H_{30}N_2O_2$	Mo{N(CHMePh)C(Ph)N(CHMePh)}(CO)$_2$(η-C$_5$H$_4$Me)		545
686	$MoC_{32}H_{46}N_2S_2$	Mo(SBu^t)$_2$(CNBu^t)$_2$(η-C$_2$Ph$_2$)		537
687	$[MoC_{34}H_{25}O]^+$	[Mo(CO)η-C$_2$Ph$_2$)$_2$(Cp)][BF$_4$]		548
688	$MoC_{34}H_{25}O_6P_3$	cis-Mo(CO)$_4${PPh$_2$P(O)PhOPPh$_2$}		549
689	$[MoC_{47}H_{69}O_2S_3]^-$	[PPh$_4$][Mo(CO)$_2${SC$_6$H$_2$(Pr$^i_3$-2,4,6}$_3$]		550
690	$[MoC_{54}H_{49}O_2P_4]^+$	[MoH(CO)$_2$(dppe)$_2$][AlCl$_4$]		551
691	$[MoC_{60}H_{77}N_5P_2]^{2+}$	[Mo(CNCy)$_5$(dppm)][PF$_6$]$_2$		552
692	$MoNb_2C_{26}H_{22}O_6$	[(Cp)$_2$Nb(CO)(μ-H)]$_2$Mo(CO)$_4$		553
693	$MoPtC_{37}H_{28}Cl_2N_2O_3P_2$	ClPt(μ-CO){μ-(P,N)-Ph$_2$PC$_5$H$_4$N}MoCl(CO)$_2$	140	554
694	$MoReC_{21}H_{17}N_2O_4$	(Cp)(OC)$_2$Mo{μ-NN(pTol)}Re(CO)$_2$(Cp)		555

No.	Formula	Structure	Details	Ref.
695	$MoRhC_{18}H_{30}IO_3P_2$	$(\eta\text{-}C_5Me_5)MeRh(\mu\text{-}PMe_2)_2MoI(CO)_3$		556
696	$MoRh_2C_{27}H_{30}O_7$	$(OC)_5MoRh_2(\mu\text{-}CO)_2(\eta\text{-}C_5Me_5)_2$	200	557
697	$MoTiC_{22}H_{23}O_4$	$(Cp)(CO)_2Mo\{\mu\text{-}(C,O)\text{-}CO)Ti(Cp)_2(THF)$		558
698	$[Mo_2C_{12}H_{12}O_{10}S_6]^{2-}$	$[NEt_4]_2[\{OMoC(CO_2Me)=C(CO_2Me)SS\}_2(\mu\text{-}S)_2]$		559
699	$Mo_2C_{14}H_{10}I_2O_4$	$[Mo(CO)_2(Cp)(\mu\text{-}I)]_2$		560
700	$Mo_2C_{14}H_{11}IO_4$	$Mo_2(CO)_4(Cp)_2(\mu\text{-}H)(\mu\text{-}I)$		560
701	$Mo_2C_{14}H_{14}S_4$	$[(Cp)Mo\{\mu\text{-}(2\mu\text{-}S)\text{-}SC_2H_2S\}]_2$		561
702	$Mo_2C_{15}H_{18}S_4$	$(\eta\text{-}C_5H_4Me)_2Mo_2\{\mu\text{-}(2\mu\text{-}S)\text{-}SC_2H_2S\}\{\mu\text{-}(2\mu\text{-}S)\text{-}S_2CH_2\}$		562
703	$Mo_2C_{15}H_{22}S_4$	$(\eta\text{-}C_5H_4Me)_2Mo_2\{\mu\text{-}(2\mu\text{-}S)\text{-}SMe\}_2\{\mu\text{-}(2\mu\text{-}S)\text{-}S_2CH_2\}$		562
704	$[Mo_2C_{16}H_{13}O_5]^+$	$[Mo_2(CO)_4(Cp)_2(\mu\text{-}\eta^2\text{-}MeCO)][BF_4]$		563
705	$Mo_2C_{16}H_{22}Cl_4$	$[Mo(\eta\text{-}C_5H_4Pr^i)(\mu\text{-}Cl)_2]_2$		522
706	$Mo_2C_{17}H_{17}NO_3$	$\{MoO(Cp)\}_2(\mu\text{-}O)\{\mu\text{-}N(pTol)\}$		564
707	$Mo_2C_{18}H_{16}O_5S$	$Mo\{\mu\text{-}(CS\eta^2,\mu\text{-}S)\text{-}SCH_2CH_2CH_2O\}(CO)_4(Cp)$		565
708	$Mo_2C_{18}H_{34}N_4S$	$(dmtc)_2Mo(\mu\text{-}S)(\mu\text{-}\eta^2\text{-}EtC_2Et)Mo(dtc)\text{-}\{(S,C)\eta^2\text{-}SCNMe_2\}$		566
709	$Mo_2C_{19}H_{19}O_4P$	$Mo_2(CO)_4(Cp)_2(\mu\text{-}\eta^2\text{-}PCBu^t)$		567
710	$Mo_2C_{21}H_{17}O_5PS$	$Mo_2(CO)_4(Cp)_2\{\mu\text{-}(SP\eta^2,\mu\text{-}P)\text{-}SPC_6H_4OMe\text{-}4\}$		568
711	$Mo_2C_{21}H_{20}N_2O_8$	$(Cp)(OC)_2Mo\{N=NC(CO_2Et)=C(OEt)O\}Mo(CO)_2(Cp)$		569
712	$[Mo_2C_{22}H_{19}O_5]^+$	$[(Cp)(OC)_2(H_2O)Mo\{\mu\text{-}(C,\eta^2)\text{-}CH=CHPh)\text{-}Mo(CO)_2(Cp)][BF_4]$		570
713	$Mo_2C_{22}H_{38}N_4$	$[Mo(pTol)(NMe_2)_2]_2$	111	571
714	$Mo_2C_{22}H_{38}N_4$	$[Mo(oTol)(NMe_2)_2]_2$	112	571
715	$Mo_2C_{22}H_{38}N_4$	$[Mo(Bz)(NMe_2)_2]_2$	108	571
716	$Mo_2C_{23}H_{18}N_2O_5$	$(Cp)(OC)_2MoNNC(Ph)C(Me)OMo(CO)_2(Cp)$		499
717	$Mo_2C_{24}H_{20}S_6$	$[(Cp)Mo(S_2CPh)(\mu\text{-}S)]_2$		561
718	$Mo_2C_{24}H_{22}S_4$	$[(Cp)Mo\{\mu\text{-}(2\mu\text{-}S)\text{-}(1,2\text{-}S_2)C_6H_3Me\text{-}4)]_2$		561
719	$Mo_2C_{24}H_{25}O_5$	$Mo(CO)_2(Cp)(\eta^3\text{-}Me_2CCHCH)C(O)C(Me)C(Me)Mo(CO)_2(Cp)$		572
720	$Mo_2C_{24}H_{28}O_3$	$(Cp)_2(OC)_2Mo_2\{\mu\text{-}(1\text{-}3\eta^3,3\text{-}4\eta^2)\text{-}Bu^tCCHCHCHC(Me)_2CO\}$		572
721	$Mo_2C_{24}H_{30}O_4$	$[Mo(CO)_2(\eta\text{-}C_5Me_5)]_2$		573
722	$Mo_2C_{24}H_{36}N_{12}O_{12}P_6$	$(OC)_3Mo[P\{N(Me)C(O)N(Me)\}_2P]_3Mo(CO)_3$		574

No.	Formula	Structure	Details	Ref.
723	$Mo_2C_{25}H_{34}N_2O_3$	$(\eta\text{-}C_5Me_5)(OC)_2Mo(\mu\text{-}N{=}CH_2)Mo(CO)(NCH_2)(\eta\text{-}C_5Me_5)$		575
724	$[Mo_2C_{26}H_{33}]^+$	$[Mo_2(Cp)_2\{\mu\text{-}(1\text{-}3,6\text{-}8{:}1,1',4,5,8)\eta\text{-}$ $CMeCMeCMeC({=}CH_2)CMeCMeCMeCMe][CF_3SO_3]$		576
725	$Mo_2C_{26}H_{33}NO_4$	$Mo_2(CO)_4(\eta\text{-}C_5Me_5)_2\{\mu\text{-}(\eta^1,\eta^2)\text{-}C{=}NMe\}$		577
726	$[Mo_2C_{26}H_{34}Cl]^+$	$[Mo_2(Cp)_2(\mu\text{-}Cl)\{\mu\text{-}(1\text{-}3,6\text{-}8{:}1,4,5,8)\eta\text{-}$ $C_8Me_8\}][SbCl_4]$	200	576
727	$Mo_2C_{27}H_{36}N_2O_4$	$(\eta\text{-}C_5Me_5)(OC)_2Mo(\mu\text{-}NCMe_2)Mo(NCO)(CO)(\eta\text{-}C_5Me_5)$		575
728	$Mo_2C_{27}H_{40}N_4O_3$	$(\eta\text{-}C_5Me_5)(OC)_2Mo\{N{=}N\text{-}N(Et)C(O)\}Mo(NEt)(\eta\text{-}C_5Me_5)$		578
729	$Mo_2C_{28}H_{52}N_2O_4$	$Mo_2(Me)_2(OBu^t)_4(py)_2$	112	579
730	$Mo_2C_{29}H_{21}O_8P$	$(OC)_3Mo\{\eta^6\text{-}Me(pTol)\}P(pTol)_2Mo(CO)_5$		542
731	$Mo_2C_{29}H_{22}O_4$	$Mo_2(CO)_4(\eta\text{-}C_5H_4Me)_2\{\mu\text{-}C(C_6H_4\text{-}2\text{-})_2\}$		580
732	$Mo_2C_{29}H_{36}N_2O_6$	$(CO)_2(\eta\text{-}C_5Me_5)Mo{=}N\text{-}N\text{-}C(COMe){=}C(Me)OMo(CO)_2\text{-}$ $(\eta\text{-}C_5Me_5)$		581
733	$Mo_2C_{29}H_{52}N_2O_7$	$Mo_2(OPr^i)_4(py)_2(\mu\text{-}OPr^i)_2(\mu\text{-}CO)$		582
734	$Mo_2C_{32}H_{48}N_2O_4P_2$	$\{Mo(CO)_2(\eta\text{-}C_5Me_5)\}_2(\mu\text{-}(PNBu^t_2)\}$		583
735	$Mo_2C_{32}H_{56}N_4O_{12}P_4$	$(Pr^i_2NPO)_4Mo_2(CO)_8$ [<u>closo</u>-$P_4O_4Mo_2$ cage]		584
736	$Mo_2C_{33}H_{36}N_4O_3$	$(Cp)(OC)_2Mo_2(NR)(Cp)\{\mu\text{-}(\mu\text{-}N,C)\text{-}NNN(R)CO\}$ $[R{=}C_6H_4Bu^t\text{-}4]$		569
737	$Mo_2C_{33}H_{72}N_2O_6$	$(Bu^tCH_2O)_3Mo(\mu\text{-}OCH_2Bu^t)\{\mu\text{-}(N,\eta^2)\text{-}N{=}CNMe_2)\text{-}$ $Mo(OCH_2Bu^t)_2$		585
738	$Mo_2C_{50}H_{40}N_2O_8P_2S_2$	$[Mo(CO)_2(py)(PPh_3)\{\mu\text{-}(\eta^2,O)\text{-}SO_2\}]_2$		539
739	$[Mo_2C_{61}H_{80}N_{13}]_3^+$	$[\{Mo(CNBu^t)_4(bipy)\}_2(\mu\text{-}CN)][PF_6]_3$		586
740	$Mo_2Ni_2C_{12}H_{10}O_2S_4$	$(Cp)_2Mo_2(\mu_3\text{-}S)_4Ni_2(CO)_2$		231
741	$Mo_2Pd_2C_{28}H_{40}O_6P_2$	$(PEt_3)_2Pd_2Mo_2(Cp)_2(\mu\text{-}CO)_4(\mu_3\text{-}CO)_2$		285
742	$Mo_2Pt_2C_{42}H_{34}O_6P_2$	$(Cp)Mo(\mu\text{-}CO)_3Pt(dppe)Pt(\mu\text{-}CO)_2Mo(CO)(Cp)$		232
743	$Mo_2Si_3C_{38}H_{60}N_2O_2$	$\{(2,6\text{-}Me_2C_6H_3O)(Me_3SiCH_2)(py)Mo\}_2(\mu\text{-}H)\text{-}$ $(\mu\text{-}CSiMe_3).PhMe$	105	587
744	$Mo_2Si_4C_{32}H_{62}O_2$	$Mo_2(CH_2SiMe_3)_4(OC_6H_3Me_2\text{-}2,6)_2$	104	587
745	$Mo_2Sn_8C_{26}H_{78}$	$[(Me_3Sn)_3SnMo(NMe_2)_2]_2$	111	588
746	$Mo_3C_{24}H_{20}O_4$	$Mo_3\{\mu_3\text{-}(1{:}1,4{:}1\text{-}4)\eta\text{-}CCHCMeCH\}(CO)_4(Cp)_3$		589

No.	Formula	Structure	Details	Ref.
747	$[Mo_4C_{12}H_4O_{12}]^{4-}$	$[Pr_4{}^nN]_4[Mo_4H_4(CO)_{12}]$		590
748	$Mo_4C_{32}H_{44}S_4$	$[Mo(\eta-C_5H_4Pr^i)(\mu_3-S)]_4$		591
749	$[Mo_4C_{32}H_{44}S_4]^+$	$[Mo(\eta-C_5H_4Pr^i)(\mu_3-S)]_4[BF_4]$		591
750	$[Mo_4C_{32}H_{44}S_4]^{2+}$	$[Mo(\eta-C_5H_4Pr^i)(\mu_3-S)]_4[I_3]_2$		591
751	$Na_4C_4H_{12}$	$(NaMe)_4$	powder-X	592
752	$Na_6Si_9C_{24}H_{72}N_6$	$[(Me_3SiNNa)_2SiMe_2]_3$		593
753	$[NbC_6O_6]^-$	$[PPN][Nb(CO)_6]$		594
754	$NbC_{19}H_{15}F_6O$	$NbO\{\eta^1$-1,2-$(CF_3)_2$bicyclo[2.2.1]hepta-1,4-dien-7-yl$\}(Cp)_2$		595
755	$NbC_{24}H_{22}$	$Nb\{(o-CH_2C_6H_4)_2\}(Cp)_2$		596
756	$[NbC_{24}H_{22}]^-$	$[Na(18\text{-crown-}6)(THF)_2][Nb\{(o-CH_2C_6H_4)_2\}(Cp)_2$		596
757	$[NbC_{24}H_{22}]^+$	$[Nb\{(o-CH_2C_6H_4)_2\}(Cp)_2][BF_4]$		596
758	$NbNiC_{14}H_{11}O_4$	$(Cp)(OC)Nb(\mu-H)Ni(CO)_3$		597
759	$NbSnC_{13}H_{14}Cl_3O$	$(\eta-C_5H_4Me)_2Nb(CO)SnCl_3$		598
760	$NbSnC_{29}H_{25}O$	$(Cp)_2Nb(CO)SnPh_3$		598
761	$Nb_2C_{30}H_{54}Cl_6N_6$	$(Bu^tNC)_4Cl_2Nb\{\eta^2-(Bu^t)NC{=}CN(Bu^t)\}NbCl_4$		599
762	$Nb_3C_{25}H_{45}Cl_8N_5$	$Nb_3Cl_5(CNBu^t)_4(\mu-Cl)_3(\mu_3-CNBu^t)$		599
763	$Nb_4C_{56}H_{40}Cl_{12}$	$\{NbCl_3(\eta-C_2Ph_2)\}_4$		600
764	$[NdSi_4C_{22}H_{42}Cl_2]^-$	$[AsPh_4][NdCl_2\{\eta-C_5H_3(SiMe_3)_2\}_2]$		54
765	$NiC_8H_{14}OS_3$	$(SacSac)Ni\{(CH_2)_2S(O)Me\}$		601
766	$NiC_{12}H_{19}I_2N_2$	$NiI_2\{C_6H_3(CH_2NMe_2)_2$-2,6$\}$		602
767	$NiC_{13}H_{17}O_4PS$	$Ni\{C(CO_2Me){=}C(CO_2Me)P(Me)_2S\}(Cp)$		603
768	$NiC_{13}H_{26}N_2O_2$	$(tmeda)Ni\{\eta^3-CH_2C(Me)C(Me)CH_2C(O)O\}$		604
769	$NiC_{19}H_{22}$	$Ni\{\eta^2-CH_2{=}CHCHPhCHCMe_2CH\}(Cp)$		605
770	$NiC_{31}H_{27}N_2P$	$Ni(bipy)\{(P,C)\eta^2-(2,6-Me_2C_6H_3)P{=}CPh_2\}$		606
771	$NiC_{32}H_{56}O_2P_2$	$Ni(dcpe)(\eta^4-C_4H_5CO_2Me)$		607
772	$[NiC_{44}H_{45}NP_3S_2]^+$	$[Ni\{(N,2P,CS)\eta^5-(Ph_2PCH_2CH_2)_2NCH_2CH_2$-$P(Ph_2)CSSMe\}][BPh_4]$		123
773	$NiC_{48}H_{44}NP_3S$	$(triphos)Ni(\eta^2-S{=}CNPh).0.5CH_2Cl_2$		608
774	$[NiC_{63}H_{57}P_4]^+$	$[Ni\{\eta^3-P(CH_2CH_2PPh_2)_3\}(\eta^3-C_3Ph_3)][BPh_4].0.5BuOH$		120

No.	Formula	Structure	Details	Ref.
775	$NiOs_3C_{14}H_8O_9$	$(Cp)NiOs_3(CO)_9(\mu\text{-}H)_3$		204
776	$NiOs_3C_{14}H_8O_9$	$(Cp)NiOs_3(CO)_9(\mu\text{-}H)_3 . 0.5Os_3(CO)_{10}(\mu\text{-}H)_2$	102	609
777	$NiOs_3C_{20}H_{16}O_9$	$(Cp)NiOs_3(\mu\text{-}H)(CO)_9\{\mu_4\text{-}(\eta^1,\eta^1,\eta^1,\eta^2)\text{-}C{=}CHBu^t\}$		610
778	$NiRu_3C_{19}H_{14}O_9$	$(Cp)NiRu_3(CO)_9(\mu\text{-}H)\{\mu_4\text{-}(\eta^1,\eta^1,\eta^1,\eta^2)\text{-}C{=}CHPr^i\}$		611
779	$NiSi_4C_{20}H_{55}P_3$	$(Me_3P)_2Ni\{CP\eta^2\text{-}(Me_3Si)_2C{=}PC(H)(SiMe_3)_2\}$		612
700	$[Ni_2C_8H_{17}]^-$	$[Na(tmeda)_2][Ni_2(\mu\text{-}H)(\eta\text{-}C_2H_4)_4]$	100	613
781	$[Ni_2C_{15}H_{15}]^+$	$[\{(Cp)Ni\}_2\{\mu\text{-}(Cp)\}][BF_4]$	190	614
782	$Ni_2C_{19}H_{36}O_3P_2$	$(OC)Ni(\mu\text{-}PBu^t{}_2)_2Ni(CO)_2$		615
783	$Ni_2C_{20}H_{10}Cl_4S_4$	$(Cp)_2Ni_2\{\mu\text{-}2(\mu\text{-}S)\text{-}C_{10}Cl_4S_4\}$		218
784	$Ni_2C_{21}H_{45}O_2P_3$	$(Me_3P)Ni(\mu\text{-}PBu^t{}_2)_2Ni(CO)_2$		615
785	$Ni_2C_{38}H_{30}P_2S_2$	$[Ni(PPh_3)\{\mu\text{-}(S,\eta^2)\text{-}CS_2\}]_2$		616
786	$Ni_2C_{38}H_{54}Cl_2F_{12}N_2P_2$	$[Ni\{C(CF_3){=}C(CF_3)CH_2C_6H_4(NMe_2\text{-}2)\}(PEt_3)(\mu\text{-}Cl)]_2$		617
787	$Ni_2Zn_4C_{30}H_{30}$	$(Cp)_2Ni_2Zn_4(Cp)_4$		618
788	$[OsC_{10}H_{18}IO_6P_2]^-$	$Na[OsI\{PO(OMe)_2\}_2(\eta\text{-}C_6H_5)].NaI.0.5Me_2CO$	258	619
789	$OsC_{10}H_{19}IO_6P_2$	$OsI\{PO(OMe)_2\}\{POH(OMe)_2\}(\eta\text{-}C_6H_6)$		619
790	$OsC_{37}H_{32}ClNOP_2$	$Os(CH_2)Cl(NO)(PPh_3)_2$		92
791	$OsC_{41}H_{35}ClP_2$	$OsCl(PPh_3)_2(Cp).CH_2Cl_2$		620
792	$OsC_{46}H_{40}Cl_2N_2P_2S$	$OsCl_2(NCS)\{C(C_6H_4NMe_2\text{-}4)\}(PPh_3)_2$		621
793	$OsC_{47}H_{37}F_3O_3P_2S$	$Os(\eta^1\text{-}O_2CCF_3)(CO)(PPh_3)_2\{\eta^2\text{-}CS(pTol)\}$		622
794	$[OsC_{50}H_{41}O_2P_4]^+$	$[Os(CHO)(CO)(dppe)_2][SbF_6].CH_2Cl_2$		623
795	$[OsC_{53}H_{47}Cl_2N_2O_2]^+$	$[OsCl_2\{CC_6H_4(NMe_2\text{-}4)\}\{(CN(pTol)\}(PPh_3)_2][ClO_4]$		621
796	$OsSbC_{22}H_{15}O_4$	$Os(CO)_4(SbPh_3)$		63
797	$OsW_2C_{33}H_{24}O_7$	$OsW_2\{\mu_3\text{-}(\eta^1,\eta^1,\eta^2)\text{-}C_2(pTol)_2\}(CO)_7(Cp)_2$		624
798	$Os_2C_6I_2O_6$	$Os_2(CO)_6(\mu\text{-}I)_2$		625
799	$Os_2C_{12}H_6O_{10}$	$Os_2(CO)_8\{\mu\text{-}(\eta^1,\eta^1)\text{-}CH_2CHCO_2Me\}$	143	626
800	$Os_3C_9O_9S_2$	$Os_3(\mu_3\text{-}S)_2(CO)_9$		627
801	$Os_3C_{10}H_2O_{10}$	$(Cp)NiOs_3(CO)_9(\mu\text{-}H)_3.0.5Os_3(CO)_{10}(\mu\text{-}H)_2$	108	609
802	$Os_3C_{11}H_2O_{10}$	$Os_3(\mu\text{-}H)(\mu_3\text{-}CH)(CO)_{10}$		628
803	$Os_3C_{11}H_2O_{10}$	$Os_3(CO)_9(\mu\text{-}H)_2(\mu_3\text{-}CCO)$		629
804	$[Os_3C_{13}H_2BrF_6O_9]^-$	$[PPN][Os_3(CO)_9(\eta\text{-}CF_3CH{=}CHCF_3)(\mu\text{-}Br)]$		630

No.	Formula	Structure	Details	Ref.
805	$Os_3C_{13}H_3BrF_6O_9$	$Os_3(CO)_9$(μ-H)(μ-Br)(η-CF_3CH=CHCF_3)		630
806	$Os_3C_{13}H_5NO_{10}S_2$	$Os_3(CO)_{10}$(μ-H)(μ-SC=NCH_2CH_2S)		631
807	$[Os_3C_{14}H_2BrF_6O_{10}]^-$	[PPN][Os_3Br$(CO)_{10}$(η-CF_3CH=CHCF_3)]		630
808	$Os_3C_{14}H_2O_{13}$	$Os_3(CO)_{12}${μ-(η^1,η^1)-CH_2CO}		632
809	$Os_3C_{14}H_6N_2$	Os_3(μ-H)$(CO)_{10}${μ-(C,N)-CNC(Me)NHCH}		633
810	$Os_3C_{14}H_6N_2O_{14}$	$Os_3(CO)_{10}${μ-O,N:O,N-MeOC(O)NNC(O)OMe}		634
811	$Os_3C_{15}H_6O_9$	$Os_3(\mu\text{-H})_2(CO)_9${μ-($\eta^1,\eta^1,\eta^2$)-$C_6H_4$}		635
812	$Os_3C_{15}H_{10}NO_9$	$Os_3(CO)_9(\mu\text{-H})_2${μ-(N,μ-C)-NHC$(CH_2)_4$C}		636
813	$Os_3C_{16}H_6FNO_9S$	$Os_3(CO)_9$(μ-H)(μ_3-S){μ-(C,N)-HC=NC_6H_4F-4}		637
814	$Os_3C_{16}H_6FNO_9S$	$Os_3(CO)_9$(μ-H){μ_3-(N,μ-S)-SCH=NC_6H_4F-4}		637
815	$Os_3C_{16}H_6N_2O_{16}$	$Os_3(CO)_{12}${μ-(N,N)-MeOC(O)NNC(O)OMe}		634
816	$Os_3C_{16}H_{11}O_8PS_2$	$Os_3(\mu_3\text{-S})_2(CO)_8$(P$Me_2$Ph)		627
817	$Os_3C_{16}H_{12}O_{10}$	$Os_3(CO)_{10}$(μ-H){μ-(η^1,η^2)-CH=CHBu^t}		638
818	$Os_3C_{17}H_6FNO_{10}S$	Os_3(μ-H)$(CO)_{10}$(μ-SCH=NC_6H_4F-4)		637
819	$Os_3C_{17}H_{20}NO_9PS$	Os_3(μ-H)$(CO)_9${μ-(P,S)-PN(H)S$Bu_2{}^t$}		639
820	$Os_3C_{19}H_{20}NO_{11}P$	$Os_3(CO)_{11}${P(NH_2)$Bu^t{}_2$}		639
821	$Os_3C_{22}H_{12}N_2O_9$	Os_3(μ-H)$(CO)_9${μ-(N,μ-N)-NPh=C(Ph)NH}		640
822	$Os_3C_{23}H_{12}O_{10}S$	Os_3(μ-H)$(CO)_{10}$(μ-SCHPh_2)		641
823	$Os_3C_{24}H_{18}NO_9PS$	$Os_3(CO)_9$(PMe_2Ph)(μ-H){μ-(N,S)-SCH=NC_6H_4F-4}		637
824	$Os_3C_{35}H_{22}O_9S_2$	Os_3(μ-H)$(CO)_9$(SCPh_2)(μ-SCHPh_2)		642
825	$Os_3C_{35}H_{22}O_9S_2$	$Os_3(\mu\text{-H})_2(CO)_8$(SC$Ph_2$){μ-(C,μ-S)-COC$Ph_2$S}		642
826	$Os_3C_{37}H_{24}O_7P_2$	$Os_3(CO)_7(\mu\text{-PPh}_2)_2${$\mu_3$-($\eta^1,\eta^1,\eta^2$)-$C_6H_4$}		643
827	$Os_3C_{38}H_{35}NO_9P$	$Os_3(CO)_9$(μ-PPh_2){μ_3-(η^1,η^1,η^2)-PhC_2C(NHBu^t)-(NHBu^n)}		644
828	$Os_3PtC_{28}H_{37}O_{10}P$	$Os_3Pt(\mu\text{-H})_4(CO)_{10}(PCy_3)$	210	645
829	$Os_3PtC_{29}H_{35}O_{11}P$	$Os_3Pt(\mu\text{-H})_2(CO)_{11}(PCy_3)$		645
830	$Os_3RhC_{16}H_{11}O_9$	(η-C_6H_5Me)Rh$Os_3(\mu\text{-H})_3(CO)_9$		646,647
831	$Os_3SiC_{15}H_{10}O_{11}$	$Os_3(CO)_{10}$(μ-CO)(μ-CHSiMe_3)		648
832	$Os_3SiC_{27}H_{18}O_9$	$Os_3(\mu\text{-H})_3(CO)_9(SiPh_3)$		649
833	$Os_3Si_2C_{12}Cl_6O_{12}$	$Os_3(CO)_{12}(SiCl_3)_2$		650

No.	Formula	Structure	Details	Ref.
834	$Os_3Si_3C_9H_3Cl_9O_9$	$Os_3(SiCl_3)_3(CO)_9(\mu\text{-}H)_3$		650
835	$Os_3WC_{20}H_{11}O_{12}S_2$	$(OC)_9Os_3W(CO)_3(PMe_2Ph)(\mu_3\text{-}S)_2$		651
836	$Os_3WC_{24}H_{12}O_{11}$	$(OC)_9Os_3W\{\mu_3\text{-}C(pTol)\}(CO)_2(Cp)$	220	624
837	$Os_3WC_{25}H_{14}O_{12}$	$(Cp)(OC)WOs_2(CO)_6\{\mu_3\text{-}(O,C,\eta^2)\text{-}OCCH_2(pTol)\}\text{-}\{\mu\text{-}Os(CO)_4\}$		652
838	$Os_3WC_{28}H_{22}O_{12}S_2$	$(OC)_9Os_3W(CO)_2(PMe_2Ph)_2(\mu_3\text{-}S)_2$		651
839	$Os_3WC_{30}H_{20}O_9$	$(Cp)HWOs_3(CO)_9\{\mu_3\text{-}C(pTol)\}_2$		653
840	$Os_3WC_{31}H_{20}O_{10}$	$(Cp)(OC)_2WOs_3(CO)_8\{\mu\text{-}(\eta^1,\eta^1,\eta^2)\text{-}C_2(pTol)_2\}$		653
841	$[Os_4C_{12}NO_{12}]^-$	$[PPN][Os_4(CO)_{12}(\mu_4\text{-}N)]$		654
842	$Os_4C_{12}O_{12}S$	$Os_4(CO)_{12}(\mu_3\text{-}S)$		655
843	$Os_4C_{12}O_{12}S_2$	$Os_4(CO)_{12}(\mu_3\text{-}S)_2$		656
844	$Os_4C_{12}H_4O_{13}$	$2[Os_4H_3(CO)_{12}(\mu\text{-}OPO_3H_2)].[Os_4H_3(CO)_{12}(\mu\text{-}OH)]$		657
845	$Os_4C_{12}H_5O_{16}P$	$2[Os_4H_3(CO)_{12}(\mu\text{-}OPO_3H_2)].[Os_4H_3(CO)_{12}(\mu\text{-}OH)]$		657
846	$Os_4C_{13}O_{13}S$	$Os_4(CO)_{13}(\mu_3\text{-}S)$		658
847	$Os_4C_{13}O_{13}S_2$	$Os_4(CO)_{13}(\mu_3\text{-}S)_2$		656
848	$[Os_4C_{16}H_9N_2O_{12}]^+$	$[Os_4H_3(CO)_{12}(NCMe)_2][BF_4]$		659
849	$Os_5C_{14}H_2O_{14}S_2$	$Os_5(\mu\text{-}H)_2(CO)_{14}(\mu_3\text{-}S)_2$		660
850	$[Os_5C_{15}O_{14}]^{2-}$	$[PPN]_2[Os_5C(CO)_{14}]$		103
851	$Os_5C_{15}O_{15}S$	$Os_5(CO)_{15}(\mu_4\text{-}S)$		658
852	$Os_5C_{17}O_6$	$Os_5C(CO)_{16}$		103
853	$Os_5C_{19}H_2O_{17}$	$Os_5(CO)_{17}\{\mu_3\text{-}(\eta^1,\eta^1,\eta^2)\text{-}HCCH\}$		661
854	$Os_5C_{20}H_{17}O_{14}P$	$Os_5(\mu\text{-}H)_2(CO)_{14}(PEt_3)$		662
855	$Os_5C_{22}H_{26}O_{16}P_2$	$Os_5(\mu\text{-}H)_2(CO)_{13}(PEt_3)P(OMe)_3$		662
856	$Os_6C_{16}O_{16}S_2$	$Os_6(CO)_{16}(\mu_4\text{-}S)(\mu_3\text{-}S)$		663
857	$Os_6C_{17}O_{17}S_2$	$Os_6(CO)_{17}(\mu_4\text{-}S)_2$		663
858	$Os_6C_{17}H_2O_{17}S_2$	$Os_6(CO)_{17}(\mu\text{-}H)_2(\mu_4\text{-}S)(\mu_3\text{-}S)$		664
859	$Os_6C_{17}H_2O_{17}S_2$	$Os_6(CO)_{17}(\mu\text{-}H)_2(\mu_4\text{-}S)(\mu_3\text{-}S)$		664
860	$Os_6C_{18}H_2O_{18}S_2$	$Os_6(CO)_{18}(\mu\text{-}H)_2(\mu_4\text{-}S)(\mu_3\text{-}S)$		664
861	$Os_6C_{19}O_{20}$	$Os_6(\mu_3\text{-}O)(\mu_3\text{-}CO)(CO)_{18}$		665
862	$Os_6C_{23}H_8O_{20}S_2$	$\{Os_3(\mu\text{-}H)(CO)_{10}\}_2\{\mu\text{-}S(CH_2)_3S\}$		666

No.	*Formula*	*Structure*	*Details*	*Ref.*
863	$Os_6C_{28}H_{18}N_2O_{14}S_2$	Os$_6$H(CO)$_{14}$(μ-H)$_5$(μ$_4$-S)(μ$_3$-S){μ-(N,C)-HC=NPh}$_2$		667
864	$Os_6C_{29}H_{12}N_2O_{15}S_2$	Os$_4$(μ$_4$-S)$_2${μ-(N,C)-HC=NPh}$_2$(CO)$_{15}$		668
865	$Os_6C_{29}H_{16}N_2O_{15}S_2$	Os$_6$(CO)$_{15}$(μ-H)$_4$(μ$_4$-S)(μ$_3$-S){μ-(N,C)-HC=NPh}$_2$		667
866	$Os_6C_{29}H_{16}N_2O_{15}S_2$	Os$_6$(CO)$_{15}$(μ-H)$_4$(μ$_4$-S)(μ$_3$-S){μ-(N,C)-HC=NPh}$_2$		667
867	$Os_6C_{30}H_{14}N_2O_{16}S_2$	Os$_6$(μ$_4$-S)$_2${μ-(N,C)-HC=NPh}$_2$(CO)$_{15}$	238	668
868	$Os_6C_{31}H_{12}F_2N_2O_{17}S_2$	(OC)$_{10}${μ-(N,C)-HC=NR}Os$_3$(μ$_4$-S)Os$_3$(μ-H)$_2$(μ$_3$-S)-{μ-(N,C)-HC=NR}(CO)$_7$ [R=C$_6$H$_4$F-4]		668
869	$Os_6C_{32}H_{14}N_2O_{18}S_2$	(OC)$_6$Os$_2${(μ$_4$-S)Os$_2$(CO)$_6${μ-(C,N)-CH=NPh)}$_2$		669
870	$Os_7C_{19}O_{19}S$	Os$_7$(CO)$_{19}$(μ$_4$-S)		655
871	$Os_7C_{20}O_{20}S_2$	Os$_7$(CO)$_{20}$(μ$_4$-S)$_2$		670
872	$Os_8C_{22}HIO_{22}$	Os$_8$H(CO)$_{22}$(μ-I)		671
873	$[Os_9C_{25}H_5O_{21}]^-$	[PPN][Os$_9$(CO)$_{21}${μ$_4$-(η^1,η^2,η^2,η^1)-CHC(Me)CH}]		672
874	$[Os_9C_{26}H_7O_{21}]^-$	[PPN][Os$_9$(CO)$_{21}${μ$_4$-(η^1,η^2,η^2,η^1)-CHC(Et)CH}]		672
875	$[Os_{10}C_{24}H_4O_{24}]^{2-}$	[PPN]$_2$[Os$_{10}$H$_4$(CO)$_{24}$]		673
876	$PbC_{26}H_{25}NS$	PbPh$_3${SC$_6$H$_4$(NMe$_2$-2)}		674
877	$PdC_{14}H_{23}ClOS_2$	PdCl{2,4-(CMe$_2$CH$_2$SMe)$_2$furan-3-yl}		675,676
878	$PdC_{15}H_{25}P$	Pd(PPh$_3$){η^1,η^3-C(C$_2$H$_4$)CH=CHCH$_2$C(C$_2$H$_4$)CHCHCH$_2$}		677
879	$PdC_{16}H_{32}P_2$	Pd(PMe$_3$)$_2${η^2-CH$_2$=C(CMe)$_3$CMe}	100	678
880	$PdC_{17}H_{10}F_6N_2O_2$	Pd(C$_6$H$_4$N=NPh)(hfacac)		679
881	$PdC_{18}H_{32}P_2$	Pd(dmpe){η^1,η^1-C(CH$_2$)$_2$CH=CHCH$_2$C(CH$_2$)$_2$CH=CHCH$_2$}		677
882	$PdC_{19}H_{31}P$	Pd(PPr$^i_3$)(η^1-C$_5$H$_5$)(Cp)	253	680
883	$PdC_{19}H_{36}ClP$	PdCl(PPr$^i_3$)(η-C$_5$Me$_5$)		681
884	$PdC_{21}H_{17}NO$	Pd{(1-MeO)naphth-8-yl}{(8-CH$_2$)quinoline}		682
885	$PdC_{22}H_{24}F_6N_2$	(NMe$_2$CH$_2$C$_6$H$_4$)Pd{C(CF$_3$)=C(CF$_3$)(C$_6$H$_4$CH$_2$NMe$_2$)}		683
886	$PdC_{24}H_{28}N_2O_8$	Pd{C(CO$_2$Me)$_2$CH$_2$CH$_2$C$_5$H$_4$N}$_2$		684
887	$PdC_{27}H_{20}F_{12}N_2$	{(NMe$_2$CH$_2$C$_6$H$_4$)C(CF$_3$)=C(CF$_3$)}Pd{C(CF$_3$)=C(CF$_3$)-(CH$_2$C$_9$H$_6$N)}		683
888	$[PdC_{31}H_{27}ClP]^+$	[Pd(Cp)(PPh$_3$)(η^2-CH$_2$=CHC$_6$H$_4$Cl-4)][BF$_4$].CH$_2$Cl$_2$		685
889	$[PdC_{31}H_{28}P]^+$	[Pd(Cp)(PPh$_3$)(η^2-CH$_2$=CHPh)][PF$_6$]		685
890	$[PdC_{32}H_{30}OP]^+$	[Pd(Cp)(PPh$_3$)(η^2-CH$_2$=CHC$_6$H$_4$OMe-4)][BF$_4$].CH$_2$Cl$_2$		685

No.	Formula	Structure	Details	Ref.
891	$PdSeC_{33}H_{28}P_2S$	Pd{o-(PPh$_2$CH$_2$)$_2$C$_6$H$_4$}{η^2-(Se,C)-SeCS}		686
892	$Pd_2C_{14}H_{16}N_2P_2S_2$	Pd$_2$(CNMe)$_2${μ-(P,S)-SPPh$_2$}$_2$.CHCl$_3$		687
893	$Pd_2C_{16}H_{34}Cl_4N_2$	[PdCl(μ-Cl){CH$_2$CH=NPr$^i{}_2$}]$_2$		688
894	$Pd_2C_{20}H_{32}Cl_4$	{Pd(μ-Cl)(1-3η-1,4,4-Me$_3$-5Cl-cycloheptenyl)		689
895	$Pd_2C_{22}H_{22}Cl_2F_{10}N_6$	(tmeda)Pd{μ-(C,N)-C(C$_6$F$_5$)=NMe}$_2$PdCl$_2$		690
896	$Pd_2C_{24}H_{48}Cl_2P_2$	[Pd{CH$_2$C(Me)=CHP(Bu^t)$_2$}(μ-Cl)]$_2$		691
897	$Pd_2C_{26}H_{20}N_4O_8$	[Pd{C$_6$H$_3$(4-NO$_2$)(2-C$_5$H$_4$N)}(μ-O$_2$CMe)]$_2$		692
898	$Pd_2C_{38}H_{26}Cl_4N_2O_4$	[Pd{C$_6$H$_3$(Cl-5)(2-C$_5$H$_3$N(5′-p-C$_6$H$_4$Cl)}(μ-O$_2$CMe)]$_2$		692
899	$Pd_2C_{38}H_{68}Cl_4N_2P_4$	[PdCl(PEt$_3$)$_2$C$_6$H$_3$(5-Cl)(2-CH=N-)}]$_2$		693
900	$Pd_2C_{46}H_{47}ClN_4P_2$	PdCl(C$_6$H$_4$CH$_2$NMe$_2$){μ-PPh$_2$CH$_2$C(NH)C(CN)PPh$_2$}-Pd(C$_6$H$_4$CH$_2$NMe$_2$).2CH$_2$Cl$_2$		694
901	$Pd_2W_2C_{28}H_{40}O_6P_2$	(PEt$_3$)$_2$Pd$_2$W$_2$(Cp)$_2$(μ-CO)$_4$(μ$_3$-CO)$_2$		285
902	$Pd_3C_{20}H_{32}Cl_6$	[Pd{η^3-CHCHCH(CH$_2$)$_2$CH(CHClMe)CH$_2$CH$_2$}(μ-Cl)$_2$]$_2$Pd		695
903	$[Pd_3C_{76}H_{66}OP_6]^{2+}$	[Pd$_3$(μ-dppm)$_3$(μ$_3$-CO)][CF$_3$CO$_2$]$_2$.3Me$_2$CO		696
904	$Pd_4C_{24}H_{28}N_8$	Pd$_4$(η-C$_3$H$_5$)$_4$(μ-2,2′-bisimidazolato)$_2$.CH$_2$Cl$_2$		697
905	$Pd_7C_{28}H_{63}O_7P_7$	Pd$_7$(CO)$_7$(PMe$_3$)$_7$ [monocapped octahedron]		698
906	$Pd_{10}C_{62}H_{108}O_{14}P_4$	Pd$_{10}$(CO)$_{14}$(PBu$^n{}_3$)$_4$ [tetracapped octahedron]		699
907	$PrC_{15}H_{15}$	[Pr(Cp)$_2$(μ-η^5,η^2-C$_5$H$_5$)]$_n$		700
908	$PtC_6H_8F_4$	Pt(η-C$_2$H$_4$)$_2$(η-C$_2$F$_4$)	X+N, 190	701
909	PtC_6H_{12}	Pt(η-C$_2$H$_4$)$_3$	X+N, 190	701
910	PtC_8H_8INO	PtI(C$_6$H$_4$CH$_2$NH$_2$)(CO)		702
911	$PtC_8H_{10}ClNO_3S$	PtCl(η^1-SMe$_2$O){η^2-(C,O)-C$_6$H$_4$NO$_2$}		703
912	$PtC_9H_{14}I_2N_4$	Pt(Me)$_2$I$_2$(pz$_2$CH$_2$)		704
913	$PtC_{12}H_{19}Cl_3N_2$	PtCl$_3${η^3-(C,2N)-C$_6$H$_3$(2,6-CH$_2$NMe$_2$)$_2$}		705
914	$PtC_{13}H_{22}I_2N_4$	Pt(Me)$_2$I$_2${(Me$_2$pz)$_2$CH$_2$}		704
915	$PtC_{14}H_{21}ClNOP$	Pt{C(Me)=CHN(=CMe$_2$)O}Cl(PMe$_2$Ph)		706
916	$PtC_{14}H_{29}Cl_2N_4$	PtCl$_2${N(NMe$_2$)=C(Me)C(Me)=NNMe$_2$}{η^2-CH$_2$=CH-(S-CHMeEt)}		707
917	$PtC_{15}H_{20}Cl_2N_2O$	PtCl$_2${CH$_2$CMe(CH$_2$OH)CH$_2$}(py)$_2$	122	708
918	$PtC_{16}H_{17}N_7$	Pt(Me){η^2-(C,N)-CCHCHNNCHpz$_2$}(py)		709

No.	Formula	Structure	Details	Ref.
919	$PtC_{17}H_{21}IN_2O_2$	$PtI(Me)_2(OOPr^i)(phen)$		710
920	$PtC_{24}H_{48}ClP_2$	$PtCl\{CH_2C(Me)=CP(Bu^t)_2\}P(Bu^t)_2(CH_2C_3H_5)$		711
921	$PtC_{24}H_{48}P_2$	<u>trans</u>-$PtH(Ph)(PPr^i_3)_2$		712
922	$PtC_{24}H_{49}ClP_2$	$PtCl\{P(Bu^t)_2CH=C(Me)CH_2\}P(Bu^t)_2(CH_2C_3H_5)$	118	713
923	$[PtC_{30}H_{30}P]^+$	$[Pt(\eta^2-H_2C=CHPh)(\eta^3-H_2CC(Me)CH_2)(PPh_3)][PF_6]$		714
924	$PtC_{31}H_{36}ClP_3S_2$	$PtCl\{\eta^2-(C,S)-CH(PPh_2S)_2\}(PEt_3)$		715
925	$PtC_{34}H_{32}N_7P$	$Pt(Me)(CCHCHNNCHpz_2)(py)(PPh_3).2py$		709
926	$PtC_{35}H_{27}ClN_4P_2$	$PtCl\{CHCH_2C(CN)_2C(CN)_2CH_2\}(dppe)$		716
927	$PtC_{40}H_{48}N_6O_4P_2$	$Pt(C{\equiv}CPh)_2\{N(R)N=NN(R)\}(PEt_3)_2$	R=<u>p</u>-$C_6H_4NO_2$	717
928	$PtC_{41}H_{35}Cl_2O_5P$	<u>cis</u>-$PtCl_2\{CC(CO_2Et)=C(Ph)C(CO_2Et)=C(Ph)O\}(PPh_3)$		718
929	$PtC_{42}H_{36}N_2O_2P_2$	$Pt\{C(CN)_2CMe_2OO\}(PPh_3)_2.CH_2Cl_2$		719
930	$PtC_{43}H_{38}O_3P_2$	$(Ph_3P)_2PtCH(COMe)C(O)CH(COMe)$		720
931	$PtC_{44}H_{39}Cl_2P_2$	$(PPh_3)_2ClPt\{\eta^2-CH=CHCHCHClCH_2CHCHCH_2\}$		721
932	$PtC_{45}H_{71}ClN_4P_2$	$PtCl\{CHCH_2C(CN)_2C(CN)_2CH_2\}(PCy_3)_2$		716
933	$PtC_{46}H_{44}O_2P_2$	<u>trans</u>-$Pt(Ph)(OOBu^t)(PPh_3)_2$		722
934	$[PtRh_{12}C_{24}O_{24}]^{4-}$	$Cs_4[PtRh_{12}(\mu-CO)_{12}(CO)_{12}].2MeOH$		723
935	$PtSi_2C_{24}H_{56}Cl_4N_4OP_2$	$PtCl(CCl=CCl_2)[\{P(NHBu^t)(NBu^tSiMe_3)\}_2O]$		724
936	$[PtWC_{21}H_{39}O_2P_2]^+$	$[(Cp)(\eta-C_2H_4)WPt(\mu-CO)_2(PEt_3)_2][BF_4]$		725
937	$PtWC_{22}H_{32}O_2P_2$	$(PMe_3)_2Pt(\mu-(\eta^1,\eta^2)-C(pTol))W(CO)_2(Cp)$		726
938	$[PtWC_{22}H_{32}P]^+$	$[(Cp)_2W(\mu-H)_2Pt(Ph)(PEt_3)][BPh_4]$		122
939	$PtWC_{26}H_{33}O_2P$	$\{(cod)PtW(CO)(PMe_3)(Cp)\}\{\mu-(\eta^1,\eta^2)-C(pTol)CO\}$		727
940	$PtWC_{28}H_{47}P_2$	$[(Cp)_2WH(\mu-H)PtPh(PEt_3)][BPh_4]$		122
941	$PtWC_{32}H_{24}O_5P_2$	$(OC)_5WPt(\mu-C=CH_2)(dppm)$		728
942	$PtWC_{39}H_{32}O_6P_2$	$(OC)PtW(CO)_4\{\mu-C(OMe)(pTol)\}(\mu-dppm).0.5CH_2Cl_2$		729
943	$PtWC_{71}H_{58}O_3P_4$	$(OC)_3W(\mu-dppm)_2Pt\{C_2(pTol)\}\{\mu-C_2(pTol)\}$		730
944	$[Pt_2C_{13}H_{37}I_4P_4X]^+$	$[(Me)_3Pt(\mu-I)(\mu-dmpm)_2PtX][I_3]$	[X=0.5(I+Me)]	731
945	$Pt_2C_{14}H_{40}P_4$	$\{Pt(Me)_2(\mu-dmpm)\}_2$		732
946	$Pt_2C_{20}H_{24}Cl_2N_2$	$[PtCl\{\mu-(\eta^2,\mu-N)-NMeC_6H_4C(Me)=CH_2\}]_2$		733
947	$Pt_2C_{25}H_{20}O$	$(Cp)Pt\{\mu-(\eta^1,\eta^1,\eta^2)-C(Ph)C(Ph)C(O)\}Pt(Cp)$		734
948	$[Pt_2C_{30}H_{67}P_4]^+$	$[Et_3P)_2Pt(\mu-H)_2Pt(Ph)(PEt_3)_2][BPh_4]$		116

No.	Formula	Structure	Details	Ref.
949	$[Pt_2C_{36}H_{61}P_4]^+$	$[(Et_3P)_2(Ph)Pt(\mu\text{-}H)Pt(Ph)(PEt_3)_2][BPh_4]$		117
950	$Pt_2C_{42}H_{36}Cl_2O_2$	$[Pt\{\underline{o}\text{-}C_6H_4P(Ph)_2CH(COMe)\}(\mu\text{-}Cl)]_2.2CDCl_3$		735
951	$Pt_2C_{53}H_{48}P_4$	$Pt_2\{\mu\text{-}(C,2P)\text{-}\underline{o}\text{-}C_6H_4P(Ph)(CH_2CH_2PPh_2)\}_2(\mu\text{-}CH_2)$		736
952	$[Pt_2C_{53}H_{49}OP_4]^+$	$[Pt_2(dppe)_2(\mu\text{-}H)(\mu\text{-}CO)][BF_4]$	115	737
953	$[Pt_2C_{54}H_{50}O_2P_4]^{2+}$	$[Pt(Me)_2(CO)_2(\mu\text{-}dppm)_2][BF_4]_2.CH_2Cl_2$		738
954	$Pt_2C_{54}H_{52}P_4$	$[Pt(Me)\{\mu\text{-}(C,2P)\text{-}\underline{o}\text{-}C_6H_4P(Ph)(CH_2CH_2PPh_2)\}]_2$		739
955	$Pt_2C_{55}H_{45}Br_2OP_3$	$(Ph_3P)BrPt(\mu\text{-}CO)PtBr(PPh_3)_2.0.5CH_2Cl_2$		740
956	$[Pt_2C_{56}H_{56}N_2P_4]^{2+}$	$[Pt_2(Me)_2(MeCN)_2(\mu\text{-}dppm)_2][BF_4]_2.MeCN$		738
957	$[Pt_2C_{72}H_{61}P_4]^+$	$[(Ph_3P)_2Pt(\mu\text{-}H)(\mu\text{-}PPh_2)PtPh(PPh_3)][BF_4]$		741
958	$[Pt_2C_{86}H_{73}P_6]^+$	$[Ph_3P)Pt(\mu\text{-}dppm)\{(\mu\text{-}(P,C)\text{-}PPh_2(CHPPh_2)\}\text{-}$ $Pt(PPh_3)][PF_6]$		742
959	$[Pt_2Rh_{11}C_{24}O_{24}]^{3-}$	$Cs_2[NEt_3Pr^n][Pt_2Rh_{11}(\mu\text{-}CO)_{12}(CO)_{12}]$		723
960	$Pt_2W_3C_{45}H_{36}O_6$	$Pt_2W_3\{\mu\text{-}C(pTol)\}_2\{\mu_3\text{-}C(pTol)\}(CO)_6(Cp)_3$		100
961	$Pt_3W_2C_{46}H_{48}O_4$	$Pt_3W_2\{\mu_3\text{-}C(pTol)\}_2(CO)_4(Cp)_2(cod)_2.CH_2Cl_2$		100
962	$Pt_7C_{108}H_{108}N_{12}$	$Pt_7\{CNC_6H_3(Me_2\text{-}2,6)\}_{12}$		743
963	ReC_5ClO_5	$ReCl(CO)_5$		744
964	$ReC_7H_5O_2S_2$	$Re(CO)_2(\eta\text{-}S_2)(Cp)$	233	745
965	$[ReC_7H_6N_2O_5S]^+$	$[Re(CO)_5(NSNMe_2)][AsF_6]$		746
966	$[ReC_{10}H_{12}N_4O_2]^+$	$\underline{cis}\text{-}[Re(CO)_2(MeCN)_4]_2[ReCl_6]$		747
967	$ReC_{14}H_{23}O_3P_2$	$Re(CO)_3(PMe_3)_2(\eta^1\text{-}C_5H_5)$		748
968	$ReC_{20}H_{29}I_2N_2$	$Re(CBu^t)(CHBu^t)I_2(py)_2$		749
969	$ReC_{21}H_{12}BrN_4O_3$	$ReBr(CO)_3\{(NN)\eta^2\text{-}C_8H_4N_2(2,7\text{-}C_5H_4N)_2\text{-}1,8\}$		750
970	$[ReC_{24}H_{22}NOPS]^+$	$[Re(NO)(PPh_3(\eta^2\text{-}H_2C{=}S)(Cp)][PF_6]$		751
971	$[ReC_{24}H_{22}NO_2P]^+$	$[Re(NO)(PPh_3)(\eta^2\text{-}H_2C{=}O)(Cp)][PF_6]$	115	751
972	$ReC_{25}H_{21}BrO_3PS_2$	$ReBr(CO)_3(S_2CCMe_2PPh_3)$	163	500
973	$[ReC_{29}H_{32}NOP]^+$	$[Re({=}CH_2)(NO)(PPh_3)(\eta\text{-}C_5Me_5)][PF_6]$		752
974	$[ReC_{29}H_{32}NO_4P]^+$	$[Re({=}CH_2)(NO)\{P(OPh)_3\}(\eta\text{-}C_5Me_5)][PF_6].C_2H_4Cl_2$		752
975	$[ReC_{33}H_{31}N_2OP]^+$	$(\underline{SS})\text{-}[Re(NO)(NCCHEtPh)(PPh_3)(Cp)][PF_6]$	115	753
976	$ReC_{38}H_{46}N_3P_4S$	$\underline{mer}\text{-}Re(\eta^1\text{-}S_2PPh_2)(\eta^1\text{-}N_2)(CNMe)(PMe_2Ph)_3$		754
977	$ReC_{41}H_{35}O_2P_2$	$Re(\eta^5\text{-}CH(O)CHCHCH_2)(PPh_3)_2(CO)$		755

No.	Formula	Structure	Details	Ref.
978	$ReC_{45}H_{35}N_2O_5P_2$	Re{OCHNC$_6$H$_4$(p-NO$_2$)}(CO)$_2$(PPh$_3$)$_2$		756
979	$ReSb_2C_{36}H_{30}Cl_2N_2O_2$	trans-ReCl$_2$(NO)$_2$(SbPh$_3$)$_2$		757
980	$ReSi_7C_{24}H_{63}N_6O_3P_2$	(OC)$_3$Re{N(SiMe$_3$)P(NSiMe$_3$)$_2$N(SiMe$_3$)P(NSiMe$_3$)-N(SiMe$_3$)$_2$}		758
981	$[Re_2C_6HI_2O_6]^-$	[NEt$_4$][Re$_2$(μ-H)(μ-I)$_2$(CO)$_6$]		759
982	$Re_2C_{12}H_4O_{10}$	(OC)$_5$ReCH$_2$CH$_2$Re(CO)$_5$	173	760
983	$Re_2C_{13}H_5NO_8$	Re$_2$(μ-H)(CO)$_8${μ-(N,C)-C$_5$H$_4$N}		761
984	$Re_2C_{13}H_8O_8$	Re$_2${μ-(η^1,η^3)-CHCHCMe$_2$}(CO)$_8$		762
985	$Re_2C_{15}H_{14}N_2O_8$	Re$_2$(μ-H)(CO)$_7$(ONMe$_3$){μ-(N,C)-C$_5$H$_4$N}		761
986	$Re_2C_{51}H_{41}Cl_3N_3P_3$	Re$_2$Cl$_3${μ-(P,N)-PPh$_2$C$_5$H$_4$N}$_2${μ-(P:N,C)-PPh(C$_6$H$_4$)(C$_5$H$_4$N)}		763
987	$Re_2Se_2C_{20}H_{14}Br_2O_6$	Re$_2$(μ-Br)$_2$(CO)$_6$(μ-PhCH$_2$SeSeCH$_2$Ph)		764
988	$Re_2SiC_{28}H_{16}O_9$	(OC)$_8$Re$_2$(μ-H){μ-C(SiPh$_3$)(CO)}.xEt$_2$O		765
989	$Re_2Si_2C_{16}H_{27}Br_2N_2O_6$	{Re(CO)$_3$}$_2$(μ-Br)$_2${μ-(N,P)-N(Bu^t)PN(SiMe$_3$)$_2$}		766
990	$Re_2Si_2C_{50}H_{40}O_{10}$	(OC)$_4$Re{μ-(μ-C,O)-C(SiPh$_3$)C(OEt)O}Re(CO)$_3$-{C(OEt)SiPh$_3$)}.xEtOH	250	765
991	$Re_2SnC_{44}H_{30}Br_2O_8P_2$	Br$_2$Sn{Re(CO)$_4$(PPh$_3$)}$_2$		767
992	$[Re_3C_{10}H_2I_2O_{10}]^-$	[NEt$_4$][Re$_3$(μ-H)$_2$(μ-I)$_2$(CO)$_{10}$]		759
993	$[Re_6C_{19}H_2O_{18}]^{2-}$	[PPh$_3$Me]$_2$[Re$_6$C(CO)$_{18}$(μ$_3$-H)$_2$]		768
994	$[RhC_2Cl_2O_2]^-$	[PtCl(PPh$_2$C$_5$H$_4$N)$_2$][RhCl$_2$(CO)$_2$].0.5CH$_2$Cl$_2$	140	769
995	$RhC_5H_4ClN_2O_2$	RhCl(CO)$_2$(pzH)		770
996	$RhC_{11}H_{16}ClN_2$	RhCl(cod)(Me$_2$pzH)		770
997	$RhC_{15}H_{24}ClN_2$	RhCl(cod){(S)-CN(CH$_2$)$_3$CHCH$_2$NMe}		193
998	$RhC_{16}H_{19}NO_2P$	Rh{OC(Ph)NCO}(PMe$_3$)(Cp)		771
999	$RhC_{16}H_{28}PS$	Rh(PPr$^i_3$){CSη^2-SCCH$_2$}(Cp)		772
1000	$RhC_{17}H_{18}NO$	Rh(oxine)(cod)		773
1001	$[RhC_{18}H_{26}S]^+$	[Rh(nbd)$_2$(SEt$_2$)][ClO$_4$]		774
1002	$RhC_{20}H_{32}P$	Rh(Me)(PMe$_3$){η^5-C$_5$H$_4$C(Me)$_2$CHC(=CMe$_2$)CHCH=CH}		775
1003	$RhC_{22}H_{32}P$	Rh(C=CHPh)(PPr$^i_3$)(Cp)		776
1004	$[RhC_{23}H_{19}F_4N]^+$	[Rh(η^4-MTFB)(η^6-indole)][ClO$_4$]		

No.	Formula	Structure	Details	Ref.
		[MTFB=Me_3F_4benzobarrelene]		777
1005	$[RhC_{23}H_{35}O_3]^+$	$[Rh_2(\eta\text{-}C_5Me_5)_2(\mu\text{-}CH_2)_2(\mu\text{-}HCO_3)][BF_4].H_2O$		778
1006	$RhC_{24}H_{44}ClP_2$	$RhHCl\{C_6H_3(CH_2PBu^t{}_2)_2\}$	115	779
1007	$RhC_{26}H_{23}F_3O_3P$	$Rh\{OC(CF_3)CHC(Pr^i)O\}(CO)(PPh_3)$		780
1008	$[RhC_{26}H_{42}]^+$	$[(cod)Rh\{\eta^5\text{-}C(Bu^t)CHC(Bu^t)CHC(=CHBu^t)\}][PF_6]$		781
1009	$RhC_{[illegible]}H_{[illegible]}P$	$(Cp)(PPr^i{}_3)Rh\{C_6H_4(2\text{-}CH=CPh)\}$		782
1010	$RhC_{31}H_{27}OP_2$	$Rh(Cp)(dppm)(CO)$		783
1011	$RhC_{32}H_{35}ClN_3S$	$(\eta\text{-}C_5Me_5)ClRh[N(Ph)C\{N(pTol)CH=N(pTol)S\}$		784
1012	$RhC_{37}H_{24}ClF_6OP_2$	$RhCl(CO)\{P(C_6H_4F\text{-}4)_3\}_2$		785
1013	$RhC_{37}H_{30}ClOP_2$	<u>trans</u>-$RhCl(CO)(PPh_3)_2$		786
1014	$[RhC_{37}H_{33}NP_2]^+$	$[Rh\{NC_5H_4CH(PPh_2)_2\text{-}2\}(nbd)][BF_4]$		787
1015	$[RhC_{37}H_{41}NOP_2]^+$	[(cod)Rh{(<u>S</u>)-$PPh_2OCH_2CH(CH_2)_3NPPh_2\}][PF_6]$		788
1016	$[RhC_{39}H_{42}P_2]^+$	[Rh{(<u>R</u>)-$PPh_2CHCyCH_2PPh_2\}(nbd)][ClO_4]$	188,173	789
1017	$RhC_{43}H_{44}P_3$	$Rh(oTol)\{PhP(CH_2CH_2CH_2PPh_2)_2\}.PhMe$		790
1018	$RhC_{44}H_{41}P_2$	$Rh(PPh_3)_2\{(1\text{-}3)\eta^3\text{-}$cycloocta-1,5-dienyl}		791
1019	$RhC_{46}H_{63}O_6P_2$	$\{P(OPr^i)_3\}_2Rh\{(1,3,4)\eta^3\text{-}CPh=CPh\text{-}$ $CPhC(H)Ph\}.0.5C_5H_{12}$	138	792
1020	$RhC_{48}H_{48}ClP_2$	$RhCl(PPh_3)_2\{(2,3)\eta^2\text{-}Me_2C=C=C=CC_2Me_4\}$		793
1021	$RhSi_4C_{21}H_{46}P$	$Rh\{CHSiMe_2CH(SiMe_3)PHCH(SiMe_3)SiMe_2\}(cod)$		794
1022	$RhWC_{27}H_{28}O_2P$	$\{(PMe_3)(\eta\text{-}ind)RhW(CO)_2(Cp)\}\{\mu\text{-}C(pTol)\}$		727
1023	$[Rh_2C_{19}H_{38}N_6O_5]^{2+}$	$[Rh_2(\mu\text{-}CO)_3[\mu\text{-}\{dien(CH_2CH_2O\text{-})_2\}_2][PF_6]_2.MeOH$		795
1024	$Rh_2C_{20}H_{28}Cl_2N_2$	$Rh_2Cl_2(cod)_2(\mu\text{-}pyz)$		796
1025	$Rh_2C_{20}H_{36}O_4$	$Rh_2(CO)_4(\mu\text{-}PPh_2)_2$		797
1026	$Rh_2C_{22}H_{30}N_4$	$[Rh(cod)(\mu\text{-}pz)]_2$		462
1027	$Rh_2C_{22}H_{30}O_2$	$Rh_2(\mu\text{-}CO)_2(\eta\text{-}C_5Me_5)_2$	200	205
1028	$Rh_2C_{24}H_{32}O_3$	$\{(\eta\text{-}C_5Me_5)(OC)Rh\}_2(\mu\text{-}CHCHO)$		499
1029	$Rh_2C_{24}H_{40}$	$Rh_2(Me)_2(\mu\text{-}CH_2)_2(\eta\text{-}C_5Me_5)_2$		798
1030	$Rh_2C_{24}H_{40}Cl_2N_2$	$\{(cod)RhCl\}_2(\mu\text{-}(N,N)\text{-}Pr^iN=CHCH=NPr^i)$		799
1031	$Rh_2C_{24}H_{42}Cl_2F_2N_2P_2$	$[(cod)RhCl]_2\{\mu\text{-}PFN(Bu^t)PFNBu^t\}$		800
1032	$Rh_2C_{24}H_{54}O_2P_4$	$Rh_2(CO)_2(PMe_3)_2(\mu\text{-}PBu^t)_2$		801

No.	Formula	Structure	Details	Ref.
1033	$Rh_2C_{26}H_{58}Cl_4O_{14}P_4$	$[RhCl(\mu\text{-}Cl)(CO)\{PO(OPr^i)_2\}\{POH(OPr^i)_2\}]_2$		802
1034	$Rh_2C_{30}H_{25}O_2P$	$(Cp)Rh(\mu\text{-}CO)_2Rh(PPh_3)(Cp)$		803
1035	$Rh_2C_{30}H_{40}O_4$	$(\eta\text{-}C_5Me_5)_2Rh_2(\mu\text{-}CO)\{\mu\text{-}(O,\eta^2)\text{-}OCHCH_2CMe_2\text{-}CH_2C(O)CCO\}$		804
1036	$[Rh_2C_{30}H_{52}S_6]^{2+}$	$[(cod)_2Rh_2\{\mu\text{-}S(CH_2)_2S(CH_2)_2SCH_2\}_3\}_2][PF_6]$		805
1037	$[Rh_2C_{39}H_{64}N_7]^+$	$[Rh_2\{\mu\text{-}CNCMe_2(CH_2)CMe_2NC\}_3\{\mu\text{-}CNCMe_2CH_2CH_2\text{-}CMe_2\}[CF_3SO_3]$		806
1038	$Rh_2C_{40}H_{92}O_{12}P_4$	$Rh_2\{P(OPr^i)_3\}_4(\mu\text{-}H)\{\mu\text{-}(C,\eta^2)\text{-}MeC{=}C(H)Me\}$	153	792
1039	$Rh_2C_{42}H_{75}O_3P_3$	$(PPr^i{}_3)_2H_2Rh(\mu\text{-}CO_3)Rh(PPr^i{}_3)(\eta\text{-}C_2Ph_2)$	113	807
1040	$Rh_2C_{52}H_{100}O_{12}P_4$	$Rh_2\{P(OPr^i)_3\}_4(\mu\text{-}H)\{\mu\text{-}(C,\eta^2)\text{-}(pTol)C{=}CH(pTol)\}$	173	792
1041	$Rh_2C_{54}H_{48}Br_6Cl_2N_6$	$[RhCl\{CNC_6H_2(Br\text{-}4)(Me_2\text{-}2,6)\}_3]_2$		808
1042	$Rh_2C_{58}H_{52}O_4P_4$	$[Rh(CO)(dppp)(\mu\text{-}CO)]_2.0.5C_6H_6$		809
1043	$Rh_2C_{59}H_{45}O_5P_3$	$(CO)(PPh_3)_2Rh(\mu\text{-}CO)_2Rh(CO)_2(PPh_3)$		810
1044	$[Rh_2C_{61}H_{49}ClN_2O_3P_4]^+$	$[(MeOH)(OC)Rh\{\mu\text{-}(P,P)\text{-}NC_5H_3(PPh_2)_2\text{-}2,6\}\text{-}RhCl(CO)][PF_6]$	140	811
1045	$[Rh_2C_{61}H_{50}N_2OP_4]^{2+}$	$[Rh_2(CO)\{\mu\text{-}(P,P,N)\text{-}NC_5H_4CH(PPh_2)_2\}_2][BPh_4]_2$		115
1046	$[Rh_2C_{62}H_{50}N_2O_2P_4]^{2+}$	$[Rh_2(CO)_2\{\mu\text{-}(P,P,N)\text{-}NC_5H_4CH(PPh_2)_2\}_2][PF_6]_2.Me_2CO$		466
1047	$[Rh_2C_{63}H_{62}O_{10}P_5]^+$	$[(MeO)_3PRhRh(O_2CMe)(\mu\text{-}CO)(\mu\text{-}dppm)_2\text{-}(\mu\text{-}MeO_2CC{=}CCO_2Me)][PF_6]$		812
1048	$[Rh_2C_{74}H_{80}F_6N_4]^{2+}$	$[Rh_2(CNBu^t)_4(\mu\text{-}dppm)_2\{\mu\text{-}(\eta^1,\eta^1)\text{-}C_2(CF_3)_2\}]\text{-}[PF_6]_2.2Me_2CO$		813
1049	$Rh_3C_{18}H_{15}O_3$	$[Rh(Cp)(\mu\text{-}CO)]_3$		814
1050	$Rh_3C_{27}H_{54}O_3P_3$	$[Rh(CO)(\mu\text{-}PBu^t{}_2)]_3$		815
1051	$Rh_3C_{32}H_{47}$	$\{Rh(\eta\text{-}C_5Me_5)\}_3(\mu_3\text{-}CH)_2$		816
1052	$Rh_3C_{40}H_{30}Cl_2O_4P_3$	$Rh_3(\mu\text{-}Cl)_2(\mu\text{-}PPh_2)_3(\mu\text{-}CO)(CO)_3.CH_2Cl_2$		817
1053	$Rh_3C_{43}H_{30}O_7P_3$	$Rh_3(\mu\text{-}PPh_2)_3(CO)_7$		818
1054	$[Rh_3C_{67}H_{58}Cl_2O_3]^+$	$[Rh_3Cl(\mu\text{-}Cl)(CO)_3\{\mu_3\text{-}(Ph_2PCH_2)_2PPh\}_2]Cl$		819
1055	$[Rh_3C_{67}H_{58}I_2O_3P_6]^+$	$[Rh_3(CO)_2(\mu\text{-}CO)(\mu\text{-}I)_2\{\mu_3\text{-}(Ph_2PCH_2)_2PPh\}_2]\text{-}[BPh_4]$	140	114

No.	Formula	Structure	Details	Ref.
1056	$Rh_4C_{44}H_{58}P_2$	$Rh_4(\mu_4\text{-}PPh)_2(cod)_4$		820
1057	$Rh_4C_{54}H_{40}O_6P_4$	$Rh_4(CO)_4(\mu\text{-}CO)_2(\mu\text{-}PPh_2)_4$		821
1058	$Rh_4C_{61}H_{46}Cl_4N_2O_3P_4$	$Rh_4Cl_2(\mu\text{-}Cl)_2(CO)_2(\mu\text{-}CO)\{\mu_3\text{-}NC_5H_3(PPh_3)_2\text{-}2,6\}_2$		822
1059	$[Rh_6C_{15}NO_{15}]^-$	$[PPN][Rh_6N(CO)_{15}]$		823
1060	$[Rh_9C_{18}O_{18}]^{3-}$	$[Cs(18\text{-}crown\text{-}6)_2][Rh_9(CO)_{18}]$		824
1061	$[Rh_{11}C_{23}O_{23}]^{3-}$	$[NMe_4]_3[Rh_{11}(CO)_{23}]$		825
1062	$[Rh_{12}C_{25}O_{23}]^{3-}$	$[NPr_4]_3[Rh_{12}C_2(CO)_{23}]$		826
1063	$[Rh_{12}C_{26}O_{24}]^{2-}$	$[PPN]_2[Rh_{12}C_2(CO)_{24}]$		827
1064	$RuC_7H_5BrO_2$	$RuBr(CO)_2(Cp)$		828
1065	$RuC_{13}H_{17}BrO_2$	$RuBr(CO)_2(\eta\text{-}C_5Me_5)$		828
1066	$RuC_{14}H_{20}O$	$Ru(CO)\{\eta^4,\eta^4\text{-}CH_2{=}CHC({=}CHMe)(CH_2)_3C({=}CHMe)CH{=}CH_2\}$		829
1067	$[RuC_{18}H_{22}ClN_4]^+$	$[RuCl(pyz)_2\{\eta\text{-}MeC_6H_4(CHMe_2\text{-}4)\}][PF_6]$		830
1068	$RuC_{19}H_{31}O_3P$	$Ru\{P(OMe)_3\}(\eta^4\text{-}cycloocta\text{-}1,3,5\text{-}triene)(cod)$		831
1069	$RuC_{20}H_{20}O_{10}$	$Ru(Cp)(\eta\text{-}C_5(CO_2Me)_5\}$		832
1070	$[RuC_{23}H_{20}OP]^+$	$[Ru(Cp)(\eta^6\text{-}C_6H_5P(O)Ph_2)][ClO_4]$		833
1071	$RuC_{23}H_{26}Cl_2O_2P_2S$	$RuCl_2(CO)\{\underline{o}\text{-}C_6H_4(PPhMe)_2\}(SOMe_2)$		72
1072	$RuC_{24}H_{28}Cl_2O_6P_2$	$RuCl\{C(CO_2Me){=}C(CO_2Me)Cl\}(CO)_2(PMe_2Ph)_2$		834
1073	$RuC_{24}H_{34}$	$Ru(\eta^6\text{-}C_6Me_6)\{\eta^4\text{-}C_6Me_4(CH_2)_2\}$		835
1074	$RuC_{28}H_{16}N_2O_2$	$\underline{cis}\text{-}Ru(CO)_2\{benzo(h)quinolin\text{-}10\text{-}yl\}_2$		836
1075	$RuC_{28}H_{32}ClOP$	$RuCl(Me)\{\underline{o}\text{-}(PPh_2)(OMe)C_6H_4\}(cod)$		837
1076	$RuC_{28}H_{36}ClNOP_2$	$RuCl(Ph)(CO)(CNBu^t)(PMe_2Ph)_2$		838
1077	$[RuC_{30}H_{28}P]^+$	$[Ru(PPh_3)(nbd)(Cp)][ClO_4]$		833
1078	$RuC_{32}H_{31}ClP_2$	$(\underline{S})\text{-}RuCl\{\underline{R}\text{-}Ph_2PCH(Me)CH_2PPh_2\}(Cp)$		839
1079	$[RuC_{32}H_{38}P]^+$	$[RuH(PPh_3)(\eta\text{-}C_2H_4)(\eta\text{-}C_6Me_6)][PF_6]$		840
1080	$RuC_{33}H_{34}P_2$	$(\underline{S})\text{-}Ru(Me)(\underline{R}\text{-}Ph_2PCH(Me)CH_2PPh_2)(Cp)$		841
1081	$RuC_{35}H_{41}NO_2P_2$	$Ru(Ph)(COPh)(CO)(CNBu^t)(PMe_2Ph)_2$		842
1082	$RuC_{37}H_{33}ClP_2$	$\underline{S}\text{-}RuCl(Cp)(\underline{R}\text{-}Ph_2PCH(Ph)CH_2PPh_2)$		843
1083	$RuC_{37}H_{39}Cl_2OP_3$	$RuCl_2(CO)(Ph\overline{PCH{=}CMeCMe{=}CH})_3$		844
1084	$RuC_{39}H_{30}F_2O_2P_2$	$Ru(CF_2)(CO)_2(PPh_3)_2.C_6H_6$		845
1085	$RuC_{40}H_{38}ClF_2O_2P_2$	$RuHCl(PF_2OPr^n)(CO)(PPh_3)_2$		846

No.	Formula	Structure	Details	Ref.
1086	$[RuC_{41}H_{33}OP_4]^+$	$[RuH(CO)\{\underline{o}\text{-}(PPhMe)_2C_6H_4\}_2][PF_6].2Me_2CO$		847
1087	$RuC_{48}H_{42}ClP_3$	$RuCl(PPh_3)(\eta^1\text{-dppm})(Cp)$		848
1088	$[RuC_{54}H_{49}O_2P_4]^+$	$\underline{trans}\text{-}[Ru(CDO)(CO)(dppe)_2][SbF_6].CH_2Cl_2$		849
1089	$RuSiC_{21}H_{19}F_6O_7P$	$Ru(PPh_2CH_2SiMe_2OH)(CO_2CF_3)_2(CO)_2.Et_2O$		850
1090	$RuSnC_{32}H_{31}Cl_3P_2$	$(\underline{R})\text{-}Ru(SnCl_3)(\underline{R}\text{-}Ph_2PCH(Me)CH_2PPh_2)(Cp)$		839
1091	$RuZrC_{21}H_{24}O_3$	$(Cp)_2(Me_3CO)ZrRu(CO)_2(Cp)$		851
1092	$Ru_2C_{14}H_8O_4$	$[Ru(CO)_2\{\mu\text{-}(\eta^1,\eta^5)\text{-}C_5H_4\}]_2$		852
1093	$Ru_2C_{14}H_8O_4$	$Ru_2(CO)_4\{\mu\text{-}(\eta^5,\eta^5)\text{-fulvalene}\}$		852
1094	$Ru_2C_{14}H_{12}O_{10}$	$(OC)_3Ru\{\mu\text{-}(\eta^1,\eta^1:\eta^4)\text{-}C_4(CH_2OH)_4\}Ru(CO)_3.H_2O$		853
1095	$Ru_2C_{15}H_{12}O_3$	$\underline{cis}\text{-}Ru_2(CO)_2(\mu\text{-}CO)(\mu\text{-}CCH_2)(Cp)_2$		854
1096	$Ru_2C_{15}H_{12}O_4$	$[(Cp)(CO)_2Ru]_2(\mu\text{-}CH_2)$	173	855
1097	$[Ru_2C_{15}H_{13}O_3]^+$	$\underline{cis}\text{-}[Ru_2(CO)_2(\mu\text{-}CO)(\mu\text{-}CCH_2)(Cp)_2][BF_4]$		854
1098	$Ru_2C_{15}H_{16}O_2$	$Ru_2(CO)_2(\mu\text{-}H)\{\mu\text{-}(\eta^1,\eta^2)\text{-}CHC(H)Me\}(Cp)_2$	200	856
1099	$Ru_2C_{16}H_{16}O_3$	$Ru_2(CO)_2(\mu\text{-}CO)(\mu\text{-}CMe_2)(Cp)_2$	220	856
1100	$Ru_2C_{16}H_{26}$	$Ru\{\eta^5\text{-}CH_2C(Me)C(Me)C(Me)CH_2\}_2$		857
1101	$Ru_2C_{17}H_{18}O_4$	$(OC)Ru\{\eta^4,\eta^4\text{-}CH_2{=}CHC({=}CHMe)(CH_2)_3C({=}CMe)\text{-}$ $CH{=}CH\}Ru(CO)_3$		829
1102	$Ru_2C_{17}H_{20}N_2O_7$	$Ru_2(CO)_5\{\mu\text{-}(\eta^2,N,\mu\text{-}N)\text{-}Pr^iN(CH)_2N(Pr^i)C(O)\text{-}$ $CH_2C(O)CH_2\}$		858
1103	$Ru_2C_{18}H_{12}O_{14}$	$(OC)_6Ru_2\{\mu\text{-}(\eta^1,\eta^1,\eta^4)\text{-}C_4(CO_2Me)_4\}$		859
1104	$Ru_2C_{18}H_{20}O_8$	$(OC)_3Ru\{\eta^4\text{-}C(CH_2CH_2OH)C(Et)C(Et)C(CH_2CH_2OH)\}\text{-}$ $Ru(CO)_3$		853
1105	$Ru_2C_{19}H_{20}O_2$	$(Cp)(CO)Ru(\mu\text{-}CO)\{\mu\text{-}(\eta^1,\eta^2)\text{-}C_4H_4CMe_2\}Ru(Cp)$		860
1106	$Ru_2C_{20}H_{18}$	$\{(Cp)Ru(\eta\text{-}C_5H_4)\}_2$		861
1107	$Ru_2C_{22}H_{20}O_6$	$(OC)_6Ru_2\{\mu\text{-}\eta^1,\eta^1,\eta^4)\text{-}H_{11}C_8C_8H_{11}\}$		315
1108	$Ru_2C_{24}H_{24}O_9$	$Ru_2(Cp)_2(\mu\text{-}CO)\{\mu\text{-}(\eta^3,\eta^2)\text{-}C_4(CO_2Me)_4CH_2\}$		860
1109	$Ru_2C_{26}H_{19}O_6P$	$Ru_2(CO)_6\{\mu\text{-}(P{:}\eta^2,\mu\text{-}C)\text{-}MeCC_6H_4PPh_2\}$		862
1110	$Ru_2C_{31}H_{26}NO_7P$	$Ru_2(CO)_6(\mu\text{-}PPh_2)\{\mu\text{-}(O,C)\text{-}OCCH{=}C(Ph)NEt_2\}$		379
1111	$[Ru_2C_{92}H_{82}P_4]^{2+}$	$[(Cp)(Ph_3P)_2RuC{=}CHC{=}CHC(CH_2CMe_2CH_2)\text{-}$ $Ru(PPh_3)_2(Cp)][PF_6]_2$		863

No.	Formula	Structure	Details	Ref.
1112	$Ru_2Si_2C_{36}H_{36}O_6P_2$	$Ru_2(\mu\text{-}PPh_2CH_2SiMe_2)_2(CO)_6$		850
1113	$[Ru_3C_{11}HO_{12}]^-$	$[PPN][Ru_3(CO)_7(\mu\text{-}CO)_3(\mu\text{-}O_2CH)]$		864
1114	$Ru_3C_{12}H_4O_{11}$	$Ru_3(CO)_{10}(\mu\text{-}H)(\mu\text{-}COMe)$		865
1115	$Ru_3C_{16}H_5NO_{10}$	$Ru_3(CO)_9(\mu_3\text{-}CO)(\mu_3\text{-}NPh)$		866
1116	$Ru_3C_{16}H_9NO_{11}$	$Ru_3(CO)_{11}(CNBu^t)$		867
1117	$Ru_3C_{16}H_{12}O_9$	$Ru_3(CO)_8(\mu\text{-}H)(\mu_3\text{-}COMe)(\eta^4\text{-cyclohexa-1,3-diene})$		865
1118	$Ru_3C_{16}H_{18}O_{16}P_2$	$Ru_3(CO)_{10}\{P(OMe)_3\}_2$		868
1119	$Ru_3C_{17}H_5ClO_9$	$(OC)_9Ru_3(\mu\text{-}Cl)\{\mu_3\text{-}(\eta^1,\mu\text{-}\eta^2)\text{-}C{=}CPh\}$		869
1120	$Ru_3C_{17}H_9NO_{10}$	$(OC)_4RuRu_2(CO)_6(\mu\text{-}H)(\mu\text{-}NHCH_2Ph)$		870
1121	$Ru_3C_{18}H_{18}O_3$	$Ru_3(\mu_2\text{-}H)(CO)_3(Cp)_3$		871
1122	$Ru_3C_{18}H_{27}O_9P_3$	$Ru_3(CO)_9(PMe_3)_3$		868
1123	$Ru_3C_{20}H_{18}N_2O_{10}$	$Ru_3(CO)_{10}(CNBu^t)_2$		867
1124	$Ru_3C_{22}H_{15}NO_9$	$Ru_3(CO)_9(\mu\text{-}H)\{\mu_3\text{-}(\eta^1,\eta^1,\eta^2)\text{-}PhC_2CNBu^t\}$		644
1125	$Ru_3C_{28}H_{17}O_8P$	$Ru_3(CO)_8(\mu\text{-}H)_2\{\mu_3\text{-}(P,\eta^1,\eta^1,\eta^2)\text{-}HC_2C_6H_4PPh_2\}$		872
1126	$Ru_3C_{29}H_{33}O_{11}P$	$Ru_3(CO)_{11}(PCy_3)$		868
1127	$Ru_3C_{30}H_{17}O_{10}P$	$Ru_3(CO)_{10}\{\mu\text{-}(P,\eta^2)\text{-}CH_2{=}CHC_6H_4PPh_2\}$		872
1128	$Ru_3C_{36}H_{50}O_6$	$Ru_3(CO)_5\{\mu_3\text{-}(\eta^3,\eta^2,\eta^1)\text{-}CH_2CBu^tCCBu^t\}\text{-}$ $(\mu_3\text{-}\eta^1,\eta^3,\eta^3\text{-}HCCBu^t\overline{CCHCBu^tCHCBu^tO})$		873
1129	$Ru_3C_{37}H_{24}O_7P_2$	$Ru_3(CO)_7(\mu\text{-}PPh_2)_2\{\mu_3\text{-}(\eta^1,\eta^1,\eta^2)\text{-}C_6H_4\}$		643
1130	$Ru_3C_{43}H_{31}O_7P_3$	$Ru_3(CO)_7(\mu\text{-}H)(\mu\text{-}PPh_2)_3$		874
1131	$[Ru_4C_{10}Cl_4O_{10}]^{2-}$	$[PPN]_2[\{(OC)_4(\mu\text{-}Cl)_2Ru_2(\mu\text{-}CO)\}_2]$		875
1132	$Ru_4C_{11}H_3NO_{11}$	$Ru_4(\mu\text{-}H)_3(CO)_{11}(\mu_4\text{-}N)$		654
1133	$[Ru_4C_{13}HO_{13}]^-$	$[PPN][Ru_4(\mu\text{-}H)(CO)_{12}(\mu\text{-}CO)]$		876
1134	$Ru_4C_{13}H_3NO_{13}$	$Ru_4(CO)_{12}(\mu\text{-}H)_3(\mu\text{-}NCO)$		877
1135	$Ru_4C_{20}H_{40}O_{20}P_4$	$Ru_4(\mu\text{-}H)_4(CO)_8\{P(OMe)_3\}_4$	X 293,N 20	878
1136	$Ru_4C_{31}H_{15}O_{11}P$	$Ru_4(CO)_{11}\{\mu_4\text{-}HC_2(\underline{o}\text{-}C_6H_4PPh_2\}$ [Ru_4C_2 cluster]		879
1137	$Ru_4C_{44}H_{47}O_{15}P_3$	$Ru_4(\mu\text{-}H)_4(CO)_9(PMe_2Ph)\{P(OpTol)_3\}\{P(OCH_2)_3CEt\}$		880
1138	$[Ru_5C_{14}NO_{14}]^-$	$[NEt_3Bz][Ru_5N(CO)_{14}]$		881
1139	$Ru_5C_{16}O_{15}$	$Ru_5C(CO)_{15}$		882
1140	$Ru_5C_{17}H_6O_{14}S$	$Ru_5C(\mu\text{-}H)(\mu\text{-}SEt)(CO)_{14}$		98

No.	Formula	Structure	Details	Ref.
1141	$Ru_5C_{18}H_3NO_{15}$	$Ru_5C(CO)_{15}(MeCN)$		882
1142	$Ru_5C_{28}H_{10}O_{16}P_2$	$Ru_5(CO)_{16}(\mu\text{-}PPh_2)(\mu_5\text{-}P).2C_6H_6$		883
1143	$Ru_5C_{33}H_{15}O_{13}P$	$Ru_5(CO)_{13}(\mu\text{-}PPh_2)\{\mu_4\text{-}C_2Ph\}$		884
1144	$Ru_5C_{33}H_{15}O_{14}P$	$Ru_5C(CO)_{14}(PPh_3)$		882
1145	$Ru_5C_{33}H_{21}O_{12}PS$	$Ru_5C(\mu\text{-}H)(\mu\text{-}SEt)(CO)_{12}(PPh_3)$		98
1146	$Ru_5C_{34}H_{15}O_{14}P$	$Ru_5(CO)_{13}(\mu\text{-}CO)(\mu\text{-}PPh_2)(\mu_5\text{-}C_2Ph)$		884
1147	$Ru_5C_{34}H_{21}O_{13}PS$	$Ru_5C(\mu\text{-}H)(\mu\text{-}SEt)(CO)_{13}(PPh_3)$		98
1148	$Ru_5C_{39}H_{26}O_{12}P_2$	$Ru_5C(\mu\text{-}H)_2(CO)_{12}(dppe)$		882
1149	$Ru_5C_{51}H_{30}O_{13}P_2$	$Ru_5C(CO)_{13}(PPh_3)_2$		882
1150	$Ru_6C_{15}N_2O_{16}$	$Ru_6C(CO)_{14}(NO)_2$		99
1151	$Ru_6C_{16}HNO_{16}$	$Ru_6C(CO)_{15}(NO)(\mu\text{-}H)$		885
1152	$Ru_6C_{22}H_9NO_{16}$	$Ru_6C(CO)_{15}(\mu\text{-}CO)(CNBu^t)$		886
1153	$[Ru_6C_{24}H_{13}N_2O_{20}]^-$	$[AsPh_4][Ru_6H(CO)_{14}(\mu\text{-}CO)_4\{\mu\text{-}(O,C)\text{-}$ $OCNMe_2\}_2].Me_2CO$	117	48
1154	$Ru_9C_{24}H_6O_{24}S_3$	$[Ru_3(\mu\text{-}H)_2(\mu_4\text{-}S)(CO)_8]_3$		887
1155	SbC_3H_9	$SbMe_3$	E	32
1156	$SbC_{12}H_8ClS$	$S(C_6H_4\text{-}2,2')_2SbCl$		888
1157	$SbC_{13}H_{18}NO$	$SbMe_4(oxine)$		889
1158	$SbC_{28}H_{28}Cl$	$SbCl(pTol)_4$		890
1159	$SeC_{16}H_{14}O_2$	$Se(CH_2COPh)_2$		891
1160	$SeC_{16}H_{22}O_4$	$Se\{C_6H_4(2\text{-}O)(6\text{-}OH)(4,4\text{-}Me_2)\}_2$		892
1161	$[SeC_{21}H_{21}]^+$	$[Se(pTol)_3][X]$ X=$Cl.H_2O,Br.H_2O,HSO_4$		893
1162	$SeC_{31}H_{25}PS$	$(PhSe)(PhS)CPPh_3$		894
1163	$SeSn_4C_{32}H_{72}$	$(Bu^t)_8Sn_4Se$		895
1164	Se_2C	CSe_2	(at 17.5,50,200K),N	896
1166	$[Se_2C_{74}H_{60}P_4]^{2+}$	$[\{(Ph_3P)_2CSe\}_2][Fe_2OCl_6].4CH_2Cl_2$		898
1167	$[Se_3C_5H_8NO]^+$	$[Se_3CNC_4H_8O][I]$		899
1168	$Se_4C_{14}H_8$	Dibenzotetraselenafulvalene		900
1169	$[Se_4C_{18}H_8]^+$	$[Se_4tetracene][Y]$ Y=Hg_mX_n;$CuBr_2$		897,901
1170	$Se_6Sn_5C_{24}H_{56}$	$Sn\{SeSn(Pr^i)_2SeSn(Pr^i)_2Se\}_2$		902

No.	Formula	Structure	Details	Ref.
1171	$[Se_8C_{20}H_{24}]^+$	$[Me_4\mathrm{tetraselenafulvalene}]_2{}^+[X]^-$		903
1172	$SiC_5H_{12}S_5$	$Me_3SiCHCH_2S_4S$		904
1173	$[SiC_8H_{14}N]^+$	$[Me_3Si(py)][I]$		905
1174	$SiC_8H_{16}O$	$(Me)(HO)SiCH_2CH(CH_2)CH_2CH_2CHCH_2$		906
1175	$[SiC_9H_{20}N_2]^+$	$[(Me)_2SiCH_2C(Me){=}NNBu^t][AlCl_4]$		907
1176	$SiC_{16}H_{24}N_2$	$Me_2SiN(Bu^t)NC(Me)CCHPh$		908
1177	$[SiC_{16}H_{35}N_2]^+$	$[(Pr^i)_2SiCH_2C(Bu^t){=}NHNBu^t]_2\{Al_4Cl_{10}O_2]$		907
1178	$SiC_{17}H_{17}NO_2$	$Ph_2SiCHC(Me)CH(NO_2)CH_2$		909
1179	$[SiC_{18}H_{38}N_2]^+$	$[Bu^t{}_2SiCH_2C(Bu^t)NN(Bu^t)][AlCl_4]$		907
1180	$SiC_{19}H_{19}Br$	(<u>RR</u>)-$Si(CHBrMe)(Me)(Ph)(C_{10}H_7)$		910
1181	$SiC_{19}H_{19}Cl$	(<u>SS</u>)-$Si(CHClMe)(Me)(Ph)(C_{10}H_7)$		910
1182	$SiC_{20}H_{27}NO_2$	$Ph_2Si(OH)(CH_2OCH_2CH_2NC_5H_{10})$		911
1183	$SiC_{22}H_{21}Br_2NO$	$(MeO)(oTol)Si\{1\text{-}C_6H_3(3\text{-}Br)\text{-}2\text{-}\}_2NEt$		417
1184	$SiC_{24}H_{18}O$	$Ph_2Si(1\text{-}C_6H_4\text{-}2\text{-})_2O$		417
1185	$SiC_{24}H_{18}S$	$Ph_2Si(1\text{-}C_6H_4\text{-}2\text{-})_2S$		417
1186	$SiC_{24}H_{20}$	$SiPh_4$	[$X\text{-}X^{ho}$ deformation density]	912
1187	$SiC_{24}H_{22}$	$\{1,4\text{-}Ph_2C_{10}H_6\}(\mu\text{-}1,4\text{-}SiMe_2)$		913
1188	$SiC_{26}H_{30}$	$\{1,4\text{-}Ph_2C_{12}H_{14}(\mu\text{-}1,4\text{-}SiMe_2)$		913
1189	$SiC_{28}H_{27}N$	$Ph_2Si\{1\text{-}C_6H_3(5\text{-}Me)\text{-}2\text{-}\}_2NEt$		417
1190	$SiTiC_6H_{15}Cl_2N$	$Me_2Si(\mu\text{-}NCMe_3)_2TiCl_2$		914
1191	$Si_2C_2H_8O$	$(SiH_2Me)_2O$	E	915
1192	$Si_2C_4H_{12}S_2$	$(Me_2Si)_2(\mu\text{-}S)_2$	153	916
1193	$Si_2C_4H_{13}O$	$(SiHMe_2)_2O$	E	915
1194	$Si_2C_{10}H_{24}N_2S_2$	$[Me_3SiN(Me)C(S)\text{-}]_2$		420
1195	$Si_2C_{10}H_{24}S_3$	$Me_3SiCHCH_2SCH_2CH(SiMe_3)SS$		904
1196	$Si_2C_{12}H_{18}O$	$\{(CH_2{=}CH)_3Si\}_2O$		917
1197	$Si_2C_{12}H_{30}O_3$	$(SiPr_2{}^iOH)_2O$		918
1198	$Si_2C_{16}H_{20}Cl_2N_2$	$(Me_2Si)_2(\mu\text{-}N(C_6H_4Cl\text{-}2)\}_2$		919
1199	$Si_2C_{16}H_{20}O_2$	$\{(Ph)(Me)SiCH_2O\}_2$		920
1200	$Si_2C_{18}H_{26}N_2$	$[Me_2Si(pTol)]_2$		921

No.	Formula	Structure	Details	Ref.
1201	$Si_2C_{18}H_{26}N_2$	$[Me_2SiN(mTol)]_2$		921
1202	$Si_2C_{19}H_{36}N_2O_2$	$Me_2SiN(Bu^t)NC(Me)CSi(Me)_2O(C_7H_9O)CH_2$		908
1203	$Si_2C_{24}H_{54}P_2$	$\{Si(Bu^t)_2(\mu\text{-}PBu^t)\}_2$		922
1204	$Si_2C_{25}H_{32}ClNP_2$	$PhP{=}C(PPhSiMe_3)\{N(SiMe_3)(C_6H_4Cl\text{-}2)\}$		923
1205	$Si_2C_{26}H_{32}F_3NP_2$	$PhP{=}C(CPhSiMe_3)\{N(SiMe_3)(C_6H_4CF_3\text{-}4)\}$		923
1206	$Si_2C_{26}H_{54}Cl_2$	<u>trans</u>-$C_2Cl_2(SiBu^t_3)_2$		924
1207	$Si_2C_{33}H_{38}$	$H_2CSi_2(C_6H_3Me_2\text{-}2,6)_4$		925
1208	$Si_2C_{36}H_{44}$	$Si_2(Mes)_2$		926
1209	$Si_2SnC_6H_{18}Cl_4N_2S$	$SnCl_4\{\eta^2\text{-}(N,N)\text{-}Me_3SiNSNSiMe_3\}$		927
1210	$Si_2SnC_{16}H_{26}$	$Sn\{\eta^5\text{-}C_5H_4P(NPr^i_2)_2\}_2$		928
1211	$Si_2ThC_{28}H_{52}$	$Th(CH_2SiMe_3)_2(\eta\text{-}C_5Me_5)_2$		929
1212	$Si_2TiC_{14}H_{20}Cl_2O$	$Cl_2Ti(\eta\text{-}C_5H_4SiMe_2)_2O$		930
1213	$Si_2W_2C_{20}H_{40}Cl_2O_6P_2$	$[WCl\{(O,C)\eta^2\text{-}COCH_2SiMe_3)(CO)_2(PMe_3)]_2$		931
1214	$Si_3C_{47}H_{58}$	$(Ph)(Me_3Si)C{=}CSi_2(Mes)_4$		932
1215	$Si_3SnC_{12}H_{33}F$	$(Me_3SiCH_2)_3SnF$		933
1216	$Si_4C_8H_{24}$	Si_4Me_8	87	934
1217	$Si_4C_{12}H_{24}$	$(Me_2Si\text{-}C{=}C\text{-}SiMe_2)_2$-(<u>SiSi</u>)		935
1218	$Si_4C_{24}H_{60}N_2P_2$	$P_2\{N(SiMe_2Bu^t)_2\}_2$		936
1219	$Si_4C_{30}H_{42}$	2,3,7,8-$(SiMe)_4$terphenylene		937
1220	$Si_4C_{32}H_{52}N_4$	$[\{SiMe_2N(Bu^t)N{=}C(Me)C\}_2\{\mu\text{-}Si(Me)(Ph)\}]_2$		938
1221	$Si_4ZrC_{18}H_{50}N_2P_2$	$Zr(CH_2SiMe_2NSiMe_3)_2(dmpe)$		939
1222	$Si_4ZrC_{20}H_{56}Cl_2N_2P_4$	$ZrCl_2\{N(SiMe_2CH_2PMe_2)_2\}_2$		940
1223	$Si_5C_{60}H_{50}O$	$(Ph)_{10}Si_5O.0.5C_6H_6$		941
1224	$Si_6C_{14}H_{38}O_3$	$HC(SiMe_2OSiMe_2)_3CH$		942
1225	$Si_6C_{16}H_{50}N_4P_2$	$\{Si(NMeSiMe_3)_2(\mu\text{-}P)\}_2$		943
1226	$Si_6C_{18}H_{54}P_2$	$(Me_3Si)_3CPPC(SiMe_3)_3$		944
1227	$Si_6C_{20}H_{50}N_6$	$[Me_3Si)_2SiN(Me)N{=}C(Me)CH]_2$		938
1228	$Si_6C_{24}H_{66}N_4P_2$	$\{Si(NMeSiMe_3)_2(\mu\text{-}PBu^t)\}_2$		922
1229	$Si_6C_{24}H_{66}N_6$	$\{Si(NMeSiMe_3)_2(\mu\text{-}NBu^t)\}_2$		945
1230	$Si_6C_{26}H_{52}N_4$	$(Ph)(Me)SiN(SiMe_3)N(SiMe_3)Si(Me)(Ph)NN(SiMe_3)_2$		946

No.	Formula	Structure	Details	Ref.
1231	$Si_6C_{42}H_{48}$	catena-$(SiMePh)_6$		947
1232	$Si_6SnC_{18}H_{54}BrN_3$	$BrSn\{N(SiMe_3)_2\}_3$		948
1233	$Si_8C_{16}H_{48}O_8$	$[Me_2SiO]_8$		949
1234	$Si_8Sn_3C_{30}H_{82}O_3$	$[(Me_3Si)_3CSnO)]_3$		950
1235	$Si_8Sn_4C_{32}H_{90}Cl_2O_4$	$[(Me_3SiCH_2)_2SnCl(\mu\text{-}OH)(\mu_3\text{-}O)Sn(CH_2SiMe_3)_2]_2$		951
1236	$Si_8Zr_2C_{24}H_{68}N_4$	$[(Me_3Si)_2NZrCHSi(Me)_2NSiMe_3]_2$		952
1237	$Si_8Zr_2C_{26}H_{78}N_4O$	$[ZrMe\{N(SiMe_3)_2\}_2]_2(\mu\text{-}O)$		953
1238	$Sm_2C_{36}H_{46}$	$\{(\eta\text{-}C_5H_4Me)_2Sm(\mu\text{-}C_2Bu^t)\}_2$		954
1239	$Sm_2C_{40}H_{62}$	$[Sm(\mu\text{-}H)(\eta\text{-}C_5Me_5)_2]_2$		955
1240	$[SnC_2H_6Cl_3]^-$	$[IrH\{P(OMe)_3\}_4\{P(OMe)_2(OSnCl_2Me_2)\}][SnMe_2Cl_3]$		460
1241	$SnC_4H_{10}S_2$	$Me_2SnSCH_2CH_2S$		956
1242	$SnC_6H_{11}Cl_3O_2$	$Cl_3Sn(CH_2)_3C(O)OEt$		957
1243	$SnC_7H_{11}N_3O$	$SnMe_3\{NCC(CN)_2\}(OH_2)$		958
1244	$SnC_7H_{13}NO_2$	$SnMe_3\{NC(O)CH_2CH_2C(O)\}$	138	959
1245	$SnC_7H_{17}NS_2$	$Me_2SnS(CH_2)_2NMe(CH_2)_2S$		960
1246	$SnC_8H_{10}ClNO_2$	$\{(Me)_2ClSnNC_5H_4(2\text{-}COO)\}_n$		961
1247	$SnC_{12}H_{14}$	$Sn(\eta\text{-}C_5H_4Me)_2$	E	415
1248	$[SnC_{12}H_{31}O_2]^+$	$[SnBu_3(OH_2)_2][C_5(CO_2Me)_5]$		962
1249	$SnC_{14}H_{14}S$	$Sn(Me)_2(1\text{-}C_6H_4\text{-}2\text{-})_2S$		963
1250	$SnC_{20}H_{26}Cl_2N_2$	$Sn(Bu^n)_2Cl_2(phen)$		964
1251	$SnC_{22}H_{34}O_6$	$Bu_2{}^nSnO_2C_7H_8O_3(OMe)(Ph)$		965
1252	$SnC_{24}H_{18}Cl_2N_4$	$SnCl_2(\underline{o}\text{-}C_6H_4N{=}NPh)_2$		966
1253	$SnC_{24}H_{20}$	$SnPh_4$		967,968
1254	$SnC_{24}H_{24}Cl_2N_4$	$SnCl_2(Et)_2\{1,2,4\text{-}N_3C_3(2\text{-}C_5H_4N)(5,6\text{-}Ph_2)\}$		969
1255	$SnC_{26}H_{21}Br_2N$	$Ph_2Sn\{1\text{-}C_6H_3(3\text{-}Br)\text{-}2\text{-}\}_2NEt$		417
1256	$SnC_{26}H_{25}NS$	$Sn(Ph_3(SC_6H_4NMe_2\text{-}2)$		970
1257	$SnC_{28}H_{25}Br_2N$	$(oTol)_2Sn\{1\text{-}C_6H_3(3\text{-}Br)\text{-}2\text{-}\}_2NEt$		417
1258	$SnC_{28}H_{27}N$	$(oTol)_2Sn(C_6H_4)_2NEt$		417
1259	$SnC_{28}H_{28}$	$Sn(oTol)_4$		967
1260	$SnC_{28}H_{35}Cl_2N_4O$	$SnCl_2(Bu^n)_2\{C_8N_2H_6(Ph)(NCHC_5H_4N)\}$		971

No.	Formula	Structure	Details	Ref.
1261	$SnC_{30}H_{28}N$	$(oTol)_2Sn\{1\text{-}C_6H_3(5\text{-}Me)\text{-}2\text{-}\}_2NEt$		417
1262	$SnC_{32}H_{36}Cl_2O_2P_2$	$Sn(Pr^n)_2Cl_2O(dppoe)$		972
1263	$SnC_{32}H_{36}N_2O_3$	$Ph_3Sn[O_2CC_6H_4\{2\text{-}N_2C_6H_3(2'\text{-}OH)(4'\text{-}Me)\}]$		973
1264	$SnC_{34}H_{40}Cl_2O_2P_2$	$Sn(Bu^n)_2Cl_2(dppoe)$		972
1265	$SnC_{34}H_{42}Cl_2O_2P_2$	$[Sn(Bu^n)_2Cl_2\{\mu\text{-}(dppoe)\}]_n$		974
1266	$SnC_{40}H_{44}$	$Sn(CPh{=}CMe_2)_4$		268
1267	$SnC_{64}H_{54}$	$Sn(CPh{=}CPh_2)_3(Bu^n)$		975
1268	$Sn_2C_{18}H_{30}Cl_4N_2O_2$	$[SnClMe_2(ONC_5H_3Me_2\text{-}2,6)(\mu\text{-}Cl)]_2$	138	976
1269	$Sn_2C_{48}H_{42}O_8P_2S_2$	$[SnPh_2\{OP(S)(OPh)_2\}(\mu\text{-}OH)]_2$	138	977
1270	$Sn_2TeC_{36}H_{30}$	$Te(SnPh_3)_2$		978
1271	$Sn_3C_9H_{27}P_7$	$P_7(SnMe_3)_3$		423
1272	$Sn_3C_{21}H_{45}N_3O_9$	$[MeSn(\mu\text{-}OCH_2CH_2)_3N]_3$		979
1273	$Sn_3C_{22}H_{22}S_3$	$\{SnPh(\mu\text{-}S)\}_3\{\mu_3\text{-}(CH_2)_3CH\}$		980
1274	$Sn_3C_{58}H_{52}$	$(Ph_3SnCH_2)_3CH$		981
1275	$Sn_3C_{60}H_{78}$	$[Sn\{C_6H_3(Et_2\text{-}2,6)\}_2]_3.0.5C_6H_6$		982
1276	$Sn_3C_{60}H_{78}O_3$	$[Sn\{C_6H_3(Et_2\text{-}2,6)\}_2(\mu\text{-}O)]_3.0.5C_6H_6$		982
1277	$Sn_4C_8H_{24}Cl_4O_2$	$[Me_2ClSn(\mu\text{-}Cl)SnMe_2(\mu_3\text{-}O)]_2$		983
1278	$Sn_4C_{24}H_{58}Cl_2O_4$	$[Pr^i{}_2SnCl(\mu\text{-}OH)(\mu_3\text{-}O)SnPr^i{}_2]_2$		951
1279	$Sn_4C_{32}H_{72}S$	$(Bu^t)_8Sn_4S$		895
1280	$Sn_4C_{50}H_{44}$	$(Ph_2SnCH_2SnPh_2)_2\text{-}(\underline{SnSn})$		984
1281	$Sn_4TeC_{32}H_{72}$	$(Bu^t)_8Sn_4Te$	188	895
1282	$[Sn_4WC_{66}H_{57}O_4]^-$	$[NMe_4][W(SnPh_3)_2\{(Ph_2Sn)_2OPr^i\}(CO)_3]$		985
1283	$Sn_5C_{24}H_{56}S_6$	$Sn\{SSn(Pr^i)_2SSn(Pr^i)_2S\}_2$		902
1284	$Sn_5C_{32}H_{72}S_6$	$Sn\{SSn(Bu^t)_2SSn(Bu^t)_2S\}_2$		902
1285	$[TaC_6O_6]^-$	$[PPN][Ta(CO)_6]$		594
1286	$TaC_{15}H_{42}P_5$	$Ta(PMe_3)_3(\eta^2\text{-}CH_2PMe_2)(\eta^2\text{-}CHPMe_2)$		986
1287	$TaC_{16}H_{37}P_2$	$TaH_4(PMe_3)_2(\eta\text{-}C_5Me_5)$		987
1288	$[Ta_2C_{20}H_{30}Cl_7]^+$	$[\{TaCl_2(\eta\text{-}C_5Me_5)\}_2(\mu\text{-}Cl)_3]_2[Ta_4Cl_{18}O_2]$		988
1289	$Ta_2C_{24}H_{39}Cl_4N$	$(\eta\text{-}C_5Me_4Et)Cl_2Ta\{\mu\text{-}(N,\eta^2)\text{-}NCHMe\}(\mu\text{-}H)(\mu\text{-}Cl)\text{-}$ $TaCl(\eta\text{-}C_5Me_4Et)$		989

No.	Formula	Structure	Details	Ref.
1290	$Ta_2C_{52}H_{42}Cl_8P_4$	{μ-(PCη^2,P)-Ph$_2$P=CHC(PPh$_2$)=C(PPh$_2$)CH=PPh$_2$}-(TaCl$_4$)$_2$.2CH$_2$Cl$_2$		990
1291	$Ta_2C_{56}H_{52}Cl_6P_4S_2$	{μ-(PCη^2,P)-Ph$_2$PCC(PPh$_2$)=C(PPh$_2$)C(PPh$_2$)}{TaCl$_3$(SMe$_2$)}$_2$		990
1292	$Ta_2ZnC_{18}H_{38}Cl_6$	{(MeOCH$_2$CH$_2$OMe)TaCl$_2$(μ-CCMe$_3$)(μ-Cl)}$_2$Zn		991
1293	TeC_2H_6	TeMe$_2$	E	32
1294	$TeC_2H_6Cl_2$	Me$_2$TeCl$_2$	151	992
1295	$[TeC_6H_5Cl_3I]^-$	[Bu$^n{}_4$N][TeCl$_3$I(Ph)]		993
1296	$TeC_{12}H_8Br_2Cl_2$	TeCl$_2$(C$_6$H$_4$Br-4)$_2$		994
1297	$TeC_{12}H_9BrCl_2$	TePh(C$_6$H$_4$Br-4)Cl$_2$		995
1298	$Te_2C_{26}H_{18}$	{Te(C$_6$H$_4$)$_2$CH}$_2$.TCNQ		996
1299	$ThC_{30}H_{50}ClO_2P$	ThCl{OC(=PMe$_3$)C(CH$_2$CMe$_3$)=O}(η-C$_5$Me$_5$)$_2$		997
1300	$TiC_9H_{13}Cl_2O$	TiCl$_2$(THF)(Cp)		998
1301	$TiC_{10}H_{10}F_6P_2$	Ti(PF$_3$)$_2$(Cp)$_2$		999
1302	$TiC_{10}H_{10}N_4S_3$	Ti{NS(NS)$_2$N}(Cp)$_2$		1000
1303	$TiC_{11}H_{10}Cl_2N_2$	TiCl$_2$(η^2-NNPh)(Cp)	238	1001
1304	$TiC_{11}H_{12}Cl_2N_2$	TiCl$_2$(η^2-NH$_2$NPh)(Cp)	231	1001
1305	$TiC_{12}H_{16}S_2$	Ti(SMe)$_2$(Cp)$_2$		1002
1306	$TiC_{13}H_{21}Cl_2O_2$	TiCl$_2$(THF)$_2$(Cp)		998
1307	$TiC_{14}H_{18}Cl_2$	TiCl$_2$(η-C$_5$H$_4$Et)$_2$		428
1308	$TiC_{15}H_{28}P_2$	Ti(Et)(η-C$_7$H$_7$)(dmpe)		1003
1309	$TiC_{17}H_{15}O_2$	Ti(O$_2$CPh)(Cp)$_2$		1004
1310	$TiC_{17}H_{25}OP$	Ti(CO)(PEt$_3$)(Cp)$_2$		999
1311	$TiC_{22}H_{20}S_4$	Ti(SPh)(SSSPh)(Cp)$_2$		1005
1312	$TiC_{22}H_{34}$	Ti(η-C$_2$H$_4$)(η-C$_5$Me$_5$)		1006
1313	$TiC_{24}H_{20}O_4$	Ti(O$_2$CPh)$_2$(Cp)$_2$		1007
1314	$TiWC_{25}H_{22}O_2$	(Cp)$_2$TiW{μ-C(pTol)}(μ-CO)(CO)(Cp)		1008
1315	$TiWC_{26}H_{24}O_2$	(Cp)(CO)W{μ-CH$_2$C(pTol)}(μ-η^2-CO)Ti(Cp)$_2$		726
1316	$Ti_2C_{22}H_{20}Cl_2N_2$	[TiCl(Cp)]$_2$(μ-NPh)$_2$		1009
1317	$Ti_2C_{22}H_{30}Cl_2O_2$	[Ti(η-C$_7$H$_7$)(THF)(μ-Cl)]$_2$		1003

No.	Formula	Structure	Details	Ref.
1318	$Ti_2C_{26}H_{26}O_4$	$(Cp)_2Ti\{O_2C(C_4H_6)CO_2\}Ti(Cp)_2$		1004
1319	$Ti_2C_{27}H_{26}O_4$	$(Cp)_2Ti\{O_2CC(CH_2)_3CCO_2\}Ti(Cp)_2$		1004
1320	$Ti_2C_{28}H_{25}Cl_2N_3$	$Ti_2Cl_2(Cp)_2(\mu\text{-}NPh)(\mu\text{-}N_2Ph_2)$		998
1321	$Ti_2C_{36}H_{30}Cl_2N_4$	$(Cp)ClTi\{\mu\text{-}(N,\eta^2)\text{-}NNCPh_2\}_2TiCl(Cp)$		998
1322	$Ti_2C_{48}H_{42}O$	$\{(Cp)_2TiC(Ph)=CH(Ph)\}_2).THF$		1010
1323	$[TlC_{22}H_{30}O_6]^+$	$[Tl(Me)_2(Cy_2\text{-}18\text{-}crown\text{-}6)][C_6H_2O(NO_2)_3]$		1011
1324	$[UC_{17}H_{15}N_2S_2]^-$	$[AsPh_4][U(NCS)_2(Cp)_3]$		49
1325	$UC_{17}H_{23}Cl_3O_2$	$UCl_3(THF)_2(\eta^5\text{-}ind)$		1012
1326	$[UC_{19}H_{21}N_2]^+$	$[U(Cp)_3(NCMe)_2]_2[UCl_4O_2].2C_4H_6$		1013
1327	$UC_{19}H_{23}O$	$U(THF)(Cp)_3$	225	1014
1328	$UC_{22}H_{36}$	$U\{\eta^3\text{-}CH_2C(Me)CH_2\}_3(\eta\text{-}C_5Me_5)$		1015
1329	$UC_{24}H_{27}P$	$U(CHPMe_2Ph)(Cp)_3$		1016
1330	$VC_{11}H_{23}Cl_2P_2$	$VCl_2(PMe_3)_2(Cp)$	110	1017
1331	$VC_{20}H_{30}S_2$	$V(\eta^2\text{-}S_2)(\eta\text{-}C_5Me_5)_2$		1018,1019
1332	$V_2C_{14}H_{10}F_6S_4$	$V_2(Cp)_2(\mu\text{-}\eta^2\text{-}S_2)\{\mu\text{-}2(\mu\text{-}S)\text{-}S_2C_2(CF_3)_2]$		1020
1333	$V_2C_{16}H_{16}$	$\{(Cp)V\}_2(\mu\text{-}\eta^6\text{-}C_6H_6)$		1021
1334	$V_2C_{16}H_{18}$	$\{(CP)V(M\text{-}H)\}_2\{M\text{-}(H^2,H^2)\text{-}C_6H_6)$		1022
1335	$V_2C_{16}H_{22}S_4$	$V_2(H\text{-}C_5H_4PR^I)_2(M\text{-}S)_2(M\text{-}S_2)$		1020
1336	$V_2C_{18}H_{18}$	$\{(Cp)V\}(\mu\text{-}\eta^5,\eta^5\text{-}C_8H_8)$		1023
1337	$V_2C_{19}H_{22}$	$\{(Cp)V\}_2(\mu\text{-}\eta^6\text{-}MesH)$		1021
1338	$V_2C_{32}H_{36}O_8$	$V_2\{\mu\text{-}(C,O)\text{-}C_6H_3(OMe)_2\text{-}2,6\}_2\{\mu\text{-}(O,\mu\text{-}C,O)\text{-}$ $C_6H_3(OMe)_2\text{-}2,6\}_2$		1024
1339	$V_2C_{58}H_{48}O_6P_4S$	$\{V(CO)_3(dppe)\}_2(\mu\text{-}S).3PhMe$		1025
1340	$WC_{11}H_{12}O_5S_3$	$W(CO)_5(SCHMe)_3$		1026
1341	$WC_{11}H_{13}NO_5$	$W(CO)_4\{NC_5H_9(2\text{-}Me)\}$		1027
1342	$WC_{12}H_6F_6O_3$	$W(CO)_2\{\eta^3\text{-}C(O)C(CF_3)=CHCF_3\}(Cp)$		1028
1343	$WC_{12}H_{28}Cl_2OP_2$	$WCl_2(=CHBu^t)(CO)(PMe_3)_2$		1029
1344	$WC_{13}H_8O_5S_3$	$W(CO)_5\{S=CHC=C(S)C(=CHS)CH_2CH_2CH_2\}$		1030
1345	$WC_{14}H_9N_3O_4$	$W(CO)_4\{(NC_5H_4\text{-}2\text{-})_2NH\}$		254
1346	$WC_{14}H_{15}F_6O_5P$	$W(CO)\{P(OMe)_3\}\{\eta^3\text{-}C(O)C(CF_3)=CHCF_3\}(Cp)$		1028

No.	Formula	Structure	Details	Ref.
1347	$[WC_{14}H_{23}O_2P]^+$	$[W(CO)(PMe_3)\{\eta^3\text{-}Me_2PC(Me){=}C{=}O\}]Cl$		1031
1348	$WC_{14}H_{30}N_4P_2$	<u>trans</u>-$W(CNMe)_4(PMe_3)_2$		1032
1349	$WC_{15}H_{12}O_3S_2$	$W\{SS(pTol)\}(CO)_3(Cp)$		1033
1350	$WC_{15}H_{18}$	$W(\eta^5\text{-}C_5H_4CMe_2)(\eta\text{-}C_6H_5Me)$		522
1351	$WC_{15}H_{19}O_2P$	$W(\eta^1\text{-}C_2Pr^i)(CO)_2(PMe_3)(Cp).CH_2Cl_2$		1034
1352	$WC_{16}H_{23}Cl$	$WCl\{\eta^3\text{-}C_3(Bu^t)_2\}(Cp)$		1035
1353	$[WC_{16}H_{29}P_2]^+$	$[W(dmpe)(\eta^3\text{-}C_3H_5)(\eta^6\text{-}C_6H_5Me)][PF_6]$		1036
1354	$WC_{17}H_{27}Cl_2$	$WCl_2(\eta^6\text{-}C_5Me_4Bu^t)(\eta\text{-}C_2Me_2)$		1037
1355	$WC_{17}H_{27}I_2N_3O_2$	$WI_2(CO)_2(CNBu^t)_3$		1038
1356	$[WC_{18}H_{22}O_2P]^+$	$[W(CO)(PMe_3)\{\eta\text{-}(HO)C_2(pTol)\}(Cp)][BF_4].Me_2CO$		1039
1357	$WC_{19}H_{15}BrN_2O_2$	$WBr(CPh)(CO)_2(py)_2$		582
1358	$[WC_{21}H_{20}O_5PS]^+$	$[W(CO)_5(Ph_2PCH_2CH_2SMe_2)][BF_4]$		532
1359	$WC_{21}H_{36}I_2N_4O$	$WI_2(CO)(CNBu^t)_4.CHCl_3$		1038
1360	$[WC_{23}H_{13}N_4O_2]^-$	$[PPN][W(CN)_2(CO)(phen)\{\eta^2\text{-}(C,C)\text{-}O{=}C{=}CPh\}$		1040
1361	$WC_{24}H_{24}Cl_2$	$WCl_2\{\eta^3\text{-}C(Ph)C(Bu^t)C(Ph)\}(Cp)$		1041
1362	$WC_{24}H_{42}N_2O_3P_2S$	$(Cp)(OC)_2W\{C(O)N(SNPBu^t{}_2)PBu^t{}_2\}$		1042
1363	$WC_{25}H_{22}$	$W(\eta^5\text{-}C_5H_4CPh_2)(\eta\text{-}C_6H_5Me)$		522
1364	$WC_{27}H_{27}P$	$W(PMe_3)(\eta\text{-}HC_2Ph)_3$		1032
1365	$WC_{33}H_{28}O_3P_2$	$W(C{=}CHCO_2Me)(CO)_3(dppe)$		1043
1366	$[WC_{34}H_{25}O]^+$	$[W(CO)(\eta\text{-}C_2Ph_2)_2(Cp)][BF_4]$		548
1367	$WZnC_{13}H_{17}NO_3$	$Me_2N(CH_2)_2CH_2ZnW(CO)_3(Cp)$		1044
1368	$WZrC_{31}H_{42}O$	$(Cp)_2W{=}CHO\text{-}ZrH(\eta\text{-}C_5Me_5)_2$		1045
1369	$W_2C_{10}H_6O_8S_2$	$W_2(CO)_8(\mu\text{-}SMe)_2$	253	1046
1370	$W_2C_{11}H_8I_2O_5S$	$(Cp)(OC)_2W(\mu\text{-}SMe)(\mu\text{-}I)W(CO)_3I$		1047
1371	$W_2C_{16}H_{12}N_2O_{10}$	$W_2(CO)_{10}(\mu\text{-}N_2C_2Me_4)$		1048
1372	$W_2C_{20}H_{24}N_4O_8$	$[W(CO)_4(\mu\text{-}N_2C_2Me_4)]_2$		1049
1373	$W_2C_{22}H_{42}O_{10}$	$(Pr^iO)_4W(\mu\text{-}OPr^i)_2W(CO)_4$		1050
1374	$W_2C_{24}H_{48}O_6$	$(Pr^iO)_3W(\mu\text{-}OPr^i)\{\mu\text{-}(\eta^1,\eta^1{:}\eta^4)\text{-}C_4H_4\}$-		
		$W(OPr^i)_2(\eta\text{-}C_2H_2)$	113	1051
1375	$W_2C_{28}H_{60}O_6$	$[W(CMe)(OBu^t)_3]_2$	113	1052

No.	Formula	Structure	Details	Ref.
1376	$W_2C_{30}H_{46}O_4$	$W_2(OBu^t)_4(\mu\text{-}CPh)_2$		1053
1377	$W_2C_{30}H_{54}N_2O_6$	$W_2(Pr^iO)_4(py)_2(\mu\text{-}OPr^i)_2(\mu\text{-}C_2H_2)$	108	1051
1378	$W_2C_{30}H_{60}O_6$	$(Pr^iO)_3W(\mu\text{-}OPr^i)\{\mu\text{-}(\eta^1,\eta^1:\eta^4)\text{-}C_4Me_4\}\text{-}W(OPr^i)_2(\eta\text{-}C_2Me_2)$	113	1051
1379	$W_2C_{30}H_{66}N_2O_6$	$[(Bu^tO)_2W(CNMe_2)(\mu\text{-}OBu^t)]_2$		585
1380	$W_2C_{44}H_{82}N_2O_6$	$(py)(Bu^tCH_2O)_3W(\mu\text{-}OCH_2Bu^t)(\mu\text{-}\eta^2\text{-}C_2Me_2)\text{-}W(OCH_2Bu^t)_2(py)$	113	1051
1381	$W_2C_{48}H_{56}O_4$	$W_2(OBu^t)_4(\mu\text{-}\eta^2\text{-}C_2Ph_2)_2$		1053
1382	$W_3C_{34}H_{20}O_{10}S_4$	$W_3(CO)_{10}(\mu\text{-}SPh)_4$	240	1046
1383	$W_4C_{48}H_{94}O_{14}$	$[(Pr^iO)_3W\{\mu_3\text{-}(O,\mu\text{-}C)\text{-}CO\}(\mu\text{-}OPr^i)W(OPr^i)_2(py)]_2$		1054
1384	$W_6C_{46}H_{100}O_{14}$	$\{W_3(OBu^t)_5(\mu\text{-}CEt)(\mu\text{-}O)\}_2(\mu\text{-}O)_2$		1055
1385	$Y_2C_{30}H_{42}N_2$	$Y_2(Cp)_2\{\mu\text{-}(C,\eta^2)\text{-}HC{=}NHBu^t\}_2$		1056
1386	$YbC_{14}H_{20}O_2$	$Yb(MeOCH_2CH_2OMe)(Cp)_2$		1057
1387	$YbC_{25}H_{44}ClP_2$	$YbCl(\eta^1\text{-}dmpm)(\eta\text{-}C_5Me_5)_2$		1058
1388	$Zn_2C_{20}H_{38}Br_2O_6$	$[\overline{ZnBr(CH_2CO}_2Bu^t)(THF)]_2$		1059
1389	$Zn_2C_{22}H_{46}N_2O_2$	$[(Et)Zn\{\mu\text{-}(N,\mu\text{-}O)\text{-}\{N(Et)(Bu^t)CH{=}C(Me)O\}]_2$		1060
1390	$Zn_2C_{26}H_{26}N_2$	$\{Zn(Me)\}_2(\mu\text{-}NPh_2)_2$		1061
1391	$ZrC_{10}H_{10}Cl_2$	$ZrCl_2(Cp)_2$		427
1392	$ZrC_{12}H_{16}$	$Zr(Me)_2(Cp)_2$		427
1393	$ZrC_{13}H_{16}ClNOS$	$ZrCl(SOCNMe_2)(Cp)_2$		1062
1394	$ZrC_{14}H_{18}Cl_2$	$ZrCl_2(\eta\text{-}C_5H_4Et)_2$		428
1395	$ZrC_{16}H_{18}F_6O_7S_2$	$Zr(OSO_2CF_3)_2(THF)(Cp)_2$		1063
1396	$ZrC_{23}H_{21}ClN_2$	$ZrCl(NN\eta^2\text{-}HN_2{=}CPh_2)(Cp)_2$		1064
1397	$ZrC_{23}H_{22}ClP$	$ZrCl(Cp)_2(PC\eta^2\text{-}CH_2PPh_2)$	140	275
1398	$ZrC_{24}H_{20}Cl_4O_2$	$Zr(OC_6H_4Cl_2\text{-}2,6)_2(\eta\text{-}C_5H_4Me)_2$		1065
1399	$ZrC_{25}H_{26}N_2$	$Zr(Me)(NN\eta^2\text{-}MeNN{=}CPh_2)(Cp)_2$		1064
1400	$ZrC_{27}H_{26}O$	$\overline{Zr(CH_2CH{=}CHCH_2C(Ph)_2O})(Cp)_2$		1066
1401	$ZrC_{27}H_{37}NO$	$Zr(CO\eta^2\text{-}H_2C{=}CO)(py)(\eta\text{-}C_5Me_5)_2$		1067
1402	$ZrC_{31}H_{34}N_2O_4$	$Zr(CH_2Ph)\{NN\eta^2\text{-}(PhCH_2)NN{=}C(CO_2Et)_2\}(Cp)_2$		1064
1403	$ZrC_{36}H_{34}P_2$	$Zr(CH_2PPh_2)_2(Cp)_2$	140	275

No.	Formula	Structure	Details	Ref.
1404	$Zr_2C_{21}H_{22}Cl_2O$	$\{ZrCl(Cp)_2\}\{\mu\text{-}(O,\eta^2)\text{-}CH_2O\}$		1068
1405	$Zr_2C_{22}H_{26}O$	$\{Zr(Me)(Cp)\}_2O$		427
1406	$Zr_2C_{24}H_{58}P_6$	$Zr_2\{(CH_2)_2PMe_2\}_2(\mu\text{-}CPMe_3)_2$		1069
1407	$Zr_2C_{28}H_{27}Cl$	$Zr_2(Cp)_4(\mu\text{-}Cl)\{\mu\text{-}(\eta^1,\eta^2)\text{-}HCCHPh\}$		1070
1408	$Zr_2C_{28}H_{36}O_2$	$[Zr(Cp)_2\{\mu\text{-}(C,\mu\text{-}O)\text{-}CH_2C(Me)CH_2O\}]_2$		1071
1409	$Zr_2C_{32}H_{30}O_2$	$\{Zr(OPh)(Cp)_2\}_2O$		1072
1410	$Zr_2C_{34}H_{34}O$	$\{Zr(oTol)(Cp)_2\}_2O$		1073
1411	$Zr_2C_{38}H_{42}O$	$\{Zr(oTol)(\eta\text{-}C_5H_4Me)_2\}_2O$		1073
1412	$Zr_3C_{33}H_{36}O_3$	$[(Cp)_2Zr\{\mu\text{-}(O,\eta^2)\text{-}CH_2O\}]_3$		1074

3 Metals Cross Reference Table

A list of mixed metal compound numbers which are listed alphabetically in the Main Table under another metal.

Metal	Compound Numbers
B	84, 85
Co	28, 47, 48, 49, 50, 61, 78, 98, 111, 146, 147, 148, 149, 150, 156, 168, 173, 186, 187
Cr	18, 19, 29, 51, 62, 151, 172, 174, 175, 218, 259
Cu	8, 122, 123
Fe	9, 20, 47, 48, 52, 53, 86, 99, 108, 109, 124, 125, 149, 152, 157, 158, 159, 160, 161, 219, 220, 221, 222, 223, 224, 225, 226, 227, 260, 261, 262, 263, 276, 277, 312, 331, 336
Ge	415, 500, 505
Hg	54, 63, 332, 466, 467, 468
Ir	176, 177, 345, 416
Li	228, 417, 418, 419, 506
Mn	30, 64, 126, 153, 222, 223, 420, 421, 469, 500, 521, 541, 545
Mo	47, 48, 49, 50, 54, 65, 76, 127, 128, 129, 130, 131, 132, 133, 154, 264, 265, 346, 347, 422, 470, 471, 472, 512, 542, 592, 623
Na	21, 22, 162
Nb	692
Ni	70, 85, 150, 155, 178, 189, 224, 227, 585, 740, 758
Np	163
Os	87, 88, 89, 100, 101, 102, 134, 169, 229, 348, 349, 522, 567, 775, 776, 777
Pd	55, 56, 72, 266, 333, 423, 546, 624, 637, 741
Pt	179, 180, 190, 350, 580, 693, 742, 828, 829
Re	135, 136, 188, 694
Rh	10, 11, 73, 90, 103, 181, 182, 183, 184, 191, 193, 230, 313, 351, 424, 425, 473, 568, 581, 695, 696, 830, 934, 959
Ru	8, 15, 50, 57, 66, 74, 91, 92, 93, 94, 95, 96, 98, 104, 105, 106, 107, 262, 267, 268, 269, 270, 278, 279, 356, 426, 427, 428, 542, 543, 569, 778
Sb	429, 796, 979
Sc	137
Se	12, 314, 430, 501, 638, 639, 891, 987
Si	18, 21, 22, 23, 24, 29, 31, 32, 51, 58, 62, 67, 137, 138, 197, 315, 334, 337, 357, 431, 474, 475, 476, 477, 478, 479, 510, 550, 570, 586, 587, 588, 589, 593, 594, 600, 625, 626, 640, 743, 744, 752, 764, 779, 831, 832, 833, 834, 935, 980, 988, 989, 990, 1021, 1089, 1112
Sn	68, 69, 156, 194, 231, 316, 480, 571, 745, 759, 760, 991, 1090, 1163, 1170, 1209, 1210, 1215, 1232, 1234, 1235
Ta	139
Te	544, 643, 1270, 1281
Th	164, 1211
Ti	627, 697, 1190, 1212
U	75, 165, 507, 508
V	166, 481
W	25, 47, 61, 97, 232, 233, 268, 317, 318, 350, 352, 432, 433, 434, 435, 482, 483, 484, 502, 572, 797, 835, 836, 837, 838, 839, 840, 901, 936, 937, 938, 939, 940, 941, 942, 943, 960, 961, 1022, 1213, 1282, 1314, 1315
Y	33
Zn	503, 787, 1292, 1367
Zr	140, 167, 319, 320, 1091, 1221, 1222, 1236, 1237, 1368

REFERENCES

1 W.C.Ho and T.C.W.Mak, J. Organomet. Chem., 1983, 241, 131.
2 J.R.DeMember, H.F.Evans, F.A.Wallace, and P.A.Tariverdian, J. Am. Chem. Soc., 1983, 105, 5647.
3 T.C.W.Mak, W.C.Ho, and N.Z.Huang, J. Organomet. Chem., 1983, 251, 413.
4 M.I.Bruce, M.L.Williams, B.W.Skelton, and A.H.White, J. Chem. Soc., Dalton Trans., 1983, 799.
5 T.C.W.Mak and W.C.Ho, J. Organomet. Chem., 1983, 243, 233.
6 C.Cohen-Addad, P.Baret, P.Chautemps, and J.-L.Pierre, Acta Crystallogr., Sect. C, 1983, 39, 1346.
7 Y.Yamamoto, K.Aoki, and H.Yamazaki, Inorg. Chim. Acta, 1983, 68, 75.
8 M.J.Freeman, M.Green, A.G.Orpen, I.D.Salter, and F.G.A.Stone, J. Chem. Soc., Chem. Commun., 1983, 1332.
9 G.N.Mott, N.J.Taylor, and A.J.Carty, Organometallics, 1983, 2, 447.
10 N.G.Connelly, A.R.Lucy, J.D.Payne, A.M.R.Galas, and W.E.Geiger, J. Chem. Soc., Dalton Trans., 1983, 1879.
11 B.T.Heaton, L.Strona, S.Martinengo, D.Strumolo, V.G.Albano, and D.Braga, J. Chem. Soc., Dalton Trans., 1983, 2175.
12 H.Hofmann, P.G.Jones, M.Noltemeyer, E.Peymann, W.Pinkert, H.W.Roesky, and G.M.Sheldrick, J. Organomet. Chem., 1983, 249, 97.
13 A.Guitard, A.Mari, A.L.Beauchamp, Y.Dartiguenave, and M.Dartiguenave, Inorg. Chem., 1983, 22, 1603.
14 T.C.W.Mak, J. Organomet. Chem., 1983, 246, 331.
15 S.Gambarotta, C.Floriani, A.Chiesi-Villa, and C.Guastini, J. Chem. Soc., Chem. Commun., 1983, 1087.
16 C.Tessier-Youngs, C.Bueno, O.T.Beachley,Jr., and M.R.Churchill, Inorg. Chem., 1983, 22, 1054.
17 C.Tessier-Youngs, C.Bueno, O.T.Beachley,Jr., and M.R.Churchill, Chem. Abs., 1983, 99, 88318n.
18 J.A.Labinger, J.N.Bonfiglio, D.L.Grimmett, S.T.Masuo, E.Shearin, and J.S.Miller, Organometallics, 1983, 2, 733.
19 L.Rösch, G.Altnau, C.Krüger, and Y.-H.Tsay, Z. Naturforsch., B, 1983, 38B, 34.
20 W.Clegg, U.Klingebiel, J.Neemann, and G.M.Sheldrick, J. Organomet. Chem., 1983, 249, 47.
21 M.R.Churchill and H.J.Wasserman, Inorg. Chem., 1983, 22, 41.
22 R.Benn, A.Rufinska, H.Lehmkuhl, E.Janssen, and C.Krüger, Angew. Chem., Int. Ed. Engl., 1983, 22, 779.
23 H.Schmidbaur, S.Lauteschläger, and B.Milewski-Mahrla, Chem. Ber., 1983, 116, 1403.
24 H.F.Klein, K.Ellrich, and K.Ackermann, J. Chem. Soc., Chem. Commun., 1983, 888.
25 C.Tessier-Youngs, W.J.Youngs, O.T.Beachley,Jr., and M.R.Churchill, Organometallics, 1983, 2, 1128.
26 D.L.Grimmett, J.A.Labinger, J.N.Bonfiglio, S.T.Masuo, E.Shearin, and J.S.Miller, Organometallics, 1983, 2, 1325.
27 U.Wannagat, T.Blumemthal, D.J.Brauer, and H.Bürger, J. Organomet. Chem., 1983, 249, 33.
28 D.W.Goebel,Jr., J.L.Hencher, and J.P.Oliver, Organometallics, 1983, 2, 746.
29 E.B.Lobkovsky, G.L.Soloveychik, B.M.Bulychev, A.B.Erofeev, A.I.Gusev, and N.I.Kirillova, J. Organomet. Chem., 1983, 254, 167.
30 J.L.Atwood and M.J.Zaworotko, J. Chem. Soc., Chem. Commun., 1983, 302.
31 J.L.Atwood, D.C.Hrncir, R.D.Priester, and R.D.Rogers, Organometallics, 1983, 2, 985.
32 R.Blom, A.Haaland, and R.Seip, Acta Chem. Scand., Ser. A, 1983, A37, 595.
33 K.Kamiya, W.B.T.Cruse, and O.Kennard, Biochem. J., 1983, 213, 217.
34 J.D.Korp, I.Bernal, L.Avens, and J.L.Mills, J. Crystallogr. Spectrosc. Res., 1983, 13, 263.
35 R.Richter, J.Sieler, A.Richter, J.Heinicke, A.Tzschach, and O.Lindqvist, Z. Anorg. Allg. Chem., 1983, 501, 146.

36 W.T.Pennington, A.W.Cordes, J.C.Graham, and Y.W.Jung, Acta Crystallogr., Sect. C, 1983, 39, 1010.
37 A.N.Sobolev, V.K.Belsky, N.Yu.Chernikova, and F.Yu.Akhmadulina, J. Organomet. Chem., 1983, 244, 129.
38 P.Jutzi, T.Wippermann, C.Krüger, and H.-J.Kraus, Angew. Chem., Int. Ed. Engl., 1983, 22, 250.
39 R.K.Brown, T.J.Bergendahl, J.S.Wood, and J.H.Waters, Inorg. Chim. Acta, 1983, 68, 79.
40 G.Hunter and T.J.R.Weakley, J. Chem. Soc., Dalton Trans., 1983, 1067.
41 M.Moll, H.Behrens, W.Popp, G.Liehr, and W.P.Fehlhammer, Z. Naturforsch., B, 1983, 38B, 1446.
42 W.Urland and E.Warkentin, Z. Naturforsch., B, 1983, 38B, 299.
43 U.Mueller, W.Kafitz, and K.Dehnicke, Z. Anorg. Allg. Chem., 1983, 501, 69.
44 A.C.McDonell, T.W.Hambley, M.R.Snow, and A.G.Wedd, Aust. J. Chem., 1983, 36, 253.
45 B.Krebs, M.Hucke, M.Hein, and A.Schäffer, Z. Naturforsch., B, 1983, 38B, 20.
46 L.Bustos, M.A.Khan, and D.G.Tuck, Can. J. Chem., 1983, 61, 1146.
47 J.M.Ball, P.M.Boorman, K.J.Moynihan, V.D.Patel, J.F.Richardson, D.Collison, and F.E.Mabbs, J. Chem. Soc., Dalton Trans., 1983, 2479.
48 N.M.Boag, C.B.Knobler, and H.D.Kaesz, Angew. Chem., Int. Ed. Engl., 1983, 22, 249.
49 G.Bombieri, F.Benetollo, K.W.Bagnall, M.J.Plews, and D.Brown, J. Chem. Soc., Dalton Trans., 1983, 45.
50 J.W.Kolis, E.M.Holt, and D.F.Shriver, J. Am. Chem. Soc., 1983, 105, 7307.
51 S.Spiliadis, A.A.Pinkerton, and D.Schwarzenbach, Inorg. Chim. Acta, 1983, 75, 115.
52 W.Urland and U.Schwaritz-Schüller, Angew. Chem., Int. Ed. Engl., 1983, 22, 1009.
53 M.Steimann, U.Nagel, R.Grenz, and W.Beck, J. Organomet. Chem., 1983, 247, 171.
54 M.F.Lappert, A.Singh, J.L.Atwood, W.E.Hunter, and H.-M.Zhang, J. Chem. Soc., Chem. Commun., 1983, 69.
55 Z.Xia and Z.Zhang, Huaxue Xuebao, 1983, 41, 577.
56 F.Richter, H.Beurich, M.Müller, N.Gärtner, and H.Vahrenkamp, Chem. Ber., 1983, 116, 3774.
57 M.Müller and H.Vahrenkamp, Chem. Ber., 1983, 116, 2748.
58 K.M.Flynn, B.D.Murray, M.M.Olmstead, and P.P.Power, J. Am. Chem. Soc., 1983, 105, 7460.
59 I.R.Butler, W.R.Cullen, F.W.B.Einstein, S.J.Rettig, and A.J.Willis, Organometallics, 1983, 2, 128.
60 A.Winter, L.Zsolnai, and G.Huttner, J. Organomet. Chem., 1983, 250, 409.
61 G.Salem, B.W.Skelton, A.H.White, and S.B.Wild, J. Chem. Soc., Dalton Trans., 1983, 2117.
62 G.L.Roberts, B.W.Skelton, A.H.White, and S.B.Wild, Aust. J. Chem., 1982, 22, 2193.
63 L.R.Martin, F.W.B.Einstein, and R.K.Pomeroy, Inorg. Chem., 1983, 22, 1959.
64 A.H.Cowley, J.G.Lasch, N.C.Norman, M.Pakulski, and B.R.Whittlesey, J. Chem. Soc., Chem. Commun., 1983, 881.
65 A.J.Ashe III, W.M.Butler, and T.R.Diephouse, Organometallics, 1983, 2, 1005.
66 G.Sennyey, F.Mathey, J.Fischer, and A.Mitschler, Organometallics, 1983, 2, 298.
67 M.Müller and H.Vahrenkamp, J. Organomet. Chem., 1983, 252, 95.
68 A.H.Cowley, J.G.Lasch, N.C,Norman, and M.Pakulski, Angew. Chem., Int. Ed. Engl., 1983, 22, 978.
69 E.C.Alyea, S.A.Dias, G.Ferguson, and P.Y.Siew, Can. J. Chem., 1983, 61, 257.
70 B.F.Hoskins and R.J.Steen, Aust. J. Chem., 1983, 36, 683.
71 G.Ferguson and A.Somogyvari, J. Crystallogr. Spectrosc. Res., 1983, 13, 49.
72 S.R.Hall, B.W.Skelton, and A.H.White, Aust. J. Chem., 1983, 36, 271.
73 A.H.Cowley, J.G.Lasch, N.C.Norman, and M.Pakulski, J. Am. Chem. Soc., 1983,

105, 5506.
74 C.Pelizzi, G.Pelizzi, and P.Tarasconi, J. Chem. Soc., Dalton Trans., 1983, 2689.
75 C.Pelizzi and G.Pelizzi, J. Chem. Soc., Dalton Trans., 1983, 847.
76 F.Cecconi, S.Midollini, and A.Orlandini, J. Chem. Soc., Dalton Trans., 1983, 2263.
77 R.Bohra, H.W.Roesky, J.Lucas, M.Noltemeyer, and G.M.Sheldrick, J. Chem. Soc., Dalton Trans., 1983, 1011.
78 L.R.Gray, D.J.Gulliver, W.Levason, and M.Webster, J. Chem. Soc., Dalton Trans., 1983, 133.
79 J.C.Calabrese, T.Herskovitz, and J.B.Kinney, J. Am. Chem. Soc., 1983, 105, 5914.
80 N.Lugan, J.-M.Savariault, G.Lavigne, and J.-J.Bonnet, J. Crystallogr. Spectrosc. Res., 1983, 13, 389.
81 D.L.Kepert, J.M.Patrick, and A.H.White, Aust. J. Chem., 1983, 36, 469.
82 A.L.Rheingold and M.R.Churchill, J. Organomet. Chem., 1983, 243, 165.
83 A.L.Rheingold and P.J.Sullivan, Organometallics, 1983, 2, 327.
84 A.L.Rheingold and P.J.Sullivan, J. Chem. Soc., Chem. Commun., 1983, 39.
85 J.Ellermann, A.Veit, E.Lindner, and S.Hoehne, J. Organomet. Chem., 1983, 252, 153.
86 R.Usón, A.Laguna, M.Laguna, E.Fernandez, M.D.Villacampa, P.G.Jones, and G.M.Sheldrick, J. Chem. Soc., Dalton Trans., 1983, 1679.
87 M.I.Bruce, J.K.Walton, B.W.Skelton, and A.H.White, J. Chem. Soc., Dalton Trans., 1983, 809.
88 R.Usón, J.Vicente, M.T.Chicote, P.G.Jones, and G.M.Sheldrick, J. Chem. Soc., Dalton Trans., 1983, 1131.
89 A.J.Canty, N.J.Minchin, J.M.Patrick, and A.H.White, Aust. J. Chem., 1983, 36, 1107.
90 G.K.Barker, N.R.Godfrey, M.Green, H.E.Parge, F.G.A.Stone, and A.J.Welch, J. Chem. Soc., Chem. Commun., 1983, 277.
91 M.I.Bruce and B.K.Nicholson, J. Organomet. Chem., 1983, 250, 627.
92 A.F.Hill, W.R.Roper, J.M.Waters, and A.H.Wright, J. Am. Chem. Soc., 1983, 105, 5939.
93 K.Burgess, B.F.G.Johnson, J.Lewis, and P.R.Raithby, J. Chem. Soc., Dalton Trans., 1983, 1661.
94 B.F.G.Johnson, J.Lewis, W.J.H.Nelson, M.D.Vargas, D.Braga, and M.McPartlin, J. Organomet. Chem., 1983, 246, C69.
95 A.Tiripicchio, M.Tiripicchio Camellini, R.Usón, L.A.Oro, and J.A.Cabeza, J. Organomet. Chem., 1983, 244, 165.
96 L.W.Bateman, M.Green, K.A.Mead, R.M.Mills, I.D.Salter, F.G.A.Stone, and P.Woodward, J. Chem. Soc., Dalton Trans., 1983, 2599.
97 B.F.G.Johnson, J.Lewis, J.N.Nicholls, J.Puga, and K.H.Whitmire, J. Chem. Soc., Dalton Trans., 1983, 787.
98 A.G.Cowie, B.F.G.Johnson, J.Lewis, J.N.Nicholls, P.R.Raithby, and M.J.Rosales, J. Chem. Soc., Dalton Trans., 1983, 2311.
99 B.F.G.Johnson, J.Lewis, W.J.H.Nelson, J.Puga, P.R.Raithby, D.Braga, M.McPartlin, and W.Clegg, J. Organomet. Chem., 1983, 243, C13.
100 M.R.Awang, G.A.Carriedo, J.A.K.Howard, K.A.Mead, I.Moore, C.M.Nunn, and F.G.A.Stone, J. Chem. Soc., Chem. Commun., 1983, 964.
101 E.Roland, K.Fischer, and H.Vahrenkamp, Angew. Chem., Int. Ed. Engl., 1983, 22, 326.
102 K.Burgess, B.F.G.Johnson, D.A.Kaner, J.Lewis, P.R.Raithby, and S.N.A.B.Syed-Mustaffa, J. Chem. Soc., Chem. Commun., 1983, 455.
103 B.F.G.Johnson, J.Lewis, W.J.H.Nelson, J.N.Nicholls, J.Puga, P.R.Raithby, M.J.Rosales, M.Shröder, and M.D.Vargas, J. Chem. Soc., Dalton Trans., 1983, 2447.
104 B.F.G.Johnson. J.Lewis, W.J.H.Nelson, P.R.Raithby, and M.D.Vargas, J. Chem. Soc., Chem. Commun., 1983, 608.
105 R.Usón, L.A.Oro, J.Gimeno, M.A.Ciriano, J.A.Cabeza, A.Tiripicchio, and M.Tiripicchio Camellini, J. Chem. Soc., Dalton Trans., 1983, 323.
106 L.J.Farrugia, M.J.Freeman, M.Green, A.G.Orpen, F.G.A.Stone, and I.D.Salter,

J. Organomet. Chem., 1983, 249, 273.
107 M.I.Bruce and B.K.Nicholson, J. Organomet. Chem., 1983, 252, 243.
108 C.E.Briant, K.P.Hall, and D.M.P.Mingos, J. Chem. Soc., Chem. Commun., 1983, 843.
109 S.Gambarotta, C.Floriani, A.Chiesi-Villa, and C.Guastini, J. Chem. Soc., Chem. Commun., 1983, 1304.
110 J.W.A. van der Velden, J.J.Bour, W.P.Bosman, and J.H.Noordik, Inorg. Chem., 1983, 22, 1913.
111 S.J.Rettig and J.Trotter, Can. J. Chem., 1983, 61, 2334.
112 M.Noltemeyer, G.M.Sheldrick, C.Habben, and A.Meller, Z. Naturforsch., B, 1983, 38B, 1182.
113 W.Kliegel, D.Nanninga, S.J.Rettig, and J.Trotter, Can. J. Chem., 1983, 61, 2493.
114 M.M.Olmstead, R.R.Guimerans, and A.L.Balch, Inorg. Chem., 1983, 22, 2473.
115 M.P.Anderson and L.H.Pignolet, Organometallics, 1983, 2, 1246.
116 F.Bachechi, G.Bracher, D.M.Grove, B.Kellenberger, P.S.Pregosin, L.M.Venanzi, and L.Zambonelli, Inorg. Chem., 1983, 22, 1031.
117 D.Carmona, R.Thouvenot, L.M.Venanzi, F.Bachechi, and L.Zambonelli, J. Organomet. Chem., 1983, 250, 589.
118 H.A.Goodwin, E.S.Kucharski, and A.H.White, Aust. J. Chem., 1983, 36, 1115.
119 B.Capelle, M.Dartiguenave, Y.Dartiguenave, and A.L.Beauchamp, J. Am. Chem. Soc., 1983, 105, 4662.
120 C.Mealli, S.Midollini, S.Moneti, and T.A.Albright, Helv. Chim. Acta, 1983, 66, 557.
121 G.Cardaci, G.Bellachioma, and P.Zanazzi, Polyhedron, 1983, 2, 967.
122 A.Albinati, R.Naegeli, A.Togni, and L.M.Venanzi, Organometallics, 1983, 2, 926.
123 C.Bianchini, C.A.Ghilardi, A.Meli, and A.Orlandini, J. Organomet. Chem., 1983, 246, C13.
124 I.A.Teslya, Z.A.Starikova, V.A.Dorokhov, and B.M.Mikhailov, Koord. Khim., 1982, 8, 1546(Engl.Ed. 847).
125 J.S.Thompson, R.L.Harlow, and J.F.Whitney, J. Am. Chem. Soc., 1983, 105, 3522.
126 G.E.Herberich, J.Hengesbach, G.Huttner, A.Frank, and U.Schubert, J. Organomet. Chem., 1983, 246, 141.
127 T.P.Fehlner, C.E.Housecroft, W.R.Scheidt, and K.S.Wong, Organometallics, 1983, 2, 825.
128 K.-B.Shiu, M.D.Curtis, and J.C.Huffman, Organometallics, 1983, 2, 936.
129 M.D.Curtis, K.-B.Shiu, and W.M.Butler, Organometallics, 1983, 2, 1475.
130 T.Desmond, F.J.Lalor, G.Ferguson, and M.Parvez, J. Chem. Soc., Chem. Commun., 1983, 457.
131 T.Desmond, F.J.Lalor, G.Ferguson, B.Ruhl, and M.Parvez, J. Chem. Soc., Chem. Commun., 1983, 55.
132 S.G.Shore, D.-Y.Jan, L.-Y.Hsu, and W.-L.Hsu, J. Am. Chem. Soc., 1983, 105, 5923.
133 A.J.Baskar and C.M.Lukehart, J. Organomet. Chem., 1983, 254, 149.
134 M.F.Lappert, A.Singh, J.L.Atwood, and W.E.Hunter, J. Chem. Soc., Chem. Commun., 1983, 206.
135 W.Clegg, Acta Crystallogr., Sect. C, 1983, 39, 387.
136 D.L.Reger, C.A.Swift, and L.Lebioda, J. Am. Chem. Soc., 1983, 105, 5343.
137 D.L.Reger, M.E.Tarquini, and L.Lebioda, Organometallics, 1983, 2, 1763.
138 J.R.Jennings, R.Snaith, M.M.Mahmoud, S.C.Wallwork, S.J.Bryan, J.Halfpenny, E.A.Petch, and K.Wade, J. Organomet. Chem., 1983, 249, C1.
139 W.Kliegel, D.Nanninga, S.J.Rettig, and J.Trotter, Can. J. Chem., 1983, 61, 2329.
140 K.Niedenzu and H.Nöth, Chem. Ber., 1983, 116, 1132.
141 S.J.Rettig and J.Trotter, Can. J. Chem., 1983, 61, 206.
142 C.J.Cardin, H.E.Parge, and J.W.Wilson, J. Chem. Research, (S), 1983, 93.
143 G.Schmid, U.Höhner, D.Kampmann, D.Zaika, and R.Boese, J. Organomet. Chem., 1983, 256, 225.
144 G.Schmid and R.Boese, Z. Naturforsch., B, 1983, 38B, 485.

145 J.Edwin, M.C.Böhm, N.Chester, D.M.Hoffman, R.Hoffmann, H.Pritzkow, W.Siebert, K.Stumpf, and H.Wadepohl, Organometallics, 1983, 2, 1666.
146 J.Edwin, M.Bochmann, M.C.Böhm, D.E.Brennan, W.E.Geiger, C.Krüger, J.Pebler, H.Pritzkow, W.Siebert, W.Swiridoff, H.Wadepohl, J.Weiss, and U.Zenneck, J. Am. Chem. Soc., 1983, 105, 2582.
147 K.Delpy, D.Schmitz, and P.Paetzold, Chem. Ber., 1983, 116, 2994.
148 G.Schmid, U.Höhner, D.Kampmann, D.Zaika, and R.Boese, Chem. Ber., 1983, 116, 951.
149 G.Müller, D.Neugebauer, W.Geike, F.H.Köhler, J.Pebler, and H.Schmidbaur, Organometallics, 1983, 2, 257.
150 H.Wadepohl, H.Pritzkow, and W.Siebert, Organometallics, 1983, 2, 1899.
151 R.B.Maynard, R.G.Swisher, and R.N.Grimes, Organometallics, 1983, 2, 500.
152 R.G.Swisher, E.Sinn, and R.N.Grimes, Organometallics, 1983, 2, 506.
153 R.P.Micciche and L.G.Sneddon, Organometallics, 1983, 2, 674.
154 N.A.Bell, H.M.M.Shearer, and C.B.Spencer, Acta Crystallogr., Sect. C, 1983, 39, 694.
155 R.Shinomoto, E.Gamp, N.M.Edelstein, D.H.Templeton, and A.Zalkin, Inorg. Chem., 1983, 22, 2351.
156 R.G.Swisher, E.Sinn, G.A.Brewer, and R.N.Grimes, J. Am. Chem. Soc., 1983, 105, 2079.
157 G.J.Zimmerman and L.G.Sneddon, Acta Crystallogr., Sect. C, 1983, 39, 856.
158 J.Bould, J.E.Crook, N.N.Greenwood, and J.D.Kennedy, J. Chem. Soc., Chem. Commun., 1983, 951.
159 M.Yalpani and R.Boese, Chem. Ber., 1983, 116, 3347.
160 R.B.Maynard, Z.-T.Wang, E.Sinn, and R.N.Grimes, Inorg. Chem., 1983, 22, 873.
161 K.A.Solntsev, L.A.Butman, I.Yu.Kuznetsov, N.T.Kuznetsov, B.Stibr, Z.Janousek, and K.Base, Koord. Khim., 1983, 9, 993.
162 J.E.Crook, N.N.Greenwood, J.D.Kennedy, and W.S.McDonald, J. Chem. Soc., Chem. Commun., 1983, 83.
163 N.W.Alcock, J.G.Taylor, and M.G.H.Wallbridge, J. Chem. Soc., Chem. Commun., 1983, 1168.
164 G.K.Barker, M.Green, F.G.A.Stone, W.C.Wolsey, and A.J.Welch, J. Chem. Soc., Dalton Trans., 1983, 2063.
165 L.Zheng, R.T.Baker, C.B.Knobler, J.A.Walker, and M.F.Hawthorne, Inorg. Chem., 1983, 22, 3350.
166 J.A.Walker, C.B.Knobler, and M.F.Hawthorne, J. Am. Chem. Soc., 1983, 105, 3370.
167 J.A.Walker, C.B.Knobler, and M.F.Hawthorne, J. Am. Chem. Soc., 1983, 105, 3368.
168 L.I.Zakharkin, M.G.Meiramov, V.A.Antonovich, A.V.Kazantsev, A.I.Yanovskii, and Yu.T.Struchkov, Zh. Obshch. Khim., 1983, 53, 90(Engl.Ed. 73).
169 R.P.Micciche, J.S.Plotkin, and L.G.Sneddon, Inorg. Chem., 1983, 22, 1765.
170 G.K.Barker, M.P.Garcia, M.Green, F.G.A.Stone, and A.J.Welch, J. Chem. Soc., Chem. Commun., 1983, 137.
171 A.I.Yanovskii, Yu.T.Struchkov, V.N.Kalinin, A.V.Usatov, and L.I.Zakharkin, Koord. Khim., 1982, 8, 1700.
172 R.E.King III, S.B.Miller, C.B.Knobler, and M.F.Hawthorne, Inorg. Chem., 1983, 22, 3548.
173 E.A.Chernyshev, L.K.Knyazeva, Z.V.Belyakova, A.V.Kisin, N.I.Kirillova, A.I.Gusev, and N.V.Alekseev, Zh. Obshch. Khim., 1983, 53, 1433(Engl.Ed. 1289).
174 P.E.Behnken, C.B.Knobler, and M.F.Hawthorne, Angew. Chem., Int. Ed. Engl., 1983, 22, 722.
175 C.Novák, V.Šubrtová, A.Línek, and J.Hašek, Acta Crystallogr., Sect. C, 1983, 39, 1393.
176 J.A.Walker, C.A.O´Con, L.Zheng, C.B.Knobler, and M.F.Hawthorne, J. Chem. Soc., Chem. Commun., 1983, 803.
177 A.I.Yanovskii, Yu.T.Struchkov, V.I.Bregadze, V.Ts.Kampel, M.V.Petriashvili, and N.N.Godovikov, Izv. Akad. Nauk SSSR, Ser. Khim., 1983, 1523(Engl.Ed. 1380).
178 A.N.Sobolev, V.K.Bel´skii, and I.P.Romm, Koord. Khim., 1983, 9, 262.

179 G.Faraglia, R.Graziani, L.Volponi, and U.Casellato, J. Organomet. Chem., 1983, 253, 317.
180 B.Murray, J.Hvoslef, H.Hope, and P.P.Power, Inorg. Chem., 1983, 22, 3421.
181 F.Calderazzo, A.Morvillo, G.Pelizzi, and R.Poli, J. Chem. Soc., Chem. Commun., 1983, 507.
182 P.J.Toscano, T.F.Swider, L.G.Marzilli, N.Bresciani-Pahor, and L.Randaccio, Inorg. Chem., 1983, 22, 3416.
183 E.J.Miller, T.B.Brill, A.L.Rheingold, and W.C.Fultz, J. Am. Chem. Soc., 1983, 105, 7580.
184 A.L.Rheingold, W.C.Fultz, T.B.Brill, and S.J.Landon, J. Crystallogr. Spectrosc. Res., 1983, 13, 317.
185 W.H.Hersh, F.J.Hollander, and R.G.Bergman, J. Am. Chem. Soc., 1983, 105, 5834.
186 K.A.Woode, J.C.J.Bart, M.Calcaterra, and G.Agnès, Organometallics, 1983, 2, 627.
187 R.P.Hughes, D.E.Samkoff, R.E.Davis, and B.B.Laird, Organometallics, 1983, 2, 195.
188 A.I.Yanovskii, G.G.Aleksandrov, Yu.T.Struchkov, I.Ya.Levitin, R.M.Bodnar, and M.E.Vol'pin, Koord. Khim., 1983, 9, 825.
189 M.F.Summers, P.J.Toscano, N.Bresciani-Pahor, G.Nardin, L.Randaccio, and L.G.Marzilli, J. Am. Chem. Soc., 1983, 105, 6259.
190 N.Dudeney, J.C.Green, P.Grebenik, and O.N.Kirchner, J. Organomet. Chem., 1983, 252, 221.
191 R.L.Harlow, R.J.McKinney, and J.F.Whitney, Organometallics, 1983, 2, 1839.
192 H.-F.Klein, J.Gross, R.Hammer, and U.Schubert, Chem. Ber., 1983, 116, 1441.
193 A.W.Coleman, P.B.Hitchcock, M.F.Lappert, R.K.Maskell, and J.H.Müller, J. Organomet. Chem., 1983, 250, C9.
194 N.Bresciani-Pahor, L.Randaccio, M.Summers, and P.J.Toscano, Inorg. Chim. Acta, 1983, 68, 69.
195 Y.Kushi, M.Kuramoto, Y.Hayashi, T.Yamamoto, A.Yamamoto, and S.Komiya, J. Chem. Soc., Chem. Commun., 1983, 1033.
196 A.A.Pasynskii, I.L.Eremenko, G.Sh.Gasanov, O.G.Ellert, V.M.Novotortsev, V.T.Kalinnikov, Yu.T.Struchkov, and V.E.Shklover, Izv. Akad. Nauk SSSR, Ser. Khim., 1983, 1446(Engl.Ed. 1315).
197 M.King, E.M.Holt, P.Radnia, and J.S.McKennis, Organometallics, 1982, 1, 1718.
198 S.Aime, D.Osella, L.Milone, and A.Tiripicchio, Polyhedron, 1983, 2, 77.
199 H.Yamazaki, K.Yasufuku, and Y.Wakatsuki, Organometallics, 1983, 2, 726.
200 M.Müller and H.Vahrenkamp, Chem. Ber., 1983, 116, 2322.
201 M.Müller and H.Vahrenkamp, Chem. Ber., 1983, 116, 2765.
202 H.Vahrenkamp, E.J.Wucherer, and D.Wolters, Chem. Ber., 1983, 116, 1219.
203 H.F.Klein, H.Witty, and U.Schubert, J. Chem. Soc., Chem. Commun., 1983, 231.
204 M.R.Churchill and C.Bueno, Inorg. Chem., 1983, 22, 1510.
205 M.Green, D.R.Hankey, J.A.K.Howard, P.Louca, and F.G.A.Stone, J. Chem. Soc., Chem. Commun., 1983, 757.
206 H.-F.Klein, K.Ellrich, D.Neugebauer, O.Orama, and K.Krüger, Z. Naturforsch., B, 1983, 38B, 303.
207 E.O.Fischer, P.Friedrich, T.L.Lindner, D.Neugebauer, F.R.Kreissl, W.Uedelhoven, N.Q.Dao, and G.Huttner, J. Organomet. Chem., 1983, 247, 239.
208 J.A.Abad, L.W.Bateman, J.C.Jeffery, K.A.Mead, H.Razay, F.G.A.Stone, and P.Woodward, J. Chem. Soc., Dalton Trans., 1983, 2075.
209 P.C.Leung and P.Coppens, Acta Crystallogr., Sect. B, 1983, B39, 532.
210 M.Arewgoda, B.H.Robinson, and J.Simpson, J. Am. Chem. Soc., 1983, 105, 1893.
211 D.Gregson and J.A.K.Howard, Acta Crystallogr., Sect. C, 1983, 39, 1024.
212 J.A.King,Jr. and K.P.C.Vollhardt, Organometallics, 1983, 2, 684.
213 K.H.Theopold and R.G.Bergman, J. Am. Chem. Soc., 1983, 105, 464.
214 K.H.Theopold and R.G.Bergman, Organometallics, 1982, 1, 1571.
215 K.D.Grande, A.J.Kunin, L.S.Stuhl, and B.M.Foxman, Inorg. Chem., 1983, 22, 1791.
216 I.T.Horváth, G.Pályi, L.Markó, and G.D.Andreetti, Inorg. Chem., 1983, 22, 1049.

217 J.A.King,Jr. and K.P.C.Vollhardt, J. Am. Chem. Soc., 1983, 105, 4846.
218 B.K.Teo, V.Bakirtzis, and P.A.Snyder-Robinson, J. Am. Chem. Soc., 1983, 105, 6330.
219 R.Zolk and H.Werner, J. Organomet. Chem., 1983, 252, C53.
220 R.A.Jones, A.L.Stuart, J.L.Atwood, and W.E.Hunter, Organometallics, 1983, 2, 1437.
221 W.A.Herrmann, C.Bauer, J.M.Huggins, H.Pfisterer, and M.L.Ziegler, J. Organomet. Chem., 1983, 258, 81.
222 A.D.Harley, R.R.Whittle, and G.L.Geoffroy, Organometallics, 1983, 2, 383.
223 P.Braunstein, I.Pruskil, G.Predieri, and A.Tiripicchio, J. Organomet. Chem., 1983, 247, 227.
224 A.D.Harley, R.R.Whittle, and G.L.Geoffroy, Organometallics, 1983, 2, 60.
225 K.Jonas, G.Koepe, L.Schieferstein, R.Mynott, C.Krüger, and Y.-H.Tsay, Angew. Chem., Int. Ed. Engl., 1983, 22, 620.
226 H.Lang, L.Zsolnai, and G.Huttner, Angew. Chem., Int. Ed. Engl., 1983, 22, 976.
227 D.E.Fjare, D.G.Keyes, and W.J.Gladfelter, J. Organomet. Chem., 1983, 250, 383.
228 B.H.Freeland, N.C.Payne, M.A.Stalteri, and H. van Leeuwen, Acta Crystallogr., Sect. C, 1983, 39, 1533.
229 E.Roland and H.Vahrenkamp, Organometallics, 1983, 2, 1048.
230 C.Mahe, H.Patin, J.-Y. Le Marouille, and A.Benolt, Organometallics, 1983, 2, 1051.
231 M.D.Curtis and P.D.Williams, Inorg. Chem., 1983, 22, 2661.
232 P.Braunstein, J.-M.Jud, Y.Dusausoy, and J.Fischer, Organometallics, 1983, 2, 180.
233 P.Braunstein, J.Rose, and O.Bars, J. Organomet. Chem., 1983, 252, C101.
234 E.Roland and H.Vahrenkamp, Organometallics, 1983, 2, 183.
235 A.Benolt, A.Darchen, J.-Y.LeMarouille, C.Mahé, and H.Patin, Organometallics, 1983, 2, 555.
236 J.Fortune, A.R.Manning, and F.S.Stephens, J. Chem. Soc., Chem. Commun., 1983, 1071.
237 K.P.C.Vollhardt and E.C.Walborsky, J. Am. Chem. Soc., 1983, 105, 5507.
238 R.B.A.Pardy, G.W.Smith, and M.E.Vickers, J. Organomet. Chem., 1983, 252, 341.
239 P.Bradamante, P.Pino, A.Stefani, G.Fachinetti, and P.F.Zanazzi, J. Organomet. Chem., 1983, 251, C47.
240 P.Brun, G.M.Dawkins, M.Green, R.M.Mills, J.-Y.Salaün, F.G.A.Stone, and P.Woodward, J. Chem. Soc., Dalton Trans., 1983, 1357.
241 S.Aime, D.Osella, L.Milone, A.M.Manotti Lanfredi, and A.Tiripicchio, Inorg. Chim. Acta, 1983, 71, 141.
242 M.Hidai, M.Orisaku, M.Ue, Y.Koyasu, T.Kodama, and Y.Uchida, Organometallics, 1983, 2, 292.
243 S.Holand, F.Mathey, J.Fischer, and A.Mitschler, Organometallics, 1983, 2, 1234.
244 G.Ciani and A.Sironi, J. Organomet. Chem., 1983, 241, 385.
245 G.Gervasio, R.Rossetti, P.L.Stanghellini, and G.Bor, J. Chem. Soc., Dalton Trans., 1983, 1613.
246 C.Burschka, F.E.Baumann, and W.A.Schenk, Z. Anorg. Allg. Chem., 1983, 502, 191.
247 D.J.Darensbourg and A.Rokicki, Organometallics, 1982, 1, 1685.
248 C.Krüger, R.Goddard, and K.H.Claus, Z. Naturforsch., B, 1983, 38B, 1431.
249 H.W.Roesky, R.Emmert, W.Isenberg, M.Schmidt, and G.M.Sheldrick, J. Chem. Soc., Dalton Trans., 1983, 183.
250 B.Lubke, F.Edelmann, and U.Behrens, Chem. Ber., 1983, 116, 11.
251 K.H.Dötz, W.Kuhn, and K.Ackermann, Z. Naturforsch., B, 1983, 38B, 1351.
252 W.S.Sheldrick, H.W.Roesky, and D.Amirzadeh-Asl, Phosphorus and Sulphur, 1982, 14, 161.
253 J.-C.Boutonnet, J.Levisalles, E.Rose, G.Precigoux, C.Courseille, and N.Platzer, J. Organomet. Chem., 1983, 255, 317.
254 R.A.Howie, G.Izquierdo, and G.P.McQuillan, Inorg. Chim. Acta, 1983, 72, 165.

255 K.Ackermann, P.Hofmann, F.H.Köhler, H.Kratzer, H.Krist, K.Öfele, and H.R.Schmidt, Z. Naturforsch., B, 1983, 38B, 1313.
256 W.Clegg, U.Klingebiel, and G.M.Sheldrick, Z. Naturforsch., B, 1983, 38B, 260.
257 G.S.Girolami, J.E.Salt, G.Wilkinson, M.Thornton-Pett, and M.B.Hursthouse, J. Am. Chem. Soc., 1983, 105, 5954.
258 G.W.Harris, M.O.Albers, J.C.A.Boeyens, and N.J.Coville, Organometallics, 1983, 2, 609.
259 R.Mercier, J.Douglade, and J.Vebrel, Acta Crystallogr., Sect. C, 1983, 39, 1177.
260 H.Schwemlein, L.Zsolnai, G.Huttner, and H.H.Brintzinger, J. Organomet. Chem., 1983, 256, 285.
261 L.Weber and R.Boese, Angew. Chem., Int. Ed. Engl., 1983, 22, 498.
262 L.Weber and R.Boese, Chem. Ber., 1983, 116, 197.
263 M.J.McGlinchey, J.L.Fletcher, B.G.Sayer, P.Bougeard, R.Faggiani, C.J.Lock, A.D.Bain, C.Rodger, E.P.Kündig, D.Astruc, J.-R.Hamon, P.LeMaux, S.Top, and G.Jaouen, J. Chem. Soc., Chem. Commun., 1983, 634.
264 R.Faggiani, N.Hao, C.J.L.Lock, B.G.Sayer, and M.J.McGlinchey, Organometallics, 1983, 2, 96.
265 A.Hengefeld, J.Kopf, and D.Rehder, Organometallics, 1983, 2, 114.
266 S.I.Khan and R.Bau, Organometallics, 1983, 2, 1896.
267 G.R.Clark and P.W.Clark, J. Organomet. Chem., 1983, 255, 205.
268 C.J.Cardin, D.J.Cardin, J.M.Kelly, R.J.Norton, A.Roy, B.J.Hathaway, and T.J.King, J. Chem. Soc., Dalton Trans., 1983, 671.
269 A.Winter, I.Jibril, and G.Huttner, J. Organomet. Chem., 1983, 247, 259.
270 R.D.Barr, M.Green, K.Marsden, F.G.A.Stone, and P.Woodward, J. Chem. Soc., Dalton Trans., 1983, 507.
271 D.Neugebauer and U.Schubert, J. Organomet. Chem., 1983, 256, 43.
272 K.M.Flynn, H.Hope, B.D.Murray, M.M.Olmstead, and P.P.Power, J. Am. Chem. Soc., 1983, 105, 7750.
273 A.Tzschach, K.Jurkschat, M.Scheer, J.Meunier-Piret, and M. van Meerssche, J. Organomet. Chem., 1983, 259, 165.
274 G.L.Hillhouse, B.L.Haymore, S.A.Bistram, and W.A.Herrmann, Inorg. Chem., 1983, 22, 314.
275 N.E.Schore, S.J.Young, and M.M.Olmstead, Organometallics, 1983, 2, 1769.
276 J.Roziere, P.Teulon, and M.D.Grillone, Inorg. Chem., 1983, 22, 557.
277 F.H.Köhler, R.de Cao, K.Ackermann, and J.Sedlmair, Z. Naturforsch., B, 1983, 38B, 1406.
278 L.Yoong Goh, T.W.Hambley, and G.B.Robertson, J. Chem. Soc., Chem. Commun., 1983, 1458.
279 C.Elschenbroich, J.Heck, W.Massa, and R.Schmidt, Angew. Chem., Int. Ed. Engl., 1983, 22, 331.
280 J.Borm, L.Zsolnai, and G.Huttner, Angew. Chem., Int. Ed. Engl., 1983, 22, 977.
281 E.Halwax and H.Voellenkle, Monatsh. Chem., 1983, 114, 687.
282 B.Bovio, J. Organomet. Chem., 1983, 252, 71.
283 I.L.Eremenko, A.A.Pasynskii, Yu.V.Rakitin, O.G.Ellert, V.M.Novotortsev, V.T.Kalinnikov, V.E.Shklover, and Yu.T.Struchkov, J. Organomet. Chem., 1983, 256, 291.
284 L.G.Kuz′mina and Yu.T.Struchkov, Koord. Khim., 1982, 8, 1542.
285 R.Bender, P.Braunstein, J.-M.Jud, and Y.Dusausoy, Inorg. Chem., 1983, 22, 3394.
286 I.L.Eremenko, A.A.Pasynskii, O.G.Ellert, O.G.Volkov, A.S.Antsyshkina, M.A.Porai-Koshits, Yu.T.Struchkov, and L.M.Dikareva, Izv. Akad. Nauk SSSR, Ser. Khim., 1983, 1444(Engl.Ed. 1313).
287 M.Hoefler, K.-F.Tebbe, H.Veit, and N.E.Weiler, J. Am. Chem. Soc., 1983, 105, 6338.
288 A.A.Pasynskii, I.L.Eremenko, Yu.V.Rakitin, V.M.Novotortsev, O.G.Ellert, V.T.Kalinnikov, V.E.Shklover, S.V.Lindeman, T.Kh.Kurbanov, and G.Sh.Gasanov, J. Organomet. Chem., 1983, 248, 309.
289 I.L.Eremenko, A.A.Pasynskii, G.Sh.Gasanov, Sh.A.Bagirov, O.G.Ellert,

V.M.Novotortsev, V.T.Kalinnikov, Yu.T.Struchkov, and V.E.Shklover, Izv. Akad. Nauk SSSR, Ser. Khim., 1983, 1445(Engl.Ed. 1314).
290 G.Doyle, K.A.Eriksen, and D. van Engen, Inorg. Chem., 1983, 22, 2892.
291 J.S.Thompson and J.F.Whitney, J. Am. Chem. Soc., 1983, 105, 5488.
292 P.Leoni, M.Pasquali, and C.A.Ghilardi, J. Chem. Soc., Chem. Commun., 1983, 240.
293 H.Schmidbaur, C.E.Zybill, G.Müller, and C.Krüger, Angew. Chem., Int. Ed. Engl., 1983, 22, 729.
294 A.T.Hutton, P.G.Pringle, and B.L.Shaw, Organometallics, 1983, 2, 1889.
295 H.Werner, J.Roll, K.Linse, and M.L.Ziegler, Angew. Chem., Int. Ed. Engl., 1983, 22, 982.
296 N.C.Payne, N.Okura, and S.Otsuka, J. Am. Chem. Soc., 1983, 105, 245.
297 D.Braga, K.Henrick, B.F.G.Johnson, J.Lewis, M.McPartlin, W.J.H.Nelson, A.Sironi, and M.D.Vargas, J. Chem. Soc., Chem. Commun., 1983, 1131.
298 G.A.Carriedo, J.A.K.Howard, and F.G.A.Stone, J. Organomet. Chem., 1983, 250, C28.
299 L.Carlton, W.E.Lindsell, K.J.McCullough, and P.N.Preston, J. Chem. Soc., Chem. Commun., 1983, 216.
300 P.Yu.Zavalii, V.S.Fundamenskii, M.G.Mys´kiv, and E.I.Gladyshevskii, Kristallografiya, 1983, 28, 662.
301 G.Doyle, K.A.Eriksen, M.Modrick and G.Ansell, Organometallics, 1982, 1, 1613.
302 M.Pasquali, P.Fiaschi, C.Floriani, and A.Gaetani-Manfredotti, J. Chem. Soc., Chem. Commun., 1983, 197.
303 R.I.Papasergio, C.L.Raston, and A.H.White, J. Chem. Soc., Chem. Commun., 1983, 1419.
304 R.L.Geerts, J.C.Huffman, K.Folting, T.H.Lemmen, and K.G.Caulton, J. Am. Chem. Soc., 1983, 105, 3503.
305 M.Pasquali, P.Fiaschi, C.Floriani, and P.F.Zanazzi, J. Chem. Soc., Chem. Commun., 1983, 613.
306 S.Gambarotta, C.Floriani, A.Chiesi-Villa, and C.Guastini, J. Chem. Soc., Chem. Commun., 1983, 1156.
307 T.C.W.Mak, H.N.C.Wong, K.H.Sze, and L.Book, J. Organomet. Chem., 1983, 255, 123.
308 G.F.Kokoszka, J.Baranowski, C.Goldstein, J.Orsini, A.D.Mighell, V.L.Himes, and A.R.Siedle, J. Am. Chem. Soc., 1983, 105, 5627.
309 K.H.Pannell, Y.-S.Chen, K.Belknap, C.C.Wu, I.Bernal, M.W.Creswick, and H.N.Huang, Inorg. Chem., 1983, 22, 418.
310 G.W.Harris, J.C.A.Boeyens, and N.J.Coville, Acta Crystallogr., Sect. C, 1983, 39, 1180.
311 R.A.Nissan, M.S.Connolly, M.G.L.Mirabelli, R.R.Whittle, and H.R.Allcock, J. Chem. Soc., Chem. Commun., 1983, 822.
312 M.Yu.Antipin, G.G.Aleksandrov, Yu.T.Struchkov, Yu.A.Belousov, V.N.Babin, and N.S.Kochetkova, Inorg. Chim. Acta, 1983, 68, 229.
313 B.W.Sullivan and B.M.Foxman, Organometallics, 1983, 2, 187.
314 A.L.Rheingold, A.D.Uhler, and A.G.Landers, Inorg. Chem., 1983, 22, 3255.
315 I.Noda, H.Yasuda, and A.Nakamura, J. Organomet. Chem., 1983, 250, 447.
316 K.H.Pannell, A.J.Mayr, and D.VanDerveer, Organometallics, 1983, 2, 560.
317 X.Lin, J.Huang, and J.Huang, Ziran Zazhi, 1982, 5, 949.
318 T.W.Leung, G.G.Christoph, and A.Wojcicki, Inorg. Chim. Acta, 1983, 76, L281.
319 M.B.Humphrey, W.M.Lamanna, M.Brookhart, and G.R.Husk, Inorg. Chem., 1983, 22, 3355.
320 J.Calabrese, S.D.Ittel, H.S.Choi, S.G.Davis, and D.A.Sweigart, Organometallics, 1983, 2, 226.
321 R.Battaglia, C.C.Frazier III, H.Kisch, and C.Krüger, Z. Naturforsch., B, 1983, 38B, 648.
322 Y.S.Yu and R.J.Angelici, Organometallics, 1983, 2, 1018.
323 Z.Fu, K.Pan, J.Lu, G.Zhang, and H.Zhu, Jiegou Huaxue, 1982, 1, 57.
324 Y.Zhang, Z.Cai, Z.Chen, K.Pan, J.Lu, G.Zhang, and H.Zhu, Jiegou Huaxue, 1982, 1, 45.
325 V.B.Rybakov, L.A.Aslanov, and S.A.Eremin, Zh. Strukt. Khim., 1983, 24(3),

183(Engl.Ed. 495).
326 B.F.Anderson and G.B.Robertson, Acta Crystallogr., Sect. C, 1983, 39, 428.
327 M.Franck-Neumann, M.P.Heitz, D.Martina, and A.de Cian, Tetrahedron Letters, 1983, 24, 1611.
328 M.E.Wright, Organometallics, 1983, 2, 558.
329 D.Sellmann, G.Lanzrath, G.Huttner, L.Zsolnai, C.Krüger, and K.H.Claus, Z. Naturforsch., B, 1983, 38B, 961.
330 N.A.Lewis, B.Patel, and P.S.White, J. Chem. Soc., Dalton Trans., 1983, 1367.
331 E.J.S.Vichi, P.R.Raithby, and M.McPartlin, J. Organomet. Chem., 1983, 256, 111.
332 M.Alini, R.Roulet, and G.Chapuis, Helv. Chim. Acta, 1983, 66, 1624.
333 A.S.Batsanov and Yu.T.Struchkov, J. Organomet. Chem., 1983, 248, 101.
334 K.-J.Jens, T.Valeri, and E.Weiss, Chem. Ber., 1983, 116, 2872.
335 P.E.Riley and R.E.Davis, Organometallics, 1983, 2, 286.
336 J.C.Dewan, Organometallics, 1983, 2, 83.
337 H.Brunner, B.Hammer, I.Bernal, and M.Draux, Organometallics, 1983, 2, 1595.
338 W.E.McEwan, C.E.Sullivan, and R.O.Day, Organometallics, 1983, 2, 420.
339 G.J.Baird, J.A.Bandy, S.G.Davies, and K.Prout, J. Chem. Soc., Chem. Commun., 1983, 1202.
340 M.W.Kokkes, D.J.Stufkens, and A.Oskam, J. Chem. Soc., Dalton Trans., 1983, 439.
341 R.L.Keiter, A.L.Rheingold, J.J.Hamerski, and C.K.Castle, Organometallics, 1983, 2, 1635.
342 D.L.Reger, K.A.Belmore, J.L.Atwood, and W.E.Hunter, J. Am. Chem. Soc., 1983, 105, 5710.
343 D.L.Reger, K.A.Belmore, E.Mintz, N.G.Charles, E.A.H.Griffith, and E.L.Amma, Organometallics, 1983, 2, 101.
344 M.Cygler, F.R.Ahmed, A.Forgues, and J.L.A.Roustan, Inorg. Chem., 1983, 22, 1026.
345 P.Conway, S.M.Grant, A.R.Manning, and F.S.Stephens, Inorg. Chem., 1983, 22, 3714.
346 A.H.Cowley, J.E.Kilduff, J.G.Lasch, N.C.Norman, M.Pakulski, F.Ando, and T.C.Wright, J. Am. Chem. Soc., 1983, 105, 7751.
347 L.K.K.LiShingMan, J.G.A.Reuvers, J.Takats, and G.Deganello, Organometallics, 1983, 2, 28.
348 D.A.Roberts, G.R.Steinmetz, M.J.Breen, P.M.Shulman, E.D.Morrison, M.R.Duttera, C.W.DeBrosse, R.R.Whittle and G.L.Geoffroy, Organometallics, 1983, 2, 846.
349 T.A.Bazhenova, R.M.Lobkovskaya, R.P.Shibaeva, A.K.Shilova, M.Gruselle, G.Leny, and E.Deschamps, J. Organomet. Chem., 1983, 244, 375.
350 T.A.Bazhenova, R.M.Lobkovskaya, R.P.Shibaeva, A.E.Shilov, A.K.Shilova, M.Gruselle, G.Leny, and B.Tchoubar, J. Organomet. Chem., 1983, 244, 265.
351 S.M.Gadol and R.E.Davis, Organometallics, 1982, 1, 1607.
352 V.B.Rybakov, L.A.Aslanov, V.M.Ionov, and S.A.Eremin, Zh. Strukt. Khim., 1983, 24(2), 74(Engl.Ed. 230).
353 K.Sünkel, U.Nagel, and W.Beck, J. Organomet. Chem., 1983, 251, 227.
354 D.Seyferth, B.W.Hames, T.G.Rucker, M.Cowie, and R.S.Dickson, Organometallics, 1983, 2, 472.
355 W.R.Cullen, T.-J.Kim, F.W.B.Einstein, and T.Jones, Organometallics, 1983, 2, 714.
356 H.R.Allcock, L.J.Wagner, and M.L.Levin, J. Am. Chem. Soc., 1983, 105, 1321.
357 D.F.Jones, P.H.Dixneuf, A.Benoit, and J.-Y.Le Marouille, Inorg. Chem., 1983, 22, 29.
358 H.-A.Kaul, D.Greissinger, W.Malisch, H.-P.Klein, and U.Thewalt, Angew. Chem., Int. Ed. Engl., 1983, 22, 60.
359 P.Deplano, E.F.Trogu, F.Bigoli, E.Leporati, M.A.Pellinghelli, D.L.Perry, R.J.Saxton, and L.J.Wilson, J. Chem. Soc., Dalton Trans., 1983, 25.
360 L.Párkányi, K.H.Pannell, and C.Hernandez, J. Organomet. Chem., 1983, 252, 127.
361 V.B.Rybakov, L.A.Aslanov, V.M.Ionov, and S.A.Eremin, Zh. Strukt. Khim., 1983, 24(3), 100(Engl.Ed. 414).

362 L.Busetto, J.C.Jeffery, R.M.Mills, F.G.A.Stone, M.J.Went, and P.Woodward, J. Chem. Soc., Dalton Trans., 1983, 101.
363 D.Seyferth, G.B.Womack, M.Cowie, and B.W.Hames, Organometallics, 1983, 2, 1696.
364 D.Seyferth, G.B.Womack, Li.-C.Song, M.Cowie, and B.W.Hames, Organometallics, 1983, 2, 928.
365 R.L.De, J. Organomet. Chem., 1983, 243, 331.
366 B.Deppisch and H.Schäffer, Acta Crystallogr., Sect. C, 1983, 39, 975.
367 M.A.El-Hinnawi, A.A.Aruffo, B.D.Santarsiero, D.R.McAlister, and V.Schomaker, Inorg. Chem., 1983, 22, 1585.
368 H.Vahrenkamp and D.Wolters, Angew. Chem., Int. Ed. Engl., 1983, 22, 154.
369 D.R.Wilson, R.D.Ernst, and T.H.Cymbaluk, Organometallics, 1983, 2, 1220.
370 A.G.Orpen, J. Chem. Soc., Dalton Trans., 1983, 1427
371 H.Umland, F.Edelmann, D.Wormsbacher, and U.Behrens, Angew. Chem., Int. Ed. Engl., 1983, 22, 153.
372 A.Hafner, J.H.Bieri, R.Prewo, W. von Philipsborn, and A.Salzer, Angew. Chem., Int. Ed. Engl., 1983, 22, 713.
373 M.Hillman and Å.Kvick, Organometallics, 1983, 2, 1780.
374 R.Aumann, H.Heinen, G.Henkel, M.Dartmann, and B.Krebs, Z. Naturforsch., B, 1983, 38B, 1325.
375 Z.Goldschmidt, Y.Bakal, D.Hoffer, A.Eisenstadt, and P.F.Lindley, J. Organomet. Chem., 1983, 258, 53.
376 E.K.Lhadi, H.Patin, A.Benoit, J.-Y. Le Marouille, and A.Darchen, J. Organomet. Chem., 1983, 259, 321.
377 G.M.Dawkins, M.Green, J.C.Jeffery, C,Sambale, and F.G.A.Stone, J. Chem. Soc., Dalton Trans., 1983, 499.
378 Y.-F.Yu, J.Gallucci, and A.Wojcicki, J. Am. Chem. Soc., 1983, 105, 4826.
379 G.N.Mott, R.Granby, S.A.MacLaughlin, N.J.Taylor, and A.J.Carty, Organometallics, 1983, 2, 189.
380 V.W.Day, M.R.Thompson, G.O.Nelson, and M.E.Wright, Organometallics, 1983, 2, 494.
381 B.A.Sosinsky, R.G.Shong, B.J.Fitzgerald, N.Norem, and C.O'Rourke, Inorg. Chem., 1983, 22, 3124.
382 T.C.W.Mak, L.Book, C.Chieh, M.K.Gallagher, L.-C.Song, and D.Seyferth, Inorg. Chim. Acta, 1983, 73, 159.
383 H.Lang, L.Zsolnai, and G.Huttner, Angew. Chem., Int. Ed. Engl., 1983, 22, 976.
384 S.Lu, N.Okura, T.Yoshida, and S.Otsuka, J. Am. Chem. Soc., 1983, 105, 7470.
385 B.Cowans, J.Noordik, and M.R.DuBois, Organometallics, 1983, 2, 931.
386 P.D.Williams, M.D.Curtis, D.N.Duffy, and W.M.Butler, Organometallics, 1983, 2, 165.
387 R.J.Haines, N.D.C.T.Steen, and R.B.English, J. Chem. Soc., Dalton Trans., 1983, 1607.
388 E.A.Chernyshev, O.V.Kuz'min, A.V.Lebedev, A.I.Gusev, M.G.Los', N.V.Alekseev, N.S.Nametkin, V.D.Tyurin, A.M.Krapivin, N.A.Kubasova, and V.G.Zaikin, J. Organomet. Chem., 1983, 252, 143.
389 M.E.Wright, T.M.Mezza, G.O.Nelson, N.R.Armstrong, V.W.Day, and M.R.Thompson, Organometallics, 1983, 2, 1711.
390 K.M.Flynn, M.M.Olmstead, and P.P.Power, J. Am. Chem. Soc., 1983, 105, 2085.
391 A.Clearfield, C.J.Simmons, H.P.Withers,Jr., and D.Seyferth, Inorg. Chim. Acta, 1983, 75, 139.
392 T.B.Rauchfuss, T.D.Weatherill, S.R.Wilson, and J.P.Zebrowski, J. Am. Chem. Soc., 1983, 105, 6508.
393 L.Busetto, M.Monari, A.Palazzi, V.Albano, and F.Demartin, J. Chem. Soc., Dalton Trans., 1983, 1849.
394 M.Green, K.Marsden, I.D.Salter, F.G.A.Stone, and P.Woodward, J. Chem. Soc., Chem. Commun., 1983, 446.
395 A.Ceriotti, L.Resconi, F.Demartin, G.Longoni, M.Manassero, and M.Sansoni, J. Organomet. Chem., 1983, 249, C35.
396 E.Iiskola, T.A.Pakkanen, T.T.Pakkanen, and T.Venäläinen, Acta Chem. Scand.,

Ser. A, 1983, A37, 125.
397 W.-K.Wong, K.W.Chiu, G.Wilkinson, A.M.R.Galas, M.Thornton-Pett, and M.B.Hursthouse, J. Chem. Soc., Dalton Trans., 1983, 1557.
398 M.Lourdichi, R.Pince, F.Dahan, and R.Mathieu, Organometallics, 1983, 2, 1417.
399 M.I.Bruce, T.W.Hambley, and B.K.Nicholson, J. Chem. Soc., Dalton Trans., 1983, 2385.
400 E.Lindner, G.A.Weiss, W.Hiller, and R.Fawzi, J. Organomet. Chem., 1983, 255, 245.
401 K.H.Pannell, A.J.Mayr, and D.VanDerveer, J. Am. Chem. Soc., 1983, 105, 6186.
402 N.S.Nametkin, V.D.Tyurin, A.I.Nekhaev, Yu.P.Sobolev, M.G.Kondrat'eva, A.S.Batsanov, and Yu.T.Struchkov, J. Organomet. Chem., 1983, 243, 323.
403 J.K.Kouba, E.L.Muetterties, M.R.Thompson, and V.W.Day, Organometallics, 1983, 2, 1065.
404 D.Melzer and E.Weiss, J. Organomet. Chem., 1983, 255, 335.
405 R.B.Petersen, J.M.Ragosta, G.E.Whitwell II, and J.M.Burlitch, Inorg. Chem., 1983, 22, 3407.
406 A.S.Batsanov, L.V.Rybin, M.I.Rybinskaya, Yu.T.Struchkov, I.M.Salimgareeva, and N.G.Bogatova, J. Organomet. Chem., 1983, 249, 319.
407 I.R.Butler, W.R.Cullen, J.Reglinski, and S.J.Rettig, J. Organomet. Chem., 1983, 249, 183.
408 R.E.Cramer, K.T.Higa, S.L.Pruskin, and J.W.Gilje, J. Am. Chem. Soc., 1983, 105, 6749.
409 J.S.Bradley, E.W.Hill, G.B.Ansell, and M.A.Modrick, Organometallics, 1982, 1, 1634.
410 R.B.Hallock, O.T.Beachley,Jr., Y.-J.Li, W.M.Sanders, M.R.Churchill, W.E.Hunter, and J.L.Atwood, Inorg. Chem., 1983, 22, 3683.
411 H.Schmidbaur, U.Thewalt, and T.Zafiropoulos, Organometallics, 1983, 2, 1550.
412 K.R.Breakall, S.J.Rettig, A.Storr, and J.Trotter, Can. J. Chem., 1983, 61, 1659.
413 H.Puff, S.Franken, W.Schuh, and W.Schwab, J. Organomet. Chem., 1983, 254, 33.
414 S.N.Gurkova, A.I.Gusev, N.V.Alekseev, R.I.Segel'man, T.K.Gar, and N.Yu.Khromova, Zh. Strukt. Khim., 1983, 24(1), 162(Engl.Ed. 155).
415 J.Almlöf, L.Fernholt, K.Faegri,Jr., A.Haaland, B.E.R.Schilling, R.Seip, and K.Taugbøl, Acta Chem. Scand., Ser. A, 1983, A37, 131.
416 C.Glidewell and D.C.Liles, J. Organomet. Chem., 1983, 243, 291.
417 V.K.Belsky, I.E.Saratov, V.O.Reikhsfeld, and A.A.Simonenko, J. Organomet. Chem., 1983, 258, 283.
418 D.J.Brauer and R.Eujen, Organometallics, 1983, 2, 263.
419 F.W.B.Einstein, R.K.Pomeroy, P.Rushman, and A.C.Willis, J. Chem. Soc., Chem. Commun., 1983, 854.
420 T.Halder, W.Schwartz, J.Weidlein, and P.Fischer, J. Organomet. Chem., 1983, 246, 29.
421 S.N.Gurkova, S.N.Tandura, A.V.Kisin, A.I.Gusev, N.V.Alekseev, T.K.Gar, N.Yu.Khromova, and I.R.Segel'man, Zh. Strukt. Khim., 1982, 23(4), 101(Engl.Ed. 574).
422 F.Brisse, M.Vanier, M.J.Olivier, Y.Gareau, and K.Steliou, Organometallics, 1983, 2, 878.
423 G.Fritz, K.D.Hoppe, W.Hönle, D.Weber, C.Mujica, V.Manriquez, and H.G.Schnering, J. Organomet. Chem., 1983, 249, 63.
424 H.Puff, S.Franken, W.Schuh, and W.Schwab, J. Organomet. Chem., 1983, 244, C41.
425 L.Ross and M.Drager, Z. Naturforsch., B, 1983, 38B, 665.
426 H.Puff, S.Franken, and W.Schuh, J. Organomet. Chem., 1983, 256, 23.
427 W.E.Hunter, D.C.Hrncir, R.V.Bynum, R.A.Penttila, and J.L.Atwood, Organometallics, 1983, 2, 750.
428 Y.Dong, S.Wu, R.Zhang, and S.Chen, Kexue Tongbao, 1982, 27, 1436.
429 R.T.Baker, J.F.Whitney, and S.S.Wreford, Organometallics, 1983, 2, 1049.
430 A.Laarif, F.Theobald, B.Gautheron, and G.Tainturier, J. Appl. Crystallogr., 1983, 16, 277.

431 A.R.Norris, S.E.Taylor, E.Buncel, F.Bélanger-Gariépy, and A.L.Beauchamp, Can. J. Chem., 1983, 61, 1536.
432 A.J.Canty, J.M.Patrick, and A.H.White, J. Chem. Soc., Dalton Trans., 1983, 1873.
433 B.Teclé, K.F.Siddiqui, C.Ceccarelli, and J.P.Oliver, J. Organomet. Chem., 1983, 255, 11.
434 J.L.Atwood, D.E.Berry, S.R.Stobart, and M.J.Zaworotko, Inorg. Chem., 1983, 22, 3480.
435 K.Onan, J.Rebek,Jr., T.Costello, and L.Marshall, J. Am. Chem. Soc., 1983, 105, 6759.
436 W.A.Herrmann, M.L.Ziegler, and O.Serhadli, Organometallics, 1983, 2, 958.
437 S.Ermer, K.King, K.I.Hardcastle, E.Rosenberg, A.M.M.Lanfredi, A.Tiripicchio, and M.T.Camellini, Inorg. Chem., 1983, 22, 1339.
438 F.W.B.Einstein, C.H.W.Jones, T.Jones, and R.D.Sharma, Inorg. Chem., 1983, 22, 3924.
439 L.G.Kuzmina, A.G.Ginzburg, Yu.T.Struchkov, and D.N.Kursanov, J. Organomet. Chem., 1983, 253, 329.
440 E.G.Mednikov, V.V.Bashilov, V.I.Sokolov, Yu.L.Slovkhotov, Yu.T.Struchkov, Polyhedron, 1983, 2, 141.
441 D.Grdenić, M.Sikirica, D.Matković-Čalogović, and A.Nagl, J. Organomet. Chem., 1983, 253, 283.
442 D.K.Breitinger, G.Petrikowski, G.Liehr, and R.Sendelbeck, Z. Naturforsch., B, 1983, 38B, 357.
443 M.Yu.Antipin, Yu.T.Struchkov, A.Yu.Volkonskii, and E.M.Rokhlin, Izv. Akad. Nauk SSSR, Ser. Khim., 1983, 452(Engl.Ed. 410).
444 T.M.Gilbert and R.G.Bergman, Organometallics, 1983, 2, 1458.
445 E.A.V.Ebsworth, R.O.Gould, N.T.McManus, D.W.H.Rankin, M.D.Walkinshaw, and J.D.Whitelock, J. Organomet. Chem., 1983, 249, 227.
446 D.L.Thorn and T.H.Tulip, Organometallics, 1982, 1, 1580.
447 E.L.Muetterties, K.D.Tau, J.F.Kirner, T.V.Harris, J.Stark, M.R.Thompson, and V.W.Day, Organometallics, 1982, 1, 1562.
448 L.Dahlenburg and F.Mirzaei, J. Organomet. Chem., 1983, 251, 103.
449 R.Usón, L.A.Oro, D.Carmona, M.A.Esteruelas, C.Foces-Foces, F.H.Cano, and S.Garcia-Blanco, J. Organomet. Chem., 1983, 254, 249.
450 L.A.Oro, D.Carmona, M.A.Esteruelas, C.Foces-Foces, and F.H.Cano, J. Organomet. Chem., 1983, 258, 357.
451 G.Nord, A.C.Hazell, R.G.Hazell, and O.Farver, Inorg. Chem., 1983, 22, 3429.
452 F.Demartin and N.Masciocchi, Acta Crystallogr., Sect. C, 1983, 39, 1225.
453 T.Hayashi, M.Tanaka, I.Ogata, T.Kodama, T.Takahashi, Y.Uchida, and T.Uchida, Bull. Chem. Soc. Jpn., 1983, 56, 1780.
454 F.W.B.Einstein, T.Jones, D.Sutton, and Z.Xiaoheng, J. Organomet. Chem., 1983, 244, 87.
455 E.Arpac and L.Dahlenburg, J. Organomet. Chem., 1983, 251, 361.
456 M.Cowie, M.D.Gauthier, S.J.Loeb, and I.R.McKeer, Organometallics, 1983, 2, 1057.
457 L.J.Farrugia, A.G.Orpen, and F.G.A.Stone, Polyhedron, 1983, 2, 171.
458 V.Schurig, W.Pille, and W.Winter, Angew. Chem., Int. Ed. Engl., 1983, 22, 327.
459 L.Dahlenburg and F.Mirzaei, J. Organomet. Chem., 1983, 251, 123.
460 P.B.Hitchcock, S.I.Klein, and J.F.Nixon, J. Organomet. Chem., 1983, 241, C9.
461 M.J.Breen, G.L.Geoffroy, A.L.Rheingold, and W.C.Fultz, J. Am. Chem. Soc., 1983, 105, 1069.
462 K.A.Beveridge, G.W.Bushnell, S.R.Stobart, J.L.Atwood, and M.J.Zaworotko, Organometallics, 1983, 2, 1447.
463 G.W.Bushnell, D.O.K.Fjeldsted, S.R.Stobart, and M.J.Zaworotko, J. Chem. Soc., Chem. Commun., 1983, 580.
464 J.Müller, C.Hansch, and J.Pickardt, J. Organomet. Chem., 1983, 259, C21.
465 C.A.Ghilardi, S.Midollini, and A.Orlandini, Angew. Chem., Int. Ed. Engl., 1983, 22, 790.
466 M.P.Anderson, C.C.Tso, B.M.Mattson, and L.H.Pignolet, Inorg. Chem., 1983, 22, 3267.

467 M.M.Harding, B.S.Nicholls, and A.K.Smith, J. Chem. Soc., Dalton Trans., 1983, 1479.
468 F.Demartin, M.Manassero, M.Sansoni, L.Garlaschelli, C.Raimondi C., and S.Martinengo, J. Organomet. Chem., 1983, 243, C10.
469 F.Demartin, M.Manassero, M.Sansoni, L.Garlaschelli, M.C.Malatesta, and U.Sartorelli, J. Organomet. Chem., 1983, 248, C17.
470 K.Pörschke, G.Wilke, and C.Krüger, Angew. Chem., Int. Ed. Engl., 1983, 22, 547.
471 P.Jutzi, E.Schlüter, C.Krüger, and S.Pohl, Angew. Chem., Int. Ed. Engl., 1983, 22, 994.
472 C.Eaborn, P.B.Hitchcock, J.D.Smith, and A.C.Sullivan, J. Chem. Soc., Chem. Commun., 1983, 1390.
473 C.Eaborn, P.B.Hitchcock, J.D.Smith, and A.C.Sullivan, J. Chem. Soc., Chem. Commun., 1983, 827.
474 B.Schubert and E.Weiss, Chem. Ber., 1983, 116, 3212.
475 J.T.B.H.Jastrzebski, G.van Koten, K.Goubitz, C.Arlen, and M.Pfeffer, J. Organomet. Chem., 1983, 246, C75.
476 J.Powell, K.S.Ng, W.W.Ng, and S.C.Nyburg, J. Organomet. Chem., 1983, 243, C1.
477 L.M.Engelhardt, A.S.May, C.L.Raston, and A.H.White, J. Chem. Soc., Dalton Trans., 1983, 1671.
478 H.Schmidbaur, A.Schier, and U.Schubert, Chem. Ber., 1983, 116, 1938.
479 G.W.Klumpp, P.J.A.Geurink, A.L.Spek, and A.J.M.Duisenberg, J. Chem. Soc., Chem. Commun., 1983, 814.
480 H.Hope and P.P.Power, J. Am. Chem. Soc., 1983, 105, 5320.
481 B.Schubert and E.Weiss, Angew. Chem., Int. Ed. Engl., 1983, 22, 496.
482 R.E.Stevens and W.L.Gladfelter, Inorg. Chem., 1983, 22, 2034.
483 M.A.Andrews, J.Eckert, J.A.Goldstone, L.Passell, and B.Swanson, J. Am. Chem. Soc., 1983, 105, 2262.
484 G.Liehr, H.-J.Seibold, and H.Behrens, J. Organomet. Chem., 1983, 248, 351.
485 A.J.Schultz, R.G.Teller, M.A.Beno, J.M.Williams, M.Brookhart, W.Lamanna, and M.B.Humphrey, Science, 1983, 220, 197.
486 C.G.Howard, G.S.Girolami, G.Wilkinson, M.Thornton-Pett, and M.B.Hursthouse, J. Chem. Soc., Chem. Commun., 1983, 1163.
487 G.W.Harris, J.C.A.Boeyens, and N.J.Coville, J. Organomet. Chem., 1983, 255, 87.
488 P.Jaitner, W.Huber, G.Huttner, and O.Scheidsteger, J. Organomet. Chem., 1983, 259, C1.
489 N.M.Loim, M.V.Laryukova, V.G.Andrianov, V.A.Tsiryapkin, Z.N.Parnes, and Yu.T.Struchkov, J. Organomet. Chem., 1983, 248, 73.
490 E.P.Kündig, P.L.Timms, B.A.Kelly, and P.Woodward, J. Chem. Soc., Dalton Trans., 1983, 901.
491 G.S.Girolami, G.Wilkinson, M.Thornton-Pett, and M.B.Hursthouse, J. Am. Chem. Soc., 1983, 105, 6752.
492 Y.K.Chung, E.D.Honig, W.T.Robinson, D.A.Sweigart, N.G.Connelly, and S.D.Ittel, Organometallics, 1983, 2, 1479.
493 E.Lindner, K.A.Starz, H.-J.Eberle, and W.Hiller, Chem. Ber., 1983, 116, 1209.
494 J.R.Bleeke and J.J.Kotyk, Organometallics, 1983, 2, 1263.
495 J.T.Weed, M.F.Rettig, and R.M.Wing, J. Am. Chem. Soc., 1983, 105, 6510.
496 E.Lindner, K.A.Starz, N.Pauls, and W.Winter, Chem. Ber., 1983, 116, 1070.
497 A.B.Antonova, S.V.Kovalenko, E.D.Korniyets, A.A.Johansson, Yu.T.Struchkov, A.I.Ahmedov, and A.I.Yanovsky, J. Organomet. Chem., 1983, 244, 35.
498 C.G.Howard, G.S.Girolami, G.Wilkinson, M.Thornton-Pett, and M.B.Hursthouse, J. Chem. Soc., Dalton Trans., 1983, 2631.
499 W.A.Herrmann, J.Plank, J.L.Hubbard, G.W.Kriechbaum, W.Kalcher, B.Koumbouris, G.Ihl, A.Schäfer, M.L.Ziegler, H.Pfisterer, C.Pahl, J.L.Atwood, and R.D.Rogers, Z. Naturforsch., B, 1983, 38B, 1392.
500 W.Winter, R.Merkel, and U.Kunze, Z. Naturforsch., B, 1983, 38B, 747.
501 M.D.Rausch, B.H.Edwards, J.L.Atwood, and R.D.Rogers, Organometallics, 1982, 1, 1567.

502 C.P.Casey, R.M.Bullock, W.C.Fultz, and A.L.Rheingold, Organometallics, 1982, 1, 1591.
503 B.F.Hoskins, R.J.Steen, and T.W.Turney, Inorg. Chim. Acta, 1983, 77, L69.
504 U.Schubert, K.Ackermann, G.Kraft, and B.Wörle, Z. Naturforsch., B, 1983, 38B, 1488.
505 M.D.Rausch, B.H.Edwards, R.D.Rogers, and J.L.Atwood, J. Am. Chem. Soc., 1983, 105, 3882.
506 E.Lindner,I.P.Butz, S.Hoehne, W.Hiller, and R.Fawzi, J. Organomet. Chem., 1983, 259, 99.
507 J.A.Iggo, M.J.Mays, P.R.Raithby, and K.Hendrick, J. Chem. Soc., Dalton Trans., 1983, 205.
508 E.Lindner, I.P.Butz, W.Hiller, R.Fawzi, and S.Hoehne, Angew. Chem., Int. Ed. Engl., 1983, 22, 996.
509 C.G.Howard, G.Wilkinson, M.Thornton-Pett, and M.B.Hursthouse, J. Chem. Soc., Dalton Trans., 1983, 2025.
510 G.Ferguson, W.J.Laws, M.Parvez, and R.J.Puddephatt, Organometallics, 1983, 2, 276.
511 P.Braunstein, J.-M.Jud, and J.Fischer, J. Chem. Soc., Chem. Commun., 1983, 5.
512 W.A.Herrmann, J.Weichmann, R.Serrano, K.Blechschmitt, H.Pfisterer, and M.L.Ziegler, Angew. Chem., Int. Ed. Engl., 1983, 22, 314.
513 J.L.Atwood, I.Bernal, F.Calderazzo, L.G.Canada, R.Poli, R.D.Rogers, C.A.Veracini, and D.Vitali, Inorg. Chem., 1983, 22, 1797.
514 M.C.Böhm, R.D.Ernst, R.Gleiter, and D.R.Wilson, Inorg. Chem., 1983, 22, 3815.
515 S.Gambarotta, C.Floriani, A.Chiesi-Villa, and C.Guastini, J. Chem. Soc., Chem. Commun., 1983, 1128.
516 M.Herberhold, D.Reiner, and D.Neugebauer, Angew. Chem., Int. Ed. Engl., 1983, 22, 59.
517 N.A.Bailey, P.L.Chell, C.P.Manuel, A.Mukhopadhyay, D.Rogers, H.E.Tabbron, and M.J.Winter, J. Chem. Soc., Dalton Trans., 1983, 2397.
518 H.Adams, N.A.Bailey, P.Cahill, D.Rogers, and M.J.Winter, J. Chem. Soc., Chem. Commun., 1983, 831.
519 H. tom Dieck, T.Mack, K.Peters, and H.-G. von Schnering, Z. Naturforsch., B, 1983, 38B, 568.
520 E.Carmona, L.Sánchez, M.L.Poveda, J.M.Marin, J.L.Atwood, and R.D.Rogers, J. Chem. Soc., Chem. Commun., 1983, 161.
521 J.W.Faller, H.H.Murray, D.L.White, and K.H.Chao, Organometallics, 1983, 2, 400.
522 M.L.H.Green, A.Izquierdo, J.J.Martin-Polo, V.S.B.Mtetwa, and K.Prout, J. Chem. Soc., Chem. Commun., 1983, 538.
523 O.Scheidsteger, G.Huttner, V.Bejenke, and W.Gartzke, Z. Naturforsch., B, 1983, 38B, 1598.
524 K.A.Mead, H.Morgan, and P.Woodward, J. Chem. Soc., Dalton Trans., 1983, 271.
525 P.Diversi, G.Ingrosso, A.Lucherini, W.Porzio, and M.Zocchi, J. Chem. Soc., Dalton Trans., 1983, 967.
526 E.Carmona, J.M.Marin, M.L.Poveda, J.L.Atwood, and R.D.Rogers, J. Am. Chem. Soc., 1983, 105, 3014.
527 A.J.Graham, D.Akrigg, and B.Sheldrick, Acta Crystallogr., Sect. C, 1983, 39, 192.
528 M.A.A.F. de C.T.Carrondo and A.M.T.S.Domingos, J. Organomet. Chem., 1983, 253, 53.
529 S.R.Allen, M.Green, N.C.Norman, K.E.Paddick, and A.G.Orpen, J. Chem. Soc., Dalton Trans., 1983, 1625.
530 G.E.Herberich, B.Hessner, and J.Okuda, J. Organomet. Chem., 1983, 254, 317.
531 R.S.Herrick, S.J.Nieter Burgmayer, and J.L.Templeton, Inorg. Chem., 1983, 22, 3275.
532 R.D.Adams, C.Blankenship, B.E.Segmüller, and M.Shiralian, J. Am. Chem. Soc., 1983, 105, 4319.
533 J.L.Davidson, W.F.Wilson, Lj.Manojlović-Muir, and K.W.Muir, J. Organomet. Chem., 1983, 254, C6.

534 R.G.Beevor, M.Green, A.G.Orpen, and I.D.Williams, J. Chem. Soc., Chem. Commun., 1983, 673.

535 S.R.Allen, P.K.Baker, S.G.Barnes, M.Bottrill, M.Green, A.G.Orpen, I.D.Williams, and A.J.Welch, J. Chem. Soc., Dalton Trans., 1983, 927.

536 G.N.Schrauzer, L.A.Hughes, N.Strampach, F.Ross, D.Ross, and E.O.Schlemper, Organometallics, 1983, 2, 481.

537 M.Kamata, K.Hirotsu, T.Higuchi, M.Kido, K.Tatsumi, T.Yoshida, and S.Otsuka, Inorg. Chem., 1983, 22, 2416.

538 P.B.Winston, S.J.Nieter Burgmeyer, and J.L.Templeton, Organometallics, 1983, 2, 167.

539 G.J.Kubas, G.D.Jarvinen, and R.R.Ryan, J. Am. Chem. Soc., 1983, 105, 1883.

540 G.N.Schrauzer, L.A.Hughes, E.O.Schlemper, F.Ross, and D.Ross, Organometallics, 1983, 2, 1163.

541 J.W.Faller and K.-H.Chao, J. Am. Chem. Soc., 1983, 105, 3893.

542 E.C.Alyea, G.Ferguson, and A.Somogyvari, Organometallics, 1983, 2, 668.

543 M.W.Creswick and I.Bernal, Inorg. Chim. Acta, 1983, 71, 41.

544 H.Scordia, R.Kergoat, M.M.Kubicki, and J.E.Guerchais, Organometallics, 1983, 2, 1681.

545 M.W.Creswick and I.Bernal, Inorg. Chim. Acta, 1983, 74, 241.

546 H.Brunner, J.Wachter, I.Bernal, G.M.Reisner, and R.Benn, J. Organomet. Chem., 1983, 243, 179.

547 I.Bernal and G.M.Reisner, Inorg. Chim. Acta, 1983, 71, 65.

548 N.E.Kolobova, V.V.Skripkin, T.V.Rozantseva, Yu.T.Struchkov, G.G.Aleksandrov, V.A.Antonovich, and V.I.Bakhmutov, Koord. Khim., 1982, 8, 1655(Engl.Ed. 910).

549 E.H.Wong, F.C.Bradley, and E.J.Gabe, J. Organomet. Chem., 1983, 244, 235.

550 J.R.Dilworth, J.Hutchinson, and J.A.Zubieta, J. Chem. Soc., Chem. Commun., 1983, 1034.

551 J.M.Hanckel and M.Y.Darensbourg, J. Am. Chem. Soc., 1983, 105, 6979.

552 P.W.Corfield, J.C.Dewan, and S.J.Lippard, Inorg. Chem., 1983, 22, 3424.

553 A.S.Antsyshkina, L.Kh.Minacheva, M.A.Porai-Koshits, V.N.Ostrikova, G.G.Sadikov, A.A.Pasynskii, Yu.V.Skripkin, and V.T.Kalinnikov, Koord. Khim., 1982, 8, 1423(Engl.Ed. 774).

554 J.P.Farr, M.M.Olmstead, N.M.Rutherford, F.E.Wood, and A.L.Balch, Organometallics, 1983, 2, 1758.

555 C.F.Barrientos-Penna, F.W.B.Einstein, T.Jones, and D.Sutton, Inorg. Chem., 1983, 22, 2614.

556 R.G.Finke, G.Gaughan, C.Pierpont, and J.H.Noordik, Organometallics, 1983, 2, 1481.

557 R.D.Barr, M.Green, J.A.K.Howard, T.B.Marder, and F.G.A.Stone, J. Chem. Soc., Chem. Commun., 1983, 759.

558 J.S.Merola, R.A.Gentile, G.B.Ansell, M.A.Modrick, and S.Zentz, Organometallics, 1982, 1, 1731.

559 T.R.Halbert, W.-H.Pan, and E.I.Stiefel, J. Am. Chem. Soc., 1983, 105, 5476.

560 M.D.Curtis, N.A.Fotinos, K.R.Han, and W.M.Butler, J. Am. Chem. Soc., 1983, 105, 2686.

561 W.K.Miller, R.C.Haltiwanger, M.C.VanDerveer, and M.Rakowski DuBois, Inorg. Chem., 1983, 22, 2973.

562 M.McKenna, L.L.Wright, D.J.Miller, L.Tanner, R.C.Haltiwanger, and M.R.DuBois, J. Am. Chem. Soc., 1983, 105, 5329.

563 K.Sünkel, K.Schloter, W.Beck, K.Ackermann, and U.Schubert, J. Organomet. Chem., 1983, -241-, 333.

564 H.Alper, J.-F.Petrignani, F.W.B.Einstein, and A.C.Willis, J. Am. Chem. Soc., 1983, 105, 1701.

565 H.Alper, F.W.B.Einstein, R.Nagai, J.-F.Petrignani, and A.C.Willis, Organometallics, 1983, 2, 1291.

566 R.S.Herrick, S.J.Nieter-Burgmayer, and J.L.Templeton, J. Am. Chem. Soc., 1983, 105, 2599.

567 G.Becker, W.A.Herrmann, W.Kalcher, G.W.Kriechbaum, C.Pahl, C.T.Wagner, and M.L.Ziegler, Angew. Chem., Int. Ed. Engl., 1983, 22, 413.

568 H.Alper, F.W.B.Einstein, J.-F.Petrignani, and A.C.Willis, Organometallics,

1983, 2, 1422.
569 J.J.D'Errico, L.Messerle, and M.D.Curtis, Inorg. Chem., 1983, 22, 849.
570 R.F.Gerlach, D.N.Duffy, and M.D.Curtis, Organometallics, 1983, 2, 1172.
571 M.J.Chetcuti, M.H.Chisholm, K.Folting, D.A.Haitko, J.C.Huffman, and J.Janos, J. Am. Chem. Soc., 1983, 105, 1163.
572 M.Green, A.G.Orpen, C.J.Schaverien, and I.D.Williams, J. Chem. Soc., Chem. Commun., 1983, 181.
573 J.-S.Huang and L.F.Dahl, J. Organomet. Chem., 1983, 243, 57.
574 H.W.Roesky, D.Amirzadeh-Asl, W.Clegg, M.Noltemeyer, and G.M.Sheldrick, J. Chem. Soc., Dalton Trans., 1983, 855.
575 W.A.Herrmann, L.K.Bell, M.L.Ziegler, H.Pfisterer, and C.Pahl, J. Organomet. Chem., 1983, -247-, 39.
576 S.G.Bott, N.G.Connelly, M.Green, N.C.Norman, A.G.Orpen, J.F.Paxton, and C.J.Schaverien, J. Chem. Soc., Chem. Commun., 1983, 378.
577 H.Brunner, H.Buchner, J.Wachter, I.Bernal, and W.H.Ries, J. Organomet. Chem., 1983, 244, 247.
578 W.A.Herrmann, G.W.Kriechbaum, R.Dammel, H.Bock, M.L.Ziegler, and H.Pfisterer, J. Organomet. Chem., 1983, 254, 219.
579 M.H.Chisholm, J.C.Huffman, and R.J.Tatz, J. Am. Chem. Soc., 1983, 105, 2075.
580 J.J.D'Errico and M.D.Curtis, J. Am. Chem. Soc., 1983, 105, 4479.
581 L.K.Bell, W.A.Herrmann, M.L.Ziegler, and H.Pfisterer, Organometallics, 1982, 1, 1673.
582 F.A.Cotton and W.Schwotzer, Inorg. Chem., 1983, 22, 387.
583 D.A.Dubois, E.N.Duesler, and R.T.Paine, Organometallics, 1983, 2, 1903.
584 E.H.Wong, M.M.Turnbull, E.J.Gabe, F.L.Lee, and Y.Le Page, J. Chem. Soc., Chem. Commun., 1983, 776.
585 M.H.Chisholm, J.C.Huffman, and N.S.Marchant, J. Am. Chem. Soc., 1983, 105, 6162.
586 A.Bell, S.J.Lippard, M.Roberts, and R.A.Walton, Organometallics, 1983, 2, 1562.
587 T.W.Coffindaffer, I.P.Rothwell, and J.C.Huffmann, J. Chem. Soc., Chem. Commun., 1983, 1249.
588 M.J.Chetcuti, M.H.Chisholm, H.T.Chiu, and J.C.Huffman, J. Am. Chem. Soc., 1983, 105, 1060.
589 M.Green. A.G.Orpen, C.J.Schaverien, and I.D.Williams, J. Chem. Soc., Chem. Commun., 1983, 583.
590 J.T.Lin and J.E.Ellis, J. Am. Chem. Soc., 1983, 105, 6252.
591 J.A.Bandy, C.E.Davies, J.C.Green, M.L.H.Green, K.Prout, and D.P.S.Rodgers, J. Chem. Soc., Chem. Commun., 1983, 1395.
592 E.Weiss, G.Sauermann, and G.Thirase, Chem. Ber., 1983, 116, 74.
593 D.J.Brauer, H.Bürger, W.Geschwandtner, G.R.Liewald, and C.Krüger, J. Organomet. Chem., 1983, 248, 1.
594 F.Calderazzo, U.Englert, G.Pampaloni, G.Pelizzi, and R.Zamboni, Inorg. Chem., 1983, 22, 1865.
595 R.Mercier, J.Douglade, J.Amaudrut, J.Sala-Pala, and J.E.Guerchais, J. Organomet. Chem., 1983, 244, 145.
596 L.M.Engelhardt, W.-P.Leung, C.L.Raston, and A.H.White, J. Chem. Soc., Chem. Commun., 1983, 386.
597 A.S.Antsyshkina, M.A.Porai-Koshits, V.N.Ostrikova, A.A.Pasynskii, Yu.V.Skripkin, and V.T.Kalinnikov, Koord. Khim., 1982, 8, 1552(Engl.Ed. 852).
598 A.A.Pasynskii, Yu.V.Skripkin, O.G.Volkov, V.T.Kalinnikov, M.A.Porai-Koshits, A.S.Antsyshkina, L.M.Dikareva, and V.N.Ostrikova, Izv. Akad. Nauk SSSR, Ser. Khim., 1983, 1207(Engl.Ed. 1093).
599 F.A.Cotton and W.J.Roth, J. Am. Chem. Soc., 1983, 105, 3734.
600 E.Hey, F.Weller, and K.Dehnicke, Naturwissenschaften, 1983, 70, 41.
601 D.S.Dudis and J.P.Fackler,Jr., J. Organomet. Chem., 1983, 249, 289; 255, C19.
602 D.M.Grove, G.van Koten, R.Zoet, N.W.Murrall, and A.J.Welch, J. Am. Chem. Soc., 1983, 105, 1379.
603 E.Lindner, F.Bouachir, and S.Hoehne, Chem. Ber., 1983, 116, 46.

604 D.Walther, E.Dinjus, J.Sieler, N.N.Thanh, W.Schade, and I.Leban, Z. Naturforsch., B, 1983, 38B, 835.
605 H.Lehmkuhl, C.Naydowski, R.Benn, A.Rufińska, G.Schroth, R.Mynott, and C.Krüger, Chem. Ber., 1983, 116, 2447.
606 T.A. van der Knaap, L.W.Jenneskens, H.J.Meeuwissen, F.Bickelhaupt, D.Walther, E.Dinjus, E.Uhlig, and A.L.Spek, J. Organomet. Chem., 1983, 254, C33.
607 H.M.Büch, P.Binger, R.Goddard, and C.Krüger, J. Chem. Soc., Chem. Commun., 1983, 648.
608 C.Bianchini, D.Masi, C.Mealli, and A.Meli, J. Organomet. Chem., 1983, 247, C29.
609 G.Lavigne, F.Papageorgiou, C.Bergounhou, and J.-J.Bonnet, Inorg. Chem., 1983, 22, 2485.
610 E.Sappa, A.Tiripicchio, and M.Tiripicchio Camellini, J. Organomet. Chem., 1983, 246, 287.
611 A.J.Carty, N.J.Taylor, E.Sappa, and A.Tiripicchio, Inorg. Chem., 1983, 22, 1871.
612 J.L.Atwood, W.E.Hunter, and H.-M.Zhang, J. Am. Chem. Soc., 1983, 105, 3737.
613 K.R.Pörschke, W.Kleimann, G.Wilke, K.H.Claus, and C.Krüger, Angew. Chem., Int. Ed. Engl., 1983, 22, 991.
614 E.Dubler, M.Textor, H.R.Oswald, and G.B.Jameson, Acta Crystallogr., Sect. B, 1983, B39, 607.
615 R.A.Jones, A.L.Stuart, J.L.Atwood, and W.E.Hunter, Organometallics, 1983, 2, 874.
616 C.Bianchini, C.A.Ghilardi, A.Meli, S.Midollini, and A.Orlandini, J. Chem. Soc., Chem. Commun., 1983, 753.
617 C.Arlen, M.Pfeffer, J.Fischer, and A.Mitschler, J. Chem. Soc., Chem. Commun., 1983, 928.
618 P.H.M.Budzelaar, J.Boersma, and G.J.M. van der Kerk. Angew. Chem., Int. Ed. Engl., 1983, 22, 329.
619 U.Schubert, R.Werner, L.Zinner, and H.Werner, J. Organomet. Chem., 1983, 253, 363.
620 M.I.Bruce, M.L.Williams, J.M.Patrick, and A.H.White, Aust. J. Chem., 1983, 36, 1353.
621 G.R.Clark, N.R.Edmonds, R.A.Pauptit, W.R.Roper, J.M.Waters, and A.H.Wright, J. Organomet. Chem., 1983, 244, C57.
622 G.R.Clark, T.J.Collins, K.Marsden, and W.R.Roper, J. Organomet. Chem., 1983, 259, 215.
623 G.Smith, D.J.Cole-Hamilton, M.Thornton-Pett, and M.B.Hursthouse, Polyhedron, 1983, 2, 1241.
624 L.Busetto, M.Green, B.Hessner, J.A.K.Howard, J.C.Jeffery, and F.G.A.Stone, J. Chem. Soc., Dalton Trans., 1983, 519.
625 E.E.Sutton, M.L.Niven, and J.R.Moss, Inorg. Chim. Acta, 1983, 70, 207.
626 M.R.Burke, J.Takats, F.-W.Grevels, and J.G.A.Reuvers, J. Am. Chem. Soc., 1983, 105, 4092.
627 R.D.Adams, I.T.Horváth, B.E.Segmüller, and L.-W.Yang, Organometallics, 1983, 2, 144.
628 J.R.Shapley, M.E.Cree-Uchiyama, G.M.St.George, M.R.Churchill, and C.Bueno, J. Am. Chem. Soc., 1983, 105, 140.
629 J.R.Shapley, D.S.Strickland, G.M.St.George, M.R.Churchill, and C.Bueno, Organometallics, 1983, 2, 185.
630 Z.Dawoodi, M.J.Mays, P.R.Raithby, K.Henrick, W.Clegg, and G.Weber, J. Organomet. Chem., 1983, 249, 149.
631 A.M.Brodie, H.D.Holden, J.Lewis, and M.J.Taylor, J. Organomet. Chem., 1983, 253, C1.
632 E.D.Morrison, G.R.Steinmetz, G.L.Geoffroy, W.C.Fultz, and A.L.Rheingold, J. Am. Chem. Soc., 1983, 105, 4104.
633 M.R.Churchill and J.R.Missert, J. Organomet. Chem., 1983, 256, 349.
634 F.W.B.Einstein, S.Nussbaum, D.Sutton, and A.C.Willis, Organometallics, 1983, 2, 1259.
635 R.J.Goudsmit, B.F.G.Johnson, J.Lewis, P.R.Raithby, and M.J.Rosales, J. Chem.

Soc., Dalton Trans., 1983, 2257.
636 G.Süss-Fink and P.R.Raithby, Inorg. Chim. Acta, 1983, 71, 109.
637 R.D.Adams, Z.Dawoodi, D.F.Foust, and B.E.Segmüller, Organometallics, 1983, 2, 315.
638 E.Sappa, A.Tiripicchio, and A.M.Manotti Lanfredi, J. Organomet. Chem., 1983, 249, 391.
639 W.Ehrenreich, M.Herberhold, G.Süss-Fink, H.-P.Klein, and U.Thewalt, J. Organomet. Chem., 1983, 248, 171.
640 K.Burgess, H.D.Holden, B.F.G.Johnson, and J.Lewis, J. Chem. Soc., Dalton Trans., 1983, 1199.
641 H.D.Holden, B.F.G.Johnson, J.Lewis, P.R.Raithby, and G.Uden, Acta Crystallogr., Sect. C, 1983, 39, 1200.
642 H.D.Holden, B.F.G.Johnson, J.Lewis, P.R.Raithby, and G.Uden, Acta Crystallogr., Sect. C, 1983, 39, 1197.
643 M.I.Bruce, J.M.Guss, R.Mason, B.W.Skelton, and A.H.White, J. Organomet. Chem., 1983, 251, 261.
644 S.A.MacLaughlin, J.P.Johnson, N.J.Taylor, and A.J.Carty, Organometallics, 1983, 2, 352.
645 L.J.Farrugia, M.Green, D.R.Hankey, A.G.Orpen, and F.G.A.Stone, J. Chem. Soc., Chem. Commun., 1983, 310.
646 S.G.Shore, W.-L.Hsu, M.R.Churchill, and C.Bueno, J. Am. Chem. Soc., 1983, 105, 655.
647 M.R.Churchill and C.Bueno, J. Organomet. Chem., 1983, 256, 357.
648 M.R.Churchill and H.J.Wasserman, J. Organomet. Chem., 1983, 248, 365.
649 A.C.Willis, F.W.B.Einstein, R.M.Ramadan, and R.K.Pomeroy, Organometallics, 1983, 2, 935.
650 A.C.Willis, G.N. van Buuren, R.K.Pomeroy, and F.W.B.Einstein, Inorg. Chem., 1983, 22, 1162.
651 R.D.Adams, I.T.Horváth, and P.Mathur, J. Am. Chem. Soc., 1983, 105, 7202.
652 J.T.Park, J.R.Shapley, M.R.Churchill, and C.Bueno, Inorg. Chem., 1983, 22, 1579.
653 J.T.Park, J.R.Shapley, M.R.Churchill, and C.Bueno, J. Am. Chem. Soc., 1983, 105, 6182.
654 M.A.Collins, B.F.G.Johnson, J.Lewis, J.M.Mace, J.Morris, M.McPartlin, W.J.H.Nelson, J.Puga, and P.R.Raithby, J. Chem. Soc., Chem. Commun., 1983, 689.
655 R.D.Adams, D.F.Foust, and P.Mathur, Organometallics, 1983, 2, 990.
656 R.D.Adams and L.-W.Yang, J. Am. Chem. Soc., 1983, 105, 235.
657 B.F.G.Johnson, J.Lewis, W.J.H.Nelson, J.Puga, K.Henrick, and M.McPartlin, J. Chem. Soc., Dalton Trans., 1983, 1203.
658 R.D.Adams, I.T.Horvàth, B.E.Segmüller, and L.-W.Yang, Organometallics, 1983, 2, 1301.
659 B.F.G.Johnson, J.Lewis, W.J.H.Nelson, J.Puga, P.R.Raithby, and K.H.Whitmire, J. Chem. Soc., Dalton Trans., 1983, 1339.
660 R.D.Adams, I.T.Horváth, and L.-W.Yang, Organometallics, 1983, 2, 1257.
661 B.F.G.Johnson, J.Lewis, P.R.Raithby, and M.J.Rosales, J. Chem. Soc., Dalton Trans., 1983, 2645.
662 B.F.G.Johnson, J.Lewis, P.R.Raithby, and M.J.Rosales, J. Organomet. Chem., 1983, 259, C9.
663 R.D.Adams, I.T.Horváth, and L.-W.Yang, J. Am. Chem. Soc., 1983, 105, 1533.
664 R.D.Adams, I.T.Horváth, P.Mathur, and B.E.Segmüller, Organometallics, 1983, 2, 996.
665 R.J.Goudsmit, B.F.G.Johnson, J.Lewis, P.R.Raithby, and K.H.Whitmire, J. Chem. Soc., Chem. Commun., 1983, 246.
666 H.D.Holden, B.F.G.Johnson, J.Lewis, P.R.Raithby, and G.Uden, Acta Crystallogr., Sect. C, 1983, 39, 1203.
667 R.D.Adams, D.F.Foust, and B.E.Segmüller, Organometallics, 1983, 2, 308.
668 R.D.Adams, Z.Dawoodi, D.F.Foust, and B.E.Segmüller, J. Am. Chem. Soc., 1983, 105, 831.
669 R.D.Adams and D.F.Foust, Organometallics, 1983, 2, 323.
670 R.D.Adams, I.T.Horváth, P.Mathur, B.E.Segmüller, and L.-W.Yang,

Organometallics, 1983, 2, 1078.
671 B.F.G.Johnson, J.Lewis, W.J.H.Nelson, M.Vargas, D.Braga, and M.McPartlin, J. Organomet. Chem., 1983, 249, C21.
672 B.F.G.Johnson, J.Lewis, M.McPartlin, W.J.H.Nelson, P.R.Raithby, A.Sironi, and M.D.Vargas, J. Chem. Soc., Chem. Commun., 1983, 1476.
673 D.Braga, J.Lewis, B.F.G.Johnson, M.McPartlin, W.J.H.Nelson, and M.D.Vargas, J. Chem. Soc., Chem. Commun., 1983, 241.
674 L.G.Kuz'mina, Yu.T.Struchkov, E.M.Rokhlina, A.S.Peregudov, and D.N.Kravtsov, Zh. Strukt. Khim., 1983, 24(1), 106(Engl.Ed. 93).
675 R.McGrindle, G.Ferguson, G.J.Arsenault, A.J.McAlees, and M.Parvez, J. Organomet. Chem., 1983, 246, C19.
676 G.Ferguson, R.McCrindle, and M.Parvez, Acta Crystallogr., Sect. C, 1983, 39, 993.
677 H.M.Büch, P.Binger, R.Benn, C.Krüger, and A.Rufińska, Angew. Chem., Int. Ed. Engl., 1983, 22, 774.
678 H.Werner, G.T.Crisp, P.W.Jolly, H.-J.Kraus, and C.Krüger, Organometallics, 1983, 2, 1369.
679 M.C.Etter and A.R.Siedle, J. Am. Chem. Soc., 1983, 105, 641.
680 H.Werner, H.-J.Kraus, U.Schubert, K.Ackermann, and P.Hofmann, J. Organomet. Chem., 1983, 250, 517.
681 H.-J.Kraus, H.Werner, and C.Krüger, Z. Naturforsch., B, 1983, 38B, 733.
682 J.Dehand, A.Mauro, H.Ossor, M.Pfeffer, R.H.de A.Santos, and J.R.Lechat, J. Organomet. Chem., 1983, 250, 537.
683 C.Arlen, M.Pfeffer, O.Bars, and D.Grandjean, J. Chem. Soc., Dalton Trans., 1983, 1535.
684 G.R.Newkome, W.E.Puckett, V.K.Gupta, and F.R.Fronczek, Organometallics, 1983, 2, 1247.
685 K.Miki, O.Shiotani, Y.Kai, N.Kasai, H.Kanatani, and H.Kurosawa, Organometallics, 1983, 2, 585.
686 H.Werner, M.Ebner, W.Bertleff, and U.Schubert, Organometallics, 1983, 2, 891.
687 B.Messbauer, H.Meyer, B.Walther, M.J.Heeg, A.F.M.Maqsudur Rahman, and J.P.Oliver, Inorg. Chem., 1983, 22, 272.
688 R.McCrindle, G.Ferguson, G.J.Arsenault, and A.J.McAlees, J. Chem. Soc., Chem. Commun., 1983, 571.
689 E.E.Björkman and J.-E.Bäckvall, Acta Chem. Scand., Ser. A, 1983, A37, 503.
690 R.Usón, J.Fornies, P.Espinet, E.Lalinde, P.G.Jones, and G.M.Sheldrick, J. Organomet. Chem., 1983, 253, C47.
691 W.J.Youngs, J.Mahood, B.L.Simms, P.N.Swepston, J.A.Ibers, S.Maoyu, H.Jinling, and L.Jiaxi, Organometallics, 1983, 2, 917.
692 J.Selbin, K.Abboud, S.F.Watkins, M.A.Gutierrez, and F.R.Fronczek, J. Organomet. Chem., 1983, 241, 259.
693 J.Granell, J.Sales, J.Vilarrasa, J.P.Declercq, G.Germain, C.Miravitlles, and X.Solans, J. Chem. Soc., Dalton Trans., 1983, 2441.
694 P.Braunstein, D.Matt, Y.Dusausoy, and J.Fischer, Organometallics, 1983, 2, 1410.
695 M.Parra-Hake, M.F.Rettig, and R.M.Wing, Organometallics, 1983, 2, 1013.
696 Lj.Manojlović-Muir, K.W.Muir, B.R.Lloyd, and R.J.Puddephatt, J. Chem. Soc., Chem. Commun., 1983, 1336.
697 R.Usón, J.Gimeno, L.A.Oro, J.M.Martínez de Ilarduya, J.A.Cabeza, A.Tiripicchio, and M.Tiripicchio Camellini, J. Chem. Soc., Dalton Trans., 1983, 1729.
698 R.Goddard, P.W.Jolly, C.Krüger, K.-P.Schick, and G.Wilke, Organometallics, 1982, 1, 1709.
699 E.G.Mednikov, N.K.Eremenko, Y.L.Slovokhotov, Yu.T.Struchkov, and S.P.Gubin, J. Organomet. Chem., 1983, 258, 247.
700 W.Hinrichs, D.Melzer, M.Rehwoldt, W.Jahn, and R.D.Fischer, J. Organomet. Chem., 1983, 251, 299.
701 J.A.K.Howard, J.L.Spencer, and S.A.Mason, Proc. R. Soc. London, Ser. A, 1983, 386, 145.
702 A.Avshu, R.D.O'Sullivan, A.W.Parkins, N.W.Alcock, and R.M.Countryman, J.

Chem. Soc., Dalton Trans., 1983, 1619.
703 E.N.Izakovich, L.M.Kachapina, R.P.Shibaeva, and M.L.Khidekel, Izv. Akad. Nauk SSSR, Ser. Khim., 1983, 1389(Engl.Ed. 1260).
704 H.C.Clark, G.Ferguson, V.K.Jain, and M.Parvez, Organometallics, 1983, 2, 806.
705 J.Terheijden, G.van Koten, J.L.de Booys, H.J.C.Ubbels, and C.H.Stam, Organometallics, 1983, 2, 1882.
706 A.T.Hutton, D.M.McEwan, B.L.Shaw, and S.W.Wilkinson, J. Chem. Soc., Dalton Trans., 1983, 2011.
707 P.Ammendola, M.R.Ciajolo, A.Panunzi, and A.Tuzi, J. Organomet. Chem., 1983, 254, 389.
708 J.T.Burton, R.J.Puddephatt, N.L.Jones, and J.A.Ibers, Organometallics, 1983, 2, 1488.
709 A.J.Canty, N.J.Minchin, J.M.Patrick, and A.H.White, J. Chem. Soc., Dalton Trans., 1983, 1253.
710 G.Ferguson, M.Parvez, P.K.Monaghan, and R.J.Puddephatt, J. Chem. Soc., Chem. Commun., 1983, 267.
711 W.J.Youngs and J.A.Ibers, J. Am. Chem. Soc., 1983, 105, 639.
712 G.B.Robertson and P.A.Tucker, Acta Crystallogr., Sect. C, 1983, 39, 1354.
713 W.J.Youngs and J.A.Ibers, Organometallics, 1983, 2, 979.
714 K.Miki, Y.Kai, N.Kasai, and H.Kurosawa, J. Am. Chem. Soc., 1983, 105, 2482.
715 J.Browning, G.W.Bushnell, K.R.Dixon, and A.Pidcock, Inorg. Chem., 1983, 22, 2226.
716 M.Calligaris, G.Carturan, G.Nardin, A.Scrivanti, and A.Wojcicki, Organometallics, 1983, 2, 865.
717 J.Geisenberger, U.Nagel, A.Sebald, and W.Beck, Chem. Ber., 1983, 116, 911.
718 F.Canziani, F.Galimberti, L.Garlaschelli, M.Carlotta Malachesta, A.Albinati, and F.Ganazzoli, J. Chem. Soc., Dalton Trans., 1983, 827.
719 G.Read, M.Urgelles, A.M.R.Galas, and M.B.Hursthouse, J. Chem. Soc., Dalton Trans., 1983, 911.
720 R.D.W.Kemmitt, P.McKenna, D.R.Russell, and L.J.S.Sherry, J. Organomet. Chem., 1983, 253, C59.
721 M.D.Waddington, J.A.Campbell, P.W.Jennings, and C.F.Campana, Organometallics, 1983, 2, 1270.
722 G.Strukul, R.A.Michelin, J.D.Orbell, and L.Randaccio, Inorg. Chem., 1983, 22, 3706.
723 A.Fumagalli, S.Martinengo, and G.Ciani, J. Chem. Soc., Chem. Commun., 1983, 1381.
724 O.J.Scherer, R.Konrad, E.Guggolz, and M.L.Ziegler, Chem. Ber., 1983, 116, 2676.
725 M.R.Awang, J.C.Jeffery, and F.G.A.Stone, J. Chem. Soc., Chem. Commun., 1983, 1426.
726 R.D.Barr, M.Green, J.A.K.Howard, T.B.Marder, I.Moore, and F.G.A.Stone, J. Chem. Soc., Chem. Commun., 1983, 746.
727 J.C.Jeffrey, C.Sambale, M.F.Schmidt, and F.G.A.Stone, Organometallics, 1982, 1, 1597.
728 M.R.Awang, J.C.Jeffery, and F.G.A.Stone, J. Chem. Soc., Dalton Trans., 1983, 2091.
729 K.A.Mead, I.Moore, F.G.A.Stone, and P.Woodward, J. Chem. Soc., Dalton Trans., 1983, 2083.
730 A.Blagg, A.T.Hutton, P.G.Pringle, and B.L.Shaw, Inorg. Chim. Acta, 1983, 76, L265.
731 S.S.M.Ling, R.J.Puddephatt, Lj.Manojlović-Muir, and K.W.Muir, J. Organomet. Chem., 1983, 255, C11.
732 S.S.M.Ling, R.J.Puddephatt, Lj.Manojlović-Muir, and K.W.Muir, Inorg. Chim. Acta, 1983, 77, L95.
733 M.K.Cooper, P.V.Stevens, and M.McPartlin, J. Chem. Soc., Dalton Trans., 1983, 553.
734 N.M.Boag, R.J.Goodfellow, M.Green, B.Hessner, J.A.K.Howard, and F.G.A.Stone, J. Chem. Soc., Dalton Trans., 1983, 2585.
735 M.L.Illingsworth, J.A.Teagle, J.L.Burmeister, W.C.Fultz, and A.L.Rheingold,

Organometallics, 1983, 2, 1364.
736 D.P.Arnold, M.A.Bennett, G.M.McLaughlin, and G.B.Robertson, J. Chem. Soc., Chem. Commun., 1983, 34.
737 G.Minghetti, A.L.Bandini, G.Banditelli, F.Bonati, R.Szostak, C.E.Strouse, C.B.Knobler, and H.D.Kaesz, Inorg. Chem., 1983, 22, 2332.
738 A.T.Hutton, B.Shabanzadeh, and B.L.Shaw, J. Chem. Soc., Chem. Commun., 1983, 1053.
739 D.P.Arnold, M.A.Bennett, G.M.McLaughlin, G.B.Robertson, and M.J.Whittaker, J. Chem. Soc., Chem. Commun., 1983, 32.
740 R.J.Goodfellow, I.R.Herbert, and A.G.Orpen, J. Chem. Soc., Chem. Commun., 1983, 1386.
741 J.Jans, R.Naegeli, L.M.Venanzi, and A.Albinati, J. Organomet. Chem., 1983, 247, C37.
742 M.P.Brown, A.Yavari, Lj. Manojlović-Muir, and K.W.Muir, J. Organomet. Chem., 1983, 256, C19.
743 Y.Yamamoto, K.Aoki, and H.Yamazaki, Organometallics, 1983, 2, 1377.
744 F.A.Cotton and L.M.Daniels, Acta Crystallogr., Sect. C, 1983, 39, 1495.
745 M.Herberhold, D.Reiner, and U.Thewalt, Angew. Chem., Int. Ed. Engl., 1983, 22, 1000.
746 G.Hartmann, R.Mews, and G.M.Sheldrick, Angew. Chem., Int. Ed. Engl., 1983, 22, 723.
747 F.A.Cotton, L.M.Daniels, and C.D.Schmulbach, Inorg. Chim. Acta, 1983, 75, 163.
748 C.P.Casey, J.M.O'Connor, W.D.Jones, and K.J.Haller, Organometallics, 1983, 2, 535.
749 D.S.Edwards, L.V.Biondi, J.W.Ziller, M.R.Churchill, and R.R.Schrock, Organometallics, 1983, 2, 1505.
750 W.Tikkanen, W.C.Kaska, S.Moya, T.Layman, R.Kane, and C.Krüger, Inorg. Chim. Acta, 1983, 76, L29.
751 W.E.Buhro, A.T.Patton, C.E.Strouse, J.A.Gladysz, F.B.McCormick, and M.C.Etter, J. Am. Chem. Soc., 1983, 105, 1056.
752 A.T.Patton, C.E.Strouse, C.B.Knobler, and J.A.Gladysz, J. Am. Chem. Soc., 1983, 105, 5804.
753 J.H.Merrifield, G.-Y.Lin, W.A.Kiel, and J.A.Gladysz, J. Am. Chem. Soc., 1983, 105, 5811.
754 A.J.L.Pombeiro, P.B.Hitchcock, and R.L.Richards, Inorg. Chim. Acta, 1983, 76, L225.
755 D.Baudry, J.-C.Daran, Y.Dromzee, M.Ephritikhine, H.Felkin, Y.Jeannin, and J.Zakrzewski, J. Chem. Soc., Chem. Commun., 1983, 813.
756 R.Rossi, A.Duatti, L.Magon, U.Casellato, R.Graziani, and L.Toniolo, Inorg. Chim. Acta, 1983, 75, 77.
757 D.Fenske, N.Mronga, and K.Dehnicke, Z. Anorg. Allg. Chem., 1983, 498, 131.
758 O.J.Scherer, J.Kerth, and M.L.Ziegler, Angew. Chem., Int. Ed. Engl., 1983, 22, 503.
759 G.Ciani, G.D'Alfonso, P.Romiti, A.Sironi, and M.Freni, Inorg. Chem., 1983, 22, 3115.
760 K.Raab, U.Nagel, and W.Beck, Z. Naturforsch., B, 1983, 38B, 1466.
761 P.O.Nubel, S.R.Wilson, and T.L.Brown, Organometallics, 1983, 2, 515.
762 M.Green, A.G.Orpen, C.J.Schaverien, and I.D.Williams, J. Chem. Soc., Chem. Commun., 1983, 1399.
763 T.J.Barder, S.M.Tetrick, R.A.Walton, F.A.Cotton, and G.L.Powell, J. Am. Chem. Soc., 1983, 105, 4090.
764 E.W.Abel, S.K.Bhargava, M.M.Bhatti, M.A.Mazid, K.G.Orrell, V.Šik, M.B.Hursthouse, and K.M.A.Malik, J. Organomet. Chem., 1983, 250, 373.
765 E.O.Fischer, P.Rustemeyer, O.Orama, D.Neugebauer, and U.Schubert, J. Organomet. Chem., 1983, 247, 7.
766 O.J.Scherer, J.Kerth, R.Anselmann, and W.S.Sheldrick, Angew. Chem., Int. Ed. Engl., 1983, 22, 984.
767 H.Preut and H.-J.Haupt, Acta Crystallogr., Sect. C, 1983, 39, 981.
768 G.Ciani, G.D'Alfonso, P.Romiti, A.Sironi, and M.Freni, J. Organomet. Chem., 1983, 244, C27.

769 J.P.Farr, M.M.Olmstead, F.E.Wood, and A.L.Balch, J. Am. Chem. Soc., 1983, 105, 792.
770 M.J.Decker, D.O.K.Fjeldsted, S.R.Stobart, and M.J.Zaworotko, J. Chem. Soc., Chem. Commun., 1983, 1525.
771 H.Werner, J.Wolf, and U.Schubert, Chem. Ber., 1983, 116, 2848.
772 H.Werner, J.Wolf, R.Zolk, and U.Schubert, Angew. Chem., Int. Ed. Engl., 1983, 22, 981.
773 J.G.Leipoldt and E.C.Grobler, Inorg. Chim. Acta, 1983, 72, 17.
774 A.Tiripicchio, M.Tiripicchio Camellini, C.Claver, A.Ruiz, and L.A.Oro, J. Organomet. Chem., 1983, 241, 77.
775 J.Müller, R.Stock, and J.Pickardt, Z. Naturforsch., B, 1983, 38B, 1454.
776 J.Wolf, H.Werner, O.Serhadli, and, M.L.Ziegler, Angew. Chem., Int. Ed. Engl., 1983, 22, 414.
777 R.Usón, L.A.Oro, J.A.Cabeza, C.Foces-Foces, F.H.Cano, and S.Garcia-Blanco, J. Organomet. Chem., 1983, 246, 73.
778 N.J.Meanwell, A.J.Smith, H.Adams, S.Okeya, and P.M.Maitlis, Organometallics, 1983, 2, 1705.
779 S.Nemeh, C.Jensen, E.Binamira-Soriaga, and W.C.Kaska, Organometallics, 1983, 2, 1442.
780 J.G.Leipoldt, S.S.Basson, and J.T.Nel, Inorg. Chim. Acta, 1983, 74, 85.
781 G.Moran, M.Green, and A.G.Orpen, J. Organomet. Chem., 1983, 250, C15.
782 H.Werner, J.Wolf, U.Schubert, and K.Ackermann, J. Organomet. Chem., 1983, 243, C63.
783 F.Faraone, G.Bruno, S.Lo Schiavo, G.Tresoldi, and G.Bombieri, J. Chem. Soc., Dalton Trans., 1983, 433.
784 P.Piraino, G.Bruno, G.Tresoldi, G.Faraone, and G.Bombieri, J. Chem. Soc., Dalton Trans., 1983, 2391.
785 A.Monge, E.Gutiérrez-Puebla, J.V.Heras, and E.Pinilla, Acta Crystallogr., Sect. C, 1983, 39, 446.
786 A.Ceriotti, G.Ciani, and A.Sironi, J. Organomet. Chem., 1983, 247, 345.
787 M.P.Anderson, B.M.Mattson, and L.H.Pignolet, Inorg. Chem., 1983, 22, 2644.
788 E.Cesarotti, A.Chiesa, G.Ciani, and A.Sironi, J. Organomet. Chem., 1983, 251, 79.
789 J.D.Oliver and D.P.Riley, Organometallics, 1983, 2, 1032.
790 E.Arpac and L.Dahlenburg, J. Organomet. Chem., 1983, 241, 27.
791 G.Vitulli, A.Raffaelli, P.A.Costantino, C.Barberini, F.Marchetti, S.Merlino, and P.S.Skell, J. Chem. Soc., Chem. Commun., 1983, 232.
792 R.R.Burch, A.J.Shusterman, E.L.Muetterties, R.G.Teller, and J.M.Williams, J. Am. Chem. Soc., 1983, 105, 3546.
793 P.J.Stang, M.R.White, and G.Maas, Organometallics, 1983, 2, 720.
794 B.D.Murray, M.M.Olmstead, and P.P.Power, Organometallics, 1983, 2, 1700.
795 J.-P.Lecomte, J.-M.Lehn, D.Parker, J.Guilhem, and C.Pascard, J. Chem. Soc., Chem. Commun., 1983, 296.
796 R.Halesha, G.K.N.Reddy, S.P.S.Rao, and H.Manohar, J. Organomet. Chem., 1983, 252, 231.
797 R.A.Jones, T.C.Wright, J.L.Atwood, and W.E.Hunter, Organometallics, 1983, 2, 470.
798 K.Isobe, A.Vázquez de Miguel, P.M.Bailey, S.Okeya, and P.M.Maitlis, J. Chem. Soc., Dalton Trans., 1983, 1441.
799 H.T.Dieck and J.Klaus, J. Organomet. Chem., 1983, 246, 301.
800 J.C.T.R.Burckett St. Laurent, P.B.Hitchcock, and J.F.Nixon, J. Organomet. Chem., 1983, 249, 243.
801 R.A.Jones and T.C.Wright, Organometallics, 1983, 2, 1842.
802 L.A.Glinskaya, E.N.Yurchenko, S.F.Solodovnikov, L.S.Gracheva, and R.F.Klevtsova, Zh. Strukt. Khim., 1982, 23(3), 79(Engl.Ed. 389).
803 F.Faraone, G.Bruno, S.Lo Schiavo, P.Piraino, and G.Bombieri, J. Chem. Soc., Dalton Trans., 1983, 1819.
804 W.A.Herrmann, C.Bauer, M.L.Ziegler, and H.Pfisterer, J. Organomet. Chem., 1983, 243, C54.
805 D.P.Riley and J.D.Oliver, Inorg. Chem., 1983, 22, 3361.
806 J.S.Najdzionek and B.D.Santarsiero, Acta Crystallogr., Sect. C, 1983, 39,

577.
807 T.Yoshida, W.J.Youngs, T.Sakaeda, T.Ueda, S.Otsuka, and J.A.Ibers, J. Am. Chem. Soc., 1983, 105, 6273.
808 Y.Yamamoto, Y.Wakatsuki, and H,Yamazaki, Organometallics, 1983, 2, 1604.
809 B.R.James, D.Mahajan, S.J.Rettig, and G.M.Williams, Organometallics, 1983, 2, 1452.
810 A.S.C.Chan, H.-S.Shieh, and J.R.Hill, J. Chem. Soc., Chem. Commun., 1983, 688.
811 F.E.Wood, J.Hvoslef, and A.L.Balch, J. Am. Chem. Soc., 1983, 105, 6986.
812 J.T.Mague, Inorg. Chem., 1983, 22, 45.
813 J.T.Mague, Inorg. Chem., 1983, 22, 1158.
814 F.Faraone, S.Lo Schiavo, G.Bruno, P.Piraino, and G.Bombieri, J. Chem. Soc., Dalton Trans., 1983, 1813.
815 J.L.Atwood, W.E.Hunter, R.A.Jones, and T.C.Wright, Inorg. Chem., 1983, 22, 993.
816 A.V.de Miguel, K.Isobe, P.M.Bailey, N.J.Meanwell, and P.M.Maitlis, Organometallics, 1982, 1, 1604.
817 R.J.Haines, N.D.C.T.Steen, and R.B.English, J. Chem. Soc., Dalton Trans., 1983, 2229.
818 R.B.English, R.J.Haines, and N.D.Steen, S. Afr. J. Chem., 1983, 36, 108.
819 R.R.Guimerans, M.M.Olmstead, and A.L.Balch, J. Am. Chem. Soc., 1983, 105, 1677.
820 E.W.Burkhardt, W.C.Mercer, G.L.Geoffroy, A.L.Rheingold, and W.C.Fultz, J. Chem. Soc., Chem. Commun., 1983, 1251.
821 C.P.Lau, C.Y.Ren, L.Book, and T.C.W.Mak, J. Organomet. Chem., 1983, 249, 429.
822 F.E.Wood, M.M.Olmstead, and A.L.Balch, J. Am. Chem. Soc., 1983, 105, 6332.
823 R.Bonfichi, G.Ciani, A.Sironi, and S.Martinengo, J. Chem. Soc., Dalton Trans., 1983, 253.
824 J.L.Vidal and R.C.Schoenig, J. Organomet. Chem., 1983, 241, 395.
825 A.Fumagalli, S.Martinengo, G.Ciani, and A.Sironi, J. Chem. Soc., Chem. Commun., 1983, 453.
826 D.Strumolo, C.Seregni, S.Martinengo, V.G.Albano, and D.Braga, J. Organomet. Chem., 1983, 252, C93.
827 V.G.Albano, D.Braga, P.Chini, D.Strumolo, and S.Martinengo, J. Chem. Soc., Dalton Trans., 1983, 249.
828 H.Adams, N.A.Bailey, and C.White, Inorg. Chem., 1983, 22, 1155.
829 I.Noda, H.Yasuda, and A.Nakamura, Organometallics, 1983, 2, 1207.
830 D.A.Tocher, R.O.Gould, T.A.Stephenson, M.A.Bennett, J.P.Ennett, T.W.Matheson, L.Sawyer, and V.K.Shah, J. Chem. Soc., Dalton Trans., 1983, 1571.
831 P.Pertici, G.Vitulli, W.Porzio, M.Zocchi, P.L.Barili, G.Deganello, J. Chem. Soc., Dalton Trans., 1983, 1553.
832 M.I.Bruce, R.C.Wallis, M.L.Williams, B.W.Skelton, and A.H.White, J. Chem. Soc., Dalton Trans., 1983, 2183.
833 R.Uson, L.A.Oro, M.A.Ciriano, M.M.Naval, M.C.Apreda, C.Foces-Foces, F.H.Cano, and S.Garcia-Blanco, J. Organomet. Chem., 1983, 256, 331.
834 P.R.Holland, B.Howard, and R.J.Mawby, J. Chem. Soc., Dalton Trans., 1983, 231.
835 J.W.Hull,Jr. and W.L.Gladfelter, Organometallics, 1982, 1, 1716.
836 J.M.Patrick, A.H.White, M.I.Bruce, M.J.Beatson, D.St.C.Black, G.B.Deacon, and N.C.Thomas, J. Chem. Soc., Dalton Trans., 1983, 2121.
837 T.V.Ashworth, E.Singleton, R.R.English, and M.M.de V.Steyn, S. Afr. J. Chem., 1983, 36, 97.
838 Z.Dauter, R.J.Mawby, C.D.Reynolds, and D.R.Saunders, Acta Crystallogr., Sect. C, 1983, 39, 1194.
839 G.Consiglio, F.Morandini, G.Ciani, A.Sironi, and M.Kretschmer, J. Am. Chem. Soc., 1983, 105, 1391.
840 H.Kletzin, H.Werner, O.Serhadli, and M.L.Ziegler, Angew. Chem., Int. Ed. Engl., 1983, 22, 46.
841 G.Consiglio, F.Morandini, G.Ciani, and A.Sironi, Angew. Chem., Int. Ed.

Engl., 1983, 22, 333.
842 S.A.Chawdhury, Z.Dauter, R.J.Mawby, C.D.Reynolds, D.R.Saunders, and M.Stephenson, Acta Crystallogr., Sect. C, 1983, 39, 985.
843 F.Morandini, G.Consiglio, B.Straub, G.Ciani, and A.Sironi, J. Chem. Soc., Dalton Trans., 1983, 2293.
844 L.M.Wilkes, J.H.Nelson, L.B.McCusker, K.Seff, and F.Mathey, Inorg. Chem., 1983, 22, 2476.
845 G.R.Clark, S.V.Hoskins, T.C.Jones, and W.R.Roper, J. Chem. Soc., Chem. Commun., 1983, 719.
846 J.W.Gilje, R.Schmutzler, W.S.Sheldrick, and V.Wray, Polyhedron, 1983, 2, 603.
847 S.C.Grocott, B.W.Skelton, and A.H.White, Aust. J. Chem., 1983, 36, 259.
848 M.I.Bruce, M.G.Humphrey, J.M.Patrick, and A.H.White, Aust. J. Chem., 1983, 36, 2065.
849 G.Smith, D.J.Cole-Hamilton, M.Thornton-Pett, and M.B.Hursthouse, J. Chem. Soc., Dalton Trans., 1983, 2501.
850 M.J.Auburn, R.D.Holmes-Smith, S.R.Stobart, M.J.Zaworotko, T.S.Cameron, and A.Kumari, J. Chem. Soc., Chem. Commun., 1983, 1523.
851 C.P.Casey and R.F.Jordan, J. Am. Chem. Soc., 1983, 105, 665.
852 K.P.C.Vollhardt and T.W.Weidman, J. Am. Chem. Soc., 1983, 105, 1676.
853 A.Astier, J.-C.Daran, Y.Jeannin, and C.Rigault, J. Organomet. Chem., 1983, 241, 53.
854 R.E.Colborn, D.L.Davies, A.F.Dyke, A.Endesfelder, S.A.R.Knox, A.G.Orpen, and D.Plaas, J. Chem. Soc., Dalton Trans., 1983, 2661.
855 Y.C.Lin, J.C.Calabrese, and S.S.Wreford, J. Am. Chem. Soc., 1983, 105, 1679.
856 R.E.Colborn, A.F.Dyke, S.A.R.Knox, K.A.Mead, and P.Woodward, J. Chem. Soc., Dalton Trans., 1983, 2099.
857 L.Stahl and R.D.Ernst, Organometallics, 1983, 2, 1229.
858 L.H.Polm, G.van Koten, K.Vrieze, C.H.Stam, and W.C.J.van Tunen, J. Chem. Soc., Chem. Commun., 1983, 1177.
859 M.I.Bruce, J.G.Matisons, B.W.Skelton, and A.H.White, J. Organomet. Chem., 1983, 251, 249.
860 P.Q.Adams, D.L.Davies, A.F.Dyke, S.A.R.Knox, K.A.Mead, and P.Woodward, J. Chem. Soc., Chem. Commun., 1983, 222.
861 D.C.Levendis, J.C.A.Boeyens, and E.W.Neuse, J. Crystallogr. Spectrosc. Res., 1982, 12, 493.
862 M.I.Bruce, M.L.Williams, and B.K.Nicholson, J. Organomet. Chem., 1983, 258, 63.
863 J.P.Selegue, J. Am. Chem. Soc., 1983, 105, 5921.
864 D.J.Darensbourg, M.Pala, and J.Waller, Organometallics, 1983, 2, 1285.
865 M.R.Churchill, L.R.Beanan, H.J.Wasserman, C.Bueno, Z.A.Rahman, and J.B.Keister, Organometallics, 1983, 2, 1179.
866 S.Bhaduri, K.S.Gopalkrishnan, G.M.Sheldrick, W.Clegg, and D.Stalke, J. Chem. Soc., Dalton Trans., 1983, 2339.
867 M.I.Bruce, J.G.Matisons, R.C.Wallis, J.M.Patrick, B.W.Skelton, and A.H.White, J. Chem. Soc., Dalton Trans., 1983, 2365.
868 M.I.Bruce, J.G.Matisons, B.W.Skelton, and A.H.White, J. Chem. Soc., Dalton Trans., 1983, 2375.
869 S.Aime, D.Osella, A.J.Deeming, A.M.Manotti Lanfredi, and A.Tiripicchio, J. Organomet. Chem., 1983, 244, C47.
870 P.Michelin Lausarot, G.Angelo Vaglio, M.Valle, A.Tiripicchio, and M.Tiripicchio Camellini, J. Chem. Soc., Chem. Commun., 1983, 1391.
871 N.J.Forrow, S.A.R.Knox, M.J.Morris, and A.G.Orpen, J. Chem. Soc., Chem. Commun., 1983, 234.
872 M.I.Bruce, B.K.Nicholson, and M.L.Williams, J. Organomet. Chem., 1983, 243, 69.
873 G.Gervasio, E.Sappa, A.M.Manotti Lanfredi, and A.Tiripicchio, Inorg. Chim. Acta, 1983, 68, 171.
874 R.P.Rosen, G.L.Geoffroy, C.Bueno, M.R.Churchill, and R.B.Ortega, J. Organomet. Chem., 1983, 254, 89.
875 P.Braunstein, J.Rose, Y.Dusausoy, and J.-P.Mangeot, J. Organomet. Chem.,

1983, 256, 125.
876 J.A.Jensen, D.E.Fjare, and W.L.Gladfelter, Inorg. Chem., 1983, 22, 1250.
877 D.E.Fjare, J.A.Jensen, and W.L.Gladfleter, Inorg. Chem., 1983, 22, 1774.
878 A.G.Orpen and R.K.McMullan, J. Chem. Soc., Dalton Trans., 1983, 463.
879 M.I.Bruce, E.Horn, M.R.Snow, and M.L.Williams, J. Organomet. Chem., 1983, 255, 255.
880 M.I.Bruce, B.K.Nicholson, J.M.Patrick, and A.H.White, J. Organomet. Chem., 1983, 254, 361.
881 M.L.Blohm, D.E.Fjare, and W.L.Gladfelter, Inorg. Chem., 1983, 22, 1004.
882 B.F.G.Johnson, J.Lewis, J.N.Nicholls, J.Puga, P.R.Raithby, M.J.Rosales, M.McPartlin, and W.Clegg, J. Chem. Soc., Dalton Trans., 1983, 277.
883 S.A.MacLaughlin, N.J.Taylor, and A.J.Carty, Inorg. Chem., 1983, 22, 1409.
884 S.A.MacLaughlin, N.J.Taylor, and A.J.Carty, Organometallics, 1983, 2, 1194.
885 B.F.G.Johnson, J.Lewis, W.J.H.Nelson, J.Puga, M.McPartlin, and A.Sironi, J. Organomet. Chem., 1983, 253, C5.
886 R.D.Adams, P.Mathur, and B.E.Segmüller, Organometallics, 1983, 2, 1258.
887 R.D.Adams, D.Mannig, and B.E.Segmüller, Organometallics, 1983, 2, 149.
888 W.T.Pennington, A.W.Cordes, J.C.Graham, and Y.W.Jung, Acta Crystallogr., Sect. C, 1983, 39, 709.
889 H.Schmidbaur, B.Milewski-Mahrla, and F.E.Wagner, Z. Naturforsch., B, 1983, 38B, 1477.
890 K.N.Akatova, R.I.Bochkova, V.A.Lebedev, V.V.Sharutin, and N.V.Belov, Dokl. Akad. Nauk SSSR, 1983, 268, 1389.
891 G.V.N.Appa Rao, M.Seshasayee, G.Aravamudan, and S.Sowrirajan, Acta Crystallogr., Sect. C, 1983, 39, 620.
892 R.Kivekäs and T.Laitalainen, Acta Chem. Scand., Ser. B, 1983, B37, 61.
893 V.K.Bel´skii, N.P.Bel´skaya, L.V.Kon´yakhina, and V.P.Syskova, Kristallografiya, 1982, 27, 1102.
894 H.Schmidbaur, C.Zybill, C.Krüger, and H.-J.Kraus, Chem. Ber., 1983, 116, 1955.
895 H.Puff, A.Bongartz, W.Schuh, and R.Zimmer, J. Organomet. Chem., 1983, 248, 61.
896 B.M.Powell and B.H.Torrie, Acta Crystallogr., Sect. C, 1983, 39, 963.
897 V.F.Kaminskii, E.E.Kostuchenko, R.P.Shibaeva, E.B.Yagubskii, and A.V.Zvarykina, Chem. Abs., 1983, 99, 205002t.
898 H.Schmidbaur, C.E.Zybill, and D.Neugebauer, Angew. Chem., Int. Ed. Engl., 1983, 22, 156.
899 F.Bigoli, E.Leporati, M.A.Pellinghelli, G.Crisponi, P.Deplano, and E.F.Trogu, J. Chem. Soc., Dalton Trans., 1983, 1763.
900 K.Lerstrup, M.Lee, F.M.Wiygul, T.J.Kistenmacher, and D.O.Cowan, J. Chem. Soc., Chem. Commun., 1983, 294.
901 R.P.Shibaeva, V.F.Kaminskii, E.B.Yagubskii, and L.A.Kulch, Kristallografiya, 1983, 28, 92.
902 H.Puff, E.Friedrichs, R.Hundt, and R.Zimmer, J. Organomet. Chem., 1983, 259, 79.
903 J.M.Williams, M.A.Beno, J.C.Sullivan, L.M.Banovetz, J.M.Braam, G.S.Blackman, C.D.Carlson, D.L.Greer, and D.M.Loesing, J. Am. Chem. Soc., 1983, 105, 643.
904 E.A.Chernyshev, O.V.Kuz´min, A.V.Lebedev, A.I.Gusev, N.I.Kirillova, N.S.Nametkin, V.D.Tyurin, A.M.Krapivin, and N.A.Kubasova, J. Organomet. Chem., 1983, 252, 133.
905 K.Hensen, T.Zengerly, P.Pickel, and G.Klebe, Angew. Chem., Int. Ed. Engl., 1983, 22, 725.
906 Mazhar-ul Haque, W.Horne, S.E.Cremer, and C.S.Blankenship, J. Chem. Soc., Perkin Trans.II, 1983, 395.
907 O.Graalmann, M.Hesse, U.Klingebiel, W.Clegg, M.Haase, and G.M.Sheldrick, Angew. Chem., Int. Ed. Engl., 1983, 22, 621.
908 W.Clegg, O.Graalmann, M.Haase, U.Klingebiel, G.M.Sheldrick, P.Werner, G.Henkel, and B.Krebs, Chem. Ber., 1983, 116, 282.
909 O.A.D´yachenko, Yu.A.Sokolova, V.M.Berestovitskaya, E.V.Trukhin, and G.A.Berkova, Izv. Akad. Nauk SSSR, Ser. Khim., 1983, 1319(Engl.Ed. 1191).
910 G.L.Larson, S.Sandoval, F.Cartledge, and F.R.Fronczek, Organometallics,

1983, 2, 810.
911 R.Tacke, H.Lange, W.S.Sheldrick, G.Lambrecht, U.Moser, and E.Mutschler, Z. Naturforsch., B, 1983, 38B, 738.
912 V.Gruhnert, A.Kirfel, G.Will, F.Wallrafen, and K.Recker, Z. Kristallogr., 1983, 163, 53.
913 H.Preut, B.Mayer, and W.P.Neumann, Acta Crystallogr., Sect. C, 1983, 39, 1118.
914 R.A.Jones, M.H.Seeberger, J.L.Atwood, and W.E.Hunter, J. Organomet. Chem., 1983, 247, 1.
915 D.W.H.Rankin and H.E.Robertson, J. Chem. Soc., Dalton Trans., 1983, 265.
916 W.E.Schklower, Yu. T. Strutschkow, L.E.Guselnikow, W.W.Wolkowa, and W.G.Awakyan, Z. Anorg. Allg. Chem., 1983, 501, 153.
917 A.I.Gusev, M.Yu.Antipin, D.S.Yufit, Yu.T.Struchkov, V.D.Sheludyakov, V.I.Zhun, and S.D.Vlasenko, Zh. Strukt. Khim., 1983, 24(3), 178(Engl.Ed. 490).
918 W.Clegg, Acta Crystallogr., Sect. C, 1983, 39, 901.
919 L.Párkányi, A.Szöllősy, L.Bihátsi, P.Hencsei, and J.Nagy, J. Organomet. Chem., 1983, 256, 235.
920 R.Tacke, H.Lange, A.Bentlage, W.S.Sheldrick, and L.Ernst, Z. Naturforsch., B, 1983, 38B, 190.
921 A.Szöllősy, L.Párkányi, L.Bihátsi, and P.Hencsei, J. Organomet. Chem., 1983, 251, 159.
922 W.Clegg, M.Haase, U.Klingebiel, and G.M.Sheldrick, Chem. Ber., 1983, 116, 146.
923 R.Appel, F.Knoch, B.Laubach, and R.Sievers, Chem. Ber., 1983, 116, 1873.
924 M.Weidenbruch, H.Flott, B.Ralle, K.Peters, and H.G. von Schnering, Z. Naturforsch., B, 1983, 38B, 1062.
925 S.Masamune, S.Murakami, H.Tobita, and D.J.Williams, J. Am. Chem. Soc., 1983, 105, 7776.
926 M.J.Fink, M.J.Michalczyk, K.J.Haller, R.West, and J.Michl, J. Chem. Soc., Chem. Commun., 1983, 1010.
927 H.W.Roesky, H.-G.Schmidt, M.Noltemeyer, and G.M.Sheldrick, Chem. Ber., 1983, 116, 1411.
928 A.H.Cowley, J.G.Lasch, N.C.Norman, C.A.Stewart, and T.C.Wright, Organometallics, 1983, 2, 1691.
929 J.W.Bruno, T.J.Marks, and V.W.Day, J. Organomet. Chem., 1983, 250, 237.
930 M.D.Curtis, J.J.D'Errico, D.N.Duffy, P.S.Epstein, and L.G.Bell, Organometallics, 1983, 2, 1808.
931 E.Carmona, J.M.Marín, M.L.Poveda, L.Sánchez, R.D.Rogers, and J.L.Atwood, J. Chem. Soc., Dalton Trans., 1983, 1003.
932 M.Ishikawa, H.Sugisawa, M.Kumada, T.Higuchi, K.Matsui, K.Hirotsu, and J.Iyoda, Organometallics, 1983, 2, 174.
933 L.N.Zakharov, Yu.T.Struchkov, E.A.Kuz'min, and B.I.Petrov, Kristallografiya, 1983, 28, 271.
934 C.Kratky, H.G.Schuster, and E.Hengge, J. Organomet. Chem., 1983, 247, 253.
935 H.Sakurai, Y.Nakadaira, A.Hosomi, Y.Eriyama, and C.Kabuto, J. Am. Chem. Soc., 1983, 105, 3359.
936 E.Niecke, R.Rüger, M.Lysek, S.Pohl, and W.Schoeller, Angew. Chem., Int. Ed. Engl., 1983, 22, 486.
937 G.H.Hovakeemian and K.P.C.Vollhardt, Angew. Chem., Int. Ed. Engl., 1983, 22, 994.
938 W.Clegg, Acta Crystallogr., Sect. C, 1983, 39, 1106.
939 R.P.Planalp and R.A.Anderson, Organometallics, 1983, 2, 1675.
940 M.D.Fryzuk, H.D.Williams, and S.J.Rettig, Inorg. Chem., 1983, 22, 863.
941 L.Párkányi, E.Hengge, and H.Stüger, J. Organomet. Chem., 1983, 251, 167.
942 C.Eaborn, P.B.Hitchcock, and P.D.Lickiss, J. Organomet. Chem., 1983, 252, 281.
943 U.Klingebiel, N.Vater, W.Clegg, M.Haase, and G.M.Sheldrick, Z. Naturforsch., B, 1983, 38B, 1557.
944 J.Jaud, C.Couret, and J.Escudie, J. Organomet. Chem., 1983, 249, C25.
945 W.Clegg, M.Haase, U.Klingebiel, J.Neemann, and G.M.Sheldrick, J. Organomet.

Chem., 1983, 251, 281.
946 W.Clegg, M.Hasse, H.Hluchy, U.Klingebiel, and G.M.Sheldrick, Chem. Ber., 1983, 116, 290.
947 S.-M.Chen, L.D.David, K.J.Haller, C.L.Wadsworth, and R.West, Organometallics, 1983, 2, 409.
948 R.D.Rogers and J.L.Atwood, J. Crystallogr. Spectrosc. Res., 1983, 13, 1.
949 N.L.Paddock, S.J.Rettig, and J.Trotter, Can. J. Chem., 1983, 61, 541.
950 V.K.Belsky, N.N.Zemlyansky, I.V.Borisova, N.D.Kolosova, and I.P.Beletskaya, J. Organomet. Chem., 1983, 254, 189.
951 H.Puff, I.Bung, E.Friedrichs, and A.Jansen, J. Organomet. Chem., 1983, 254, 23.
952 R.P.Planalp, R.A.Andersen, and A.Zalkin, Organometallics, 1983, 2, 16.
953 R.P.Planalp and R.A.Anderson, J. Am. Chem. Soc., 1983, 105, 7774.
954 W.J.Evans, I.Bloom, W.E.Hunter, and J.L.Atwood, Organometallics, 1983, 2, 709.
955 W.J.Evans, I.Bloom, W.E.Hunter, and J.L.Atwood, J. Am. Chem. Soc., 1983, 105, 1401.
956 A.S.Secco and J.Trotter, Acta Crystallogr., Sect. C, 1983, 39, 451.
957 R.A.Howie, E.S.Paterson, J.L.Wardell, and J.W.Burley, J. Organomet. Chem., 1983, 259, 71.
958 D.Britton and Y.M.Chow, Acta Crystallogr., Sect. C, 1983, 39, 1539.
959 F.E.Hahn, T.S.Dory, C.L.Barnes, M.B.Hossain, D. van der Helm, and J.J.Zuckerman, Organometallics, 1983, 2, 969.
960 M.Dräger, J. Organomet. Chem., 1983, 251, 209.
961 I.W.Nowell, J.S.Brooks, G.Beech, and R.Hill, J. Organomet. Chem., 1983, 244, 119.
962 A.G.Davies, J.P.Goddard, M.B.Hursthouse, and N.P.C.Walker, J. Chem. Soc., Chem. Commun., 1983, 597.
963 W.T.Pennington, A.W.Cordes, J.C.Graham, and Y.W.Jung, Acta Crystallogr., Sect. C, 1983, 39, 712.
964 P.Ganis, V.Peruzzo, and G.Valle, J. Organomet. Chem., 1983, 256, 245.
965 C.W.Holzapfel, J.M.Koekemoer, C.F.Marais, G.J.Kruger, and J.A.Pretorius, S. Afr. J. Chem., 1982, 35, 80.
966 J.L.Brianső, X.Solans, and J.Vicente, J. Chem. Soc., Dalton Trans., 1983, 169.
967 V.K.Belsky, A.A.Simonenko, V.O.Reikhsfeld, and I.E.Saratov, J. Organomet. Chem., 1983, 244, 125.
968 L.M.Engelhardt, W.-P.Leung, C.L.Raston, and A.H.White, Aust. J. Chem., 1982, 22, 2383.
969 L.Prasad, Y.Le Page, and F.E.Smith, Inorg. Chim. Acta, 1983, 68, 45.
970 L.G.Kuz'mina, Yu.T.Struchkov, E.M.Rokhlina, A.S.Peregudov, and D.N.Kravtsov, Zh. Strukt. Khim., 1982, 23(6), 108(Engl.Ed. 914).
971 C.Pelizzi, G.Pelizzi, and P.Tarasconi, Polyhedron, 1983, 2, 145.
972 P.G.Harrison, N.W.Sharpe, C.Pelizzi, G.Pelizzi, and P.Tarasconi, J. Chem. Soc., Dalton Trans., 1983, 1687.
973 P.G.Harrison, K.Lambert, T.J.King, and B.Majee, J. Chem. Soc., Dalton Trans., 1983, 363.
974 P.G.Harrison, N.W.Sharpe, C.Pelizzi, G.Pelizzi, and P.Tarasconi, J. Chem. Soc., Dalton Trans., 1983, 921.
975 C.J.Cardin, D.J.Cardin, R.J.Norton, H.E.Parge, and K.W.Muir, J. Chem. Soc., Dalton Trans., 1983, 665.
976 S.-W.Ng, C.L.Barnes, D. van der Helm, and J.J.Zuckerman, Organometallics, 1983, 2, 600.
977 F.A.K.Nasser, M.B.Hossain, D. van der Helm, and J.J.Zuckerman, Inorg. Chem., 1983, 22, 3107.
978 F.W.B.Einstein, C.H.W.Jones, T.Jones, and R.D.Sharma, Can. J. Chem., 1983, 61, 2611.
979 R.G.Swisher, R.O.Day, and R.R.Holmes, Inorg. Chem., 1983, 22, 3692.
980 A.L.Beauchamp, S.Latour, M.J.Olivier, and J.D.West, J. Am. Chem. Soc., 1983, 105, 7778.
981 A.L.Beauchamp, S.Latour, M.J.Olivier, and J.D.Wuest, J. Organomet. Chem.,

1983, 254, 283.

982 S.Masamune, L.R.Sita, and D.J.Williams, J. Am. Chem. Soc., 1983, 105, 630.

983 R.Graziani, U.Casellato, and G.Plazzogna, Acta Crystallogr., Sect. C, 1983, 39, 1188.

984 J.Meunier-Piret, M. van Meerssche, M.Gielen, and K.Jurkschat, J. Organomet. Chem., 1983, 252, 289.

985 G.L.Rochfort and J.E.Ellis, J. Organomet. Chem., 1983, 250, 277.

986 V.C.Gibson, P.D.Grebenik, and M.L.H.Green, J. Chem. Soc., Chem. Commun., 1983, 1101.

987 J.M.Mayer, P.T.Wolczanski, B.D.Santarsiero, W.A.Olson, and J.E.Bercaw, Inorg. Chem., 1983, 22, 1150.

988 J.C.Green, C.P.Overton, K.Prout, and J.M.Marin, J. Organomet. Chem., 1983, 241, C21.

989 M.R.Churchill and H.J.Wasserman, J. Organomet. Chem., 1983, 247, 175.

990 F.A.Cotton, L.R.Falvello, and R.C.Najjar, Organometallics, 1982, 1, 1640.

991 A.W.Gal and H. van der Heijden, J. Chem. Soc., Chem. Commun., 1983, 420.

992 R.F.Ziolo and J.M.Troup, J. Am. Chem. Soc., 1983, 105, 229.

993 N.W.Alcock and W.D.Harrison, J. Chem. Soc., Dalton Trans., 1983, 2015.

994 R.K.Chadha, J.E.Drake, and J.L.Hencher, Can. J. Chem., 1983, 61, 1222.

995 R.K.Chadha, J.E.Drake, and M.A.Khan, Acta Crystallogr., Sect. C, 1983, 39, 45.

996 R.M.Lobkovskaya, R.P.Shibaeva, and O.N.Eremenko, Kristallografiya, 1983, 28, 276.

997 K.G.Moloy, T.J.Marks, and V.W.Day, J. Am. Chem. Soc., 1983, 105, 5696.

998 S.Gambarotta, C.Floriani, A.Chiesi-Villa, and C.Guastini, J. Am. Chem. Soc., 1983, 105, 7295.

999 B.H.Edwards, R.D.Rogers, D.J.Sikora, J.L.Atwood, and M.D.Rausch, J. Am. Chem. Soc., 1983, 105, 416.

1000 C.G.Marcellus, R.T.Oakley, A.W.Cordes, and W.T.Pennington, J. Chem. Soc., Chem. Commun., 1983, 1451.

1001 J.R.Dilworth, I.A.Latham, G.J.Leigh, G.Huttner, and I.Jibril, J. Chem. Soc., Chem. Commun., 1983, 1368.

1002 M.A.A.F. de C.T.Carrondo and G.A.Jeffrey, Acta Crystallogr., Sect. C, 1983, 39, 42.

1003 M.L.H.Green, N.J.Hazel, P.D.Grebenik, V.S.B.Mtetwa, and K.Prout, J. Chem. Soc., Chem. Commun., 1983, 356.

1004 A.W.Clauss, S.R.Wilson, R.M.Buchanan, C.G.Pierpont, and D.N.Hendrickson, Inorg. Chem., 1983, 22, 628.

1005 A.Shaver, J.M.McCall, P.H.Bird, and N.Ansari, Organometallics, 1983, 2, 1894.

1006 S.A.Cohen, P.R.Auburn, and J.E.Bercaw, J. Am. Chem. Soc., 1983, 105, 1136.

1007 D.M.Hoffman, N.D.Chester, and R.C.Fay, Organometallics, 1983, 2, 48.

1008 G.M.Dawkins, M.Green, K.A.Mead, J.-Y.Salaün, F.G.A.Stone, and P.Woodward, J. Chem. Soc., Dalton Trans., 1983, 527.

1009 C.T.Vroegop, J.H.Teuben, F. van Bolhuis, and J.G.M. van der Linden, J. Chem. Soc., Chem. Commun., 1983, 550.

1010 V.B.Shur, S.Z.Bernadyuk, V.V.Burlakov, V.G.Andrianov, A.I.Yanovsky, Yu.T.Struchkov, and M.E.Vol'pin, J. Organomet. Chem., 1983, 243, 157.

1011 J.Crowder, K.Hendrick, R.W.Matthews, and B.L.Podejma, J. Chem. Research, (S), 1983, 82.

1012 J.Rebizant, M.R.Spirlet, and J.Goffart, Acta Crystallogr., Sect. C, 1983, 39, 1041.

1013 G.Bombieri, F.Benetollo, E.Klähne, and R.D.Fischer, J. Chem. Soc., Dalton Trans., 1983, 1115.

1014 H.J.Wasserman, A.J.Zozulin, D.C.Moody, R.R.Ryan, and K.V.Salazar, J. Organomet. Chem., 1983, 254, 305.

1015 T.H.Cymbaluk, R.D.Ernst, and V.W.Day, Organometallics, 1983, 2, 963.

1016 R.E.Cramer, R.B.Maynard, J.C.Paw, and J.W.Gilje, Organometallics, 1983, 2, 1336.

1017 J.Nieman, J.H.Teuben, J.C.Huffman, and K.G.Caulton, J. Organomet. Chem., 1983, 255, 193.

1018 S.A.Koch and V.Chebolu, Organometallics, 1983, 2, 350.
1019 S.Gambarotta, C.Floriani, A.Chiesi-Villa, and C.Guastini, J. Chem. Soc., Chem. Commun., 1983, 184.
1020 C.M.Bolinger, T.B.Rauchfuss, and A.L.Rheingold, J. Am. Chem. Soc., 1983, 105, 6321.
1021 A.W.Duff, K.Jonas, R.Goddard, H.-J.Kraus, and C.Krüger, J. Am. Chem. Soc., 1983, 105, 5479.
1022 K.Jonas, V.Wiskamp, Y.-H.Tsay, and C.Krüger, J. Am. Chem. Soc., 1983, 105, 5480.
1023 Ch.Elschenbroich, J.Heck, W.Massa, E.Nun, and R.Schmidt, J. Am. Chem. Soc., 1983, 105, 2905.
1024 F.A.Cotton, G.E.Lewis, and G.N.Mott, Inorg. Chem., 1983, 22, 560.
1025 J.Schiemann, P,Hübener, and E.Weiss, Angew. Chem., Int. Ed. Engl., 1983, 22, 980.
1026 E.W.Abel, G.D.King, K.G.Orrell, G.M.Pring, V.Šik, and T.S.Cameron, Polyhedron, 1983, 2, 1117.
1027 J.D.Korp, I.Bernal, J.L.Mills, and H.T.Weaver,Jr., Inorg. Chim. Acta, 1983, 75, 173.
1028 F.Y.Pétillon, J.-L. Le Quéré, F. Le Floch-Pérennou, J.-E.Guerchais, M.-B.Gomas de Lima, Lj. Manojlović-Muir, K.W.Muir, and D.W.A.Sharp, J. Organomet. Chem., 1983, 255, 231.
1029 M.R.Churchill and H.J.Wasserman, Inorg. Chem., 1983, 22, 1574.
1030 C.Glidewell, D.C.Liles, and P.J.Pogorzelec, Acta Crystallogr., Sect. C, 1983, 39, 542.
1031 F.R.Kreissl, M.Wolfgruber, W.Sieber, and K.Ackermann, J. Organomet. Chem., 1983, 252, C39.
1032 K.W.Chiu, D.Lyons, G.Wilkinson, M.Thornton-Pett, and M.B.Hursthouse, Polyhedron, 1983, 2, 803.
1033 A.Shaver, J.Hartgerink, R.D.Lai, P.Bird, and N.Ansari, Organometallics, 1983, 2, 938.
1034 W.Sieber, M.Wolfgruber, D.Neugebauer, O.Orama, and F.R.Kreissl, Z. Naturforsch., B, 1983, 38B, 67.
1035 L.G.McCullough, M.L.Listemann, R.R.Schrock, M.R.Churchill, and J.W.Ziller, J. Am. Chem. Soc., 1983, 105, 6729.
1036 A.Gourdon and K.Prout, Acta Crystallogr., Sect. C, 1983, 39, 865.
1037 M.R.Churchill and H.J.Wasserman, Organometallics, 1983, 2, 755.
1038 J.C.Dewan, M.M.Roberts, and S.J.Lippard, Inorg. Chem., 1983, 22, 1529.
1039 J.C.Jeffery, J.C.V.Laurie, I.Moore, and F.G.A.Stone, J. Organomet. Chem., 1983, 258, C37.
1040 E.O.Fischer, A.C.Fillippou, H.G.Alt, and K.Ackermann, J. Organomet. Chem., 1983, 254, C21.
1041 M.R.Churchill, J.W.Ziller, L.McCullough, S.F.Pedersen, and R.R.Schrock, Organometallics, 1983, 2, 1046.
1042 M.Herberhold, W.Ehrenreich, K.Guldner, W.Jellen, U.Thewalt, and H.-P.Klein, Z. Naturforsch., B, 1983, 38B, 1383.
1043 K.R.Birdwhistell, S.J.N.Burgmayer, and J.L.Templeton, J. Am. Chem. Soc., 1983, 105, 7789.
1044 P.H.M.Budzelaar, H.J.Alberts-Jansen, K.Mollema, J.Boersma, G.J.M. van der Kerk, A.L.Spek, and A.J.M.Duisenberg, J. Organomet. Chem., 1983, 243, 137.
1045 P.T.Wolczanski, R.S.Threlkel, and B.D.Santarsiero, Acta Crystallogr., Sect. C, 1983, 39, 1330.
1046 A.Winter, O.Scheidsteger, and G.Huttner, Z. Naturforsch., B, 1983, 38B, 1525.
1047 J.L.Le Quéré, F.Y.Pétillon, J.E.Guerchais, Lj.Manojlović-Muir, K.W.Muir, and D.W.A.Sharp, J. Organomet. Chem., 1983, 249, 127.
1048 B.Bovio, J. Organomet. Chem., 1983, 244, 257.
1049 B.Bovio, J. Organomet. Chem., 1983, 241, 363.
1050 F.A.Cotton and W.Schwotzer, J. Am. Chem. Soc., 1983, 105, 5639.
1051 M.H.Chisholm, K.Folting, D.M.Hoffman, J.C.Huffman, and J.Leonelli, J. Chem. Soc., Chem. Commun., 1983, 589.
1052 M.H.Chisholm, D.M.Hoffman, and J.C.Huffman, Inorg. Chem., 1983, 22, 2903.

1053 F.A.Cotton, W.Schwotzer, and E.S.Shamshoum, Organometallics, 1983, 2, 1167.
1054 F.A.Cotton and W.Schwotzer, J. Am. Chem. Soc., 1983, 105, 4955.
1055 F.A.Cotton, W.Schwotzer, and E.S.Shamshoum, Organometallics, 1983, 2, 1340.
1056 W.J.Evans, J.H.Meadows, W.E.Hunter, and J.L.Atwood, Organometallics, 1983, 2, 1252.
1057 G.B.Deacon, P.I.MacKinnon, T.W.Hambley, and J.C.Taylor, J. Organomet. Chem., 1983, 259, 91.
1058 T.D.Tilley, R.A.Andersen, and A.Zalkin, Inorg. Chem., 1983, 22, 856.
1059 J.Dekker, J.Boersma, and G.J.M. van der Kerk, J. Chem. Soc., Chem. Commun., 1983, 553.
1060 M.R.P. van Vliet, J.T.B.H.Jastrzebski, G. van Koten, K.Vrieze, and A.L.Spek, J. Organomet. Chem., 1983, 251, C17.
1061 N.A.Bell, H.M.M.Shearer, and C.B.Spencer, Acta Crystallogr., Sect. C, 1983, 39, 1182.
1062 M.E.Silver and R.C.Fay, Organometallics, 1983, 2, 44.
1063 U.Thewalt and W.Lasser, Z. Naturforsch., B, 1983, 38B, 1501.
1064 S.Gamborotta, C.Floriani, A.Chiesi-Villa, and C.Guastini, Inorg. Chem., 1983, 22, 2029.
1065 J.Dai, M.Lou, J.Zhang, and S.Chen, Jiegou Huaxue, 1982, 1, 63.
1066 G.Erker, K.Engel, J.L.Atwood, and W.E.Hunter, Angew. Chem., Int. Ed. Engl., 1983, 22, 494.
1067 E.J.Moore, D.A.Straus, J.Armantrout, B.D.Santarsiero, R.H.Grubbs, and J.E.Bercaw, J. Am. Chem. Soc., 1983, 105, 2068.
1068 S.Gambarotta, C.Floriani, A.Chiesi-Villa, and C.Guastini, J. Am. Chem. Soc., 1983, 105, 1690.
1069 G.W.Rice, G.B.Ansell, M.A.Modrick, and S.Zentz, Organometallics, 1983, 2, 154.
1070 G.Erker, K.Kropp, J.L.Atwood, and W.E.Hunter, Organometallics, 1983, 2, 1555.
1071 H.Takaya, M.Yamakawa, and K.Mashima, J. Chem. Soc., Chem. Commun., 1983, 1283.
1072 W.Chang, J.Dai, and S.Chen, Jiegou Huaxue, 1982, 1, 73.
1073 Z.Wang, S.Chen, X.Yao, Y.Dong, S.Wu, J.Zhang, A.Wei, and R.Zhang, Sci. Sin. Ser. B(Engl. Ed.), 1982, 25, 1133.
1074 K.Kropp, V.Skibbe, G.Erker and C.Krüger, J. Am. Chem. Soc., 1983, 105, 3353.